Environmental Calculations

Environmental Calculations
A Multimedia Approach

Robert G. Kunz

WILEY

A JOHN WILEY & SONS, INC., PUBLICATION

Copyright © 2009 by John Wiley & Sons, Inc. All rights reserved

Published by John Wiley & Sons, Inc., Hoboken, New Jersey
Published simultaneously in Canada

No part of this publication may be reproduced, stored in a retrieval system, or transmitted in any form or by any means, electronic, mechanical, photocopying, recording, scanning, or otherwise, except as permitted under Section 107 or 108 of the 1976 United States Copyright Act, without either the prior written permission of the Publisher, or authorization through payment of the appropriate per-copy fee to the Copyright Clearance Center, Inc., 222 Rosewood Drive, Danvers, MA 01923, (978) 750-8400, fax (978) 750-4470, or on the web at www.copyright.com. Requests to the Publisher for permission should be addressed to the Permissions Department, John Wiley & Sons, Inc., 111 River Street, Hoboken, NJ 07030, (201) 748-6011, fax (201) 748-6008, or online at http://www.wiley.com/go/permission.

Limit of Liability/Disclaimer of Warranty: While the publisher and author have used their best efforts in preparing this book, they make no representations or warranties with respect to the accuracy or completeness of the contents of this book and specifically disclaim any implied warranties of merchantability or fitness for a particular purpose. No warranty may be created or extended by sales representatives or written sales materials. The advice and strategies contained herein may not be suitable for your situation. You should consult with a professional where appropriate. Neither the publisher nor author shall be liable for any loss of profit or any other commercial damages, including but not limited to special, incidental, consequential, or other damages.

For general information on our other products and services or for technical support, please contact our Customer Care Department within the United States at (800) 762-2974, outside the United States at (317) 572-3993 or fax (317) 572-4002.

Wiley also publishes its books in a variety of electronic formats. Some content that appears in print may not be available in electronic formats. For more information about Wiley products, visit our web site at www.wiley.com.

Library of Congress Cataloging-in-Publication Data:

Kunz, Robert G.
  Environmental calculations : a multimedia approach / Robert G. Kunz.
    p. cm.
  Includes index.
  ISBN 978-0-470-13985-1 (cloth)
  1. Sanitary engineering–Equipment and supplies–Problems, exercises, etc. 2. Pollution control equipment–Problems, exercises, etc. 3. Environmental sciences–Mathematics. 4. Engineering mathematics–Formulae. I. Title.
TD192.K86 2009
628.5–dc22

                                                                  2009009712

Printed in the United States of America

10 9 8 7 6 5 4 3 2 1

To my lovely wife, Maureen, who has had the patience to put up with me during the writing of this book, who accompanied me on the reconnaissance missions outlined in Appendices G and I, and who played the role of laboratory assistant for the experiment included in Appendix H. To my children, their spouses, and the grandchildren as well, at whose homes much of the book was written during family gatherings while the rest of the family attended to family matters.

A special dedication goes out to my classmates, living and deceased, of the Regis High School, New York City, Class of 1957, where I learned all my Greek letters, not just those appearing in this volume. Dedication also goes to the Dominican Academy, New York City, Class of 1958, the alma mater of my helpmate, trip companion, personal laboratory assistant, and indeed best friend.

"It's Not Easy Being Green."
— Kermit the Frog

Contents

Preface	xxi

Acknowledgments	xxiii

1. Introduction   1

    1.1   Who, What, Why, and How of This Book   1
    1.2   Potential Users   2
    1.3   Arrangement   2
    1.4   Truths and Myths About Environmental Control   3
    1.5   Adequate Preparation Is the Key   4
    1.6   A List of Do's   4
    1.7   And At Least One Don't   5
    1.8   Author's Supplemental Disclaimer   5
           References   5

2. Basic Concepts   6

    2.1   Basic Chemistry   6
             2.1.1   The Atom   6
             2.1.2   The Periodic Table   7
             2.1.3   Valence and Balanced Chemical Equations   9
             2.1.4   Summary of Basic Chemistry   13
    2.2   The Ideal Gas   13
             2.2.1   Dalton's Law   15
             2.2.2   Amagat's Law   16
             2.2.3   Example Problems for the Ideal Gas Law   16
    2.3   Concentrations and Mass Flow Rates   19
             2.3.1   Concentrations in Mass Per Unit Volume   19
             2.3.2   Emission Factors in Mass Per Unit of Fuel Firing Rate   21
             2.3.3   Higher and Lower Heating Values   22
             2.3.4   Mass Flow Rates   24
    2.4   Vapor Pressure   25
             2.4.1   What It Is   25
             2.4.2   How to Calculate   26
    2.5   Henry's Law   31
             2.5.1   Gases   31

viii Contents

        2.5.2   Volatile Liquids   33
        2.5.3   Ionizing Materials: Ionic Equilibrium   34
        2.5.4   Henry's Law Constants for Ionizing Species   36
2.6   Vapor–Liquid Equilibrium   41
        2.6.1   Raoult's Law   41
        2.6.2   Systems Where Raoult's Law Does Not Apply   46
        2.6.3   Azeotropic Systems   49
        2.6.4   Activity Coefficients Versus Mole Fraction Plots   52
        2.6.5   Henry's Law Constants from Vapor–Liquid Equilibrium Activity Coefficients   52
        2.6.6   Rigorous Vapor–Liquid Equilibria for the Air–Water System   58
2.7   Energy Balances and Heat Transfer   60
        2.7.1   Addition or Subtraction of Energy with No Change of Phase   60
        2.7.2   Latent Heat   61
        2.7.3   Heat Transfer   62
2.8   A Smattering of Statistics   72
        2.8.1   Characterizations of the Center of a Distribution   73
        2.8.2   Characterization of the Spread of a Distribution   74
        2.8.3   The Rolling Average   76
        2.8.4   Probability Density Functions: The Normal Distribution   77
        2.8.5   Test for a Normal Distribution   78
        2.8.6   The Log-Normal Distribution   80
        2.8.7   Other Statistical Applications   81
References   81

3. Air Combustion   83

3.1   Introduction   83
3.2   Combustion   83
        3.2.1   Combustion Inputs and Outputs   84
3.3   Fuel   84
        3.3.1   Fuel Heating Values   85
3.4   Air   87
3.5   Water   91
        3.5.1   Estimation of Combustion Air Moisture   91
3.6   Combustion Calculations from Basic Principles   98
        3.6.1   Pure O_2: Stoichiometric Case   98
        3.6.2   Pure O_2: Excess O_2 Case   99
        3.6.3   Case Using Pure O_2 to Obtain a Specified O_2 Concentration in the Flue Gas   100
        3.6.4   Stoichiometric Case Using Dry Air   101
        3.6.5   Percent Excess Air Case Using Dry Air   101
        3.6.6   The Case of a Specific Excess O_2 Concentration in the Flue Gas Using Dry Atmospheric Air   103
        3.6.7   Stoichiometric Case Using Moist Air   104
        3.6.8   Excess Air Case Using Moist Air   104
        3.6.9   The Case of a Specific Excess O_2 in the Flue Gas   105

3.7	Flue Gas	107	
3.8	Combustion Calculations Based on EPA Method 19 F Factors		110
3.9	Combustion Problems: Major Species in the Flue Gas		115
	3.9.1	Relationship Between Percent Excess Air and Percent O_2 in the Flue Gas 131	
3.10	Adiabatic Flame Temperature	131	
3.11	Estimation of Pollutant Emissions	137	
3.12	The Stack Test	166	
	3.12.1	Test Methods and Calculations	168
	3.12.2	Air Infiltration	180
3.13	Continuous Emission Monitoring Systems		182
3.14	Miscellaneous Sources of Air Emissions		186
	3.14.1	Fugitive Emissions	186
	3.14.2	Cooling Towers	190
	3.14.3	Other Sources of Fugitive Emissions	191
	3.14.4	Organic Liquid Storage Tanks	192
	3.14.5	Emergency Diesel Generator and Fuel Oil Tank	192
	3.14.6	Organic Chemical Vent Stream	195
	3.14.7	Monoethanolamine (MEA) Emissions	195
	References	202	

4. Air Control Devices 207

4.1	Overview	207	
4.2	Flares	207	
	4.2.1	Flare Operation	208
	4.2.2	Flare Emissions	208
4.3	Selective Catalytic Reduction		211
	4.3.1	SCR Equations	211
	4.3.2	SCR Operation	212
4.4	Selective Noncatalytic Reduction		214
	4.4.1	SNCR Design and Operation	215
	4.4.2	SNCR Performance	215
	4.4.3	SNCR plus SCR	217
4.5	Flue Gas Recirculation	218	
	4.5.1	FGR Performance	219
	4.5.2	FGR Data Analysis	219
4.6	Water/Steam Injection into Combustion Gas Turbine		223
	4.6.1	Emissions and Abatement	223
	4.6.2	Permitting	224
4.7	Low-Temperature Oxidation (Ozone Reaction/Scrubbing)		226
	4.7.1	Background	226
	4.7.2	Process Description	227
	4.7.3	Chemical Reactions	228
	4.7.4	Ozone to NO_x Ratio	229
	4.7.5	Possible Problem Areas	229
	4.7.6	Demonstration on Natural-Gas Fired Boiler	229
4.8	Particulate Removal	232	

x   Contents

4.9	Flue Gas Scrubbing	237
4.10	Atmospheric Dispersion	238
	4.10.1 Atmospheric Conditions—Stability	239
	4.10.2 Atmospheric Conditions—Wind Speed and Air Temperature	239
	4.10.3 Characteristics of the Emission	239
	4.10.4 Good Engineering Practice (GEP) Stack Height	240
	4.10.5 Dispersion Modeling Calculations	241
	4.10.6 Stack Draft	246
	References	247

5. Water/Wastewater Composition   252

5.1	Introduction	252
5.2	Concentrations Expressed as mg/L as $CaCO_3$	252
	5.2.1 Balancing a Typical Water Analysis	253
	5.2.2 Calculation of Conductivity	257
5.3	Dissolved Oxygen	257
5.4	Nonspecific Indicators of Water Pollution	259
	5.4.1 Biochemical Oxygen Demand	259
	5.4.2 Chemical Oxygen Demand	260
	5.4.3 Permanganate Oxygen Demand Test	260
	5.4.4 Total Organic Carbon	261
5.5	BOD, COD, and TOC in Industrial Wastewater	262
5.6	Domestic Wastewater (Also known as Sewage)	268
	5.6.1 Septic Tanks	268
5.7	Dissolved Oxygen Concentration in a Receiving Stream	269
5.8	Alkalinity	271
	5.8.1 Indicators Used in Alkalinity Titrations	272
	5.8.2 Alkalinity of Natural Waters	272
5.9	The Nitrogen Cycle	273
	5.9.1 Kinetic Analysis of Organic Nitrogen Decomposition	274
	5.9.2 Ammonia-Nitrogen	275
	5.9.3 Ammonia Ionization	276
5.10	Chlorination/Dechlorination	277
	5.10.1 Chlorine in Solution	278
	5.10.2 Oxidizing Power of Chlorine-Containing Chemicals	280
	5.10.3 Breakpoint Chlorination	281
	5.10.4 Dechlorination	283
	5.10.5 The Sulfurous Acid System	285
5.11	Petroleum Oil	288
5.12	Cooling Water Operations	291
5.13	Boiler Operations	291
	5.13.1 Dissolved Solids Removal from Boiler Feedwater	292
	5.13.2 Deaeration	292
	5.13.3 Boiler Blowdown	293
	5.13.4 Addition of Treatment Chemicals	293
	5.13.5 Boiler Wastewater Composition	294
	References	300

6. Water/Wastewater Hydraulics   303

- 6.1   Measurement of Effluent Flow   303
  - 6.1.1   The Weir   303
  - 6.1.2   Flumes   308
- 6.2   Flow in Rivers and Streams   313
  - 6.2.1   Pitot Tube   314
  - 6.2.2   Current Meter   314
  - 6.2.3   Tracer Techniques   315
  - 6.2.4   Calculation of Open-Channel Flow   318
- 6.3   Meeting Water Quality Limits   322
  - 6.3.1   Low-Flow Conditions   322
  - 6.3.2   Discharge Under No-Flow or Low-Flow Conditions   324
- 6.4   Groundwater Flow   324
- 6.5   Storm Water Calculations   326
  - 6.5.1   The Rational Method   326
  - 6.5.2   Coefficient   327
  - 6.5.3   Intensity   328
  - 6.5.4   Criteria for Storm Water Design   328
  - 6.5.5   Time of Concentration   329
- 6.6   Back to the Manning Equation   331
  - 6.6.1   Rectangular Channels   331
  - 6.6.2   Circular Channels/Pipe   332
  - References   337

7. Water/Wastewater Draining of Tanks   339

- 7.1   Introduction   339
- 7.2   Time to Drain Tanks   340
  - 7.2.1   Vertical Tank of Constant Cross Section   340
  - 7.2.2   (Vertical) Cone-Shaped Tank   343
  - 7.2.3   Spherical Tank   345
  - 7.2.4   Spheroidal or Ellipsoidal Tank   348
  - 7.2.5   Vertical Cylindrical Tank with Cone Bottom   349
  - 7.2.6   Vertical Cylindrical Tank with Hemispherical Bottom   350
  - 7.2.7   Vertical Cylindrical Tank with Ellipsoidal Bottom   351
  - 7.2.8   Vertical Cylindrical Tank with Dished Bottom   352
  - 7.2.9   Vertical Tank of Elliptical Cross Section with Ellipsoidal Bottom   354
  - 7.2.10   Horizontal Cylindrical Tank with Flat Ends   354
  - 7.2.11   Horizontal Cylindrical Tank with Hemispherical Ends   357
  - 7.2.12   Horizontal Cylindrical Tank with Elliptical Ends   359
  - 7.2.13   Horizontal Cylindrical Tank with Dished Ends   361
  - 7.2.14   Horizontal Tank with Elliptical Cross Section and Flat Ends   364
  - 7.2.15   Horizontal Tank of Elliptical Cross Section with Elliptical Ends   366
- 7.3   Trajectory of the Jet from a Leaking Tank   367
  - References   374

xii   Contents

8. Solid Waste   375

    8.1   Introduction   375
    8.2   Selected Waste Designations/Definitions   376
            8.2.1   Coal Wastes   376
            8.2.2   Construction/Demolition Waste/Debris   376
            8.2.3   Garbage   377
            8.2.4   Hazardous Waste   377
            8.2.5   Hazardous Household Waste   378
            8.2.6   Industrial (Solid) Waste   379
            8.2.7   Infectious/Medical Waste   379
            8.2.8   Liquid Waste   380
            8.2.9   Municipal Solid Waste (MSW)   380
            8.2.10   Nuclear/Radioactive Waste   381
            8.2.11   Refuse/Rubbish/Trash   381
            8.2.12   Residual Waste   382
            8.2.13   Solid Waste   383
            8.2.14   Special Waste   383
            8.2.15   Toxic Waste   384
            8.2.16   Universal Waste   385
            8.2.17   Waste Oil   385
    8.3   Waste Analysis   386
    8.4   Calculations for Solid/Hazardous Waste Permitting   386
            8.4.1   Conversion of Units   386
            8.4.2   More Unit Conversions   387
            8.4.3   Conversion of Volume to Weight   389
            8.4.4   Sampling of Solid Waste   390
    8.5   Waste Incineration   393
            8.5.1   Principal Organic Hazardous Constituents   393
            8.5.2   Significant Figures for DRE   398
            8.5.3   DRE Determination for POHC via a Surrogate Compound   399
            8.5.4   MSW Incineration with AP-42 Estimated Emissions   400
            8.5.5   Wrap-Up of MSW Problem   412
        References   412

9. Noise   415

    9.1   General   415
    9.2   Sound Versus Noise   415
            9.2.1   Sound Characteristics   416
    9.3   Sound Properties   417
            9.3.1   Loudness   417
            9.3.2   Pitch   417
            9.3.3   Timbre   418
    9.4   Frequency Spectrum   418
    9.5   The Octave   419
            9.5.1   Octave Definition—the Musical Scale   419
            9.5.2   Octave Band Analysis—Sound Meter   421
    9.6   Combining Decibels   422
            9.6.1   Decibel Addition   422

	9.6.2	Averaging Decibels 423
	9.6.3	Subtraction of Decibels 424
	9.6.4	"Shortcut" Graphical Procedure 425
9.7	Composite Sound Level 428
	9.7.1	Flat Weighting 428
	9.7.2	Weighting Networks 428
	9.7.3	A-Weighting—The dBA Scale 428
9.8	Speech Interference 430
9.9	A-Weighting Statistics 431
	9.9.1	Statistical Definitions 431
	9.9.2	Determination of L_{eq} 432
	9.9.3	Time Periods for L_{eq} 433
	9.9.4	Adjustments to L_{eq}—L_{dn} and CNEL 435
9.10	Noise Regulations 438
	9.10.1	General 438
	9.10.2	Two Approaches—Common Law Nuisance/ Qualitative Versus Quantitative 439
	9.10.3	Adjustments to Permissible Noise Levels 439
	9.10.4	Impulse Noises 440
	9.10.5	Pure Tones 440
	9.10.6	Night Versus Day 441
9.11	Industrial Noise 441
9.12	Sound Propagation from Point Source to Receptor 441
	9.12.1	Sound Power Versus Sound Pressure 441
	9.12.2	Sound Radiation Models 442
	9.12.3	Attenuation of Sound over Distance 443
9.13	Excess Attenuation of Noise over Distance 448
	9.13.1	Attenuation by Atmospheric Air 448
9.14	Highway Noise (A Line Source) 454
9.15	Noise Control 456
9.16	The Community Noise Survey 459
	9.16.1	Data Sheets 462
	9.16.2	Effect of Wind 462
	9.16.3	Precipitation 463
	9.16.4	Temperature 463
	9.16.5	Sound Meter Setting 463
	9.16.6	Other Phenomena Encountered during Noise Surveys 464
	References 466

10. Radioactive Decay   468

10.1	Definitions and Units 468
	10.1.1 Definitions 468
	10.1.2 Units of Radioactive Decay and Exposure 468
	10.1.3 Other Measures of Dosage 469
10.2	Some Sources of Radioactivity 469
10.3	Types of Radioactive Decay 470
	10.3.1 Alpha Particle Decay 470

xiv Contents

        10.3.2   Beta Particle Decay   471
        10.3.3   Gamma Radiation   471
        10.3.4   Other Types   472
10.4   Pathways of Radioactive Decay   472
        10.4.1   Radioactive Decay—Single Reaction   472
        10.4.2   Single-Reaction Rate of Decay   477
        10.4.3   Parallel Decay   481
        10.4.4   Independent, Unrelated Simultaneous Decay Reactions   483
10.5   Decay Series   486
        10.5.1   Maxima and Inflection Points   489
10.6   Longer Decay Chains   491
        10.6.1   Example—A Five-Member Chain   491
10.7   Rate of Decay in a Radioactive Series   492
        10.7.1   Secular Equilibrium   494
        10.7.2   Transient Equilibrium   494
        10.7.3   No Equilibrium   494
10.8   A Transition from Science to the Realm of Regulatory Control   495
10.9   Some Notable Accidents Involving Nuclear Materials   495
10.10   Governmental Regulations and Licensing Procedures   497
        10.10.1   Nuclear Regulatory Commission   497
        10.10.2   Regulations Governing Nuclear Material   497
        10.10.3   Waste Management Consists of Licensing   497
        10.10.4   Licensing   498
        10.10.5   Requirements   498
        10.10.6   General Licenses and Generally Licensed Devices   499
        10.10.7   Further Comments on General Licenses   499
        10.10.8   Periodic Testing for Contamination—Wipe Tests   501
10.11   Radioactive Waste Disposal   501
        10.11.1   NRC Limits   503
        10.11.2   Radioactive Waste Disposal—Conclusion   504
References   504

Appendix A. Suggested Undergraduate Environmental Curriculum   507

        Reference   510

Appendix B. Relationship among Expressions for Atmospheric Contaminants as Concentrations (ppm), Mass Flow Rates (lb/h), and Emission Factors (lb/MMBtu)   511

        B.1   Summary   511
        B.2   Concentration Limits   512
        B.3   Mass Flow Rate Limits   514
        B.4   Emission Factor   514
        B.5   Conclusion   515
        References   515

Appendix C. Burner NO_x from Ethylene Cracking Furnaces — 517

- C.1 General 517
- C.2 Summary 517
- C.3 Introduction 518
- C.4 Disclaimer 518
- C.5 Regulatory Considerations 519
- C.6 Technical Considerations 520
- C.7 NO_x Correlation for SMR Furnace Burners 520
- C.8 Extension of the Correlation to Ethylene Cracking Furnaces 522
  - C.8.1 Processes Are Different—A Review 522
  - C.8.2 But Combustion Is Similar 524
  - C.8.3 Firebox Temperatures Are Different 524
  - C.8.4 Adiabatic Flame Temperatures Are Both Different and Similar 525
- C.9 NO_x Correlating Equations for Ethylene Furnaces 526
- C.10 Influence of the Variables 527
  - C.10.1 Oxygen Dependence 527
  - C.10.2 Combustion Air Temperature 531
  - C.10.3 Fuel Temperature 533
  - C.10.4 Ambient Air Humidity 533
  - C.10.5 Hydrogen Content of Methane/Hydrogen Mixtures 534
  - C.10.6 Acetylene Content of Methane and Hydrogen Fuels 536
- C.11 Experimental Verification of NO_x Predictions 537
- C.12 Opportunities for Improvement 538
- C.13 Conclusions 540
- References 540

Appendix D. What Is BOD and How Is It Measured? — 543

- D.1 Summary 543
- D.2 Dissolved Oxygen (DO) and Its Measurement 543
  - D.2.1 Overview 543
  - D.2.2 DO Measurement 545
  - D.2.3 Reproducibility 547
  - D.2.4 Wrap-Up 547
- D.3 Biochemical Oxygen Demand (BOD) and Its Measurement 548
  - D.3.1 Theoretical Considerations 548
  - D.3.2 Mechanics of the BOD Test 551
  - D.3.3 BOD Experiment 552
  - D.3.4 Reproducibility 553
  - D.3.5 Wrap-Up 553
- D.4 Chemical Oxygen Demand (COD) and Its Measurement 554
  - D.4.1 The COD Test 554
  - D.4.2 Reproducibility 556
  - D.4.3 Wrap-Up 557
- References 557

xvi   Contents

Appendix E. Cooling Water Calculations — 559

- E.1 Summary   559
- E.2 Tower Parameters   559
- E.3 Water Parameters   561
  - E.3.1 Impurities in Cooling Water   561
  - E.3.2 pH, Alkalinity, and Hardness   562
  - E.3.3 Milligrams Per Liter as Calcium Carbonate   565
  - E.3.4 Electroneutrality   566
- E.4 Control of Cycles of Concentration   567
- E.5 pH Effects   568
  - E.5.1 The pH Increases with Cycles of Concentration   568
  - E.5.2 Prediction of Circulating Water pH   568
  - E.5.3 Dissolved CO_2 Is Not Exactly Constant   571
  - E.5.4 Quantity of Acid Required   572
- E.6 Total Dissolved Solids and Conductivity   573
  - E.6.1 TDS   573
  - E.6.2 Prediction of Conductivity   573
- E.7 Allowable Cycles of Concentration   574
  - E.7.1 Langelier Saturation Index   575
  - E.7.2 Ryznar Stability Index   576
  - E.7.3 Corrosion Ratio   577
  - E.7.4 Allowable pH Range   577
  - E.7.5 Calcium Salts   577
  - E.7.6 Calcium and Zinc Phosphate Saturation Indexes   578
  - E.7.7 Silica   579
  - E.7.8 Other Dissolved Materials   579
  - E.7.9 TDS and Conductivity   580
  - E.7.10 Suspended Solids   580
- E.8 Example Problem   581
- E.9 Computerized Calculations   586
- E.10 Case Studies   586
- References   588

Appendix F. Increase in Runoff from Industrial/Commercial/Urban Development: The Telltale Bridge — 591

- F.1 Summary   591
- F.2 Introduction   591
- F.3 The Case in Point   592
- F.4 The Bridge   593
- F.5 Storm Water Flow in the Swale   594
- F.6 Bottlenecks   595
  - F.6.1 At the Bridge   595
  - F.6.2 At the Culverts under Road No. 1   597
- F.7 Continued Flooding   597
- References   598

Appendix G. Water Quality Improvement for a Small River — 599

- G.1 Summary and Conclusions 599
- G.2 Introduction 600
- G.3 Flow of Whippany River 600
- G.4 Water Quality of River 601
  - G.4.1 Dissolved Oxygen 605
  - G.4.2 Biochemical Oxygen Demand 611
- G.5 Waste Treatment at the Whippany Paper Board Company 615
  - G.5.1 Nature of the Waste 615
  - G.5.2 Treatment Before 1964–1965 615
  - G.5.3 Effect of Early Treatment on Water Quality 616
  - G.5.4 Upgrading of Treatment Facilities 617
  - G.5.5 Effect of Improvements on Water Quality 619
  - G.5.6 Summary 621
- G.6 Modeling the Whippany River 621
  - G.6.1 Selection of Constants 622
  - G.6.2 Theoretical Calculation of Stream Velocity Impractical 622
  - G.6.3 Velocity—Time of Travel Measurements 624
  - G.6.4 Assembling the Model 626
  - G.6.5 Results and Discussion 627
- G.7 Postscript 628
- References 630
- Supplemental References for Postscript to Appendix G 631

Appendix H. Experimental Determination of Coefficient for Draining of Tank — 632

- H.1 Summary 632
- H.2 Description of Experiment—Equipment 632
  - H.2.1 Equipment Used 633
- H.3 Description of Experiment—Procedure 633
- H.4 Experimental Data 634
- H.5 Data Analysis—Falling Head/Unsteady-State Experiment 638
- H.6 Time to Drain the Tank 642
- H.7 Steady-State Experiment 643
  - H.7.1 Transient Solution to Fill and Drain Problem 644
- References 645

Appendix I. Noise Case Studies — 647

- I.1 Case 1—Sound Meter Readings behind a Highway Noise Barrier 647
  - I.1.1 Theory 647
  - I.1.2 Synopsis of the Measurements 647
  - I.1.3 Turnpike Noise Barriers 648

xviii   Contents

              I.1.4   Turnpike Noise Measurements   650
              I.1.5   Neighborhood Side of the Noise Barrier   650
              I.1.6   Noise Measurements behind the Barrier   653
    I.2   Case 2—Another Noise Barrier Study   654
              I.2.1   Highway Noise Data   655
              I.2.2   Summary of Results   656
    I.3   Case 3—Successful Noise Permitting Procedure   656
              I.3.1   The Project   658
              I.3.2   Method of Execution   658
              I.3.3   The Background Noise Survey   658
              I.3.4   Estimate of Noise from New Equipment   659
              I.3.5   Postconstruction Noise Survey Confirmed Compliance   659
    References   659

Appendix J. Air Pollution Aspects of the Fluid Catalytic Cracking Process   661

    J.1   Summary   661
    J.2   Introduction   661
    J.3   FCC Process Description   662
              J.3.1   FCC Cracking Catalyst   662
              J.3.2   The Reactor   664
              J.3.3   The Regenerator   667
    J.4   Atmospheric Contaminants from the Regenerator—Origin and Treatment   667
              J.4.1   Carbon Monoxide (CO)   668
              J.4.2   Particulate Matter (PM)   669
              J.4.3   Sulfur Oxides (SO_x)   671
              J.4.4   Oxides of Nitrogen (NO_x)   672
    J.5   Summary   676
    References   676

Appendix K. Case Studies in Air Emissions Control   681

    K.1   Summary   681
    K.2   Addition of Steam to Reduce Burner NO_x 681
    K.3   Addition of SCR to Reduce Burner Emissions   682
    K.4   Integration of a Furnace with Gas Turbine Exhaust   683
    References   684

Appendix L. Combustion of Refinery Fuel Gas   685

    L.1   Refinery Fuel Gas   685
    L.2   Combustion Calculations for Refinery Fuel Gas   686
    L.3   EPA Method 19 Combustion Calculations   686

Contents **xix**

    L.4 F Factors for Refinery Fuel Gas   687
                L.4.1   Calculated F Factors for Published Natural Gas Compositions   689
    L.5   Example—Use of F Factors in Combustion Calculations   691
                L.5.1   Stoichiometric Case   691
                L.5.2   3% Excess O_2 Case   694
    L.6   Conclusions   696
    References   697

Index   **699**

Preface

This preface has the advantage of hindsight, having been written after the completion of the book. Looking back, it is a summary of author's many years of experience in industry as a permit getter and permit-getter supervisor. The explanations, example problems, and solutions presented here are considered to represent the most typical situations encountered when performing calculations in support of environmental permit applications. Nevertheless, the material presented involves nothing of a proprietary nature.

Following a brief description of the nature of the environmental permitting process in Chapter 1 and the desired education, training, and experience for its practice (Appendix A), fundamental principles are reviewed in Chapter 2. The remaining subject matter includes the areas of air pollution control (Chapters 3 and 4; Appendices B, C, J, K, and L), water/wastewater (Chapters 5, 6, and 7; Appendices D, E, F, G, and H), solid/hazardous waste (Chapter 8), and noise generation, propagation, and control (Chapter 9 and Appendix I), with a little radioactive decay provided in Chapter 10. For the most part, the appendices explore case studies from the author's experience and offer a thorough explanation without burdening the main text with excessive details.

Criteria concerning which problems/subject matter to emphasize, to mention merely in passing, or to omit entirely are based solely on the author's judgment of what are the most meaningful topics to present in a volume of finite length. The length is however sufficient to give rise to yet another of Kunz's Maxims, "Never (again) write a book by yourself."

This does not imply that the book has been written in a vacuum. Literature citations are listed at the end of each chapter and appendix, and the essential contributions to the book have been acknowledged. The author apologizes to those individuals whose contributions may not have been duly noted. Thanks are also due to the numerous characters whom the author met along the way and whose amusing behavior provides the basis for the real-life anecdotes sprinkled throughout the text.

So, buckle up for an exciting ride. It is a real "page turner."

ROBERT G. KUNZ
Hillsborough, NC
2009

Acknowledgments

The author acknowledges Roberta Kunz Fox of fox^2 design and Suzanne Roth for typing the vast majority of the text, and Roberta Fox for enhancing the photographs, producing the more complicated figures, for the cover art, and for overall assistance in preparing the manuscript for publication. The author also acknowledges the unsung computer guru who retrieved the lost file of Chapter 3 from cyberspace never-never land when "the computer ate my homework."

All photographs are by the author, except for the photographs of the storm drain provided in Chapter 6. They were taken by Laurie Kunz, and the first of those photos was retouched by Andrew Fox of fox^2 design to remove unsightly weeds from the dry channel.

Chapter 1

Introduction

You have to start somewhere.

1.1 WHO, WHAT, WHY, AND HOW OF THIS BOOK

This work is an outgrowth of the author's many years in industry as a member of or supervising a group of engineers charged with securing, maintaining, and/or negotiating environmental quality permits issued by government agencies. (On the author's watch, no capital project was ever delayed on account of not having been issued an environmental permit on time.) Demonstrating compliance and, when necessary, troubleshooting processes and equipment to ensure such compliance go with the territory. To "fill out the form" as per the printed instructions, one would often have to fall back on fundamental principles or develop one's own methods to estimate emissions to the environment, in addition to or in spite of using a prescribed methodology that came with no explanation or perhaps made no sense at all. This book attempts to capture many of those calculations and procedures in one place to serve as a reference for interested parties.

If one is looking for a book on environmental regulations, perhaps translated from legalese into plain English, this is *not* the right book. Although some discussion of regulations is necessary to clarify the motivation for certain calculations, the presentation strives to avoid being too specific lest it become dated as regulations, subject to change, indeed do change. In contrast, the underlying scientific principles, and the calculations derived from them, are timeless.

The presentation presumes at least a general background in science. One can accept on faith the equations and calculation procedures presented and proceed from there. However, to understand the origin of some of the mathematical relationships to the fullest, knowledge of algebra, calculus, and differential equations is required. The author's suggestions for a course curriculum useful in preparing practitioners in the environmental control function are explored in Appendix A.

Environmental Calculations: A Multimedia Approach, by Robert G. Kunz
Copyright © 2009 John Wiley & Sons, Inc.

There may be some overlap between sections as common elements are perhaps repeated to make the example problems as independent as possible, so that the reader does not have to refer to the entire book while solving a particular problem. The book, however, contains a minimal amount of "handbook information." Sufficient chemical and physical property data, basic constants, and conversion factors are included to enable an understanding of the subject matter and solve the example problems. This document may be used along with other references as necessary when modifying the problems for one's own use and/or for other materials. It cites such sources, rather than incorporating vast amounts of supplemental material in the text. (Besides, where better to find handbook data than in a handbook?)

1.2 POTENTIAL USERS

One title that came to mind was *Dr. Bob's Handy Household Guide to Environmental Permitting Calculations*, but the grown-ups at Wiley prevailed; hence, the present title. The book is designed to help someone in industry to complete environmental permit applications. It is also intended for an Environmental Regulatory Permit Engineer in a government agency to evaluate the technical content of such an application and write a meaningful permit for a new project, or ascertain compliance/noncompliance with an existing permit. This compendium of calculation procedures could also be used as a primary or supplemental text in a college course in environmental stoichiometry or as a reference source to prepare for professional engineering or other professional licensing examinations. It may serve as a useful guide for government officials and mass media personnel when called upon to educate the public in the midst of an environmental disaster, such as a chemical fire, explosion, or release of hazardous/toxic materials. A recent local example comes to mind [1]. Since the book is oriented more toward technical issues than to specific environmental regulations, the basic principles discussed could have international appeal beyond the United States, especially in developing countries where environmental control efforts may be just getting started.

1.3 ARRANGEMENT

The book is arranged by environmental media, with several chapters on air, water and wastewater, solid waste, noise, and radioactive decay, following an overview of basic concepts. (The chapter on solid waste includes a discussion of potentially hazardous materials stored on site, which may result in the aforementioned environmental disaster.) The format is an explanation of a given topic (and possibly a related anecdote) followed by a worked example problem to illustrate the concept. Where appropriate, a general discussion of environmental laws and regulations driving the need for a permit is included. Each succeeding topic builds upon others that have preceded it. When a more extensive discussion is warranted, original, real-world case study material is presented in an appendix. Finally, one of Kunz's Maxims, truisms/snippets of philosophy accumulated over the years, appears after each chapter heading. For example, "If you are at all capable of something no one else wants to do, you've got the job." Enjoy!

1.4 TRUTHS AND MYTHS ABOUT ENVIRONMENTAL CONTROL

Environmental control is based upon what the author calls the *First Three Laws of the Environmental Material Balance*, to wit

- What enters either stays there or leaves.
- It's all got to go somewhere.
- That somewhere is usually the most inconvenient place.

The last statement is a corollary to Murphy's law. Since the universe does not come with an owner's manual and new discoveries are still being made, there may well be other laws of which the author is not yet aware.

In addition to the self-evident truths enumerated above, there is a series of accompanying environmental myths commonly held by people not directly involved in obtaining or issuing environmental permits. These are listed in Table 1.1. The

Table 1.1 List of Environmental Myths

Myth no. 1	Permits? Permits? We don't need no stinking permits!
Myth no. 2	Environmental regulations have to make sense.
Myth no. 3	These rules don't apply to us:
	Small plants don't require permits.
	The plant is out in the middle of nowhere.
	It's only water/steam.
	The agency will listen to reason.
	We'll get an exemption.
	We can always change the law.
	Political influence works wonders.
Myth no. 4	The contractor, vendor, project engineer, consultant, or someone else gets the permits, certainly not us.
Myth no. 5	Permit expenses are not in the budget.
Myth no. 6	Requirements are getting easier.
Myth no. 7	There is no need:
	No need to explain: environmental engineers (ours and theirs) are psychic.
	No need to be straightforward: slip it in (or leave it out); maybe they won't notice.
	No need to plan ahead: agency personnel work nights, weekends on our permit applications; the really dedicated ones give up their vacation time.
Myth no. 8	Plant managers never go to jail.
Myth no. 9	No problem! The plant's neighbors will gladly wear earplugs, gas masks, and dark glasses.
Myth no. 10	All environmentally concerned persons are "tree huggers."
Myth no. 11	If only our environmental people were more clever ...
Myth no. 12	The Operations Department will fix it in the field.
Myth no. 13	There is a magic universal permitting handbook.

tabulation is limited here to a baker's dozen (13) to avoid devoting the entire book to this one topic. It is hoped that this book will help to remove the existence of a universal permitting handbook from the list of myths.

1.5 ADEQUATE PREPARATION IS THE KEY

One last item of interest: It would seem that the role of a "permit getter" in industry is caught in the middle—representing the "big, bad, predatory, price-gouging, mean-spirited, polluting" corporation before a regulatory agency and then having to explain the rules of a seemingly intransigent agency to company management just interested in moving things along.

However, lest agency personnel be overlooked, their job is no picnic either. Many on the front line in the permitting process are overworked and underpaid. They too have deadlines to meet. At least initially, they know less about the applicant's manufacturing process than the applicant and must learn quickly, one process after another, to perform each new evaluation and write a meaningful permit. Routine permitting often takes several to many months; more complicated permitting can take considerably longer.

1.6 A LIST OF DO'S

Some advice about expediting the permitting process has been discussed previously [2], as summarized in Table 1.2. This involves the so-called *preapplication meeting*. In the author's experience, some steps may be omitted/modified, and the order of activities may vary on a case-by-case basis. The keys are to make it a

Table 1.2 Suggested Procedures for Obtaining Environmental Permits

Start early and devote adequate resources to develop a complete application.
Define the scope thoroughly (emissions/discharges, emission/discharge points, control equipment and monitoring systems, environmental effects/impacts (both onsite and offsite)) to avoid the application's being placed on the permit writer's "back burner." Perform emission and modeling calculations for routine and nonroutine operations. Follow agency's checklist, if available.
Determine permits required and develop a permit list for each and every agency having jurisdiction.
Review previously issued permits and background information, as necessary.
If agency rules allow, arrange for and attend a preapplication meeting with the agency, and possibly a site visit. Modify application, as necessary, as a result of agency input.
Obtain buy-in from Operations Department. Allow sufficient operational flexibility to meet requirements at all times without the need for (a) permit amendment(s).
Be proactive and provide required information before being asked, but do not submit an application prematurely.

downright pleasure for the agency permit engineer to work on one's permit application in preference to the many others that may be on his/her desk at the same time and to develop a positive track record of cooperation, which may help to dispel any sense of suspicion and replace that feeling with one of trust.

1.7 AND AT LEAST ONE DON'T

A final caution: Never, never, never obtain the agency's commitment to expedite the permitting process and then fail to provide the necessary information correctly, completely, and in a timely manner. This might not be the worst thing that one can do to shoot oneself in the foot, but it certainly ranks right up there with those that are.

1.8 AUTHOR'S SUPPLEMENTAL DISCLAIMER

No animals were harmed in the writing of this book; trees may be quite another matter. And now on with the show ...

REFERENCES

1. R. Hall, P.E., Supervisory Investigator, Case study: fire and community evacuation in Apex, North Carolina, Report No. 2007-01-NC, U.S. Chemical Safety and Hazard Investigation Board (USCSB), Washington, DC, April 16, 2008, 14 pp.
2. M. Peters and G. Scappatura, Expediting issuance of cracking furnace permits, Proceedings of the 6th Ethylene Producers' Conference, American Institute of Chemical Engineers (AIChE), New York, NY, 1994, pp. 491–497.

Chapter 2

Basic Concepts

When on the road, eat now; you don't know when your next meal is coming.

This chapter sets the stage for the specific permitting applications in the ensuing chapters. It introduces several fundamental principles upon which many of those calculations depend. For some readers, this will be merely a review of what they already know. In any event, the topics introduced here are followed up, as necessary, with the more advanced topics in the subsequent material. Each problem stands in its own and/or is cross-referenced to related issues.

2.1 BASIC CHEMISTRY

This section on basic chemistry is prompted by a meeting the author once had with a state environmental agency permit writer. The explanation of a chemical process in which 1 molecular weight (MW) (mole) of something plus two moles of something else results in two moles of a product.

$$A + 2B \rightarrow 2C$$

was met with the response, "Right, I understand . . . but what's a mole?" Readers who are familiar with small furry, blind, underground dwelling creatures or double-agent spies may wish to skip this section. All you others, hang on.

2.1.1 The Atom

Here on Earth, human beings continue to struggle to understand—and continue to try to control—many aspects of their environment. Experience is often gained by the costly process of trial and error. To be glib, those who ate the wrong nuts or berries were quickly weeded out of the gene pool.

Environmental Calculations: A Multimedia Approach, by Robert G. Kunz
Copyright © 2009 John Wiley & Sons, Inc.

In the beginning, knowledge of chemistry was inferred from circumstantial evidence of how the Earth's materials behave. The ancient Greeks postulated that all matter is composed of four basic elements: earth, air, fire, and water, which when combined in the correct proportions make up all of the "stuff" on earth [1, pp. 12–13].

Although the concept of an element has changed, today one has reason to believe that there are 109 such building blocks from which all the matter in the world is assembled. Each element is composed of only one type of so-called *atom*, which cannot be further divided and still retain the unique chemical properties of that element.

In 1807, an English schoolteacher, John Dalton, proposed an atomic theory based on the scientific evidence at the time [1, pp. 12–13]:

- Elements are made up of atoms (thought at the time not to be further divisible; see below).
- All atoms of a given element are identical (this turned out not to be totally true because of the later discovery of isotopes).
- Individual atoms combine with one another to form molecules of chemical compounds, each of which always contains the same types of atoms in the same ratios.
- The atoms participating in a chemical reaction are simply rearranged from one combination to another but are otherwise unchanged.

Dalton's postulates still hold true in essence despite the further discovery of subatomic particles, such as protons, neutrons, electrons, and others, and nuclear fission reactions, which were first reported in 1939 [1, pp. 12–13].

The identity of an element and its atomic number is determined by the number of positively charged protons in the nucleus of the atom. Those protons are balanced by an equal number of orbiting electrons, which have equal and opposite charge but negligible mass compared to the protons. A number of additional particles termed *neutrons* may also appear in the nucleus; the neutron is similar in mass to the proton but carries no charge.

The atomic weight of an element is determined by the total number of protons and neutrons. Atoms that have the same number of protons—thereby constituting the same element—but a different number of neutrons are known as *isotopes*.

2.1.2 The Periodic Table

Around 1868–1870, Dmitri Mendeleev in Russia and Lothar Meyer in Germany independently succeeded in rearranging the then known elements into a two-dimensional matrix called the Periodic Table [2, pp. 46–49]. In a modern version of the Periodic Table (Table 2.1), the elements are arranged by atomic number (number of protons, upper number inside each box). Approximate atomic weights are shown at the bottom of each box.

Table 2.1 Periodic Table of the Elements

Periodic Table of the Elements

Source: Reprinted by permission of John Wiley & Sons.

Elements in the vertical columns, known as *groups*, exhibit similar but gradually changing properties as one proceeds from top to bottom. For example, the column at the extreme left of the Periodic Table consists of hydrogen and the alkali metals—a highly reactive set of elements. Except for hydrogen, all of these elements are soft metals that are good conductors of heat and electricity.

In contrast, the elements occupying the extreme right-hand column of the Periodic Table are called the *inert, rare,* or *noble* gases. As such, these elements are relatively unreactive and rarely react or combine with other elements.

To their left lie the halogens, a highly reactive set of nonmetals that are poor conductors of heat and electricity. Each of the groups occupying other vertical columns in between these extremes shows a progressive change in metallic character and exhibits its own unique set of properties.

Meanwhile, each horizontal row in the Periodic Table is called a *period*. The first period contains only hydrogen and helium. Within a given group (vertical column), the size of the atom increases from the top period to the bottom, and with this changing atom size come changes in other properties that are related to atomic size. For example, the halogens proceed from gas to liquid to solid as one moves down the table.

In general, size increases from right to left in a given period [1, p. 24]. Ionization energy, electron affinity, and electronegativity increase from the lower left-hand corner of the table to the upper right; atomic and ionic radii increase from upper to lower left [3, p. 7]. Atomic or ionic radii differ considerably according to the environment of neighboring atoms and ions and the type of chemical bond involved [1, p. 24].

When the atomic theory of matter was first postulated, the existence of the atom had to be taken on faith because it was impossible to see an individual atom. However, today it is no longer just a theory since individual atoms and molecules can now be analyzed and even manipulated, thanks to the latest scientific advances. Fundamentals are contained in Ref. [4, pp. 66–68] from which this material was drawn.

Since the mass of protons, neutrons, and electrons is so small, the atomic weight is stated on a relative basis, which reflects the proportion of atoms of elements combining with other atoms to form *molecules* of chemical *compounds*. When the weight of a compound in grams is equal to the molecular weight, the amount of material present is said to be a *gram mole*, or a *gram molecular weight* (*GMW*). A *lb mole* is similarly defined when the weight of material in pounds is equal to the molecular weight.

2.1.3 Valence and Balanced Chemical Equations

Atoms of elements want to have a stable number of electrons and will gain, lose, or share electrons with other elements when forming the molecules of compounds. This concept is captured in the idea of valence, the degree of combining power of an atom or group of atoms with other atoms or groups. Items of positive (+) valence combine with items of negative (−) valence. Valence is related to the column the element

10 Chapter 2 Basic Concepts

occupies in the Periodic Table. It is the essence of preparing the correct formulas from which a balanced equation is written.

As an example, consider the reaction of hydrochloric acid (HCl) on limestone ($CaCO_3$) to liberate carbon dioxide (CO_2) gas. The coefficients of the equation below represent the number of moles of each species taking part in the reaction.

$$2HCl + CaCO_3 \rightarrow CaCl_2 + H_2O + CO_2 \uparrow$$

In this case, two moles of HCl react with one mole of $CaCO_3$ to create one mole each of the products. Carbon dioxide escapes to the atmosphere from a water solution of the salt, calcium chloride ($CaCl_2$).

Hydrogen is said to have a valence of $+1$, calcium $+2$, chloride -1, oxygen -2, and the carbonate radical (CO_3^{2-}) -2. This makes the valence of carbon here $+4$. The charges, $+$ and $-$, are balanced within each molecule, and the total number of each atom is the same on both sides of the equation.

Several other examples follow, starting with balanced equations for different types of chemical reactions. The objective is to calculate the correct weights from the number of moles participating in the reactions.

PROBLEM 2.1 *Acid–Base Reaction*

How much sodium hydroxide (NaOH) is required to neutralize 100 lb of sulfuric acid (H_2SO_4)?

SOLUTION The balanced equation is written as

$$\underset{(2)(40.00)}{\overset{x}{2NaOH}} + \underset{(1)(98.08)}{\overset{100}{H_2SO_4}} \rightarrow \underset{(1)(142.04)}{\overset{y}{Na_2SO_4}} + \underset{(2)(18.0153)}{\overset{z}{2H_2O}}$$

Hydrogen (H) and sodium (Na) have a valence of $+1$ each, the hydroxyl (OH^-) radical a valence of -1, and the sulfate radical (SO_4^{2-}) a valence of -2.

(a) The solution assumes that 100 lb of H_2SO_4 is to be neutralized by x lb of pure NaOH. To calculate the amount of NaOH solution to be used on a less than 100% H_2SO_4 solution would require further steps [part (b)]. The molecular weight of each species multiplied by its coefficient in the equation is written below that species in the equation. The weight of each species is shown above that component. Molecular weights added up from individual atomic weights come from a standard reference [5, pp. B-85 to B-178].

For the number of moles to balance

$$\frac{x}{(2)(40.00)} = \frac{100}{(1)(98.08)} = \frac{y}{(1)(142.04)} = \frac{z}{(2)(18.0153)}$$

from which

$$x = (100)(2)(40)/(98.08) = 81.57 \text{ lb NaOH required}$$
$$y = (100)(142.04)/(98.08) = 144.82 \text{ Na}_2\text{SO}_4 \text{ formed, and}$$
$$z = (100)(2)(18.0153)/(98.08) = 36.74 \text{ lb H}_2\text{O formed}$$

	lb of Reactants		lb of Products
H_2SO_4	100.00	Na_2SO_4	144.82
NaOH	81.57	H_2O	36.74
Total	181.57	Total	181.56

For a reaction, such as this, going to completion, the reaction products must equal the reactants. The slight difference calculated here is caused by numerical roundoff.

(b) Sixty-six degree Baumé (Bé) (93.19 wt%) sulfuric acid, the industrial strength most generally used [6, p. 284], has a freezing point of $-29°F$ [7, p. 10–114], requiring insulation/heating only in the coldest outdoor areas. Its solution density is 114.47 lb/ft^3. A 20 wt% NaOH (26.1 Bé) solution has a similar freezing point [8, p. 849]; this solution contains 15.22 lb NaOH per cubic foot, or 2.035 lb NaOH per U.S. gal [7, p. 10–106].

For the sulfuric acid solution, 1 ft^3 corresponds to 114.47 lb of solution, or $(0.9319)(114.47) = 106.67$ lb H_2SO_4 and $(1 - 0.9319)(114.47) = 7.80$ lb H_2O. Therefore, 100 lb of H_2SO_4 translates to $(1)(100)/106.67 = 0.9375$ ft^3.

$$0.9375 \text{ ft}^3 \times 7.48 \text{ gal/ft}^3 = 7.01 \text{ gal}$$

For the NaOH, 81.57 lb NaOH calculated above represents $81.57/15.22 = 5.3594$ ft^3 or

$$(81.57)/(2.035) = 40.08 \text{ gal}$$

About 40 gal of 20% NaOH are required to neutralize some 7 gal of 93.19 wt% H_2SO_4. Water balance

36.74 lb formed in reaction

7.80 lb from sulfuric acid solution

$(80/20)(81.57) = 326.28$ lb from sodium hydroxide solution

370.82 lb water total

144.82 lb from sodium sulfate

515.64 lb sodium sulfate solution

$$(144.82)/(515.64) = 0.2809 \times 100 = 28.09 \text{ wt\% } Na_2SO_4 \text{ solution} \rightarrow \text{call } 28\%$$

Unlike weights, volumes are not additive. Sodium sulfate (Na_2SO_4) has a solubility of up to 48.8% at 40°C [5, p. 4–123]. Specific gravity (20°C/4°C) is not listed beyond 24 wt% (1.2336) [7, p. 10–139]. By approximate rule of thumb, specific gravity of an inorganic salt solution is $[1 + (\text{wt\%}/100)] = 1 + (28/100) \rightarrow 1.28$ (see Section 2.3, Equation (2.21) and Problem 2.10).

$$\text{Specific gravity} \times \text{water density} = \text{solution density}$$
$$1.28 \times 8.34 \text{ lb/gal} \cong 10.68 \text{ lb/gal}$$

$(515.64)/(10.68) = 48.28$ gal of solution, compared to a reagent volume of $40 + 7 = 47$ gal

Volume of pure water would be $(370.82)/(8.34) = 44.5$ gal. (This is also *not* the way to calculate the volume of Na_2SO_4 solution for this problem). ∎

PROBLEM 2.2 Oxidation–Reduction Reaction

(One in which the valences of certain participants change through gain or loss of electrons)

Five hundred milligrams of potassium dichromate is reacted with 1000 mg of ferrous ($+2$) sulfate. How much ferric ($+3$) sulfate is formed?

SOLUTION The balanced equation is as follows:

$$\underset{(6)(151.91)}{\overset{1000}{6FeSO_4}} + \underset{(1)(294.19)}{\overset{500}{K_2Cr_2O_7}} + 7H_2SO_4 \rightarrow \underset{(3)(399.27)}{\overset{x}{3Fe_2(SO_4)_3}} + Cr_2(SO_4)_3 + K_2SO_4 + 7H_2O$$

$$Cr_2^{12+} O_7^{14-} = Cr_2O_7^{2-}$$

Each Cr^{6+} gains 3 electrons (e^-) to go to Cr^{3+}

$$2Cr^{6+} + 6e^- \rightarrow 2Cr^{3+}$$

Reduction is gain of electrons, and in so doing Cr^{6+} is the oxidizing agent. Each Fe^{2+} ion loses an electron to go to Fe^{3+}.

$$6Fe^{2+} - 6e^- \rightarrow 6Fe^{3+}$$

Oxidation is loss of electrons, and Fe^{2+} is the reducing agent.
Ferric sulfate formation based on ferrous sulfate:

$$\frac{x}{(3)(399.27)} = \frac{1000}{(6)(151.91)}$$

$$x = (1000)(3)(399.27)/[(6)(151.91)] = 1314.17\,g$$

Ferric sulfate formation based on potassium dichromate:

$$\frac{x}{(3)(399.27)} = \frac{500}{(1)(294.19)}$$

$$x = (500)(3)(399.27)/[(1)(294.19)] = 2035.78$$

Only the lesser amount, 1314.17 g of $Fe_2(SO_4)_3$, is possible. The amount of $FeSO_4$ present does not match the overabundance of $K_2Cr_2O_7$. The $K_2Cr_2O_7$ is in excess, and the $FeSO_4$ here is the limiting reagent.

Potassium dichromate consumed by the available $FeSO_4$:

$$\frac{y}{(1)(294.19)} = \frac{1000}{(6)(151.91)}$$

$$y = (1000)(1)(294.19)/[(6)(151.91)] = 322.77\,g\,of\,K_2Cr_2O_7\,consumed$$

$$500 - 322.77 = 177.23\,g\,of\,K_2Cr_2O_7\,in\,excess$$

■

PROBLEM 2.3 Combustion Reaction

(Another form of oxidation–reduction reaction)

How much CO_2 and H_2O result from the combustion of 100 lb of propane (C_3H_8) with atmospheric oxygen? Molecular weight of propane is 44.11.

SOLUTION The applicable balanced equation is

$$\underset{44.11}{\overset{100}{C_3H_8}} + \underset{(5)(31.9988\cong 32.00)}{\overset{x}{5O_2}} \rightarrow \underset{(3)(44.01)}{\overset{y}{3CO_2}} + \underset{(4)(18.0153)}{\overset{z}{4H_2O}}$$

$x = (5)(32)(100)/44.11 = 362.73$ lb oxygen consumed

$y = (3)(44.01)(100)/44.11 = 299.32$ lb CO_2 formed

$z = (4)(18.0153)(100)/44.11 = 163.37$ lb H_2O formed

Total reactants $= 100 + 362.73 = 462.73$

Total products $= 299.32 + 163.37 = 462.69$

Again, the difference between total reactants and total products is roundoff. ∎

In combustion with the atmospheric air, the products of combustion (known as *flue gas* (FG)) include the inert components of air such as nitrogen, argon, atmospheric CO_2, and moisture (H_2O) carried through the combustion device. Moreover, an excess of atmospheric air and the oxygen therein is used to minimize the formation of carbon monoxide (CO) and other products of incomplete combustion. Additional contaminants/pollutants can also be produced. These matters will be discussed in Chapter 3.

2.1.4 Summary of Basic Chemistry

It is impossible to compress the entirety of general chemistry within a few pages here. The reader is advised to refer to any good chemistry text to fill in the blanks as needed.

Another important branch of chemistry is concerned with compounds of carbon, their reactions, and their relationships with the cell structure of living beings. Propane in the example above is one simple example of an organic chemical, namely, a hydrocarbon. This subject also deserves a more complete treatment than can be provided here. An excellent summary of organic chemistry from an environmental perspective is contained in 70 pages or less in Ref. [9, pp. 86–150] and in the subsequent edition [10, pp. 94–163].

2.2 THE IDEAL GAS

The kinetic theory of gases postulates that gases are made up of a multitude of tiny particles called molecules, all continuously in random motion. Perfectly elastic collisions between the individual "billiard-ball" molecules and between the molecules and the walls of the containment vessel account for the observed pressure. Increased molecular motion, and hence pressure, accompanies an increase in temperature.

The ideal gas model further assumes that the molecules possess a negligible volume and are not subject to intermolecular forces of attraction and repulsion. These assumptions are best approximated under dilute conditions of low pressure and high temperature, where the molecules constitute a small fraction of the overall gas volume and spend most of their time far apart. Combustion air and furnace flue gas can be described to a high degree of accuracy by the ideal gas law.

Table 2.2 Relationships among Temperature Scales

°K = °C + 273.15
°R = °F + 459.67
°R = 1.8°K
°F = 1.8°C + 32
°C = (°F − 32)/(1.8), or the familiar
°C = (5/9)(°F − 32)
Alternatively, by the "Rule of 40"
　°F = (1.8)(°C + 40) − 40
　°C = (°F + 40)/(1.8) − 40

An ideal, or perfect, gas obeys the equation of state

$$PV = nRT \tag{2.1}$$

where P and V are the pressure and volume of the gas, n is the amount of gas present, expressed in moles,[1] T is its absolute temperature [Kelvin (K) or Rankine (R)] (Table 2.2), and R is the universal gas constant, applicable to all such gases. The numerical value of R changes only to be consistent with the units of other factors in the equation (Table 2.3).

This ideal gas law relationship is a combination of Boyle's law

$$PV = \text{a constant at constant temperature} \tag{2.2}$$

describing pressure and volume as inversely proportional for a given mass of gas, and Charles' law

$$V/T = \text{constant at constant pressure} \tag{2.3}$$

showing volume and absolute thermodynamic temperature to be directly proportional, again for a given mass of gas.

For volumetric flow calculations, the volume of a given mass of gas is needed. Since gas volume depends on temperature and pressure, the volume of an ideal gas is often expressed as the volume that it would occupy at some "standard" condition of temperature and pressure. The adjustment from the actual condition (a) to the standard condition (s) is as follows:

$$V_s = V_a(P_a/P_s)(T_s/T_a) \tag{2.4}$$

This equation is derived by applying Equation (2.1) to each condition, dividing through, canceling the constants n and R, and rearranging.

---

[1] One gram molecular weight of gas contains approximately 6.022×10^{23} molecules. Count Amedeo Avogadro (1776–1856) was a professor of Physics at the University of Turin, who in 1811 first postulated what is known as *Avogadro's law* [11, Vol. 3, p. 62]. This states that the number of molecules present in a GMW of any substance is the same.

2.2 The Ideal Gas

Table 2.3 Some Values of the Ideal Gas Law Constant R

0.08205 (L)(atm)/(g mol)(°K)
0.7302 (ft^3)(atm)/(lb mol)(°R)
8.315 (m^3)(kP$_a$)/(kg mol)(°K)
10.731 (ft^3)(psia)(lb mol)(°R)
82.06 (cm^3)(atm)/(g mol)(°K)
18,540 (in.^{3})(psia)/(lb mol)(°R)
62,360 (cm^3)(mmHg)/(g mol)(°K)
1.986 (Btu)/(lb mol)(°R)
1.986 (cal)/(g mol)(°K)
8.314 (J)/(g mol)(°K)

See Ref. 5, p. F-241 and elsewhere for a more complete listing of R, for various combinations of English and metric units of temperature, pressure, and volume.

There are multiple conventions for standard conditions, and those conditions should be defined whenever the term "standard" is used. Standard conditions in common use employ 1 atm as the pressure standard, but the standard temperature varies.

The molar volumes in Table 2.4 are calculated by rearranging Equation (2.1) for the ideal gas

$$V/n = (RT)/P \qquad (2.5)$$

and substituting the values of the standard temperature as listed and the standard pressure of 1 atm, along with the gas constant $R = 0.7302$ (ft^3 atm/lb mol R) from Table 2.3 and the definition of absolute temperature (R) from Table 2.2.

2.2.1 Dalton's Law

The ideal gas law applies to both a single ideal gas and to mixtures of ideal gases. By Dalton's law, each component exerts its own pure component pressure, while occupying the total volume (V_T) of the mixture. The sum of these partial pressures

Table 2.4 Ideal Gas Molar Volume at Various Standard Temperatures

Standard temperature (°F)[a]	Molar volume (V/n) (ft^3/lb mol)	Where used
32°F = 0°C	359.0	Scientific applications
60°F	379.5	Petroleum industry heating values
68°F = 20°C	385.3	EPA procedures, some other environmental agencies
70°F	386.8	HVAC,[b] some environmental agencies

[a] With standard pressure = 1 atm.
[b] Heating, ventilating, and air conditioning.

equals the total pressure (P_T), and the sum of the moles of gas equals the total moles (n_T).

$$P_1 V_T = n_1 RT, \text{ etc.} \tag{2.6}$$
$$P_1 + P_2 + \cdots P_N = P_T \tag{2.7}$$
$$P_T V_T = n_T RT \tag{2.8}$$

Therefore, $P_N/P_T = n_N/n_T$ with P_N and n_N the partial pressure and number of moles, respectively, of the Nth species.

2.2.2 Amagat's Law

By Amagat's law, each component is considered to occupy its proportional share of the volume of the mixture at total pressure. The sum of these partial volumes equals the total volume.

$$P_T V_1 = n_1 RT, \text{ etc.} \tag{2.9}$$
$$V_1 + V_2 + \cdots V_N = V_T \tag{2.10}$$
$$P_T V_T = n_T RT \tag{2.8}$$

Therefore,

$$V_N/V_T = n_N/n_T \tag{2.11}$$

where V_N is the partial volume of the Nth species.

Thus, in an ideal gas,

$$P_N/P_T = V_N/V_T = n_N/n_T \tag{2.12}$$

and a pressure ratio, volume ratio, and molar ratio are synonymous. The ratio n_N/n_T is denoted as y_N (mole fraction (m.f.) of component N in the vapor phase).

Additional information on the ideal gas equation of state can be found in basic textbooks in chemistry, physics, and engineering.

2.2.3 Example Problems for the Ideal Gas Law

We now turn to a series of example problems dealing with the ideal/perfect gas.

PROBLEM 2.4

Verify the earlier statement that combustion air and combustion flue gas both conform to the equation of state for the ideal/perfect gas.

SOLUTION This requires a little research in a standard handbook [12, pp. 2–140 to 2–149] to back up the statement above. At conditions widely removed from liquefaction, that is, relatively high temperatures and low pressures, the equation-of-state relationship approaches the shape of an equilateral hyperbola on a pressure–volume (PV) diagram [13, p. 195].

Table 2.5 Compressibility Factor $Z = PV/RT$ for 1 mol of Gas at 1 atm Pressure[a]

T (°K)	T (°F)	Air	N_2	O_2	Ar	CO_2	CO	H_2O
300	80.3	0.9999	0.9998	0.9994	0.9995	0.9950	0.9997	—
450	350.33	1.0003	1.0003	1.0002	1.00015	0.9988	1.0003	0.993
600	620.33	1.0004	1.0004	1.0003	1.0003	0.9997	1.0005	0.998
1000	1340.33	1.0004	1.0003	1.0003	1.0002	1.0003	1.0004	1.000

[a] Excerpted and interpolated (where necessary) from Compressibility Factor Tables, *Perry's Chemical Engineers' Handbook*, 7th edition, pp. 2–140 to 2–149 [12]. All tables used were stated in terms of temperature in K and pressure in bars (1.01325 bars = 14.696 psia = 1 atm; 1 bar = 14.5038 psia), except for carbon monoxide (CO) that is stated in °C and bars and CO that is stated in °F and atmospheres. The two tables for water, one in °F and psia and one in K and bars, are not completely consistent; the one using K and bars was used.

Table 2.5 shows the compressibility factor Z, the ratio of PV to nRT, for air, its principal constituents, and the primary components of flue gas from combustion of hydrocarbon fuel. At atmospheric pressure and typical temperatures for combustion air and for flue gas, the figures in the table clearly show that these gases behave as ideal, that is, PV/nRT is extremely close to 1.0.

The poorest agreement is for water ($Z \geq 0.99$) at combustion temperatures, but even so, water vapor is only a minority constituent (~10–25%) of flue gas. Combustion air normally contains on the order of only 1% moisture or less.

The compressibility factor for CO makes little difference for a properly operating combustion device, where CO exists at parts per million (ppm) concentrations. ∎

PROBLEM 2.5

Calculate the number of gram moles of carbon dioxide occupying 0.759 mL at 0°C and 760 mmHg.

SOLUTION From the ideal gas law,

$$n = PV/RT$$

Here use $R = 0.08205$ (L)(atm)/(g mol)(°K).

1 mL = L/1000, 760 mmHg = 1 atm, and 0°C = 0 + 273.15 = 273.15°K

Substituting

$$n = (1)(0.759/1000)/[(0.08205)(273.15)]$$
$$n = 3.39 \times 10^{-5} \text{ g mol}$$

∎

PROBLEM 2.6

Calculate the volume in cubic feet occupied by 28.01 lb of nitrogen at atmospheric pressure and 68°F, based on the ideal gas law.

SOLUTION

$$V = nRT/P \qquad (2.13)$$

where n = number of moles = weight/molecular weight, T = absolute temperature (°K or °R), p = absolute pressure, R = universal gas constant (Table 2.3) in appropriate units, $n = 28.01$ (weight in pounds)/28.01 (molecular weight of nitrogen) = 1 lb mol, $T = 68°F + 459.67 = 527.67°R$, and $P = (1 \text{ atm})(14.696 \text{ lb}_{force}/\text{in.}^2 \text{ per atm}) = 14.696$ psia (pounds per square inch absolute).

$$V(\text{ft}^3) = 1(\text{lb mol})(10.731 \text{ psia ft}^3/\text{lb mol R})(527.67°\text{R})/14.696 \text{ psia}$$
$$V(\text{ft}^3) = 385.3 \text{ ft}^3$$

This is the molar volume (at 68°F) used by the U.S. EPA in 40 CFR 60 Appendix A, Method 19 [14]. This and other commonly employed standard molar volumes are listed in Table 2.4. ∎

PROBLEM 2.7

(a) Convert the volumetric flow of 100,000 actual cubic feet per minute (ACFM) of flue gas at 350°F and atmospheric pressure to the standard condition of 60°F and 1 atm. A similar problem, containing an actual flue gas pressure other than standard atmospheric, is treated in the next chapter (Chapter 3).

(b) Express this flow in lb moles/h.

SOLUTION

(a) By writing the ideal gas law twice and canceling terms as appropriate, this calculation can be handled quickly and without even having to know the gas constant. The subscript 2 denotes the standard condition; the subscript 1 denotes actual flue gas conditions

$$\frac{V_2}{V_1} = \frac{n_2 R T_2/P_2}{n_1 R T_1/P_1} \text{ (see Equation (2.4))}$$

from which

$$V_2 = V_1(T_2/T_1)$$

Factors cancel (same mass of gas, same R, same pressure)

$$V_2 = 100,000(60 + 459.67°\text{R})/(350 + 459.67°\text{R})$$
$$= 100,000(519.67)/(809.67)$$
$$= 64,200 \text{ standard cubic feet per minute(SCFM) (at 60 °F, 1 atm)}$$

(b) SCFM × (60 min/h)/(379.5 SCF/lb mol) = 10,150 lb mol/h

This is another basis for a standard condition (different from EPA's 68°F, 1 atm), and it is important to specify it clearly. This basis is commonly used in industry (Table 2.4), especially in conjunction with fuel heating values (Btu/lb mol) in the ideal gas state at 60°F. ∎

PROBLEM 2.8

Calculate the density in lb/ft³ of nitrogen gas at a laboratory temperature of 25°C and atmospheric pressure, assuming that the ideal gas law applies (refer again to Table 2.5 for a Z value $\cong 1.0$.)

SOLUTION

$$V = nRT/P \qquad (2.13)$$
$$\rho(\text{density}) = m/V = nM/V \qquad (2.14)$$

where m is the mass of gas and M is its molecular weight.

By combining the two equations above,

$$\rho = PM/(RT) \qquad (2.15)$$
$$\rho = (14.696)(28.01)/[(10.731)(536.67)]$$
$$\rho = 0.07148 \text{ lb}_{\text{mass}}/\text{ft}^3$$

in which $T\,(°F) = (25+40)(1.8)-40 = 77°F$ and $T\,(°R) = (77+459.67) = 536.67°F$

∎

2.3 CONCENTRATIONS AND MASS FLOW RATES

2.3.1 Concentrations in Mass Per Unit Volume

Concentration (the amount of "stuff" per unit of carrier fluid) is expressed differently for liquids and gases. A common unit for liquids is the milligram of constituent per liter of solution (mg/L). For dilute concentrations in water, this is equivalent to parts per million by weight (ppmw).

$$\frac{\text{mg of constituent}}{\text{liter of solution}} \frac{\text{liter of solution}}{1000\,\text{gm of solution}} \frac{\text{gm of constituent}}{1000\,\text{mg of constituent}} \frac{10^6\,\text{grams}}{\text{million grams}} = \text{ppmw}$$
$$(2.16)$$

The equivalence is not exact when the concentration of the nonwater constituent is high enough to change the density of the solution from 1000 g/L or 1 g/mL. A good rule of thumb one may choose to use when the experimental density is not readily available is $[1 + (\text{wt}\%/100)]$. The weight percent (wt%) is, of course, 100 times the weight or mass (m) of a constituent divided by the total mass of the N constituents, including the carrier fluid/solvent.

$$\text{wt}\% = 100\left(m_i / \sum_{i=1}^{N} m_i\right) \qquad (2.17)$$

Another measure of concentration often used in wastewater analysis is grains (gr) per gallon [6, p. 415].

$$\frac{1 \text{ gr}}{\text{U.S.gal}} \frac{\text{lb}}{7000 \text{ gr}} \frac{453.59 \text{ g}}{\text{lb}} \frac{1000 \text{ mg}}{\text{g}} \frac{\text{U.S.gal}}{231 \text{ in.}^3} \frac{\text{in.}^3}{2.543 \text{ cm}^3} \frac{1000 \text{ cm}^3}{\text{L}} = 17.1 \text{ mg/L}$$

(2.18)

Since the Imperial gallon contains 4.8 quarts versus the 4 quarts in the U.S. gallon,

$$\text{Grain/Imperial gallon} = 14.3 \text{ mg/L}$$

A typical situation that arises is the requirement to report the mass flow rate of a contaminant in lb/day when the contaminant concentration is given in mg/L and the wastewater flow is in gallons per minute (gpm).

$$\text{lb/day} = (\text{mg/L})(\text{gal/min})(\text{gm}/1000 \text{ mg})(\text{lb}/453.59 \text{ gm})(3.785 \text{ L/gal})$$
$$\times (1440 \text{ min/day})$$
$$\text{lb/day} = 1.202 \times 10^{-2} \, (\text{mg/L})(\text{gal/min}) \tag{2.19}$$

Conversion factors for other concentration and flow situations can be derived in like manner. ∎

PROBLEM 2.9

Calculate lb/day of pollutant in a 150-gpm wastewater effluent containing 60 mg/L of that pollutant.

SOLUTION From Equation (2.19),

$$(1.202 \times 10^{-2})(60)(150) = 108 \text{ lb/day}$$

In gases, the concentration is expressed as volume percent (vol%), which we have seen in Section 2.2.2 is equivalent to mole percent (mol%) for an ideal gas. Smaller concentrations are expressed as parts per million by volume (ppmv). Parts per million by volume is further differentiated between ppm by volume on a wet (w) basis (which includes the moisture in the gas) and ppm by volume on a dry (d), or moisture-free, basis. This pertains to a combustion effluent (flue gas), among others.

Wet-to-dry basis : $\quad \text{ppmvd} = \text{ppmvw}[100/(100-\text{vol}\%\text{H}_2\text{O})]$ (2.20a)

Dry-to-wet basis : $\quad \text{ppmvw} = \text{ppmvd}[(100-\text{vol}\% \text{ H}_2\text{O})/100]$ (2.20b)

The concentration of particulate matter in a gas is often expressed as grains per dry standard cubic foot. There are 7000 grains per pound. (This unit is based on the average weight of a grain of wheat).

Removal of particulates from the gas by an aqueous medium in a wet scrubber would result in a different measurement of concentration. In a liquid, the concentration of particulate matter, also known as suspended solids, is given in wt%, as is the concentration of salts (dissolved solids). Substantial concentrations of dissolved solids change the density (ρ) of an aqueous solution. As mentioned in Sections 2.1.3 and 2.3.1, an approximate rule of thumb one may choose to employ when an experimental density is not available is

$$\rho \text{ solution} \cong \rho \text{ water}[1 + (\text{wt\% dissolved solids}/100)] \quad (2.21)$$

However, reliable experimental density data should always be used, whenever available.

With flue gas, the concentration (usually on a dry basis) is standardized from what it actually is to what it would be at a specific value of oxygen concentration present in excess of stoichiometric combustion requirements. This oxygen concentration is also usually on a dry basis. This is to prevent an operator of a combustion device from achieving compliance with a ppm limit simply by dilution. Typical standards are 3% O_2 by volume on a dry basis (3% O_2, dry) for boilers, heaters, and furnaces; 15% O_2 (dry) for gas turbines; and 7% O_2 (dry) for incinerators. The author has also seen 0% O_2 in use in standards.

$$\text{ppmvd } X \text{ @ 3\% } O_2 \text{ (dry)} = [\text{ppmvd @ another } O_2 \text{ (dry)}][(20.95-3)/(20.95-O_2)] \quad (2.22)$$

By substituting 15, 7, and 0 in the place of 3 in the equation above, similar equations can be written for 15% O_2 (dry), 7% O_2 (dry), and 0% O_2. ∎

PROBLEM 2.10

Show that a dissolved solids concentration of 1 wt% is nearly equivalent to a concentration of 10,000 mg/L.

SOLUTION

$$\frac{10,000 \text{ mg}}{L} \frac{\text{gm}}{1000 \text{ mg}} \frac{\text{lb solids}}{453.59 \text{ gm}} \frac{3.785 \text{ L}}{\text{gal}} \frac{\text{gal}}{8.34 \text{ lb water}} \times 100 = 1.00 \text{ wt\%}$$

when the density of the water solution is 1 g/cc, 62.4 lb/ft^3, or 8.34 lb/gallon.

However, for a 1 wt% solution, the density can be estimated to be approximately $[1 + (\text{wt\%}/100)] \times$ the density of water, from Equation (2.21). Therefore, 10,000 mg/L $\cong$ (1.00 wt%)/(1.01) $\cong$ 0.99 wt%.

For example, density at 20°C of several common salt solutions at 1 wt% are calculated below in units of g/cm^3 from data contained in the cited handbook [7, pp. 10–128 to 10–141].

CaSO$_4$	1.009	NaCl	1.0053
FeCl$_3$	1.009	NaNO$_3$	1.0049
KCl	1.0046	Na$_2$S	1.0098
Na$_2$CO$_3$	1.0086	Na$_2$SO$_4$	1.0073

Densities of other 1 wt% salt solutions from the handbook table are in general somewhat lower. ∎

2.3.2 Emission Factors in Mass Per Unit of Fuel Firing Rate

Another measure of concentration is in the form of the so-called *emission factor* in units of pounds per million Btu fired (lb/MMBtu) or nanograms/Joule (ng/J). The amount of Btu fired is based on the heating value of the fuel in Btu/lb or in Btu/standard cubic foot (SCF) in the ideal gas state, normally at 60°F and 1 atm. This means that a factor of 379.5 SCF/lb mol from Section 2.2 (Table 2.4) is used along with the fuel's MW to convert from Btu/lb to Btu/SCF, and vice versa: Btu/SCF (60°F, 1 atm) equals (Btu/lb) (MW, lb/lb mol)/(379.5 SCF/lb mol). This topic is discussed further in Chapter 3.

2.3.3 Higher and Lower Heating Values

In the regulatory community, the heating value is based on the *higher heating value* (*HHV*) of the fuel by convention; burner manufacturers use the *lower heating value* (*LHV*) actually realized in a practical case. When water (H_2O) is one of the products of combustion, a different amount of heat is realized from the combustion process, depending on whether the resulting water vapor is condensed and its heat of evaporation is recovered, or whether the water vapor is allowed to escape as a gas. In a normal case of combustion with commercial equipment, the water of combustion is not condensed.

An example is provided by the combustion (c) of methane gas (g) to liquid (l) water versus water vapor or gas (g).

$$CH_4(g) + 2O_2(g) \rightarrow CO_2(g) + 2H_2O(l) \quad \Delta H_{c\ 60°F} = 1010\ Btu/SCF(HHV)$$
$$CH_4(g) + 2O_2(g) \rightarrow CO_2(g) + 2H_2O(g) \quad \Delta H_{c\ 60°F} = 909.4\ Btu/SCF(LHV)$$

The difference between the HHV and the LHV above is due to the heat of vaporization (v) (ΔH_v) of water, 1059.7 Btu/lb H_2O at 60°F [15, p. 88].

An emission factor in lb/MMBtu (LHV) is greater than one employing lb/MMBtu (HHV); for typical hydrocarbon fuels, this difference is on the order of 10%. Again, one must be careful.

PROBLEM 2.11

Confirm that the difference between the HHV and the LHV for methane combustion is the heat of vaporization of the product water. The heat of vaporization (ΔH_v) of water (H_2O) at 60°F is 1059.7 Btu/lb H_2O [15, p. 88]; its MW is 18.0153 [5, p. B-11].

SOLUTION

$$CH_4 + 2O_2 \rightarrow CO_2 + H_2O$$

Basis 1 SCF of CH_4 combusted

$$\frac{2\text{SCF } H_2O}{\text{SCF } CH_4} \frac{\text{lb mol } H_2O}{379.5 \text{SCF } H_2O} \frac{18.0153 \text{ lb } H_2O}{\text{lb mol } H_2O} \frac{1059.7 \text{ Btu}}{\text{lb } H_2O} = \frac{100.6 \text{ Btu}}{\text{SCF } CH_4}$$

HHV of CH_4 at 60°F = 1010 Btu/SCF (60°F, 1 atm)

LHV of CH_4 at 60°F = 909.4 Btu/SCF(60°F, 1 atm)

The higher or gross heating value (ΔH_v) here is evolved at 60°F by complete combustion of a standard cubic foot of gas in the ideal gas state with all water of combustion condensed to the liquid state; the lower or net heating value is the same except that the water of combustion remains in the vapor state [16]. Fuel, combustion air, and products of combustion are all at 60°F.

$$HHV - \Delta H_v \text{ of water (in the proper units)} = LHV$$
$$1010 - 100.6 = 909.4$$
$$909.4 = 909.4 \text{ Q.E.D.}$$ ■

PROBLEM 2.12

Given the HHV of propane (C_3H_8), hydrogen (H_2), and carbon monoxide (CO), find the LHV for each at the same conditions.

Fuel Gas	MW	HHV at 60°F (Btu/SCF (60°F, 1 atm)
C_3H_8	44.4	2516.1
H_2	2.0158	324.2
CO	28.01	320.5

(See Problem 2.11 for definitions of HHV and LHV at 60°F.)

ΔH_v of H_2O at 60°F = 1059.7 Btu/lb H_2O [15, p. 88].

SOLUTION

(a) C_3H_8

Basis 1 SCF of C_3H_8 combusted:

$$C_3H_8 + 5O_2 \rightarrow 3CO_2 + 4H_2O$$

$$\frac{4 \text{ SCF } H_2O}{\text{SCF } C_3H_8} \frac{\text{lb mol } H_2O}{379.5 \text{ SCF } H_2O} \frac{18.0153 \text{ lb } H_2O}{\text{lb mol } H_2O} \frac{1059.7 \text{ Btu}}{\text{lb } H_2O} = \frac{201.2 \text{ Btu}}{\text{SCF } C_3H_8}$$

$$2516.1 - 201.2 = 2314.9 \text{ Btu/SCF } (60°F, 1 \text{ atm})$$

(b) H_2

Basis 1 SCF of H_2:
In this case, H_2O is the entirety of the combustion flue gas. As in part (a),

$$H_2 + \frac{1}{2}O_2 \rightarrow H_2O$$

$$\frac{1 \text{ SCF } H_2O}{\text{SCF } H_2} \frac{\text{lb mol } H_2O}{379.5 \text{ SCF } H_2O} \frac{18.0153 \text{ lb } H_2O}{\text{lb mol } H_2O} \frac{1059.7 \text{ Btu}}{\text{lb } H_2O} = \frac{50.3 \text{ Btu}}{\text{SCF } H_2}$$

$$324.2 - 50.3 = 273.9 \text{ Btu/SCF } (60°F, 1 \text{ atm})$$

(c) Similarly, for CO,
Basis 1 SCF of CO

$$CO + \frac{1}{2}O_2 \rightarrow CO_2$$

$$\frac{0 \text{ SCF } H_2O}{\text{SCF CO}} \text{ multiplied by the rest of the factors} = 0 \text{ Btu/SCF CO}$$

Since burning of carbon monoxide as fuel results in no water of combustion (gaseous, liquid, or otherwise), there is no heat of vaporization to be subtracted from the HHV or added to the LHV. The HHV and LHV of CO are therefore equal, namely, 320.5 Btu/SCF (60°F, 1 atm). ∎

2.3.4 Mass Flow Rates

Allowable concentrations in ppm and emission factors in lb/MMBtu show up as limits on environmental regulatory permits. A third type of limit is the mass flow rate of a pollutant. This is found by multiplying the flow rate of the total gas times the concentration of pollutant, using appropriate conversion factors as necessary to make the units consistent. For example, ppm on a dry basis must be multiplied by total flow on a dry basis, and both must be on the same basis of excess O_2. For the same O_2 basis, ppm wet times wet total flow also works.

Mass flow rate can also be calculated by multiplying the firing rate in Btu/h by lb of pollutant/MMBtu, thereby obtaining lb/h of pollutant. Care must be taken because heating values in Btu/SCF may be based on a 60°F, 1 atm SCF (379.5 SCF/lb mol) and regulatory flow rates on 68°F, 1 atm (385.3 SCF/lb mol) (Table 2.4).

The author has seen at least one environmental air quality permit stating limits in all three units: ppmd @ 3% O_2 (dry), lb/MMBtu (HHV) and lb/h. As shown in Appendix B, these values are proportional to one another across the entire spectrum of O_2 in the combustion flue gas. If the limits are consistent, they are redundant. If they are not, the strictest limit applies, and the others are meaningless. Therefore, caution must be exercised here also when interpreting such permit limits.

PROBLEM 2.13

A flue gas from firing 10 MMBtu/h of fuel and containing 2.6% O_2, 18% H_2O, and 100 ppmv of pollutant X (all on a wet basis) amounts to 12,500 SCFM (corrected to 68°F, 1 atm). Determine the emission of pollutant X in (a) ppmvd at 3% O_2 (dry), (b) lb/h, and (c) lb/MMBtu (HHV). Pollutant X has a molecular weight of 55 lb/lb mol.

SOLUTION First step is to correct the % O_2 to a dry basis

$$O_2 \text{ (dry)} = O_2 \text{ (wet)}[100/(100-\% H_2O)] = (2.6)(100/82) = 3.17\% O_2 \text{ (dry)}$$

and 100 ppmvw of X to ppmvd of X at flue gas conditions

$$\text{ppmvd} = \text{ppmvw}[100/(100-\% H_2O)] = (100)(100/82) = 121 \text{ ppmvd of } X$$

and then the flue gas flow to a dry basis

$$\text{SCFM dry} = \text{SCFM wet}[(100-\% H_2O)/100] = (12,500)(82/100) = 10,250 \text{ SCFM (dry)}$$

(a) Convert ppmvd of X at conditions to ppmvd of X at 3% O_2 (dry)

$$\text{ppmvd at } 3\% O_2 \text{ (dry)} = (\text{ppmvd})[(20.95-3)/(20.95-3.17)]$$
$$= (121)(17.95/17.78) = 122 \text{ ppmvd at } 3\% O_2 \text{ (dry)}$$

(b) Convert 10,250 SCFM dry to SCFM dry at 3% O_2 (dry).

$$\text{SCFM dry at 3\%O}_2 \text{ (dry)} = (\text{SCFM dry})[(20.95-3.17)/(20.95-3)]$$
$$= (10,250)(17.78/17.95) = 10,153 \text{ SCFM dry at 3\%O}_2 \text{ (dry)}$$

For everything on a 3% O_2 (dry) basis,

$$\text{lb/h} = \frac{122 \text{ mol } X}{10^6 \text{ mol FG}} \frac{55 \text{ lb } X}{\text{mol } X} \frac{\text{mol FG}}{385.3 \text{ SCF FG}} \frac{10,153 \text{ SCF FG}}{\text{min}} \frac{60 \text{ min}}{\text{h}}$$
$$= 10.6 \text{ lb/h of pollutant } X$$

ppm = moles of pollutant per 10^6 mol of flue gas

The calculation is also valid using the dry, uncorrected basis

$$\frac{121}{10^6} \frac{55}{1} \frac{1}{385.3} (12,500)(60) = 10.6 \text{ lb/h}$$

as well as for the original wet basis

$$\frac{100}{10^6} \frac{55}{1} \frac{1}{385.3} (12,500)(60) = 10.7 \text{ lb/h}$$

the difference being roundoff. (The 10.7 figure is more precise because its factors are the given values without the accumulated roundoff of the intermediate calculations.)

(c) Calculate on the basis of lb/MMBtu (HHV) fired

$$10.7 \text{ lb}/10 \text{ MMBtu (HHV)} = 1.07 \text{ lb/MMBtu (HHV)}$$

The emission factor in lb/MMBtu (LHV) can be estimated here on the rule of thumb stated in the text (10% greater than when using HHV).

$$\text{lb/MMBtu (LHV)} \cong [(1+(10/100)] \text{lb/MMBtu (HHV)}$$
$$= (1.1)(1.07) = 1.18 \text{ lb/MMBtu (LHV)}$$

∎

2.4 VAPOR PRESSURE

2.4.1 What It Is

A condensable vapor in equilibrium with its own liquid exerts a pressure that depends on temperature alone. This pressure, termed the *vapor pressure*, applies to a pure substance along the phase boundary between liquid and vapor. As long as two phases are involved, functional dependence is that of a single variable; fix the temperature and the pressure is defined and vice versa. When a solid such as naphthalene [17, p. 112], iodine crystals [17, p. 112], snow/water ice [17, p. 112], or carbon dioxide dry ice [2, p. 201] changes directly from solid to vapor without passing through the liquid state, the process is called *sublimation*, and the pressure is known as the *sublimation pressure* [17, p. 112].

2.4.2 How to Calculate

Although vapor pressure data are presented in many forms (Tables [5, pp. D-179 to D-215; 18], graphs/charts (including the so-called *Cox Chart*) [19, p. 564], and equations [7, p. 10–31]), vapor pressure is often correlated by a functional form known as the *Antoine equation*. The Antoine equation [7, p. 10–31]

$$\log_{10}[P\,(\text{mmHg})] = A - B/[t(^\circ C) + C] \tag{2.23}$$

is simple and easy to use. It can be programmed into spreadsheets much more easily than a table lookup with an algorithm for interpolation. For hand calculations, it is simpler to employ than a formulation containing more terms with additional constants, as tabulated, for example, in Ref. 20. The greater accuracy to be achieved over a wider temperature range for such more complicated functions may not be necessary for our purposes and not worth the additional effort.

The Antoine constants A, B, and C compiled from several sources [22] are listed in Table 2.6 for a number of chemical substances, with synonyms shown in italic type. The constants are derived from experimental data, and care should be taken not to use those constants outside the stated temperature range. In some cases, more than one set of constants is shown in the table for a given material. The original references should be consulted to determine the vapor pressure of other materials not listed here.

PROBLEM 2.14

Calculate the vapor pressure of methanol at 64.7°C using the Antoine equation and each of three sets of constants listed in Table 2.6.

SOLUTION Note that all three sets of constants are valid at 64.7°C:

First	-20 to $+140^\circ C$
Second	-14 to $+65^\circ C$
Third	$+64$ to $+110^\circ C$

Substituting and exponentiating (to the base 10)

$$P(\text{mmHg}) = 10^\wedge \{A - B/[t\,(^\circ C) + C]\}$$

First $P = 10^\wedge \{7.87863 - 1473.11/[64.7 + 230.0]\} = 758.50\,\text{mmHg}$
Second $P = 10^\wedge \{7.89750 - 1474.08/[64.7 + 229.13]\} = 759.84\,\text{mmHg}$
Third $P = 10^\wedge \{7.97328 - 1515.4/[64.7 + 232.85]\} = 760.73\,\text{mmHg}$

None of these sets of constants reproduces exactly the normal boiling point temperature of 67.4°C listed in Table 2.6. At the normal boiling point, the pressure should equal 760 mmHg (see also Problem 2.15). Deviations range from 0.01% to 0.2%. Such deviations are thought to be typical (see Problem 2.16). While each set of constants is said to be valid at 67.4°C, they were obviously not derived from a data fit at this point. Temperatures corresponding exactly to 760 mmHg are calculated as 64.75, 64.705, and 64.68°C, respectively, using the three sets of Antoine constants listed. (again, see Problem 2.15). ■

Table 2.6 Selected Antoine Equation Constants[a]

Chemical name and synonyms	Formula	MW[b]	BP (°C)[c]	Applicable range (°C)	A	B	C
Acetone	H_3CCOCH_3	58.08	56.2	—	7.02447	1161.0	224
2-Propanone, Dimethyl Ketone				Liquid	7.11714	1210.595	229.664
Aniline	$C_6H_5NH_2$	92.13	184.13	—	7.24179	1675.3	200
Aminobenzene, phenylamine				+102 to +185	7.32010	1731.515	206.049
Benzene	C_6H_6	78.12	80.1	+8 to +103	6.90565	1211.033	220.790
				−12 to +3	9.1064	1885.9	244.2
Chloroform	$CHCl_3$	119.38	61.7	−30 to +150	6.90328	1163.03	227.4
Trichloromethane				−35 to +61	6.4934	929.44	196.03
Ethanol	C_2H_5OH	46.07	78.4	—	8.04494	1554.3	222.65
Ethyl alcohol, grain alcohol, methyl carbinol				−26 to +100	8.32109	1718.10	237.52
Monoethanolamine (MEA)	$H_2N(CH_2)_2OH$	61.09	171.0[d]	+65 to +171	7.4568	1577.67	173.37
2-Aminoethanol, colamine				—	8.52557[d]	2303.41[d]	237.08[d]
(Di)ethyl ether	$H_5C_2OC_2H_5$	74.12	34.6	—	6.78574	994.195	220.0
Ether, ethoxyethane				−61 to +20	6.92032	1064.07	228.08
n-Hexane	C_6H_{14}	86.18	68.7	—	6.87776	1171.530	224.336
Dipropyl				−25 to +92	6.87601	1171.17	224.41
Methanol	CH_3OH	32.04	64.7	−20 to +140	7.87863	1473.11	230.0
Methyl alcohol				−14 to +65	7.89750	1474.08	229.13
Wood alcohol, carbinol				+64 to +110	7.97328	1515.14	232.85

(*continued*)

Table 2.6 (*Continued*)

Chemical name and synonyms	Formula	MW[b]	BP (°C)[c]	Applicable range (°C)	A	B	C
Methyl ethyl ketone (MEK)	H₃COC₂H₅	72.12	79.6	—	6.97421	1209.6	216
2-Butanone				+43 to +88	7.06356	1261.34	221.97
1-Propanol	C₃H₇OH	60.11	97.2	—	7.99733	1569.70	209.5
n-Propyl alcohol				+2 to +120	7.84767	1499.21	204.64
2-Propanol	H₃CC(OH)CH₃	60.11	82.4	0 to +113	6.66040	813.055	132.93
Isopropyl alcohol				0 to +101	8.11778	1580.92	219.61
Toluene	C₆H₅CH₃	92.14	110.6	+6 to +137	6.95464	1344.800	219.482
Methyl benzene, phenyl methane							
Water	H₂O	18.02	100.00	0 to +60	8.10765	1750.286	235.0
				+60 to +150	7.96681	1668.21	228.0

[a] Antoine constants from Ref. 21, except as noted.
[b] Molecular weight from Ref. 5, rounded for water to two decimal places.
[c] Boiling point at 760 mmHg from Ref. 7 or Ref. 5, chosen to be the most consistent with boiling point calculated using Antoine constants.
[d] From manufacturer's literature [22].

2.4 Vapor Pressure

PROBLEM 2.15

Calculate the normal boiling point of benzene from the Antoine equation.

SOLUTION The normal boiling point is defined as the temperature at which the vapor pressure exerted equals the absolute pressure of a standard atmospheric, itself defined as 760 mmHg, 29.921 in. Hg, 14.696 lb force per square inch (psia), or 1.01325 bars. (One bar is equal to $10 \, N/m^2$, and therefore 0.9869 atm.)

Rearrangement of the Antoine equation to solve for temperature yields

$$t \, (°C) = B/[A - \log_{10} P \, (mmHg)] - C \qquad (2.24)$$

For $P = 760$ mmHg, $\log_{10} P = 2.8880814$.

Substituting A, B, and C for the function valid from $+8$ to $+103°C$ from Table 2.6,

$$t \, (°C) = (1211.033)/[6.90565 - 2.880814] - 220.79 = 80.1°C$$

Compare with the boiling point of 80.1°C listed in Table 2.6 for benzene. ∎

PROBLEM 2.16

Water vapor obeys the Antoine relationship, with constants given in Table 2.6. Calculate the vapor pressure of water from 0 to 150°C (32–302°F) using the Antoine equation with those constants. Compare with steam table entries [15, pp. 85–88; 23, pp. 28–31] from two standard references commonly consulted for the properties of water and steam.

SOLUTION Vapor pressure calculations performed using Equation (2.23) at every 10°C from 0 to 150°C are listed in Table 2.7. The resulting vapor pressures in mmHg (column 3) are converted to $lb_f/in.^2$ (psia) (column 4) for comparison with the steam tables as follows:

$$\text{Vapor pressure (psia)} = (14.696/760) \text{Vapor pressure (mmHg)} \qquad (2.25)$$

Temperatures in °C (column 1) are converted to °F (column 2) as follows (Table 2.2):

$$T \, (°F) = (1.8)(T \, (°C) + 40) - 40 \qquad (2.26)$$

Vapor pressures in psia at the selected temperatures in °F as read from the steam tables appear in columns 5 and 7 of Table 2.7. Percent deviations in columns 6 and 8 are calculated from

$$\%\text{Deviation} = (100)(\text{Antoine} - \text{steam tables})/\text{steam tables} \qquad (2.27)$$

As shown in the table, deviations are less than 0.04% except for both ends of the variation in temperature, where deviations range approximately from $\pm 0.1\%$ to 0.3%. The basis for the steam table entries and the smoothing formulas used are discussed therein [15, pp. 13–27; 23, p. 14]. ∎

PROBLEM 2.17

The final problem in this section is a simple humidity calculation for the saturated vapor pressure of water vapor in standard atmospheric air at 60°F. Humidity at less than saturated conditions is taken up in the combustion calculations of Chapter 3.

30   Chapter 2   Basic Concepts

Table 2.7 Vapor Pressure of Water: Comparison of Antoine Equation with Steam Tables

		Equation		Steam tables, with source as noted			
T (°C)	T (°F)	mmHg	psia	psia [15]	% Deviation	psia [23]	% Deviation
0	32	4.5669	0.08831	0.08859	−0.32	0.08854	−0.26
10	50	9.1966	0.17783	0.17796	−0.07	0.17811	−0.16
20	68	17.5301	0.3390	0.33889	+0.03	0.3390	0.00
30	86	31.8271	0.6154	0.61518	+0.04	0.6152	+0.03
40	104	55.3317	1.0699	1.06965	+0.02	1.0695	+0.04
50	122	92.5328	1.7893	1.7891	+0.01	1.7888	+0.03
60	140	149.4435	2.8898	2.8892	+0.02	2.8886	+0.04
60	140	149.4219	2.8893	2.8892	+0.00	2.8886	+0.02
70	158	233.7706	4.520	4.5197	+0.01	4.519	+0.02
80	176	355.2579	6.870	6.8690	+0.01	6.868	+0.03
90	194	525.8557	10.168	10.168	0.00	10.168	0.00
100	212	759.9830	14.696	14.696	0.00	14.696	0.00
110	230	1074.6758	20.781	20.779	+0.01	20.789	0.00
120	248	1489.7147	28.806	28.796	+0.03	28.797	+0.03
130	266	2027.7085	39.210	39.179	+0.08	39.182	+0.07
140	284	2714.1300	52.483	52.414	+0.13	52.418	+0.12
150	302	3577.3055	69.174	69.038	+0.20	69.046	+0.19

SOLUTION   Using the methods of Problem 2.16,

$$T\,(°\mathrm{C}) = [T\,(°\mathrm{F}) + 40]/1.8 - 40$$
$$T = 15.555\ldots°\mathrm{C} \tag{2.28}$$

$$P\,(\mathrm{mmHg}) = 10^{\wedge}[8.10765 - (1750.286)/(t\,(°\mathrm{C}) + 235.0)] = 13.244$$
$$P\,(\mathrm{psia}) = (14.696/760)(13.244) = 0.25610$$

P (psia) from steam tables is 0.25611 [15] and 0.2563 [23], respectively.

There is a small correction to the vapor pressure of water in contact with air [7, p. 10–25], but it is not significant for this calculation. [19, p. 293].

From the ideal gas law,

$$\frac{n\ \text{water}}{n\ \text{dry air}} = \frac{p_{\text{vap water}}}{(P_\mathrm{T} - p_{\text{vap water}})} \tag{2.29}$$
$$= (13.244)/(760 - 13.244) = 0.25610/(14.696 - 0.25610)$$
$$= 0.01774\ \text{mol of water/mol of dry air}$$

Steam table values for the vapor pressure of water could also have been used here. ∎

2.5 HENRY'S LAW

2.5.1 Gases

The pressure (p) exerted by a volatile species, dissolved in a liquid, typically water, is described by Henry's law.

$$p = Hx \qquad (2.30)$$

where x is the mole fraction of the solute and H is the Henry's law constant in units of pressure per mole fraction (m.f.).

Conversely, x is the equilibrium solubility in the liquid in response to the partial pressure of that component in the gas phase. Note that this describes an equilibrium condition without regard to how long it would take to achieve that condition from either side, from low pressure to high pressure or vice versa.

Equations to calculate Henry's law constants for a number of gases in water are listed in Table 2.8 [24]. Handbook data referenced in the table were fitted as a function of temperature by the author. Either the functions or the tabulated entries can be used, as one prefers, in Henry's law pressure or solubility calculations.

The Henry's law constant also varies somewhat with the concentration of dissolved solids (salts) present in the liquid; the functions in the table and the data to which they were fitted apply to pure water.

Table 2.8 Temperature Functions for Henry's Law Constants for Selected Materials[a]

Component	a	b	c	Range of temperature	Data source
H_2	39.677	−1722.91	−11.7406	0–100°C (32–212°F)	[68–71]
CO	72.332	−3500.5	−22.5621	0–100°C (32–212°F)	[68–70]
CH_4	79.369	−3931.52	−24.8819	0–157°C (32–315°F)	[68–75]
N_2	70.4753	−3398.67	−21.8794	0–100°C (32–212°F)	[68–71, 76]
C_2H_6	104.3	−5321.9	−33.1278	0–100°C (32–212°F)	[68–70, 75]

[a] Form of equation is $\log_{10}[H(\text{atm/m.f.})] = a + b/T\,(°K) + c\log_{10}[T\,(°K)]$ (1 atm = 14.696 psia = 1.01325 bar = 101.325 kP$_a$).

PROBLEM 2.18

Calculate the equilibrium solubility of oxygen in pure water at 20°C (68°F) and 1 atm pressure under the following conditions:

(a) The gas phase is pure dry 100% oxygen (O_2).
(b) The gas phase is dry atmospheric air at 14.696 psia = 1 atm and 20.9% O_2 by volume.
(c) The gas phase is the same as above but saturated with water vapor.

32 Chapter 2 Basic Concepts

SOLUTION

(a) Solving for x in Equation (2.30),

$$x = p/H \qquad (2.31)$$

At 20°C from Ref. [19, p. 675],

$$H = 4.01 \times 10^4 \text{ atm/m.f. } O_2 \text{ for } O_2$$

partial pressures up to 1 atm.
Substituting in Equation (2.31),

$$x = (1)/(4.01 \times 10^4) = 2.494 \times 10^{-5} \text{ m.f. (mol } O_2/\text{mol total)}$$

$$O_2(\text{mol/L}) = \frac{2.494 \times 10^{-5} \text{ mol } O_2}{(1-2.494 \times 10^{-5}) \text{ mol } H_2O} \frac{\text{mol } H_2O}{18.0153 \text{ g } H_2O} \frac{1000 \text{ g } H_2O}{\text{L } H_2O}$$

$$= 0.001385 \text{ g mol } O_2/\text{L}$$

(Here the 2.494×10^{-5} term in the denominator can be neglected, if desired.)

$$O_2 \text{ in mg/L} = (0.001385 \text{ g mol } O_2/\text{L})(32 \text{ g } O_2/\text{g mol } O_2)(1000 \text{ mg/g})$$

$$= 44.32 \text{ rounded to } 44.3 \text{ mg/L} \cong \text{ppmw}$$

(b) For dry atmospheric air, p_{O_2} is reduced to $20.9\%/100 = 0.209$

and O_2 in mg/L or ppmw $= (0.209)(44.32 \text{ from part(a)})$

$$= 9.3 \text{ mg/L, or ppmw}$$

(c) For air saturated with water vapor (moisture, H_2O), the partial pressure of O_2 is reduced by the presence of H_2O.
From Table 2.7, at 20°C $p_{\text{vap water}} = 17.5301$ mmHg.

$$\frac{n \text{ water}}{n \text{ dry air}} = \frac{p_{\text{vap water}}}{(P_T - p_{\text{vap water}})} = 17.5301/(760-17.5301) = 0.0236$$

Here, the calculation for the dimensionless moles of water/mol dry air is carried out with all pressures expressed in mmHg.
Basis 1 mol and 1 atm total pressure:

Component	mol/mol of dry air	÷ 1.0236 = mol/mol wet air	×100 = % wet
O_2	0.209	0.2042	20.42
Other gases	0.791	0.7728	77.28
H_2O	0.0236	0.0230	2.30
Total	1.0236	1.0000	100.00

p of the dry gases $= 760 - 7.5301 = 742.4699 \div 760$

$$= 0.9769 \text{ atm} \times 0.209$$

$$= 0.2042 \text{ atm} = p_{O_2}$$

$$= (0.2042)(44.32 \text{ part(a) with } p_{O_2} = 1 \text{ atm}) \ 9.1 \text{ mg/L or ppmw}$$

Calculations of dissolved oxygen at several temperatures are explored more fully later in Table D.1 of Appendix D. ∎

2.5.2 Volatile Liquids

Henry's law can also be used to calculate the volatility of other trace constituents dissolved in wastewater. Henry's law constants are tabulated for a substantial number of organic compounds in water at ambient temperature (mostly 25°C) [25]. These calculated values agree well with experimental data for the examples shown. They are applicable for low solute/high solvent concentrations.

PROBLEM 2.19 *(adapted from Ref. 25)*

Estimate the concentration of carbon tetrachloride (CCl_4) in atmospheric air above water containing 85.4 ppmw (mg/L) CCl_4. The value of H for CCl_4 at 25°C is 1634 atm/m.f.

SOLUTION On a molar basis,

$$\text{mol/L } CCl_4 = \frac{CCl_4 (mg/L)}{(MW \text{ of } CCl_4 = 153.82 \text{ g/g mol})(1000 \text{ mg/g})} = 5.5519 \times 10^{-4}$$

$$\text{mol/L } H_2O = \frac{1000 \text{ g } H_2O}{L} \frac{\text{g mol } H_2O}{18.0153 \text{ g } H_2O} = 55.5084$$

$$(5.5519 \times 10^{-4})/(55.5084 + 5.5519 \times 10^{-4}) = 10 \times 10^{-6} \text{ m.f. of } CCl_4$$

$$P_{CCl_4} = (H_{@25°C})(x) = (1634 \text{ atm/m.f.})(10 \times 10^{-6} \text{ m.f.}) = 0.01634 \text{ atm}$$

$$y_{CCl_4} = P_{CCl_4}/P_T = (0.01634 \text{ atm}/1 \text{ atm}) \times 10^6 = 16,340 \text{ ppmv} \times 10^{-4} = 1.634 \text{ vol\%}$$

Note: The maximum water solubility of carbon tetrachloride is listed as 0.097 parts per 100 parts of water at 0°C and 0.08 parts per 100 parts of water at 20°C [7, pp. 7–122 to 7–123]. This translates to 970 and 800 ppmw, respectively. ∎

PROBLEM 2.20 *(adapted from Ref. 20)*

Estimate the concentration of benzene (C_6H_6) in water below air (at standard atmospheric pressure) containing 1000 ppmv C_6H_6. The H value for C_6H_6 at 25°C is 309.2 atm/m.f.

SOLUTION

$$P_{C_6H_6} = (y_{C_6H_6})(P_T)$$
$$y_{C_6H_6} = 1000 \times 10^{-6} \text{ (mole fraction of } C_6H_6 \text{ in the vapor phase)}$$
$$P_{C_6H_6} = (1000 \times 10^{-6})(1) = 10^{-3} \text{ atm}$$
$$x = (P_{C_6H_6})/(H_{@25°C}) = (10^{-3})/(309.2) = 3.23 \times 10^{-6} \text{ m.f. of } C_6H_6$$

$$C_6H_6 \text{ in mg/L} = (MW_{C_6H_6}/MW_{H_2O})(\rho_{H_2O} = 1000 \text{ g/L})(m.f.)(10^3 \text{ mg/g})$$
$$= (78.12/18.0153)(1)(3.23 \times 10^{-6})(10^{-6}) = 14.0 \text{ mg/L or ppmw}$$

Here the 3.23×10^{-6} in the $1 - 3.23 \times 10^{-6}$ factor has been neglected.

Note: The maximum solubility of benzene in water is listed as 0.07 parts per 100 parts of water at 22°C [7, pp. 7–98 to 7–99]. This translates to 700 ppmw. ∎

2.5.3 Ionizing Materials: Ionic Equilibrium

It is sometimes said that Henry's law is not applicable to solutes that ionize in solution to form a nonvolatile species. The fact is, however, that Henry's law continues to apply, but only to the volatile fraction. Two examples are in order—the carbon dioxide–bicarbonate–carbonate system and the ammonia–ammonium ion system. Both are important in wastewater treatment and in some process condensates. The first system is the basis of buffering in natural waters.

Ionization occurs among carbon dioxide and the bicarbonate (HCO_3^-) and carbonate (CO_3^{2-}) ions according to the following equilibria:

$$CO_2 + H_2O \xrightleftharpoons{K_1} HCO_3^- + H^+ \tag{2.32}$$

$$HCO_3^- \xrightleftharpoons{K_2} CO_3^{2-} + H^+ \tag{2.33}$$

Ionization constants are defined by

$$K_1 = \frac{[HCO_3^-][H^+]}{[CO_2]} \tag{2.34}$$

$$K_2 = \frac{[CO_3^{2-}][H^+]}{[HCO_3^-]} \tag{2.35}$$

where the bracketed quantities [ ] are thermodynamic activities, approximated as concentrations in g mol/L in dilute solution.

The total species concentration $[T_{CO_2}]$ is the sum of CO_2, bicarbonate, and carbonate.

$$[T_{CO_2}] = [CO_2] + [HCO_3^-] + [CO_3^{2-}] \tag{2.36}$$

The volatile fraction is molecular CO_2.

$$[CO_2]/[T_{CO_2}] = 1/\{1 + K_1/[H^+] + K_1 K_2/[H^+]^2\} \tag{2.37}$$

The nonvolatile fraction is made up of ionic species, HCO_3^- and CO_3^{2-}.

$$[HCO_3^-]/[T_{CO_2}] = 1/\{1 + [H^+]/K_1 + K_2/[H^+]\} \tag{2.38}$$

$$[CO_3^{2-}]/[T_{CO_2}] = 1/\{1 + [H^+]/K_2 + [H^+]^2/(K_1 K_2)\} \tag{2.39}$$

Ionization equilibrium curves for this system in dilute solution at 25°C (77°F) are plotted versus pH in Figure 2.1. The pH is defined as

$$pH = -\log_{10}[H^+] \tag{2.40}$$

For this plot, ionization constants at 25°C, $K_1 = 4.47 \times 10^{-7}$ and $K_2 = 4.68 \times 10^{-11}$ are taken from a standard reference [7, p. 5–40].

At a pH below about 4, all of the CO_2 essentially exists as dissolved nonionized gas. As pH is gradually increased to about 8.3, more and more of the CO_2 ionizes to HCO_3^-. Beyond pH 8.3, where HCO_3^- peaks, equilibrium between HCO_3^- and CO_3^{2-} governs, with virtually no molecular CO_2 present. Beyond a pH of 13, CO_3^{2-} concentration dominates and HCO_3^- becomes negligible.

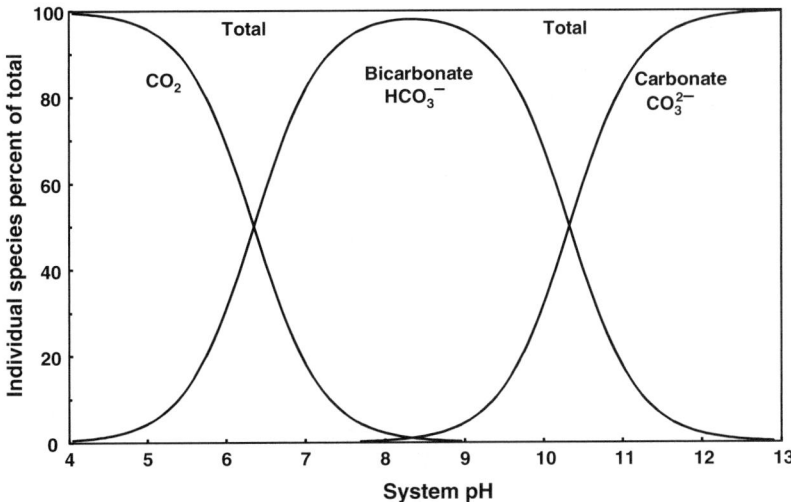

Figure 2.1 Distribution of species in the system CO_2–bicarbonate–carbonate at 25°C.

Similarly, for ammonia ionization

$$NH_4^+ \leftrightarrow H^+ + NH_3 \qquad (2.41)$$

$$K_a = \frac{[H^+][NH_3]}{[NH_4^+]} \qquad (2.42)$$

The total species concentration $[T_{NH_3}]$ is the sum of molecular ammonia and ammonium ion

$$[T_{NH_3}] = [NH_3] + [NH_4^+] \qquad (2.43)$$

The volatile fraction is nonionized molecular ammonia

$$[NH_3]/[T_{NH_3}] = 1/\{1 + [H^+]/K_a\} \qquad (2.44)$$

and the nonvolatile component is the ammonium ion (NH_4^+).

$$[NH_4^+]/[T_{NH_4}] = 1/\{1 + K_a/[H^+]\} \qquad (2.45)$$

Ionization equilibrium curves for NH_3 and NH_4^+ versus pH are shown in Figure 2.2, using $K_a = 5.69 \times 10^{-10}$ at 25°C (77°F) [7, p. 5–40]. Ammonium ion dominates at low pH, while nonionized molecular ammonia at high pH. Both occur together in whole number percentage compositions in the pH range from 7.2 to 11.2.

Since ionization constants are functions of temperature, the curves in Figures 2.1 and 2.2 shift somewhat as the temperature changes. Temperature functions fitting the ionization constants for the CO_2 and the ammonia system as well as those for various other ionic equilibria have been obtained by least squares regression [24]. They are listed in Table 2.9 in the form

$$pK = a + bT \,(°K) + c/T \,(°K) \qquad (2.46)$$

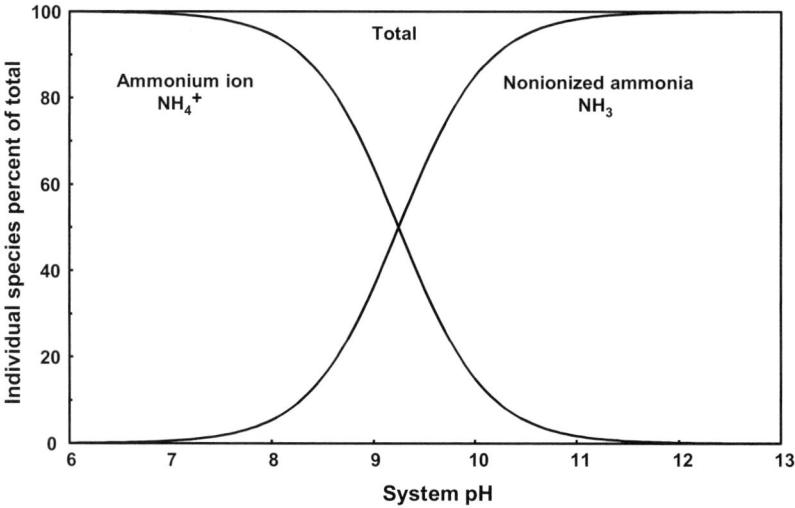

Figure 2.2 Distribution of species in the ammonia–ammonium ion system at 25°C.

where

$$pK = -\log_{10} K \tag{2.47}$$

similar to the definition of pH in Equation (2.40).

2.5.4 Henry's Law Constants for Ionizing Species

2.5.4.1 CO_2

Carbon dioxide follows Henry's law (i.e., P versus x is approximately linear) up to CO_2 partial pressures of 500 kP_a (~5 atm) [27; 28, p. 168]. However, the liquid mole fraction to be used is that of nonionized dissolved CO_2 gas or, equivalently, the total dissolved CO_2 species multiplied by the fraction of nonionized CO_2. The following expression for H to be used with the mole fraction of total CO_2 species was obtained by regression of Henry's law constants [19, pp. 673–676; 29, pp. 14–2 to 14–7; 30] as a function of temperature between 0 and 60°C (32 and 140°F); this result was then combined with the temperature expressions for the ionization constants K_1 and K_2 from Table 2.9 [24]. The temperature function for the CO_2 Henry's law constant agrees with data from another source [31] in the range of 50–200°C (122–392°F) to within 0% to 7%.

$$H(\text{atm/m.f.}) = 10^{\wedge}\{59.921 - 3404.2/T(\text{K}) - 18.3037\log_{10} T(\text{K})\}/\{1 + K_1/10^{\wedge}(-\text{pH})\} \tag{2.48}$$

For a pH of 8.3 and above, CO_2 becomes nonvolatile as H approaches zero, regardless of temperature, and all of the CO_2 in the gas phase wants to take up residence in the liquid phase. It does so in the form of the nonvolatile ions, bicarbonate

Table 2.9 Ionization Constants as Functions of Temperature[a]

Constant	Reaction	a	b	c	Temperature range (°C)	Data source
K_1 from Equation (2.34)	CO_2–HCO_3^- equilibrium	−14.8441	0.0327726	3406.14	0–50	[7, p. 5–40], [5, pp. D-151 to D-152], [26, pp. 204–206]
K_2 from Equation (2.35)	HCO_3^-–CO_3^{2-} equilibrium	−6.5476	0.0238628	2910.82	0–50	[7, p. 5–40], [5, pp. D-151 to D-152], [26, pp. 204–206]
K_a from Equation (2.42)	NH_4^+–NH_3 equilibrium	0.091468	0	2729.14	0–100	[7, p. 5–40], [5, pp. D-151 to D-152]
K_w from Equation (2.50) $K_w = [H^+][OH^-]$	Dissociation of water $H_2O \leftrightarrow H^+ + OH^-$	−6.1665	0.0171777	4484.53	0–50	[5, pp. D-151 to D-152]
$K_{HF} = \dfrac{[\text{formate}][H^+]}{\text{formic acid}}$	Dissociation of formic acid $HCOOH \leftrightarrow H^+ + HCOO^-$	−5.4144	0.0154103	1363.07	0–50	[5, pp. D-151 to D-152]
$K_{HAc} = \dfrac{[\text{acetate}][H^+]}{[\text{acetic acid}]}$	Dissociation of acetic acid $CH_3COOH \leftrightarrow H^+ + CH_3COO^-$	−3.2427	0.0135300	1181.97	0–50	[7, p. 5–40], [5, pp. D-151 to D-152]

[a] Constants a, b, and c from curve fits [24] of the data indicated.
$pK = a + b(T\,°K) + c/T\,(°K)$
$pK = -\log_{10} K$

and carbonate. At the other extreme of low pH, H is a function of temperature only. Carbon dioxide is more volatile (less soluble in water) with rising temperature, with an increasing value of H. In between, pH and temperature are both important. The numerical values of H at low pH are in the range of about 700 atm/m.f. at 0°C to about 3000 atm/m.f. at 50°C (1 atm = 101.325 kP$_a$).

2.5.4.2 NH$_3$

For ammonia, an equation combining the effects of volatility and ionization was taken directly from Ref. 32.

$$H \text{ (atm/(m.f.)} = \{1.441 \times 10^5 \exp[-3513/T \text{ (°K)}]\}/\{1 + 2.528 \times 10^{\wedge}(-\text{pH})$$
$$\times \exp[6054/T \text{ (°K)}] \quad (2.49)$$

It is also a function of pH and temperature. The numerator relates H and temperature for nonionized NH$_3$ in solution, and the denominator accounts for ionization with pH and temperature.

A plot of Equation (2.49) for ammonia would be a reversal of that for CO$_2$. Ammonia is nonvolatile below about pH 6, regardless of temperature. Volatility becomes independent of pH above pH 10. In the mid-range, H increases with both pH and temperature. At high pH, H for ammonia calculates to about 0.4 atm/m.f. at 0°C, about 3 atm/m.f. at 50°C, and about 9 atm/m.f. at 90°C. This implies that ammonia in the gas phase at a ppm to hundreds of ppm concentration (by volume) can exist in equilibrium with a ppm by weight of similar magnitude in solution.

PROBLEM 2.21

Carbon dioxide dissolved in water is called carbonic acid [5, pp. B72, B102].

(a) For a saturated solution at 25°C (68°F) and a pH of 3.8 [5, p. D-135], estimate the percentage of nonionized molecular CO$_2$ in solution.
(b) Solubility of CO$_2$ at 25°C and 760 mmHg (1 atm) is 0.759 mL (measured at 0°C and 760 mmHg) dissolved in 1 mL of water [33, pp. 5.3–5.5]. This is assumed to be total CO$_2$ in solution. Solubility of CO$_2$ increases with increasing pressure and decreases with increasing temperature [19, p. 674]. For this case, estimate the concentration of nonionized CO$_2$ in solution if the pH is 3.8.

SOLUTION

(a)
$$[\text{CO}_2]/[T_{\text{CO}_2}] = 1/\{1 + K_1/[\text{H}^+] + K_1K_2/[\text{H}^+]^2\} \quad (2.37)$$

For pH = 3.8,
$[\text{H}^+] = 1.5849 \times 10^{-4}$ and $[\text{H}^+]^2 = 2.5119 \times 10^{-8}$
$K_1 = 4.47 \times 10^{-7}$ and $K_2 = 4.68 \times 10^{-4}$
$= 1/\{1 + (4.47 \times 10^{-7})/(1.5849 \times 10^{-4}) + (4.47)(4.68)(2.5119) \times 10^{-19}\}$
$= 1/\{1 + 2.8204 \times 10^{-3} + 52.5479 \times 10^{-19}\}$
$= 0.9972 = 99.72\%$

(b) From Problem 2.5 in Section 2.2, 0.759 mL of CO_2 at $0°C$ and 760 mmHg corresponds to

3.39×10^{-5} g mol of CO_2

3.39×10^{-5} g mol $CO_2/1$ mL $\times 1000 = 0.0339$ g mol/L of water $= [T_{CO_2}]$

In other common units (g $CO_2/100$ g of water),

$$\frac{3.39 \times 10^{-2} \text{ g mol}}{\text{liter of water}} \frac{\text{liter of water}}{1000 \text{ g water}} \frac{44.01 \text{ g } CO_2}{\text{g mol } CO_2} = 0.1490 \text{ g } CO_2/100 \text{ g of water}$$

For mg CO_2/L of water solution,

$$(3.39 \times 10^{-2} \text{ g mol/L})(44.01 \text{ g } CO_2/\text{g mol } CO_2)(1000 \text{ mg/g})$$
$$= 1490 \text{ mg/L nonionized } CO_2 \text{ in solution at pH 2.8}$$
$$= (0.9972 \text{ from Part(a)})(3.39 \times 10^{-2} \text{ g mol/L})$$
$$= 0.0338 \text{ g mol/L})$$

Note: The ionized species in solution, HCO_3^- and CO_3^{2-} (total here $= 0.28\%$), are balanced by the $[H^+]$ and $[OH^-]$ from the ionization of water.

$$K_w = [H^+][OH^-] \cong 10^{-14} \text{ at } 25°C \text{ [7, p. 5–7]} \quad (2.50)$$

according to the principle of electroneutrality

$$[H^+] = [OH^-] + [HCO_3^-] + 2[CO_3^{2-}] \quad (2.51)$$

It is this $[H^+]$ that accounts for the pH of the carbonic acid solution. ∎

PROBLEM 2.22

Let us revisit Problem 2.21 and calculate the pH from the constituents in solution and their interaction, rather than accepting a general statement from a handbook that the pH of a carbonic acid solution is 3.8.

SOLUTION Estimate the pH from electroneutrality of the solution. Total positive charges must equal total negative charges.

$$[H^+] = [OH^-] + [HCO_3^-] + 2[CO_3^{2-}] \quad (2.51)$$

$$[H^+] = \frac{K_w}{[H^+]} + \frac{[T_{CO_2}]}{1 + [H^+]/K_1 + K_2/[H^+]} + \frac{2[T_{CO_2}]}{1 + [H^+]/K_2 + [H^+]^2/(K_1 K_2)} \quad (2.52)$$

from Equations (2.50), (2.38), and (2.39).
Fraction of molecular CO_2 to total CO_2–bicarbonate–carbonated species

$$\frac{[CO_2]}{[T_{CO_2}]} = \frac{1}{1 + K_2/[H^+] + (K_1 K_2)/[H^+]^2} \quad (2.53)$$

from Equation (2.37).
Rearranging Equation (2.53),

$$[T_{CO_2}] = [CO_2]\{1 + K_1/[H^+] + (K_1 K_2)/[H^+]^2\} \quad (2.54)$$

$$pH = -\log_{10}[H^+] \quad (2.40)$$

40 Chapter 2 Basic Concepts

Solving for $[H^+]$,

$$[H^+] = 10^{\wedge}(-pH) \tag{2.55}$$

At 25°C,

$$pK_w = 13.996 \quad K_w = 1.0093 \times 10^{-14} \quad [7, \text{p. 5--7}]$$
$$pK_1 = 6.35 \quad K_1 = 4.47 \times 10^{-7} \quad [7, \text{p. 5--40}]$$
$$pK_2 = 10.33 \quad K_2 = 4.68 \times 10^{-11} \quad [7, \text{p. 5--40}]$$

Find $[CO_2]$ from Henry's law

$$\text{at } 25°C \quad H = 1640 \text{ atm/m.f.} \ [19, \text{p. 674}]$$
$$x = p/H = 1/1640 = 0.0006098 \text{ m.f.} \tag{2.31}$$

$$(0.0006098) \frac{\text{g mol } CO_2}{\text{g mol total}} \frac{\text{g mol total}}{(1 - 0.0006098 \text{ g mol } H_2O)} \frac{\text{g mol } H_2O}{18.02 \text{ g } H_2O} \frac{1000 \text{ g } H_2O}{\text{liter } H_2O \text{ total}}$$

For small moles of CO_2 as above, the 0.0006098 in the denominator could be neglected.

$$= 0.03388 \text{ mol } CO_2/L = [CO_2]$$
$$= (0.03388)(44.01 \text{ g } CO_2/\text{g mol } CO_2)$$
$$= 1.490 \text{ g } CO_2/L$$
$$\times 1000 = 1490 \text{ mg/L}$$

Solution for $[H^+]$ is by trial and error. Guess the pH and therefore $[H^+]$. This then determines $[T_{CO_2}]$ from $[CO_2]$ and the $[H^+]$ function. Continue to iterate until the left-hand side (LHS) equals the right-hand side (RHS) of the electroneutrality equation.

The final trial results in

$$[T_{CO_2}] = [0.03388]\{1 + 4.47 \times 10^{-7}/1.23 \times 10^{-4} + (4.47)(4.68) \times 10^{-18}/(1.23 \times 10^{-4})^2\} \tag{2.54}$$

$$[T_{CO_2}] = [0.03388]\{1 + 3.631 \times 10^{-3}/1.4 \times 10^{-9}\} = [0.003388]\{1.0036\} = 0.03399$$

$$[H^+] = \frac{1.0093 \times 10^{-14}}{[H^+]} + \frac{0.03399}{1 + 1.23 \times 10^{-4}/4.68 \times 10^{-11} + 4.68 \times 10^{-11}/1.23 \times 10^{-4}}$$
$$+ \frac{(2)(0.03399)}{1 + 1.23 \times 10^{-4}/4.68 \times 10^{-11} + (1.23 \times 10^{-4})^2/[(4.47)(4.68) \times 10^{-18}]} \tag{2.52}$$

$$\underset{[\text{LHS}]}{[H^+]} \cong 1.23 \times 10^{-4} = \underset{[\text{RHS}]}{8.21 \times 10^{-11}} + 1.23 \times 10^{-4} + 3.08 \times 10^{-12} = 1.23 \times 10^{-4}$$

At the pH shown below, the right-hand side and the left-hand side agree to within 1.0×10^{-16}.

$$\overset{\text{more exact}}{pH = -\log_{10}[H^+] = -\log_{10}[1.229957 \times 10^{-4}] = 3.910110075793} \tag{2.40}$$

In realistic terms, pH = 3.91 from which

$$[CO_2]/[T_{CO_2}] = (0.03388)/(0.03399) = 0.9964 \tag{2.53}$$

Compare pH = 3.91 with the pH of 3.8 from Problem 2.21 and the fraction of nonionized molecular CO_2. This is excellent agreement for data and inputs from different sources. ∎

Where CO_2 is being dissolved into pure water solution, it does not make much difference whether $[CO_2]$ or $[T_{CO_2}]$ is used in conjunction with the Henry's law constant. However, at a higher pH in other systems, where the pH is set by the combination of other ionic species in solution, it is appropriate to use the volatile fraction $[CO_2]$ in the Henry's law relationship.

2.6 VAPOR–LIQUID EQUILIBRIUM

We now move on to systems of two liquid components, in which the composition must be specified to define the vapor pressure versus temperature relationship. These systems are tractable mathematically, but the situation, in general, becomes much more complicated upon the introduction of a third component or more.

This will not be the most elegant treatment of the thermodynamics of solution equilibrium. Pressures rather than fugacities and concentrations instead of activities will be employed. The resulting relationships are valid at the low pressures typical of environmental concerns.

2.6.1 Raoult's Law

The simplest vapor–liquid systems are described by Raoult's law.

$$p_1 = p_1^{vap} x_1 \tag{2.56}$$

and

$$p_2 = p_2^{vap} x_2 \tag{2.57}$$

where p is the partial pressure exerted by the given component, p^{vap} is that pure component's vapor pressure (a function of temperature only), and x is the liquid-phase mole fraction of that particular component, with

$$x_1 + x_2 = 1 \tag{2.58}$$

and

$$\text{total pressure } P_T = p_1 + p_2 \tag{2.59}$$

Therefore,

$$P_T = p_1^{vap} x_1 + p_2^{vap} x_2 \tag{2.60}$$
$$P_T = p_1^{vap} x_1 + p_2^{vap}(1 - x_1) \tag{2.61}$$
$$P_T = (p_1^{vap} - p_2^{vap}) x_1 + p_2^{vap} \tag{2.62}$$

By convention, x_1 is chosen as the more volatile component (higher vapor pressure, lower boiling temperature).

42 Chapter 2 Basic Concepts

The vapor-phase mole fraction is denoted as y.

$$y_1 = p_1/P_T = p_1^{vap} x_1/P_T \quad (2.63)$$

$$y_2 = p_2/P_T = p_2^{vap}(1-x_1)/P_T \quad (2.64)$$

Combining Equations (2.63) and (2.62),

$$y_1 = p_1^{vap} x_1 / [(p_1^{vap} - p_2^{vap})x_1 + p_2^{vap}] \quad (2.65)$$

Since

$$y_1 + y_2 = 1 \quad (2.66)$$

$$y_2 = (1 - y_1) \quad (2.67)$$

The *vapor–liquid distribution ratio* (K) is defined as

$$K = (y_1/x_1) \quad (2.68)$$

for all systems, not only those for which Raoult's law applies.

Likewise, the relative volatility (α)

$$\alpha = [y_1/x_1]/[y_2/x_2] = [y_1/x_1]/[(1-y_1)/(1-x_1)] \quad (2.69)$$

Relative volatility is a term used in distillation to characterize the ease of separation of one component from another.

For systems in which Raoult's law applies,

$$K = p_1^{vap} / [(p_1^{vap} - p_2^{vap})x_1 + p_2^{vap}] \quad (2.70)$$

and

$$\alpha = p_1^{vap}/p_2^{vap} \quad (2.71)$$

The classic system used to illustrate Raoult's law is benzene–toluene. Because of the similarity of the two molecules, the unsaturated benzene ring (C_6H_6) and the methyl ($-CH_3$)-substituted benzene ring ($C_6H_5CH_3$), interactions between unlike molecules are pretty much the same as interactions between like molecules, leading to a situation where Raoult's law applies.

2.6.1.1 Varying Total Pressure at Constant Temperature

PROBLEM 2.23

At a temperature of 120°C, calculate the total pressure exerted by a solution of benzene and toluene containing a benzene liquid mole fraction (x_1) of 0.2. Calculate the corresponding mole fraction of benzene in the vapor (y_1).

SOLUTION Vapor pressures at 120°C are computed for the pure components using the Antoine equation, with constants from Table 2.6 and a slight change in nomenclature.

Benzene : p_1^{vap} (mmHg) $= 10^\wedge[6.90565-1211.033/(120°C+220.790)] = 2249.3$ mmHg
Toluene : p_2^{vap} (mmHg) $= 10^\wedge[6.95464-1344.800/(120°C+219.482)] = 984.7$ mmHg

$$P_T = p_1^{\text{vap}} x_1 + p_2^{\text{vap}}(1-x_1) \\
= (2249.3)(0.2) + (984.7)(0.8) = 1237.62 \text{ mmHg} \tag{2.61}$$

$y_1 = p_1^{\text{vap}} x / P_T = (2249.3)(0.2)/(1237.62) = 0.3635$ from Equation (2.63)

In a similar manner, the total pressure and vapor mole fraction are worked out at constant temperature for a number of other liquid mole fraction (x_1) points. The mole fractions x_1 and y_1 are plotted against P_T as the smooth curves in Figure 2.3. The x_1 pressure curve is known as the *bubble point* curve; the y_1 pressure curve is called the *dew point* curve, although the horizontal and vertical coordinates are switched from the way they are usually plotted. The bubble point occurs when the first vapor forms from a large mass of liquid, whereas the dew point occurs when the first drop of liquid condenses from a large mass of vapor. The discreet points shown in the figure are plotted from Ref. 34.

Typically, the composition of the liquid and the condensed vapor phase is experimentally determined by density, refractive index, or gas chromatograph measurements.

For the pure components, toluene without benzene ($x_1 = 0$ and $y_1 = 0$) and benzene without toluene ($x_1 = 1$ and $y_1 = 1$), the equations reduce to the pure component vapor pressures.

At $x_1 = 0$, $\quad P_{T_{\text{toluene}}} = p_2^{\text{vap}}$ (at $x_1 = 0$ and temperature $= 120°C$) $= 984.7$ mmHg
and at $x_1 = 1$, $\quad P_{T_{\text{benzene}}} = p_2^{\text{vap}}$ (at $x_1 = 1$ and temperature $= 120°C$) $= 2249.3$ mmHg ∎

These provide the end points of the curves in Figure 2.3. (A temperature of 120°C is slightly beyond the range of applicability of benzene's Antoine constants in Table 2.6, but it is not a serious problem here).

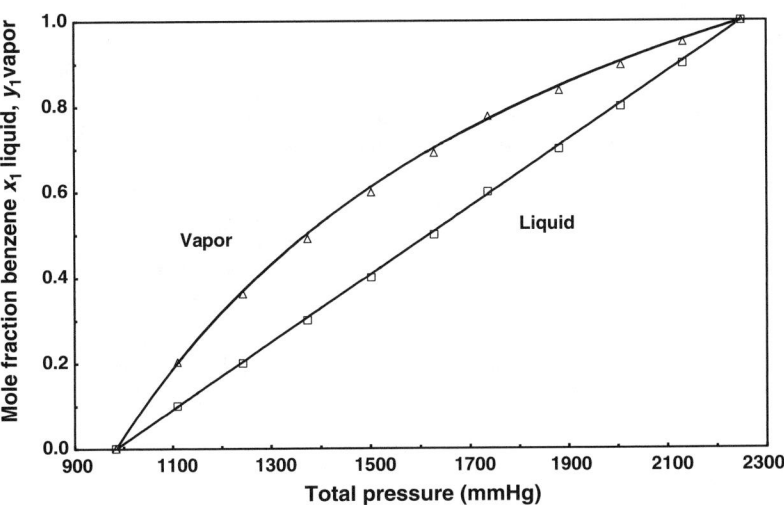

Figure 2.3 Benzene–toluene vapor–liquid equilibrium at 120°C (*P–x* and *P–y* curves).

2.6.1.2 Varying Temperature at Constant Total Pressure

Keeping the pressure constant and allowing the temperature to vary as the mole fraction changes is more useful for practical cases, especially for atmospheric pressure situations. Calculations at constant pressure are illustrated by the following example problem.

PROBLEM 2.24

Calculate the bubble point curve (x_1 versus temperature) and the dew point curve (y_1 versus temperature) for benzene–toluene at 1 atm (760 mmHg) total pressure.

SOLUTION

Step 1 Pick a temperature between the atmospheric boiling point of benzene (80.1°C) and toluene (110.6°C) (Table 2.6). Let the temperature be 92.1°C.

Step 2 Calculate p_1^{vap} and p_2^{vap} at this temperature from the Antoine equation, as in Problem 2.23.

$$p_1^{vap} = 1084.3632 \text{ mmHg}, \quad p_2^{vap} = 435.0844 \text{ mmHg}$$

Step 3 Set $P_T = 1$ atm (760 mmHg)

Step 4 By rearrangement of Equation (2.62), calculate

$$x_1 = (P_T - p_2^{vap})/(p_1^{vap} - p_2^{vap}) \tag{2.72}$$

With rounded vapor pressures,

$$x_1 = (760 - 435.1)/(1084.4 - 435.1) = 0.5004 \cong 0.500$$

Step 5 Calculate $y_1 = p_1^{vap} x_1 / P_T$ from Equation (2.63).

$$y_1 = (1084.4)(0.5004)/(760) = 0.7140 \cong 0.714$$

Step 6 Plot points x_1 versus T and y_1 versus T.

Step 7 Repeat for additional temperatures.

Curves are shown in Figure 2.4, again plotted in an unconventional manner for convenience, along with experimental data from several sources [34,35]. ∎

Another useful relationship is the curve of y_1 versus x_1 at constant pressure (Figure 2.5). This is a replot of Figure 2.4, in which each individual x_1 and y_1 point on the curve of Figure 2.5 corresponds to a unique temperature. It is helpful in determining $K = y_1/x_1$ and $\alpha = [y_1/x_1]/[(1 - y_1)/(1 - x_1)]$. The straight line in the figure, at 45° from both the horizontal and vertical, is the $y_1 = x_1$ line. The distance between the curve and this line marks the deviation between the vapor-phase and the liquid phase compositions. The discrete data points in the figure are from a number of literature sources [34,35].

PROBLEM 2.25

At $x_1 = 0.5$, estimate K and α graphically from Figure 2.6, without having to resort to vapor pressure calculations.

2.6 Vapor–Liquid Equilibrium   45

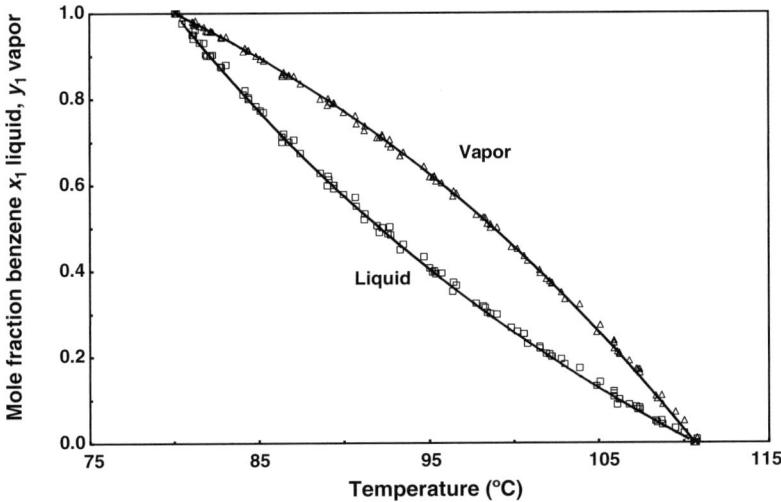

Figure 2.4 Benzene–toluene vapor–liquid equilibrium at 1 atm (T–x and T–y curves).

SOLUTION From Problem 2.24, $y_1 = 0.714$ at $x_1 = 0.500$. Then $(1 - x_1) = 0.5$ and $(1 - y_1) = 0.286$.

$$K = (y_1/x_1) = (0.714)/(0.500) = 1.4340 \cong 1.43 \text{ from Equation (2.68)}$$

Equation (2.69):

$$\alpha = [y_1/x_1]/[(1-y_1)/(1-x_1)] = [(0.714)/(0.500)]/[(0.286)/(0.500)] \cong 2.50 \qquad \blacksquare$$

A further ideal system obeying Raoult's law is methanol–ethanol, where both the constituents are soluble in water. These short-chain alcohols appear, among other

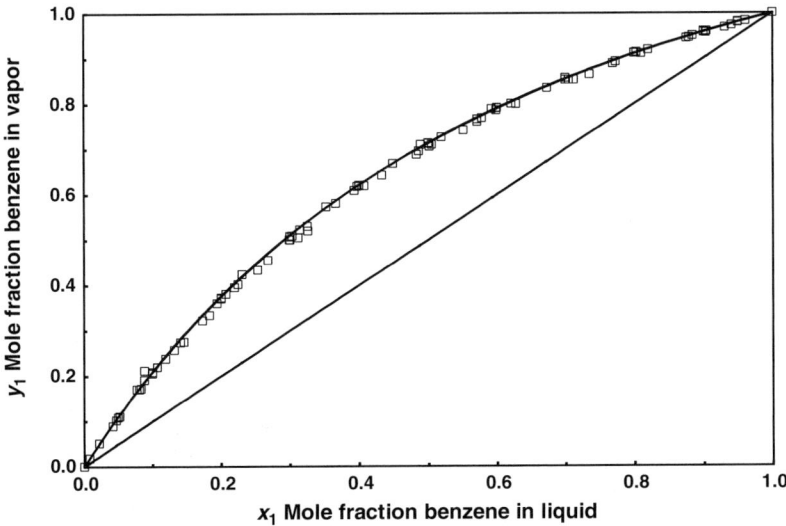

Figure 2.5 Benzene–toluene vapor–liquid equilibrium at 1 atm (x–y curve).

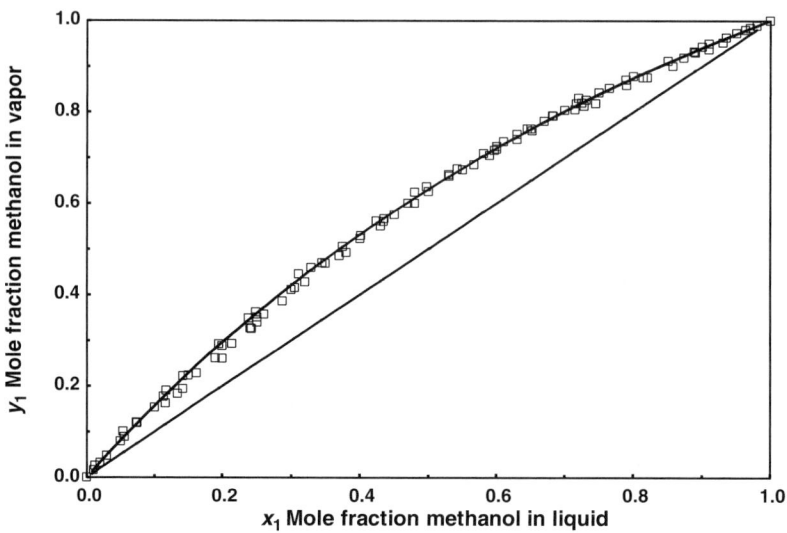

Figure 2.6 Methanol–ethanol vapor–liquid equilibrium at 1 atm (x–y curve).

environmental contaminants, in some process condensates. The x–y curve calculated from Raoult's law, the 45° $x = y$ line, and experimental data from several different investigators [34,35] are all shown in Figure 2.6.

2.6.2 Systems Where Raoult's Law Does Not Apply

Raoult's law is an exception rather than a rule in vapor–liquid equilibrium (VLE). It is limited to similar molecules, for example, benzene–toluene, methane–ethanol, ethylene dibromide–propylene dibromide, and chlorobenzene–bromobenzene [17, p.156; 19, p.526]; to isomers and members of homologous series of hydrocarbons like propane–butane and butane–isobutane [19, p. 526]; and to some special cases such as carbon tetrachloride–toluene and ethylene dichloride (1,2-dichloroethane) with either benzene or toluene [17, p. 156; 19, p. 526; 36].

For the nonideal systems, a factor γ, called an *activity coefficient*, is introduced into the Raoult's law formulations as follows:

$$p_1 = p_1^{vap} \gamma_1 x_1 \tag{2.73}$$

$$p_2 = p_2^{vap} \gamma_2 x_2 = p_2^{vap} \gamma_2 (1 - x_1) \tag{2.74}$$

$$P_T = p_1 + p_2 \tag{2.75}$$

$$y_1 = p_1/P_T = p_1^{vap} \gamma_1 x_1 / P_T \tag{2.76}$$

$$y_2 = p_2/P_T = p_2^{vap} \gamma_2 x_2 / P_T = p_2^{vap} \gamma_2 (1 - x_1) / P_T \tag{2.77}$$

Some would say that this is simply a "fudge factor." However, it exhibits its own unique properties within the mathematical framework of thermodynamics. Some of these include an approach of γ to 1.0 as the mole fraction of that component approaches 1.0, an approach to Henry's law as that component approaches zero at the other end of the concentration spectrum, and a temperature relationship to heat of mixing data. In addition, thermodynamically consistent activity coefficients will produce a net zero area (positive area = negative area) under the curve of $\ln(\gamma_1/\gamma_2)$ or $\log_{10}(\gamma_1/\gamma_2)$ versus x, that is,

$$\int_{x_1=0}^{x_1=1} \ln(\gamma_1/\gamma_2) \mathrm{d}x = \int_{x_1=0}^{x_1=1} \log_{10}(\gamma_1/\gamma_2) \mathrm{d}x = 0 \qquad (2.78)$$

Empirical formulations such as the van Laar equations [19, p. 527; 37]

$$\log_{10} \gamma_1 = \frac{A_{1-2}}{[1+(A_{1-2}x_1)/(A_{2-1}x_2)]^2} \qquad (2.79)$$

$$\log_{10} \gamma_2 = \frac{A_{2-1}}{[1+(A_{2-1}x_2)/(A_{1-2}x_1)]^2} \qquad (2.80)$$

and the Margules equations [19, p. 527; 37]

$$\log_{10} \gamma_1 = x_2^2[A_{1-2} + 2x_1(A_{2-1} - A_{1-2})] \qquad (2.81)$$

$$\log_{10} \gamma_2 = x_1^2[A_{2-1} + 2x_2(A_{1-2} - A_{2-1})] \qquad (2.82)$$

have been in use for many years to fit activity coefficients derived from experimental vapor–liquid equilibrium data. Expressions for the constants can be found by solving the activity coefficient equations simultaneously [19, p. 538; 37].

van Laar:

$$A_{1-2} = (\log_{10}\gamma_1)\{1+[(x_2\log_{10}\gamma_2)/(x_1\log_{10}\gamma_1)]\}^2 \qquad (2.83)$$

$$A_{2-1} = (\log_{10}\gamma_2)\{1+[(x_1\log_{10}\gamma_1)/(x_2\log_{10}\gamma_2)]\}^2 \qquad (2.84)$$

Margules:

$$A_{1-2} = [(\log_{10}\gamma_1)/(x_2^2)] + 2(x_1)[(\log_{10}\gamma_2)/(x_1^2) - (\log_{10}\gamma_1)/(x_2^2)] \qquad (2.85)$$

$$A_{2-1} = [(\log_{10}\gamma_2)/(x_1^2)] + 2(x_2)[(\log_{10}\gamma_1)/(x_2^2) - (\log_{10}\gamma_2)/(x_1^2)] \qquad (2.86)$$

Constants (a and b) for the above formulations on a natural logarithm basis (ln γ instead of $\log_{10}\gamma$) are (ln (10) = 2.3026...) $\times A_{1-2}$ and A_{2-1}, respectively.

Other more recent formulations for the activity coefficient, such as Wilson, NRTL (Nonrandom Two Liquid) and UNIQUAC (Universal Quasi-Chemical) [38] are also employed, and some are often much better than others in reproducing VLE data for a particular system. Nonetheless, the van Laar equations are adequate for the present discussion to illustrate the underlying principles.

2.6.2.1 Methanol–Water

The first such system considered here is methanol–water at 1 atm pressure (Figure 2.7). It has been studied by a number of investigators in various different types of experimental distillation equipment, and constants for the van Laar equation are readily available as handbook values [19, p. 528]. Values of $A_{1-2} = 0.36$ and

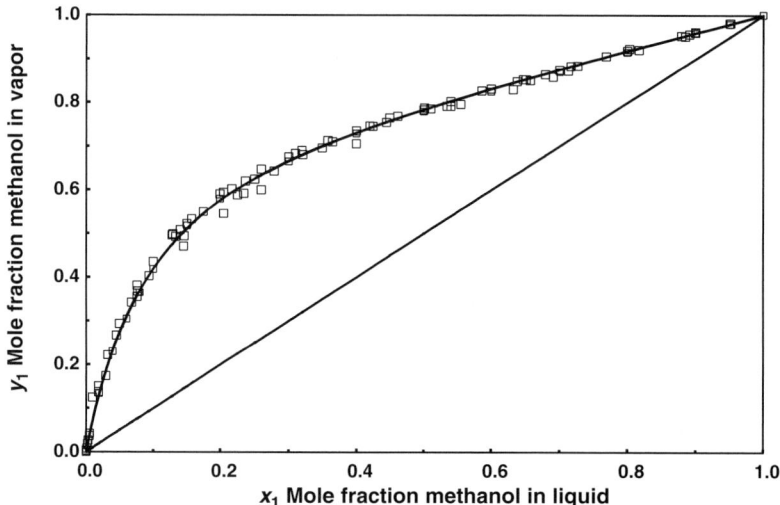

Figure 2.7 Methanol–water vapor–liquid equilibrium at 1 atm (x–y curve).

$A_{2-1} = 0.22$ ($a = 0.83$, $b = 0.51$) for the system at 1 atm pressure and a temperature range of 64.6°C–100°C have been used to draw the curve in the figure. Solutions for pairs of temperatures and mole fractions that result in a total pressure of 1 atm is by trial and error, as exemplified in Problem 2.26. Data points are from Refs [34,35].

PROBLEM 2.26

Using the Antoine constants for methanol and for water from Table 2.6 and the van Laar constants from the text above, calculate the liquid-phase mole fraction of methanol (x_1) corresponding to a temperature of 81.5°C to produce a total pressure of 1 atm (760 mmHg). Calculate the vapor-phase mole fraction (y_1) that goes along with this value of x_1, using Equation (2.76), assuming that the ideal gas law applies to the vapor phase.

SOLUTION From the Antoine equations with constants from Table 2.6, vapor pressures are calculated to be

$$\text{Methanol}: \quad p_1^{\text{vap}} = 10^{\wedge}[7.87863 - (1473.11)/(81.5 + 230.0)] = 1411.1 \text{ mmHg}$$

$$\text{Water}: \quad p_2^{\text{vap}} = 10^{\wedge}[7.96681 - 1668.21/(81.5 + 228.0)] = 377.4 \text{ mmHg}$$

The value of $x_1 = 0.21308885$ is obtained by trial and error in a spreadsheet program to produce a total pressure of 760 mmHg at a temperature of 81.5°C. The final trial is shown below with x_1 rounded to 0.21309:

$$\gamma_1 = \exp\left\{\frac{(0.83)}{[1 + (0.83)(0.21309)/(0.51)(0.78691)]^2}\right\} = 1.4916 \quad (2.87)$$

$$\gamma_2 = \exp\left\{\frac{(0.51)}{[1+(0.51)(0.78691)/(0.83)(0.21309)]^2}\right\} = 1.0489 \qquad (2.88)$$

$$P_T = \gamma_1 p_1^{\text{vap}} x_1 + \gamma_2 p_2^{\text{vap}} (1-x_1) \qquad (2.89)$$

from a combination of Equations (2.74) and (2.75).

$$P_T = (1.4916)(1411.1)(0.21309) + (1.0489)(377.4)(0.78691) = 760 \text{ mmHg}$$

The use of exact values from the spreadsheet program results in a P_T of 760.0000.

$$y_1 = p_1/P_T = p_1^{\text{vap}} \gamma_1 x_1 / P_T = (1411.1)(1.4916)(0.21309)/(760) = 0.5901$$

Similar calculations with values from $x_1 = 0$ to $x_1 = 1$ produce the curve in Figure 2.7. ∎

2.6.3 Azeotropic Systems

As systems become less and less ideal, the total pressure at constant temperature or the system temperature at constant pressure goes through a maximum or minimum. When this occurs, a situation known as an *azeotrope* occurs. For an azeotropic point, the composition of the liquid becomes identically equal to the composition of the vapor. At this point, the two components cannot be separated by distillation.

2.6.3.1 Ethanol–Water

The classic example frequently cited to illustrate an azeotrope is the ethanol–water system. It is well investigated in the literature and serves as a student exercise in college/university laboratories. This system forms an azeotrope composed of 95.6 wt% ethanol and 4.4 wt% water (probably rounded) at 1 atm pressure [19, p. 631] and 78.15°C [9, p. 633]. This corresponds to 89.43 mol% ethanol and 10.57 mol% water [19, p.633]. The azeotropic point is different at other total pressure levels [19, p. 631; 39].

PROBLEM 2.27

For the ethanol–water system, convert weight percent to mole percent necessary for VLE calculations.

SOLUTION Basis 100 lb of solution with 95.58 wt% ethanol and 4.42 wt% water.

Constituent	MW	lb	lb mol	mol%
Ethanol	46.07	95.58	2.0747	89.43
Water	18.0153	4.42	0.2453	10.57
Total	—	100.00	2.3200	100.00

lb mol (column 4) = lb (column 3) × MW (column 2)

mol% (column 5) = 100 × [lb mol of constituent (column 4)/lb mol total] ∎

PROBLEM 2.28

Determine the van Laar constants for the ethanol–water system at 1 atm.

SOLUTION In the special case of an azeotrope, one can obtain the van Laar constants by setting $y_1 = x_1$ and $y_2 = x_2$ in Equations (2.76) and (2.77) and solving the equations for the γ's at the azeotropic point.

$$\gamma_1 = (y_1/x_1)P_T/p_1^{\text{vap}} = (1)P_T/p_1^{\text{vap}} \qquad (2.90)$$

$$\gamma_2 = (y_2/x_2)P_T/p_2^{\text{vap}} = (1)P_T/p_2^{\text{vap}} \qquad (2.91)$$

Then the van Laar constants are given by Equations (2.83) and (2.84).

At 78.15°C for the azeotrope (89.43 mol% ethanol and 10.57 mol% water), vapor pressures are p_1^{vap} (ethanol) = 754.60 mmHg and p_2^{vap} (water) = 329.47 mmHg. They are calculated from the Antoine equation with the constants from Table 2.6.

	A	B	C
Ethanol	8.04494	1554.3	222.65
Water	7.96681	1668.21	228.0

From Equations (2.90) and (2.91),

$$\gamma_1 = P_T/p_1^{\text{vap}} = (760)/(754.60) = 1.00715; \quad \gamma_2 = P_T/p_2^{\text{vap}} = (760)/(329.47) = 2.30674$$

On a natural logarithm basis,

$$a = (\ln \gamma_1)[1 + (x_2 \ln \gamma_2)/(x_1 \ln \gamma_1)]^2$$
$$= [\ln(1.00715)]\{1 + [(0.1057)\ln(2.30674)]/[(0.8943)\ln(1.00715)]\}^2 = 1.5745$$
$$b = (\ln \gamma_2)[1 + (x_1 \ln \gamma_1)/(x_2 \ln \gamma_2)]^2$$
$$= [\ln(2.30674)]\{1 + [(0.8943)\ln(1.00715)]/[(0.1057)\ln(2.30674)]\}^2 = 0.9607$$

On a common logarithm basis, $A_{1-2} = 0.68381$ and $A_{2-1} = 0.41724$.

Figure 2.8 depicts the x–y diagram for ethanol–water at 1 atm total pressure. It uses the van Laar constants already determined, the Antoine constants from Table 2.6, and the same trial and error procedure employed in a previous example to draw the continuous curve. This curve crosses the $y = x$ line at the azeotropic point, as expected.

Data shown in the diagram are taken from a myriad of independent investigations. Data points are so numerous that they obscure the curve in some places. This may be a bit of overkill since the calculations have filled up the underlying computer file in generating the figure. Data were not selectively chosen to show agreement with the curve. The few stray points are from no one single investigation.

All in all, the van Laar equations with constants fitted at the published azeotrope reproduce the x–y data in the figure very well. Those x–y data were not used to obtain the constants of fit. The fitting procedure involves only the azeotropic point, and agreement with experimental values at other places along the curve is not guaranteed. The fact of such agreement testifies to the adequacy of the van Laar functional form for this system. Nevertheless, there may indeed be other pairs of constants that do not match the published azeotropic point exactly but fit the data overall as well or better. Different activity coefficient functional forms may also prove better, but that is another story. ∎

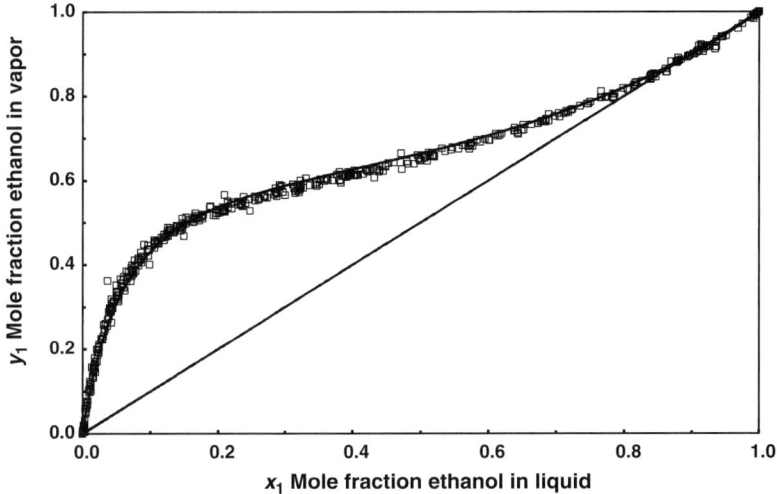

Figure 2.8 Ethanol–water vapor–liquid equilibrium at 1 atm (x–y curve).

2.6.3.2 Acetone–Chloroform

The boiling temperature for the ethanol–water azeotrope at 1 atm pressure (78.15°C) lies below the boiling points of the pure components (78.4°C for ethanol and 100°C for water). This is an example of a *minimum-boiling* azeotrope. Although less common, maximum-boiling azeotropes are found in other systems. One such case is acetone–chloroform at 1 atm pressure (Figure 2.9). Its azeotrope at 34.5 mol%

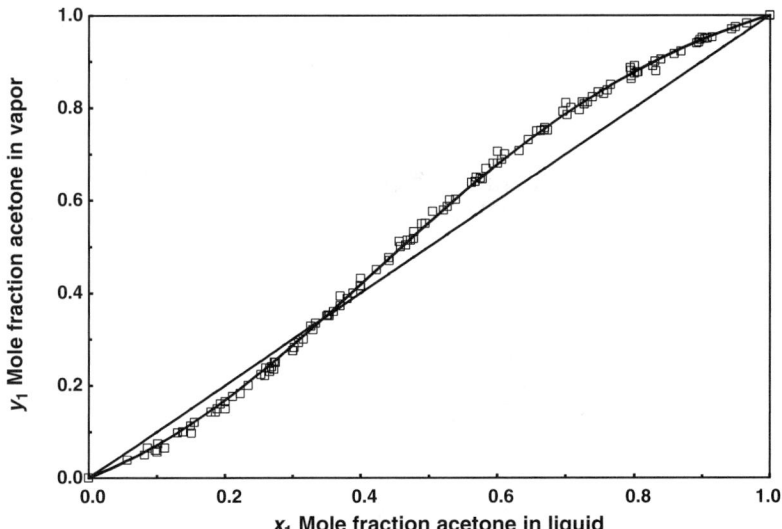

Figure 2.9 Acetone–chloroform vapor–liquid equilibrium at 1 atm (x–y curve).

acetone, 65.6 mol% chloroform, and 64.5°C [19, p. 634] occurs at a temperature above that of the pure components (56.2°C for acetone and 61.7°C for chloroform). The x–y diagram in the figure employs the mole fraction of acetone (the more volatile component) as x_1. However, treatment of this system in the literature has been mixed, and sometimes the mole fraction of chloroform is plotted as x_1 instead.

The curve is generated in the same manner as that for ethanol–water, already described. Antoine constants from the first entry in Table 2.6 are used for each of the components. As usual, experimental data points come from a number of literature sources [35,36]. The van Laar constants obtained from a fit at the azeotropic point are $A_{1-2} = -0.3080$ and $A_{2-1} = -0.2677$ on a common logarithm basis, and $a = -0.7091$ and $b = -0.6164$ using natural logarithms; verification is left to the reader.

2.6.4 Activity Coefficients Versus Mole Fraction Plots

The last graph of this section is a standard plot depicting the activity coefficients calculated for the VLE systems discussed (Figure 2.10). It shows, from top to bottom, $\ln \gamma$'s generated here for ethanol–water, methanol–water, and acetone–chloroform. The logarithm of the activity coefficients for benzene–toluene and methanol–ethanol, where Raoult's law applies throughout the entire range of x values, would fall on a straight horizontal line coincident with the x-axis.

The activity coefficient (γ) approaches 1.0 ($\ln \gamma \rightarrow 0$) for the constituent in greater abundance and actually reaches 1.0 for the pure material. In other words, Raoult's law holds for the solvent at infinite dilution of the solute material. The other end of each curve, as its $x \rightarrow 0$, approaches the natural logarithm of what is known as the *activity coefficient at infinite dilution*, γ^∞. Infinite dilution is approximated by mg/L or ppm concentrations of the species in lesser abundance (perhaps, an environmental contaminant).

A plot of the logarithm of the ratio of γ's (not shown) proves their thermodynamic consistency when the positive and negative areas under the curve are numerically equal.

2.6.5 Henry's Law Constants from Vapor–Liquid Equilibrium Activity Coefficients

In this section we will see how the Henry's law constant can be estimated from vapor–liquid equilibrium data. The Henry's law constant (H) for a volatile liquid can be calculated from the pure component vapor pressure (P^{vap}) and the activity coefficient at infinite dilution (γ^∞). The activity coefficient at infinite dilution, in turn, can be obtained from heat of mixing/heat of solution data[2] [19, pp. 529–530].

---

[2]When two liquids mix, they evolve or absorb heat. An extreme example is the dilution of concentrated sulfuric acid with water. One is cautioned to wear proper personal protective equipment (PPE) in the chemistry lab and to pour the more dense acid into the water to prevent splattering and contact with hot acid.

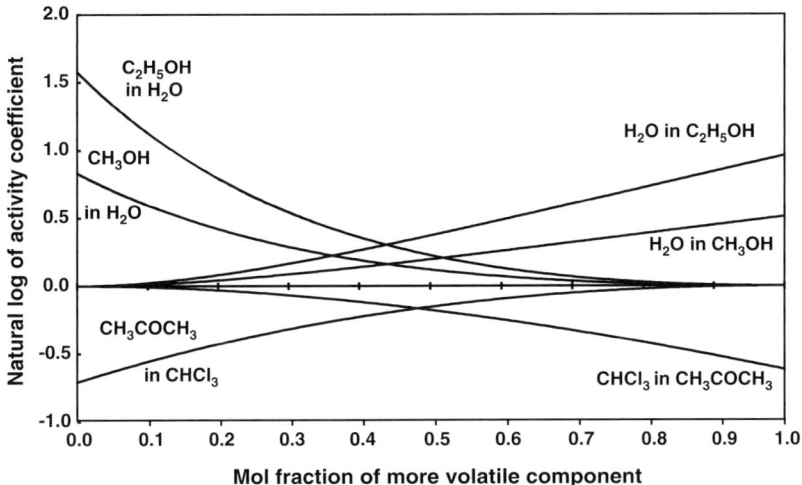

Figure 2.10 Activity coefficient plots for various systems at 1 atm.

A data search was conducted and such calculations were performed for methanol and ethanol in water because no Henry's law information could be found as a function of temperature [24,40,41]. This same technique can be extended to other chemicals as well, provided that the requisite data are available.

From Henry's law, recall that the partial pressure of a component is given by

$$p = Hx \tag{2.30}$$

From vapor–liquid equilibrium,

$$p = p^{\text{vap}} \gamma x \tag{2.73}$$

For each value of x, there is a corresponding γ, as in the previous figures showing vapor–liquid equilibrium relationships. For increasingly dilute solutions, the worst case (highest) activity coefficient occurs at zero mole fraction of the nonaqueous solute, so-called *infinite dilution* (see, for example, Figure 2.10).

Therefore,

$$Hx = \gamma^\infty p^{\text{vap}} x \tag{2.92}$$

and

$$H = \gamma^\infty p^{\text{vap}} \tag{2.93}$$

Equation (2.93) is assumed valid for all dilute concentrations.

Activity coefficients at infinite dilution and the resulting Henry's law constants determined for methanol and ethanol in water [24,40,41] are listed as a function of temperature in Table 2.10 along with the equations used to compute them. In these equations, terms are combined and the equations are simplified mathematically compared to those reported in the author's original publications. Component vapor pressures in the table are calculated from the Antoine equation with the proper constants from Table 2.6.

Table 2.10 Gamma Infinity, Vapor Pressure, and Henry's Law Constant for Methanol and Ethanol in Water

		Methanol (CH$_3$OH)			Ethanol (C$_2$H$_5$OH)			Water
T (°C)	T (K)	γ^∞	P^{vap} (atm)	H (atm/m.f.)	γ^∞	P^{vap} (atm)	H (atm/m.f.)	P^{vap} (atm)
25	298.15	1.711	0.1663	0.2846	4.095	0.0773	0.3164	0.0318
50	323.15	2.081	0.5454	1.135	5.429	0.2907	1.578	0.1218
75	348.15	2.310	1.4721	3.401	6.240	0.8745	5.463	0.3805
100	373.15	2.395	3.4186	8.818	6.444	2.2225	14.32	1.0000
125	198.15	2.356	7.0504	16.61	6.135	4.9346	30.27	2.2919
150	423.15	2.227	13.2194	29.44	5.487	9.8444	54.02	4.7070

Graphs calculated from the equations are presented in the original references, along with activity coefficients derived from published vapor–liquid equilibrium data. Each of the activity coefficient curves passes through a maximum point corresponding to the temperature at which the heat of solution at infinite dilution equals zero. Calculated activity coefficients at infinite dilution agree within the scatter of experimental values [34,41], although the functional form would not have been obvious had such an indirect method not been employed.

Activity coefficients in Table 2.10 are computed from

Methanol: $\ln \gamma^\infty = -4270.6967/T(°K) - 11.331862 \ln[T(°K)] + 79.42543$

(2.94)

Ethanol: $\ln \gamma^\infty = -6787.3992/T(°K) - 18.37086 \ln[T(°K)] + 128.84455$

(2.95)

and

$$\gamma^\infty = \exp[\ln \gamma^\infty]$$

Vapor pressures are obtained from the Antoine equation, Equation (2.23), with the following specific constants from Table 2.6:

	A	B	C	Range (°C)
Methanol	7.87863	1473.110	230	−20 to +140
Ethanol	8.04494	1554.3	222.65	—
Water	8.10765	1750.286	235	0 to +60
	7.96681	1668.21	228	+60 to +150

The Henry's law constant is calculated from the product of $\gamma^\infty \times P^{vap}$. For water, γ at infinite dilution of the alcohol species is that of pure water, that is, 1.0. Its Henry's law constant at this condition is, therefore, equal to its vapor pressure. Henry's law constants at 25°C (77°F) agree within the same order of magnitude with those in the cited tabulation [25], 0.2846 and 0.3164 (Table 2.10) versus 0.3858 and 0.4515 [25] for methanol and ethanol, respectively.

2.6 Vapor–Liquid Equilibrium

Use of Henry's law to estimate how much methanol and ethanol can be expected to condense with the water in a process gas is illustrated in Problem 2.29. In this example, it is assumed that the various components vaporize or condense independently; that is, each forms its own separate binary system with water, and nonaqueous component-to-component interactions are not important. It is further assumed that the thermodynamic constants used in the estimation are independent of pressure and of the concentration of their own or other species. All of these assumptions become better with increasingly dilute solutions.

PROBLEM 2.29

A hypothetical process gas at 500°F and 100 psig (pounds per square inch gauge) consisting mainly of inerts contains 15% moisture and 500 and 100 ppmw of methanol and ethanol, respectively, on a total or wet basis. This amounts to approximately 588 and 118 ppmw on a dry or moisture- (water-) free basis, 500 (100/85) and 100 (100/85). It is cooled to 50°C and let down to atmospheric pressure. Estimate how much gas condenses and with it how much methanol and ethanol end up in the condensate. How much water, methanol, and ethanol remain in the uncondensed gas?

SOLUTION The first step is to see whether the initial conditions of the problem are possible. From steam tables [15, p. 84], at 500°F, the vapor pressure of water is 680.86 psia. This is more than the total pressure of the process gas stream.

A total pressure of 100 psig is the pressure of the stream in addition to atmospheric pressure. For a standard atmosphere, the absolute pressure (psia) equals psig + 14.696. Hence,

$$\text{psia} = \text{psig} + 14.696; \text{ here psia} = 100 + 14.696 = 114.696$$
$$\div 14.696 = 7.8, \text{ or almost 8 atm} \quad (2.96)$$

Therefore, 5% moisture is possible. Since methanol and ethanol are more volatile (have higher vapor pressures) and exist at lower concentrations than the water in the process gas, the given concentrations for these contaminants are also possible.

Basis for calculations is 100 lb mol (total) of process gas at the initial conditions.

$$\text{Water }(H_2O) = (15 \text{ lb mol}/100 \text{ lb mol})(100) = 15 \text{ mol}$$
$$\text{Methanol }(CH_3OH) = (500 \text{ mol}/10^6 \text{ mol})(100) = 0.05 \text{ mol}$$
$$\text{Ethanol }(C_2H_5OH) = (100 \text{ mol}/10^6 \text{ mol})(100) = 0.01 \text{ mol}$$

mol H_2O/total gas = 15/100 = 0.15; mol H_2O/mol dry gas = 15/(100−15) = 0.1765
mol H_2O/mole of noncondensible gas = 15/(100−15−0.05−0.01) = 0.1766

It is assumed that the water will condense independent of the contaminants as the temperature is lowered and the pressure is let down. The alcohols will follow, distributing themselves between liquid and vapor according to Henry's law. It is also assumed that there is no interaction between methanol and ethanol and that each forms its own separate binary system with water. It is further assumed that the various thermodynamic constants used in the estimation are independent of pressure and concentration. All of these assumptions improve with increasingly dilute solutions [41].

At 50°C (122°F), the vapor pressure of water is given by the Antoine equation, Equation (2.23), with the appropriate constants chosen from Table 2.6.

$$\log_{10}[P(\text{mmHg})] = 8.10765 - 170.286/(50°C + 235.0) = 1.9663$$
$$P\,(\text{psia}) = (14.696/760)10^{\wedge}\log_{10}[P(\text{mmHg})] = 1.7893 \text{ psia}$$

Steam table value at 122°F [15, p. 87] is 1.7891 psia.

$$(n_{H_2O}/n_{\text{Total}}) = p_{\text{water}}/P_T = 1.7891/14.696 = 0.1217$$
$$n_{H_2O}/n_{\text{dry gas}}) = p_{\text{water}}/(P_T - p_{\text{water}}) = 0.1386$$

If the noncondensible gas is approximately the dry gas,

$$\text{Moles of water remaining in the gas} = (0.1386)(85) = 11.781$$
$$\text{and water condensed} = 15 - 11.781 = 3.219 \text{ mol } H_2O$$

The problem can be solved by an iterative technique. In the first iteration, the moles of CH_3OH and C_2H_5OH are neglected when being added to the noncondensibles and the water in both liquid and vapor. If no methanol and ethanol condense, this would add only 0.06 mol to the final vapor. If all the alcohols condense, this adds only 0.06 mol to the liquid. The CH_3OH and C_2H_5OH must material balance; that is, final moles of these contaminants in the liquid plus the gas must equal their initial values in the process gas. In addition, the moles in the gas are related to moles in the liquid by Henry's law.

First Trial

$$\text{moles in gas} = \text{moles in gas}$$

$$CH_3OH \text{ total initial moles} - n_{CH_3OH} \text{ in liquid} = \frac{(\text{total moles})(\text{mole fraction in vapor})}{(\text{total moles in gas})(y_{CH_3OH} \text{ in gas})}$$

Employing Henry's law:

$$\text{moles of } CH_3OH \text{ in gas} = (\text{total moles in gas})(H_{CH_3OH})(x_{CH_3OH} \text{ in liquid})/P_T$$
$$= (\text{total moles in gas})(H_{CH_3OH})(n_{CH_3OH} \text{ in liquid})/$$
$$\div [(P_T)(\text{total moles in liquid})]$$

Substituting values:

$$0.05 - n_{CH_3OH} \text{ in liquid} = (85 + 11.781)(1.135)(n_{CH_3OH} \text{ in liquid})/[(1)(3.219)]$$
$$0.05 - n_{CH_3OH} \text{ in liquid} = (34.1244)(n_{CH_3OH} \text{ in liquid})$$

Solving:

$$n_{CH_3OH} \text{ in liquid} = (0.05)/(35.1244) = 0.0014 \text{ mol of } CH_3OH \text{ in liquid}$$
$$n_{CH_3OH} \text{ in gas} = (34.1244)(0.0014) = 0.0486 \text{ mol of } CH_3OH \text{ in gas}$$

Similarly, for C_2H_5OH,

$$0.01 - n_{C_2H_5OH} \text{ in liquid} = \frac{(85 + 11.781)(1.578)}{(1)(3.219)} (n_{C_2H_5OH} \text{ in liquid})$$
$$= (47.4434)(n_{C_2H_5OH} \text{ in liquid})$$
$$n_{C_2H_5OH} = (0.01)/(48.4434) = 0.0002 \text{ mol } C_2H_5OH \text{ in liquid}$$
$$(47.4434)(0.0002) = 0.0098 \text{ mol } C_2H_5OH \text{ in gas}$$

Second Iteration

For the next iteration, the moles of the alcohols just calculated for liquid and gas are added to the water and inerts in both phases.

$$n_{H_2O}/n_{\text{non-}H_2O \text{ constituents}} = p_{H_2O}/(P_T - p_{H_2O} - p_{CH_3OH} - p_{C_2H_5OH})$$

$$p_{CH_3OH} = (H)(x_{CH_3OH}) = (H_{CH_3OH})(n_{CH_3OH}\text{in liquid})/(\text{total moles in liquid})$$

$$p_{CH_3OH} = (1.135)[(0.0014)/(3.219 + 0.0014 + 0.0002)]$$

$$p_{CH_3OH} = (1.135)(0.0014)/(3.2206) = 0.0005 \text{ atm} \times 14.696 = 0.0073 \text{ psia}$$

$$p_{C_2H_5OH} = (1.578)(0.0002)/(3.2206) = 0.0001 \text{ atm} \times 14.696 = 0.0014 \text{ psia}$$

$$n_{H_2O}/n_{\text{non-}H_2O \text{ constituents}} = (1.7891)/(14.696 - 1.7891 - 0.0073 - 0.0014)$$

$$= (1.7891)/(12.8982) = 0.1387$$

$$(0.1387)(85) = 11.7895 \text{ mol } H_2O \text{ in gas}$$

$$15 - 11.7895 = 3.2105 \text{ mol } H_2O \text{ condensed in liquid}$$

$$CH_3OH \quad 0.05 - n_{CH_3OH} \text{ in liquid} = \frac{(85 + 11.7895)(1.135)}{(1)(3.2105 + 0.0014 + 0.0002)}(n_{CH_3OH})$$

$$0.05 - n_{CH_3OH} = (34.2007)(n_{CH_3OH})$$

$$n_{CH_3OH} = (0.05)/(35.2007) = 0.0014 \text{ in liquid} \times 34.2007 = 0.0486 \text{ in gas}$$

$$C_2H_5OH \quad 0.05 - n_{C_2H_5OH} = \frac{(85 + 11.7895)(1.578)}{(1)(3.2105 + 0.0014 + 0.0002)}(n_{C_2H_5OH})$$

$$0.01 - n_{C_2H_5OH} = (47.6022)(n_{C_2H_5OH})$$

$$n_{C_2H_5OH} = (0.01)/(48.6022) = 0.0002 \text{ in liquid}$$

$$n_{C_2H_5OH} = (47.6022)(0.0002) = 0.0098 \text{ in gas}$$

Results are essentially the same as from the first trial. Stop!

Wrap-up:

$$(100)(3.2105/15) = 21.4\% \text{ of initial water condenses}$$

11.7895 mol H_2O, 0.0486 mol CH_3OH, and 0.0098 mol C_2H_5OH remain in vapor. ∎

Constituent in gas	mol	mol or vol% (wet)	mol or vol% (dry)
Inerts	85	87.77	99.93
H_2O	11.7895	12.17	—
CH_3OH	0.0486	0.05 = 502 ppmv	0.06 = 571 ppmv
C_2H_5OH	0.0098	0.01 = 101 ppmv	0.01 = 115 ppmv
Total	96.8479	100.00	100.00

In the liquid:

$$(3.2105 \text{ lb mol } H_2O)\frac{18.0153 \text{ lb } H_2O}{\text{lb mol } H_2O}\frac{\text{kg}}{2.20462 \text{ lb}}\frac{L}{1 \text{ kg}} = 26.2350 \text{ L round to } 26.2 \text{ L}$$

Methanol:

$$\frac{0.0014 \text{ lb mol}[(32.04 \text{ lb } CH_3OH)/(\text{lb mol } CH_3OH)][(\text{kg})/(2.20462 \text{ lb})][(10^6 \text{ g})/(\text{kg})]}{(3.2105 \text{ lb mol } H_2O)[(18.0153 \text{ lb } H_2O)/(\text{lb mol } H_2O)][(1 \text{ kg})/(2.20462 \text{ lb})][(1 \text{ L})/(\text{kg})]}$$

$$= 780 \text{ mg } CH_3OH/L \text{ in condensate}$$

Ethanol:

$$\frac{[(0.0002 \times 46.07 \times 10^6)/(2.20462)]}{[(3.2105 \times 18.0153 \times 1)/(2.20462)]} = 159 \text{ mg/L, properly rounded to } 200 \text{ mg/L}$$

Note that the conversion factor 2.20462 lb/kg cancels out in these calculations and could, therefore, have any value.

Because of the presence of methanol and ethanol, it is unlikely that the process gas would be allowed to be vented directly to atmosphere or that the raw condensate could be legally discharged untreated directly to the environment. Permit limits will vary, depending on the governing jurisdiction.

2.6.6 Rigorous Vapor–Liquid Equilibria for the Air–Water System

In the previous examples in this chapter, we have considered the concentration of water vapor in air at saturation and the maximum solubility of atmospheric gases in water as separate calculations. While these are extremely good approximations, the phenomenon is in fact an example of a single vapor–liquid equilibrium for the air–water system. To demonstrate this approach, an illustrative example in Ref. [42, pp. 46–56] is reworked in Problem 2.30. The very minor differences between the answers here and those of the cited example are caused by roundoff.

PROBLEM 2.30

Compare the normal simplified calculations with the more rigorous relationships for the air–water vapor–liquid equilibrium. Use an approximate composition of atmospheric air of 79% N_2 and 21% O_2, neglecting the argon (Ar) and the trace components.

SOLUTION At 20°C [19, p. 675], Henry's law constant for

$$N_2 = 8.04 \times 10^4 \text{ atm/m.f.}$$
$$O_2 = 4.01 \times 10^4 \text{ atm/m.f.}$$

From Table 2.7,

vapor pressure of H_2O at 20°C = 0.3390 psia

$$\div 14.696 = 0.023 \text{ atm}$$

$$n_{H_2O}/n_{\text{dry gas}} = p_{\text{water}}/(P_T - p_{\text{water}})$$
$$= (0.3390)/(14.696 - 0.3390) = 0.0236$$

This calculation assumes evaporation of water vapor from a pure water phase.

2.6 Vapor–Liquid Equilibrium

Basis 100 mol of dry air in the gas phase

Constituent	MW	mol% (dry)	mol (wet)	mol% (wet)
N_2	28.01	79	79	77.18
O_2	16.00	21	21	20.52
H_2O	18.0153	—	2.36	2.30
Total	—	100	102.36	100.00

From Henry's law, calculate the mole fraction (x) in the liquid.

$$x = p/H \qquad (2.31)$$

for O_2 $\quad x = (0.7718)(1)/(8.04 \times 10^4) = 9.5995 \times 10^{-6} \rightarrow$ rounded to 10×10^{-6}

for N_2 $\quad x = (0.2052)(1)/(4.01 \times 10^4) = 5.1172 \times 10^{-6} \rightarrow$ rounded to 5×10^{-6}

By difference, $\quad x_{H_2O} = 1 - 10 \times 10^{-6} - 5 \times 10^{-6} = 0.999985$

This is the normal way of performing these calculations and allows a decoupling of the thermodynamic equilibrium equations.

In actuality, the liquid is not pure water but water containing dissolved nitrogen and oxygen, albeit at miniscule concentrations.

The following equations apply:

$p_{H_2O} = (y_{H_2O})P_T = (p_{H_2O}^{vap})(x_{H_2O})$ from Raoult's law and an ideal gas phase

$p_{N_2} = (y_{N_2})P_T = H_{N_2}(x_{N_2})$ from Henry's law and an ideal gas phase

$p_{O_2} = (y_{O_2})P_T = H_{O_2}(x_{O_2})$ from Henry's law and an ideal gas phase

$$x_{H_2O} + x_{N_2} + x_{O_2} = 1 \quad (\Sigma x = 1)$$
$$y_{H_2O} + y_{N_2} + y_{O_2} = 1 \quad (\Sigma y = 1)$$

The final equation is an assumption put forth in the cited example that states that the ratio of $N_2/O_2 = 79/21$ regardless of water vapor content.

$$y_{O_2} = (21/79)(y_{N_2}) = 0.266(y_{N_2})$$

From $\Sigma x = 1$,

$$y_{H_2O} + (1 + 0.266)(y_{N_2}) = 1$$
$$y_{H_2O} = 1 - 1.266(y_{N_2})$$

From $\Sigma y = 1$,

$$x_{H_2O} + x_{N_2} + x_{O_2} = 1$$

Substituting for the x's,

$$(y_{H_2O}/p_{H_2O}^{vap})(P_T) + (y_{N_2}/H_{N_2})(P_T) + (y_{O_2}/H_{O_2})(P_T) = 1$$
$$(y_{H_2O}/p_{H_2O}^{vap})(P_T) + (y_{N_2})[(1/80,400) + (0.266)/(40,100)](1) = 1$$
$$(y_{H_2O}/0.023)(1) + (y_{N_2}/H_{N_2})(P_T) + (0.266)(y_{O_2}/H_{N_2})(P_T) = 1$$
$$(43.47826)(y_{H_2O}) + (0.0000190712)(y_{N_2}) = 1$$

60 Chapter 2 Basic Concepts

Substitution in $\Sigma = 1$ equation to solve simultaneously

$$(43.47826)[1-(1.266)(y_{N_2})] + (0.0000190712)(y_{N_2}) = 1$$
$$(55.0434579)(y_{N_2}) = 42.47826$$
$$y_{N_2} = 0.771722$$

To summarize,

Results from this method	From previous simplified method
$y_{N_2} = 0.771722$	0.7718
$y_{O_2} = 0.205278$	0.2052
$y_{H_2O} = 0.02299966$	0.023
$x_{N_2} = 9.5985 \times 10^{-6}$	10×10^{-6}
$y_{O_2} = 5.1192 \times 10^{-6}$	5×10^{-6}
$y_{H_2O} = 0.9999852$	0.999985
$(y_{O_2})/(y_{N_2}) = 0.2660$	0.26587

∎

The two sets of results are extremely close, and the additional refinement is not worth the added complications. As stated by the cited author [42, p. 56], not all of the digits are significant. They are carried along to show the small difference in results between the two techniques. No one would ever solve this particular problem the hard way except perhaps as an illustrative example of how more complex equilibria would be handled.

Nonetheless, any reader desiring additional practice is invited to work out a similar though somewhat easier exercise considering air as a single component, and resulting in only four simultaneous equations and four unknowns. The Henry's law constant for "air" at 20°C is 6.64×10^4 atm/m.f. [19, p. 674]; the vapor pressure of water remains the same.

2.7 ENERGY BALANCES AND HEAT TRANSFER

There will be occasions where an energy (heat) balance will be necessary in addition to or in lieu of a material (mass) balance. This section outlines the basics of such balances and discusses the very basics of heat transfer as well. Additional topics in these areas will be brought up as necessary in the application areas later in this book.

2.7.1 Addition or Subtraction of Energy with No Change of Phase

For a solid remaining a solid, a liquid remaining a liquid, or a gas/vapor remaining as such, a change in temperature (ΔT) by whatever means involves the *heat capacity* or *specific heat* of the material.

$$\text{Energy charge} = (m \text{ or } n)(C_p)\Delta T \tag{2.97}$$

where m is the mass or n is the number of moles of material and C_p is the heat capacity in units of energy per mass or mole units per temperature unit. Typically, C_p is given in calories per gram or gram mole per °C change or Btu per lb or lb mol per °F change. Because of the combination of conversion factors involved, these values are numerically equal.

More appropriately,

$$\text{Energy change} = (m \text{ or } n) \int_{T_1}^{T_2} C_p dt \qquad (2.98)$$

in cases where the heat capacity cannot be assumed constant over the applicable temperature range. Experimental values of heat capacity are correlated as empirical functions of temperature. The uncertainty of the correlated C_p decreases with the number of temperature terms included in the function.

For gases, there are two heat capacities, the heat capacity at constant pressure (C_p) and the heat capacity at constant volume (C_v). C_p is greater than C_v because the gas expands at constant pressure as the temperature is raised and thereby does work ($P\Delta V$).

For an ideal gas,

$$C_p - C_v = R \qquad (2.99)$$

where R is the universal gas constant (with a value of 1.986) in molar and thermal units from the ideal gas equation. The ratio of C_p to C_v is γ (not an activity coefficient, despite the unfortunate duplication of nomenclature), the dimensionless exponent of ideal gas expansion in an adiabatic process (i.e., one without heat transfer to/from the surroundings, or outside environment).

$$P_1 V_1^{\gamma} = P_2 V_2^{\gamma} \qquad (2.100)$$

The distinction in heat capacity for a liquid or a solid makes little difference because of their minor changes in volume with temperature.

2.7.2 Latent Heat

In addition to a change in temperature with no change in phase, the so-called *sensible heat*, the heat added to or removed from a material during a phase change must be considered. This is termed *latent heat*. Included are the *latent heat of fusion* (ΔH_f) or its converse, the *latent heat of melting*, equal in magnitude but opposite in sign, for the transition from the liquid to the solid phase, or vice versa. The latent heat added to cause passage from the liquid to the gaseous state is the *heat of evaporation* (ΔH_v), it is called the *heat of condensation* when heat is removed from a vapor to allow it to form a liquid. When a solid passes directly to a vapor, the heat effect is called the heat of *sublimation*. These heat effects are much greater in magnitude than the value of heat capacity, and a large change in temperature is needed to equal the effect of latent heat.

2.7.3 Heat Transfer

Heat transfer takes place by the process of conduction, convection, and/or radiation. *Conduction* is the transfer of energy from one molecule to another without gross movement of the medium. It occurs in solids and in liquids and gases as well. An important parameter in conduction is the thermal conductivity (k), a property of the material expressed in units of Btu/h per ft per °F.

Convection is the transfer of energy by the macroscopic circulation of a fluid, that is, a liquid or gas/vapor. Transfer of heat to or from a fluid through a hypothetical thin film of fluid can be calculated by means of a heat transfer coefficient (h) typically in units of Btu/h per ft^2 per °F.

Radiation is the transfer of energy by emission or absorption of electromagnetic waves. It occurs primarily at high temperatures and is proportional to the fourth power of the absolute temperature.

The type of heat transfer arguably most important in environmental matters involves some form of convection.

Now, we move on to the problems, the first of which is a throwback to Section 2.2.

PROBLEM 2.31

An ideal gas is expanded adiabatically (heat exchange with the surroundings = zero) from 300°F and 100 psig to standard atmospheric pressure. Calculate the final temperature and the relative volume of the gas. For heat capacities, take $C_p = 7$ Btu/lb mol per °F and $C_v = 5$ Btu/lb mol per °F, both constant over the range of this problem.

SOLUTION $C_p - C_v = R = 7 - 5 = 2$ Btu/lb mol per R, compared to the more exact value of 1.986.

$$R \text{ in thermal units} = \frac{10.731 \text{ ft}^3 \text{ lb}_f}{\text{lb mol R in.}^2} \frac{144 \text{ in.}^2}{\text{ft}^2} \frac{\text{Btu}}{778.17 \text{ ft lb}_f} = \frac{1.986 \text{ Btu}}{\text{lb mol R}}$$

$$C_p/C_v = 7/5 = 1.4$$

$$P_1 = 100 \text{ psig} + 14.696 = 114.696 \text{ psia} \tag{2.96}$$

$T_1 = 300°\text{F} + 459.67 = 759.67°\text{R}$ (see Table 2.2)

$$P_1 V_1^\gamma = P_2 V_2^\gamma \tag{2.100}$$
$$(V_2/V_1)^\gamma = (P_1/P_2)$$
$$\ln(V_2/V_1) = (1/\gamma)\ln(P_1/P_2) = (1/1.4)[\ln(114.696/14.696)]$$
$$= (0.7143)[\ln(7.8046)] = (0.74143)(2.0547) = 1.4677$$
$$V_2/V_1 = \exp[1.4677] = 4.3390 = 4.34 \text{ rounded to three significant figures}$$

From the ideal gas law (Section 2.2),

$$T_2 = (P_2/P_1)(V_2/V_1)(T_1)$$
$$(1/7.8046)(4.3390)(759.67) = 422.34°\text{R}$$
$$422.34°\text{R} - 459.67 = -37.33°\text{F} \qquad\blacksquare$$

PROBLEM 2.32

Prove that a heat capacity expressed in units of cal/g mole per ($\Delta°C$ or ΔK) is numerically equivalent in units of Btu/lb mol per ($\Delta°F$ or ΔR).

SOLUTION First, a temperature difference in °C or °K and a temperature difference in °F or °R are the same because the Celsius (C) and the Kelvin (K) degrees are the same in size; the size of the Fahrenheit (F) and the Rankine (R) degrees are also the same. The Celsius or the Kelvin degree is 1.8 times the size of the Fahrenheit or the Rankine degree.

$$\frac{1 \text{ cal}}{\text{g mol }\Delta°C \text{ or }\Delta°K} \frac{\text{Btu}}{252 \text{ cal}} \frac{453.59 \text{ g mol}}{\text{lb mol}} \frac{(\Delta°C \text{ or }\Delta°K)}{(1.8)(\Delta°F \text{ or }\Delta°R)} = 1$$

Since the molecular weight in g/g mol or lb/lb mol is the same, heat capacities in cal/g per $\Delta°C$ or $\Delta°K$ = heat capacities in Btu/lb per $\Delta°F$ or $\Delta°R$.

The heat capacity of liquid water is approximately 1.0 in both systems of units. ∎

PROBLEM 2.33

Calculate the mean heat capacity in the gaseous state of the materials listed below between 300°R and various higher temperatures. Units of C_p are cal/(g mol °C) or Btu/(lb mol °F).

N_2	$C_p = 6.50 + 0.00100T$ (°K)	range (300–3000K)
O_2	$C_p = 8.27 + .000258T$ (°K) $- 187{,}700/[T\,(°K)]^2$	range (300–3500K)
CO_2	$C_p = 10.34 + 0.00274T$ (°K) $- 195{,}500/[T\,(°K)]^2$	range (273–1200K)
H_2O	$C_p = 8.22 + 0.00015T$ (°K) $+ 0.00000134[T\,(°K)]^2$	range (300–3000K)

These particular equations are tabulated among many others for other materials in Ref. [19, pp. 220–226]. Maximum uncertainty ranges from 1% for O_2 through 1.5% for CO_2 to 3% for N_2. Uncertainty for H_2O is not specified. We shall have occasion to use other heat capacity equations containing a greater number of temperature terms for these same materials elsewhere in this book.

SOLUTION Starting from Equation (2.98),

$$\bar{C}_p = (1/\Delta T) \int_{T_{\text{base}}}^{T_{\text{final}}} C_p \, dt \qquad (2.101)$$

where $\Delta T = T_{\text{final}} - T_{\text{base}} = T_{\text{final}} - 300°K$ for this problem, and $\bar{C}_p$ is the mean heat capacity over this temperature range.

For N_2,

$$\bar{C}_p = [1/(T_{\text{final}} - 300)][6.50T + (0.00100/2)T^2]_{300}^{T_{\text{final}}}$$

$$\bar{C}_p = (1/(\Delta T)[6.50(\Delta T) + (0.0005)T_{\text{final}}^2 - (0.0005)(300)$$

Similarly, for

O_2 $\bar{C}_p = (1/\Delta T)\{(8.27)(\Delta T) + 0.000258/2)(T_{\text{final}}^2 - 300^2)$
$\qquad + (187{,}700)[(1/T_{\text{final}}) - (1/300)]\}$

CO_2 $\bar{C}_p = (1/\Delta T)\{(10.34)(\Delta T) + 0.00274/2)(T_{\text{final}}^2 - 300^2)$
$\qquad + (195{,}500)[(1/T_{\text{final}}) - (1/300)]\}$

H_2O $\bar{C}_p = (1/\Delta T)[(8.22)(\Delta T) + (0.00015/2)(T_{\text{final}}^2 - 300^2)$
$\qquad + (0.00000134/3)[(T_{\text{final}}^3 - 300^3)]$

64 Chapter 2 Basic Concepts

Readers who may be a bit rusty in the mathematics of integration should consult a calculus book or table of integrals.

Results are given in Table 2.11 for a final temperature from 300°K (approximately 25°C or 77°F) up to 3000°K. The value for 300°K is found simply by plugging into the original heat capacity equation without integration since the integrated equation gives rise to the mathematically indeterminate 0/0 for $T_{final} = T_{base}$ (here, 300°F). ∎

PROBLEM 2.34

It is desired to cool 100 SCF (60°F, 1 atm) of nitrogen (N_2) from 500°F to 50°F. How much heat in Btu must be removed? As in Problem 2.33, the heat capacity for gaseous N_2 at constant pressure (C_p) in cal/g mol/°K is given as $6.50 + 0.00100T$ (°K) between 300 and 3000°K, with an uncertainty of 3% [19, p. 222].

SOLUTION Since the temperature in the heat capacity equation is in K, calculations will be performed in the metric system and converted at the end to Btu.

$$\text{The amount of gas} = 100\ \text{SCF}\ (60°F, 1\ \text{atm}) \frac{\text{lb mol}}{379.5\ \text{SCF}\ (60°F, 1\ \text{atm})} = 0.263505\ \text{lb mol}$$

$$\times\ 453.59\ \text{g mol/lb mol} = 119.52\ \text{g mol}$$

The molar volume used (379.5 SCF/lb mol) comes from Table 2.4.

Convert temperatures (see Table 2.2)

$$(500°F + 459.67)/(1.80) = 533.15°K$$
$$533.15°K - 273.15 = 260°C$$
$$(50°F + 459.67)/(1.8) = 283.15°K$$
$$283.15°K - 273.15 = 10°C$$

$$\text{Energy transferred} = n \int_{T_1}^{T_2} C_p dt \tag{2.98}$$

$$= (119.52\ \text{g mol}) \int_{T_1}^{T_2} (6.5 + 0.001000T) dt$$
$$= 119.52[6.5T + 0.00100(T^2/2)]_{T_1=283.15°K}^{T_2=538.15°K}$$
$$= 119.52[(6.5)(533.15 - 283.15) + (0.00050)(533.15^2 - 283.15^2)]$$
$$= 119.52[1625 + 102.0375] = (119.52)[1727.0375] = 206{,}415.5220\ \text{cal}$$
$$\div\ 252\ \text{cal/Btu} = 819\ \text{Btu}$$

An alternate solution comes from the definition of the mean heat capacity such that

$$\bar{C}_p = (1/\Delta T) \int_{T_1}^{T_2} C_p dt \tag{2.101}$$

$$\text{Energy transferred} = n\bar{C}_p \Delta T \tag{2.97}$$

The value of the integral in metric units (from above) is 1727.0375.

$$\bar{C}_p = (1727.0375)/(\Delta T = 533.15 - 283.15) = 6.9082/\text{g mol }°K$$

This is equivalent to 6.9082 Btu/lb mol/°F.
Energy transferred = (0.263505 lb mol)(6.9082 Btu/lb mol/°F)(500–50)°F = 819 Btu. ∎

Table 2.11 Mean Heat Capacities of Various Gases Between 300 K and the Indicated Temperature

				N$_2$		O$_2$		CO$_2$		H$_2$O	
T (K)	T (°F)	T (°C)	ΔT (°K)	$\bar{C}_p$ molar	$\bar{C}_p$ mass	$\bar{C}_p$ molar	$\bar{C}_p$ mass	$\bar{C}_p$ molar	$\bar{C}_p$ mass	$\bar{C}_p$ molar	$\bar{C}_p$ mass
300	80.33	26.85	0	6.80	0.243	6.26	0.196	8.99	0.204	8.39	0.465
500	440.33	226.85	200	6.90	0.246	7.12	0.223	10.13	0.230	8.50	0.472
1000	1340.33	726.85	700	7.15	0.255	7.81	0.244	11.47	0.261	8.94	0.496
1200	1700.33	926.85	900	7.25	0.259	7.94	0.248	11.85	0.269	9.18	0.509
1500	2240.33	1226.85	1200	7.40	0.264	8.09	0.253	—	—	9.60	0.533
2000	3140.33	1726.85	1700	7.65	0.273	8.25	0.258	—	—	10.49	0.582
2500	4040.33	2226.85	2200	7.90	0.282	8.38	0.262	—	—	11.60	0.644
3000	4940.33	2726.85	2700	8.15	0.291	8.49	0.265	—	—	12.93	0.718

$\bar{C}_p$ molar in units of Btu/lb mol °F; $\bar{C}_p$ mass in units of Btu/lb °F. Molecular weights: N$_2$ 28.01, O$_2$ 32.00, CO$_2$ 44.01, and H$_2$O 18.0153. Base temperature = 300°K.

PROBLEM 2.35

An ice water slurry of 100 lb is combined with 40 gal of water at 150°F in an open vat. The ice melts and the resulting water achieves a temperature of 114°F.

(a) How much ice was present in the original mixture? Assume that no evaporation occurs and no heat is lost to the surroundings.

(b) What is the final temperature without ice originally present?

SOLUTION Because ice and water are present at atmospheric pressure, the temperature can be assumed to be 0°C (32°F). The heat of fusion at this temperature is 1.436 kcal/g mol [7, p. 9–97; 19, p. 210]. The specific heat of liquid water is about 1.007 cal/g per °C (Btu/lb per °F) at both ends of the temperature range of 0–100°C, and it passes through a minimum of 0.99735 at 34–35°C [5, p. D-158]. It will be considered constant here at 1 Btu/lb per °F. (The heat capacity of ice is approximately constant at 0.5 Btu/lb per °F, certainly from 0°C to −10°C, and the heat capacity of subcooled water from 0°C to −6°C is approximately 1 in the same units [5, p. D-159]. The density of liquid water as a function of temperature is readily available as the reciprocal of the specific volume (ft^3/lb) in steam tables [15, pp. 83–88]. Specific volume or density of liquid water is not a strong function of temperature.

The conversion factor between cubic feet and U.S. gallons can be derived from the definitions of 231 in.^{2} = 1 gal and 12 in. = 1 ft.

(a) Final Temperature 114°F—Conversion of Units:

$$\frac{(1.436 \text{ kcal/g mol})(1000 \text{ cal/kcal})(453.59 \text{ g mol/lb mol})}{(252 \text{ cal/Btu})(18.0153 \text{ lb H}_2\text{O/lb mol H}_2\text{O})} = 143.5 \text{ Btu/lb } \Delta H_f$$

$$(1 \text{ gal})(1728 \text{ in.}^3/\text{ft}^3)(1 \text{ gal}/231 \text{ in.}^3) = 7.48 \text{ gal/ft}^3$$

Density of liquid water at 150°F: $1/(0.0116343 \text{ ft}^3/\text{lb}) = 61.188 \text{ lb/ft}^3$

Water added: $(40 \text{ gal})(61.188 \text{ lb/ft}^3)(1 \text{ ft}^3/7.48 \text{ gal}) = 327.2 \text{ lb}$

Raise temperature of ice water mixture = cool hot water.
Let x = lb of ice, 100 = lb of ice water mixture
Then, (100 − x) = lb of cold water initially.

$$\underbrace{(x \text{ lb})(143.5 \text{ Btu/lb})}_{\text{Melt ice}} + \underbrace{(100 \text{ lb})(1 \text{ Btu/lb/°F})(T_{\text{final}}°\text{F}-32°\text{F})}_{\text{Heat original water and melted ice}}$$

$$\underbrace{(327.2 \text{ lb})(150°\text{F}-T_{\text{final}}°\text{F})(1)}_{\text{Cool hot water added}}$$

Substituting 114°F for the final temperature (T_{final}) (given)

$$143.5x + (100)(1)(114-32) = (327.2)(150-114)(1)$$
$$143.5x + (100)(1)(82) = (327.2)(36)(1)$$
$$x = 24.9 \text{ lb of ice in mixture}$$

(b) For no ice present, $x = 0$. Solve for T_{final}.

$$0 + (100)(T_{\text{final}}-32) = (327.2)(150-T_{\text{final}})$$

$$T_{\text{final}} = (52,220)/(427.2) = 122.4°\text{F final temperature without ice present initially} \quad \blacksquare$$

2.7 Energy Balances and Heat Transfer

PROBLEM 2.36 Flashing Factor

When a liquid such as water is let down from a higher pressure to a lower pressure, some of the liquid vaporizes while the rest remains in the liquid state. The fraction of vapor formed is known as the *flashing factor*.

Consider here an example of the flashing factor as 1 lb of condensate water from 600 psig saturated steam is let down to atmospheric pressure. Using steam tables, determine how much steam will flash to atmosphere and how much liquid will find its way to wastewater treatment?

SOLUTION Blowdown across a valve or steam trap is a process that occurs at constant enthalpy (pronounced (en-THAL-py). The enthalpy (H) is a thermodynamic function defined as the internal energy (E) + [the pressure (P) × the volume (V)].

$$H = E + PV \tag{2.102}$$

It is a function of convenience because the group $E + PV$ occurs so often as a thermodynamic variable. Refer to any thermodynamics text for details.

At constant enthalpy ($\Delta H = 0$), the enthalpy of the final state must equal the enthalpy of the initial state. For this to happen, some of the condensate must evaporate.

From Equation (2.96), initial condition = 600 psig + 14.696 = 614.696 psia.
From steam tables [15, p. 92],

Pressure (psia)	Temperature (°F)	Enthalpy of water (Btu/lb)
620	489.74	475.8
614.696	488.81 (Interpolated)	474.74 (Interpolated)
610	487.98	473.8

Linear interpolation for temperature (°F):

$[(614.696-610)/(620-610)](489.7-487.98) + 487.98 = 488.8065 \cong 488.81°F$

Linear interpolation for enthalpy (Btu/lb):

$[(614.696-610)/(620-610)](475.8-473.8) + 473.8 = 474.7392 \cong 474.74\,\text{Btu/lb}$

Final condition [15, pp. 86, 95]

Pressure (psia)	Temperature (°F)	Enthalpy (Btu/lb)	
(Given)	(From steam tables)	Water	Steam
14.696	212.00	180.17	1150.5

$$H_{\text{initial}} = H_{\text{final}}\,(\Delta H = 0)$$

Hot water under pressure = Vapor at atm pressure + Liquid at atm pressure

$(1)(474.74) \qquad\qquad (x)(1150.5) \qquad\qquad (1-x)(180.17)$

Solving for x, the number of lb of water vapor (steam), is 0.30358 and $(1-x)$, the number of lb of liquid water, is 0.69642. Approximately, 0.3 or 30% flashes and about 0.7 or 70% remains as liquid and becomes wastewater unless used elsewhere in the plant.

68 Chapter 2 Basic Concepts

A similar phenomenon occurs when high-pressure steam is let down to a lower pressure; the fraction of vapor in the resulting two-phase mixture is known as the *quality*. If the process occurs in a steam turbine or a flow nozzle, calculations would be performed as above, but at constant entropy (EN-tro-py) values from the steam tables ($\Delta S = 0$). ■

PROBLEM 2.37

A flue gas at 1000°F, consisting of 72% N_2, 2% O_2, 16% CO_2, and 10% H_2O by volume, is scrubbed with water for pollution control. Calculate the adiabatic saturation temperature.

SOLUTION When a hot flue gas is scrubbed with water, some water is evaporated, cooling the gas. The water will evaporate to the extent governed by the saturation vapor pressure at the final temperature achieved.

The heat given up by the gas is balanced by the heat necessary for the evaporative change of phase of the scrubbing water. At steady state, the scrubbing water and the gas exit at the same temperature. This is known as the theoretical *adiabatic saturation temperature*.

Neglecting Sensible Heat from Makeup Water

To keep the problem simple, without having to specify much details of operation, the sensible heat in raising the liquid from the supply temperature to the steady-state temperature in the scrubber will first be neglected. In a second case, sensible heat for the water is considered but only the sensible heat increase of the makeup water. The results will be compared. Solution is by trial and error, in which the temperature must first be guessed to calculate physical properties dependent on temperature.

Heat capacity equations from Problem 2.33 are being used. Mean heat capacities between 1000°F and the temperature of the final trial are as follows:

Constituent	$\bar{C}_p$ (molar)	$\bar{C}_p$ (mass)
N_2	7.08	0.253
O_2	7.75	0.242
CO_2	11.23	0.255
H_2O	8.78	0.487

See Problem 2.33 for calculation of mean heat capacity.

The vapor pressure of water is calculated from Equation (2.23), the Antoine equation, for the appropriate temperature range (Table 2.6). Preliminary calculations have indicated that the final temperature is within the range of 160°F and 170°F. This determines which set of constants from Table 2.6 to use (60–150°C, 140–302°F).

The heat of evaporation from steam tables [15, p. 87] between 160°F and 170°F and slightly beyond is linear and is fitted by the equation

$$\Delta H_v \text{ (Btu/lb)} = 1098.2 - (0.6)[T \text{ (°F)}]$$

Basis is 100 mol/h of flue gas.

$$\left[\sum n\bar{C}_p\right][1000°F - T_{\text{final}} (°F)] = (\text{max allowable moles } H_2O - n_{H_2O} \text{ in the gas})(\Delta H_v)$$

2.7 Energy Balances and Heat Transfer 69

Trial and error solution via spreadsheet calculation yields a temperature of 162.4736417802°F for the final trial, at which point, the difference (ΔBtu/h) between the left-hand side of the above equation minus its right-hand side $= 0.000000$ Btu/h. See the following.

Saturation water at temperature:

$$(n_{H_2O}/n_{dry\ gas}) = p^{vap}_{H_2O}/(P_T - p^{vap}_{H_2O}) = (0.3422\ atm)/(1 - 0.3422)$$
$$= 0.5202\ mol\ H_2O\ per\ mole\ of\ dry\ gas$$
$$\times (100-10)\ mol/h\ of\ dry\ gas = 46.8194 \cong 46.82\ mol\ H_2O/h$$

$46.8194 - 10\ mol/h\ H_2O$ present in gas $= 36.8194\ mol/h\ H_2O$ to be evaporated

ΔH_v at temperature $= 1000.715\ Btu/lb \times 18.0153\ lb/lb\ mol = 18{,}028.18\ Btu/lb\ mol$

$$\underbrace{[(72)(7.08) + (2)(7.75) + (16)(11.23) + (10)(8.78)](1000 - 162.47+)}_{\text{Heat from the gas}}$$
$$= \underbrace{(36.8194)(18{,}028.18)}_{\text{Evaporative cooling}}$$
$$= 663{,}787.7\ Btu/h$$

Considering Sensible Heat of Makeup Water

An added degree of complexity is to consider the sensible heat that must be added to raise the temperature of the makeup water replacing the scrubber water being evaporated. In reality, makeup water would be added to replace the evaporation plus whatever blowdown is taken to control the concentration of some contaminant among those being removed from the gas. That situation would require a complete material balance, which is beyond the scope intended for a supposedly simple example problem.

Now we return to the single added degree of complexity, namely, heating the makeup water that replaces the evaporation. This introduces an additional term on the RHS of the heat balance equation

$$(n_{H_2O})(C_{p_{H_2O}})(T_{final} - 60°F)$$

where the temperature of the makeup water is assumed to be 60°F.

$$(33.50172\ lb\ mol/h\ H_2O)(1\ Btu/lb°F)(18.0153\ lb/lb\ mol\ H_2O) \times$$
$$(T_{final} - 60°F) = Btu/h$$

The final temperature, again by trial and error, works out to be 160.41042383735. The mean molal heat capacities are virtually the same as before, and the heat of evaporation at this new temperature by linear interpolation of steam table values is 1001.153 Btu/lb. The vapor pressure is 0.325851 atm, making the allowable maximum moisture content in the gas 0.483352 mol H_2O/mol dry gas. Multiplying this last figure by 90 mol/h of dry gas per 100 mol/h of wet flue gas gives a value of 43.50172 mol/h. Subtracting the 10 mol/h H_2O already in the gas gives the 33.50172 lb mol/h in the equation immediately above.

The final heat balance is calculated as

From cooling the gas   From heating and evaporating the water
665,324.8 Btu/h   = 665,324.8 Btu/h
with ΔBtu/h   = 0.000000

The Btu/h contributions to the heat balance equation readjust to make the final temperature only about 2°F less than that of the first case. In round numbers, the gas provides an additional 1600 Btu/h by being cooled an extra 2°F. Less water, corresponding to a heat of evaporation of 59,000 Btu/h, is capable of evaporation at the lower temperature. This heat duty is replaced by the new sensible heat term, 60,600 Btu/h, from heating the makeup water from 60°F to approximately 160.4°F.

A Quicker But Approximate Calculation

A good rule of thumb for more rapid approximate calculations with flue gas constituents is to use a heat capacity of 0.5 Btu/lb °F for H_2O and 0.25 Btu/lb °F for the non-H_2O components. An approximate value for ΔH_v is a constant 1000 Btu/lb. The results are quite good compared to the exact values above. Often, a "ballpark" temperature will be all that is necessary.

The LHS of the heat balance equation becomes

$$\left(\sum mC_{p_{\text{mass}}}\right)(1000-T_f)$$

Mass flow rates are calculated as follows:

Constituent	MW	mol/h	lb/h = mol/h × MW
N_2	28.01	72	2016.72
O_2	32.00	2	64.00
CO_2	44.01	16	704.16
H_2O	18.0153	10	180.153
Total	—	100	2784.880 (dry, non-H_2O components); 2965.033 (all components, including H_2O)

$$(\text{lb/h } H_2O)(\text{Btu/lb°F}) + (\text{lb/h non-}H_2O)(\text{Btu/lb°F})$$
$$(180.153)(0.5) + (2784.880)(0.25) = 786.3 \text{ Btu/h/°F}$$

Results are summarized below and are a few tenths of a degree F different from the answers for the respective rigorous cases.

Case	T_{final} (°F)	LHS (Btu/h)	(RHS (Btu/h)	Δ(Btu/h)
No sensible heat	162.3197492503	658,665.0	658,665.0	0.000000
Heat makeup water	160.2851703985	660,264.8	660,264.8	0.000000

There are times when heating or cooling is desired not by direct mixing of two or more liquid or gaseous streams. In such cases, indirect heat exchange is used. Not only is the final temperature important, but the rate of heat transfer must also be addressed. Consider the following example. ∎

PROBLEM 2.38

You are an environmental engineer in a small chemical process plant, where a pesky wastewater stream is always on the verge of violating the plant's wastewater discharge temperature limits, even when blended with other wastewaters in the final effluent. The plant manager "wants to know your thoughts" on this matter but is unwilling to spend a lot of money in coming up with a fix.

The stream is only 20 gal/min (gpm), but its temperature is 100°F. You would like to lower that to a comfortable 80°F. A stream of fresh water is available at 40°F, and 10 gpm of the resulting hot water can be used in the process if uncontaminated by wastewater.

This may be your lucky day! You find a perfectly good but used shell-and-tube heat exchange sitting in storage in the warehouse. A tag on it reads, "Heat exchange area 20 ft²—overall heat transfer coefficient for water-to-water service 500 Btu/h ft² °F." The plant maintenance department can install it for virtually no out-of-pocket cost.

Will this exchanger do the job?

SOLUTION

$(20 \text{ gpm})(500 \text{ lb/h/gpm}) = 10,000 \text{ lb/h}$ (note that 1 gpm ≅ 500 lb/h for water)

Heat to be removed $= Q = mC_p\Delta T$

$= (10,000 \text{ lb/h})(1 \text{ Btu/lb°F})(100-80)°F = 200,000 \text{ Btu/h} = 0.2 \text{ MMBtu/h}$

Cooling Water Heat Balance

Q = heat to be removed = heat to be picked up by the cooling water

$200,000 \text{ Btu}/h = (10)(500)(1)(x-40)$

$x = 80°F$

There is an equation for the heat exchange rate of a parallel-flow or counter-flow heat exchanger.

$$Q = UA\Delta T_{lm} \quad (2.103)$$

where U is the overall heat transfer coefficient (Btu/h ft² °F), A is the area available for heat exchange, and ΔT_{lm} is the log-mean temperature difference, defined as follows:

$$\Delta T_{lm} = (\Delta T_2 - \Delta T_1)/[\ln(\Delta T_2/\Delta T_1)] \quad (2.104)$$

ΔT_2 is the temperature difference between the fluids at one end of the exchanger, and ΔT_1 is the same temperature difference at the other end.

A derivation of the expression for ΔT_{lm} is provided elsewhere [43,44].

Counter-flow is more efficient than parallel flow because it maintains the greatest temperature difference throughout the exchanger.

Hot stream 100°F → 80°F $\Delta T_1 = 100-80 = 20°F$
Cold stream 80°F ← 40°F $\Delta T_2 = 80-40 = 40°F$

From Equation (2.104),

$$\Delta T_{lm} = (40-20)/[\ln(40/20)] = 20/\ln(2) = 28.85°F$$

72 Chapter 2 Basic Concepts

From Equation (2.103), the design Q for this installation

$$Q = (500 \text{ Btu/h ft}^2 \, {}^\circ\text{F})(20 \text{ ft}^2)(28.85 {}^\circ\text{F}) = 288{,}500 \text{ Btu/h}$$

Since this number is greater than the Q to be processed (200,000 Btu/h), the exchanger in the warehouse is capable of doing the job. As a precaution, it would be a good idea to verify the basis for the value of U on the tag, to check that the exchanger is capable of handling the required flow and to recompute the heat transfer area from the number and dimensions of the tubes.

Installation of the wastewater stream on the tube side could facilitate cleaning during turnarounds to remove any accumulated scale buildup. Monitoring the clean stream for contamination from leaking tubes during operation would be advisable. ∎

2.8 A SMATTERING OF STATISTICS

Statistics is the mathematical science concerned with the collection and treatment of large amounts of numerical data. Some uses of statistics are as follows:

- Determining which data to collect
- Understanding a collection of data
- Extending the usefulness of such data
- Deciphering patterns not otherwise apparent
- Drawing inferences and forming generalizations
- Predicting missing or future values or other outcomes
- Reaching conclusions and making decisions from imperfect data
- Summarizing vast quantities of data with a few key parameters
- Presenting and displaying data in a clear, effective manner

Like two snowflakes, no two measurements of the same variable will ever be exactly the same. Even if this were not so intrinsically, there are always errors introduced during measurement. Sometimes, the differences are less than the ability to measure, but they are still there, all the same.

A series of measurements, therefore, leads to a distribution of results, as shown in Table 2.12, characterizing the daily temperature measurements, reported to the nearest °F, of an industrial plant's liquid effluent during a given month.

This distribution can be plotted as a special kind of bar graph or histogram known as a *stem plot*, or with broader ranges of data, a *stem-and-leaf plot* [45, pp. 194–195; 46, pp. 31–33; 47, pp. 20–23]. This provides a quick method of plotting and viewing the distribution while retaining the actual data, as shown in Table 2.13.

Now that we have the distribution plotted, what can we do with it?

Table 2.12 Plant Effluent Temperatures

Day no.	Temperature (°F)	Day no.	Temperature (°F)
1	78	16	84
2	74	17	79
3	76	18	84
4	78	19	77
5	81	20	80
6	79	21	74
7	82	22	72
8	85	23	81
9	88	24	78
10	79	25	82
11	84	26	79
12	70	27	76
13	80	28	80
14	81	29	79
15	85	30	78

2.8.1 Characterizations of the Center of a Distribution

The mean ($\bar{x}$) is the unweighted arithmetic average of the number (n) of x-values,

$$\bar{x} = (1/n) \sum_{i=1}^{n} x_i \qquad (2.105)$$

Simply add up the values and divide by the total number.

The median is the midpoint of the distribution; no more than half the values lie above it, and no more than half below it. For an odd number of values, it is the center point. For an even number of values, it is the arithmetic average of the two central (c) values, the lower value (l) and the higher value (h).

Table 2.13 Distribution of Plant Effluent Temperature Measurements (°F) (Number of Measurements at Stated Value)

Number																					
								79													
							78	79													
							78	79	80	81			84								
				74		76		78	79	80	81	82		84	85						
70		72		74		76	77	78	79	80	81	82		84	85			88			
70	71	72	73	74	75	76	77	78	79	80	81	82	83	84	85	86	87	88	89	90	
								Measured temperature (°F)													

74 Chapter 2 Basic Concepts

$$\text{Median} = (Xcl + Xch)/2 \tag{2.106}$$

Simply add them up and divide by 2.

In addition to the median, other statistical measurements can be used to subdivide the data. The *quartiles* are the data values lying about one-fourth of the distance from either end of the data set. The highest one-fourth of the data lies between the upper quartile (Q_3, the 75th percentile) and the highest data value. The next one-fourth of the data falls between the upper quartile and the median. Then the next one-fourth of the data falls between the median and the lower quartile (Q_1, the 25th percentile). Finally, the lowest one-fourth of the data values are contained between the lower quartile and the minimum data value. The interquartile range lies between the upper quartile and the lower quartile and contains the middle half of the data. The set of values consisting of the median, the quartiles, and the extremes of the data is known as the *five-number summary* [45, p. 216].

Percentiles are similarly defined, for example, the 90th or the 99th percentile. However, division of the data into such narrowly defined percentiles is more appropriate for very large data sets, much larger than that of the example discussed above.

The *mode* is the most frequent value, at the maximum point of the distribution. The maximum point itself, representing the maximum number or frequency of measurements, defines the vertical height of the distribution.

In cases where it is important to address the relative importance of the various data entries, a *weighted mean* is used [46, pp. 45–46]. It is defined as

$$\sum_{i=1}^{n} w_i x_i / \sum_{i=1}^{n} w_i \tag{2.107}$$

When all of the weighting factors are equal, this becomes

$$w_i \sum_{i=1}^{n} x_i/(nw_i) \tag{2.108}$$

The factor w_i cancels in numerator and denominator, and the expression reduces to Equation (2.105).

2.8.2 Characterization of the Spread of a Distribution

The variance (s^2) is defined as

$$s^2 = [1/(n-1)] \sum_{i=1}^{n} (x_i - \bar{x})^2 \tag{2.109}$$

The square is used to keep positive and negative deviations from canceling out. The term (n – 1) is used in the denominator rather than (n) to provide an unbiased estimate since only (n – 1) deviations from the mean are independent. The mean ($\bar{x}$) is calculated from the xi's by means of Equation (2.105).

2.8 A Smattering of Statistics

The formula for the variance is mathematically equivalent to

$$s^2 = \{1/[n(n-1)]\}\left[n\sum_{i=1}^{n} x_i^2 - \left(\sum_{i=1}^{n} x_i\right)^2\right] \quad (2.110)$$

and

$$s^2 = \left[\sum_{i=1}^{n} x_i^2 - n\bar{x}^2\right]/(n-1) \quad (2.111)$$

One form might be preferable over the others, depending on the method of computation being performed.

The standard deviation (s) is the square root of the variance

$$s = (s^2)^{1/2} \quad (2.112)$$

PROBLEM 2.39

Calculate the mean, mode, median, the quartiles, the variance, and the standard deviation of the data set in Table 2.12. Do these numbers appear reasonable when compared with the histogram of Table 2.13?

SOLUTION *Mean*: The sum of the 30 daily temperature readings is 2382. From Equation (2.105),

$$\text{mean} = 2383/30 = 79.4333\ldots \rightarrow \text{call } 79.4°F$$

Mode: By inspection, the mode is the most frequent value = 79°F.
Median: For the median, there are 15 values below a value of 79°F, and 15 values below another value of 79°F. From Equation (2.106),

$$\text{median} = (79+79)/2 = 79°F$$

Quartiles: The lower quartile (Q_1) is the median of the 15 data points lying below the overall median. This will be located between the seventh and eighth data points in the lower half of the data (78 + 78)/2 = 78°F Q_1. The upper quartile (Q_3) lies between the seventh and eighth data points in the upper half of the data (82 + 82)/2. Central range of the data lies between 78°F and 82°F.
Variance:

$$\sum_{i=1}^{n} x_i^2 = 189{,}751,\ n\sum x_i^2 = (30)(189{,}751) = 5{,}692{,}530$$

From a previous step,

$$\sum_{i=1}^{n} x_i = 2383,\ (\Sigma x_i)^2 = (2383)^2 = 5{,}678{,}689$$

76 Chapter 2 Basic Concepts

From Equation (2.110),

$$s^2 = \{(1)/[(30)(29)]\}[5{,}692{,}530 - 5{,}678{,}689]$$
$$s^2 = [(1)/(870)][13{,}841] = 15.9092$$

Standard Deviation: Equation (2.112),

$$(s^2)^{1/2} = (15.9092)^{1/2} = 3.9886°F \rightarrow \text{call } 4°F \qquad \blacksquare$$

These numbers are quite consistent with the pictorial representation of Table 2.13.

2.8.3 The Rolling Average

A concept that comes up in environmental permits is the rolling or moving average. The average is computed in the normal manner, but the oldest value drops off as each new value is added.

PROBLEM 2.40

Consider the previous problem of plant effluent temperatures (Problem 2.39). Assume instead that 90°F constitutes a violation on the basis of a 30-day rolling average and that the plant has been chugging along with a constant value of 89°F for the past 29 days. On the last day of the month, however, there is a massive excursion that nobody notices until the effluent temperature is recorded at 140°F. If the problem is fixed, and the effluent temperature returns to a constant 89°F for the next 30 days, what happens to the rolling average?

SOLUTION

$$30\text{-day average} = [(89°F)(29 \text{ days}) + (140°F)(1 \text{ day})]/(30 \text{ days})$$
$$= [2581 + 140]/(30) = (2721)/(30) = 90.7°F$$

This violation will continue for the next 29 days (total of 30 days) until the 140°F upset value washes out of the system. This is not a good position for the plant operator to be in since substantial fines are assessed per day, with each day constituting a separate violation.

If the temperature could be immediately decreased to 70°F, the lowest recorded value in Table 2.12, and held at that level for only a few days, there would be only one more day of violation before getting back into compliance.

Day 1 : Rolling average (RA) = 90.7 (violation)
Day 2 : RA = $[(28)(89) + (1)(140) + (1)(70)]/(30) = 90.06$ (presumed violation)
Day 3 : RA = $[(27)(89) + (1)(140) + (2)(70)]/(30) = 89.43$ (OK)

After Day 3, the temperature could be allowed to return to 89°F, if desired, and still remain in compliance. In that case,

Day 4 : RA = $[(26)(89) + (1)(140) + (2)(70) + (1)(89)]/(30) = 89.43$ (OK)

Other combinations are possible; you do the math.

2.8.4 Probability Density Functions: The Normal Distribution

For a really large number of data points with a continuous x-variable, the histogram tends toward a continuous function, called a probability density function. Often, this function is the familiar Gaussian[3] bell-shaped curve of the normal distribution [47, p.213; 48, p. 72].

Because there are an infinite number of such curves, each with a different combination of means and standard deviations, a mathematical transformation is used to collapse them into a single curve. It is actually this function, the so-called *normal curve of error*, *standard normal curve*, or simply the *probability curve*, that is widely tabulated.

Mathematically,

$$f(x, \mu, \sigma) = \{1/[\sigma(2\pi)^{1/2}]\}\exp\{-\frac{1}{2}[(x-\mu)/\sigma]^2\} \qquad (2.113)$$

where μ and σ represent the mean and standard deviation, respectively, of the particular continuous function.

In any normal distribution, the mean, the median, and the mode are identical and occur at the midpoint of the distribution. The frequency of occurrence between any two x-values is the value of the integral of the function. This is the area under the curve between those values. This integral cannot be evaluated by traditional methods in closed form. Rather, tabulated values found in statistics textbooks and mathematics handbooks are used.

$$F(z) = \{1/[(2\pi)^{1/2}]\} \int_{-z}^{+z} \exp[-t^2/2]dt \qquad (2.114)$$

The standardized random variable

$$z = (x-\mu)/\sigma \qquad (2.115)$$

is used to transform a particular normal distribution to the standard normal distribution [49, p. 144]. (The definition of the function's integration limits and, therefore, the tabulated values vary from table to table; so be careful.)

This curve of Figure 2.11 is symmetrical, centered around the y-axis, with a mean (μ) of zero and a standard deviation (σ) of 1.0. Key points of the distribution are listed in Table 2.14.

---

[3]Karl Friedrich Gauss (1777–1855) was a German mathematician and scientist. In addition to the mathematical function and statistics curve that bear his name, many other important discoveries in physics, astronomy, mathematics, and statistics are credited to Dr. Gauss. The unit of magnetic field intensity is also named after him [11, Vol. 11, p. 36; 47, p. 39].

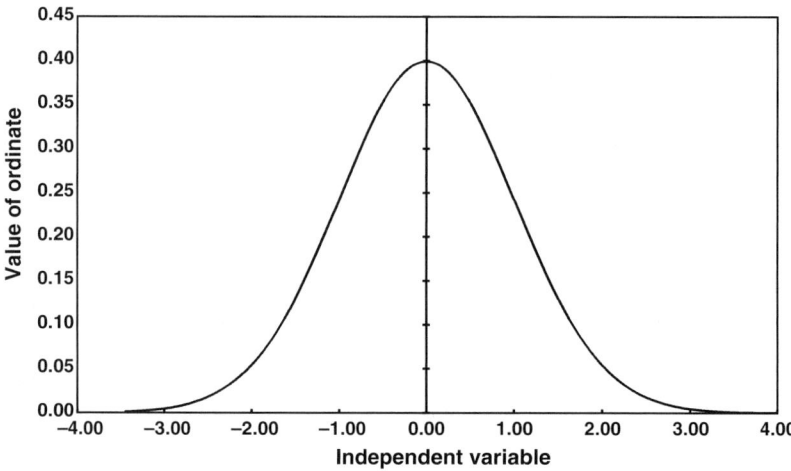

Figure 2.11 Standard normal curve of error.

Its ordinate at $x = 0$ is 0.3989. Its σ is unity, and the area under the curve from $-\infty$ to $+\infty$ is 1.0, or 100%. In practical terms, however, approximately 68% of the area is contained between $-z$ and $+z$, approximately 95% between $-2z$ and $+2z$, and approximately 99.7% between $-3z$ and $+3z$, as shown in Table 2.14, excerpted, interpolated, and modified from Refs [46,49,50]. These values are referred to as 1, 2, and 3 standard deviations of the standard normal distribution. The curve is essentially complete after four to five standard deviations.

2.8.5 Test for a Normal Distribution

Even smaller data sets can approach and can be fitted to a normal distribution by plotting the cumulative percent greater or less than each data point entry as the ordinate on special "probability" graph paper [51, pp. 100–101], shown in conjunction with

Table 2.14 Table of the Standard-Normal Curve at Key Points

z	Fraction of area between $-z$ and $+z$		Ordinate at $-z$ or $+z$	
0.00	0.0	(exact)	0.3989	
0.9946 (interpolated)	0.6800		0.2433	(calculated)
1.00	0.6826		0.2420	
1.96	0.9500		0.0584	
2.00	0.9546		0.0540	
2.96	0.9970		0.0050	
3.00	0.9974		0.0044	
4.00	0.99994		0.0001	
5.00	0.9999994		0.0000015	(calculated)
∞	1.0	(exact)	0.0	(exact)

Problem 2.41. The nonlinear abscissa of this graph is calculated to correspond to the area under the probability curve. A truly normal distribution plots as a perfect straight line. Others, not so perfect but reasonably normal, exhibit some scatter.

The slope of this line is indicative of the spread of the distribution, the steeper the line, the broader the spread; a horizontal line indicates a uniform distribution. The mean is found from the value of the variable at 50% point. The standard deviation is the arithmetic difference between the values at the $(50 + 68.26/2) = 84.13\%$ point and the 50% point or between the 50% point and the $(50 - 68.26/2) = 15.87\%$ point.

PROBLEM 2.41

Determine whether the plant effluent temperature (Tables 2.12 and 2.13), previously investigated as a discrete distribution, can be approximated by a continuous normal distribution.

SOLUTION   The data are plotted on the aforementioned probability coordinates in Figure 2.12. The temperature measurements are shown on the vertical axis, and the cumulative percent of the number of data points "less than" each measured temperature is plotted along the horizontal. Because of the discrete nature of the measured temperatures compared to a continuous random variable, each reported temperature value represents the range from $-0.5°F$ to plus $0.5°F$ of the reported central value. Points are plotted at the lower boundary ($-0.5°F$) since "less than" includes everything up to that point [46, p. 26]. This is known as the *continuity correction* [48, pp. 73–74; 49, pp. 146–147].

As seen in the figure, a straight line with some scatter results, meaning that the distribution is not perfectly normal. The straight line was fitted by eye, giving more weight to the center of the graph rather than the tail areas at its ends. Based on this fit, the mean (50% point) is 79.4°F

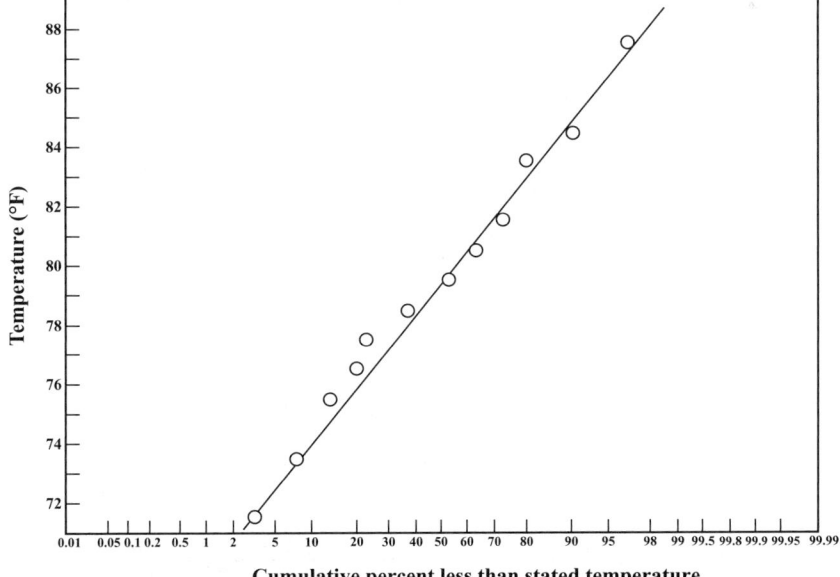

Figure 2.12   Plant effluent temperatures on probability coordinates.

80   Chapter 2   Basic Concepts

compared to the same 79.4°F from the discrete distribution. The 16% point and the 84% point plot as 75.2°F and 83.6°F, respectively, making the standard deviation (79.4–75.2°F) or (83.6–79.4°F) = 4.2°F. This compares to the 4.0°F calculated in Problem 2.39.

Finally, the 25th percentile (Q_1) determined from the line is 76.5°F, and the 75th percentile is 82.2°F, versus their respective values of 78°F and 82°F previously calculated. Agreement for this example turns out to be quite good.

In summary, regardless of the exact values obtained, it is important to understand and apply the principles involved. ∎

2.8.6   The Log-Normal Distribution

When circumstances occur that prohibit negative numbers, such as certain physical measurements, the distribution becomes skewed. In that case, sometimes the logarithm of the value rather than the value itself follows a normal distribution. This happens with the distribution of sizes of fine powders [51, p. 102] and the typical fluidized catalytic cracking catalyst distribution [51, p. 251] depicted in Figure 2.13. The logarithm of particle size plots as a straight line or nearly straight line on probability graph paper. One can either take logarithms and plot them on probability graph paper with an arithmetic vertical scale or use another type of commercially available probability paper whose vertical scale is logarithmic [52, p. 30].

On a logarithmic probability plot, the mean is the 50% point of the distribution, as usual. The standard deviation, however, is the ratio of the 84.3% point to the 50% point or of the 50% point to the 15.87% point. When the size distribution of small particles measured by weight plots as a straight line on log-probability coordinates, the distribution on a number- or surface-area-basis shows a parallel straight line if plotted on the same coordinates [52, p. 30, p. 39].

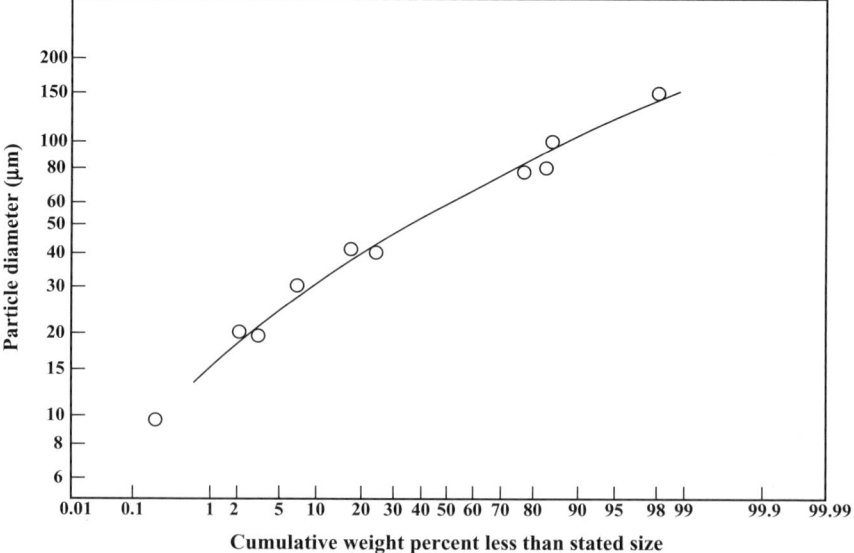

Figure 2.13   Particle size distribution by weight for a typical fluidized catalytic cracking catalyst.

2.8.7 Other Statistical Applications

One cannot possibly hope to cover the entire field of statistics in the few pages of this section. Additional applications of statistics related to environmental permitting will be addressed as necessary later in this book, and Ref. 53, for example, is devoted entirely to the application of statistics to environmental situations.

REFERENCES

1. M.K. Wilson, G. Kannangara, G. Smith, M. Simmons, and B. Raguse, *Nanotechnology Basic Science and Emerging Technologies*, Chapman & Hall/CRC, Boca Raton, FL, 2002, 271 pp. (reprinted 2004)
2. M.J. Sienko and R.A. Plane, *Chemistry*, 4th edn, McGraw-Hill, New York, 1971, 638 pp.
3. B.S. Mitchell, *An Introduction to Materials Engineering and Science for Chemical and Materials Engineers*, Wiley–Interscience, Hoboken, NJ, 2004, 953 pp.
4. L. Theodore and R.G. Kunz, *Nanotechnology: Environmental Implications and Solutions*, Wiley–Interscience, Hoboken, NJ, 2005, 378 pp.
5. R.C. Weast, editor, *CRC Handbook of Chemistry and Physics*, 58th edn, CRC Press, Inc., Cleveland, OH, 1977.
6. *Betz Handbook of Industrial Water Conditioning*, 6th edn, Betz Laboratories, Inc., Trevose, PA, 1962.
7. J.A. Dean, editor, *Lange's Handbook of Chemistry*, 11th edn, McGraw-Hill, New York, 1973.
8. R.E. Kirk, et al., editors, *Kirk–Othmer Encyclopedia of Chemical Technology*, 3rd edn, Vol. **1**, Wiley, New York, 1978, pp. 849–850.
9. C.N. Sawyer and P.L. McCarty, *Chemistry for Sanitary Engineers*, 2nd edn, McGraw-Hill, New York, 1978, 518 pp.
10. C.N. Sawyer and P.L. McCarty, *Chemistry for Environmental Engineering*, 3rd edn, McGraw-Hill, New York, 1978, 532 pp.
11. *Funk & Wagnalls New Encyclopedia*, Funk & Wagnalls, Inc., New York, 1971
12. D.W. Green, and J.O. Maloney, editors, *Perry's Chemical Engineers' Handbook*, 7th edn, McGraw-Hill, New York, 1997, pp. 2–140 to 2–149.
13. M.W. Zemansky and H.C. Van Ness, *Basic Engineering Thermodynamics*, McGraw-Hill, New York, 1966, 380 pp.
14. 40 CFR 60, Appendix A, Method 19, U.S. Government Printing Office, Washington, DC (various editions).
15. *1967 ASME Steam Tables: Thermodynamic and Transport Properties of Steam*, 2nd edn, American Society of Mechanical Engineers, New York 1968.
16. ASTM Method D 3588–81, Standard Method of Calculating Calorific Value and Specific Gravity (Relative Density) of Gaseous Fuels, American Society for Testing and Materials, Philadelphia, PA, January 1982.
17. C.F. Prutton and S.H. Maron, *Fundamental Principles of Physical Chemistry*, Revised Edition, Macmillan, New York, 1951, 803 pp.
18. T.E. Jordan, *Vapor Pressure of Organic Compounds*, Interscience Publishers, New York, 1954.
19. J.H. Perry, editor, *Chemical Engineers' Handbook*, 3rd edn, McGraw-Hill, New York, 1950, pp. 673–676.
20. C.L. Yaws, *Handbook of Vapor Pressure*, Vols 1–3, Gulf Publishing Company, Houston, TX, 1994.
21. *Lange's Handbook of Chemistry*, 11–16 edns, McGraw-Hill, New York, 1973–2005.
22. Union Carbide Chemicals Company, Physical Property Data, January 16, 1956.
23. J.H. Keenan and F.G. Keyes, *Thermodynamic Properties of Steam Including Data for the Liquid and Solid Phases*, 1st edn, Wiley, New York, 1936.
24. R.G. Kunz and W.F. Baade, Predict methanol and ammonia in hydrogen-plant process condensate, Paper ENV-00-171, presented at the NPRA 2000 Environmental Conference, San Antonio, TX (September 10–12, 2000).
25. C. Yaws, H.C. Yang, and X. Pan, Henry's law constants for 362 organic compounds in water, *Chemical Engineering*, **98**(11), 179–185, 1991.

26. W. Stumm and J.J. Morgan, *Aquatic Chemistry*, 2nd edn, Wiley, New York, 1981, pp. 204–206.
27. K. Forlich, E.J. Tauch, J.J. Hogan, and A.A. Peer, Solubilities of gases in liquids at high pressure, *Industrial & engineering chemistry*, **23**(5), 548–550, 1931.
28. W. Gerhartz, et al., editors, *Ulmann's Encyclopedia of Industrial Chemistry*, 5th edn, Vol. **A5**, VCH Publishers, Deerfield Beach, Florida, 1986, p. 169.
29. R.H. Perry, C.H. Chilton, and S.D. Kirkpatrick, editors, *Chemical Engineers' Handbook*, 4th edn, McGraw-Hill, New York, 1963.
30. E.W. Washburn, editor, *International Critical Tables of Numerical Data: Physics, Chemistry, and Technology*, 1st edn, 4th impression, Vol. III, McGraw-Hill, New York, 1928, pp. 255–261.
31. A. Zawisa and B. Malesinska, Solubility of carbon dioxide in liquid water and of water in gaseous carbon dioxide in the range 0.2–5 MPa and at temperatures up to 473°K, *Journal of Chemical & Engineering Data*, **26**, 388–391, 1981.
32. G. Saracco and G. Genon, High temperature ammonia stripping and recovery from process liquid wastes, *Journal of Hazardous Materials*, **37**, 191–206, 1994.
33. J.A. Dean, editor, *Lange's Handbook of Chemistry*, 14th edn, McGraw-Hill, New York, 1992.
34. J. Gmehling, et al., *Vapor–Liquid Equilibrium Data Collection* (including Supplements), DECHEMA Chemistry Data Series, Springer-Verlag, Frankfurt, Germany, 1977–1988.
35. J.C. Chu, R.J. Getty, L.F. Brennecke, and R. Paul, *Distillation Equilibrium Data*, Reinhold, New York, 1950.
36. C.A. Jones, E.M. Schoenborn, and A.P. Colburn, Equilibrium still for miscible liquids: data on ethylene dichloride–toluene and ethanol–water, *Industrial &Engineering Chemistry*, **35**(6), 666–672, 1943.
37. H.C. Carlson and A.P. Colburn, Vapor–liquid equilibria of nonideal solutions: utilization of theoretical methods to extend data, *Industrial & Engineering Chemistry*, **34**(5), 581–589, 1942.
38. J.M. Smith, H.C. Van Ness, and M.M. Abbott, *Introduction to Chemical Engineering Thermodynamics*, 5th edn, McGraw-Hill, New York, 1996, pp. 388–389.
39. H. Otsuki and F.C. Williams, Effect of pressure on vapor–liquid equilibria for the system ethyl alcohol–water, *Chemical Engineering Progress Symposium Series*, No. **6**(49), 1953, pp. 55–67.
40. R.G. Kunz and W.F. Baade, Predict contaminant concentrations in deaerator-vent emissions, *Hydrocarbon Processing (International Edition)*, **80**(6), 100-A–100-O, 2001.
41. R.G. Kunz and W.F. Baade, Vapor–liquid activity coefficients for methanol and ethanol from heat of solution data: application to steam–methane reforming, *Journal of Hazardous Materials*, **88**(1), 53–62, 2001.
42. N. de Nevers, *Physical and Chemical Equilibrium for Chemical Engineers*, Wiley–Interscience, New York, 2002, pp. 46–56.
43. E.R.G. Eckert and R.M. Drake Jr., *Heat and Mass Transfer*, 2nd edn, McGraw-Hill, New York, 1959, pp. 14–16.
44. W.L. McCabe, J.C. Smith, and P. Harriott, *Unit Operations in Chemical Engineering*, 5th edn, McGraw-Hill, New York, 1993, pp. 316–319.
45. D.S. Moore, *Statistics: Concepts and Controversies*, 3rd edn, Freeman, New York, 1991, 439 pp.
46. J.E. Freund and R.M. Smith, *Statistics: A First Course*, 4th edn, Prentice-Hall, Englewood Cliffs, NJ, 1986, 557 pp.
47. A.F. Siegel, *Statistics and Data Analysis: An Introduction*, Wiley, New York, 1988, 518 pp.
48. I. Miller and J.E. Freund, *Probability and Statistics for Engineers*, Prentice-Hall, Englewood Cliffs, NJ, 1965, 432 pp.
49. I. Miller, J.E. Freund, and R.A. Johnson, *Probability and Statistics for Engineers*, 4th edn, Prentice-Hall, Englewood Cliffs, NJ, 1990, 612 pp.
50. C.D. Hodgman,editor, *Mathematical Tables from Handbook of Chemistry and Physics*, 10th edn, Chemical Rubber Publishing Co, Cleveland, OH, 1954, pp. 209–213.
51. F.A. Zenz and D.F. Othmer, *Fluidization and Fluid-Particle Systems*, Reinhold, New York, 1960, 513 pp.
52. R.R. Irani and C.F. Callis, *Particle Size: Measurement, Interpretation, and Application*, Wiley, New York, 1963, 165 pp.
53. P.M. Berthouex and L.C. Brown, *Statistics for Environmental Engineers*, Lewis Publishers/CRC Press, Boca Raton, FL, 1994, 335 pp.

Chapter 3

Air Combustion

*Nothing is ever easy; yet somehow, everything is easy for
people who don't have to do it themselves.*

3.1 INTRODUCTION

This chapter presents a treatment of combustion calculations as employed in environmental permitting for the construction of a combustion source and its subsequent emission testing. Typically, much information is available on the fuel and its firing rate. However, additional information is needed on the combustion flue gas (FG) to complete the assessment of pollution potential. The link between fuel and flue gas is the combustion calculations.

The approach is both from the basic principles and from the EPA shortcut equations used in permitting. Their origin is explained in relation to those basic principles, and any simplifying assumptions are identified.

Although many of the equations presented may be used for other purposes, the focus is on environmental permitting and follow-up activities. The treatment is not concerned *per se* with process design, equipment selection, or optimizing design or operation, to name a few. However, some of those issues will be addressed as they come up in the context of environmental control/permitting.

In addition, some miscellaneous sources of air emissions not resulting from combustion are included at the end of the chapter.

3.2 COMBUSTION

Combustion may be defined as a rapid chemical reaction involving oxygen and resulting in the evolution of light and heat. This oxidation process requires a *fuel* (one that is burned to produce the light or heat) as well as *oxygen*, normally provided by atmospheric air. The atmospheric combustion air, unless absolutely bone dry, carries its water vapor (H_2O), or *humidity*, into the combustion process; this moisture is in

---

Environmental Calculations: A Multimedia Approach, by Robert G. Kunz
Copyright © 2009 John Wiley & Sons, Inc.

addition to any that may be formed chemically as a combustion product or added in the combustion process. Fuel and air combine to form *flue gas*—a mixture of the reaction products, unconsumed reactants, and undesired by-products plus inert materials, which enter with the fuel or air and pass through unchanged.[1]

The material balance equations that account for the conversion of reactants into various oxide products and the disposition of inerts are known as *combustion calculations*. The calculations are most simply carried out in terms of molecular weight units, or *moles*, the actual weight of substance in grams or pounds divided by its molecular, or formula, weight. Furthermore, conventional practice expresses the fractional composition of a gas mixture by volume and the composition of a solid or liquid by weight. The volumetric relationships of atmospheric combustion air and flue gas at atmospheric pressure are adequately described for this purpose by the ideal gas law (Chapter 2).

3.2.1 Combustion Inputs and Outputs

The combustion calculations may be thought of as a procedure in which inputs of fuel, air, and atmospheric moisture are manipulated to produce a flue gas as output. This output includes the flow rate and composition of major constituents in percentage concentrations to which an estimate of lesser amounts of atmospheric contaminants/pollutants in the parts per million (ppm) range may be added.

In the field, fuel flow would normally be measured, and its analysis performed at least periodically. The makeup of the combustion air is assumed constant at some standard value with atmospheric moisture measured indirectly. The required airflow is not usually measured but falls out of the calculations, based on the concentration of unused oxygen (excess O_2) in the flue gas. The calculated flue gas flow is verified in the field from differential pressure measurements by pitot tube in the exhaust stack to obtain velocity times the cross-sectional area of the stack (EPA Method 2 [2], or equivalent).

3.3 FUEL

The fuel may be gaseous, liquid, or solid; it may consist of a single component or multiple components. Examples are natural gas, a No. 2 petroleum-distillate fuel oil, and bituminous or anthracite coal. Natural gas contains primarily methane (CH_4) with lesser quantities of higher hydrocarbons, such as ethane (C_2H_6) and propane (C_3H_8); inerts, such as nitrogen (N_2), carbon dioxide (CO_2), and water (H_2O); and possibly some sulfur compounds, such as hydrogen sulfide (H_2S) and mercaptans (RSH).

---

[1] The furnace, heater, boiler, incinerator, or other combustion device may function with a natural draft [1, pp. 121–122], a forced draft, or an induced draft. In many cases, the combustion chamber and flue gas train operate at a slight vacuum, inches of water gauge (in. w.g.), below atmospheric pressure. Such operation is susceptible to the in-leakage of ambient atmospheric air, and this so-called *air infiltration*, or *tramp air*, joins the flue gas downstream of combustion. This is normally a minor factor but must always be considered in the evaluation of flue gas properties, as noted later in this chapter.

Table 3.1 Analysis of Typical Natural Gas[a]

Constituent	Formula	Molecular weight	Composition (vol%)
Methane	CH_4	16.04	93.14
Ethane	C_2H_6	30.07	2.50
Propane	C_3H_8	44.11	0.67
Butanes	C_4H_{10}	58.12	0.32
Pentanes	C_5H_{12}	72.15	0.12
Hexanes[+]	C_6H_{14}	86.18	0.05
Carbon dioxide	CO_2	44.01	1.06
Nitrogen	N_2	28.01	2.14
Total	—	—	100.00

HHV* at 60°F, 30 in. Hg (dry) (14.73 psia) = 1024 Btu/ft^3 (corrected to 60°F, 14.696 psia = 1021.7). Specific gravity[†] = 0.599.

[a]Analysis for Birmingham, Alabama, natural gas (1954 data) from the utility company serving the city as reported in Table 9.15 on p. 9–12 of Ref. [3]. Molecular weights from tabulations in Ref. [5], "Physical Properties of Organic Compounds" and "Physical Properties of Inorganic Compounds."

An example of a sulfur-free natural gas [3, p. 9–12] is provided in Table 3.1. Petroleum and coal are made up of a complex mixture of many different hydrocarbons and mineral matter, which forms ash upon combustion.

Gaseous fuels are characterized by the volumetric composition of chemical compounds in the blend, whereas liquid and solid fuels are characterized by weight using an overall percentage composition element by element. An empirical molar chemical formula, such as $C_xH_yO_zN_wS_v$, may also be used for liquids and solids. Combustion calculations are performed for a multiconstituent fuel by treating each constituent separately and summing the results.

3.3.1 Fuel Heating Values

When a fuel is combusted, it liberates heat. This *firing rate*, *fired duty*, or *heat release* depends on the composition of the fuel and the physical state assumed for the products of combustion.

The gross or higher heating value (HHV) is the maximum heat that would be realized, including the condensation of any water of combustion. EPA calls this the gross calorific value (GCV) [4]. The net or lower heating value (LHV) occurs in the more practical case when the combustion moisture remains in the vapor state. Heating values for selected fuels are listed in Table 3.2. Data for other materials can be found in standard references.

*HHV calculated for this natural gas composition using the heating values of Table 3.2 for the individual constituents is 1019.4 Btu/SCF (Problem 3.1).

†This is ratio of density (ρ) of fuel to that of air. For ideal gases, $\rho = PM/RT$, and this becomes a ratio of molecular weights. Molecular weight of this fuel is 17.37; MW of dry air is 28.96. Ratio = 17.37/28.96 = 0.5998.

Table 3.2 Heating Values of Selected Fuels[a]

Fuel	Formula	Molecular weight	Heating values [Btu/SCF (60°F, 1 atm)]	
			HHV[b]	LHV
Hydrogen	H_2	2.02	324.2	273.8
Carbon monoxide	CO	28.01	320.5	320.5
Methane	CH_4	16.04	1010.0	909.4
Ethane	C_2H_6	30.07	1769.6	1618.7
Propane	C_3H_8	44.11	2516.1	2314.9
n-Butane	C_4H_{10}	58.12	3262.3	3010.8
i-Butane	C_4H_{10}	58.12	3251.9	3000.4
(average C_4H_{10})		58.12	3257.1	3005.6
n-Pentane	C_5H_{12}	72.15	4008.9	3706.9
i-Pentane	C_5H_{12}	72.15	4000.9	3699.0
Neo-pentane	C_5H_{12}	72.15	3984.7	3682.9
(average C_5H_{12})		72.15	3998.2	3696.3
n-Hexane	C_6H_{14}	86.18	4755.9	4403.0

[a] Heating value of a fuel gas is obtained when 1 ft³ (at 60°F, 30 in. Hg = 14.73 psia) is combusted in air and the combustion products are returned to 60°F. As specified, the fuel gas may either be saturated with water vapor or may be dry, as is the natural gas of Table 3.1 [3, p. 9–12]. The HHV results when all the water vapor after combustion is condensed to a liquid, whereas the LHV results when the moisture is allowed to remain as a gas. The difference between HHV and LHV is the heat of vaporization of water.

[b] Heating values in this table are based on dry fuel gas in the ideal gas state at 60°F, 1 atm = 29.92 in. Hg, 760 mmHg, 14.696 psia. At this condition, 1 lb mol of gas occupies 379.5 ft versus 378.6 ft at 30 mmHg (14.73 psia).

The difference between the HHV and the LHV is simply the enthalpy of condensation of water vapor, at 60°F (1059.7 Btu/lb H_2O [6], 19,091 Btu/lb mol H_2O, or 50.3 Btu/SCF of H_2O as an ideal gas). For example, the HHV for methane is 1010 Btu/SCF, and its LHV is 1010−2(50.3) = 909.4 Btu/SCF, since 2 mol (or SCFs) of H_2O are formed per mole (or SCF) of CH_4 combusted.

By rule of thumb, the HHV and the LHV differ by about 10% for typical fuel gases, except those containing substantial amounts of hydrogen (H_2) or carbon monoxide (CO). The HHV of H_2 is some 18% higher than its LHV, and the HHV and LHV for CO are identical because no combustion water is produced when burning CO to form CO_2. Higher and lower heating values for a composite fuel, the natural gas of Table 3.1, are calculated in Problem 3.1.

PROBLEM 3.1

Calculate the HHV and LHV of the natural gas of Table 3.1 using the component heating values of Table 3.2. Compare with the HHV given in Table 3.1.

SOLUTION The composite heating value is determined by multiplying the component heating value by the fractional composition of that constituent and summing, as follows:

Basis: 100 SCF of natural gas (60°F, 1 atm = 14.696 psia, 22.92 in. Hg) (ideal gas state)

Constituent	SCF	HHV	LHV	HHV × SCF	LHV × SCF
CH_4	93.14	1010.0	909.4	94,071.4000	84,701.5160
C_2H_6	2.50	1769.6	1618.7	4424.0000	4046.7500
C_3H_8	0.67	2516.1	2314.9	1685.7870	15,550.9830
C_4H_{10}	0.32	3257.1	3005.6	1042.2720	961.7920
C_5H_{12}	0.12	3998.2	3696.3	479.7840	443.5560
$C_6H_{14}^+$	0.05	4755.9	4403.0	237.7950	220.1500
CO_2	1.06	0	0	0	0
N_2	2.14	0	0	0	0
Total	100.00	—	—	101,941.0380	91,924.7470

$$HHV = (101,941.0380)/(100) = 1019.4 \text{ versus } 1024 \text{ reported } (14.73 \text{ psia})$$
$$\text{and } 1021.7 \text{ corrected to } 14.696 \text{ psia}$$
$$LHV = (91,924.7470)/(100) = 919.2 \text{ (not reported)}$$
$$HHV/LHV = (1019.4)/(919.2) = 1.109 \times 100 = 110+\% \text{ (HHV > LHV by } \sim 10\%)$$

Since the hydrocarbon isomers for the butanes (C_4) and the pentanes (C_5) have not been specified, the heating values of average butanes and pentanes have been used in the calculations. The heating value of *n*-hexane has been used for the 0.05% C_6^+ fraction.

The reported HHV (corrected to 14.696 psia) for this natural gas is 1021.7. The calculated HHV of 1019.4 is within 0.23% of the reported HHV.

Heating values of noncombustible materials such as CO_2 and N_2 are identically zero. ∎

3.4 AIR

Except for unusual cases, fuels are burned in air, not pure oxygen. Atmospheric air consists of approximately 79% nitrogen (N_2), +20.9% oxygen (O_2), about 1% argon (Ar), some carbon dioxide, water vapor, plus trace quantities of the rare gases, pollutants, and miscellaneous other constituents. A typical composition of dry air including trace components is tabulated as handbook data [5, p. F-210].

How to account for the trace ingredients in combustion calculations has brought about different approaches. For hand calculations, the dry air composition is sometimes assumed to be 79.1% N_2/20.9% O_2 (or even 79% N_2/21% O_2). In that case, the argon and the minor constituents are lumped in with the nitrogen. EPA Method 19 and stack-testing methods use a composition of 79.1% N_2 and 20.9% O_2 [4].

The Babcox & Wilcox Steam Manual [7] uses 79.05% N_2 and 20.95% O_2[8]. The nitrogen including argon and other non-oxygen constituents is known as *atmospheric nitrogen*. The molecular weight (MW) of this so-called atmospheric nitrogen is adjusted from 28.01 to 28.16 to end up with the same molecular weight of 28.96 for dry atmospheric air. A molecular weight of 28.01 is still employed for nitrogen originating from the fuel.

This simplification and arbitrary definition, which complicate the nitrogen balance, are no longer necessary with the advent of high-speed computers to assist in the calculations. It is possible, if desired, to carry along absolutely every trace component in the air. This tends, however, to obscure the calculated results with unnecessary detail.

As a compromise, the abbreviated and normalized composition of Table 3.3 in use here is drawn from the complete handbook analysis. On a dry basis, this model air contains O_2 at 20.947%, 20.95% (rounded), or simply 20.9% (truncated) along with about 78% "real" N_2 (MW = 28.01), argon, and CO_2. As calculated in Problem 3.2, the average molecular weight (AMW) of this dry air remains at 28.96, or approximately 29, and somewhat lower for moist air. Estimation of atmospheric water vapor content will be addressed in a subsequent section.

Table 3.3 Dry Basis Composition of Atmospheric Air[a,b]

Constituent	Formula	Molecular weight	Composition (vol%)
Nitrogen	N_2	28.01	78.086
Oxygen	O_2	32.00	20.947
Argon	Ar	39.95	0.934
Carbon dioxide	CO_2	44.01	0.033

[a]Abbreviated and normalized from the composition listed in Ref. [5, p. F-210]. Trace components in the ppm range consist of other rare gases [neon (Ne), helium (He), krypton (Kr), and xenon (Xe)], hydrogen (H_2), methane (CH_4), and nitrous oxide (N_2O). Unless otherwise noted, this atmospheric air composition, including a variable water vapor content dependent on atmospheric humidity (H), is the basis for combustion calculations in this book.

[b]Table 3.3 is derived from the dry basis content of atmospheric air reported in the 58th edition of the *CRC Handbook of Chemistry and Physics* [5, p. F-210]. Until the 72nd edition (1991–1992) [9, p. 14–19], subsequent annual editions of this handbook carry this same atmospheric air composition.

In the 71st edition (1990–1991) [10, p. 14–11], reference begins to the "U.S. Standard Atmosphere, 1976," a book attributed to several U.S. governmental agencies jointly. Sea-level values from this reference for average temperature, pressure, density, and molecular weight have been standard for many decades [11, p. 14–19]. Atmospheric composition (from the Standard Atmosphere) reported in the *Handbook of Chemistry and Physics* is now somewhat different. For example, volume percent of nitrogen (N_2) and argon are exactly the same as in Table 3.3, with only a slight difference for oxygen (20.9476% versus 20.947%). Carbon dioxide is shown as 314 ppm.

A history for annual average CO_2 concentration measured at Mauna Loa Observatory in Hawaii begins in the 72nd edition (1991–1992) [9, p. 14–20] and continues to the present day [11, pp. 14–27 to 14–28]. (Mauna Loa contains the longest continuous record of direct measurements of atmospheric CO_2, dating back to 1958.) This is accompanied in the handbook by a graph of the increased combustion of fossil fuels. Annual average CO_2 has steadily increased from 330 ppm in 1974 to 377 ppm in 2004. Data for later years are available on the Internet.

This phenomenon, although of interest for other reasons, makes very little difference in calculated flue gas composition, and results obtained using Table 3.3 are considered to remain valid. In fact, as noted in the text, some accepted bases for combustion calculations assume atmospheric air to be totally free of carbon dioxide altogether.

PROBLEM 3.2

(a) Calculate the average molecular weight of dry combustion air having the air composition shown in Table 3.3. AMW is a useful intermediate in some calculations.

(b) Repeat for *complete air, atmospheric nitrogen,* and *air containing atmospheric nitrogen.*

SOLUTION

(a) To illustrate the principle, this calculation is also best performed in spreadsheet form.
    Basis: 100 mol (dry) of Table 3.3 air

Constituent	MW	vol or mol%	Moles	wt = MW × mol	wt%
Nitrogen (N_2)	28.01	78.086	78.086	2187.45	75.52
Oxygen (O_2)	32.00	20.947	20.947	670.28	23.14
Argon	39.95	0.934	0.934	37.31	1.29
Carbon dioxide	44.01	0.033	0.033	1.45	0.05
Total	—	100.000	100.000	2896.49	100.00

$$wt\% = 100(wt/total\ wt)$$
$$\text{For nitrogen} = 100[(2187.45/2896.49)] = 75.52\%$$

Weight (wt) for the other constituents is calculated in a similar manner.

$$\text{Average molecular weight} = total\ wt/total\ mole$$
$$= (2896.49)/(100) = 28.96 \cong 29$$

The molecular weight calculation for dry air can be simplified from the spreadsheet form into

$$MW_{dry\ air} = (0.2801)(\%N_2) + (0.3200)(\%O_2) + (0.3995)(\%Ar) + (0.4401)(\%CO_2) \quad (3.1)$$
$$= (0.2801)(78.086) + (0.3200)(20.947) + (0.3995)(0.934) + (0.4401)(0.033)$$
$$= 28.96 \cong 29$$

(b) Additional average molecular weights are calculated below using more precise values for component molecular weights to show differences.
    Calculation of AMW for *complete air*
    Basis: 100 mol of complete air (dry)

Component	vol%	Adjusted vol%	MW	wt	wt%
N_2	78.084	78.08418	28.0134	2187.403	75.51961
O_2	20.946	20.94604	31.9988	670.2484	23.14017
CO_2	0.033	0.033000	44.01	1.452333	0.050141
Ar	0.934	0.934002	39.948	37.31152	1.288171
Ne	0.001818	0.001818	20.183	0.036692	0.001266

90 Chapter 3 Air Combustion

He	0.000524	0.000524	4.0026	0.002097	0.000072
Kr	0.000114	0.000114	83.80	0.009553	0.000329
Xe	0.000008	0.000008	131.30	0.001050	0.000036
H_2	0.00005	0.000050	2.0158	0.000100	0.000003
CH_4	0.0002	0.000200	16.04	0.003208	0.000110
N_2O	0.00005	0.000050	44.01	0.002200	0.000075
Total	99.99976	100.000	—	2896.470	100.000000

$$AMW = (2896.470)/(100) = 28.96470$$

Calculation of AMW for "air" nitrogen
Basis: 100 mol of complete air less moles of O_2

Component	vol%	Adjusted vol%	MW	wt	wt%
N_2	78.084	78.08418	28.0134	2187.403	98.25629
O_2	0	0	31.9988	0	0
CO_2	0.033	0.033000	44.01	1.452333	0.065237
Ar	0.934	0.934002	39.948	37.31152	1.676001
Ne	0.001818	0.001818	20.183	0.036692	0.001648
He	0.000524	0.000524	4.0026	0.002097	0.000094
Kr	0.000114	0.000114	83.80	0.009553	0.000429
Xe	0.000008	0.000008	131.30	0.001050	0.000047
H_2	0.00005	0.000050	2.0158	0.000100	0.000004
CH_4	0.0002	0.000200	16.04	0.003208	0.000144
N_2O	0.00005	0.000050	44.01	0.002200	0.000098
Total	79.05376	79.0540	—	2226.222	100.0000

$$AMW = (2226.222)/(79.0540) = 28.1608 \cong 28.16$$

Calculation of AMW for air with "air" nitrogen
Basis: 100 mol of air (dry)

Component	vol%	MW	wt	wt%	AMW
N_2	79	28.16	2224.640	76.8007	—
O_2	21	32.00	672.000	23.1993	—
Total	100	—	2896.640	100.0000	28.96640
N_2	79.05	28.16	2226.048	76.8544	—
O_2	20.95	32.00	670.400	23.1456	—
Total	100	—	2896.448	100.0000	28.96448
N_2	79.1	28.16	2227.456	76.9081	—
O_2	20.9	32.00	668.800	23.0919	—
Total	100	—	2896.256	100.0000	28.96256

The molecular weight of moist air is calculated in Problem 3.6. ■

3.5 WATER

Water is an important part of the combustion process, either being formed as a product of combustion or entering the flue gas from somewhere upstream. The product "water of combustion" is by far the greatest source of moisture in the flue gas. (The water of combustion is really an output of the combustion process but is addressed here for convenience).

Water also enters as constituent of the fuel and as humidity in the combustion air. Other sources of water include steam for atomization of a liquid fuel, water introduced into the flue gas from a wet scrubber, steam injected for control of nitrogen oxides (NO_x), and any steam used for cooling, quenching, or miscellaneous purposes. Flue gas moisture is the sum of all of these sources.

> *Water of Combustion.* As it turns out, for combustion of a hydrocarbon such as methane, the volume of product water, in general, is of the same order of magnitude as the oxygen used in combustion. It is accounted for in the subsequent combustion reaction equations.
>
> *Water in the Fuel.* Any water present as a constituent of the fuel is assumed to pass through combustion and appear unchanged in the flue gas, and it is accounted for accordingly.
>
> *Steam Injection.* The presence of steam from sources extraneous to the combustion process must also be accounted for. However, some of the equations presented here may not apply directly to certain cases involving steam injection, and caution should be exercised in those situations.
>
> *Water in Combustion Air.* Unless air is perfectly dry, vapor-phase moisture is carried into the combustion process.[2] The concentration of water vapor in the air being used in combustion is determined by the temperature, pressure, and degree of moisture saturation of the ambient atmosphere. Even if the combustion air is preheated for heat recovery, its water vapor content still depends on conditions existing in the ambient atmosphere. Procedures to estimate combustion air moisture are described in the following section.

3.5.1 Estimation of Combustion Air Moisture

There are several methods for estimating the water vapor entering a combustion process with the incoming air. The techniques described here employ vapor pressure and relative humidity (RH), dew point temperature, and wet- and dry bulb temperatures. The choice of method depends on the available data.

3.5.1.1 Vapor Pressure and Relative Humidity

This procedure starts with given values of air temperature, barometric pressure, and relative humidity. The temperature-dependent vapor pressure of water is used as an intermediate in the calculation.

---

[2] Entrainment of droplets of liquid water during precipitation in the atmosphere is controlled by proper design of the air intake and is not being considered here.

A condensable vapor in equilibrium with its own liquid exerts a pressure that depends on temperature alone. This so-called *vapor pressure* (p_w) is often correlated by a functional form known as the Antoine equation (Chapter 2).

$$\log_{10}[p_w(\text{mmHg})] = A - B/[C + t\,(^\circ\text{C})] \tag{2.23}$$

The vapor pressure of water vapor (p_w) obeys this relationship, with constants given below [12]:

Temperature range	A	B	C
0–60°C (32–180°F)	8.10765	1750.286	235.0
60–150°C (180–302°F)	7.96681	1668.21	228.0

In the atmosphere, the maximum amount of water vapor that can be present at any given temperature and atmospheric pressure (P_T) is termed the *saturation humidity*. This occurs when the partial pressure of water (p') equals its vapor pressure (p_w). Then the maximum absolute molal humidity (H_s) (mol H$_2$O/mol dry air) at saturated conditions is calculated as follows:

$$H_s = (n_{w_{\max}})/(n_{\text{dry air}}) = p_w/(P_T - p_w) \tag{3.2}$$

For lesser amounts of water vapor below the maximum, the percent *relative humidity* (% RH) is defined as 100 times the ratio of the actual partial pressure (p') to the vapor pressure (p_w) at that temperature:

$$\%\,\text{RH} = 100(p'/p_w) \tag{3.3}$$

Therefore, the actual absolute molal humidity (H)

$$H = (n_w)/(n_{\text{dry gas}}) = p'/(P_T - p') \tag{3.4}$$

can be expressed in terms of the vapor pressure and the % RH.

$$H = p_w(\%\,\text{RH}/100)/[P_T - p_w(\%\,\text{RH}/100)] \tag{3.5}$$

The value of p_w is calculated via Equation (2.23) using constants A, B, and C from the table that immediately follows it.

This equation allows one to estimate the amount of water vapor present in the incoming combustion air. Values will lie typically in the range of 0 to 0.1 mol H$_2$O/mol dry air (Figure 3.1). The corresponding mass ratio (mass of H$_2$O/mass of dry air, constituent units) can be obtained by multiplying the absolute molal humidity by the ratio of molecular weights for water and dry air (18.02/28.96 = 0.62). Relative-humidity curves in mass units are typically found on psychrometric charts.

3.5.1.2 Dew Point Temperature

However, the relative humidity may not be given directly. Other ways of expressing atmospheric moisture-content include the dew point and the wet-bulb temperature.

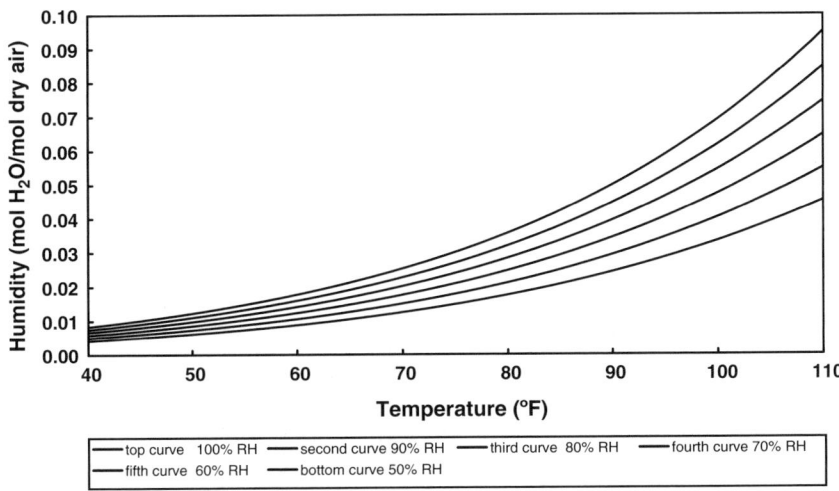

Figure 3.1 Absolute humidity (H) versus temperature with RH as parameter.

The dew point is the temperature at which liquid water just begins to condense from air saturated with water vapor. Hence, the partial pressure at the actual air temperature is the vapor pressure at the dew point (from Equation (2.23), using the dew-point temperature). The humidity is then calculated from Equation (3.4), and the % RH is calculated by rearrangement of Equation (3.5) to give

$$\%\text{RH} = (100)(P_T)(H)/[p_w(1+H)] \tag{3.6}$$

PROBLEM 3.3

Find the absolute humidity (H) for 60°F air at 60% RH and a total atmospheric pressure of 1 atm = 760 mmHg = 29.92 in. Hg = 14.696 psia.

SOLUTION Convert 0°F to °C

$$(60°F + 40)/(1.8) - 40 = 15.5555 \,°C$$

$$\log_{10}[p_w] \text{ at } (60°F = 15.5555\,°C) = 8.10765 - 1750.286/(235.0 + 15.5555) = 1.1220$$

$$p_w = 10^{1.1220} = 13.2433 \text{ mmHg}$$

$$\times (14.696/760) = 0.25610 \text{ psia}$$

From Equation (3.5),

$$H = (0.25610)(60/100)/[14.696 - (0.25610)(60/100)]$$
$$= 0.010566 \text{ mol } H_2O/\text{mole of dry air}$$

Alternatively, p_w can be read directly from steam tables [6] at 60°F as 0.25611 psia. ∎

PROBLEM 3.4

Find the absolute humidity (H) and the percent relative humidity for 69°F air at a barometric pressure of 30.13 in. Hg (14.80 psia) if the dew point is 60°F.

SOLUTION   From Problem 3.3, p_w at 60°F = 0.25610 psia.
In a similar manner, p_w at 69°F = 0.35082 psia.
Since the (saturation) dew point is 60°F,

$$\text{partial pressure } (p') \text{ at } 69°F = \text{vapor pressure } (p_w) \text{ at } 60°F$$

From Equation (3.4),

$$H = (0.25610)/(14.80 - 0.25610) = 0.01761 \text{ mol } H_2O/\text{mol of dry air}$$

From Equation (3.6),

$$\% \text{ RH} = (100)(14.80)(0.01761)/[(0.35082)(1.01761)]$$
$$= 73.0055 \cong 73\% \text{RH}$$

■

3.5.1.3 Wet Bulb and Dry Bulb Temperatures

The wet bulb temperature is the equilibrium temperature attained by a small amount of liquid water during its adiabatic vaporization (without gain or loss of heat) into atmospheric air at a given (dry bulb) temperature. The designations "wet bulb" and "dry bulb" arise from the two thermometers, the bulb of one being kept wet, on a sling psychrometer used to determine the amount of water vapor in the atmosphere. Humidity, absolute or relative, may then be obtained from wet and dry bulb temperatures and barometric pressure by reading a psychrometric chart or by interpolating tabulated values [5, p. E-45].

There is, however, a straightforward analytical method of calculating humidity from wet and dry bulb temperatures without recourse to a psychrometric chart or humidity table [13]. Multiple steps are required but no iterations. The following relationship, obtained by rearrangement of the energy balance equation [14, pp. 285–286], is useful for programmed calculations.

$$H = \frac{\lambda H_w - 6.95(t_d - t_w)}{\lambda + 8.04(t_d - t_w)} \tag{3.7}$$

where is the absolute molal humidity at dry bulb temperature, H_w is the absolute molal humidity at wet bulb temperature (saturated) [calculated using total pressure and vapor pressure at the wet bulb temperature, Equation (3.2)], t_w and t_d are the wet and dry bulb temperatures (°F), respectively, and λ is the molal heat of evaporation of water at the wet bulb temperature (Btu/lb mol = Btu/lb × molecular weight of water).

Between 32 and 150°F,

$$\lambda(\text{Btu/lb mol}) = 19{,}705.0 - 10.4041 t\,(°F) + 0.0039230[t\,(°F)]^2 - 0.00002086[t\,(°F)]^3 \tag{3.8}$$

3.5 Water 95

by regression of steam table entries [13]. The constants 6.95 and 8.04 are the molal heat capacities at constant pressure for air and water vapor, respectively, over the temperature range from 32–200°F. An example using some atmospheric data gathered during stack testing [13] is as follows.

PROBLEM 3.5

Calculate H and % RH using the procedure above and compare with table lookup/interpolation.

$$\text{Dry bulb temperature } (t_d) = 83.0°F = 28.3°C$$
$$\text{Wet bulb temperature } (t_w) = 68.0°F = 20.0°C$$
$$(t_d - t_w) = 15°F = 8.3°C$$
$$\text{Barometric pressure} = 14.707 \text{ psia}$$

SOLUTION Molal heat of evaporation at the wet bulb temperature (λ) is calculated from the regression equation, Equation (3.8). Vapor pressure of water at the dry bulb temperature from Equation (2.23) is substituted into Equation (3.2) to calculate H_w. These inputs are used in Equation (3.7) to calculate H.

Results are as follows:

$$\lambda = 19{,}009.10 \text{ Btu/lb mol}$$
$$p_w \text{ at wet bulb temperature} = 0.33898 \text{ psia}$$
$$H_w \text{ at wet bulb temperature} = 0.023592 \text{ mol } H_2O/\text{mol of dry air}$$

Then from Equation (3.7) (divided by λ for a more accurate calculation)

$$H = \frac{[(19{,}009.10)/(19{,}009.10)](0.023592) - [(6.95)(15)/(19{,}009.10)]}{(19{,}009.10/19{,}009.10) + [(8.04)(15)/(19{,}009.10)]}$$

$$H = \frac{0.023592 - 0.0054842}{1 + 0.0063443} = 0.017994 \text{ mol } H_2O/\text{mol dry air}$$

To check the % RH from this method with tabulated values, p_w at the dry bulb temperature is calculated from Equation (2.23).

$$p_w = 0.55896 \text{ psia}$$

From Equation (3.6)

$$\% \text{ RH} = \frac{(100)(14.707)(0.017994)}{(0.55896)(1.017994)} = 46.51$$

Entering humidity table [5, p. E-45] with

$$t_d = 28.3°C \text{ and } (t_d - t_w) = 8.3°C, \% \text{ RH} = 46.5$$

Agreement between the values of % RH is quite good. A more extensive table of comparative values is contained in Ref. 13.

To recapitulate, one may arrive at the requisite absolute molal humidity (H) for use in combustion calculations in several ways, depending on the given starting point. Some suggested pathways are listed below. Dry bulb temperature and total atmospheric pressure are assumed to be known in all cases.

Starting point	Suggested methodology
H	Use directly
% RH	Use Equation (3.5) with p_w and P_T to calculate H (Problem 3.3)
Dew point	See Problem 3.4
Wet bulb	See Problem 3.5 or consult psychrometric chart/table

■

3.5.1.4 Moist Air Composition

Absolute humidity (H) in mol H_2O per mol dry air is added to the dry air composition to obtain the composition of the moist air to be used in combustion calculations. A worked example is shown below, using the value of H calculated in Problem 3.3. The MW of the moist air is seen to be less than that of the dry air because the MW of water is less than that of dry air.

PROBLEM 3.6

Starting with the standard dry air composition in use here, add humidity (H) at 0.010566 mol H_2O/mol dry air to obtain the wet air composition. Calculate the average molecular weight of the moist air and compare with that of dry air. The AMW of a gas is a useful intermediate for many further calculations.

SOLUTION   Basis: 100 mol of dry air

Constituent (formula)	Molecular weight	vol or mol% (dry)	Moles	mol% (wet)	wt = MW × mol	wt%
Nitrogen (N_2)	28.01	78.086	78.086	77.269	2187.45	75.03
Oxygen (O_2)	32.00	20.947	20.947	20.728	670.28	22.99
Argon (Ar)	39.95	0.934	0.934	0.924	37.31	1.28
Carbon dioxide	44.01	0.033	0.033	0.033	1.45	0.05
Water vapor	18.02	—	1.057*	1.046	19.04	0.65
Total	—	100.000	101.057	100.000	2915.54	100.00

$$\text{Mole } H_2O = (0.010566 \text{ mol } H_2O/\text{mol dry air})(100 \text{ mol dry air}) = 1.057$$

$$\text{Mol\% (wet)} = (100)(\text{moles of each constituent})/(\text{total moles including } H_2O)$$

$$\text{AMW} = \text{total weight/total moles} = (2915.54)/(101.057) = 28.85$$

Here also the molecular weight calculation can be simplified into

$$MW_{air\ wet} = \frac{(0.2801)(\%N_2) + (0.3200)(\%O_2) + (0.3995)(\%Ar) + (0.4401)(\%CO_2) + (18.02)(H)}{(1+H)}$$

(3.9)

$$\frac{(0.2801)(78.086) + (0.3200)(20.947) + (0.3995)(0.934) + (0.4401)(0.033) + (18.02)(0.010566)}{(1+0.010566)} = 28.85$$

Combining this equation with Equation (3.1) for the molecular weight of dry air yields

$$MW_{air\ wet} = \frac{MW_{air\ dry} + (18.02)(H)}{[1+H]}$$

(3.10)

Values derived from Equation (3.10) are as follows:

H	$MW_{air\ dry}$	$MW_{air\ wet}$
0	28.96	28.96
0.01	28.96	28.85
0.02	28.96	28.75
0.03	28.96	28.64
0.04	28.96	28.54
0.05	28.96	28.44
0.06	28.96	28.34
0.07	28.96	28.24

There is a slight curvature brought about by the $[1 + H]$ factor in the denominator of Equation (3.10), but the molecular weight of moist air is effectively linear over the practical range of atmospheric humidity. The molecular weight of combustion air decreases from 28.96 at zero H to 28.24 at 0.07 mol H_2O/mol dry air corresponding to 100% RH at 100.4°F (38.0°C) (Figure 3.2). Molecular weights for several values of H beyond 0.07 (not shown) continue the linear trend.

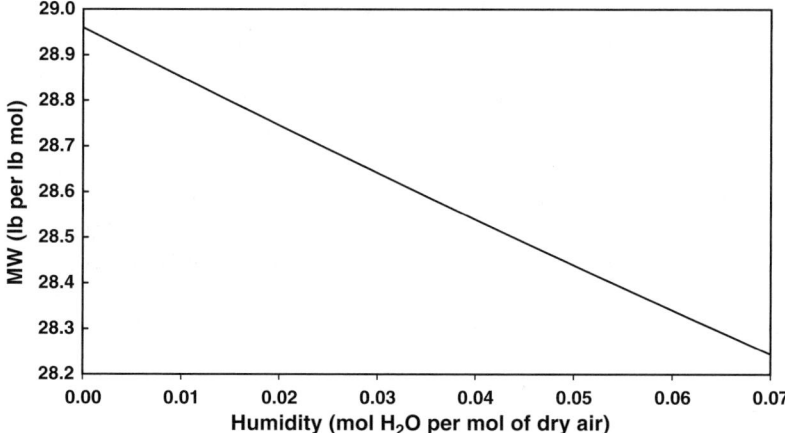

Figure 3.2 Molecular weight of moist air.

98 Chapter 3 Air Combustion

A truly linear fit forced through MW = 28.96 at $H = 0$

$$MW = 28.96 - (10.39239)(H) \tag{3.11}$$

describes the function very well in the range $H = 0$–0.07, with only a few deviations in the hundredths decimal place.

To reiterate a previous concept, sometimes the combustion air is heated before being introduced into the combustion process. Nevertheless, the combustion air moisture is determined by the conditions of the ambient air and not the preheated air. ∎

3.6 COMBUSTION CALCULATIONS FROM BASIC PRINCIPLES

These calculations relate the volume, composition, and mass flow rate of the combustion flue gas to those of the fuel and air streams. Flue gas properties are needed to assess the pollution potential of the combustion process. In the following paragraphs, we will perform the calculations from basic principles. A shortcut procedure derived from EPA Method 19 [4] will be illustrated in a subsequent section.

3.6.1 Pure O_2: Stoichiometric Case

Combustion calculations begin with a balanced chemical equation for each combustible species. Combustion of methane (CH_4) in pure, dry oxygen (O_2), for example, is represented as follows:

$$CH_4 + 2O_2 \rightarrow CO_2 + 2H_2O \tag{3.12}$$

Complete combustion to carbon dioxide and water is assumed, and there is no parallel decomposition of the fuel by any other reaction.

The balanced equation shows that 1 mol of methane requires 2 mol of oxygen (O_2) for complete combustion, producing 1 mol of carbon dioxide and 2 mol of water. At a 2:1 molar ratio of O_2 to CH_4, all of the methane fuel and oxygen are consumed, and the total flue gas (3 mol per mol of methane) consists of 2 mol of H_2O and 1 mol of CO_2. Flue gas composition is expressed as $(100)(1/3) = 33.33\%$ CO_2 by volume and $100(2/3) = 66.67\%$ H_2O by volume on a wet basis (counting H_2O), or $100(1/1) = 100\%$ CO_2 on a dry basis (excluding H_2O). This balance, in which only as much O_2 is added as is necessary to consume the fuel completely, is known as the *stoichiometric case*. The stoichiometric case will be different for other fuels and other oxygen sources; however, this is the stoichiometric case for this example using pure oxygen. This treatment starts off with a single fuel species using the exact amount of dry O_2 needed for combustion. As more information is provided, the situation becomes increasingly more complex until calculations are performed for an excess of moist combustion air with a fuel containing multiple constituents.

Inputs and results in tabular form are as follows:

	Methane combustion (stoichiometric case with pure O_2)							
Component	Moles in fuel	Moles in O_2	Moles O_2 consumed	Moles CO_2 produced	Moles H_2O produced	Moles in flue gas	mol% (wet)	mol% (dry)
CH_4	1	0	2	1	2	0	0.00	0.00
CO_2	0	0	0	0	0	1	33.33	100.00
O_2	0	2	0	0	0	0	0.00	0.00
H_2O	0	0	0	0	0	2	66.67	—
Total (dry)	1	2	2	1	0	1	—	100.00
Total (wet)	1	2	2	1	2	3	100.00	—

Balanced equations for other fuels lead to similar conclusions, but with different stoichiometric ratios:

$$\text{Ethane:} \quad C_2H_6 + 7/2 O_2 \rightarrow 2CO_2 + 3H_2O \qquad (3.13)$$

$$\text{Propane:} \quad C_3H_8 + 5O_2 \rightarrow 3O_2 + 4H_2O \qquad (3.14)$$

In general, for a hydrocarbon fuel,

$$C_nH_m + (n + m/4)O_2 \rightarrow nCO_2 + (m/2)H_2O \qquad (3.15)$$

3.6.2 Pure O_2: Excess O_2 Case

In reality, a significant amount of undesirable CO would be formed from incomplete combustion of any of these fuels at idealized stoichiometric conditions since an equilibrium exists between CO, CO_2, and O_2.

$$CO + \frac{1}{2}O_2 \leftrightarrow CO_2 \qquad (3.16)$$

To drive this equilibrium toward CO_2, an excess of oxygen must be added. The combustion summary for methane using a 10% excess of oxygen above stoichiometric then becomes as follows:

	Methane combustion (using a 10% excess of pure O_2)							
Component	Moles in fuel	Moles in O_2	Moles O_2 consumed	Moles CO_2 produced	Moles H_2O produced	Moles in flue gas	mol% (wet)	mol% (dry)
CH_4	1	0	2	1	2	0	0.00	0.00
CO_2	0	0	0	0	0	1	31.25	83.33
O_2	0	2.2	0.2	0	0	0.2	6.25	16.67

H$_2$O	0	0	0	0	0	2	62.50	—
Total (dry)	1	2.2	2.2	1	0	1.2	100.00	100.00
Total (wet)	1	2.2	2.2	1	2	3.2	100.00	—

Here,

$$10\% \text{ excess } O_2 = 2 \text{ mol } O_2 \text{ stoichiometric} + (0.1)(2 \text{ mol } O_2 \text{ for stoichiometric})$$
$$= 2.2 \text{ mol } O_2$$

The denominator for mol% wet is 3.2 whereas for mol% dry, it is 1.2.

In this example, the excess oxygen added to the combustion mixture results in an (excess) O_2 concentration of 6.25% (wet) or 16.67% (dry) in the flue gas, which now totals 3.2 mol O_2 per mole of methane fuel. Total moles of CO_2 and H_2O remain the same, but their percentage contributions decrease in the larger flue gas volume.

3.6.3 Case Using Pure O$_2$ to Obtain a Specified O$_2$ Concentration in the Flue Gas

Excess oxygen is also commonly specified by percentage composition (wet or dry) in the flue gas rather than by a percentage of the theoretical oxygen required for combustion. For this simple example, 2% O_2 (wet) in the flue gas would mean

$$\% \, O_2/100 = O_2/(O_2 + CO_2 + H_2O) = 2/100$$

Solving for O_2,

$$O_2 = [(\% \, O_2 \, (\text{wet}))/(100 - \% \, O_2 \, (\text{wet}))][CO_2 + H_2O]$$
$$O_2 = [(2)/(100-2)][CO_2 + H_2O]$$
$$= (2/98)(1+2) = 0.061 \text{ mol } O_2 \text{ in flue gas}$$

Likewise, % O_2 (dry) would result in

$$O_2/(O_2 + CO_2) = \% \, O_2 \, (\text{dry})/100$$
$$O_2 = [(\% \, O_2 \, (\text{dry}))/(100 - \% \, O_2 \, (\text{dry}))][CO_2]$$

Now suppose we cleverly choose a value of 5.75% O_2 (dry) for illustrative purposes:

$$O_2 = (5.75/94.75)(1) = 0.061 \text{ mol } O_2 \text{ in flue gas}$$

The combustion summary for these cases (collapsed into one) appears as follows:

Component	Moles in fuel	Moles in O$_2$	Moles O$_2$ consumed	Moles CO$_2$ produced	Moles H$_2$O produced	Moles in flue gas	mol% (wet)	mol% (dry)
CH$_4$	1	0	2	1	2	0	0.00	0.00
CO$_2$	0	0	0	0	0	1	32.67	94.25
O$_2$	0	2.061	0.061	0	0	0.061	**2.00**	**5.75**

H_2O	0	0	0	0	0	0	65.33	—
Total (dry)	1	2.061	2.061	1	0	1.061	—	100.00
Total (wet)	1	2.061	2.061	1	2	3.061	100.00	—

Values of % O_2 wet and dry (in bold) were selected to illustrate the point in a single table.

3.6.4 Stoichiometric Case Using Dry Air

Except for special circumstances, pure oxygen is not used for combustion, and combustion takes place with atmospheric air. Stoichiometric combustion for the methane example proceeds as before, with zero remaining oxygen, but with the other components of air passing through into the flue gas. As above, each mole of methane requires 2 mol of oxygen for combustion, producing 1 mol of CO_2 and 2 mol of water. Enough air is used to provide 2 mol of oxygen. For the standard air of Table 3.3, this brings $(78.086/20.947)(2) = 7.456$ mol of N_2, $(0.934/20.947)(2) = 0.089$ mol of Ar, and $(0.033/20.947)(2) = 0.003$ mol of CO_2 along for the ride.

The familiar combustion summary is tabulated as follows:

			Methane combustion (stoichiometric case with dry air)					
Component	Moles in fuel	Moles in air	Moles O_2 consumed	Moles CO_2 produced	Moles H_2O produced	Moles in flue gas	mol% (wet)	mol% (dry)
CH_4	1	0	2	1	2	0	0.00	0.00
CO_2	0	0.003	0	0.003	0	1.003	9.51	11.73
N_2	0	7.456	0	0	0	7.456	70.69	87.23
Ar	0	0.089	0	0	0	0.089	0.84	1.04
O_2	0	2	0	0	0	0	0	0.00
H_2O	0	0	0	0	0	2	18.96	—
Total (dry)	1	**9.548**	2	1.003	2	*8.548*	—	100.00
Total (wet)	1	**9.548**	2	1.003	0	*10.548*	100.00	—

The CO_2 in the flue gas is the sum of what is formed from combustion of CH_4 plus the minor amount in the incoming air; there is no CO_2 contained in the fuel in this example. Methane, again assumed to react completely with oxygen to CO_2 plus H_2O, is completely used up and does not appear in the flue gas. Total air used (in bold) is 9.548 mol/mol of CH_4 burned. Note that the flue gas (in bold italics) amounts to about 10 times the amount of fuel on a molar basis.

3.6.5 Percent Excess Air Case Using Dry Air

As before, excess air is defined as the amount of air that is used in combustion beyond the stoichiometric amount necessary for all of the oxygen to be consumed exactly.

102 Chapter 3 Air Combustion

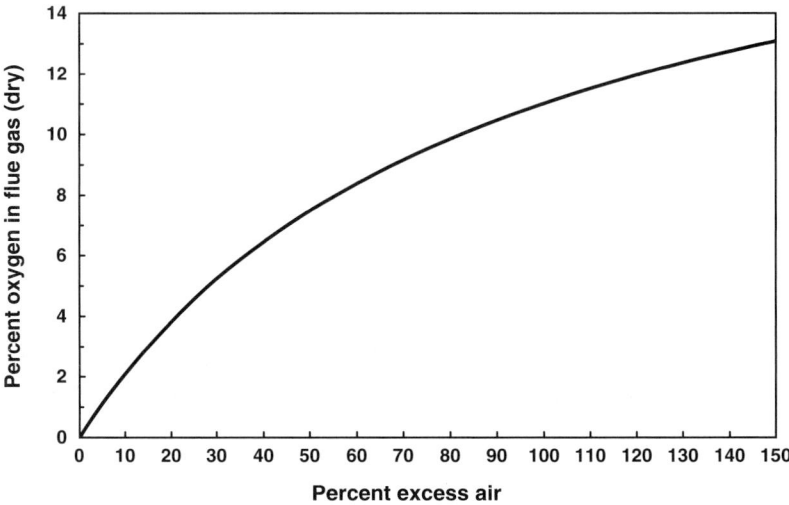

Figure 3.3 Percent oxygen in flue gas versus percent excess air for methane fuel using EPA F factors.

This includes any oxygen that may be contained as O_2 or oxygenated species in the fuel.

The relationship between excess combustion air and percent oxygen in the flue gas on a dry basis (% O_2 dry) for combustion of methane is depicted in Figure 3.3. It is equally valid regardless of the moisture content in the combustion air. The curve is nonlinear, showing values of 0, 2.1, 3, and 13% O_2 dry at a percent excess air of 0, 10, 15, and 150, respectively. Later in this chapter, we shall see how this curve is calculated.

The example continues for methane combustion with an excess of dry air. For 10% excess air, for example, the oxygen required is 2.2 mol as in the % excess O_2 pure oxygen case. The combustion summary now includes the excess O_2 in the flue gas plus its accompanying nitrogen and other constituents from the incoming air. These other constituents of air are prorated upward according to the composition of the dry air (from Table 3.3) and summed.

	Methane combustion (using 10% dry excess air)							
Component	Moles in fuel	Moles in air	Moles O_2 consumed	Moles CO_2 produced	Moles H_2O produced	Moles in flue gas	mol% (wet)	mol% (dry)
CH_4	1	0	2	1	2	0	0.00	0.00
CO_2	0	0.003	0	0.003	0	1.003	8.72	10.56
N_2	0	8.201	0	0	0	8.201	71.30	86.31
Ar	0	0.098	0	0	0	0.098	0.85	1.03
O_2	0	2.2	0	0	0	0.2	*1.74*	*2.10*
H_2O	0	0	0	0	0	2	17.39	—

| Total (dry) | 1 | 10.502 | 2 | 1.003 | 0 | 9.502 | — | 100.00 |
| Total (wet) | 1 | 10.502 | 2 | 1.003 | 0 | **11.502** | 100.00 | — |

With no oxygen in the fuel, all of the excess O_2 comes from the air and the new amount of air is (within roundoff) 1.1 × the air from the stoichiometric case. The flue gas is now 11.502 mol/mol of methane, and oxygen in the flue gas is 1.74% (wet) or 2.10% (dry).

3.6.6 The Case of a Specific Excess O_2 Concentration in the Flue Gas Using Dry Atmospheric Air

This calculation is sometimes performed by trial and error by increasing the amount of combustion air until the desired flue gas O_2 concentration is achieved. However, calculation to a specific value of excess O_2 wet or dry need not be by trial and error if the stoichiometric flue gas (SFG) resulting from the stoichiometric case is first used to calculate the moles of O_2 in the flue gas and the incoming air necessary to produce the desired O_2 concentration in the flue gas. From an oxygen balance on the air and flue gas

$$\text{mole } O_2 \text{ in the flue gas with excess } O_2 = \frac{[Q @ 0\% \, O_2 \, (\text{SFG})](\% \, O_2/100)}{1-(\% \, O_2/20.947)} \quad (3.17)$$

where Q @ 0% O_2 is the total moles in the SFG. With dry combustion air, the desired % O_2 in the flue gas can be either on a wet or on a dry basis. If wet, the wet SFG is used; if dry, the dry SFG is used. *Note:* If moisture is present in the combustion air, this equation works only for dry basis O_2 in the flue gas (see Section 3.6.8).

Here, 2.5% O_2 wet and 3.0% O_2 dry are inserted separately into Equation (3.17). From the stoichiometric case of Section 3.6.4, SFG = 10.548 mol (wet)/mol CH_4 and 8.54 mol (dry)/mol CH_4. Flue gas O_2 = 10.548 (2.5/100)/(1 − 2.5/20.947) = 0.299 for the wet case, and "coincidentally," moles of O_2 = 18.548 (3/100)/(1 − 3/20.947) = 0.299 for the dry case. This value is entered in bold in the combustion summary that follows. It is then added to the 2 mol of O_2 in the air from the stoichiometric case (bold italics). The entries for the other constituents of air are prorated up from the stoichiometric case by the factor (2.299/2) to yield the other entries in italics. These are summed in the flue gas with the calculated moles of O_2, and the percentage compositions (wet and dry) are calculated in the last two columns. In actual practice, one would use either a desired % O_2 wet or a % O_2 dry, not both. The numbers were chosen for this illustrative example to result in the same moles of O_2 and therefore a single table.

	Methane combustion (dry excess air case to a desired flue gas O_2)							
Component	Moles in fuel	Moles in air	Moles O_2 consumed	Moles CO_2 produced	Moles H_2O produced	Moles in flue gas	mol% (wet)	mol% (dry)
CH_4	1	0	2	1	2	0	0.00	0.00
CO_2	0	*0.003*	0	0.003	0	1.003	8.38	10.06

N_2	0	8.571	0	0	0	8.571	71.57	85.92
Ar	0	0.102	0	0	0	0.102	0.85	1.02
O_2	0	**2.299**	0	0	0	**0.299**	2.50	3.00
H_2O	0	0	0	0	0	2	16.70	—
Total	1	10.975	2	1.003	2	11.975	100.00	—
Total (dry)	1	10.975	2	1.003	0	9.975	—	100.00

Percent excess air $= 100(\text{mol air for excess air case/stoichiometric moles air}) - 100$
$= 100(10.975/9.548) - 100 = 115 - 100 = 15\%$ excess air

3.6.7 Stoichiometric Case Using Moist Air

The next step is to add atmospheric water vapor (humidity) to the combustion air. The air from Problem 3.3 with RH of 60% at 60°F and 1 atm pressure contains 0.010566 mol H_2O/mol of dry air. For the stoichiometric case of the combustion of 1 mol of methane, this results in $(9.548)(0.010566) = 0.101$ mol of H_2O in the combustion summary table that follows. The figure of 9.548 is the same as the stoichiometric air for the dry air case of Section 3.6.4.

	Methane combustion (stoichiometric case with 60°F, 60% RH, 1 atm air)							
Component	Moles in fuel	Moles in air	Moles O_2 consumed	Moles CO_2 produced	Moles H_2O produced	Moles in flue gas	mol% (wet)	mol% (dry)
CH_4	1	0	2	1	2	0	0.00	0.00
CO_2	0	0.003	0	0.003	0	1.003	9.42	11.73
N_2	0	7.456	0	0	0	7.456	70.01	87.23
Ar	0	0.089	0	0	0	0.089	0.84	1.04
O_2	0	2	0	0	0	0	0	0
H_2O	0	**0.101**	0	0	0.101	2.101	19.73	—
Total (dry)	1	9.548	2	1.003	0	8.548	—	100.00
Total (wet)	1	9.649	2	1.003	2.101	10.649	100.00	—

3.6.8 Excess Air Case Using Moist Air

Calculations proceed as before except that the humidity in the combustion air must be considered.

The 10% excess air case with moist air: For 60% RH, 60°F, 1 atm air supply, $H = 0.010655$ mol H_2O/mol of dry air. The dry air required for the corresponding stoichiometric case is 9.548. Adding 10% results in 10.502 mol of dry air, the accompanying mol $H_2O = (10.502)(0.010566) = 0.111$ is shown in bold in the table that follows. Other constituents of the combustion are increased by 10% over stoichiometric as well. The entries are summed to obtain a new flue gas composition and divided by wet and dry totals to provide mole percents.

	Methane combustion with a 10% excess of 60°F, 60% RH, 1 atm air							
Component	Moles in fuel	Moles in air	Moles O_2 consumed	Moles CO_2 produced	Moles H_2O produced	Moles in flue gas	mol% (wet)	mol% (dry)
CH_4	1	0	2	1	2	0	0.00	0.00
CO_2	0	0.003	0	0.003	0	1.003	8.64	10.56
N_2	0	8.201	0	0	0	8.201	70.62	86.31
Ar	0	0.098	0	0	0	0.098	0.84	1.03
O_2	0	2.2	0	0	0	0.2	1.72	2.10
H_2O	0	**0.111**	0	0	0	2.111	18.18	—
Total (dry)	1	10.502	2	1.003	0	9.502	—	100.00
Total (wet)	1	10.613	2	1.003	2	11.613	100.00	—

3.6.9 The Case of a Specific Excess O_2 in the Flue Gas

When atmospheric humidity (H) is contained in the incoming air, Equation (3.17) works only to find the dry O_2 concentration in the flue gas. This equation is modified and repeated below. To obtain the wet O_2 concentration corresponding to a desired % O_2 (wet), a factor of $[1 + H]$ must be added in the denominator, resulting in Equation (3.18).

$$\text{mole } O_2 \text{ in the flue gas with excess } O_2 = \frac{[Q_{dry}@0\%O_2(SFG_{dry})](\% O_2 \text{ dry}/100)}{1-(\% O_2 \text{ dry}/20.947)}$$

(3.17)

$$\text{mole } O_2 \text{ in the flue gas with excess } O_2 = \frac{[Q_{wet}@0\% O_2(SFG_{wet})](\% O_2 \text{ wet}/100)}{1-[1+H](\% O_2 \text{ wet}/20.947)}$$

(3.18)

From Equation (3.17), 3% O_2 dry results in

$(8.548)(3/100)/(1-3/20.947) = 0.299$ mol of O_2 in the flue gas as before

But Equation (3.18)

$$(10.649)(2.5/100)/\{1-[1+0.010566](2.5/20.947)\}$$
$$= 0.303 \text{ mol of } O_2 \text{ in the flue gas}$$

must be used to achieve 2.5% O_2 wet.

106 Chapter 3 Air Combustion

The 8.548 and 10.649 factors are the dry and wet total moles of flue gas, respectively, from the corresponding stoichiometric case containing atmospheric humidity at 0.010566 mol H_2O/mol of dry air (Section 3.6.7).

The 0.299 mol of O_2 results in 2.47% O_2 (wet), and the 0.303 mol of O_2 results in 3.03% O_2 (dry), as shown in the following combustion summary tables. The cases of 3% O_2 dry and of 2.5% O_2 wet are no longer the same, and the two separate cases are presented in individual combustion summary tables.

	The 3% O_2 (dry) case with 60°F, 60% RH, 1 atm air							
Component	Moles in fuel	Moles in air	Moles O_2 consumed	Moles CO_2 produced	Moles H_2O produced	Moles in flue gas	mol% (wet)	mol% (dry)
CH_4	1	0	2	1	2	0	0.00	0.00
CO_2	0	*0.003*	0	0.003	0	1.003	8.30	10.06
N_2	0	*8.571*	0	0	0	8.571	70.89	85.92
Ar	0	*0.102*	0	0	0	0.102	0.84	1.02
O_2	0	***2.299***	0	0	0	**0.299**	2.47	3.00
H_2O	0	**0.116**	0	0	0.116	2.116	17.50	—
Total (dry)	1	*10.975*	2	1.003	0	9.975	—	100.00
Total (wet)	1	**11.091**	2	1.003	2.116	12.091	100.00	—

As before, the starting point is the moles of O_2 in the flue gas (bold). This plus the 2 mol consumed by reaction with methane means that 2.299 mol O_2 must have been present in the incoming air (bold italics). The total dry air is $(2.299)/(0.20947) = 10.975$, and the other constituents of dry air (in italics) are found by multiplying total dry air by their fractional composition from Table 3.3.

Excess air $= (100)(11.091/9.649) - 100 = 114.9 - 100 = 14.9\%$ excess air

The 9.649 for this calculation and the one below is the total wet air needed in the corresponding stoichiometric case. The humidity (H) of 0.010566 is multiplied by total moles of dry air to yield the water vapor content of the air (also in bold). The rest of the table above is then filled out in a straightforward manner. Note the % O_2 (wet and dry) in the table and the % excess air calculated below the table.

The computation procedure for the 2.5% O_2 (wet) case is the same as for the 3% O_2 (dry) case, starting with the moles of flue gas O_2 calculated to provide the desired mol% O_2 (wet) in the flue gas.

	The 2.5% O_2 (wet) case with 60°F, 60% RH, 1 atm air							
Component	Moles in fuel	Moles in air	Moles O_2 consumed	Moles CO_2 produced	Moles H_2O produced	Moles in flue gas	mol% (wet)	mol% (dry)
CH_4	1	0	2	1	2	0	0.00	0.00
CO_2	0	*0.004*	0	0.004	0	1.004	8.29	10.05

N_2	0	8.585	0	0	0	8.585	70.89	85.89
Ar	0	0.103	0	0	0	0.103	0.85	1.03
O_2	0	2.303	0	0	0	0.303	2.50	3.03
H_2O	0	0.116	0	0	0	2.116	17.47	—
Total (dry)	1	10.995	2	1.0036	0	9.995	—	100.00
Total (wet)	1	11.111	2	1.0036	2	12.111	100.00	—

Excess air = $100(11.111/9.649) - 100 = 115.2 - 100 = 15.2\%$ excess air

The additional increment of excess air results in the somewhat higher mol% O_2 (wet and dry) in the flue gas, compared to the 3% O_2 (dry) case. For both of these cases, the dry flue gas is about 10 times the fuel, and the wet flue gas about 12 times.

3.7 FLUE GAS

For the general case, the flue gas contains products of combustion, inerts passing through from the fuel and air, and unreacted fuel and oxygen. Pollutants in the flue gas arise from undesirable combustion by-products, noxious inerts, unconsumed fuel constituents, and possibly materials from the chemical or mechanical degradation of the combustion train such as particles of rust or refractory insulation.

The mass flows and concentrations of the major constituents in the flue gas (N_2, CO_2, H_2O, O_2) are determined by the material balance of fuel and atmospheric air (i.e., the combustion calculations). Pollutants/contaminants are estimated by a variety of techniques, including empirical emission factors (EFs), to be explained later.

Flue gas flow and composition is especially sensitive to how much air is used or how much excess O_2 is present. Concentration of the major species as well as the lesser amount of contaminants all depend on how much excess air is present to dilute the combustion mixture. The flue gas flow (Q) can be shown to obey the following relationships:

$$Q_{dry}/(Q_{dry} @ 0\% O_2) = [20.9/(20.9 - \% O_2 \text{ dry})] \quad (3.19)$$
$$Q_{wet}/(Q_{wet} @ 0\% O_2) = \{20.9/(20.9 - \% O_2 \text{ wet}[1 + H])\} \quad (3.20)$$

where H is the previously defined humidity of the ambient air (mol H_2O/mol dry air), and Q_{dry} or Q_{wet} @ 0% O_2 was previously denoted as the stoichiometric flue gas (dry or wet). These theoretically rigorous relationships are derived from ideal gas combustion using ambient air with a dry basis O_2 concentration of 20.9%. They are plotted in Figures 3.4 and 3.5, with H as a parameter in Figure 3.6.

Likewise, a CO_2 balance under various excess air conditions yields the following:

$$Q_2/Q_1 = (CO_2 @ \text{ condition } 1 - CO_2 \text{ in air})/(CO_2 @ \text{ condition } 2 - CO_2 \text{ in air}) \quad (3.21)$$

and

$$Q_2/Q_1 = (CO_2 @ \text{ condition } 1)/(CO_2 @ \text{ condition } 2) \quad (3.22)$$

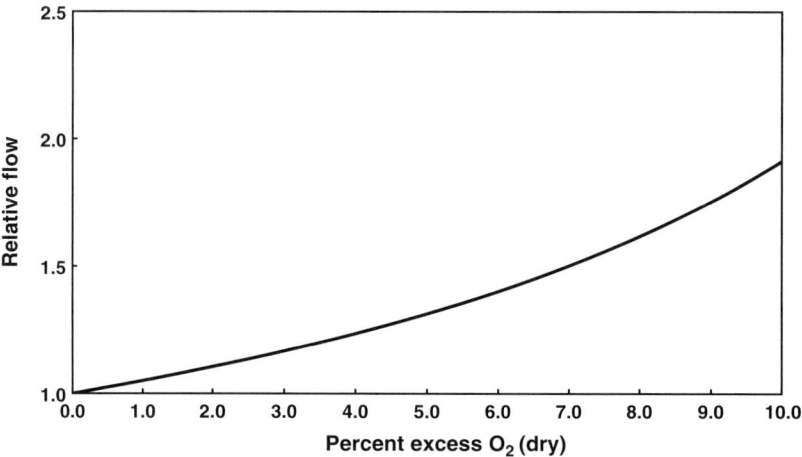

Figure 3.4 Ratio of flue gas flow to flue gas flow at stoichiometric conditions (dry basis).

The second equation results if one ignores the small amount of CO_2 in the combustion air (0.033%), compared to CO_2 on the order of 10% found in flue gas. These equations work equally well for either the wet basis or dry basis calculations, provided that consistent inputs are used, all wet or all dry.

Conversion from a basis of wet to dry or from dry to wet is accomplished by

$$Q_{wet} = Q_{dry}[100/(100 - \% \, H_2O)] \qquad (3.23)$$

and

$$Q_{dry} = Q_{wet}[(100 - \% \, H_2O)/100] \qquad (3.24)$$

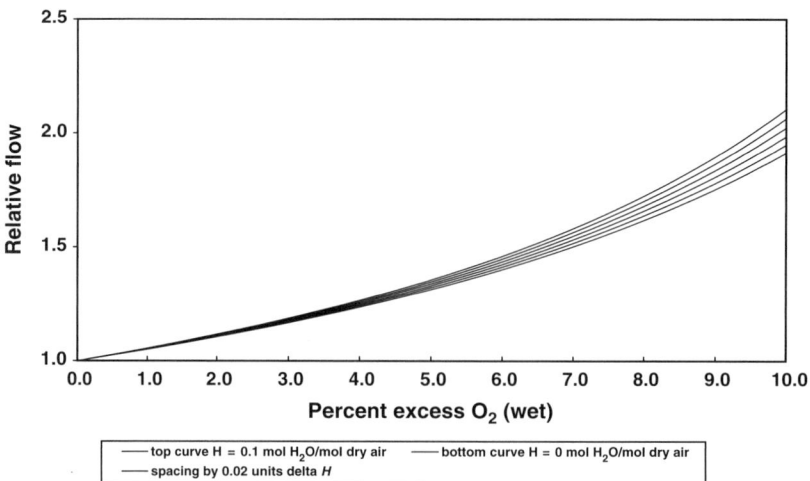

Figure 3.5 Ratio of flue gas flow to flue gas flow at stoichiometric conditions (wet basis) with humidity as parameter.

3.7 Flue Gas

Frequently, it is important to convert pollutant concentration from one condition of excess air to another, expressed as excess O_2 or flue gas CO_2 concentrations. Allowable concentrations of pollutants are often stated in regulations, standards, or permits at some specified condition of excess air to prevent meeting the prescribed limit simply by dilution. Typical conditions are 3% O_2 for boilers and heaters, 7% O_2 for municipal waste combustion, 15% O_2 for stationary gas turbines, and 12% CO_2 for incinerators, all on a dry basis. The author has seen 0% O_2 used as a basis as well.

The key to such calculations is the fact that the mass emission of pollutant is the same, regardless of the dilution. Hence,

$$(C_1)(Q_1) = (C_2)(Q_2) = \text{mass/time} \tag{3.25}$$

The concentration and flow must be on a consistent basis, both wet or both dry. For adjustments using O_2,

$$C_2 \text{ dry} = C_1 \text{ dry}(20.9\% - \% \, O_2 \text{ dry}_{(2)})/(20.9 - \% \, O_2 \text{ dry}_{(1)}) \tag{3.26}$$

where C_1 and C_2 are pollutant concentrations on a dry basis.

For example, when it is required to convert concentration to 3% O_2 (dry), Equation (3.26) becomes

$$C_{\text{dry}} \text{ @ } 3\% \, O_2 \text{ (dry)} = C_{\text{dry}} \text{ @ conditions } (20.9-3 = 17.9)/(20.9-\% \, O_2 \text{ (dry) @ conditions}) \tag{3.27}$$

An equation similar to Equation (3.26) can be derived relating wet basis pollutant and O_2 concentrations.

$$C_{2\text{wet}} = \frac{(C_{1\text{wet}})(20.9-\% \, O_2 \text{ wet}_{(2)}[1+H])}{(20.9-\% \, O_2 \text{ wet}_{(1)}[1+H])} \tag{3.28}$$

This equation is also exact and based on theory. However, to make use of it to convert to a final dry basis condition, as is often required, both the moisture content at the final condition and the ambient humidity must be known. Although it is possible to derive equations for moisture content (and hence the factor $100/(100-\% \, H_2O)$ to convert from ppm wet to ppm dry) as a function of the stoichiometric moisture, the humidity, and the dry basis % O_2, these equations are somewhat complicated (see Equations (B.3) and (B.4)).

There is less opportunity for error if concentrations on a wet basis are first changed to a dry basis using the moisture content and then Equation (3.26) is employed to convert to the final condition.

$$C_{\text{dry}} = C_{\text{wet}}[100/(100-\% \, H_2O)] \tag{3.29}$$
$$C_{\text{wet}} = C_{\text{dry}}[(100-\% \, H_2O)/100] \tag{3.30}$$

where C is the concentration of any flue gas species other than H_2O.

For adjustments based on CO_2,

$$C_2 = C_1(CO_2 \text{ @ condition 2})/(CO_2 \text{ @ condition 1}) \tag{3.31}$$

where all concentrations are consistent, either all wet or all dry.

For example, to convert to 12% CO_2 dry,

$$C_{2_{dry}} @ 12\% \ CO_{2_{dry}} = C_{1_{dry}} @ \text{ original conditions} (12/CO_2 \text{ dry} @ \text{ original conditions})$$
(3.32)

3.8 COMBUSTION CALCULATIONS BASED ON EPA METHOD 19 F FACTORS

Method 19 in 40 CFR 60, Appendix A, provides data reduction procedures for the results of stack testing. In addition, one can make use of Method 19's F factors as a shortcut in performing combustion calculations.

Method 19 relates the volume of combustion products to the HHV of the fuel, a quantity that is denoted there as the *gross calorific value*. The factor F_d, for example, is defined as the ratio of the dry stoichiometric flue gas volume at standard conditions of 68°F (20°C) and 1 atm (760 mmHg, 29.92 in. Hg, or 14.696 psia) to the GCV (HHV).

Some F factor values are listed in Table 3.4. In English units for natural gas, the dry basis F_d factor is 8710 SCF (dry)/(10^6 or MMBtu). The same F_d factor is listed for propane and for butane. At these conditions, the molar volume is 385.3 SCF/lb mol (Table 2.4). This dry stoichiometric flue gas volume can be converted to other conditions of temperature and pressure via the ideal gas law (Chapter 2). Conversion to other conditions of flue gas oxygen or excess air is accomplished by means of Equation (3.19), repeated here for convenience.

$$Q_{dry}/(Q_{dry} @ 0\% \ O_2) = [20.9/(20.9 - \% \ O_2 \text{ dry})]$$
(3.19)

Table 3.4 Selected Values of F Factors for Various Fuels (English Units)[a,b]

Fuel type	F_d SCF (dry) per 10^6 Btu (MMBtu)	F_w SCF (wet) per 10^6 Btu (MMBtu)	F_c SCF per 10^6 Btu (MMBtu)
Gas			
Natural gas	8710	10,610	1040
Propane	8710	10,200	1190
Butane	8710	10,390	1250
Oil			
Crude, residual or distillate	9190	10,320	1420
Coal			
Anthracite	10,100	10,540	1970
Bituminous	9,780	10,640	1800
Lignite	9,860	11,950	1910

[a] Abridged from Table 19-1 of 40 CFR 60, Appendix A, Method 19 [4].
[b] Basis for standard cubic foot (SCF) is 68°F (20°C) and 1 atm pressure (760 mmHg, 29.92 in. Hg, or 14.696 psia).

3.8 Combustion Calculations Based on EPA Method 19 F Factors   111

Table 3.5 Combustion Relationships [per MMBtu (HHV) Fired] Using Method 19 Factors: Stoichiometric Case

Constituent	Combustion air	Flue gas
CO_2	0	F_c
N_2	$(F_d - F_c)$	$(F_d - F_c)$
O_2	$(20.9/79.1)(F_d - F_c)$	0
Total dry	$(100/79.1)(F_d - F_c)$	F_d
H_2O in air	$(100/79.1)(F_d - F_c) \times [B_{wa}/(1 - B_{wa})]$	—
H_2O of combustion	0	$(F_w - F_d)$
Total H_2O	$(100/79.1)(F_d - F_c) \times [B_{wa}/(1 - B_{wa})]$	$(F_w - F_d) + (100/79.1)(F_d - F_c) \times [B_{wa}/(1 - B_{wa})]$
Total wet	$(100/79.1)(F_d - F_c) \times \{1 + B_{wa}/(1 - B_{wa})\}$	$F_w + (100/79.1)(F_d - F_c) \times [B_{wa}/(1 - B_{wa})]$

Other F factors, F_w and F_c, are also defined. These have different numerical values for different fuels. F_w serves the same purpose as F_d, except relating the wet basis stoichiometric flue gas flow to the heating value. F_w reflects only the water from combustion of the hydrogen in the fuel and does not include moisture in the combustion air or water added by a wet scrubber, steam injection, and so on. F_c is the amount of carbon dioxide produced per quantity of heat input, regardless of the excess O_2 concentration in the flue gas. This assumes that there is no CO_2 present in the combustion air.

Step-by-step procedures developed here using F factors for combustion calculations are summarized in Tables 3.5–3.8. Separate tables are given for the stoichiometric condition, in terms of excess air, and for percent excess oxygen (O_2) in the flue gas, both wet and dry. Each of these procedures makes use of what Williams and Johnson [15, pp. 7, 66, 215–216] call a *key component*. In this case, it is the identical mass of nitrogen in the flue gas and in the atmospheric air.

This F factor procedure assumes that there is no nitrogen in the fuel and no CO_2 in the air. Since F_w reflects only water of combustion, the total flue gas moisture must be computed from the sum of the water of combustion and the moisture in the combustion air. The combustion air moisture is determined from the moisture content of the air and the total amount of air used. That air is proportional to the quantity of flue gas nitrogen or its equivalent, combustion air nitrogen.

Combustion calculations making use of the Method 19 relationships are illustrated in several example problems appearing later in this chapter.

The F factor method is a refinement of a general relationship noted between the stoichiometric flue gas volume generated by a fossil fuel or a waste fuel mixture and the heating value of the fuel [16]. That relationship is said to be similar to an old rule of thumb that one cubic foot of air releases 100 Btu of (gross) heat [17].

$$\text{SCF (@ 32°F, 1 atm) per lb of fuel} = [\text{HHV of fuel (Btu/lb)}]/100 \quad (3.33)$$

Table 3.6 Combustion Relationships [per MMBtu (HHV) Fired] Using Method 19 Factors: Percent Excess Air Case

Constituent	Combustion air	Flue gas
CO_2	0	F_c
N_2	$(F_d - F_c)[1 + (\% \text{ XS air}/100)]$	$(F_d - F_c)[1 + (\% \text{ XS air}/100)]$
O_2	$(20.9/79.1)(F_d - F_c) \times [1 + (\% \text{ XS air}/100)]$	$(20.9/79.1)(F_d - F_c) \times (\% \text{ XS air}/100)$
Total dry	$(100/79.1)(F_d - F_c) \times [1 + (\% \text{ XS air}/100)]$	$F_d + (100/79.1)(F_d - F_c) \times (\% \text{ XS air}/100)$
H_2O in air	$(100/79.1)(F_d - F_c) \times [B_{wa}/(1 - B_{wa})] \times [1 + (\% \text{ XS air}/100)]$	—
H_2O of combustion	0	$(F_w - F_d)$
Total H_2O	$(100/79.1)(F_d - F_c) \times [B_{wa}/(1 - B_{wa})] \times [1 + (\% \text{ XS air}/100)]$	$(F_w - F_d) + (100/79.1)(F_d - F_c) \times [B_{wa}/(1 - B_{wa})] \times [1 + (\% \text{ XS air}/100)]$
Total wet	$(100/79.1)(F_d - F_c) \times \{1 + B_{wa}/(1 - B_{wa})\} \times [1 + (\% \text{ XS air}/100)]$	$F_w + (100/79.1)(F_d - F_c) \times [B_{wa}/(1 - B_{wa})] \times [1 + (\% \text{ XS air}/100)] + (100/79.1) \times (F_d - F_c) \times (\% \text{ XS air}/100)$

% Excess air = (100)(air at conditions/stoichiometric air) − 100.

Air may be wet or dry but must be consistent with excess air and stoichiometric conditions.

3.8 Combustion Calculations Based on EPA Method 19 F Factors   113

Table 3.7 Combustion Relationships [per MMBtu (HHV) Fired] Using Method 19 Factors: Percent O_2 Dry in Flue Gas

Constituent	Combustion air	Flue gas
CO_2	0	F_c
N_2	N_2 in flue gas	$[1 - (\% \, O_2 \, \text{dry}/100)] \times$ (total dry flue gas) $- F_c$
O_2	$(20.9/79.1)(N_2$ in flue gas$)$	$(\% \, O_2 \, \text{dry}/100) \times$ (total dry flue gas)
Total dry	$(100/79.1)(N_2$ in flue gas$)$	$F_d \, [20.9/(20.9 - \% \, O_2 \, \text{dry})]$
H_2O in air	$(100/79.1)(N_2$ in flue gas$) \times [B_{wa}/(1 - B_{wa})]$	—
H_2O of combustion	0	$(F_w - F_d)$
Total H_2O	$(100/79.1)(N_2$ in flue gas$) \times [B_{wa}/(1 - B_{wa})]$	$(F_w - F_d) + H_2O$ in air
Total wet	$(100/79.1)(N_2$ in flue gas$)/(1 - B_{wa})$	Total dry flue gas $+$ total H_2O in flue gas

Per MMBtu combusted and correcting by the ratio of absolute temperatures for the difference in volume at EPA's standard condition of 68°F, 1 atm and the 32°F, 1 atm basis of Equation (3.33), upon rearrangement this relationship becomes

$$\text{SCF}(@68°F, 1 \, \text{atm})/\text{MMBtu} = (10^6/100)[(68 + 459.67)/(32 + 459.67)]$$

$$= 10{,}732 \, \text{SCF/MMBtu} \quad (3.34)$$

Table 3.8 Combustion Relationships [per MMBtu (HHV) Fired] Using Method 19 Factors: Percent O_2 Wet in Flue Gas

Constituent	Combustion air	Flue gas
CO_2	0	F_c
N_2	N_2 in flue gas	$[(1 - B_{ws}) - (\% \, O_2 \, \text{wet}/100)] \times$ (total wet flue gas) $- F_c$
O_2	$(20.9/79.1)(N_2$ in flue gas$)$	$(\% \, O_2 \, \text{wet}/100)$(total wet flue gas)
Total dry	$(100/79.1)(N_2$ in flue gas$)$	$(1 - B_{ws})$(total wet flue gas)
H_2O in air	$(100/79.1)(N_2$ in flue gas$) \times [B_{wa}/(1 - B_{wa})]$	—
H_2O of combustion	0	$F_w - F_d$
Total H_2O	$(100/79.1)(N_2$ in flue gas$) \times [B_{wa}/(1 - B_{wa})]$	(B_{ws})(total wet flue gas)
Total wet	$(100/79.1)(N_2$ in flue gas$)/(1 - B_{wa})$	$[20.9/[20.9 - \% \, O_2 \, \text{wet}/(1 - B_{wa})]] \times [F_w + (100/79.1)(F_d - F_c) \times [B_{wa}/(1 - B_{wa})]]$

From the example given in the cited reference, it is obvious that the volume of combustion products is on a wet basis and includes the water of combustion. This single figure is in the same "ballpark" as the values of F_w (10,200–11,950) that are listed in Table 3.4. However, rather than simply representing an overall average relationship, the F factors in the table are tailored to the specific fuel or class of fuels listed.

With the use of these tables, percentage composition is obtained by dividing an individual entry by the corresponding total air or total flue gas (wet or dry) and multiplying by 100.

EPA Method 19 uses moisture terms B_{ws} and B_{wa}. The term B_{ws} is equivalent to % H_2O/100 in the flue gas, and $(1 - B_{ws})$ then equals $(100-\% H_2O)/100$.

B_{wa} is the moisture fraction of ambient air.

From its usage in the Method 19 equations, the unstated units of B_{wa} must be mol H_2O/mol of total or wet air. In that case, the following identities apply between the absolute humidity $[H]$ (mol H_2O/mol dry air) defined elsewhere in this book and Method 19's B_{wa} (mol H_2O/mol total wet air).[3]

$$B_{wa} = \frac{H}{1+H} \tag{3.35}$$

$$H = \frac{B_{wa}}{1-B_{wa}} \tag{3.36}$$

$$(1+H) = \frac{1}{1-B_{wa}} \tag{3.37}$$

$$(1-B_{wa}) = \frac{1}{1+H} \tag{3.38}$$

Method 19 also contains guidelines regarding combustion air moisture (B_{wa}) to be used in reducing stack test data:

- Actual measurement
- Estimation
  - Using highest monthly average that occurred within the previous calendar year at the nearest Weather Service Station—determined annually and used as an estimate for the entire current calendar year.
  - Using highest daily average that occurred within a calendar month at the nearest Weather Service Station—a value calculated for each month from the data of the past 3 years.
- A default value, $B_{wa} = 0.027$, which may be used at any location at all times.

---

[3] Method 19 mentions a default humidity of 0.027, explained below, for data reduction. A factor of 0.027 in moles H_2O/wet air is equivalent to 0.027749 mol H_2O/mol dry air, less than a 3% difference. This corresponds roughly to air at 80F with a relative humidity between 75 and 80%, but other combinations of temperature and RH that produce the same value are, or course, equally valid. Agreement is even better at lower values of H.

3.9 COMBUSTION PROBLEMS: MAJOR SPECIES IN THE FLUE GAS

With the basics behind us, we now embark on a series of problems to wrap up combustion calculations for methane and to show the results for combustion of propane (C_3H_8) and the natural gas of Table 3.1. Results from the method employing the chemical reaction combustion equations and the F factor method are compared. Agreement found here in flue gas compositions between the two approaches is certainly reasonable. The basis for Figure 3.3 describing the relationship between percent air and percent O_2 in the flue gas from methane combustion is explained at the end of this section.

PROBLEM 3.7

By way of review, revisit the combustion of pure methane from basic principles (the long way), this time for a flue gas oxygen content of 2.0% O_2 wet. Combustion air (Table 3.3) is at 1 atm, 60°F, 60% RH. Prepare a combustion summary table for this condition.

SOLUTION Once again Equation (3.18) will be employed to calculate the moles of O_2 in the flue gas. The stoichiometric flue gas for methane with this combustion air is contained in Section 3.6.7. Humidity (H) equals 0.010566 mol H_2O/mol dry air.

$$\text{Mole } O_2 \text{ in flue gas} = \frac{[10.649 \text{ (wet)}](2/100)}{1-(1.010566)(2/20.947)} = 0.2357$$

This is shown in bold in the combustion summary table that follows. One then adds this figure to the 2 mol O_2 required for stoichiometric combustion (bold italics) (Section 3.6.7). From this, the moles of N_2, Ar, CO_2, and H_2O in the air (italics) are obtained. The rest of the table can be filled out by material balance with the help of Equation (3.12) for the methane combustion reaction. The N_2 and Ar are exactly the same in the flue gas as in the combustion air. The methane completely disappears and is replaced by its combustion products. The CO_2 and H_2O in the flue gas are the respective sums from combustion products and from the air. The moles of each constituent in the flue gas are summed and percentages (wet and dry) are computed. The mol% O_2 wet checks at 2.0%.

Combustion of pure methane 2.0% O_2 (wet) case with 60°F, 60% RH, 1 atm air

Component	Moles in fuel	Moles in air	Moles O_2 consumed	Moles CO_2 produced	Moles H_2O produced	Moles in flue gas	mol% (wet)	mol% (dry)
CH_4	1	0	2	1	2	0	0.00	0.00
CO_2	0	0.0035	0	0.0035	0	1.0035	8.51	10.37

N_2	0	*8.3342*	0	0	0	8.3342	70.71	86.16
Ar	0	*0.0997*	0	0	0	0.0997	0.85	1.03
O_2	0	*2.2357*	0	0	0	**0.2357**	2.00	2.44
H_2O	0	*0.1128*	0	0	0	2.1128	17.93	—
Total (dry)	1	10.6731	2	1.0035	0	9.6731	—	100.00
Total (wet)	1	10.7859	2	1.0035	2	11.7859	100.00	—

■

PROBLEM 3.8

Compute F_d, F_w, and F_o when firing 1 mol/h of pure methane using Table 3.3 air.

SOLUTION For the stoichiometric case (Section 3.6.7), 1 mol/h of CH_4 produces 8.545 mol/h of dry flue gas excluding CO_2 in the ambient air and 1 mol/h of CO_2 excluding the same ambient air CO_2. Wet flue gas composed of dry flue gas plus only the water of combustion amounts to 10.545 mol/h. One lb mol/h releases

$$\frac{1010 \text{ Btu}}{\text{SCF}} \frac{1 \text{ lb mol}}{\text{h}} \frac{379.5 \text{ SCF}}{\text{lb mol}} = 0.3833 \times 10^6 \frac{\text{Btu}}{\text{h}}$$

CO_2 1 lb mol/h   ×385.3 SCF (68°F, 1 atm)   = 385.3 SCFH
Dry flue gas   (8.5450)   (385.3)   = 3292 SCFH
Wet flue gas   (10.545)   (385.3)   = 4063 SCFH

$F_d = (3292)/(0.3833) = 8590$
$F_w = (4103)/(0.3833) = 10{,}600$ $\Big\}$ SCF (68°F, 1 atm) per 10^6 Btu (all rounded as in Table 3.4)
$F_c = (385.3)/(0.3833) = 1010$

See Problem 3.9 for a further explanation of the calculation of F factors. ■

PROBLEM 3.9

Using the chemical equations for the oxidation reactions occurring during combustion, calculate the following when firing 1 lb/mol of propane (C_3H_8) at 2% excess O_2 (wet) in the flue gas using 60°F, 60% RH combustion air of Table 3.3 dry-basis composition.

- flue gas composition
- flue gas flow rate
- molecular weight of flue gas
- combustion air flow rate
- percent excess air
- heat input
- Method 19 F factors

3.9 Combustion Problems: Major Species in the Flue Gas

SOLUTION Solution of this problem follows the same methodology employed for the previous problems in which methane was fired, except that the chemical reaction for propane combustion is different.

$$C_3H_8 + 5O_2 \rightarrow 3CO_2 + 4H_2O \tag{3.14}$$

For this reaction, 1 mol of propane generates 3 mol of CO_2 and 4 mol of water of combustion, and consumes 3 mol of oxygen from the combustion air. This brings along nitrogen plus other constituents, including atmospheric moisture, present along with the oxygen in the air.

At stoichiometric conditions, enough air is introduced to use up all the oxygen exactly. When an excess of air is added to minimize formation of pollutants such as CO, particulate matter (also known as soot), and unburned hydrocarbons, oxygen appears in the products of combustion, accompanied by an additional quantity of those same components of atmospheric air. Estimation of pollutant concentrations and emissions is handled in Section 3.11.

Heat is generated during combustion, and the hot flue gas is released to atmosphere through a chimney or stack.

The combustion table follows for the stoichiometric case.

Basis: 1 lb mol of C_3H_8 fuel; 60°F, 60% RH, 1 atm air

Component	Moles in fuel	Moles in air	Moles O_2 consumed	Moles CO_2 produced	Moles H_2O produced	Moles in flue gas	mol% (wet)	mol% (dry)
C_3H_8	1	0	5	3	4	0	0.00	0.00
CO_2	0	0.0079	0	0.0079	0	3.0079	11.52	13.75
N_2	0	18.6389	0	0	0	18.6389	71.35	85.23
Ar	0	0.2229	0	0	0	0.2229	0.85	1.02
O_2	0	5	0	0	0	0	0.00	0.00
H_2O	0	0.2522	0	0	0	4.2522	16.28	—
Total (dry)	1	23.8698	5	3.0079	0	21.8697	—	100.00
Total (wet)	1	24.1220	5	3.0079	4	26.1219	100.00	—

Moles N_2 in air = (moles O_2 consumed)(78.086/20.947) = (5)(3.7278) = 18.6389

Total dry air = (18.6389)(100/78.096) = 23.8698

Moles Ar in air = (23.8698)(0.934/100) = 0.2229

Moles CO_2 in air = (23.8698)(0.033/100) = 0.0079

Moles H_2O in air = (23.8698)(0.010566 lb mol H_2O/lb mol dry air) = 0.2522

Figures for air composition are from Table 3.3.

Total air = 23.8698 (dry), 24.1220 (wet)

Total flue gas = 21.8697 (dry), 26.1219 (wet)

Molecular weight of dry flue gas

= [(3.0079)(44.01) + (18.6389)(28.01) + (0.2229)(39.95)]/(21.8697) = 30.3 lb/lb mol (dry)

For wet flue gas, add (4.2522)(18.02) and divide the total by 26.1219 = 28.3 lb/lb mol (wet).

118 Chapter 3 Air Combustion

Heating value of propane = 2516.1 (HHV) and 2314.9 (LHV) Btu/SCF (60°F, 1 atm).

HHV : (2516.1 Btu/SCF)(379.5 SCF/lb mol)(1 lb mol/h)/10^6
   = 0.9549 MMBtu/h per lb mol/h of C_3H_8
   = 95.49 MMBtu/h per 100 lb mol/h, etc.

LHV : (2314.9)(379.5)(1)/10^6
   = 0.8785 MMBtu/h per 1 lb mol/h of C_3H_8
   = 87.85 MMBtu/h per 100 lb mol/h, etc.

For the case of 2% excess O_2 (wet) in the flue gas, use Equation (3.18) to solve for moles of O_2 in the flue gas. For this equation, the wet SFG is necessary. From the combustion table above, this equals 26.1219 lb mol and $H = 0.010566$. Then,

$$\text{Moles } O_2 = \frac{[26.1219][(2/100)]}{1-(1+0.010566)(2/20.947)} = 0.5782$$

This entry starts off the following combustion table for the excess air case, along with the C_3H_8 in the fuel column.

Basis: 1 lb mol of C_3H_8 fuel; 60°F, 60% RH, 1 atm air (2% O_2 wet)

Component	Moles in fuel	Moles in air	Moles O_2 consumed	Moles CO_2 produced	Moles H_2O produced	Moles in flue gas	mol% (wet)	mol% (dry)
C_3H_8	1	0	5	3	4	0	0.00	0.00
CO_2	0	0.0088	0	0.0088	0	3.0088	10.41	12.21
N_2	0	20.7945	0	0	0	20.7945	71.92	84.43
Ar	0	0.2487	0	0	0	0.2487	0.86	1.01
O_2	0	5.5782	0	0	0	0.5782	2.00	2.35
H_2O	0	0.2814	0	0	0	4.2814	14.81	—
Total (dry)	0	26.6302	5	3.0088	0	24.6302	—	100.00
Total (wet)	0	26.9116	5	3.0088	4	28.9116	100.00	—

Moles O_2 needed in the combustion air is the sum of 0.5782 and 5 from the stoichiometric case. The moles of the other constituents in the combustion air are computed as above for the stoichiometric case but with 5.5782 instead of 5. Results are shown above in the air column. The moles O_2 column and the moles H_2O column are exactly the same as for the stoichiometric case. The moles CO_2 column and the remaining entries in the flue gas column change to reflect the additional moles in the incoming air. The components in the flue gas column are summed, and the mole percents are calculated as before.

For the excess air case, as read from the table,

Total air = 26.6302 (dry), 26.9116 (wet)

Total flue gas = 24.6302 (dry), 28.9116 (wet)

Excess air = (100)(26.9116/24.1220)−100 = 11.6%

3.9 Combustion Problems: Major Species in the Flue Gas   119

The figure 24.1220 comes from the stoichiometric case.

As in the stoichiometric case, flue gas average molecular weights are calculated as follows. Ensure that the oxygen is included in the flue gas.

$$\text{AMW dry} = \frac{[(3.0088)(44.01) + (20.7945)(28.01) + (0.2487)(39.95) + (0.5782)(32.00)]}{24.6302} = 30.2$$

(Note that the rule of thumb that the dry molecular weight of flue gas is about 30 works well here for combustion of propane.)

$$\text{AMW wet} = \frac{[(3.0088)(44.01) + (20.7945)(28.01) + (0.2487)(39.95) + (0.5782)(32.00) + (4.2814)(18.02)]}{28.9116}$$

$$= 28.4$$

The Method 19 F_d factor is defined as the SCF (68°F, 1 atm) of flue gas for the stoichiometric case divided by the gross (HHV) heat input. (CO_2 in air ignored here.)

$$\frac{(21.8697 \text{ lb mol})(385.3 \text{ SCF/lb mol})}{(2516.1 \text{ Btu/SCF})(379.5 \text{ SCF/lb mol})(1 \text{ lb mol/h})(1/10^6)} = \frac{8820 \text{ SCF}(68°F, 1 \text{ atm})}{10^6 \text{ Btu}}$$

The F_w factor is the ratio of the wet flue gas, excluding atmospheric moisture, to the GCV or HHV. (CO_2 in air ignored here also.)

$$\frac{(26.1219 - 0.2522) \text{lb mol}(385.3 \text{ SCF/lb mol})}{(2516.1 \text{ Btu/SCF})(379.5 \text{ SCF/lb mol})(1 \text{ lb mol/h})(1/10^6)} = \frac{10,440 \text{ SCF}(68°F, 1 \text{ atm})}{10^6 \text{ Btu}}$$

$$F_w = \frac{10,440 \text{ SCF } (68°F, 1 \text{ atm}) \text{ wet flue gas (excluding combustion air moisture)}}{10^6 \text{ Btu}}$$

The F_c factor is the ratio of the CO_2 produced, excluding CO_2 in the combustion air, to the GCV or HHV.

$$\frac{(3 \text{ lb mol})(385.3 \text{ SCF/lb mol})}{(2516.1 \text{ Btu/SCF})(379.5 \text{ SCF/lb mol})(1 \text{ lb mol/h})(1/10^6)} = \frac{1210 \text{ SCF } (68°F, 1 \text{ atm})}{10^6 \text{ Btu}}$$

$$F_c = \frac{1210 \text{ SCF } (68°F, 1 \text{ atm}) CO_2 (\text{excluding } CO_2 \text{ in combustion air})}{10^6 \text{ Btu}}$$

These values calculated here average about 2% higher than the F factors published in Method 19 [4]. They will be used to compare the combustion calculations in this problem, done the long way, with the EPA Method 19 shortcut procedure explained in Problem 3.11. Estimation of emissions from a small propane-fired heater will be illustrated in Problem 3.23. ■

PROBLEM 3.10   *Natural Gas the Long Way*

Using the chemical equations for the oxidation reactions occurring during combustion, calculate the following when firing 1 lb mol/h of the natural gas of Table 3.1 at (a) the stoichiometric condition, (b) 10% excess air, and (c) 2% excess O_2 (wet) in the flue gas using 60°F, 60% RH atmospheric combustion air of Table 3.3 dry-basis composition.

- flue gas composition
- flue gas flow rate

- molecular weight of flue gas
- combustion air flow rate
- percent excess air
- heat input
- Method 19 F factors (compare with the F factors for pure methane)

SOLUTION

(a) Solution follows the same methodology employed for methane and propane with the addition of the chemical reactions for the other combustible components. The F factors will be employed in a shortcut combustion calculation procedure based on 40 CFR 60, Appendix A, Method 19 [4] and explained here in a subsequent section. Combustion equations are listed below. Equations for methane, Equation (3.12); ethane, Equation (3.13); and propane, Equation (3.14) have already been presented. Equations for C_4, C_5, and C_6 hydrocarbons are derived from the general equation, Equation (3.15).

$$\text{Methane} \quad CH_4 + 2O_2 \rightarrow CO_2 + 2H_2O \tag{3.12}$$
$$\text{Ethane} \quad C_2H_6 + 7/2 O_2 \rightarrow 2CO_2 + 3H_2O \tag{3.13}$$
$$\text{Propane} \quad C_3H_8 + 5O_2 \rightarrow 3CO_2 + 4H_2O \tag{3.14}$$
$$\text{Butane} \quad C_4H_{10} + 13/2 O_2 \rightarrow 4CO_2 + 5H_2O \tag{3.39}$$
$$\text{Pentane} \quad C_5H_{12} + 8O_2 \rightarrow 5CO_2 + 6H_2O \tag{3.15}$$
$$\text{Hexanes} \quad C_6H_{14} + 19/2 O_2 \rightarrow 6CO_2 + 7H_2O \tag{3.16}$$

Each combustible component generates the indicated number of moles of CO_2 and H_2O, and consumes the number of moles of O_2 from the combustion air. As before, this brings along nitrogen plus other constituents, including other atmospheric moisture. At stoichiometric conditions, all of the O_2 in the air admitted to the combustion process is used up, leaving zero oxygen in the flue gas. The combustion table follows:

Stoichiometric Case
Basis: 1 lb mol/h of natural gas fuel; 60°F, 60% RH, 1 atm air

Component	Moles in fuel	Moles in air	Moles O_2 consumed	Moles CO_2 produced	Moles H_2O produced	Moles in flue gas	mol% (wet)	mol% (dry)
CH_4	0.9314	0	1.8628	0.9314	1.8628	0	0	0
C_2H_6	0.0250	0	0.0875	0.0500	0.0750	0	0	0
C_3H_8	0.0067	0	0.0335	0.0201	0.0268	0	0	0
C_4H_{10}	0.0032	0	0.0208	0.0128	0.0160	0	0	0
C_5H_{12}	0.0012	0	0.0096	0.0060	0.0072	0	0	0
C_6H_{14}	0.0005	0	0.0048	0.0030	0.0035	0	0	0
CO_2	0.0106	0.0032	0	0.0138	0	1.0371	9.63	11.96
N_2	0.0214	7.5264	0	0	0	7.5478	70.09	87.00
Ar	0	0.0900	0	0	0	0.0900	0.84	1.04

O$_2$	0	2.0190	0	0	0	0	0	0
H$_2$O	0	0.1018	0	0	0	2.0931	19.44	—
Total (dry)	1.0000	9.6386	2.0190	1.0371	1.9913	8.6749	—	100.00
Total (wet)	1.0000	9.7404	2.0190	1.0371	1.9913	10.7680	100.00	—

$$\text{Mol N}_2 \text{ in air} = (\text{mol O}_2 \text{ consumed})[(\text{N}_2/\text{O}_2) \text{ in air}] = (2.0190)(78.086/20.947) = 7.5264$$

$$\text{Total dry air} = (2.0190)(100/20.947) = 9.6386$$

$$\text{Mole Ar in air} = (2.0190)(0.934/20.947) = 0.0900$$

$$\text{Mole CO}_2 \text{ in air} = (2.0190)(0.033/20.947) = 0.0032$$

$$\text{Mole H}_2\text{O in air} = (9.6386)(0.010566) = 0.1018$$

Mol N$_2$ in flue gas = Mol N$_2$ in air + Mol N$_2$ in fuel = 7.5264 + 0.0214 = 7.5478

Other components are summed, inserted into the table, and percentages in flue gas are determined. Figures for combustion air composition are from Table 3.3.

The Method 19 F factors are determined from the stoichiometric condition.

$$F_d = \frac{(8.6749 \text{ lb mol dry flue gas})(385.3 \text{ SCF/lb mol})}{(1019.4 \text{ Btu/SCF})(379.5 \text{ SCF/lb mol})(1 \text{ lb mol/h})(1/10^6)} = \frac{8640 \text{ SCF}(68°\text{F, 1 atm})}{10^6 \text{ Btu}}$$

Btu/SCF is computed in Problem 3.1 and recorded in Table 3.1.

$$F_w = \frac{[(10.7680 - 0.1018) \text{ lb mol wet flue gas})](385.3)}{(1019.4)(379.5)(1/10^6)} = \frac{10,620 \text{ SCF}(68°\text{F, 1 atm})}{10^6 \text{ Btu}}$$

F_w includes only water of combustion in the total wet flue gas.

$$F_c = \frac{[(1.0371 - 0.0032) \text{ lb mol CO}_2 \text{ in flue gas})](385.3)}{(1019.4)(379.5)(1/10^6)} = \frac{1030 \text{ SCF}(68°\text{F, 1 atm})}{10^6 \text{ Btu}}$$

Moles of CO$_2$ in air have been excluded from the F_c calculation. Negligible effect of CO$_2$ from air on calculated F_d and F_w is ignored. F factors for pure methane are

$$F_d = 8590$$
$$F_w = 10,600$$
$$F_c = 1010 \text{ SCF } (68°\text{F, 1 atm})/10^6 \text{ Btu (all from Problem 3.8)}$$

EPA F factors for the "generic" natural gas are listed in Table 3.4.

Molecular weight of flue gas (dry)
$$= (0.1196)(44.01) + (0.8700)(28.01) + (0.0104)(39.95) = 30.0$$

Molecular weight of flue gas (wet)
$$= (0.0963)(44.01) + (0.7009)(28.01) + (0.0084)(39.95) + (0.1944)(18.02) = 27.7$$

(b) For 10% excess air, the amount of combustion air is increased by 10%, to 110% of that for the stoichiometric case. Oxygen in the flue gas is 10% of the O$_2$ consumed in the stoichiometric case. The combustion table follows:

10% Excess Air Case
Basis: 1 lb mol/h of natural gas fuel; 60°F, 60% RH, 1 atm air

Component	Moles in fuel	Moles in air	Moles O_2 consumed	Moles CO_2 produced	Moles H_2O produced	Moles in flue gas	mol% (wet)	mol% (dry)
CH_4	0.9314	0	1.8628	0.9314	1.8628	0	0	0
C_2H_6	0.0250	0	0.0875	0.0500	0.0750	0	0	0
C_3H_8	0.0067	0	0.0335	0.0201	0.0268	0	0	0
C_4H_{10}	0.0032	0	0.0208	0.0128	0.0160	0	0	0
C_5H_{12}	0.0012	0	0.0096	0.0060	0.0072	0	0	0
C_6H_{14}	0.0005	0	0.0048	0.0030	0.0035	0	0	0
CO_2	0.0106	0.0035	0	0.0141	0	1.0374	8.84	10.76
N_2	0.0214	8.2790	0	0	0	8.3004	70.69	86.12
Ar	0	0.0990	0	0	0	0.0990	0.84	1.03
O_2	0	2.2209	0	0	0	0.2019	1.72	2.09
H_2O	0	0.1120	0	0	0	2.1033	17.91	—
Total (dry)	1.0000	10.6024	2.0190	1.0374	1.9913	9.6387	—	100.00
Total (wet)	1.0000	10.7144	2.0190	1.0374	1.9913	11.7420	100.00	—

$$\text{Moles } O_2 \text{ in air} = (1.1)(2.0190) = 2.2209$$

Moles of atmospheric components are then calculated as in the stoichiometric case, but with this higher amount of O_2.

$$\text{Moles } O_2 \text{ in flue gas} = 2.2209 - 2.0190 = 0.2019$$

Molecular weight of flue gas (dry)

$$= (0.0.1076)(44.01) + (0.8612)(28.01) + (0.0103)(39.95) + (0.0209)(32.00) = 29.9$$

Molecular weight of flue gas (wet)

$$= (0.0884)(44.01) + (0.7069)(28.01) + (0.0084)(39.95)$$
$$+ (0.0172)(32.00) + (0.1791)(18.02) = 27.8$$

(c) Moles O_2 in flue gas

$$= \frac{[Q_{\text{wet}} @ 0\% O_2](\% O_2 \text{ wet}/100)}{1-[1+H](\% O_2 \text{ wet}/20.947)} = \frac{(10.7680)(2.00/100)}{1-(1.010566)(2.00/20.947)} = 0.2384 \quad (3.18)$$

Moles O_2 in air $= 0.2384 + 2.0190$ (from stoichiometric case) $= 2.2574$.
With these points of entry (in bold), the following combustion table is filled out, as done previously:

3.9 Combustion Problems: Major Species in the Flue Gas   123

2% O_2 (Wet) in Flue Gas
Basis: 1 lb mol/h of natural gas fuel; 60°F, 60% RH, 1 atm air

Component	Moles in fuel	Moles in air	Moles O_2 consumed	Moles CO_2 produced	Moles H_2O produced	Moles in flue gas	mol% (wet)	mol% (dry)
CH_4	0.9314	0	1.8628	0.9314	1.8628	0	0	0
C_2H_6	0.0250	0	0.0875	0.0500	0.0750	0	0	0
C_3H_8	0.0067	0	0.0335	0.0201	0.0268	0	0	0
C_4H_{10}	0.0032	0	0.0208	0.0128	0.0160	0	0	0
C_5H_{12}	0.0012	0	0.0096	0.0060	0.0072	0	0	0
C_6H_{14}	0.0005	0	0.0048	0.0030	0.0035	0	0	0
CO_2	0.0106	0.0036	0	0.0142	0	1.0375	8.71	10.57
N_2	0.0214	8.4151	0	0	0	8.4365	70.79	85.97
Ar	0	0.1007	0	0	0	0.1007	0.84	1.03
O_2	0	**2.2574**	0	0	0	**0.2384**	2.00	2.43
H_2O	0	0.1139	0	0	0	2.1052	17.66	—
Total (dry)	1.0000	10.7768	2.0190	1.0375	1.9913	9.8131	—	100.00
Total (wet)	1.0000	10.8907	2.0190	1.0375	1.9913	11.9183	100.00	—

%Excess air = (100)(10.8907/9.7407 from stoichiometric case)−100 = 11.8%
         = (100)(10.7768/9.6386 from stoichiometric case)−100 = 11.8%

Molecular weight of flue gas (dry)

$$= (0.1057)(44.01) + (0.8597)(28.01) + (0.0103)(39.95) + (0.0243)(32.00) = 29.9$$

Molecular weight of flue gas (wet)

$$= (0.0871)(44.01) + (0.7079)(28.01) + (0.0084)(39.95) + (0.0200)(32.00)$$
$$+ (0.1766)(18.02) = 27.8$$ ■

PROBLEM 3.11

Using the *F* factor method, calculate the flue gas composition from combustion of propane (C_3H_8) at 2% excess O_2 wet (2.35% excess O_2 dry).

SOLUTION This illustrates another aspect of the *F* factor method. Here, knowing the excess O_2 on both a wet basis and a dry basis is equivalent to having the moisture fraction (% H_2O/100) of the effluent gas. Method 19 calls this quantity B_{ws}.

$$(2.35)(1-B_{ws}) = 2.0$$
$$B_{ws} = (2.35-2)/2.35 = 0.1489$$

For propane, the Method 19 F_d (Table 3.4) is 8710 dry SCF @ 0% O_2/10^6 Btu. At 2.35% excess O_2 dry,

$$(8710)[20.9/(20.9-2.35)] = 9813 \text{ SCF total dry flue gas}$$

Total wet flue gas at this condition is then

$$(9813)/(1-B_{ws}) = 9813/(0.8511) = 11{,}530 \text{ SCF total wet flue gas}$$

These last two equations taken together are a variation of Method 19's Equation 19.4 that accomplishes the transformation of F_d to total wet flue gas at the excess oxygen condition in a single step. Flue gas moisture cannot be obtained directly from F_w (which includes only the water resulting from combustion of the hydrogen constituent in the fuel) because of the presence of humidity in the atmospheric combustion air.

Actual moisture in the flue gas is determined by difference.

$$H_2O = 11{,}530 - 9813 = 1717$$

Oxygen is 2.35% of the dry flue gas or 2% of the wet flue gas.

$$(9813)(2.35/100) = (11{,}530)(2/100) = 230.6$$

CO_2 is derived from the F_c factor, 1190. This is the same for any and all values of excess O_2. It assumes that there is zero CO_2 in the combustion air, an assumption that is largely correct.

Flue gas "nitrogen" is obtained by difference.

$$\underset{\text{Total}}{11{,}530} - \underset{H_2O}{1717} - \underset{O_2}{230.6} - \underset{CO_2}{1190} = 8392.4$$

For this amount of "nitrogen" in the flue gas, $(8392.4)(100/79.1) = 10{,}610$ units of dry combustion air must have been used, and $(8392.4)(20.9/79.1) - 230.6 \cong 1987$ units of oxygen must have been consumed.

To be fancy, one can go on to estimate the split between N_2 and argon in the flue gas. For a fuel containing zero nitrogen, the ratio in the flue gas of nitrogen to argon, the other constituent in air of major proportion, is the same as that in dry atmospheric air (approximated by entries in Table 3.3).

$$N_2/Ar = 78.086/0.934$$

and

$$\begin{aligned} N_2/(N_2 + Ar) &= 0.9822 \\ Ar/(N_2 + Ar) &= 0.0118 \\ (8392.4)(0.9882) &= 8293.4 \; N_2 \\ (8392.4)(0.0118) &= \dfrac{99.0 \; Ar}{8392.4 \; (N_2 + Ar)} \end{aligned}$$

To summarize,

Constituent	SCF/MMBtu	% Wet	% Dry
CO_2	1190.0	10.32	12.13
N_2	8293.4	71.93	84.51
Ar	99.0	0.86	1.01
O_2	230.6	2.00	2.35
H_2O	1717.0	14.89	—
Total (dry)	9813.0	—	100.00
Total (wet)	11,530.0	100.00	—

Correction for N_2 in the fuel is illustrated in Problem 3.14.

Compare this flue gas composition with that calculated in Problem 3.9 using the equations for the chemical reactions occurring during combustion. The slight differences are caused by a somewhat different F_d determined from the calculations done the long way, 8820 versus Method 19's 8710 from Table 3.4, the minor amount of CO_2 in the combustion air used in those calculations, and roundoff in the % O_2 wet and dry used to determine the % H_2O in the flue gas. The atmospheric humidity that ripples through both problems is 0.010566 lb mol H_2O/lb mol dry air from 60°F, 60% relative humidity air.

Determining the composition of major constituents in the flue gas facilitates expression of pollutant concentrations on whatever basis is required.

Estimation of emissions from a small propane-fired heater will be illustrated in a subsequent example problem. ∎

PROBLEM 3.12   *Natural Gas F Factors*

(a) Characterize the flue gas from firing the natural gas of Table 3.1 at 10% excess air using the atmospheric air of Table 3.8 at 60°F, 1 atm, and 60% RH. Use the Method 19 *F* factor technique and compare with the combustion calculations of Problem 3.7 done the long way. (b) Repeat at 2% O_2 wet in the flue gas.

SOLUTION This example is similar to Problem 3.11 for propane, but with different values for the *F* factors, F_w and F_c. The F_d factor is the same. From Table 3.4,

$$F_d = 8710$$
$$F_w = 10,610$$
$$F_c = 1040$$

Then,

$$(F_w - F_d) = 1900$$
$$(F_d - F_c) = 7670$$

F factors have units of SCF (68°F, 1 atm)/MMBtu. 1 MMBtu = 10^6 Btu.

This problem will be solved using the equations/entries in Tables 3.5–3.8, as appropriate. Those tables summarize the relationships among the constituents of the flue gas and the combustion air for the stoichiometric condition, firing with a fixed percent excess air, and for a given percent oxygen (dry or wet basis) in the flue gas. Part (a) confirms that the flue gas oxygen content from firing a typical natural gas at 10% excess is about 1.7% O_2 wet and 2.1% O_2 dry.

(a) Procedure for the 10% excess air case

From Table 3.6, total dry flue gas is

$$8710 + (100/79.1)(8710 - 1040)(10/100) = 9679.6587$$

O_2 in flue gas = $(20.9)(79.1)(8710-1040)(10/100) = 202.6587$

N_2 in flue gas = $(8710-1040)[1 + (10/100)] = 8437.0000$

CO_2 in flue gas = 1040

N_2 in air = N_2 in flue gas = 8437.0000

Total dry air = $(100/79.1)(8710-1040)[1 + (10/100)] = 10,666.2453$

O_2 in air = $(20.9/79.1)(8710-1040)[1 + (10/100)] = 2229.2453$

H_2O in air = $(100/79.1)(8710-1040)(0.010566)[1 + (10/100)] = 112.6995$

H_2O of combustion = $(10,610 - 8710) = 1900$

H_2O in flue gas = $1900 + 112.6995 = 2012.6995$

Total wet flue gas = $9679.6587 + 2012.6995 = 11,692.3582$

Flue gas composition is summarized in the table that follows, along with total flue gas flow when firing 38.69 MMBtu/h. The nitrogen in the table represents the sum of the actual N_2 and the other inerts, mainly argon, originating from the combustion air. The firing rate is chosen to facilitate comparison with the detailed calculations of Problem 3.10.

Flue Gas at 10% Excess Air

Constituent	SCF/MMBtu	% Wet	% Dry
CO_2	1040.0000	8.40	10.75
N_2	8437.0000	72.16	87.16
O_2	202.6587	1.73	2.09
Total (dry)	9679.6587	—	100.00
Total H_2O	2012.6995	17.21	—
Total (wet)	11,692.3582	100.00	—

$$(38.69 \text{ MMBtu/h})(9679.6587) = 374,500 \text{ SCFH dry flue gas}$$
$$(38.69 \text{ MMBtu/h})(11,692.3582) = 452,400 \text{ SCFH wet flue gas}$$

(b) Procedure for the 2% O_2 (wet) case

From Table 3.5, total flue gas moisture at stoichiometric conditions is

$$(10,610 - 8710) + (100/79.1)(8710 - 1040)(0.010566) = 2002.4541$$

since $B_{wa}/(1 - B_{wa}) = H = 0.010566$ for this problem.

From Table 3.5, total wet flue gas at stoichiometric conditions is

$$(10,610) + (100/79.1)(8710 - 1040)(0.010566) = 10,712.4541$$

% H_2O/100 at stoichiometric conditions is then

$$(2002.4541)/(10,712.4541) = 0.1869$$

From Equation (B.3), (% H_2O/100) at 2% O_2 wet is

$$0.1869 - (2.00/20.9)[(1.010566)(0.1869) - 0.010566] = 0.1698$$

This is B_{ws}.

From Table 3.8, total wet flue gas at 2% O_2 wet is

$$\frac{20.9}{20.9 - (2.00)(1.010566)}[10,610 + (100/79.1)(8710 - 1040)(0.010566)] = 11,859.3070$$

From Table 3.8,

Total $CO_2 = 1040$

N_2 in flue gas $= [(1 - 0.1698) - (2.00/100)](11,859.3070) - 1040 = 8568.4105$

O_2 in flue gas $= (2.00/100)(11,859.3070) = 237.1861$

Total dry flue gas $= (1 - 0.1698)(11,859.3070) = 9845.5967$

Total H_2O in flue gas $= (0.1698)(11,859.3070) = 2013.7103$

3.9 Combustion Problems: Major Species in the Flue Gas

N_2 in air = 8568.4105

O_2 in air = $(20.9/79.1)(8568.4105) = 2263.9669$

Total dry air = $(100/79.1)(8568.4105) = 10,832.3774$

H_2O in air = $(100/79.1)(8568.4105)(0.010566) = 114.4549$

Check:

H_2O in air + H_2O of combustion = total H_2O in flue gas

$114.4549 + (10,610 - 8710) = 2014.4549$ versus 2013.7104

(Agreement is close but inexact because of roundoff in input quantities calculated along the way.)

Flue gas composition is summarized in the following table, along with total flue gas flow when firing 38.69 MMBtu/h.

Flue Gas at 2% O_2 Wet

Constituent	SCF/MMBtu	% Wet	% Dry
CO_2	1040.0000	8.77	10.56
N_2	8568.4105	72.25	87.03
O_2	237.1861	2.00	2.41
Total (dry)	9845.5967	—	100.00
Total H_2O	2013.7103	16.98	—
Total (wet)	11,859.3070	100.00	—

For a firing rate of 38.69 MMBtu/h,

Total dry flue gas = $(38.69)(9845.5967) = 380,900$ SCFH

Total wet flue gas = $(38.69)(11,859.3070) = 458,800$ SCFH

Flue gas flows are close to those of the 10% excess air case because the two conditions themselves are close together, 2.00% O_2 wet (2.41% O_2 dry) versus 10% excess air (1.73% O_2 wet, 2.09% O_2 dry).

In some cases, it may be more convenient to convert % O_2 wet in the flue gas to % O_2 dry and proceed with the % O_2 dry table. This can be done once the % H_2O or B_{ws} in the flue gas is determined. For the case at hand,

% O_2 dry = % O_2 wet $[100/(100-\% H_2O)]$ = % O_2 wet $[1/(1-B_{ws})]$

$= (2.00)[1/(1-0.1698)] = 2.41$

Then,

Total dry flue gas flow = $(8710)[(20.9)/(20.9-2.41)] = 9845.2677$ versus 9845.5967

(Here again, roundoff causes a slight difference.)

The total dry flue gas flow with the percent moisture yields total wet flue gas flow, and with the % O_2 dry, the O_2 in the flue gas. Adding in the CO_2, which is the same in all cases (F_c), and subtracting from total flue gas, one finds the flue gas N_2, the N_2 in the combustion air, and so on. That exercise is illustrated in Problem 3.13. ∎

128   Chapter 3   Air Combustion

PROBLEM 3.13

(a) Using the Method 19 F factor technique, characterize the flue gas from firing natural gas with air at 60°F, 60% RH, 1 atm, and 2.41% O_2 dry in the flue gas.
(b) Compare with Problem 3.12 for another flue gas condition using Method 19 and with Problem 3.10 done the long way for the natural gas of Table 3.3.
(c) Repeat part (a) with the Method 19 default moisture fraction of ambient air ($B_{wa} = 0.027$).

SOLUTION   As in Problem 3.12, the Method 19 F factors for natural gas combustion are as follows:

$$F_d = 8710$$
$$F_w = 10,610$$
$$F_c = 1040$$

Then,

$$(F_w - F_d) = 1900$$
$$(F_d - F_c) = 7670$$

(a) Procedure for 2.41% O_2 dry in the flue gas

From Table 3.7, total dry flue gas is

$$(8710)[20.9/(20.9 - 2.41)] = 9845.2677$$

O_2 in flue gas $= (2.41/100)(9845.2677) = 237.2710$
CO_2 in flue gas $= 1040$
N_2 in air $= [1 - (2.41/100)][9845.2677] - 1040 = 8567.9968$

N_2 in air	=	8567.9968
O_2 in air	=	$(20.9/79.1)(8567.9968) = 2263.8576$
Total dry air	=	$(100/79.1)(8567.9968) = 10,831.8543$
H_2O in air	=	$(100/79.1)(8567.9968)(0.010566) = 114.4494$
Total wet air	=	$(100/79.1)(8567.9968)(1.010566) = 10,946.3037$
H_2O in flue gas	=	$(10,610 - 8710) + 114.4494 = 2014.4494$
Total wet flue gas	=	$9845.2677 + 2014.4494 = 11,859.7171$

For the stoichiometric case,

Total dry air $= (100/79.1)(8710 - 1040) = 9696.5866$
Total wet air $= (100/79.1)(8710 - 1040)(1.010566) = 9799.0407$
%Excess air $= 100(\text{total air}/\text{stoichiometric air}) - 100$
$\phantom{\%\text{Excess air}}= 100(10,831.8534/9696.5866) - 100 = 11.7\%$
$\phantom{\%\text{Excess air}}= 100(10,946.3037/9799.0407) - 100 = 11.7\%$

Flue gas composition is summarized below, along with total flue gas flow when firing 38.69 MMBtu/h to facilitate comparison with the other problems cited.

3.9 Combustion Problems: Major Species in the Flue Gas

Flue Gas at 2.41% O_2 Dry in the Flue Gas

Constituent	SCF/MMBtu	% Wet	% Dry
CO_2	1040.0000	8.77	10.56
N_2	8567.9968	72.24	87.03
O_2	237.2710	2.00	2.41
Total (dry)	9845.2677	—	100.00
Total H_2O	2014.4494	16.99	—
Total (wet)	11,859.7171	100.00	—

$$(38.69\,\text{MMBtu/h})(9845.2677) = 380{,}900\,\text{SCFH dry flue gas}$$
$$(38.69\,\text{MMBtu/h})(11{,}859.7171) = 458{,}900\,\text{SCFH wet flue gas}$$

(b) This is essentially the same case as that of Problem 3.12. The Method 19 results differ only because of roundoff. Problem 3.10 shows this case done the long way using the natural gas composition of Table 3.3. For proper comparison, it must be remembered that the Method 19 N_2 is the sum of N_2 and Ar. There are slight differences in the two problems because the F factors computed for the natural gas of Table 3.3 are slightly different; for example, $F_d = 8640$ versus 8710. Other differences are caused by the small percentage of N_2 in the fuel of Table 3.3 and the tiny amount of CO_2 in the combustion air, both not accounted for in the procedure outlined here for use with the Method 19 F factors. Nitrogen (N_2) in the natural gas fuel is considered in Problem 3.14. ■

PROBLEM 3.14

The natural gas of Table 3.1 contains 2.14 vol% N_2. Combustion of this gas at 10% excess air results in a dry flue gas composition of 10.76% CO_2 and 2.09% O_2, the rest being N_2 + argon. One hundred lb mol/h of this fuel results in 963.86 lb mol/h of dry flue gas and 1174.19 mol/h of wet flue gas. Calculate the flue gas composition including Ar and the flue gas molecular weight, both wet and dry.

SOLUTION The first step is to find how much of the N_2 in the flue gas is coming from the fuel and to subtract it out. To perform this calculation with experimental data requires a good fuel analysis and fuel flow rate in addition to the flue gas analysis and its flow rate.

$$N_2 \text{ in fuel} = (2.14/100)(100) = 2.14\,\text{lb mol/h } N_2 \text{ from the fuel}$$
$$(N_2 + Ar) \text{ in the flue gas (dry)} = [(100-10.76-2.09)/100](963.86)$$
$$= 840.00\,\text{lb mol/h } (N_2 + Ar)$$
$$840.00 - 2.14 = 837.86\,\text{lb mol/h of } (N_2 + Ar) \text{ from the combustion air}$$

Of this total, $78.086/(78.086 + 0.934) = 0.9882\,N_2$ fraction (Table 3.3) and

$$0.934/(78.086 + 0.934) = 0.0118\,\text{Ar fraction}$$
$$(837.86)(0.9882) = 827.97\,\text{lb mol/h } N_2 \text{ from air}$$
$$(837.86)(0.0118) = 9.89\,\text{lb mol/h Ar from air}$$

130 Chapter 3 Air Combustion

$$\text{Total lb mol/h } N_2 = 2.14 \text{ from fuel} + 827.97 \text{ from air}$$
$$= 830.11$$
$$\div 963.86 = 0.8612$$
$$\times 100 = 86.12 \text{ mol\% } N_2 \text{ (dry)}$$
$$100(9.89/963.86) = 1.03 \text{ mol\% Ar (dry)}$$

$$H_2O = \text{wet flue gas} - \text{dry flue gas}$$
$$= 1174.19 - 963.86 = 210.33 \text{ lb mol/h } H_2O$$
$$\% H_2O \text{ (wet basis)} = (100)(210.33/1174.19) = 17.91\%$$

In summary, flue gas analysis is as follows:

Constituent	% Wet	% Dry
Carbon dioxide	8.83	10.76
Nitrogen (N_2)	70.70	86.12
Argon (Ar)	0.84	1.03
Water (H_2O)	17.91	—
Oxygen (O_2)	1.72	2.09
Total	100.00	100.00

In this particular example, there is not enough nitrogen in the fuel to affect the results significantly.

In the table above,

$$\text{mol\% (wet)} = \text{mol\% (dry)} [(100 - \% H_2O)/100]$$
$$\text{for } CO_2 = 10.76(100 - 17.91)/100 = 8.83$$

and using EPA's $B_{ws} = 10.76 (1 - B_{ws}) = 10.76 (1 - 0.1791) = 8.83$.

For the other flue gas constituents, mol% (wet) is calculated in a similar manner.

Flue Gas Molecular Weight

By modification of EPA's Method 3, Equation (3.1),

$$M_d = (0.440)(\% CO_2) + (0.320)(\% O_2) + (0.280)(\% N_2 + \% CO) + (0.399)(\% Ar)$$
$$= (0.440)(10.76) + 0.320(2.09) + 0.280(86.12) + 0.399(1.03)$$
$$= 29.93 = 29.9 \text{ rounded to nearest 0.1 lb/lb mol} \tag{3.42}$$

The dry molecular weight (M_d) continues to come out close to the rule of thumb 30.

The wet molecular weight (M_w) $= 0.440(\% CO_2 \text{ wet}) + 0.320(\% O_2 \text{ wet})$
$$+ 0.280(\% N_2 \text{ wet} + \% CO \text{ wet}) + 0.399(\% Ar \text{ wet})$$
$$+ 0.180(\% H_2O)$$
$$= (0.440)(8.83) + (0.320)(1.72) + (0.280)(70.70 + 0)$$
$$+ (0.399)(0.84) + (0.180)(17.91)$$
$$= 27.79 = 27.8 \text{ rounded to the nearest 0.1 lb/lb mol}$$
$$\tag{3.43}$$

Similar to the derivation of Equation (3.10) for the molecular weight of combustion air as a function of atmospheric humidity, one may generalize from Equations (3.42) and (3.43) dry and wet molecular weights of flue gas to obtain

$$M_w = M_d[(100-\% H_2O)/100] + 18(\% H_2O/100) \quad (3.44)$$

or

$$M_w = M_d(1-B_{ws}) + 18B_{ws} \quad (3.45)$$

and

$$M_d = \frac{M_w - 18(\% H_2O/100)}{[(100-\% H_2O)/100]} \quad (3.46)$$

or

$$M_d = \frac{M_w - 18B_{ws}}{(1-B_{ws})} \quad (3.47)$$

∎

3.9.1 Relationship Between Percent Excess Air and Percent O_2 in the Flue Gas

Figure 3.3, the relationship between percent excess air and % O_2 dry in the flue gas, can be derived by equating the entries for dry flue gas in Table 3.6 (the % excess air case) and Table 3.7 (the % O_2 dry case) of the F factor technique.

$$F_d + (100/79.1)(F_d - F_c)(\%XS\ air/100) = F_d[20.9/(20.9 - \% O_2\ dry)] \quad (3.48)$$

When simplified, the resulting equation for % O_2 dry as a function of percent excess air (% XS air) is parametric in the ratio of F factors (F_c/F_d).

$$\% O_2\ dry = 20.9 \left\{ 1 - \frac{1}{1 + [(1/79.1) - (F_c/F_d)(1/79.1)(\%XS\ air)]} \right\} \quad (3.49)$$

Using the values of $F_c = 1010$ and $F_d = 8590$ calculated for methane in Problem 3.8, one obtains the dimensionless ratio $F_c/F_d = 0.1176$ to substitute into Equation (3.49) and the basis for the curve.

Because the ratio of F_c to F_d for other fuels, especially natural gas at 0.1194, is close to 0.1176, curves for those other fuels closely approximate that of methane. One such curve for an unspecified fuel appears in Ref. [18]. The curve of Figure 3.3 can be altered to reflect another fuel or class of fuels simply by changing the F factors.

3.10 ADIABATIC FLAME TEMPERATURE

The theoretical adiabatic flame temperature (AFT) assuming complete combustion is the temperature attained when a fuel is burned, without mechanical work or gain or loss of heat to the theoretical end products such as CO_2 and H_2O, regardless of any equilibrium condition that may apply [14, pp. 354–357; 19]. It depends on the

Table 3.9 Calculated Adiabatic Flame Temperatures for Several Pure Components

	AFT (°F)	
Component	Stoichiometric	10% Excess air
Carbon monoxide	4311	4059
Hydrogen (H_2)	4056	3807
Propane (C_3H_8)	3818	3556
Ethane (C_2H_6)	3795	3536
Methane (CH_4)	3698	3450

composition of the fuel and combustion "air" (atmospheric air, oxygen, or mixtures thereof), their temperatures, the amount of excess oxidant, and to a certain extent the humidity in the combustion air. It captures much information about the combustion mixture, including a broad spectrum of fuel gas compositions. Once agreement is reached on the requisite input thermodynamic data, the AFT can be computed objectively in a straightforward manner.

Adiabatic flame temperatures calculated for combustion of several pure components with atmospheric air are listed in Table 3.9 [19]. Fuel and atmospheric air are at 60°F, 60 % RH in air.

The entries in the table are arranged in decreasing order of AFT from CO and hydrogen (H_2) at the top to methane (CH_4) at the bottom. Because the inherent molar heating value of CO is larger than that of hydrogen and the total heat capacity of its combustion gas is lower relative to that of hydrogen, flame temperature of CO ranks first. When CO is fired, no water of combustion (H_2O) is generated. Water in the flue gas, from whatever source, lowers the AFT because more heat is required to raise the temperature of the water compared to the other flue gas constituents to the final AFT achieved. The heat capacity of water is $\sim$0.5 Btu/(lb °F) versus $\sim$0.25 Btu/(lb °F) for other typical flue gas constituents.

The AFT is highest for all fuels at stoichiometric conditions (combustion at 0% excess air) because less nitrogen (N_2) and other inerts from the combustion air are present in the flue gas. Likewise, the AFT achieved in pure oxygen (O_2) would be higher than air across the board for all fuels because of this same absence of nitrogen.

The AFT for methane is depicted in Figure 3.6 as a function of flue gas excess oxygen (wet) from 0% to 6% for several values of combustion air temperature. The curves are linear in excess O_2 up to about 5–6%, although a slightly better fit can be obtained through the addition of a quadratic term. Slope of these lines is on the order of 150°F per % O_2 (83°C per % O_2), and the AFT increases about 2°F for every 3°F of increased air preheat, while keeping % O_2 constant. They are essentially the same as those calculated for a typical natural gas in a previous presentation [19].

AFT has been demonstrated to agree closely with peak temperature measurements in flames, as documented in Ref. [19]. Agreement is not exact, however, since the AFT assuming complete combustion is always higher than actual temperatures because of heat loss from the flame, equilibrium preventing complete combustion,

3.10 Adiabatic Flame Temperature

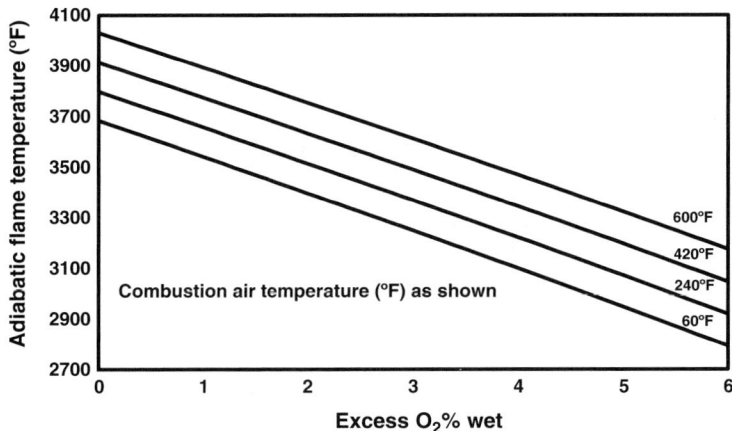

Figure 3.6 Adiabatic flame temperature for methane fuel at 60°F, ambient air at 60°F, 60% relative humidity.

and the generation of free radicals and such not accounted for in the calculation [14, pp. 354–357]. Regardless, the AFT turns out to be a useful surrogate for the actual peak flame temperature and a convenient correlating variable for NO_x (Appendix C) whether or not it is actually attained in the flame. Similarly, the average O_2 concentration measured at the furnace outlet substitutes for point values of oxygen reacting in the flame.

PROBLEM 3.15

Calculate the theoretical AFT assuming complete combustion for methane gas at a flue gas oxygen of 2% wet. Fuel and combustion air (Table 3.3) are at 60°F. RH in the ambient air is 60%. Flue gas composition is given in Problem 3.7.

SOLUTION This problem involves combustion from an initial condition to the final condition in the flame. The result is independent of the thermodynamic path chosen. A convenient path is to heat the fuel and combustion air from 60°F (15.6°C) to 25°C (77°F); combust the fuel at 25°C, for which the standard heat of combustion is tabulated; and heat the combustion products from 25°C (77°F) to the final, unknown flame temperature. There is no gain or loss of heat, and no mechanical work is performed or extracted.

The solution is by trial and error in which the sum of the heat effects is set equal to zero. Furthermore, since such a large temperature change is encountered for the flue gas, heat capacities as a function of temperature must be employed. This is most easily accomplished by using mean heat capacities over the interval from 77°F to the final flame temperature.

$$C_{p_{\text{mean}}} = (1/\Delta T) \int_{T_{\text{initial}}}^{T_{\text{final}}} C_p dt \qquad (2.101)$$

When heat capacity (C_p) equations as a function of temperature (T) in the form

$$C_p = a + bT + cT^2 + dT^3 \qquad (3.50)$$

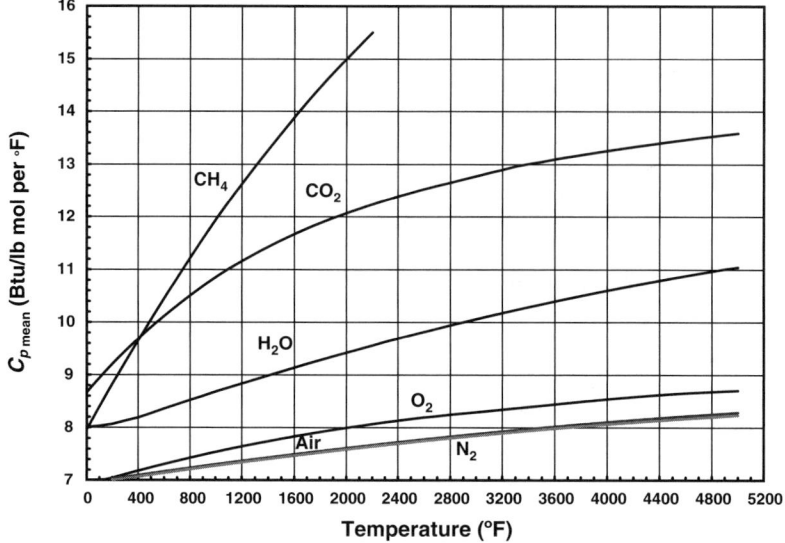

Figure 3.7 Mean molal heat capacities for selected gases at constant pressure from 77°F to indicated temperature.

are available, $C_{p_{mean}}$ can be obtained by integration over the interval from T_1 to T_2.

$$C_{p_{mean}} = \frac{1}{(T_2 - T_1)} \int_{T_1}^{T_2} (a + bT + cT^2 + dT^3) dT$$
$$= a + (b/2)(T_2 + T_1) + (c/3)(T_2^2 + T_1 T_2 + T_1^2) + (d/4)(T_2^2 + T_1^2)(T_2 + T_1) \tag{3.51}$$

The equations simplify when any of the constants equal zero, and Equation (3.51) reduces to Equation (3.50) when T_1 and T_2 are the same. Other functional forms can be integrated in a similar manner. For all such equations, care must be taken to note the units of heat capacity and to employ a temperature in the proper units.

Mean molal heat capacities (i.e., per mole of gas) calculated from a base temperature of $T_1 = 25°C$ (77°F) have been tabulated [14, p. 258] and plotted [14, p. 259; 20, pp. 126–127] as a function of T_2 for numerous gases. The curves of Figure 3.7 have been newly computed here from heat capacity data and equations for methane fuel and common combustion gases. Such charts are convenient for hand calculations. Use of the integrated equations is preferred for computer applications. For this particular problem, the graphical technique will be used.

The applicable heat balance equation is

Heat fuel + heat combustion air + combust fuel + heat flue gas = 0   (3.52)

Btu from fuel combustion = Btu needed to heat combustion air + fuel + flue gas

Heat Combustion Air and Fuel to 77°F

Referring to Problem 3.7, 1 lb mol of methane uses 10.6731 lb mol of dry air plus 0.1128 lb mol of ambient humidity/moisture/water vapor and produces the flue gas of composition listed below.

Reading the curves of Figure 3.7 at 60°F gives the following values for $C_{p_{mean}}$:

$C_{p_{mean}}$ for dry air = 7.0

$C_{p_{mean}}$ for water vapor/moisture = 8.0

$C_{p_{mean}}$ for methane fuel = 8.2

3.10 Adiabatic Flame Temperature   135

Air	$(10.6731)(7.0)(77-60) = +1270.0989$
Moisture	$(0.1128)(8.0)(77-60) = +\ \ \ 15.3048$
Fuel	$(1.0)(8.2)(77-60) = +\ \ 139.4000$
Total	$+1424.8037$ Btu/lb mol

React Methane at 25°C (77°F)

$$\text{LHV} = (-21,502 \text{ Btu/lb})(16.04 \text{ lb/lb mol}) = -344,892 \text{ Btu/lb mol}$$

LHV at 25°C (77°F) is from Ref. [21, p. 244].

Heat Flue Gas. Try 3400°F for first attempt. Final trial is at 3408°F.

lb mol of flue gas constituents		$C_{p_{\text{mean}}}$ at 3400°F read from Figure 3.7	Heat transferred (Btu/lb mol)	
			3400°F	3408°F[a]
CO_2	1.0035	13.0	43,350.1965	43,454.5605
N_2	8.3342	7.9	218,786.9181	219,313.6396
Ar	0.0997	4.97[b]	1646.5764	1650.5405
O_2	0.2357	8.4	6579.1412	6594.9803
H_2O	2.1128	10.3	72,314.5943	72,488.6890
		Total from heating flue gas	342,677.4265	343,502.4099
	Combustion air and fuel (from above)		1428.8037	1428.8037
		Total sensible heat versus 344,892 (above)	344,106.2302	344,931.2136 (0.011% difference at 3408°F)

[a] Cannot read a change in $C_{p_{\text{mean}}}$ at 3408°F versus $C_{p_{\text{mean}}}$ at 3400°F.
[b] Like helium, argon at low pressure is an ideal monatomic gas; its $C_{p_{\text{mean}}}$ (not shown in Figure 3.7) is 5/2 $R = 4.97$, independent of temperature [14, pp. 251–252].

AFT is closest to 3408°F. A temperature of 3407°F is too low, and 3409°F is even higher. Trying to calculate it with greater precision via the graphical technique is pointless because the curves cannot be read any more closely. ∎

PROBLEM 3.16

Repeat the previous problem with the same ambient combustion air preheated to 400°F.

SOLUTION   Since the same ambient air is used, the material balance does not change; only the heat balance differs. In this case, heat is provided to the combustion heat balance at 25°C (77°F) by the preheated combustion air. Fuel temperature remains the same at 60°F.

Cool the Preheated Combustion Air and Heat the Fuel. Figure 3.7 at 400°F gives $C_{p_{\text{mean}}}$ for dry air = 7.1 and $C_{p_{\text{mean}}}$ for water vapor/moisture = 8.2. As read from Figure 3.7 for Problem 3.15, $C_{p_{\text{mean}}}$ for methane fuel at 60°F is 8.2.

136 Chapter 3 Air Combustion

Air	$(10.6731)(7.1)(77 - 400) = -24{,}476.6202$
Moisture	$(0.1128)(8.2)(77 - 400) = -298.7621$
Fuel	$(1.0)(8.2)(77 - 60) = +139.4000$
Net total	$-24{,}635.9823$ Btu/lb mol

React Methane at $25°C$ *(*$77°F$*).* Heat of combustion $= -344{,}892$ Btu/lb mol (from Problem 3.15).

Total heat provided by cooling combustion air + heat of reaction − heating fuel

In this case, heat supplied =	344,892	from heat of reaction
	+ 24,635.9823	net from reactants
	369,527.9827	total available to heat flue gas

Heat the Flue Gas. Try $3700°F$ as the first attempt to determine the final temperature.

lb mol of flue gas constituents		$C_{p\mathrm{mean}}$ at $3700°F$ (Btu/lb mol per $°F$)	Heat transferred (Btu/lb mol)
CO_2	1.0035	13.15	47,809.1986
N_2	8.3342	8.00	241,558.4528
Ar	0.0997	4.97	1795.2291
O_2	0.2357	8.45	7215.8023
H_2O	2.1128	10.45	79,991.3475
		Total sensible heat to raise flue gas temperature:	378,370.0303

$3700°F$ is too high (378,370.0303 versus 369,527.9827 Btu/lb mol)

Next try $3612°F$ (too low) (368,628.8951 versus 369,527.9827 Btu/lb mol) and settle on $3620°F$, the closest whole-number degree, for the final trial.

lb mole of flue gas constituents		$C_{p\mathrm{mean}}$ at $3612°F$ (Btu/lb mole per $°F$)	Heat transferred (Btu/lb mol)	
			$3612°F$	$3620°F^a$
CO_2	1.0035	13.10	46,470.5798	46,575.7466
N_2	8.3342	8.00	235,691.1760	236,224.5648
Ar	0.0997	4.97	1751.6243	1755.5884
O_2	0.2357	8.45	7040.5358	7056.4691
H_2O	2.1128	10.40	77,674.9792	77,850.7642
		Total sensible heat to raise flue gas temperature:	368,628.8951	369,463.1331
		versus 369,527.9827 (above)		(0.018% difference at $3620°F$)

[a] Cannot read a change in $C_{p\mathrm{mean}}$ at $3620°F$ versus $C_{p\mathrm{mean}}$ at $3612°F$.

∴ AFT = $3620°F$ with $400°F$ air preheat. ∎

3.11 ESTIMATION OF POLLUTANT EMISSIONS

The percentage concentrations of the major flue gas constituents (CO_2, N_2, O_2, H_2O, etc.) are calculated from the combustion reactions. However, environmental contaminants in the ppm range are obtained in several different ways. These employ vendor data/guarantees, correlations, emission factors, mass balances when possible, and allowable exit concentrations and mass flow rates. The resulting changes to O_2, CO_2, and other flue gas constituents in the combustion material balance are usually negligible and therefore ignored.

Use of equilibrium relationships may not be appropriate because certain processes are kinetically limited rather than equilibrium controlled. Use of an equilibrium constant may be adequate to show directional trends, provide "ballpark" values, and explain the situation after the fact, but it is not necessarily good enough to produce a quantitative estimate of emissions by itself.

Vendor Input
Estimated or guaranteed emissions can be obtained from manufacturers of burners used in furnaces and boilers, flares, and other control devices. As a rule, the permitting agency would like to ensure that the manufacturer has guaranteed that mandated emission limits can and will be met. Manufacturers' data for burners include nitrogen oxides (NO_x), CO, and particulate mater (PM). Control devices such as flares, are discussed in Chapter 4.

Correlations
Emissions may be estimated from a correlation of one's own or published data. This approach has proven successful in negotiating permit limits in lieu of using agency factors that were not really applicable for the equipment under consideration. Correlations for certain types of furnace burners are discussed in Appendix C and its references.

Emission Factors
One technique for estimating emissions employs empirical *emission factors*, which incorporate the effects of incompletely understood mixing, equilibria, and reaction kinetics. These tabulated factors, in mass of pollutant emitted per mass or volume of fuel fired (e.g., lb/ton, lb/gal, or equivalent metric units) or in mass of pollutant emitted per heat input rate (lb/MMBtu, ng/J), are derived from experience with similar equipment at comparable conditions. They are provided by EPA [22] or listed in permitting instructions/guidance documents from various individual state or local air quality control agencies.

Multiplying the heat input, for example, by the corresponding emission factor results in the mass flow rate of pollutant to the atmosphere.

$$(\text{lb/h}) = (\text{MMBtu/h})(\text{lb/MMBtu}) \tag{3.53}$$

The basis for the heating value in the emission factor and the heat input rate must be consistent, both HHV or LHV. Although the LHV is more meaningful for heat balance calculations, the HHV has been adopted by many (but not all) regulatory agencies as a convention in defining emission factors. However, emission estimates or guarantees from burner manufacturers are commonly written in terms of the LHV. Consequently, caution should be duly exercised in interpreting emission factors when the heating value basis is not specified. As noted earlier in the text, the HHV for hydrocarbon fuels not containing significant amounts of H_2 and/or CO is ~110% of the LHV.

Emission factors are tabulated for oxides of nitrogen (NO_x), CO, volatile organic compounds (VOCs), particulate matter (also known as PM_{10}), total sulfur oxides (SO_x), and others. A few comments are in order concerning each type of pollutant.

- NO_x represents the sum of nitric oxide (NO) and nitrogen dioxide (NO_2) in a combustion effluent. NO_x formed from the reaction of molecular N_2 and O_2 in the flame at elevated temperature is called *thermal* NO_x. *Fuel* NO_x arises from the oxidation of ammonia (NH_3) or organic nitrogen contained in the fuel; however, the conversion of the fuel-bound nitrogen to NO_x is not quantitative, and can be considerably less than 100%. A small amount of *prompt* NO_x is formed in the flame by reactions of hydrocarbon radicals from the fuel with nitrogen (N_2) molecules in the combustion air. The total NO_x measured is the sum of NO_x from all three mechanisms. A further discussion of NO_x is contained in Appendix C and its associated references.

- Some CO occurs in the flue gas because of incomplete combustion, even when excess O_2 is present. In many cases, there is a sharp CO *emissions breakpoint*, the excess O_2 below which CO increases steeply and climbs rapidly out of range [23]. Likewise, PM and VOCs are products of incomplete combustion. The control technique is to maintain excess O_2 consistently at an adequate level to keep all of these contaminants from being even higher than estimated.

- Particulate matter includes various subdivisions by size such as PM_{10}, respirable particulate matter with an aerodynamic diameter of 10 µm (microns) or less. Total suspended particulate (TSP) consists of filterable particulate, the material captured by the filter assembly in a stack test, and condensible (or condensable) particulate, material showing up in the back-end impinger catch downstream of the particulate filter. Separate emission factors are given in AP-42 for filterable and condensible particulate [24,25]. Additional discussion of condensible particulate matter is contained in Section 3.12.
  - VOCs are precursors of the criteria pollutant ozone, which is formed from them by photochemical reactions in the atmosphere. VOCs have been defined by EPA [22] as "any compound of carbon, excluding CO, carbon dioxide, carbonic acid (H_2CO_3), metallic carbides or carbonates, and ammonium carbonate ((NH_4)$_2CO_3$), which participates in atmospheric chemical reactions." A number of compounds deemed to have "negligible photochemical reactivity" are exempt from the definition of VOC, namely, methane (CH_4), ethane (C_2H_6), methylene chloride (also known as dichloromethane) (CH_2Cl_2), 1,1,1-trichloroethane (CH_3CCl_3) (also known as methyl chloroform), many chlorofluorocarbons, and certain classes of perfluorocarbons. Additional compounds may be added to the exempt list in the future.

- The term *total organic compounds* (TOC) is used in EPA's compilation of air pollutant emission factors, AP-42 [22] to indicate all VOCs and all exempted organic compounds, including methane (CH_4), ethane (C_2H_6), chlorofluorocarbons, toxics and hazardous air pollutants (HAPs), aldehydes, and semivolatile compounds. The term *total nonmethane organic compound* (TNMOC) is TOC excluding methane.
- In certain jurisdictions, organics from incomplete combustion are called *reactive organic compounds (ROCs), reactive organic gases (ROGs),* or *precursor organic compounds (POCs)* instead of VOCs. Definitions vary from place to place, and it is important to understand which chemical species are included and excluded in each definition, as well as what molecular weight is to be used for a mixture of undetermined organics. Molecular weights of 12.01 (expressed as carbon (C)), 16.04 (expressed as methane (CH_4)), 14.70 (expressed as propane divided by three ($C_3H_8/3$)), 14.36 (expressed as hexane divided by six ($C_6H_{14}/6$)), and possibly others have been employed, and the basis should be clearly stated.

- In the general case, SO_x consists of the sum of sulfur dioxide (SO_2) and a few percent total of sulfur trioxide (SO_3), sulfates in the flue gas and ash, and sulfuric acid mist formed with the flue gas moisture. Emission factors for these separate entities are all proportional to the amount of sulfur in the fuel. The particulate matter emission factor may also show an explicit dependence on amount of ash as well. For natural gas combustion, the EPA AP-42 emission factor for SO_2 assumes complete combustion to SO_2 and an average fuel sulfur content, rather than showing an explicit functional dependence on fuel sulfur; as noted, the SO_2 emission factor given can be prorated using the site-specific natural gas sulfur content [24].

Material Balance
In the combustion of sweet natural gas and low-sulfur distillate oil, at most only a few percent of the fuel sulfur is oxidized beyond SO_2, and it is customary to compute the SO_2 or SO_x emission rate directly by material balance using the sulfur analysis of the fuel.

The regulatory definition of *natural gas* and *sweet natural gas*, however, varies somewhat from jurisdiction to jurisdiction. The State of Texas, for example, has defined *sweet natural gas* as containing per 100 ft^3 1.5 gr or less of hydrogen sulfide (H_2S) or 30 gr or less of total sulfur [26]. For H_2S, this translates into roughly 24 ppm in a natural gas and 2–3 ppm SO_x in the flue gas at a 10:1 ratio of flue gas to fuel.

EPA has defined *natural gas* as a gaseous fuel either containing more than 70% methane by volume or having a gross calorific value between 950 and 1100 Btu/SCF, with a hydrogen sulfide (H_2S) content of less than 1.0 gr per 100 SCF and H_2S constituting more than 50% by weight of the total sulfur in the fuel. A similar (but not the same) EPA definition of *pipeline natural gas* limits H_2S to less than 0.3 gr/100 SCF, with H_2S constituting at least 50% by weight of the total sulfur in the fuel [27]. The AP-42 natural gas emission factor is based on 2000 gr/100 SCF [24].

140 Chapter 3 Air Combustion

The (San Francisco) Bay Area Air Quality Management District (BAAQMD), the Sacramento Metropolitan Air Quality Management District (SMAQMD), and the South Coast Air Quality Management District (SCAQMD) of California have all defined *natural gas* as a mixture of gaseous hydrocarbons containing at least 80% methane by volume [28–30]. The BAAQMD and the SCAQMD rules cite a standard test method for this determination, and the SCAQMD rule specifies *pipeline quality* for this mixture "such as gas sold or distributed by any utility company regulated by the California Public Utilities Commission." BAAQMD is the California air quality agency responsible for the San Francisco Bay Area, and SCAQMD is the air quality agency serving the counties of Los Angeles, Riverside, and San Bernadino in the Los Angeles area.

Exit Concentrations and Mass Flow Rates

Frequently, allowable emissions are specified in terms of mass flow rate or concentration, especially after the application of some pollution control device such as a wet gas scrubber (WGS), filter (baghouse), electrostatic precipitator (ESP), or NO_x reduction process. The pollutant mass flow rate derived from a mass balance or an emission factor and fuel flow rate (and removal efficiency, if applicable) can be converted to a concentration by dividing by the calculated gas flow. Conversely, when the outlet concentration is known or can be measured or estimated, concentrations can be multiplied by flue gas volume to obtain mass flow rate. Concentration and flow must both be on the same basis, wet or dry, and expressed at the same conditions of excess air. The pollutant MW, flue gas molar volume, and unit conversion factors are also needed for these calculations. The general formula is as follows:

$$\text{lb/h of } X = \frac{\text{ppm}}{10^6} \frac{\text{(lb mole of } X\text{)}}{\text{lb mole of flue gas}} \times \frac{1/\text{molar volume}}{\text{SCF of flue gas}} \times \frac{\text{flue gas rate}}{\text{h}} \times \frac{\text{MW}}{\text{lb mole of } X} \frac{\text{lb of } X}{}$$

(3.54)

The relationships among ppm concentrations, lb/MMBtu, and lb/h are explored further in Appendix B.

For comparison with concentration-based regulatory standards, conversion of ppm at actual conditions of excess O_2 to conditions specified in the standard can be accomplished using Equation (3.26) or Equation (3.28). If necessary, ppm (wet) can first be converted to ppm (dry) by means of Equation (3.29).

Example Problems

Several example problems illustrating the calculation/estimation of atmospheric contaminants follow.

PROBLEM 3.17

The term *oxides of nitrogen* (NO_x) consists of nitric oxide (NO) and nitrogen dioxide (NO_2). The commonly accepted split between NO and NO_2 emitted from a combustion source is 95% NO and 5% NO_2[25]. From a gas turbine, the NO_2 to NO_x ratio is 0.25, that is, 75% NO and 25% NO_2[31].

3.11 Estimation of Pollutant Emissions   141

For each of these cases, calculate the average molecular weight of NO_x using the procedure from previous problems. The molecular weight of NO is 30.01 and of NO_2 is 46.01.

SOLUTION This is a trick question! Although NO_x represents the sum of nitric oxide and nitrogen dioxide in the ratios given in the problem statement, the molecular weight of NO_x is always taken by convention as 46 (or more precisely, 46.01), the molecular weight of NO_2, as for example in AP-42 [22]. The reason is that NO slowly but relentlessly transforms into NO_2 in the atmosphere, and NO_2 is the culprit triggering adverse health effects.

Likewise, the molecular weight of compounds jointly referred to as oxides of sulfur (SO_x) is reported on the basis of the molecular weight of sulfur dioxide (SO_2) (64.06, or simply 64) [22]. ∎

PROBLEM 3.18

Combustion of 100 MMBtu/h of the natural gas in Table 3.1 at 10% excess air produces 1168 MSCFH (68°F, 1 atm) of total, or wet, flue gas. Note its % O_2, % CO_2, and % H_2O.

	Partial flue gas composition	
Component in flue gas	Wet	Dry
O_2	1.72	2.10
CO_2	8.84	10.77
H_2O	17.91	—

Calculate (a) lb/h NO_x; (b) verify % O_2 (dry) and % CO_2 (dry) from % O_2 (wet) and % CO_2 (wet); (c) calculate ppm NO_x (wet) and ppm NO_x (dry) at flue gas conditions; and (d) calculate ppm NO_x (dry) at 3% O_2 (dry) and at 12% CO_2 (dry). $NO_x = 0.1$ lb/MMBtu.

SOLUTION

(a) lb/h $NO_x = (100\,\text{MMBtu/h})(0.1\,\text{lb/MMBtu}) = 10\,\text{lb/h} = 240\,\text{lb/day} = 43.8\,\text{ton/year}$

(b) % O_2 (dry) $= (1.72)[100/(100-17.91)] = 2.10$ from Equation (3.29)
 % CO_2 (dry) $= (8.84)[100/(100-17.91)] = 10.77$ from Equation (3.29)

(c) ppm NO_x (wet) $= \dfrac{(10^6)(10\,\text{lb NO}_x/\text{h})(385.3\,\text{SCF/lb mol})}{[(1168 \times 10^3\,\text{SCF/h})(46.01\,\text{lb NO}_x/\text{lb mol NO}_x)]} = 71.7$

by rearrangement of Equation (3.54).

Dry flue gas $= (1168)[(100-17.91)/100] = 958.9$ MMSCFH from Equation (3.24)

ppm NO_x (dry) $= \dfrac{(10^6)(10\,\text{lb NO}_x/\text{h})(385.3\,\text{SCF/lb mol})}{[(958.9 \times 10^3\,\text{SCF/h})(46.01\,\text{lb NO}_x/\text{lb mol NO}_x)]} = 87.3$

Alternatively,

ppm NO_x (dry) $= (71.7)[100/(100-17.91)] = 87.3$ from Equation (3.54)

(d) ppm NO_x (dry) at 3% O_2 (dry) $= (87.3)[(20.9-3)/(20.9-2.10)] = 83.1$
 from Equation (3.27)
 ppm NO_x (dry) at 12% CO_2 (dry) $= 87.3(12/10.77) = 97.4$ from Equation (3.32)

142   Chapter 3   Air Combustion

A concentration of 12% CO_2 (dry) is essentially the stoichiometric condition for this gaseous fuel. However, it represents a condition of approximately 50% excess air for coal combustion or refuse incineration [1, p. 147] and provides a convenient basis to standardize reporting of pollutant concentrations for such applications. ∎

PROBLEM 3.19   *Natural Gas F Factor Problem*

(a) Calculate the dry flue gas flow for a boiler firing 100 MMBtu/h (HHV) of natural gas at 10% excess air using the *F* factor method. (b) Calculate the emission rate in lb/MMBtu for NO_x at 9 ppm (dry) and CO at 50 ppm (dry) in the flue gas corrected to 3% O_2 (dry).[4,5] (c) Compare with the results of rigorous material balance combustion calculations for the natural gas of Table 3.1. Combustion with 10% excess air for natural gas or pure methane results in an excess O_2 in the flue gas of approximately 1.7% (wet) and 2.1% (dry).

SOLUTION

(a) For natural gas, $F_d = 8710$ SCF (dry)/MMBtu at the stoichiometric condition.

$$(8710)(100) = 871,000 \text{ SCF (dry)/h (SCFH, dry)}$$

At 3% O_2 (dry),

$$(871,000)[20.9/(20.9-3)] = 1017,000 \text{ SCFH (dry)}$$

(b) Pollutant mass flow rate of a pollutant can then be found by multiplying this flow at actual conditions by a pollutant concentration (dry) at those same conditions. This results in Equation 19.1 in the EPA procedure [4].

$$E = C_d F_d [20.9/(20.9 - \% O_2 \text{ dry})] \tag{3.55}$$

To obtain *E* in units of lb/MMBtu, C_d must be in lb/SCF, or pollutant concentration expressed in ppm must be converted to lb/SCF. Conversion factors of 1.660×10^{-7} and 1.194×10^{-7} are provided in Method 19 for SO_2 and NO_x, respectively.

Other equations involving various combinations of wet basis and dry basis quantities are also provided in Method 19 to facilitate data reduction.

From Equation (3.55), using the conversion factor for NO_x,

$$E = (9)(1.194 \times 10^{-7})(8710)[20.9/(20.9-3)] = 0.01 \text{ lb/MMBtu (HHV)}$$

---

[4] These are Best Available Control Technology (BACT) guidelines as of 2003 applied on a case-by-case basis by the Texas Commission on Environmental Quality (TCEQ) for boilers greater than 40 MMBtu/h . NO_x emission levels of this magnitude have become common in many environmental jurisdictions. One method of control to achieve such a level is selective catalytic reduction (SCR), discussed in Chapter 4.

[5] No. 2 fuel oil with 0.05 wt% sulfur is allowed as a backup fuel for up to 750 h per year, not subject to the 9 ppmd NO_x guideline . No. 2 fuel oil is a distillate fuel boiling below about 600°F (316°C) and is used in home heating and in some industrial applications; no preheating is necessary in combustion or handling [3, pp. 9-8 to 9-9].

3.11 Estimation of Pollutant Emissions   143

Alternatively, showing conversion of units,

$$\frac{(9 \text{ lb mol NO}_x)}{(10^6 \text{ lb mol flue gas})} \frac{(1)}{(100 \text{ MMBtu/h})} (871,000)[20.9/(20.9-3)]$$

$$\times (\text{SCF/h flue gas at } 3\% \text{ O}_2 \text{ (dry)}) \frac{(\text{lb mol})}{385.3 \text{ SCF}} \frac{(46.01 \text{ lb NO}_x)}{(\text{lb mol NO}_x)}$$

$$= (1.194 \times 10^{-7})(9)(8710)[20.9/(20.9-3)] = 0.01 \text{ lb NO}_x/\text{MMBtu (HHV)}$$

where 1.194×10^{-7} is the combination of the factors $(46.01)/[(385.3)(10^6)]$, and

$$\frac{(50 \text{ lb mol CO})}{(10^6 \text{ lb mol flue gas})} \frac{(1)}{(100 \text{ MMBtu/h})} (871,000)[20.9/(20.9-3)]$$

$$\times (\text{SCF/h flue gas at } 3\% \text{ O}_2 \text{ (dry)}) \frac{(\text{lb mol})}{385.3 \text{ SCF}} \frac{(28.01 \text{ lb CO}_2)}{(\text{lb mol CO}_2)}$$

$$= 0.037 \text{ lb CO/MMBtu (HHV)}$$

(c) For the particular natural gas in question, 100 mol/h fuel with a HHV of 1019.4 Btu/SCF (60°F, 1 atm) generates 38.69 MMBtu/h (HHV) heat input. This produces 867.47 mol/h = 334,240 SCF (60°F, 1 atm) dry flue gas at 0% O_2. Dividing the dry flue gas volume by the heat input yields a specific F_d for this particular natural gas.

$$F_d = 8640 \text{ SCF of dry flue gas @ } 0\% \text{ O}_2/10^6 \text{ Btu (HHV)}$$

Prorating,

$$\text{NO}_x \ E = (8640/8710)(0.01) = 0.0099 \text{ lb/MMBtu (HHV)}$$
$$\text{CO} \ E = (8640/8710)(0.037) = 0.0367 \text{ lb/MMBtu (HHV)}$$

within roundoff of the values of 0.01 lb NO_x/MMBtu (HHV) and 0.037 lb CO/MMBtu (HHV) previously computed. ∎

PROBLEM 3.20   *NO_x from a Correlation (Adapted from Ref. [23])*

A certain natural gas has a HHV of 1020.8 and a LHV of 919.7, both in units of Btu/SCF (60°F, 1 atm). Firing this fuel at 60°F at a rate of 80 MMBtu/h (LHV) in a furnace with 10% excess ambient air produces 1021 MSCFH of wet flue gas containing 1.72% O_2 (wet) and 18.1% H_2O. The adiabatic flame temperature determined by trial and error calculation is 3451°F.

Estimate the NO_x concentration, mass flow rate, and the emission factor using the low-NO_x burner correlations given in Appendix C for (a) a steam–methane reformer (SMR) furnace and (b) an ethylene pyrolysis cracking furnace.

SOLUTION   Calculate dry basis O_2 concentration.

$$\% \text{ O}_2 \text{ (dry)} = \% \text{ O}_2 \text{ (wet)}(100)/(100-\%H_2O)$$
$$= (1.72)(100)/(100-18.1) = 2.10$$

Calculate $(10,000)/\text{AFT (R)} = (10,000)/(3451 + 459.67) = 2.557$

(a) Calculate ppm NO_x for SMR.

From Appendix C,

$$\ln[NO_x \text{ ppm dry}/(\% O_2 \text{ dry})] = 12.2 - (3.6)[(10,000)/AFT\ (^\circ R)]$$
$$\ln[(NO_x \text{ ppm dry}/(\% O_2 \text{ dry})] = 12.2 - (3.6)[2.557] = 2.9948$$
$$NO_x \text{ ppm dry} = (\% O_2 \text{ dry}) \exp[2.9948] = (2.10)(19.9814) = 42.0$$
$$NO_x \text{ ppmd @ 3\% } O_2 \text{ (dry)} = NO_x \text{ ppm dry}[(20.9-3)/(20.9-\%O_2\text{dry})]$$
$$= (42.0)(20.9-3)/(20.9-2.10) = 40.0$$

Calculate NO_x (lb/h)

$$NO_x \text{ (lb/h)} = (Q \text{ dry})(NO_x \text{ ppm dry})(\text{conversion factors})$$
$$= (1021 \times 10^3 \text{ SCF/h wet FG}) \frac{[(100-18.1)\text{ dry FG}]}{(100)\text{ wet FG}}$$
$$\times \frac{(42.0 \text{ SCF } NO_x)}{(10^6 \text{ SCF dry FG})} \frac{(\text{lb mol } NO_x)}{(379.5 \text{ SCF } NO_x)} \frac{(46.01 \text{ lb } NO_x)}{(\text{lb mol } NO_x)}$$
$$= 4.26 \text{ lb/h } NO_x$$

Calculate firing rate based on HHV

$$\text{Firing rate (HHV)} = \text{firing rate (LHV)(HHV/LHV)}$$
$$= (80 \times 10^6 \text{ Btu/h})(1020.8/919.7) = 88.79 \times 10^6 \text{ Btu/h}$$
$$= 88.79 \text{ MMBtu/h}$$

Calculate EF (lb/MMBtu (HHV)).

$$EF = (4.26 \text{ lb/h } NO_x)/(88.79 \text{ MMBtu/h}) = 0.048 \text{ lb/MMBtu (HHV)}$$

(b) Calculate ppm NO_x dry for ethylene furnace.

Also from Appendix C,

$$\ln[(NO_x \text{ ppm dry}/(\% O_2 \text{ dry})] = 12.6 - (3.6)[(10,000)/AFT\ (R)]$$
$$\ln[(NO_x \text{ ppm dry}/(\% O_2 \text{ dry})] = 12.6 - (3.6)[2.557] = 3.395$$
$$NO_x \text{ ppm dry} = (\% O_2 \text{ dry})\exp[3.395] = (2.10)(29.8147) = 62.6$$

Similarly,

$$NO_x \text{ ppmd @ 3\% } O_2 \text{ (dry)} = 62.6[(20.9-3)/(20.9-2.10)] = 59.6$$
$$NO_x \text{ (lb/h)} = 1021 \times 10^3 \frac{(100-18.1)}{(100)} \frac{(62.6)}{(10^6)} \frac{(46.01)}{(379.5)} = 6.35 \text{ lb/h } NO_x$$
$$EF = (6.35 \text{ lb/h } NO_x)/(88.79 \text{ MMBtu/h}) = 0.072 \text{ lb/MMBtu (HHV)}$$

NO_x is higher from the burners in the ethylene furnace because ethylene furnaces tend to run with higher firebox temperatures than SMR furnaces (Appendix C).

Be advised that better lower NO_x-emission burners are available at this time to meet lower mandated emission limits. For example, the NO_x limit for ethylene pyrolysis furnaces in the Houston–Galveston–Brazonia (HGB) Nonattainment Area of Texas is 0.036 lb/MMBtu (HHV) [33] corresponding to 30 ppmd NO_x @ 3% O_2 (dry) for natural gas/methane. ∎

PROBLEM 3.21

(a) Using AP-42 for natural gas combustion with low-NO_x burners, estimate the NO_x from a 100 MMBtu/h process furnace/fired heater (analogous to a large wall-fired boiler). (b) Reevaluate the NO_x prediction when the combustion air temperature is increased from ambient to 400°F.

SOLUTION

(a) Table 1.4-1 of AP-42 [24] lists a NO_x emission factor of 140 lb/10^6 SCF of natural gas fired. This emission factor has an "A" rating (excellent). According to a footnote to the table, the emission factor can be converted to lb/MMBtu (HHV) by dividing by a natural gas HHV of 1020 Btu/SCF.

$$\frac{140 \text{ lb}}{10^6 \text{ SCF}} \frac{\text{SCF}}{1020 \text{ Btu}} = 0.137 \text{ lb/MMBtu}$$

(b) NO_x is known to increase when the combustion air is heated above ambient temperatures. However, the AP-42 natural gas boiler emission factors do not include the increase in NO_x expected from preheated combustion air. One way to correct the effect of air preheat is to use a curve such as Figure C.10. This is similar to the performance curves available from burner manufacturers for their burners. Using the curve for methane in the figure at a 400°F combustion air temperature, one reads a factor for NO_x with air preheat/NO_x with ambient air of about 1.6.

Then,

(0.137 from AP-42)(1.6) = 0.22 lb/MMBtu (HHV) using 400°F combustion air

(In an actual case, one would use the burner vendor's curve for that particular burner, rather than Figure C.10).

Other techniques for estimating NO_x emissions when permitting this furnace include the following:

- A burner manufacturer's guarantee incorporating the effect of air preheat.
- Data from a similar furnace already in operation.
- Non-EPA agency emission factors for use when submitting a permit application for a source within that particular agency's jurisdiction.
- A correlation applicable to specific types of process furnaces such as those discussed in Appendix C. Those correlations also address fuels other than natural gas, as well as air preheat. ■

PROBLEM 3.22

The air preheat factor was read from Figure C.10 as part of Problem 3.21. Calculate this factor analytically from Equation C.4 using the results of Problems 3.15 and 3.16.

SOLUTION From Appendix C,

$$\frac{NO_x \text{ with air preheat}}{NO_x \text{ ambient}} = \exp[B\{(10{,}000/\text{AFT (°R)})_{\text{ambient}} - (10{,}000/\text{AFT (°R)})_{\text{air preheat}}\}]$$

$$= \exp[3.6\{10{,}000/(3413 + 459.67) - 10{,}000/(3617 + 459.67)\}]$$

$$= \exp[0.4652] = 1.5923 \text{ versus } 1.6 \text{ read}$$

from the graph for 100% methane ■

PROBLEM 3.23

Using AP-42 emission factors, estimate the atmospheric emissions from a small commercial heater firing commercial-grade propane at 2 MMBtu/h. Sulfur content of the fuel is 15 gr/100 SCF, as used in Ref. [34].

SOLUTION It is convenient first to convert the volume basis emission factors of AP-42 (lb/10^3 gal of liquefied fuel) to an energy basis (lb/MMBtu of vaporized gas fired) by using the typical heating value of 90,500 Btu/gal (after vaporization) given in AP-42 for commercial propane [35].

AP-42 emission factors in lb/10^3 gal for propane fired in commercial boilers (generally between 0.3 and 10 MMBtu/h) along with factors in lb/MMBtu calculated from them are as follows:

$$\text{EF (lb/MMBtu)} = \frac{[\text{EF (lb/}10^3\text{ gal)}](10^6 \text{ Btu/MMBtu})}{\dfrac{(90,500 \text{ Btu})}{\text{gal}} \dfrac{(1000 \text{ gal})}{(10^3 \text{ gal})}}$$

$$\frac{(1000)}{(90,500)}[\text{EF (lb/}10^3\text{ gal)}] = (0.01105)[\text{EF (lb/gal)}]$$

As an example, using the conversion factor (0.01105) just calculated, for PM,

$$\text{EF (lb/MMBtu)} = (0.01105)(0.4) = 0.004 \text{ (rounded)}$$

Summary of emission factors:

Pollutant	Propane emission factor	
	lb/10^3 gal	lb/MMBtu
PM/PM$_{10}$[a]	0.4	0.004
SO$_2$	0.10 S[b]	0.0011 S[a]
NO$_x$	14	0.155
CO	1.9	0.020
TOC	0.5	0.0055
CH$_4$	0.2	0.0022

[a]Filterable particulate matter is that collected on or prior to the filter of an EPA Method 5 (or equivalent) sampling train. It is assumed here to be all PM$_{10}$, like the particulate generated from combustion of natural gas, a fuel with similar combustion characteristics. Condensible particulate is calculated using the same split given in the AP-42 table for combustion of natural gas.

[b]S in the above table equals the sulfur content expressed in grains/100 SCF of vaporized fuel gas. This will be verified by material balance, assuming 100% of fuel sulfur is converted to SO$_2$ in the flue gas.

3.11 Estimation of Pollutant Emissions

Emission Calculations

$$(PM_{10_{filterable}})(2\text{ MMBtu/h})(0.004\text{ lb/MMBtu}) = 0.008\text{ lb/h}$$
$$\times\ 4.38 = 0.035\text{ ton/year}$$

For natural gas combustion [24],

$$PM_{total} = PM_{filterable} + PM_{condensible}$$
$$PM_{filterable} = (0.25)(PM_{total})$$
$$PM_{condensible} = (0.75)(PM_{total})$$
$$\therefore PM_{condensible} = (3)(PM_{filterable})$$

$$PM_{10_{condensible}} = (3)(0.008) = 0.024\text{ lb/h}$$
$$= (3)(0.035) = 0.11\text{ ton/year}$$

Using SO_2 emission factor,

$$SO_2 = (0.0011)(15\text{ gr}/100\text{ SCF}) = 0.0166\text{ lb/MMBtu}$$
$$(0.0166)(2) = 0.033\text{ lb/h}$$
$$\times\ 4.38 = 0.145\text{ ton/year}$$

Verification by Material Balance

The HHV of gaseous propane is 2516.1 Btu/SCF (60°F, 1 atm)

$$2\times10^6\text{ Btu/h (HHV)}\frac{(\text{SCF 60°F, 1 atm})}{2516.1\text{ Btu (HHV)}}\frac{(\text{lb mol})}{379.5\text{ SCF}} = 2.0945\text{ lb mol of propane fuel gas/h}$$

$$\frac{(15\text{ gr S})}{(100\text{ SCF fuel gas})}\frac{(1\text{ lb S})}{(7000\text{ gr S})}\frac{(385.3\text{ SCF fuel gas})}{(\text{lb mol fuel gas})}\frac{(\text{lb mol S})}{(32.06\text{ lb S})}(2.0945\text{ lb mol fuel gas/h})$$
$$= 0.00054\text{ lb mol S/h in fuel}$$

1 lb mol S in fuel corresponds to 1 lb mol SO_2 in flue gas.

$$S + O_2 \rightarrow SO_2$$
$$= 0.00054\text{ lb mol }SO_2/\text{h in flue gas}$$
$$(0.00054)(64.06\text{ lb }SO_2/\text{lb mol }SO_2) = 0.035\text{ lb/h }SO_2$$

$$NO_x = (0.155)(2) = 0.31\text{ lb/h}$$
$$\times\ 4.38 = 1.4\text{ ton/year}$$
$$CO = (0.020)(2) = 0.040\text{ lb/h}$$
$$\times\ 4.38 = 0.18\text{ ton/year}$$
$$TOC = (0.0055)(2) = 0.0110\text{ lb/h}$$
$$4.38 = 0.048\text{ ton/year}$$
$$CH_4 = (0.0022)(2) = 0.0044\text{ lb/h}$$
$$4.38 = 0.019\text{ ton/year}$$

148 Chapter 3 Air Combustion

Pollutant Concentrations in Flue Gas
From combustion calculations,

$$F_d = 8710 \text{ SCF (dry) per MMBtu from Method 19 [4] (Table 3.4)}$$
$$= 8820 \text{ SCF (dry) per MMBtu from Problem 3.9}$$

Flue gas volume at 3% O_2 (dry)

$(2 \text{ MMBtu/h})(8710 \text{ SCF (dry)/MMBtu})[20.9/(20.9-3)] = 20,340 \text{ SCFH (dry) for } F_d = 8710$
$= 20,600 \text{ SCFH for } F_d = 8820$

PM_{10}:

$$PM_{10_{filterable}} = \frac{(0.008 \text{ lb/h})(7000 \text{ gr/lb})}{20,340 \text{ or } 20,600 \text{ SCF (dry)/h}} = 0.003 \text{ gr per dry standard cubic foot}$$

$$PM_{10_{condensible}} = (3)(0.003) = 0.009$$

$$PM_{total} = \underset{(25\%)}{0.003} + \underset{(75\%)}{0.009} = 0.012 \text{ gr per dry standard cubic foot}$$

SO_2:

$$\frac{10^6 \ (0.033 \text{ lb/hr})(385.3 \text{ SCF/lb mole})/(64.06 \text{ lb/lb mole})}{20,340 \text{ to } 20,600 \text{ SCF flue gas (dry)/hr}}$$
$$= 9.6-9.8 \text{ ppmd at 3\% } O_2 \text{ (dry)}$$

NO_x:

$$\frac{10^6 (0.31 \text{ lb/h})(385.3 \text{ SCF/lb mol})/(46.01 \text{ lb/lb mol})}{20,340 \text{ to } 20,600 \text{ SCF flue gas (dry)/h}}$$
$$= 126.0-127.6 \text{ ppmd at 3\% } O_2 \text{ (dry)}$$

CO:

$$\frac{10^6 (0.040 \text{ lb/h})(385.3 \text{ SCF/lb mol})/(28.01 \text{ lb/lb mol})}{20,340 \text{ to } 20,600 \text{ SCF flue gas (dry)/h}}$$
$$= 26.7-27.1 \text{ ppmd at 3\% } O_2 \text{ (dry)}$$

A number of air quality management districts in California require that emissions from new heaters in this size range must not exceed 30 ppm (0.036 lb/MMBtu) NO_x and 400 ppm CO, corrected to 3% O_2 (dry) in the flue gas. Lower limits are projected for the future. A requirement to install new burner kits to retrofit existing units emitting NO_x and CO in excess of these limits is on the table as well.

TOC (expressed as methane):

$$\frac{10^6 (0.0110 \text{ lb/h})(385.3 \text{ SCF/lb mol})/(16.04 \text{ lb/lb mol})}{20,340 \text{ to } 20,600 \text{ SCF flue gas(dry)/h}}$$
$$= 12.8-13.0 \text{ ppmd at 3\% } O_2 \text{ (dry)}$$

CH_4:

$$\frac{10^6 (0.0044 \text{ lb/h})(385.3 \text{ SCF/lb mol})/(16.04 \text{ lb/lb mol})}{20,340 \text{ to } 20,600 \text{ SCF flue gas(dry)/h}}$$
$$= 5.1-5.2 \text{ ppmd at 3\% } O_2 \text{ (dry)}$$

3.11 Estimation of Pollutant Emissions   149

Summary of pollutants from 2 MMBtu/h propane-fired heater:

Pollutant	Calculated emissions					Permit limit	
	ppm (dry)	ppmd at 3% O_2	lb/h	tons/year	lb/MMBtu	ppmd at 3% O_2	lb/MMBtu
PM/PM$_{10}$	—	a	0.032	0.14	0.016	—	—
SO_2	—	9.6–9.8	0.033	0.145	0.0165	—	—
NO_x	—	126–127.6	0.31	1.4	0.155	30	0.036
CO	—	26.7–27.6	0.040	0.18	0.020	400	—
TOC	—	12.8–13.0	0.0110	0.048	0.0055	—	—
CH4	—	5.1–5.2	0.0044	0.019	0.0022	—	—

1. Based on EPA AP-42 emission factors.
2. PM/PM$_{10}$ EF for total particulate = 4 × filterable particulate EF.
3. Molecular weight of TOC and methane hydrocarbons = 16.04.
4. Based on commercial propane with 15 gr S/100 dry standard cubic foot (dscf) (68°F).
5. Burner manufacturer will guarantee NO_x to comply with limit.
^{a}0.012 gr/(dscf) (68°F), calculated at 3% O_2 (dry) in the flue gas.

Problem 3.24 is concerned with combustion of a refinery fuel gas containing the EPA sulfur limit. With certain exceptions, EPA considers as *fuel gas* any gas generated and combusted at a petroleum refinery, including natural gas combined with it in any proportion [36]. ∎

PROBLEM 3.24

A refinery fuel gas [37] of highly simplified composition, 25% CH_4, 46% C_2H_6, 26% H_2, and 3% N_2, is burned in a refinery heater at 10% excess air. This gas also contains hydrogen sulfide (H_2S) at 230 mg/dscm, or 0.10 gr/dscf, the limit contained in 40 CFR 60, Subpart J—Standards of Performance for Petroleum Refineries [36].

SOLUTION   This sulfur limit corresponds to approximately 162 ppm H_2S by volume on a dry basis.

$$10^6 \frac{230 \text{ mg}}{\text{m}^3} \frac{\text{g}}{1000 \text{ mg}} \frac{\text{lb}}{453.59 \text{g}} \frac{\text{lb mol } H_2S}{34.08 \text{ lb}} \frac{0.02832 \text{ m}^3}{\text{ft}^3} \frac{385.3 \text{ ft}^3}{\text{lb mol fuel gas}}$$

$$= 162.35 \cong 162 \text{ ppmv } H_2S \text{ (dry)} \text{ (based on a 68°F SCF)}$$

The following combustion table results from the calculations shown below it.
Basis 100 mol of fuel gas and combustion air at 60°F, 60% RH, and 1 atm.

Component	Moles in fuel	Moles in air	Moles O_2 consumed	Moles CO_2 produced	Moles H_2O produced	Moles in flue gas	mol% (wet)	mol% (dry)
H_2	26	0	13	0	26	0	0	0
CH_4	25	0	50	25	50	0	0	0

C_2H_6	46	0		161	92	138	0	0	0
CO_2	0	0.3882		0	0.3882	0	117.3882	9.04	10.95
N_2	3	918.5273		0	0	0	921.5273	70.96	85.94
Ar	0	10.9867		0	0	0	10.9867	0.85	1.02
H_2O	0	12.4288		0	0	0	226.4288	17.43	—
O_2	0	22.4000		0	0	0	22.4000	1.72	2.09
Total (wet)	100	964.7310	224		117.3882	214	1298.7310	100.00	—
Total (dry)	100	952.3022	224		117.3882	0	1072.3012	—	100.00

$$H_2 + \frac{1}{2}O_2 \rightarrow H_2O \tag{3.56}$$

26 mol H_2

13 mol O_2 consumed

0 mol CO_2 produced

26 mol H_2O produced

$$CH_4 + 2O_2 \rightarrow CO_2 + 2H_2O \tag{3.12}$$

25 mol CH_4

50 mol O_2 consumed

25 mol CO_2 produced

50 mol H_2O produced

$$C_2H_6 + 7/2 O_2 \rightarrow 2CO_2 + 3H_2O \tag{3.13}$$

46 mol C_2H_6

161 mol O_2 consumed

92 mol CO_2 produced

138 mol H_2O produced

$$N_2 \text{ in fuel} \rightarrow N_2 \text{ in flue gas} \tag{3.57}$$

3 mol N_2

$$\text{Total } O_2 \text{ consumed} = 13 + 50 + 161 = 224$$

$$10\% \text{ excess}: (1.10)(224) = 246.4$$

$$(246.4 - 224 = 22.4) \, O_2 \text{ shows up in the flue gas}$$

The 246.4 mol of O_2 corresponds to $(100/20.947)(246.4) = 1176.3021$ mol of air and brings along $(1176.3021)(78.086/100) = 918.5273$ mol of N_2, $(1176.3021)(0.934/100) = 10.9867$ mol of Ar, $(1176.3021)(0.033/100) = 0.3882$ mol of CO_2, and $(1176.3021)(0.010566) = 12.4288$ mol of H_2O using the dry air composition of Table 3.3 and the moisture content of 60°F, 60% RH, 1 atm air.

3.11 Estimation of Pollutant Emissions

When the smoke clears and the dust settles, 100 mol of refinery fuel gas produces approximately 1299 mol of total, or wet, flue gas and 1072 mol of dry flue gas. These figures correspond to 500 and 413 MSCFH (68°F, 1 atm). Flue gas molecular weight is 30.0 (dry), as expected, and 27.9 (wet).

$$H_2S + \frac{3}{2}O_2 \rightarrow SO_2 + H_2O \qquad (3.58)$$

1 mol of H_2S produces 1 mol of SO_2 or SO_x

(100 mol of fuel gas)(162 ppm H_2S) = (1299 or 1072 mol of flue gas)(X ppm)

$$X = \frac{(100)(162)}{1299} = 12.5 \text{ ppm } SO_2 \text{ (wet)}$$

$$X = \frac{(100)(162)}{1072} = 15.1 \text{ ppm } SO_2 \text{ (dry)}$$

Mass flow rate of SO_2 (or SO_x expressed as SO_2) per 100 mol/h of refinery fuel gas

$$(100 \text{ mol/h fuel})(162 \text{ mol } H_2S/10^6 \text{ mol fuel})$$
$$\times (1 \text{ mol } SO_2/\text{mol } H_2S)(64.06 \text{ lb } SO_2/\text{mol } SO_2)$$
$$= 1.04 \text{ lb/h } SO_2$$

Flue gas flow computed from the complete published composition of the refinery fuel gas [37] is 1669 mol/h (643 MSCFH) wet and 1402 mol/h (540 MSCFH) dry. This would make the SO_2 concentration ~30% lower, 9.7 ppm (wet) and 11.6 ppm dry. ∎

PROBLEM 3.25

This problem goes through the chemical equations for the combustion of a No. 2 fuel oil[6] containing 0.22 wt% S (Table 3.10). This sulfur level may be higher than what is permitted and available for sale in some air quality jurisdictions nowadays, but the problem is a useful exercise to illustrate the chemical reactions and material balances involved. Calculations are compared with results from a published nomograph for fuel oil combustion [38].

(a) Calculate the flue gas composition including SO_x concentration from firing the following typical No. 2 fuel oil at a condition of 3% O_2 (dry) in the flue gas.
(b) How much excess air does this require?
(c) If 100 MMBtu/h (HHV) of this fuel is fired, what is the mass flow rate of SO_x expressed as SO_2 in lb/h and in ton/year?
(d) Calculate an emission factor for SO_x in lb/MMBtu, based on the HHV.
(e) Without recomputing the flue gas, estimate ppm (dry) of SO_3 for a range of fuel sulfur (S) from 0 to 0.5 wt% as SO_3/SO_x varies from 0 to 5%. Express result to the nearest 0.1 ppm.

---

[6] No. 2 fuel oil is a distillate oil boiling below about 600°F (316°C) and is used in home heating and some industrial applications. No preheating is necessary in combustion or handling [3, pp. 9-8 to 9-9].

152 Chapter 3 Air Combustion

Table 3.10 Typical Analysis of No. 2 Fuel Oil (wt%)[a]

Component	As listed	Normalized[b]
Carbon	87.3	87.155
Hydrogen	12.6	12.579
Oxygen	0.04	0.040
Nitrogen	0.006	0.006
Sulfur	0.22	0.220
Ash	<0.01	0
C/H ratio	6.93	6.93

API gravity: 33°, HHV: 141,000 Btu/gal[a], LHV: 133,000 Btu/gal[a].

[a] Analysis as listed, C/H ratio, and API Gravity from Ref. 3, p. 9-10. Heating values read from graph on the same page.

[b] Composition normalized to 100%. Ash is assumed to be zero, and oxygen, nitrogen, and sulfur remain the same. The C/H ratio also remains the same for the normalized composition.

SOLUTION

(a) This problem will be solved through the use of combustion reaction equations, starting with the normalized composition and conversion to lb moles.

Combustion calculations on fuel oil will be performed in a later example (Problem 3.26), using the EPA shortcut method with F factors.

$$C + O_2 \rightarrow CO_2 \tag{3.59}$$

7.2569 mol C
7.2569 mol O_2 consumed
7.2569 mol CO_2 formed
0 mol H_2O formed

$$H_2 + \frac{1}{2}O_2 \rightarrow H_2O \tag{3.56}$$

6.2272 mol H_2
3.1136 mol O_2 consumed
0 mol CO_2 formed
6.2272 mol H_2O formed

$$S + O_2 \rightarrow SO_2 \tag{3.60}$$

0.0069 mol S
0.0069 mol O_2 consumed
0.0069 mol SO_2 formed
0 mol CO_2 formed
0 mol H_2O formed

Oxygen in fuel will be consumed by other compounds.

0.0013 mol O_2

3.11 Estimation of Pollutant Emissions   153

Stoichiometric Case

$$\text{Moles of } O_2 \text{ required} = 7.2569 + 3.1136 + 0.0069 - (0.040/32.00 = 0.0013)$$
$$= 10.3761$$
$$\text{Moles of dry air required} = (10.3761)(100/20.947) = 49.5350$$
$$\text{Moles of } N_2 \text{ along for the ride} = (49.5350)(78.086/100) = 38.6790$$
$$\text{Moles of Ar in air} = (49.5350)(0.934/100) = 0.4627$$
$$\text{Moles of } CO_2 \text{ in air} = (49.5350)(0.033/100) = 0.0163$$
$$\text{Moles of } H_2O \text{ in air} = (49.5350)(0.0209) = 1.0353$$

These figures enable the rest of the following stoichiometric combustion table to be filled out.

Combustion Summary Table
Basis: 100 lb of fuel

Moles of excess O_2 necessary for 3% O_2 (dry) in the flue gas can be calculated from the stoichiometric dry flue gas in the following table.

$$\text{moles } O_2 = (46.4229)(3/100)/[1 - 3/20.947] = 1.6255 \text{ from Equation}(3.17)$$

Calculations are repeated with mol O_2 required = 10.3761 (in bold) + 1.6255

Flue gas composition (wet and dry) is listed in the last two columns of the table.

Combustion Summary Table
Basis: 100 lb of fuel

(b) Excess air is calculated from the total air requirement compared to the air for the stoichiometric case.

$$\% \text{ Excess air} = (100)(58.4924/50.5703) - 100 = (100)(57.2946/49.5350) - 100 = 15.7\%$$

Molecular weight of dry flue gas = (1647.1994)/(54.1829) = 30.4 with pounds of dry flue gas = Σ(component $\times$ MW of component) = 1647.1994 and mol of dry flue gas = 54.1829 given in table above.

(c) $(100 \times 10^6 \text{ Btu/h})/(133{,}000 \text{ Btu/gal}) = 752 \text{ gal/h}$

$$\text{Degrees API} = (141.5/s) - 131.5 \qquad (3.61)$$

from Ref. [3, p. 9–10] where s is the specific gravity of the oil, the density of the oil at 60°F divided by the density of water at 60°F.
Solving for s,

$$s = \frac{141.5}{\text{degrees API} + 131.5}$$
$$= \frac{141.5}{33 + 131.5} = 0.8602$$

Density of liquid water at 60°F from steam tables [6, p. 88]

$$= 1/(0.016033 \text{ ft}^3/\text{lb}) = 62.3714 \text{ lb/ft}^3$$
$$\div 7.48052 \text{ gal/ft}^3 = 8.3378 \text{ lb/gal}$$

Combustion Summery Table Basis: 100 lb of fuel oil

Stoichiometric case, air at 80°F, 60% RH (0.0209 mol of water/mol of dry air)

Component	Pounds in fuel	MW	Moles in fuel	Moles in air	Moles O_2 consumed	Moles CO_2 produced	Moles H_2O produced	Moles in flue gas	mol% (wet)	mol% (dry)
Carbon	87.155	12.01	7.2569	0	7.2569	7.2569	0	0	0	0
Hydrogen	12.579	2.02	6.2272	0	3.1176	0	6.2272	0	0	0
CO_2	0	44.01	0	0.0163	0	0.0163	0	7.2732	13.55	15.67
Sulfur	0.220	32.06	0.0069	0	0.0069	0.0069 SO_2	0	0	0	0
SO_2	30	64.06	0	0	0	0	0	0.0069	127.8 ppm	147.8 ppm
Nitrogen	0.006	28.01	0.0002	38.6799	0	0	0	38.6801	72.05	83.32
Argon	0	39.95	0	0.4627	0	0	0	0.4627	0.86	1.00
Oxygen	0.040	32.00	0.00125	**10.3761**	−0.00125	0	0	0	0	0
Water	0	18.02	0	1.0353	0	0	1.0353	7.2625	13.53	—
Total (dry)	100.00	—	13.4924	49.5350	10.3761	7.2732	0	46.4229	—	100.00
Total (wet)	100.00	—	13.4924	50.5703	10.3761	7.2732	7.2625	53.6854	100.00	—

Combustion Summery Table Basis: 100 lb of fuel oil

					Excess air case					
Component	Pounds in fuel	MW	Moles in fuel	Moles in air	Moles O_2 consumed	Moles CO_2 produced	Moles H_2O produced	Moles in flue gas	mol% (wet)	mol% (dry)
Carbon	87.155	12.01	7.2569	0	7.2569	7.2569	0	0	0	0
Hydrogen	12.579	2.02	6.2272	0	3.1176	0	6.2272	0	0	0
CO_2	0	44.01	0	0.0189	0	0.0189	0	7.2758	11.81	13.43
Sulfur	0.220	32.06	0.0069	0	0.0069	0.0069 SO_2	0	0	0	0
SO_2	0	64.06	0	0	0	0	0	0.0069	111.4 ppm	126.6 ppm
Nitrogen	0.006	28.01	0.0002	44.7394	0	0	0	44.7396	72.62	82.57
Argon	0	39.95	0	0.5351	0	0	0	0.5351	0.87	0.99
Oxygen	0.040	32.00	0.00125	12.0016	−0.00125	0	0	1.6255	2.64	3.00
Water	0	18.02	0	1.1975	0	0	1.1975	7.4247	12.05	—
Total (dry)	100.00	—	13.4924	57.2946	10.3761	7.2750	0	54.1829	—	100.00
Total (wet)	100.00	—	13.4924	58.4924	10.3761	7.2750	7.4247	61.6076	100.00	—

Density of oil is, therefore,

$$(0.8602)(62.3714) = 53.7 \text{ lb/ft}^3$$
$$(0.8602)(8.3378) = 7.17 \text{ lb/gal}$$
$$(752 \text{ gal/h, from above})(7.17 \text{ lb/gal}) = 5392 \text{ lb/h of fuel}$$

Combustion calculations in this problem are on a per 100 lb of fuel basis

$$SO_2 \quad \frac{(0.0069 \text{ lb mol } SO_2)}{100 \text{ lb of fuel}} (5392 \text{ lb/h of fuel})(64.06 \text{ lb } SO_2/\text{lb mol } SO_2)$$
$$= 23.8 \text{ lb/h } SO_2$$
$$\times 4.38 \text{ tons/year per lb/h} = 104.4 \text{ ton/year}$$

(d) 100 MMBtu/h based on the LHV corresponds to

$$(141{,}000/133{,}000)(100) = 106 \text{ MMBtu/h (HHV)}$$
$$23.8 \text{ lb/h}/106 \text{ MMBtu/h} = 0.225 \text{ lb/MMBtu (HHV)}$$

(e) The SO_2 calculated in part (a) is in fact SO_x expressed as SO_2. Neglecting the minor change in the O_2 used and its effect on the flue gas volume, 0.22 wt% S corresponds to 126.6 ppm SO_x at 3% O_2 (dry). Prorating this value to other values of fuel sulfur and taking various percentages of SO_3/SO_x results in the following table.

For total $SO_x = 126.6$ ppm for fuel $S = 0.22$:

Estimated ppm SO_3 for Oil Burning Case

	% SO_3/SO_x					
wt% S↓	0	1	2	3	4	5
0	0	0	0	0	0	0
0.1	0	0.6	1.2	1.7	2.3	2.9
0.2	0	1.2	2.3	3.5	4.6	5.8
0.22	0	1.3	2.5	3.8	5.1	6.3
0.3	0	1.7	3.5	5.2	6.9	8.6
0.4	0	2.3	4.6	6.9	9.2	11.5
0.5	0	2.9	5.8	8.6	11.5	14.4

See also Problem 3.26 for an estimation of percentage composition of sulfur oxides in the flue gas explained by rule of thumb and thermodynamic equilibrium. ∎

POSTSCRIPT TO PROBLEM 3.25

The SO_2 in flue gas from combustion of a hydrocarbon fuel oil can be found from a nomograph published many years ago in Ref. 38. There must certainly be others. The nomograph can be used to check the results of this problem.

3.11 Estimation of Pollutant Emissions

Entering this particular nomograph with 15% excess air and a *C/H* weight ratio of 6.93 (as well as can be read), one obtains a combustion air requirement of 17 lb dry air/lb fuel and an SO_2 flue gas concentration of 125 ppm, for 0.22% sulfur in the fuel. Both of these numbers are interpolated by supplying one's own graduated scale.

Rigorous calculations above for this fuel oil lead to

$$\frac{(57.2946 \text{ mol dry air})}{(100 \text{ lb of fuel})} \frac{(28.96 \text{ lb dry air})}{(\text{mole of dry air})} = \frac{16.6 \text{ lb dry air}}{\text{lb of fuel}}$$

and an SO_2 concentration of 126.6 ppm (dry) in the flue gas at 3.0% O_2 (dry) (corresponding to $\sim$15% excess air).

Sulfur oxides are not only an environmental contaminant but they also constitute a corrosion issue in the flue. The next several problems lead to the calculation of the sulfuric acid dew point for a flue gas. This arises from the sulfur trioxide (SO_3) contained in the sulfur oxides (SO_x) generated by combustion of the sulfur content in a fuel. When the flue gas temperature falls below this dew point, sulfuric acid (SO_3 + flue gas $H_2O \rightarrow H_2SO_4$) condenses and causes corrosion to a carbon steel stack and other unprotected metallic components of the flue gas train. A brick stack can also be destroyed by the formation of concentrated sulfuric acid [39]. For this exercise, a generic liquid fuel oil with 0.1–0.5 wt% sulfur will be considered. ∎

PROBLEM 3.26

Estimate the flue gas moisture content when this fuel oil is combined with sufficient excess air to achieve 3% O_2 on a dry basis in the flue gas. The combustion air is at 80°F with a 60% RH (0.0209 mol of water/mol of dry air) [8].

SOLUTION With a complete analysis of the fuel, one could make use of the equations for combustion of its chemical constituents as illustrated in Problem 3.25. However, with the limited information provided here, we will proceed via a different route. Let us revisit Table 3.5, derived from Table 19.1 in EPA Method 19 [4], and see what additional information can be extracted from it. This problem could easily involve combustion of coal, simply by changing the *F* factors.

(The treatment may be somewhat repetitive of the gas combustion problems using *F* factors, but this problem is meant to be as self-contained as possible.)

The entries for fuel oil in Table 3.5 show the volume of flue gas generated at stoichiometric conditions (0% O_2) per 10^6 Btu (MMBtu) of fuel fired.

$$F_w = 10,320 \text{ wet (or total) SCF (68°F, 1 atm)/MMBtu}$$

$$F_d = 9190 \text{ dry SCF (68°F, 1 atm)/MMBtu}$$

$$\therefore H_2O = 1130 \text{ SCF (68°F, 1 atm)/MMBtu}$$

(calculated by difference)

$$\% H_2O = (100)(1130/10,320) = 10.95\%$$

However, the F_w factor includes only moisture from the combustion of fuel. It therefore does not include the effect of combustion air humidity.

Also from the *F* factor table,

$$F_c = 1420 \text{ SCF } CO_2 \text{ (68°F, 1 atm)/MMBtu}$$

158 Chapter 3 Air Combustion

The factor F_c denotes the amount of CO_2 that is liberated in the products of combustion from carbon-based materials. It is independent of the excess O_2 concentration. The difference $(F_d - F_c) = (9190 - 1420) = 7770 =$ the amount of non-H_2O constituents, primarily N_2, in the flue gas at 0% O_2. Assuming for the moment that this is all N_2 and there is no nitrogen in the fuel oil, $(7770)(100/79.1) = 9823$ SCF/MMBtu, the amount of dry combustion air. (The N_2 in dry combustion air is taken here as 79.1%.)

$$(9823)(0.0209) = 205.3 \text{ SCF } H_2O/\text{MMBtu}$$

This is the moisture from the humidity in the air.

Total flue gas moisture is $1130 + 205.3 = 1335.3$ SCF H_2O/MMBtu and total wet flue gas is

$$9190 + 1335.3 = 10,525.3 \text{ wet SCF/MMBtu}$$

$$\% H_2O = (100)(1335.3/10,525.3) = 12.69\%$$

The ratio of flue gas moisture with humidity to flue gas moisture without humidity is

$$(1335.3)/(1130) = 1.18$$

Flue gas moisture at 3% O_2 (dry) in the flue gas is given by Equation B.4.

$$\frac{\% H_2O \text{ at } 3\% O_2}{100} = \frac{(12.69/100) - (3/20.9)[(1+0.0209)(12.69/100) - 0.0209]}{1 - (3/20.9)[(1+0.0209)(12.69/100) - 0.0209]}$$

$$= 0.1131 = 11.31\%$$

$$\frac{\% O_2 \text{ wet}}{100} = \frac{(3)(100-11.31)}{(100)} = 2.66\%$$

Use this value to check Equation (B.3).

$$\% H_2O \text{ at } 2.66\% O_2 \text{ wet} = (12.69/100) - (2.66/20.9)[(1+0.0209)(12.69/100) - 0.0209]$$
$$= 0.1131 = 11.31\%$$

Flue gas moisture at 10% by volume is said to be typical for fuel oil combustion [40]. ∎

PROBLEM 3.27

The HHV of the fuel oil is 19,000 Btu/lb. For the conditions of Problem 3.26, estimate (a) the amount of flue gas, both wet and dry, produced by firing 2000 lb/h of this fuel, and (b) the concentration (wet and dry) of total sulfur oxides (SO_x) in the flue gas.

SOLUTION

(a) Firing rate $= (2000 \text{ lb/h})(19,000 \text{ Btu/lb}) = 38$ MMBtu/h at the HHV

Dry flue gas @ 0% $O_2 = (38 \text{ MMBtu/h})(F_d = 9190 \text{ SCF dry/MMBtu})$
$= 349,220$ SCF dry (68°F, 1 atm)

Wet flue gas @ 0% $O_2 = (38 \text{ MMBtu/h})(10,525.3 \text{ SCF wet/MMBtu})$
$= 399,960$ SCF wet (68°F, 1 atm)

Dry flue gas @ 3% O_2 (dry) $= (349,220)[(20.9-0)/(20.9-3)] = 407,750$ SCF dry

Wet flue gas @ 3% O_2 (dry) $= (407,750)[100/(100-11.31)] = 459,750$ SCF wet

(b) At 1 wt% S in the fuel oil: (2000 lb/h fuel)(0.1 lb S/100 lb fuel) = 2 lb/h S

$$S + O_2 \rightarrow SO_2 \tag{3.60}$$

$$S + \frac{3}{2}O_2 \rightarrow SO_3 \tag{3.61}$$

(2 lb/h S)/(MW S = 32.06) = 0.06238 lb mol S
Since 1 mol of S produces 1 mol of SO_x, 0.06238 lb mol S = 0.06238 lb mol SO_x.

(0.06238 lb mol SO_x)[385.3 SCF (68°F, 1 atm)/lb mol] = 24.04 SCF of SO_x
@ 0% O_2 = (24.04/349,220)(10^6) = 68.8 ppmv SO_x (dry)
@ 0% O_2 = (24.04/399,960)(10^6) = 60.1 ppmv SO_x (wet)
@ 3% O_2 dry = (24.04/407,750)(10^6) = 58.9 ppmv SOx (dry)
@ 3% O_2 dry = (24.04/461,150)(10^6) = 52.1 ppmv SO_x (wet)

Results at 0.1, 0.25, 0.5, and 1.0 wt% S are listed in Table 3.11.

Table 3.11 Calculation of Sulfuric Acid Dew Point at 2% SO_3/SO_x and Slightly Over 10% H_2O in Fuel Gas

Volumetric flow		Dry	Wet
Flue gas @ 0% O_2 (SCF/MMBtu)		9190.0	10,525.3
Flue Gas @ 3% O_2 dry (SCF/MMBtu)		10,730.2	12,135.5

Humidity of ambient air = 0.0209 mol H_2O/mol dry air (80°F, 60% RH)				
S in fuel wt% →	0.1	0.25	0.5	1.0
Concentration @ 0% O_2				
% H_2O = 12.69				
FG SO_x (ppmv, dry)	68.8	172.1	344.1	688.3
FG SO_x (ppmv, wet)	60.1	150.2	300.5	601.0
for SO_3/SO_x = 2%				
FG SO_3 (ppmv, dry)	1.4	3.4	6.9	13.8
FG SO_3 (ppmv, wet)	1.2	3.0	6.0	12.0
H_2SO_4 dew point				
°K	399	407	414	421
°C	126	134	141	148
°F	258	273	286	298
Concentration @ 3% O_2(dry)				
% H_2O = 11.58				
FG SO_x (ppmv, dry)	58.9	147.4	**294.7**	589.5
FG SO_x (ppmv, wet)	52.1	130.3	260.6	521.2
for SO_3/SO_x = 2%				
FG SO_3 (ppmv, dry)	1.2	2.9	5.9	11.8
FG SO_3 (ppmv, wet)	1.0	2.6	5.2	10.4
H_2SO_4 dew point				
°K	396	405	412	419
°C	123	132	138	145
°F	254	269	281	294

160 Chapter 3 Air Combustion

SO_x from the 0.05 wt% S allowable backup fuel [32] cited in Problem 3.25 is less than 30 ppmd @ 3% O_2 (dry), 1/10 of the result in Table 3.11 for 0.5 wt% S (in bold italics). How much of this SO_x is SO_2 and how much is SO_3 will be explored in a subsequent problem. Whatever be the breakdown, the molecular weight for SO_x is taken as 64.06, the molecular weight of SO_2, whenever a molecular weight is needed in calculations. ∎

PROBLEM 3.28

Complete the estimate for the flue gas composition of the previous problem.

SOLUTION The composition in terms of SCF/MMBtu is listed in the tables below, first for the stoichiometric case and then at 3% excess O_2 (dry). As in a stack test, nitrogen is obtained by difference.

If one wants to be fancy, argon can be estimated easily provided the assumption is made that all of the flue gas nitrogen comes from the combustion air. With an actual analysis of the fuel, nitrogen can be subtracted out first as in Problem 3.14 before the argon is estimated from the nitrogen–argon split in standard atmospheric air [5, p. F-210].

$$N_2 = 78.084\% \text{ and } Ar = 0.934\%, \ (N_2 + Ar) = 79.018\%$$

Factors $N_2/(N_2 + Ar) = (78.084/79.018) = 0.9882$

$Ar/(N_2 + Ar) = (0.934/79.018) = 0.0118$

For 0% O_2 $(0.9882)(7770) = 7678.3 \ N_2, \ (0.0118)(7770) = 91.7 \ Ar$

For 3% O_2 (dry) $(0.9882)(9130.2) = 9200.3 \ N_2, \ (0.0118)(9310.2) = 109.9 \ Ar$

Combustion Summary Tables

Constituent	SCF/MMBtu	vol% (wet)	vol% (dry)	Remarks
Estimate of flue gas composition: stoichiometric case				
Carbon dioxide	1420.0	13.49	15.45	From F factor table
Nitrogen + argon (N_2 + Ar)	7770.0	73.82	84.55	(F_d–F_c) (by difference)
Nitrogen (N_2)	7678.3	72.95	83.55	Proportionate in
Argon (Ar)	91.7	0.87	1.00	atmospheric air
Oxygen (O_2)	0.0	0.00	0.00	Stoichiometric case
Sulfur oxides (SO_x)	—	—	—	See Table 3.11
Total dry	9190.0	87.31	100.00	F_d
Water	1335.3	12.69	—	From previous problem
Total wet	10,525.3	100.00	—	By addition
Estimate of flue gas composition: 3% excess O_2 (dry)				
Carbon dioxide	1420.0	11.74	13.23	From F factor table
Nitrogen + argon (N_2 + Ar)	8988.3	74.29	83.77	Total dry–F_c by difference
Nitrogen (N_2)	8882.2	73.41	82.78	Proportionate in
Argon (Ar)	106.1	0.88	0.99	atmospheric air

Oxygen (O_2)	321.9	**2.66**	**3.00**	% Dry by definition, % wet previous problem
Sulfur oxides (SO_x)	—	—	—	See Table 3.11
Total dry	10,730.2	**88.42**	**100.00**	See below
Water	1368.3	**11.31**	—	From previous problem
Total wet	12,098.5	**100.00**	—	By addition

Total flue gas in the excess air case $= (9190)[(20.9-0)/(20.9-3)] = 10,730.2$

$$\frac{H_2O}{\text{Dry flue gas} + H_2O} = (11.31)/(100) = 11.31\%$$

Solve for H_2O:

$$H_2O = (0.1131)(\text{dry flue gas})/(1-0.1131)$$

$$H_2O = (0.1131)(10,730.2)/(0.8869) = 1368.3O_2$$

$$\%O_2 \text{ (wet)} = (100)[(321.9)/(10,730.2+1368.3)] = 2.66\%$$

$$(N_2 + Ar) = \text{Total dry} - CO_2 - O_2 = 10,730.2 - 1420 - 321.9 = 8988.3$$

Flue gas flows from firing 38 MMBtu/h (HHV) are summarized below.

Flue gas flow at condition	SCFH (dry)	SCFH (wet)
At 0% O_2	349,220	399,960
At 3% O_2	407,750	459,740

■

PROBLEM 3.29

Estimate the percentage of SO_3 in total sulfur oxides ($SO_2 + SO_3 = SO_x$).

SOLUTION By rule of thumb, SO_3 constitutes 1–3% of total sulfur oxides (SO_x). According to one source [41], up to 5% of the total fuel sulfur is converted to SO_3 in the flue gas. Why this should be so can be explained after the fact by equilibrium and kinetic arguments. However, accurate prediction of SO_3 *a priori* in any given case is more problematic.

The equilibrium between SO_2 and SO_3 with flue gas O_2 is defined by the following equations:

$$2SO_2 + O_2 \rightarrow 2SO_3 \qquad (3.63)$$

$$\frac{P_{SO_3}^2}{P_{SO_2}^2 P_{O_2}} = \frac{y_{SO_3}^2 P_T^2}{y_{SO_2}^2 P_T^2 y_{O_2} P_T} = \frac{y_{SO_3}^2}{y_{SO_2}^2 y_{O_2} P_T} = K_p \qquad (3.64)$$

where the P's represent partial pressures, P_T is the total pressure, and the y's are mole fractions in the vapor phase.

At 1 atm total pressure,

$$K_p = \frac{y_{SO_3}^2}{y_{SO_2}^2 y_{O_2}} \tag{3.65}$$

$$y_{SO_3}/y_{SO_2} = (K_p y_{O_2})^{1/2} \tag{3.66}$$

and

$$\%(SO_3/SO_x) = \frac{100[(K_p y_{O_2})^{1/2}]}{[1 + (K_p y_{O_2})^{1/2}]} \tag{3.67}$$

The equilibrium constant is given as a function of temperature in Ref. [42].

$$\log_{10}[K_p \text{ (atm)}] = (10,373)/T(°K) + 1.222 \log_{10}[T(°K)] - 13.5 \tag{3.68}$$

Tabulated values of $\log_{10} [K_p]$ [43] are consistent with this equation.

The % (SO_3/SO_x) calculated from 100°F to 3100°F (approximately 300–2000°K) for the chemical reaction of Equation (3.63), the equilibrium relationships of Equations (3.65) and (3.68), and a y_{O_2} value of 0.0266 (2.66% O_2 on a wet basis, corresponding to 3% O_2 on a dry basis) are listed in Table 3.12. Over this broad range, a plot of these values produces an s-shaped curve (not shown). Over a narrower range of interest (1500–3000°F, ~1100–1900°K), the shape is hyperbolic (Figure 3.8).

At low temperatures, the potential generation of SO_3 is huge; at high temperatures, it becomes tiny indeed. This is where kinetic theory comes in. Oxidation of SO_2 to SO_3 in the gaseous phase without the influence of a catalyst takes place only in a relatively narrow temperature range of 900–1050°C (very approximately 1200–1350°K, or 1700–2000°F); no significant oxidation has been observed below 900°C (~1173°K, ~1650°F) [44]. At high temperatures, the oxidation rate makes little difference since the equilibrium SO_3 is so small. This corresponds to an SO_3/SO_x ratio of 4.43–1.47% in Table 3.12. The temperature range of 1800–2100°F corresponds to the aforementioned rule of thumb.

The oxidation reaction is catalyzed by such agents as vanadium [45,46], high flue gas moisture content [45], and by nitrogen oxides (NO_x) in the flue gas [44]. Therefore, the range of significant oxidation can be somewhat broader.

The SO_3 concentration is a necessary input for prediction of the sulfuric acid dew point, and this will be treated in Problem 3.30. See also Problem 3.27 for an estimation of SO_3 concentration for various values of fuel sulfur of percentage composition of sulfuric oxides in the flue gas. ■

PROBLEM 3.30

Estimate the sulfuric acid dew point (T_{DP}) in the flue gas for the case of oil firing under discussion.

SOLUTION The acid dew point will be estimated from an empirical correlation of experimental data [40].

$$(1000)/T_{DP}(°K) = 1.7842 + 0.0269 \log_{10} [p_{H_2O}, \text{ atm}]$$
$$- 0.1029 \log_{10} [p_{SO_3}, \text{ atm}] + 0.0329 \log_{10} [p_{H_2O}, \text{ atm}] \times \log_{10} [p_{SO_3}, \text{ atm}] \tag{3.69}$$

Table 3.12 SO_2–SO_3 Equilibrium[a]

T (°F)	T (°K)	$\log_{10} [K_p]$[b]	Tabulated[c] $\log_{10} [K_p]$	K_p (atm)	SO_3/SO_2	SO_3/SO_x	% SO_3/SO_x
100.0	310.9	22.9080	—	8.1E + 22	4.6E + 10	1.000000	100.00
200.0	366.5	17.9379	—	8.7E + 17	1.5E + 08	0.999999	100.00
300.0	422.0	14.2870	—	1.9E + 14	2266464	0.999999	100.00
400.0	477.6	11.4936	—	3.1E + 11	90917.89	0.999989	100.00
500.0	533.2	9.2888	—	1.9E + 09	7182.129	0.999860	99.99
600.0	588.7	7.5054	—	32017194	921.5856	0.998916	99.89
700.0	644.3	6.0339	—	1081061	169.3437	0.994129	99.41
800.0	699.8	4.7996	—	63036.67	40.89222	0.976129	97.61
900.0	755.4	3.7500	—	5623.297	12.21348	0.924319	92.43
982.1	801.0	2.9989	2.9920	997.4292	5.143812	0.837234	83.72
1000.0	810.9	2.8469	—	702.8770	4.318011	0.811959	81.20
1073.0	852.0	2.2565	2.2800	180.4953	2.188149	0.686338	68.63
1100.0	866.5	2.0619	—	115.3211	1.749035	0.636236	63.62
1160.0	900.0	1.6362	1.6320	43.27436	1.071419	0.517239	51.72
1200.0	922.0	1.3736	—	23.63656	0.791837	0.441913	44.19
1255.7	953.0	1.0256	1.0220	10.60772	0.530463	0.346603	34.66
1300.0	977.6	0.7653	—	5.825208	0.393097	0.282174	28.22
1340.3	1000.0	0.5396	0.5400	3.464176	0.303140	0.232623	23.26
1400.0	1033.2	0.2241	—	1.675247	0.210806	0.174104	17.41
1451.9	1062.0	−0.0341	−0.0360	0.924590	0.156609	0.135404	13.54
1500.0	1088.7	−0.2605	—	0.548959	0.120674	0.107679	10.77
1515.0	1097.0	−0.3288	—	0.469043	0.111545	0.100351	10.04
1529.3	1105.0	−0.3931	−0.4040	0.404508	0.103587	0.093864	9.39
1600.0	1144.3	−0.6966	—	0.201079	0.073034	0.068063	6.81

(*continued*)

Table 3.12 (Continued)

T (°F)	T (°K)	$\text{Log}_{10} [K_p^b]$	Tabulated[c] $\log_{10} [K_p]$	K_p (atm)	SO_3/SO_2	SO_3/SO_x	% SO_3/SO_x
1646.3	1170.0	−0.8843	−0.8880	0.130541	0.058846	0.055575	5.56
1700.0	1199.8	−1.0912	—	0.081055	0.046369	0.044314	4.43
1800.0	1255.4	−1.4498	—	0.035498	0.030686	0.029772	2.98
1900.0	1310.9	−1.7770	—	0.016711	0.021054	0.020620	2.06
2000.0	1366.5	−2.0766	—	0.008382	0.014911	0.014692	1.47
2100.0	1422.0	−2.3521	—	0.004445	0.010859	0.010742	1.07
2200.0	1477.6	−2.6060	—	0.002477	0.008106	0.008041	0.80
2300.0	1533.1	−2.8408	—	0.001442	0.006186	0.006148	0.61
2400.0	1588.7	−3.0585	—	0.000874	0.004815	0.004792	0.48
2500.0	1644.3	−3.2608	—	0.000548	0.003814	0.003799	0.38
2600.0	1699.8	−3.4494	—	0.000355	0.003070	0.003060	0.31
2700.0	1755.4	−3.6254	—	0.000236	0.002506	0.002500	0.25
2800.0	1810.9	−3.7902	—	0.000162	0.002073	0.002069	0.21
2900.0	1866.5	−3.9446	—	0.000113	0.001735	0.001732	0.17
3000.0	1922.0	−4.0897	—	0.000081	0.001468	0.001466	0.15
3100.0	1977.6	−4.2262	—	0.000059	0.001255	0.001253	0.13

[a] Equilibrium relationship is based on $2SO_2 + O_2 \leftrightarrow 2SO_3$ from Equation (3.63).
[b] Equation for equilibrium constant is from Ref. [42].

$$\log_{10}[K_p \text{ (atm)}] = (10{,}373)/T\,(°K) + 1.2222\log_{10}[T\,(°K)] - 13.5 \quad (3.68)$$

[c] Tabulated values are from Ref. 43 for the reaction $SO_2 + 1/2\,O_2 \leftrightarrow SO_3$, multiplied by 2.

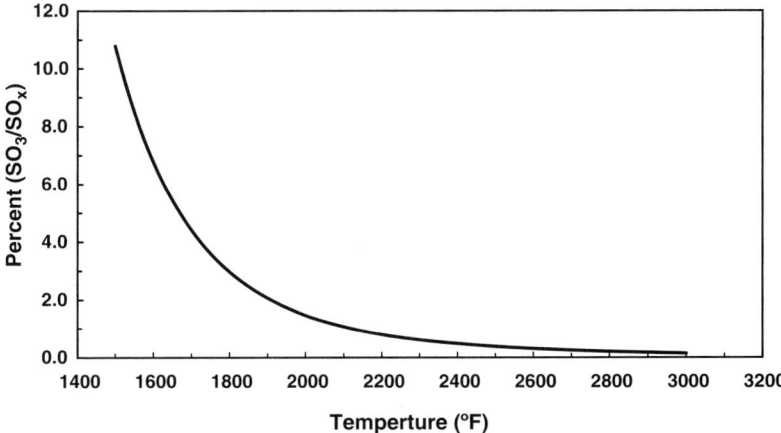

Figure 3.8 Equilibrium percent (SO_3/SO_x) versus temperature (°F) in fuel oil flue gas at 2.66% O_2 wet (3% O_2 dry).

It is based on a previous equation developed by other researchers [47].

The entries are partial pressures (mole fractions times the total pressure in atmospheres) to produce a temperature in °K. This means that % H_2O is entered as (% H_2O/100) (1 atm) and ppm SO_3 as (ppm $SO_3/10^6$) (1 atm). Graphs and an example problem are given in the cited reference and serve as a check that one has plugged in the proper numbers in a proper manner. Dew point as a function of temperature is also shown in Refs [41,48,49].

Results of this exercise, with dew point temperatures to the nearest degree, are listed in Table 3.11. The concentration of SO_3 is estimated as 2% of SO_x calculated from the wt% sulfur in the fuel. Wet basis concentrations for H_2O and SO_3 have been employed.

For example, for the excess air flue gas at 3% O_2 (dry) in Table 3.11,

$$\% H_2O = 11.58, \text{ and ppm } SO_3 \text{ wet for } 0.1 \text{ wt\% S} = (52.1)(0.02) = 1.0$$

and

$$(1000)/T_{DP}(°K) = 1.7482 + 0.0269 \log_{10}[(11.58/100)(1)]$$

$$0.1029 \log_{10}[(52.1)(0.02)(1)/(10^6)] + 0.0329 \log_{10}[(11.58/100)(1)]$$
$$\times \log_{10}[(52.1)(0.02)(1)/(10^6)]$$

$$(1000)/T_{DP}(°K) = 1.7482 - 0.0251862 + 0.0615542 + 0.184296 = 2.522851$$

and

$$T = (1000/(2.522851) = 396.38°K \rightarrow 396°K$$
$$T(°C) = 396.38 + 273.15 = 123.23°C \rightarrow 123°C$$
$$T(°F) = (396.38)(1.8) - 459.67 = 253.81°F \rightarrow 254°F$$

The author of the correlation quotes people as saying that they design to maintain the temperature of a steel stack from 275°F (135°C) to 300°F (149°C) all the way to the top. The upper temperature would be necessary to stay above the dew points calculated for this exercise. Dew points in Table 3.11 can be seen to increase with both the percent moisture and the SO_3

concentration. For a case involving 10 wt% S in the fuel and SO_3/SO_x as high as 5% and the same flue gas moisture, the design stack temperature would have to be about 350°F to keep above the calculated value (not shown). That exercise is left to the reader.

That author also states that in cases where hydrochloric acid (HCl) is present along with SO_3/H_2SO_4, the sulfuric acid controls since the HCl dew point will be lower. Equations to predict the dew point of hydrochloric acid, hydrobromic acid (HBr), nitric acid (HNO_3), and sulfurous acid (H_2SO_3) are provided in Ref. [50].

There is a balance that must be struck between having a low flue gas temperature in the interest of combustion efficiency and a high enough temperature for stack gas buoyancy and to remain above the sulfuric acid dew point. The prudent designer will consider all of these (and possibly other) factors. However, staying above the sulfuric acid dew point should become easier as the push continues to burn lower-sulfur fuels.

For the flue gas scrubbing example (Problem 2.37) of Chapter 2, the adiabatic saturation temperature is well below the sulfuric acid dew point and the proper corrosion resistant materials must be used [45,49]. Condensation of the typical blue-gray sulfuric acid mist outside the stack during certain weather conditions is also of concern [42,45]. ■

3.12 THE STACK TEST

Sometime within a specified period after start-up, mandatory stack testing will be conducted to demonstrate compliance with permit conditions and/or to verify the relative accuracy (RA) of a continuous emission monitoring system (CEMS), if one has been installed (Section 3.13). Repeat compliance testing can be expected at least annually, more frequently in special circumstances. Testing consists of a series of runs, on the order of one half-hour to one hour each, whose outcomes are averaged to obtain the final result. A typical standard for compliance testing is three 1 h runs. With proper planning, the test data can be used to satisfy both compliance demonstration and CEMS requirements. For a process operating at true steady state, the averages after 5 min do not change significantly.

An excellent schematic representation of stack-sampling facilities is available from the Texas Air Control Board (TACB) and its successor agencies—the Texas Natural Resource Conservation Commission (TNRCC, affectionately known as *Train Wreck*) and the Texas Commission on Environmental Quality (TCEQ). This diagram [51] depicts size, location, and design of stack sampling ports; work platform access, specifications, dimensions, clearances, guardrails, and support requirements; and the location of electrical power sources. It is a handy reference for illustrative purposes for any location, but the latest information/requirements for the jurisdiction in which an actual source to be tested is located should always be consulted before proceeding.

The following EPA methods [52] or other authorized equivalent test procedures for state or other air quality management/control agencies are commonly employed:

Method 1	Location of velocity-traverse and sampling points for stationary sources
Method 2	Determination of stack gas velocity and volumetric flow rate by pitot tube traverse

Method 3	Determination of O_2, CO_2, CO, and dry molecular weight (including Orsat analysis)
Method 4	Determination of moisture content in stack gas by gravimetric or volumetric analysis of condensate
Method 5	Determination of particulate emissions from stationary sources by isokinetic sampling
Method 6	Determination of sulfur dioxide emissions from stationary sources
Method 7	Determination of nitrogen oxide emissions (NO_x) from stationary sources[7]
Method 8	Determination of sulfuric acid mist, sulfur trioxide, and sulfur dioxide from stationary sources
Method 9	Visual determination of stack-gas opacity by a qualified observer
Method 10	Determination of carbon monoxide (CO) by means of a nondispersive infrared analyzer
Method 18	Measurement of gaseous organic compound emissions by gas chromatography
Method 20	Determination of nitrogen oxides, sulfur dioxide, and diluent emissions from stationary gas turbines
Method 25	Determination of total gaseous nonmethane hydrocarbons as carbon. Method 25A uses a flame ionization analyzer; Method 25B a nondispersive infrared analyzer

Certain approved procedures for more specialized sources, such as those emitting lead, fluoride, reduced sulfur species, and so on, are not listed here.

Most of the methods involve a sampling train with several impinger vessels in an ice bath. Initial contents of the impingers are specific to the test method. More than one constituent can be determined by appropriate modifications to the impinger train. The reader is encouraged to consult the latest official test procedures before making use of any of the methods enumerated above.

Test protocols are normally conducted by a professional stack-testing contractor hired specifically for this purpose. Stack sampling is a skill developed with practice [53], and some people are more adept than others. (And then there is the true story of the stack testing technician who showed up for the tests but was afraid of heights.)

It is best to verify ahead of time that someone on the test crew is trained as a *qualified observer*, if opacity must be determined (Method 9).

One can also have the stack-testing personnel measure the ambient wet-bulb and dry-bulb temperatures periodically during the test program for use in combustion calculations to check the consistency of the data [13].

It is good policy not to let the stack tests run themselves. Invariably, there seems to be at least one situation where an on-the-spot decision must be made immediately lest a test run be aborted or the entire test program grind to an untimely halt. What has

---

[7] In the author's experience, continuous extraction and analysis by Method 7E results in a smaller standard deviation among replicate runs than wet-chemical analysis of grab samples by the Method 7 procedure.

worked well in the author's experience is to have an individual on hand who does not have any duties in running the plant and who will not be called away to attend to other matters. An additional person from the plant responsible for plant liaison is always helpful.

It is best to send someone who can keep a straight face when announcing to plant personnel, "I'm from Headquarters; I'm here to help you." It is also a good idea to send someone to the plant who is familiar with the chemical manufacturing process.

Most of the cost of stack testing by a third party contractor is in mobilization, travel to/from the plant site, set up, and breakdown. It is, therefore, possible for a relatively small incremental cost to make additional measurements or conduct a few additional test runs beyond the bare minimum for compliance. These often prove useful in understanding perturbations around the design test point (for this or subsequent design) or in establishing a parameter monitoring system. A "wish list" for stack test measurements is contained at the end of Appendix C.

3.12.1 Test Methods and Calculations

This section illustrates how stack test measurements are turned into flows and concentrations used to assess compliance with permit conditions. We start first with Methods 3 and 4 because results for flue gas MW and percent moisture (H_2O) are necessary in the calculation of stack gas velocity and volumetric flow rate (Method 2). Sampling for individual pollutant concentrations can then be combined with volumetric flow rate to obtain pollutant mass flow rates.

Even if the required calculations are performed by the stack test contractor, one should be able to check those calculations and verify their accuracy. For any real case, one should always consult the latest version of the applicable regulations of the controlling jurisdiction and perform the calculations accordingly.

EPA Method 3 describes procedures for sampling and analysis of a flue gas stream of a fossil fuel combustion process. A sample, filtered to remove particulate matter and cooled to condense moisture, is removed from the stack into a leak-free flexible plastic bag made, for example, of tedlar, mylar, teflon, or aluminized mylar, and so on. A typical capacity of the bag is in the range of 55–90 L consistent with the flow rate and length of time selected for a test run. The sample is obtained by single-point grab sampling, single-point integrated sampling, or multiple-point integrated sampling. The contents of the bag are analyzed using either an Orsat analyzer or a Fyrite-type combustion gas analyzer. These are discussed below.

Method 3A uses an instrumental analysis procedure standardized with calibration gases. This method can be used to determine excess air or to obtain an emission rate correction factor, normally by integrated sampling and passing a dried gas sample directly through the instrument in real time during a pollutant emission testing run.

Method 3B is similar to Method 3 but with the same purpose as Method 3A, namely, to determine CO_2, O_2, and CO for excess air or emission rate correction factor calculations. This method notes that a Fyrite-type combustion gas analyzer is not acceptable for excess air or emission rate correction factor determinations unless approved by the EPA Administrator, and it goes on to mention the Orsat analysis.

A formula for percent excess air (% EA) is listed as Equation 3B-3 in Method 3B.

$$\% \text{ EA} = \frac{\%O_2 - 0.5\% \text{ CO}}{0.264\%N_2 - (\% O_2 - 0.5\% \text{ CO})} \times 100 \qquad (3.70)$$

This equation assumes that ambient air is used as the source of O_2 and that the fuel does not contain appreciable amounts of N_2. Coal, oil, and natural gas are not considered to contain appreciable amounts of N_2 in contrast to coke oven or blast furnace gases.

3.12.1.1 Orsat Analysis

The Orsat analyzer determines carbon dioxide, oxygen (O_2), and CO in the sample. Nitrogen (N_2) is calculated by difference. With the primary constituents of the flue gas thus determined, the molecular weight of the gas can be calculated.

In an Orsat analyzer, a gas sample of known volume is passed in turn through a series of absorption pipettes filled with prepared solutions. The first contains potassium hydroxide, which absorbs all of the CO_2 but none of the O_2 and CO. The volume of the gas sample decreases when the gas is passed through alkaline pyrogallol to remove O_2; it decreases once again after cuprous chloride removes the CO [15, pp. 52–53]. A photograph of an Orsat analyzer can be found on the Internet [54], and a simplified Orsat diagram is shown in Perry's Handbook [21, p. 1304].

An Orsat analysis is capable of determining CO_2 and O_2 in the percentage range typical of combustion effluents. CO can be determined by Orsat in the whole number percent range down to the tenths of a percent range [15, pp. 52–53]. For CO_2 less than 4.0% or O_2 greater than 15.0%, EPA Method 3B states that the Orsat measuring burette must be marked with at least 0.1% subdivisions.

EPA Method 10 using a Luft-type nondispersive infrared analyzer (NDIR) or equivalent with a range of 0–1000 ppm (0.1%) and a minimum detectable concentration of 20 ppm CO for a 0–1000 ppm span is more appropriate for determining ppm concentrations of CO. A Method 10 sample is likewise filtered for removal of particulate matter and treated for removal of water and CO_2, which interfere in the instrumental analysis of CO. Certified calibration gases containing known concentrations of CO in nitrogen (N_2) are used to calibrate the instrument.

3.12.1.2 Fyrite Analysis

The Fyrite-type volumetric gas analyzer absorbs carbon dioxide or oxygen (O_2), each separately in a special reagent contained within a closed plastic canister [55,56]. The solution for CO_2 absorption is potassium hydroxide with a red dye, while the

170 Chapter 3 Air Combustion

solution for O_2 is chromous chloride and is blue. These captive fluids may be used many times until they become exhausted and need replacement. They are, however, poisonous, corrosive, and otherwise hazardous, and care must be exercised in their use.

When a gas sample is admitted into a Fyrite canister, its volume change is related to the amount of CO_2 or O_2 absorbed from the sample. Concentration is read directly on a scale mounted on the side of the canister. The Fyrite reagents are each available in three different strengths; the range of 0–20% CO_2 and 0–21% O_2 is appropriate for analysis of combustion effluents. They are calibrated for analysis of gases saturated with moisture. Accuracy is said to be within ±0.5% CO_2 or O_2 [56].

Fyrite does not measure CO. This is not important in the determination of flue gas molecular weight since the balance of the flue gas composition is assumed to be nitrogen (N_2). This balance is obtained by difference of 100% and the sum of CO_2 and O_2, and the molecular weights of N_2 and CO to be used in the calculation of flue gas molecular weight are the same at 28.01.

Details of construction and operation [55], schematics [55], and pictures [56] are available on the Internet.

PROBLEM 3.31

(Using "data" from Ref. [15, pp. 52–53])

Calculate the dry basis percentage of CO_2, O_2, and CO in a (wet) flue gas sample from the following information obtained in an Orsat analysis:

Sample temperature (°F)	75
Sample pressure (mmHg)	758
Sample volumes (cm³)	
Original volume	100.0
After CO_2 removal	87.9
After O_2 removal	80.7
After CO removal	80.5

SOLUTION Vapor pressure of water at 75°F is 0.42964 psia [6, p. 88] or 22.2 mmHg. Partial pressure of the dry gases is then

$$758 - 22.2 = 735.8 \text{ mmHg}$$

This correction is important for determining the water vapor content of the original sample on a wet gas basis. However, it cancels out in the calculation of the dry basis percentages of the other constituents since some of the water vapor condenses at each step in the Orsat analysis, and the remaining sample continues to be saturated with water vapor. If, as expected, the sample behaves as an ideal gas mixture, percent by volume on a dry basis can be found by dividing the

3.12 The Stack Test

volume of each constituent by the total volume and multiplying by 100. The volume of the water after condensation is negligible.[8]

Volume occupied

$$\begin{aligned}
\text{by } CO_2 &= 100 - 87.9 = 12.1 \text{ cm}^3 \\
\text{by } O_2 &= 87.9 - 80.7 = 7.2 \text{ cm}^3 \\
\text{by } CO &= 80.7 - 80.5 = 0.2 \text{ cm}^3
\end{aligned}$$

$$\begin{aligned}
&= (100)(12.1/100) &&= 12.1\% \; CO_2 \text{ (dry)} \\
&= (100)(7.2/100) &&= 7.2\% \; O_2 \text{ (dry)} \\
&= (100)(0.2/100) &&= 0.2\% \; CO \text{ (dry)} \\
&\text{Total } (CO_2 + O_2 + CO) &&= 19.5\% \text{ (dry)}
\end{aligned}$$

The remainder of the dry gas is normally assumed to be N_2.

$$N_2 = 100 - 19.5 = 80.5\% \; N_2 \text{(dry)}$$

This point will be explored in Problem 3.32.

$$\text{mol } H_2O/\text{mol dry gas at sample conditions} = p_w/(P_T - p_w) \quad (3.2)$$
$$= (22.2)/(758 - 22.2 = 735.8) = 0.0302$$

Caution: This is not the water vapor content of the flue gas since the gas has been cooled from flue gas conditions to sample conditions without accounting for condensed moisture. Flue gas moisture would be measured using an EPA Method 4, 5, 6, or 8 sampling train. ■

PROBLEM 3.32

Calculate the dry molecular weight of the flue gas whose dry basis composition was determined by Orsat analysis in Problem 3.31, assuming (a) the concentration calculated by difference is nitrogen (N_2), and (b) that it is a mixture of nitrogen (N_2) and argon but that there is no nitrogen contained in the fuel.

SOLUTION

(a) Argon is ignored.
    Basis: 100 mol (dry) of flue gas

Component	MW	vol or mol%	Moles	wt = MW × mol	wt%
Carbon dioxide	44.01	12.1	12.1	532.521	17.61
Oxygen (O_2)	32.00	7.2	7.2	230.400	7.62

---

[8] By the ideal gas law, moles of dry gas is given by

$$n = (PV)/(RT)$$
$$n = [(735.8/760) \text{ atm}(100/1000)L]/[(0.08205)(297K)] = 0.0040 \text{ mol dry gas}$$
$$(0.0040 \text{ mol dry gas})(0.0302 \text{ mol } H_2O/\text{mol dry gas})(18.02 \text{ g/g mol}) = 0.0022 \text{ g } H_2O$$
$$0.0022 \text{ g } H_2O \times 1 \text{ g/cm}^3 = 0.0022 \text{ cm}^3 \text{ condensed, compared to } 100 \text{ cm}^3.$$

Carbon monoxide	28.01	0.2	0.2	5.602	0.19
Nitrogen (N$_2$)	28.01	80.5	80.5	2254.805	74.58
Total	—	100.0	100.0	3023.328	100.00

MW = total wt/total mol = (3023.328)/(100) = 30.23 = 30.2, to the nearest 0.1 g/g mol or lb/lb mol as specified in EPA Method 3.

The above spreadsheet simplifies to EPA Method 3, Equation 3.1, using molecular weights rounded to the nearest whole number with argon missing.

$$M_d = 0.440(\%CO_2) + 0.320(\%O_2) + 0.280(\%N_2 + CO)$$
$$= 0.440(12.1) + 0.320(7.2) + 0.280(80.5 + 0.2) \quad (3.1)$$
$$= 30.2240 = 30.2 \text{ rounded to the nearest } 0.1 \text{ g/g mol or lb/lb mol}$$

(b) Argon is included but no nitrogen in fuel.

If there were no nitrogen in the fuel, all of the N$_2$ in the flue gas would have originated from atmospheric air. In that case, using the dry air analysis of Table 3.3,

$$N_2 = [(78.086)/(78.086 + 0.934)](\text{Total of } N_2 + Ar)$$
$$= [78.086/79.02](80.5) = (0.9882)(80.5) = 79.55\% \text{ versus } 80.5\%$$

and

$$Ar = [(0.934)/(78.086 + 0.934)](\text{Total of } N_2 + Ar)$$
$$= [0.934/79.02](80.5) = (0.0118)(80.5) = 0.95\% \text{ versus } 0\%$$

Back to the spreadsheet,
Basis: 100 mol (dry) of flue gas

Component	MW	vol or mol%	Moles	wt = MW × mol	wt%
Carbon dioxide	44.01	12.1	12.1	532.521	17.30
Oxygen (O$_2$)	32.00	7.2	7.2	230.400	7.62
Carbon monoxide	28.01	0.2	0.2	5.602	0.19
Nitrogen (N$_2$)	28.01	79.55	79.55	2228.196	73.64
Argon	39.95	0.95	0.95	37.953	1.25
Total	—	100.0	100.0	3025.672	100.00

MW = total wt/total mol = (3025.672)/(100) = 30.26
= 30.3 to the nearest 0.1 g/g mol or lb/lb mol as specified in EPA Method 3

By modification of EPA Method 3, Equation 3.1, using rounded molecular weights and EPA's molecular weight of 39.9 for argon,

$$M_d = 0.440(\% CO_2) + 0.320(\% O_2) + 0.280(\% N_2 + CO_2) + 0.399(\%Ar)$$
$$= (0.440)(12.1) + (0.320)(7.2) + 0.280(79.55 + 0.2) + 0.399(0.95) \quad (3.1)$$
$$= 30.3371 = 30.3 \text{ rounded to the nearest } 0.1 \text{ g/g mol or lb/lb mol}$$

Method 3 indicates that the tester may choose to include argon in the analysis subject to approval of the EPA Administrator.

If the molecular weight of dry flue gas (M_d) is approximated using $M_d = 30.0$ by rule of thumb, the molecular weight on a wet basis (M_w) needed in EPA Method 2, Equation 2, can then be estimated by substituting a value of 30.0 for M_d in Equation 3.46 or Equation 3.47 to yield

$$M_w = 0.300(100 - \% \, H_2O) + 0.180(\% \, H_2O)$$
$$= 30.0 - 0.120\% \, H_2O \tag{3.71}$$

or

$$M_w = 30.0 - 30.0 B_{ws} + 18 B_{ws}$$
$$= 30.0 - 12.0 B_{ws} \tag{3.72}$$

Local values of the velocity in a stack or duct are measured sequentially at a series of points traversed in a plane across the stack or duct. The traverse points are located according to EPA Method 1. The velocity is determined from differential pressure measurements by pitot tube (EPA Method 2).

The pitot tube is named after a French physicist, Henri de Pitot, who originated the device and pioneered the use of a bent glass tube in 1730 for measurement of velocities in the River Seine [57–59]. It measures the difference between the stagnation pressure facing upstream in a flowing stream and the static pressure at another point out of the direct flow.

Velocity in a gas below 200 ft/s is proportional to the square root of the pressure difference (Δp) [3, p. 5–8]. The lower limit of velocity measurement in air at 1 atm pressure is about 15 ft/s because of the very small differential pressures involved (on the order of 0.045 in. H_2O) [3, p. 5–8].

The individual velocity determinations are averaged over the cross section of the flue, and the volumetric flow rate is computed by multiplying by the cross-sectional area. A good stack tester will verify by measurement the cross section of each stack for every stack testing campaign.

In the author's experience, experimental flow rate by pitot tube traverse is usually less than the value obtained from combustion calculations. Agreement in flow to a maximum of 15–20% might be considered good with an accurate fuel flow rate and analysis; as a rule, agreement for concentrations of major flue gas constituents is much better. ■

PROBLEM 3.33

Compute the average value of velocity head (Δp) for the flow traverses listed in the facsimile of the Method 2 data sheet below. The Δp values are given in the Ref. [60, p. 27]. (These values have been listed here by increasing numerical value, rather than by location in the stack.)

Traverse point number	Δp (in. H_2O)	$(\Delta p)^{1/2}$ (in. $H_2O)^{1/2}$
1	0.18	0.4243
2	0.18	0.4243
3	0.19	0.4359
4	0.19	0.4359
5	0.21	0.4583
6	0.22	0.4690

174 Chapter 3 Air Combustion

7	0.22	0.4690
8	0.23	0.4796
9	0.26	0.5099
10	0.26	0.5099
11	0.27	0.5196
12	0.28	0.5292
13	0.28	0.5292
14	0.29	0.5385
15	0.29	0.5385
16	0.30	0.5477
Sum	—	7.8188
Average	0.239	0.4887 → 0.489

SOLUTION The square root of Δp is averaged for the 16 traverse points, not Δp itself.

$$\text{Average } (\Delta p)^{1/2} = (1/n)\sum_{i=1}^{n}(\Delta p_i)^{1/2} \qquad (3.73)$$
$$= (1/16)(7.8188) = 4.887 \to 0.489$$

The average Δp is then obtained by squaring the average square root.

$$\text{Average } \Delta p = (\text{Average } \Delta p^{1/2})^2 \qquad (3.74)$$
$$= 0.489^2 = 0.239$$

∎

PROBLEM 3.34

(Based on data given in Problem 3.33 adapted from Ref. [60, pp. 26–28])

Determine the average effluent velocity and volumetric flow from the following additional measurements:

Method 2

Stack diameter 4.5 ft

Cross-sectional area 15.9 ft^2

Barometric pressure 27.8 in. Hg

Pressure differential between barometric pressure and stack pressure 0 in. H$_2$O (To convert from in. H$_2$O to in. Hg, multiply by 0.0735539.)

Average stack temperature 240°F

Pitot tube $(\Delta p)^{1/2}$ (from Problem 3.33) (arithmetic average from 16 traverse points)[9]
= 0.489 (in. H$_2$O)$^{1/2}$

---

[9] From EPA Method 1, the minimum number of traverse points for stacks greater than 24 in. (2 ft) in diameter is 12 when sampling is performed at a site located at least 8 diameters downstream and 2 diameters upstream of any flow disturbance such as a bend, expansion or contraction in the stack, or from a visible flame. This includes the stack exit. For sampling points not meeting the above criteria, the minimum number of traverse points is to be determined from charts in Method 1.

(Note once again that the square root of Δp is averaged from the traverse point data, not Δp itself. The average Δp is computed by squaring the average square root. Here, $0.489^2 = 0.239$ in. H_2O.)

Method 3
The specific gravity of the wet flue gas is stated to be 1.0 with respect to (dry) air, making its molecular weight equal to 28.96.

Method 4/Method 5
Moisture (H_2O) content 18% by volume.

SOLUTION From Equation (2.9) in EPA Method 2,

$$v_s(\text{ft/s}) = K_p C_p (\Delta p_{avg})^{1/2} \{(T_{S_{avg}})/[(P_s)(M_s)]\}^{1/2} \quad (3.75)$$

where

v_s = average stack velocity, $K_p = 85.49 \text{ ft/s} \left[\frac{(\text{lb/lb mol})(\text{in. Hg})}{(°R)(\text{in. } H_2O)}\right]^{1/2}$

C_p = dimensionless pitot tube coefficient (nominally 0.84 for a Type S pitot tube [EPA Method 2] and 0.99 for a standard pitot tube design [EPA Method 2]. Other references give a range of 0.98–0.995 [59] or 0.98–1.00 [3, p. 5–8] for the standard design.

Δp = velocity head of stack gas (in. H_2O)

T_s = absolute temperature of stack gas (°R)

P_s = absolute stack pressure (in. Hg)

M_s = molecular weight of stack gas (wet basis)

The pitot tube constant (K_p) is a combination of the gravitational constant (g), the gas constant in the ideal gas equation, and conversion factors to enable measurements in the selected units to be employed.

Substituting numbers in Equation (3.75),

$$v_s = (85.49)(C_p)(0.239)^{1/2} \left[\frac{(240+460)\,°R}{(27.8 \text{ in. Hg})(28.96)}\right]^{1/2}$$

$$v_s = (85.49)(C_p)(0.489)(0.9325) = 39.0 C_p \text{ft/s}$$

$$= (39.0)(0.99) = 38.6 \text{ ft/s}$$

if these data were obtained using a standard pitot tube.

$$(39.0)(0.84) = 32.8 \text{ ft/s}$$

for data from Type-S pitot tube compared to the 38.9 ft/s calculated from the slightly different formula given in Ref. [60, p. 15] for a standard design pitot tube.

The velocity calculated above is on a wet basis (including moisture). To find wet basis volumetric flow, multiply by the cross-sectional area of the stack. Using the velocity from the type-S pitot tube, $(32.8 \text{ ft/s})(15.9 \text{ ft}^2)(60\text{s/min}) = 31,300$ actual cubic feet per minute (ACFM).

To convert to EPA Method 19 standard conditions of 68°F (528°R) and 29.92 mmHg pressure, use the relationship for an ideal gas from Chapter 2:

31,300 ACFM $(528/700)(27.8/29.92) = 21,900$ standard cubic feet per minute (SCFM), still on a wet basis.

Since pollutant concentrations emitted from the stack are measured on a dry basis after flue gas moisture is removed, dry basis flow is also needed. This is calculated using Equation (3.24).

176   Chapter 3   Air Combustion

$$Q_{dry} = Q_{wet}[(100-\% \text{ H}_2\text{O})/100]$$
$$= 31,300 \text{ ACFM}_{(wet)}[(100-18)/100] = 25,700 \text{ ACFM}_{(dry)} \quad (3.24)$$
$$= 21,900 \text{ SCFM}_{(wet)}[(100-18)/100] = 18,000 \text{ SCFM}_{(dry)}$$

EPA denotes the factor [(100 − % H$_2$O)/100] as (1 − B_{ws}) in both Method 2 and Method 19. This makes B_{ws} equal to the moisture content as a decimal, in this case, 0.18.

EPA Reference Methods 4 or 5 can be used to determine the moisture content of a flue gas. Method 5 allows simultaneous determination of particulate matter and moisture. Moisture may also be determined along with particulate matter and sulfur dioxide (SO$_2$) by modifying a Method 5 train per Method 6 or Method 8. These methods include a series of impingers contained in an ice bath.

In the Method 5 train, each of the first two impingers contains 100 mL of water, the third is initially empty, and the fourth contains approximately 200–300 g of preweighed silica gel. The purpose of the silica gel is to remove the residual water vapor not condensed in the impinger upstream. A gas sample is extracted from the stack by a vacuum pump located downstream of the impingers, and temperature, pressure, and flow at locations in the sampling line are measured.

In the author's experience, moisture determination via stack test compares favorably with that obtained from combustion calculations. However, extraction of a moisture sample, handling, and manual determination of liquid weight or volume has the potential to result in a moisture concentration as much as one percentage point above or below the calculated value, in what is hoped to be the worst case. ∎

PROBLEM 3.35

Calculate the moisture content of flue gas according to the Reference Method of EPA Method 4 or 5 from the following hypothetical stack test measurements.

In this test run, final volume of water collected in the impingers was recorded as 337 mL, starting with an initial charge of 100 mL in each of the first two impingers. Silica gel in the fourth impinger picked up 17.2 g of water weight. Temperature and pressure at the exit of the impingers/at the meter were measured at 65°F and 660 mmHg, respectively. A total dry gas sample of 1.0840 m^3 (uncorrected) was collected over a 1 h period of sampling time. The sampling train met the criteria of the mandatory leak test procedure.

SOLUTION

1. Calculate volume of water vapor condensed ($V_{wc\ (std)}$).

$$(V_{wc(std)}) = \text{volume of water vapor condensed corrected to standard conditions}$$

where

V_f, V_i = total final and initial volumes, respectively, of water in the impingers (mL)

ρ_w = density of liquid water (0.9982 g/mL)

R = ideal gas constant (0.06236 mmHg m^3/(g mol °K))

T_{std}, P_{std} = standard absolute temperature and pressure, respectively, (293°K, 528°R) and (760 mmHg, 29.92 mmHg)

M_w = molecular weight of water (18.0 g/g mol or lb/lb mol)

(3.76)

$$(V_{wc(std)}) = \frac{(V_f - V_i)\rho_w RT_{std}}{P_{std}M_w} \quad \text{(Method 4 Equation 4-1)}$$
$$= (337-200)(0.9982)(0.06236)(293)/[(760)(18)]$$
$$= 0.1827 \text{ m}^3 \text{ of } H_2O \text{ collected in impingers (vapor equivalent)} \quad (3.76)$$

2. Calculate volume of water collected on silica gel ($V_{wsg\ (std)}$).

$(V_{wsg\ (std)})$ = volume of H_2O collected on silica gel corrected to standard conditions

and

W_f, W_i = final and initial weights, respectively, of silica gel or silica gel plus impinger

Other factors have been defined above.

$$(V_{wsg\ (std)}) = \frac{(W_f - W_i)RT_{std}}{P_{std}M_w} \quad \text{(Method 4 Equation 4-2)}$$
$$= (17.2)(0.06236)(293)/[(760)(18)] = 0.0230 \quad (3.77)$$

3. Calculate dry gas volume collected by dry gas meter at standard conditions ($V_{m\ (std)}$), where

$(V_{m\ (std)})$ = (dry gas volume collected by dry gas meter corrected to standard conditions)

V_m, P_m, T_m = actual volume, pressure, and temperature, respectively, at the dry gas meter

Y = meter calibration factor (dimensionless)

Other factors have been defined above.

$$(V_{m(std)}) = \frac{V_m Y P_m T_{std}}{P_{std} T_m} \quad \text{(Method 4 Equation 4-4)}$$
$$= (1.0840)(1)(660)(528)/[(760)(525)] = 0.9467 \text{ m}^3 \quad (3.78)$$

As long as the same pressure units are used for P_m and P_{std} and the same temperature units are used for T_{std} and T_m, it makes no difference what units are employed. The units of $V_{m\ (std)}$ will be the same as those of V_m. Here the units of $V_{m\ (std)}$ are cubic meters at standard conditions. Typically, Y is close to but not identically equal to 1.0.

4. Calculate fractional moisture content (B_{ws}), with factors as defined above.

$$B_{ws} = \frac{V_{wc\ (std)} + V_{wsg\ (std)}}{V_{wc\ (std)} + V_{wsg\ (std)} + V_{m\ (std)}} \quad \text{(Method 4 Equation 4-4)}$$
$$= \frac{0.1827 + 0.0230}{0.1827 + 0.0230 + 0.9467} = 0.1785 = 17.85\% \text{ H}_2\text{O} \quad (3.79)$$

EPA Method 5 determines particulate emissions. A multipoint sample is drawn from the stack through a heated filter (248°F (±25°F), 120°C (±14°C)) and an impinger train. Sampling flow is measured by a dry gas meter after the bulk of the stack gas moisture is condensed out in the impinger train, and then the residual water is removed by silica gel in its final step. It is also acceptable to calculate the moisture leaving the condenser from the vapor pressure of water and the temperature and pressure at the exit. Even then, Method 5 recommends that silica gel or equivalent be used between the condenser system and the

178 Chapter 3 Air Combustion

pump and metering devices to prevent condensation therein and to avoid the need to make corrections for moisture in the metered volume.

Sampling for determination of particulate matter is to be achieved at isokinetic rates±10%, and a pitot tube traverse (Method 2) is conducted simultaneously. *Isokinetic sampling* means that the gas sample is drawn from a stack traverse point at the same velocity that exists there in the absence of sampling to obtain a true measure of the particulate concentration at that point.

The procedure of Method 5 with the Method 5 train can be used as a reference method for moisture determination. The Method 5 train is also employed in the sampling and instrumental analyses of various gaseous pollutants on a dry basis after moisture removal.

Matter existing as particulate at the filter temperature and caught on the filter is the difference between the final and tare weights of the filter or of the filter assembly. Elaborate cleaning/wash procedures are specified for sample handling to prevent contamination or loss of sample.

EPA Method 5 considers only the material existing as a particulate at filter temperature and caught by the filter assembly in the sampling train to be a particulate emission. This is termed *filterable particulate matter*. However, the impinger catch at lower temperature may contain material in addition to condensed moisture. If a determination of the particulate matter collected in the impingers is desired in addition to moisture content, EPA Method 5 notes that the individual states or control agencies requiring this information shall be contacted concerning methods for sample recovery and analysis of the impinger contents.

This is the so-called *condensible particulate*, vaporous matter at the filter temperature that behaves as particulate matter only after condensation at collection train temperatures [22], picked up by those other methods, including the more recently promulgated EPA Method 202 [61]. Method 202 is a collection of test procedures that allows several options to accommodate state/local test methods that were in existence at the time that this method was proposed and promulgated as a final rule. Each of the allowable options may change the collected mass that would be counted as condensible particulate. Therefore, which of the optional procedures that are to be employed in a given case should be clearly defined so as to achieve results consistent with the basis of the specified emission limits. The condensible particulate emitted from boilers fueled on coal or oil is primarily inorganic in nature, with the split shown in AP-42 as 65:35 inorganic to organic for firing no. 2 oil and 85:15 for no. 6 oil firing [25]. *Condensible* is spelled *condensable* in some of the references. ■

PROBLEM 3.36

If the source of sampling in Problem 3.34 or Problem 3.35 took place using a Method 5 sampling train, calculate the particulate concentration if 21.4 g of solids were recovered at a sampling point upstream of a particulate removal device. Sample flow of $0.9354\,\text{m}^3$ (dry) at EPA's standard temperature and pressure (STP) (68°F, 29.92 in. Hg) was measured, and its water vapor content before moisture removal was determined.

SOLUTION The particulate catch was measured along with the sample flow noted above. To express concentration in conventional units of lb or grams per cubic ft or grams per cubic meter multiply test data by the appropriate conversion factors.

$$21.4\,\text{g}\frac{\text{lb}}{453.59\text{g}} = 0.04718\,\text{lb}$$

$$\times 7000\frac{\text{gr}}{\text{lb}} = 330.3\,\text{gr}$$

$$0.9354\,\text{m}^3\frac{\text{ft}^3}{0.02832\,\text{m}^3} = 33.023\,\text{ft}^3$$

$$(0.04718 \text{ lb})/(33.023 \text{ ft}^3) = 0.00143 \text{ lb/SCF (dry)}$$
$$(330.3 \text{ gr})/(33.023 \text{ ft}^3) = 10.00 \text{ gr/SCF (dry)}$$
$$(21.4 \text{ g})/(0.9354 \text{ m}^3) = 22.9 \text{ g/m}^3 \text{ (STP) (dry)}$$

$$(0.00143 \text{ lb/SCF (dry)})(100-17.85/100) = 0.00117 \text{ lb/SCF (wet)}$$
$$(10.00 \text{ gr/SCF (dry)})(100-17.85/100) = 8.21 \text{ gr/SCF (wet)}$$
$$(22.9 \text{ g/m}^3 \text{ (STP) (dry)})(100-17.85/100) = 18.8 \text{ g/m}^3 \text{ (STP) (wet)}$$

This concentration in the appropriate units would then be multiplied by the volumetric flow (wet or dry) from the Method 2 stack test to obtain the corresponding mass flow rate of particulate in the flue gas. The composition of the gas sampled prior to particulate and moisture removal is taken to be the same as that of the full-scale flue gas.

A particulate removal efficiency on the order of $99 + \%$ for the collection device would be needed to reduce the particulate concentration to the range of typical particulate emission standards. ∎

PROBLEM 3.37

In this hypothetical case, the furnace of a controversial project has started up at last. Despite delays, mandatory stack testing has finally been performed to verify compliance with permit conditions. In the permit, there is a pollutant limit in the flue gas not to exceed 100 ppmvd corrected to 3% oxygen on a dry basis, as determined by the average of three 1 h stack tests. Test results for the three back-to-back tests are listed below:

Test no.	Pollutant concentration (ppmvd)	vol% O_2 (dry)
1	99.9	4.4
2	90.0	3.1
3	94.8	4.2
Average	94.9	3.9

Further stack testing was curtailed because of severe weather, and the stack testing crew hurriedly demobilized and left the site. Although test results are close to 100 ppm, it was thought that the limit had been met, at least at first. (Can you see why?) What is the story?

SOLUTION The limit is written in the permit as ppmd @ 3% O_2 (dry), and it is this metric that is to be averaged on the basis of three stack tests, not the raw data. Correcting the results of each test separately via Equation (3.27) to 3% O_2 (dry) and averaging, and then rounding off to the same number of places as the uncorrected pollutant concentration,

$$(99.9)(20.9-3)/(20.9-4.4) = 108.38$$
$$(90.0)(20.9-3)/(20.9-3.1) = 90.51$$
$$(94.8)(20.9-3)/(20.9-4.2) = 101.61$$
$$\text{Average} = 100.17 \rightarrow 100.2 > 100$$

180 Chapter 3 Air Combustion

The source is technically out of compliance. Unfortunately, the allowable period for testing stated in the permit has expired (not a good idea to let happen). To continue, an extension of time must be requested from the agency.

In the meantime, the public, armed with binoculars and aware of the testing, demanded and got the interim results. In the rather vociferous public meeting that ensued, Friends of the Sauerbraten, an ethnic but equal opportunity social club based at a local firehouse, grilled the plant personnel and called for an explanation of such outrageous behavior. The NIMBY (Not In My Back Yard) organization, a local chapter of BANANA (Build Absolutely Nothing Anywhere Near Anybody), cried, "Shut the polluters down! Make them pay!" The Director of the Home for Tubeless Tires, a local charitable residence facility, voiced great concerns about health effects with regard to the neighboring population. Only the representative of the Save Our Werewolves Federation was willing to discuss a compromise; after all, the majority of their work took place at night when one could not really see what was happening.

As a result, the company was fined and the project forced to shut down. However, the offending furnace was allowed to restart many months later after repairs and adjustments had been made and the paperwork straightened out. A retest showed compliance, borne out every day now by the newly installed continuous emission monitoring system. In the end, no one was completely satisfied. ∎

3.12.2 Air Infiltration

There will be times when measurements of O_2 or CO_2 at the stack are not consistent with their process-control counterparts upstream. This is often caused by infiltration of so-called *tramp air* into the flue gas when the combustion train operates at a negative pressure slightly below atmospheric. Air infiltration can be estimated by comparing oxygen readings at the furnace and at the stack or O_2 and CO_2 readings measured into and out of a control device and at the stack as well.

A furnace box consists of a large amount of external surface area with seams between adjacent parts. "Big-box" control devices, such as electrostatic precipitators (ESPs) or selective catalytic reduction (SCR) units, add a large amount of surface area with seams also having the potential for leakage.

From an oxygen balance,

$$(Q_a/Q_s) = (O_{2_s} - O_{2_f})/(20.9 - O_{2_f}) \tag{3.80}$$

where Q_a is the flow of infiltrated air, Q_s is the flue gas flow at the stack, and O_{2_s} and O_{2_f} are the oxygen concentrations (% dry) at the stack and the furnace, respectively.

A CO_2 balance employing measured values into and out of a control device external to the furnace leads to the following equation:

$$(Q_a/Q_{out}) = (CO_{2_{in}} - CO_{2_{out}})/(CO_{2_{in}} - CO_{2_{air}}) \tag{3.81}$$

where, again, all CO_2 measurements are on a dry basis. The CO_2 in ambient air can be ignored without appreciable error.

Estimation of air infiltration is illustrated in the example problems that follow.

PROBLEM 3.38

A certain furnace is regularly controlled at 2.0% O_2 wet at the firebox exit. Combustion calculations indicate that flue gas moisture leaving the furnace is 20%. Estimate the amount of air infiltration between the furnace firebox and the stack if the oxygen (O_2) measured at the stack is 2.7% (dry) in one instance and 6.0% (dry) on a later occasion.

SOLUTION This estimate can be made by means of an oxygen balance in the flows of three streams, the flow from the furnace (Q_f), the flow at the stack (Q_s), and the flow of infiltrated air (Q_a). To apply Equation (3.80), values of O_2 must be on a dry basis. The dry basis O_2 is measured during stack testing. We take advantage of the plant's combustion control O_2 monitor, properly calibrated, but note that it measures O_2 on a wet basis in the combustion milieu. One way to correct the difference is to use the calculated moisture content of the flue gas at the furnace exit. Combustion calculations have shown this value to be 20%.

At the furnace

$$O_{2_{dry}} = O_{2_{wet}}[100/(100-\%H_2O)] \quad (3.29)$$
$$O_{2_{dry}} = O_{2_{dry}} = (2.0)[100/(100-20)] = 2.5\% \, O_{2_{dry}}$$

From Equation (3.80),

$$(Q_a/Q_s) = (O_{2_s} - O_{2_f})/(20.9 - O_{2_f})$$

All O_2 readings are on a dry basis.

(a) Then, $100 \, (Q_a/Q_s) = (2.7 - 2.5)/(20.9 - 2.5) \cong 1\%$ air infiltration
(b) and $100 \, (Q_a/Q_s) = (6.0 - 2.5)/(20.9 - 2.5) \cong 19\%$ air infiltration on the later occasion.

Although case (a) may be reasonable, case (b) appears to be somewhat extreme and would warrant further investigation. ∎

PROBLEM 3.39

The flue gas from a furnace followed by a "big-box" pollution control device was tested to verify compliance with its air quality permit. (This device could be an electrostatic precipitator, a fabric filter installation (baghouse), a SCR unit, etc.) Average values from compliance demonstration runs were as follows:

Process control value at furnace	2.1% O_2 (wet)	19.2% H_2O by calculation
Pollution control device inlet	2.8% O_2 (dry)	11.4% CO_2 (dry)
Device outlet/stack	3.3% O_2 (dry)	11.1% CO_2 (dry)

Estimate the percent air infiltration between the furnace and the control device, across the control device, and overall between the furnace and the stack.

SOLUTION As before,

$$O_{2_f}(\text{dry}) = O_{2_f}(\text{wet})[100/(100 - \% H_2O)] \quad (3.29)$$
$$= (2.1)[100/(100 - 19.2)]$$
$$= 2.6\% \, O_2(\text{dry})$$

Overall infiltration between the furnace and the stack:

$$(Q_a/Q_s) = (O_{2_s} - O_{2_f})/(20.9 - O_{2_f}) \quad (3.80)$$
$$(3.3 - 2.6)/(20.9 - 2.6) = 3.83\%$$

Across the pollution control device:
From O_2 measurements—

$$(Q_a/Q_{out}) = (O_{2_{out}} - O_{2_{in}})/(20.9 - O_{2_{in}}) \quad (3.82)$$
$$(3.3 - 2.8)/(20.9 - 2.8)$$
$$= 0.0276 = 2.76\%$$

From CO_2 measurements—

$$(Q_a/Q_{out}) = (CO_{2_{in}} - CO_{2_{out}})/(CO_{2_{in}} - CO_{2_{air}}) \quad (3.81)$$
$$= (11.4 - 11.1)/(11.4 - 0.033)$$
$$= 0.0264 = 2.64\%$$

The O_2 and CO_2 measurements provide a decent check on each other, and the air infiltration estimates are simply averaged here.

$$\text{Average} = (2.76 + 2.64)/2 = 2.70\%$$

Between the furnace and the control device:

$$(Q_a/Q_{in}) = (2.8 - 2.6)/(20.9 - 2.6) = 1.09\% \cong 1.1\%$$
$$(\text{by difference}) \, 3.83\% - 2.70\% = 1.13\% \cong 1.1\%$$

Both sets of computations lead to a calculated infiltration rate of about 1%.
 The extent of air infiltration and the split at various locations along the flue gas train may well be different in an actual case. ∎

3.13 CONTINUOUS EMISSION MONITORING SYSTEMS

CEMS are used to ascertain compliance of certain sources of air emissions with their permit limits. They are mandated for some sources by EPA, and the requirement can be implemented for others by a state/local air pollution control agency under its discretionary authority. Unlike manual stack testing, monitoring is on a continuous basis, rather than a snapshot of operations every so often. CEMS consist of an automatic sampling system coupled with a computer and software programmed to carry out the necessary calculations and generate the required reports.[10]

A CEMS unit installed on the stack of the primary reformer furnace in an ammonia plant is described Ref. [62]. Background, requirements, CEMS selection criteria, specifications, analyzers, data acquisition, emission rate determinations, and various calibration checks are discussed in some detail. Analyzer instruments measure NO_x, CO, and O_2 in the furnace flue gas. Fuel flow is measured along with a chemical analysis of the fuel(s). Flue gas flow rate is calculated via EPA Method 19 F_d factors.

A summary article [63] gives a good overview of CEMS technology existing at the time. In an interesting sidebar, it discusses continuous flow monitors using EPA Method 2 with a type S pitot tube and how these CEMS are indicating stack flow rates significantly higher than the actual (greater than 20% in some cases). There is an excellent text available that may answer any question on the subject of CEMS [64].

CEMS measurements must be calibrated against a standard and must be demonstrated in an initial test to hold that calibration within 2.5% of span for a week (168 h). The CEMS must pass an initial test (plus repeats at periodic intervals) where they are compared to manual stack test measurements. The CEMS must be recalibrated daily using certified gases and must be online and available to generate valid data on some mandated high percentage of the time. System availability greater than 98% can be achieved [64, p. 23].

Minor lapses typically result in various levels of "nastygrams" from the governing regulatory agency culminating in a notice of violation (NOV) when the regulatory threshold for an unacceptable amount of downtime or invalid data is exceeded [64, p.23].

Procedures are contained in 40 CFR 75 [67] on how to handle missing data. Other data from various sources are substituted for the missing data, as prescribed, to enable the calculation of mass flow rate and other averages. A properly designed CEMS will contain well-tested software customized for the given application, and these and other calculations will be done automatically, leaving a minimal amount of calculations to be done by hand.

One set of manual calculations that comes to mind are those needed for the Relative Accuracy Test Audit (RATA). This is performed for initial operation and as often as once per year thereafter. In the RATA, the absolute mean difference between CEMS readings and the results of simultaneous manual stack testing using EPA reference methods are compared along with a 2.5% statistical confidence coefficient. Test runs (9 at minimum) lasting for 30–60 min each are required [64, p. 311]. With some judicious planning, it is possible to satisfy an initial or recurring permit compliance-testing requirement at the same time [68].

---

[10] Another form of continuous monitoring is a predictive emission monitoring system (PEMS), in which process parameter data correlated to emissions are recorded, tracked, and controlled as surrogates for the emissions themselves. Examples include keeping flue gas O_2 high enough to minimize CO formation and low enough to limit NO_x. The author's experience with one such PEMS is noted briefly in Appendix K. Two separate journal articles discuss the PEMS installed on different ethylene plant furnaces [65,66].

184 Chapter 3 Air Combustion

The RATA calculations will be illustrated using an example adapted from an EPA reference [69]. Stack testing calculations for determination of permit compliance follow in a second example.

PROBLEM 3.40

The following data for pollutant concentration were obtained during nine 30 min tests. Measured SO_2 concentrations have already been corrected for flue gas % O_2 and expressed in the units of the applicable standard, as shown below.

	SO_2 (ng/J)		
RATA test no.	Stack test	CEMS	Difference
1	422.4	403.5	18.9
2	426.6	407.7	18.9
3	430.8	405.4	25.4
4	428.4	403.2	25.2
5	420.0	405.4	14.6
6	422.4	424.7	−2.3
7	430.8	433.3	−2.5
8	424.2	426.6	−2.4
9	428.4	439.3	−10.9
Sum	3834.0	3749.1	+84.9
Average	426.0	416.6	9.43

SO_2 (mg/J) = 2.66×10^6 SO_2 (ppm) F [20.9/(20.9 − % O_2 dry)] with F = 2.72×10^{-7} dry standard cubic meters (dsm^3/J), from EPA 40 CFR 60, Appendix A, Method 19, Equation 19.1 [4].
Calculate the Relative Accuracy (RA).

SOLUTION Applicable equations are also contained in 40 CFR 60, Appendix B, Performance Specification 2 [68] and Ref. [64, pp. 312–313].

Step 1: The average difference, d, is calculated for the SO_2 CEMS monitor.

$$\bar{d} = (1/n)\sum_{i=1}^{n}(X_i - Y_i) = (1/n)\sum_{i=1}^{n} d_i \qquad (3.83)$$

where n is the number of test points and X_i and Y_i are the individual entries in columns 2 and 3 of the table above. $\bar{d}$ is the sum of the individual differences calculated in the last column (84.9) divided by the number of tests (9) = 84.9/9 = 9.43 mg/J, the final entry in the last column.

The mathematical signs of the individual differences d_i (last column) between the individual stack test results (column 2) and the CEMS readings (column 3) are retained to calculate the net result.

Step 2: Calculate the standard deviation S_d.

$$S_d = \left\{[1/(n-1)]\left[\sum_{i=1}^{n} d_i^2 - (1/n)\left(\sum_{i=1}^{n} d_i\right)^2\right]\right\}^{1/2} \quad (3.84)$$

$$\sum_{i=1}^{n} d_i = 2343.89$$

$$S_d = \{(1/8)[2343.89-(1/9)(84.9)^2]\}^{1/2} = 13.9 \text{ ng/J}$$

Step 3: Calculate the 2.5% error confidence coefficient (cc).

$$cc = t_{0.975}S_d/(n)^{1/2} \quad (3.85)$$

where $t_{0.975}$ is the value from the statistical t-distribution for a one-sided t-test corresponding to the probability that a measurement will be biased low at the 95% confidence level [64, p. 313].

Its value for nine test points $(9-1=8$ degrees of freedom) $= 2.306$ [64,67–69].

$$cc = (2.306)(13.9)/(9)^{1/2} = 10.68$$

Step 4: Calculate RA.

Here, the absolute values of $\bar{d}$ and cc are used in the calculations, and RM is the average of the stack test reference method results.

$$RA = \frac{100[|\bar{d}|+|cc|]}{RM} \quad (3.86)$$

$$= \frac{(100)[|9.43|+|10.68|]}{426.0} = 4.72\%$$

This value of RA compares favorably with a figure on the order of 5% quoted for CEMS operating today [64, p. 23]. For a performance specification for RA on the order of 10–15% [64, p. 314] up to 20% [64, p. 23], this CEMS would pass its RA test. ∎

PROBLEM 3.41

From the stack testing results contained in Problem 3.40, structure the data for a compliance demonstration report.

SOLUTION For a typical compliance testing, three 1 h stack test runs using EPA reference methods are required. Using the first six entries from the table of Problem 3.40, one obtains

Compliance test number	SO_2 (ng/J)
1	$(422.4+426.6)/2 = 424.5$
2	$(430.8+428.4)/2 = 429.6$
3	$(420.0+422.4)/2 = 421.2$

Sum	1275.3
Average	425.1

It makes little difference which three contiguous 1 h periods are chosen. Selection of RATA test with 3 and 4, 6 and 7, and 8 and 9 results in the highest average emission of 427.5 mg/J, with numbers 1 and 2, 5 and 6, and 7 and 8 the lowest, 424.0 mg/J. Regardless of which data sets are selected for the compliance determination, one may have to present all of the data in the report submitted to the agency, as required for a set of RATA test data when more than nine RATA runs are conducted [68]. ∎

3.14 MISCELLANEOUS SOURCES OF AIR EMISSIONS

This section brings together emission sources that typically accompany the emission of flue gas from large furnaces, heaters, and boilers in an industrial process plant. They are discussed here to provide information necessary to complete the permit application process.

The first three sections are concerned with several forms of fugitive emissions, although air emissions from cooling towers, strictly speaking, are considered by some to constitute point sources. The remaining emission sources discussed are truly point sources, including the hydrocarbon emissions from the fuel tank of an emergency backup electrical generator, those emissions being negligible for the size range in the example. The emissions from the generator's diesel engine itself are combustion emissions.

3.14.1 Fugitive Emissions

Fugitive emissions are unintended emissions that arise from leakage from piping components and associated equipment. These emissions are estimated by counting the numbers of each type of piping component (valves, flanges, etc.), multiplying each by a published leak rate factor for that component, and summing to obtain the total emission rate.

When there is a mixture of chemical species flowing inside the process piping and equipment, the total emission calculated from a given line is multiplied by the weight fraction (wt%/100) of each contaminant of interest to obtain the emission specific to that contaminant. A separate fugitive emission calculation must be performed for each section of a chemical manufacturing plant to reflect the changing composition that takes place from one process unit to the next.

EPA leak testing was originally conducted at petroleum refineries [70–72]. This led to the development of the so-called *refinery factors* [73], more recently updated in a later document [74] and noted in AP-42 [75]. Subsequently, testing was extended to the synthetic organic chemicals manufacturing industry (SOCMI), and somewhat different factors [76] for chemical plants were generated from the data [77,78].

The latest refinery average emission factors and SOCMI average emission factors from Ref. [74] are reproduced here in Tables 3.13 and 3.14 for convenience. In the

Table 3.13 Refinery Average Fugitive Emission Factors [74][a]

Equipment type	Service	Emission factor (kg/h/source)[b]
Valves	Gas	0.0268
	Light liquid	0.0109
	Heavy liquid	0.00023
Pump seals[c]	Light liquid	0.114
	Heavy liquid	0.021
Compressor seals	Gas	0.636
Pressure relief valves	Gas	0.16
Connectors	All	0.00025
Open-ended lines	All	0.0023
Sampling connections	All	0.0150

[a]*Source*: Ref. [80].
[b] These factors are for nonmethane organic compound emission rates.
[c] The light liquid pump seal factor can be used to estimate the leak rate from agitator seals.

tables, *connectors* include flanges and threaded connections. The refinery factors are on a nonmethane hydrocarbon basis and must be adjusted to account for this fact. The original document [74] should be consulted for details.

Because there are differences among various types of chemical plants, different fugitive emission factors are in use where the process fluids contain more than 85% ethylene (C_2H_4 or $H_2C=CH_2$, abbreviated $C_2^=$) or less than 11% ethylene [79]. These factors are, respectively, termed *SOCMI with ethylene* (w $C_2^=$) and *SOCMI without ethylene* (w/o $C_2^=$). The average SOCMI factors are applied between 11% and 85%.

Table 3.14 SOCMI Average Fugitive Emission Factors [74][a]

Equipment Type	Service	Emission factor (kg/h/source)[b]
Valves	Gas	0.00597
	Light liquid	0.00403
	Heavy liquid	0.00023
Pump seals	Light liquid	0.0199
	Heavy liquid	0.00862
Compressor seals	Gas	0.228
Pressure relief valves	Gas	0.104
Connectors	All	0.00183
Open-ended lines	All	0.0017
Sampling connections	All	0.0150

[a] These factors are for total organic compound emission rates.
[b] The light liquid pump seal factor can be used to estimate the leak rate from agitator seals.

188 Chapter 3 Air Combustion

In addition, other factors have evolved for components in service with specific chemicals, for natural gas processing, for oil and gas production, and for bulk petroleum marketing terminals.

The illustrative example that follows employs the original average SOCMI factors [77] for gas/vapor service on a natural gas chemical feedstock containing a high percent of heavy constituents. The number and type of equipment components are taken from a sample calculation prepared by the State of Texas using a different mixture of organic chemicals as the process fluid [79]. The fluid here is a natural gas with a content of higher hydrocarbons somewhat greater than that of the natural gas of Table 3.1.

The example shown in Problem 3.42 is intended to demonstrate the calculation procedure only; specific guidance on an actual project should be sought from the governing environmental agency or agencies regarding the applicable factors to use and any and all other conditions.

PROBLEM 3.42

Prepare an estimate of fugitive emissions from the given equipment and natural gas composition, both listed in the following solution. Constituents in the natural gas of specific interest are the propane and heavier fraction (C_3^+). Use the original SOCMI factors, unmodified, for this estimate (Table 3.15).

Light liquids are defined as having a vapor pressure greater than 0.044 psia, whereas heavy liquids are defined as having a vapor pressure of 0.044 psia or less [79].

SOLUTION The first step is to calculate the weight percent of the C_3^+ fraction, as shown in the table below.

Weight Percent Calculation
Basis: 100 mol.

Component	vol%	MW	lb = vol% × MW	wt%	wt% C_3^+
CH_4	77.68	16.04	1245.9872	60.18	0.00
C_2H_6	13.18	30.07	396.3226	19.14	0.00
C_3H_8	3.37	44.11	148.6507	7.18	7.18
C_4H_{10}	0.68	58.12	39.5216	1.91	1.91
C_5H_{12}	0.44	72.15	31.7460	1.53	1.53
C_6H_{14}	0.07	86.18	6.0326	0.29	0.29
C_7H_{16}	0.04	100.21	4.0084	0.19	0.19
C_8H_{18}	0.02	114.23	2.2846	0.11	0.11
C_9H_{20}	0.01	128.26	1.2826	0.06	0.06
$C_{10}H_{22}$	0.01	142.29	1.4229	0.07	0.07
CO_2	4.20	44.01	184.8420	8.93	0.00
N_2	0.30	28.01	8.4030	0.41	0.00
Total	100.00	—	2070.5042	100.00	11.34

Table 3.15 SOCMI Average Emission Factors for Fugitive Emissions (for Historical Reference Only—These Factors are Updated in Table 3.14)

Type of equipment	Service	Emission factor (kg/h/source)
Valves	Gas	0.0056
	Light liquid	0.0071
	Heavy liquid	0.00023
Pump seals	Light liquid	0.0494
	Heavy liquid	0.0214
Compressor seals	Gas/vapor	0.228
Pressure relief seals	Gas/vapor	0.104
Connectors	All	0.00083
Open-ended lines	All	0.0017
Sampling connections	All	0.0150

The next step is to apply the fugitive emission factors.

Emission Calculation with $C_3^+ = 11.3474\ wt\%$

	SOCMI factors[a]					
Type of equipment	Count	kg/h per source	lb/h per source	Uncontrolled[b] emission rates		Control[b] efficiency
				lb/h	TPY	
Valves	10919	0.0056	0.01235	1.428	6.25	
Flanges	1435	0.00083	0.00183	0.298	1.31	
Compressors	1	0.228	0.10342	0.012	0.05	
Relief valves[c]	12	0.104	0.22928	0.312	1.37	
Open-ended lines[d]	3	0.0017	0.00375	0.001	0.00	
Sampling connectors	0	0.0150	0.03307	0	0	
			Total:	2.051	8.98	

[a] For this example, the SOCMI factors for liquid service area are not required.

[b] After start-up, periodic monitoring of equipment components will be required. One may take credit for a certain degree of control efficiency and may, therefore, use a lower emission factor by monitoring and fixing leaks more frequently than absolutely required.

[c] Routing relief valves to a flare constitutes a 100% control credit. Accordingly, one may subtract their fugitive emission contribution for this configuration.

[d] Open-ended lines are to be plugged, capped, or equipped with a second valve, and some credit can be taken. Sampling valves are to be backed up by a second shutoff valve immediately upstream, and both valves are to be shut off except when sampling is actually taking place.

In this example problem, the fugitive emissions of VOCs are just under 9 ton/year. In an actual case, be prepared to submit detailed drawings so that agency personnel can verify the number and type of equipment components used in the calculations. (One may wish to include a slightly greater number of equipment components as a safety factor against miscounting, just as long as this does not inadvertently and unnecessarily trigger some limit and put the project into a more severe regulatory category. Also, once the plant has begun operation, be prepared to retain plant records of fugitive emission leak monitoring and maintenance for regulatory inspection.

One final note: The estimation procedures for fugitive emissions were developed for and apply to VOCs. They do not really apply to the fugitive emissions of inorganic pollutants such as CO and H_2S. Nevertheless, they can be and have been used for such purpose on the basis that they are the "best tool available" [79]. ∎

3.14.2 Cooling Towers

Cooling towers can be a source of pollution. These are really point sources but somewhat behave like fugitive emissions. Cooling towers are considered sources of particulate emissions (PM_{10}) when liquid droplets are carried over to the atmosphere along with evaporated vapor-phase water. The term for this liquid carryover is *drift*. It is assumed to have the same composition, including total dissolved solids (TDS), as the recirculating cooling water. When the drift eventually evaporates in the atmosphere, it leaves behind its TDS as a particulate emission. AP-42 [81] indicates that a *conservatively high* PM_{10} emission factor can be obtained by multiplying the total liquid drift factor by the TDS in the recirculating water and assuming that once the water evaporates, all remaining solid particles are within the PM_{10} size range.

When the fluid being cooled by noncontact cooling water is a volatile organic compound at a higher pressure, there is a potential for leakage into the cooling water. Once again, when the cooling tower drift evaporates, a VOC emission occurs. Leaking VOCs can also emanate directly from cooling towers without the need for the drift mechanism.

Emissions of PM_{10} and VOCs are estimated in the next two example problems.

PROBLEM 3.43

Estimate particulate emissions from a 50,000 gpm recirculating flow, induced-draft cooling tower in which the makeup water contains 250 ppm by weight of dissolved solids and operates at 10 cycles of concentration based on conductivity. (See Appendix E regarding cooling tower operation.)

SOLUTION The drift is always the most difficult to estimate. Perhaps, a vendor guarantee for the cooling tower's drift eliminators can be obtained and used in calculations. In this illustrative example, we will use the AP-42 value of 0.02% of circulation [81]. The drift factor can be as low as 0.001% [82] or less (0.0005%) [83]. It will vary with the initial design, age, and degree of maintenance.

$$\text{Drift} = (0.02/100)(50,000 \text{ gpm recirculation rate})$$
$$= 10 \text{ gpm}$$
$$10 \text{ (gpm)} [500 \text{ (lb/h)/gpm}] = 5000 \text{ lb/h of drift}$$

TDS in recirculating water and drift = (250 ppm in MU)(10 cycles) = 2500 ppm
[5000(lb/h)][2500 × 10⁻⁶ lb solids/lb drift] = 12.5 lb/h of solids
× 4.38 = 54.8 ton/year (TPY) of solids

∎

PROBLEM 3.44

Estimate possible VOC emissions from the cooling tower of Problem 3.43. The organic liquid being cooled has a solubility of 50 ppm [84].

SOLUTION In the absence of better information, the factor in Section 5.1, Petroleum Refining, of AP-42 [75] for uncontrolled emissions of VOCs from cooling towers is often used. From Table 5.1-2 of the cited document, this factor is 6 lb/10^6 gal of cooling water. A factor of 0.7 lb/10^6 gal of cooling water is listed for controlled emissions as well. For the cooling tower at hand, 6 lb/10^6 gal amounts to an estimated VOC emission of

$$\frac{(6 \text{ lb})}{10^6 \text{ gal})} \frac{(50,000 \text{ gal})}{(\text{min})} \frac{(60 \text{ min})}{(\text{h})} = 18 \text{ lb/h}$$

$$\times 4.38 = 78.8 \text{ ton/year (TPY) VOCs}$$

The Texas Council of Environmental Quality (TCEQ) will *not* allow one to use the EPA factor for *controlled* emissions (0.7 lb/10^6 gal of cooling water) in lieu of the uncontrolled factor for determination of cooling tower emissions [85]. The controlled factor is contingent upon the use of control and monitoring technology to reduce VOC leakage into the cooling water. Such data and procedures would then be detailed to generate a specific emission rate applicable to a given cooling tower; this approach is allowed.

It may be possible to reduce the estimated emissions by means of specific measurements from a given cooling system. For example, at a measured hydrocarbon solubility of 50 ppm [84],

$$(5000)(50 \times 10^{-6}) = 0.25 \text{ lb/h of VOC in drift}$$

$$\times 4.38 = 1.1 \text{ ton/year}$$

The calculation above assumes that only *dissolved* VOC is emitted up to its solubility limit. However, bear in mind that leakage of VOCs may also occur without dissolving in the cooling water.

∎

3.14.3 Other Sources of Fugitive Emissions

How many people can recall the thrill of tooling down a rural dirt road and being followed by a magnificent cloud of dust? One would venture to say that there are fewer such dirt roads in the United States at the present time than in the past. The remaining roads are a source of fugitive emissions, and methods exist for their estimation [86]. A predictive equation for unpaved road emissions is also given in AP-42 [87]. Important variables are vehicle miles traveled, mean vehicle speed, weight, number of wheels, silt content and size distribution of the road material, and the amount of precipitation that the dirt road receives.

In some geographical areas, ambient air quality cannot tolerate any further degradation. This can occur because air quality is really good, and it is desired/mandated to keep it that way, or really poor and one cannot afford to make it any worse. In such cases, it is common to have to provide *emission reduction credits* by lowering one's own or someone else's emissions before being allowed to introduce a new source of atmospheric contamination.

A favorite stratagem to realize a credit for reduction of particulate emissions is to pave dirt roads within a certain distance from the new source. For hydrocarbon sources, possible credits can be obtained by buying out and shutting down a number of commercial dry cleaning establishments, also within an allowable distance from the proposed source. More conventional ways are to provide improved air pollution control equipment and/or fuels with a lower sulfur content to other emitters.

3.14.4 Organic Liquid Storage Tanks

AP-42 contains a chapter [88] on the estimation of emissions from storage tanks such as these employed in the following industries:

- Petroleum producing and refining
- Petrochemical and chemical manufacturing
- Bulk storage and transfer operations
- Other industries consuming or producing organic liquids

The AP-42 chapter presents models for estimating air emissions from organic liquid storage tanks. It also contains detailed descriptions of typical varieties of such tanks, including horizontal, vertical, and underground fixed roof tanks.

The emission estimation equations have been developed by the American Petroleum Institute (API), which retains the rights to these equations. API has granted EPA permission for the nonexclusive, noncommercial distribution of this material to government and regulatory agencies. However, API reserves its rights regarding all commercial duplication and distribution of its material. Hence, the material presented is available for public use, but it cannot be sold without written permission from both the American Petroleum Institute and the U.S. Environmental Protection Agency.

The reader is therefore referred directly to Chapter 7 in AP-42 for estimation of emissions from organic liquid storage tanks.

3.14.5 Emergency Diesel Generator and Fuel Oil Tank

An emergency generator operates to provide electrical power in the event of commercial power failure. This allows at least critical plant equipment to continue to operate during a power failure to avoid a forced shutdown. It does not include peaking units at electric utilities, generators at industrial facilities that typically operate at low rates but are not confined to emergency purposes, and any standby

3.14 Miscellaneous Sources of Air Emissions

generator that is used during time periods when power is readily available from the utility [89]. Calculation of emissions is illustrated in Problem 3.45.

PROBLEM 3.45

A diesel engine/generator set with an integral fuel tank is being installed to produce emergency backup power in the event of a power failure. The generator is rated at 35–40 KW, the engine fires diesel fuel (no. 2 oil) at 25.9 lb/h, and the tank volume is 75 gal (U.S. gallons). This unit will be operated only when commercial power is lost, except for periodic testing to ensure operability when it is needed in an emergency. Compute the pollutant emissions from both (a) the engine and (b) its tank.

SOLUTION

(a) Emission factors for the engine and the heating value of the diesel fuel are given in EPA AP-42, Table 3.3-1 (10/96) [90].

Firing rate is $(25.9 \text{ lb/h})(19,300 \text{ Btu/lb})/10^6 = 0.500 \text{ MMBtu/h}$

Calculations are as follows:

Estimated Emissions from Diesel Engine

Pollutant	EPA factor (lb/MMBtu)	× 0.500 MMBtu/h lb/h	×500 h/year[a] lb/year	÷ 2000 ton/year
NO_x	4.41	2.20	1102.21	0.55
CO	0.95	0.47	237.44	0.12
SO_x[b]	0.29	0.15	72.78	0.04
PM_{10}	0.31	0.16	77.48	0.04
TOC				
Exhaust	0.35	0.175	87.48	0.0437
Crankcase	0.01	0.005	2.50	0.0012
Total TOC	0.36	0.180	89.98	0.0450

[a] Default value from Ref. [89].
[b] The AP-42 factor for SO_x appears to be based on a fuel oil sulfur content at 0.3 wt%.

EPA has addressed the maximum potential to emit (PTE) for emergency generators in the cited guidance memorandum [89]. As expressed in that memorandum, it was felt that full year operation (8760 h) should not be used to calculate the PTE for emergency generators, but that a maximum of 500 h/year is an appropriate default assumption for such equipment because of the inherent constraints on their operation. For owners and operators who install, modify, or reconstruct their emergency generators after July 11, 2005, maintenance checks and readiness testing of such units is now limited to 100 h/year, absent some compelling reason to the contrary [91]. This new Standard of Performance was promulgated in the *Federal Register* of July 11, 2006. The testing limitation, as originally proposed in the July 11, 2005 *Federal Register* (70 FR 39870), was 30 h. The owner/operator must install a nonresettable counter to keep track of total hours of operation and

194 Chapter 3 Air Combustion

keep records of the time of operation for maintenance checks and readiness testing. There is no time limit on the use of the emergency generator in emergency situations.

Fuel sulfur content for such newly constructed or modified units is limited to 0.05 wt% S (500 ppm) as of October 1, 2007 and 15 ppm S as of October 1, 2010. The AP-42 emission factor SO_x as SO_2, current as of 10/96 [90], should therefore be regarded with extreme caution.

(b) Fuel Oil Tank

Instead of using a complicated procedure applicable to the much larger tanks, emissions from the 75 gal fuel oil tank have been estimated by means of Table 5.2-7 found in EPA publication AP-42 [92] to quantify filling, breathing, and working losses for gasoline tanks at service stations. The losses thus calculated are then prorated on vapor pressure (AP-42 Table 7.1-2) [88] from gasoline to diesel fuel.

Throughput: (25.9 lb/h/7.1 lb/gal [88, Table 7.1-2]) × 500 h/year = 1810 gal/year
Usage/refilling of tank: (25.9 lb/h/7.1 lb/gal) = 3.62 gal/h
(75 gal/3.62 gal/h) = almost 21 h of operation per filling
(1810 gal/year/75 gal/tankful) = 24+ tankfuls per year

Estimated Emissions from 75 gal Diesel Fuel Tank

Vapor pressure[a]	Distillate fuel oil no. 2 (diesel fuel)	0.012 psi at 80°F
	Gasoline RVP7	5.2 psi at 80°F
	Ratio diesel to gasoline	(0.012/5.2) = 0.0023

Emission source	Gasoline		Diesel fuel
	Emission rate[b] lb/1000 gal throughput	Annual emissions [(1810/1000) × rate = lb]	Annual emissions[c] lb = 0.0023 × lb for gasoline
Splash filling of tank (Displacement losses)	11.5	20.8	0.048
Tank breathing and emptying	1.0	1.8	0.004
"Spillage"	0.7	1.3	0.003
Total	13.2	23.9	0.055

[a] From Table 7.1-2, EPA AP-42 (11/06) [88].
[b] From Table 5.2-7, EPA AP-42 (6/08) [92].
[c] Prorated from gasoline emissions using diesel fuel to gasoline vapor pressures calculated above. This technique produces emissions for diesel fuel (no. 2 oil) in the same ratio to those of gasoline in Table 5.2-5 of EPA AP-42 (6/08) [92] for rail tank cars and tank trucks.

The resulting figure of 0.0055 lb/year for diesel fuel hydrocarbon emissions is negligible. Even the figure 23.9 lb/year calculated for gasoline is extremely small. The technique of using the ratio of vapor pressures to estimate emissions for diesel fuel (no. 2 oil) is consistent with emission factors for gasoline and no. 2 oil for rail tank cars and tank trucks in another table in AP-42, where such entries can be compared directly. (See footnote c to the table above.)

This estimation technique for the fuel tank emissions has been accepted by environmental agencies, most likely because the hydrocarbon vapor emissions for diesel fuel turn out to be negligible. However, there is no guarantee that it will be universally accepted. You are on your own. ∎

3.14.6 Organic Chemical Vent Stream

PROBLEM 3.46

In 40 CFR 265.1034, a formula is presented to calculate E_h, the total organic mass flow rate (kg/h) of a vent stream.

$$E_h = Q_{sd} \left[\sum_{i=1}^{n} C_i \, MW_i \right] [0.0416][10^{-6}]$$

where Q_{sd} is the volumetric flow rate of gas, C_i is the organic concentration in ppm (dry basis) of component i in the stream, MW_i is the molecular weight of component i in vent gas (kg/kg mol), n is the number of organic compounds in vent gas, and 10^6 is the conversion factor from ppm to mole fraction.

Verify that the other conversion factor, 0.0416 kg mol/m at 293°K and 760 mmHg, is equivalent to 385.3 SCF/lb mol (68°F and 760 mmHg = 1 atm = 14.696 psia).

SOLUTION This is an exercise in conversion of units. Temperature of 293°K is equivalent to 20°C = 68°F. Pressure basis is the same. Molar volume at 68°F and 1 atm is 385.3 SCF/lb mol.

$$\frac{(lb\ mol)}{(385.3\ ft^3)} \frac{(ft^3)}{(2.831684 \times 10^{-2}\ m^3)} \frac{(kg\ mol)}{(2.20462\ lb\ mol)} = \frac{0.0416\ kg\ mol}{m^3}$$

QED: The conversion factor is verified. ∎

3.14.7 Monoethanolamine (MEA) Emissions

Monoethanolamine ($H_2NCH_2CH_2OH$), also known as *MEA, ethanolamine, 2-amino ethanol*, or *colamine*, is a liquid material used in gas treating to remove acid gases such as carbon dioxide and hydrogen sulfide (H_2S). It is similar to ammonia (NH_3) in this respect but more easily regenerated to spring the acid gas and enable reuse of the absorbent solution. Higher alkanolamine homologues, for example, diethanolamine (DEA), triethanolamine (TEA), methyldiethanolamine (MDEA), diisopropanolamine (DIPA), and diglycolamine (DGA) are also employed for this purpose (other CO_2 removal processes with different types of chemicals are used as well [93]).

The MEA in water solution circulates continuously between two gas–liquid contacting vessels. In one, the *absorber vessel*, the so-called *lean MEA solution*, is brought in contact with a gas containing CO_2 and/or H_2S. Acid gas components move

196 Chapter 3 Air Combustion

into the liquid phase, now termed the *rich MEA solution*, and the cleaned gas exits either to further processing or is vented to atmosphere, if unwanted.

The MEA solution, now containing the acid gases, enters the regenerator, or stripper, where hot steam is introduced to shift the equilibrium and free the acid gases to flow overhead. These go on to further processing or may be vented, depending on the purpose for their removal. Some water and MEA vaporize from solution in either vessel and leave with the overhead gas. When the gas is vented, the MEA constitutes a VOC contaminant and must be accounted for in environmental permits.

The MEA solution itself is extremely corrosive to the equipment, and a corrosion inhibitor is added [94]. MEA solution is an organic water pollutant owing to the MEA and its degradation products, and contamination also arises from the other ingredients added from the corrosion inhibitor. Water pollution aspects of MEA are discussed in Chapter 5.

This section concludes with Problems 3.47 and 3.48 inspired by a combination of the process conditions given in Refs [95–97]. Problem 3.47 shows how to estimate the MEA emission from the absorber vessel in a process concentrating the CO_2 generated from various combustion gases for injection into oil-production reservoirs for enhanced oil recovery. Problem 3.48 shows what happens when materials other than CO_2 are the desired species contained in the feed gas being treated and the concentrated CO_2 removed in the absorber is not recovered from the overhead of the regenerator/stripper. Since MEA is more volatile than its alkanolamine cousins, estimation of MEA emission represents a sort of worst case.

PROBLEM 3.47

Estimate the MEA in the overhead of the absorber vessel when the unwanted material in combustion gases is vented after the desired CO_2 product is removed by the MEA solution. Process conditions are as follows: Gas vented to atmosphere: 78.8 MMSCFD containing 84% N_2, 14% H_2O, and traces of H_2, CO, CO_2, and CH_4. The absorber vessel operates at atmospheric pressure (or just slightly above to enable venting). Typical temperature is approximately 130°F. Lean solution entering the absorber contains 0.15 mol CO_2/mol of amine, whereas rich solution contains 0.46 mol CO_2/mol of amine. MEA concentration is a typical 15.3 wt% MEA.

SOLUTION

(a) The solution to the problem is not as easy as it might first appear. One could use a simple vapor pressure approach and wind up with a conservative answer. If this makes no difference in permitting, the higher number obtained could provide a margin of safety.

$$n_c/n_{dg} = (p_{vap_c})/(P_T - p_{vap_c}) \quad (3.2)$$

where n_c and n_{dg} are the moles of a volatile component and moles of dry gas, respectively, p_{vap_c} is the vapor pressure of the volatile component, and P_T is the total pressure.

Vapor pressure constants for the Antoine vapor pressure equation are given in Chapter 2.

$$\log_{10}[P\,(\text{mmHg})] = A - B/[t\,(°C) + C] \quad (2.23)$$

3.14 Miscellaneous Sources of Air Emissions

This is summarized below for water and MEA.

Constituent	A	B	C
H$_2$O (0–60°C, 32–140°F)	8.10765	1750.286	235.0
H$_2$O (60–150°C, 140–302°F)	7.96681	1668.21	228.0
MEA	8.52557	2303.41	237.08

$$\text{Vapor pressure of water at } 130°F \, (54.4°C)$$
$$= 114.97 \, \text{mmHg} \times (14.696/760) = 2.2232 \, \text{psia}$$

Rearranging Equation (3.2) to solve for P_T with $n_c = n_{H_2O}$, one obtains

$$P_T = p_{vap_{H_2O}}/(n_{H_2O}/n_{dg}) + p_{vap_{H_2O}} \tag{3.87}$$

From the problem statement,

$$n_{H_2O}/n_{dry \, gas} = (14)/(100-14) = 0.1628 \tag{3.2}$$
$$P_T = 15.8794 \, \text{psia}$$

Vapor pressure of MEA at 130°F (54.4°C)
$$= 4.21 \, \text{mmHg} \times (14.696/760) = 0.08141 \, \text{psia}$$

and

$$n_{MEA}/n_{dry \, gas} = p_{vap_{MEA}}/(P_T - p_{vap_{H_2O}}) \tag{3.2}$$
$$= (0.08141)/(15.8794 - 2.2232) = 0.0060$$
$$= 0.60 \, \text{mol\%} \times 10^4 = 6000 \, \text{ppmv (dry)}$$

Subtracting $p_{vap_{MEA}}$ along with $p_{vap_{H_2O}}$ from P_T still results in 6000 ppmv (dry). This approach predicts an extremely high mass flow rate of MEA to the atmosphere and makeup rate to replenish lost MEA in solution.

$$\frac{78.8 \times 10^6 \, \text{SCF}}{\text{day}} \frac{(100-14)}{(100)} \frac{\text{day}}{(24 \, \text{h})} \frac{\text{lb mol}}{379.5 \, \text{SCF}} = 7440.5 \, \text{lb mol/h of dry gas}$$

$$(7440.5)(0.60/100)(61.09 \, \text{lb MEA/lb mol MEA}) = 2727.24 \, \text{lb/h MEA}$$
$$\times 4.38 = 11,945 \, \text{ton/year of MEA}$$

This would translate into a loss of MEA chemical of 32.7 ton everyday, and a required makeup of almost 214 ton/day of 15.3 wt% MEA solution.

Clearly, something is wrong here and this approach is too conservative. Better techniques to estimate MEA emissions are explored in parts (b) and (c).

(b) The answer is that this is not a situation involving the vapor pressure of a pure component, but rather a case of vapor–liquid equilibrium (VLE) of a species in solution. As long as there

198 Chapter 3 Air Combustion

is a liquid phase present along with a vapor/gas phase, water and MEA constitute a binary VLE system in the presence of N_2 and the other permanent gases. As we shall see below, this does not make much difference in the estimation of the water component in the gas/vapor phase, but it makes a huge difference for the MEA.

From Chapter 2,

$$p_{partial_{H_2O}} = \gamma_{H_2O} p_{vap_{H_2O}} X_{H_2O} \qquad (2.56)$$

$$p_{partial_{MEA}} = \gamma_{MEA} p_{vap_{MEA}} X_{MEA} \qquad (2.57)$$

$$p_{MEA_{system}} = p_{partial_{H_2O}} + p_{partial_{MEA}} \qquad (2.59)$$

where the quantities denoted by X are mole fractions in the liquid state.
For 15.3 wt% MEA (Basis: 100 lb),

Component	MW	wt%	mol = wt/MW	mol%
H_2O	18.02	84.70	4.7003	94.9402
MEA	61.09	15.30	0.2505	5.0598
Total	—	100.00	4.9508	100.0000

This calculates approximately to 5 mol% MEA and 95 mol% H_2O, or to mole fractions of 0.05 and 0.95.

A typical range for MEA solutions is approximately 15–20 wt% MEA, or about 5–7% on a molar basis.

The approach here in part (b), much less conservative than in part (a), is to assume that Raoult's law is valid, meaning that the activity coefficients of both components are unity for all values of x. Then,

$$p_{partial_{H_2O}} = (1) p_{vap_{H_2O}} (0.95) = (2.2232 \text{ psia})(0.95) = 2.1120 \text{ psia}$$

and

$$p_{partial_{MEA}} = (1) p_{vap_{MEA}} (0.05) = (0.08141)(0.05) = 0.0040705 \text{ psia}$$

In essence, the partial pressure of MEA is reduced to 5% of its pure component vapor pressure.

For H_2O,

$$P_T = p_{partial_{H_2O}} / (n_{H_2O}/n_{dry\ gas}) + p_{partial\ H_2O} \qquad (3.87)$$
$$= (2.1120/0.1628) + 2.1120 = 15.0850 \text{ psia}$$

and for MEA

$$n_{MEA}/n_{dry\ gas} = p_{partial_{MEA}} / (P_T - p_{partial_{H_2O}}) \qquad (3.2)$$
$$= (0.0040705)(15.0850 - 2.1120)$$
$$= 0.000314$$
$$= 0.0314 \text{ mol}\% \times 10^4 = 314 \text{ ppmv (dry)}$$

3.14 Miscellaneous Sources of Air Emissions

The new result for mass flow rate is 142.7 lb/h or 625.1 ton/year of MEA released to the atmosphere and only 11.2 ton/day of solution makeup.

(c) This next approach is the suggested one and accounts for the nonideal nature of the water–MEA system. This system has been studied by a number of investigators, and their data are summarized by DECHEMA [98,99]. Water–MEA shows negative deviations from Raoult's law like the acetone–chloroform system, discussed in Chapter 2, as shown by a plot of negative ln γ's in one of the original sources [100]. At $X_{H_2O} = 0.95$, γ_{H_2O} is 0.99 or higher from slightly over 100°C (212°F) down to 25°C (77°F). This is to be expected since γ approaches 1.0 as concentration approaches a pure component.

The activity coefficient for MEA, γ_{MEA}, at $X_{MEA} = 0.05$ is plotted in Figure 3.9. The point values were computed here from arithmetic averages of the van Laar and Margules equations whose constants were determined from the data of various investigations and documented by DECHEMA. The computed values for γ_{MEA} at $X_{MEA} = 0.05$ range from about 0.75 at the higher end temperature down to about 0.25 at 25°C (77°F).

The two straight lines in the figure were then fitted to the points by least squares.

$$\gamma_{MEA} \text{ (at } X_{MEA} = 0.05) = (0.001582)T \text{ (°F)} + 0.124373 \quad (3.88)$$
$$(T = 77°F \text{ to } 172.4°F)$$

and

$$\gamma_{MEA} \text{ (at } X_{MEA} = 0.05) = (0.008442)T \text{ (°F)} - 1.05833 \quad (3.89)$$
$$(T = 172.4°F \text{ to } 214.5°F)$$

Ideally, one would expect a nice smooth curve for physical property data, but intersecting straight lines are good enough to use here for estimation of MEA partial pressures.

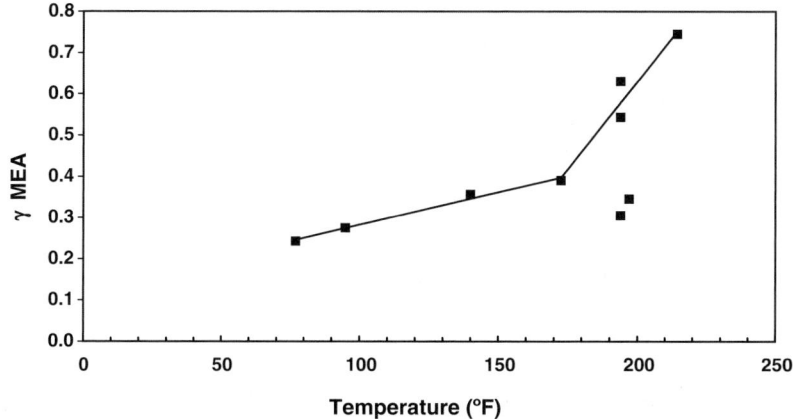

Figure 3.9 Activity coefficient at 0.05 MEA mole fraction in water–MEA system at atmospheric pressure.

200   Chapter 3   Air Combustion

For this case,
$$p_{\text{partial}_{H_2O}} = (0.99)(0.95)p_{\text{vap}_{H_2O}} \tag{2.56}$$
for $\gamma_{H_2O} = 0.99$ and $X_{H_2O} = 0.95$
$$p_{\text{partial}_{H_2O}} = (0.99)(0.95)(2.2232) = 2.0909 \text{ psia}$$

From Equation (3.87),
$$P_T = (2.0909/0.1628) + 2.0909 = 14.9343 \text{ psia}$$

From Equation (3.88), at 130°F γ_{MEA} (at $X_{MEA} = 0.05$) = 0.3300,
$$p_{\text{partial}_{MEA}} = (0.33)(0.05)(0.08141) \tag{2.57}$$
$$= 0.001343 \text{ psia}$$

$$n_{MEA}/n_{\text{dry gas}} = (0.001343)/(14.9343 - 2.0909) \tag{3.2}$$
$$= 0.00010457$$
$$\times 10^6 = 104.6 \text{ ppmd}$$
(from part(a))

MEA mass flow rate = (7440.5 lb mol/h of dry gas)$(104.6/10^6)(61.09) = 47.54$ lb/h
$\times 4.38 = 208.2$ ton/year
$= 0.57$ ton/day $= 1141$ lb/day of MEA

for a 15.3 wt% MEA solution with a density of 62.74 lb/ft^3 (8.39 lb/gal) at 20°C [101],

$$\frac{(1141)}{(0.153)} \frac{(1)}{(8.39)} = 889 \text{ gal/day MEA solution makeup}$$

∎

PROBLEM 3.48

This problem is related to Problem 3.47, inspired by Ref. [96,97]. Here, the overhead stream from the MEA absorber vessel is the desired product, and the unwanted CO_2-rich stream leaving the MEA regenerator/stripper is to be vented to atmosphere. Temperature at the top of the regenerator is 200°F and pressure is 23 psia. It is followed by a cooling water condenser and a condensate drum at 150°F. The underflow from the drum is pumped back to the MEA regenerator, and the vent is located on the drum. Estimate the atmospheric emission of MEA. The mass flow rate of CO_2 is 8.5 MMSCFD.

SOLUTION The CO_2 will remain in the gas phase as it passes through the condenser. It is the primary permanent gas constituent leaving the regenerator vessel and the condenser/condensate drum. Drawing a sketch may be helpful.

At 200°F, γ_{MEA} (at $X_{MEA} = 0.05$) = 0.6301 from Equation (3.89). Vapor pressure of MEA at 200°F (93.3°C) = 35.8320 mmHg = 0.6929 psia calculated from the Antoine equation with the constants from Problem 3.47.

$$p_{\text{partial}_{MEA}} = (0.6301)(0.05)(0.6929) = 0.021830 \text{ psia}$$

Similarly,

vapor pressure of H_2O at 200°F = 596.0533 mmHg = 11.5258 psia

3.14 Miscellaneous Sources of Air Emissions

$$p_{partial_{H_2O}} = (0.99)(0.95)(11.5258) = 10.8400 \text{ psia}$$

From the problem statement, total pressure of the 200°F gas is 23 psia. It is assumed to consist only of CO_2, H_2O, and MEA. Otherwise, the term CO_2 includes the actual CO_2 plus any other permanent gases present.

$$23.0000 \text{ psia total}$$
$$10.8400 \text{ psia } H_2O$$
$$0.0218 \text{ psia MEA}$$
$$12.1382 \text{ psia } CO_2 \text{ (by difference)}$$

Cooling to 150°F will condense out some water and MEA.
At 150°F, γ_{MEA} @ $X_{MEA} = 0.05 = 0.3617$ from Equation (3.88). Vapor pressure of MEA at 150°F (65.6°C) = 8.2111 mmHg = 0.1588 psia calculated from the Antoine equation with the constants from Problem 3.47.

$$p_{partial_{MEA}} = (0.3617)(0.05)(0.1588) = 0.0028719 \text{ psia}$$

In like manner,

$$\text{vapor pressure of } H_2O \text{ at } 150°F = 192.3249 \text{ mmHg} = 3.7190 \text{ psia}$$
$$p_{partial_{H_2O}} = (0.99)(0.95)(3.7190) = 3.4977 \text{ psia}$$

At 150°F,

$$p_{CO_2} = (12.1383)(150 + 459.67)/(200 + 459.67) = 11.2182 \text{ psia } CO_2$$
(from above and correcting for lower temperature)   3.4977 psia H_2O
                                                     0.0029 psia MEA
                                                     14.7188 psia total

On the basis of 100 mol CO_2 at either temperature

	Moles at 200°F		Moles at 150°F		Moles condensed
CO_2		100.00		100.00	0.0000
H_2O	$\dfrac{(10.8400)(100)}{(12.1382)} =$	89.3048	$\dfrac{(3.4977)(100)}{(11.2182)} =$	31.1788	58.1260
MEA	$\dfrac{(0.021830)(100)}{(12.1382)} =$	0.1798	$\dfrac{(0.0028719)(100)}{(11.2182)} =$	0.0256	0.1542
Total		189.4846		131.2044	58.2802

Check:

$$H_2O : \frac{(31.1788)}{(131.2044)}(14.7188) = 3.4977 \text{ psia}$$

$$MEA : \frac{(0.0256)}{(131.2044)}(14.7188) = 0.0029 \text{ psia}$$

$$MEA = (0.0256/100)(10^6) = 256 \text{ ppmd}$$

202 Chapter 3 Air Combustion

$$\frac{256 \text{ lb mol MEA}}{10^6 \text{ lb mol CO}_{2\,(\text{dry basis})}} 8.5 \times 10^6 \text{ SCF/day CO}_2 \frac{\text{lb mol CO}_2}{379.5 \text{ SCF CO}_2} \frac{\text{day}}{24 \text{ h}} \frac{69.01 \text{ lb MEA}}{\text{lb mol MEA}}$$

$$= 16.49 \text{ lb/h} \times 4.38 = 72.2 \text{ ton/year}$$

$$\times 24 = 395.7 \text{ lb/day}$$

$$\frac{(395.7)}{(0.153)} \frac{(1)}{(8.39)} = 308 \text{ gal/day MEA solution makeup}$$

The MEA concentration in the vented gas is higher here than in part (c) of Problem 3.47. In contrast, the mass flow rate of MEA to atmosphere is considerably less because the mass flow rate of dry vented gas is much smaller here than in part (c).

Verification of estimated MEA emissions can be achieved via stack testing after start-up. To a lesser extent, good long-term records of MEA solution usage and material balance provide a consistency check on calculated emissions.

A final note: Operating conditions for the problems were determined by a combination of inputs from more than one source, supplemented by some changes and assumptions. Therefore, the calculations presented here for these hypothetical cases are not intended to represent the actual emissions from any particular MEA system at any specific location. ∎

REFERENCES

1. R.C. Corey, editor, *Principles and Practices of Incineration*, Wiley–Interscience, New York, 1969, 297 pp.
2. U.S. Environmental Protection Agency, Determination of Stack Gas Velocity and Volumetric Flow, 40 CFR 60, Appendix A, Method 2, U.S. Government Printing Office, Washington, DC, 2002 (and later editions).
3. R.H. Perry and C.H. Chilton, *Chemical Engineers' Handbook*, 5th edn, McGraw-Hill, New York, 1973.
4. U.S. Environmental Protection Agency, Determination of Sulfur Dioxide Removal Efficiency and Particulate Matter, Sulfur Dioxide, and Nitrogen Oxide Emission Rates, 40 CFR 60, Appendix A, Method 19, July 1, 2007.
5. R.C. Weast, editor, *CRC Handbook of Chemistry and Physics*, 58th edn, CRC Press, Inc., Cleveland, OH, 1977.
6. C.A. Meyer, R.B. McClintock, G.J. Silvestri, and R.C. Spencer Jr., *1967 ASME Steam Tables*, 2nd edn, American Society of Mechanical Engineers (ASME), New York, 1968, pp. 88.
7. *Steam: Its Generation and Use*, 40th edn, The Babcock & Wilcox Company, Baberton, OH, 1992.
8. R. Kneille, Anatomy of combustion calculations, *Hydrocarbon Processing*, **74**(5), 87–96, 1995.
9. D.R. Lide, editor, *CRC Handbook of Chemistry and Physics*, 72nd edn, CRC Press, Inc., Boca Raton, FL, 1991.
10. D.R. Lide, editor, *CRC Handbook of Chemistry and Physics*, 71st edn, CRC Press, Inc., Boca Raton, FL, 1990.
11. D.R. Lide, editor, *CRC Handbook of Chemistry and Physics*, 88th edn, CRC Press, Boca Raton, FL, 2008.
12. J.A. Dean, editor, *Lange's Handbook of Chemistry*, 11th edn, McGraw-Hill, New York, 1973, p. 10–45.
13. R.G. Kunz, Calculate water in combustion air, *Hydrocarbon Processing*, **77**(9), 119–122, 1998.
14. O.A. Hougen, K.M. Watson, and R.A. Ragatz, *Chemical Process Principles: Part I—Material and Energy Balances*, 2nd edn, Wiley, New York, 1954, 504 pp.
15. E.T. Williams and R.C. Johnson, *Stoichiometry for Chemical Engineers*, McGraw-Hill, New York, 1958, 350 pp.
16. I. Frankel, Shortcut calculation for fluegas volume, *Chemical Engineering*, **88**(11), 88–89, 1981.

References

17. North American Combustion Handbook, North American Manufacturing Co., Cleveland, OH (1st edn, 2nd printing, p.16, 1957; 2nd edn, p.18, 1978).
18. O.P. Goyal, Guidelines help combustion engineers, *Hydrocarbon Processing*, **59**(11), 205–211, 1980.
19. R.G. Kunz, D.D. Smith, N.M. Patel, G.P. Thompson, and G.S. Patrick, Control NO_x from gas-fired hydrogen reformer furnaces, Paper AM 92-56 presented at the 1992 NPRA Annual Meeting, New Orleans, LA, March 22–24, 1992.
20. J.M. Smith and H. C. Van Ness, *Introduction to Chemical Engineering Thermodynamics*, 2nd edn, McGraw-Hill, New York, 1959.
21. J.H. Perry, editor, *Chemical Engineers' Handbook*, 3rd edn, McGraw-Hill, New York, 1950, 1941 pp.
22. *Compilation of Air Pollutant Emission Factors*, AP-42, Vol. I, 5th edn, Introduction, U.S. EPA, Washington, DC, January 1995.
23. R.G. Kunz, D.D. Smith, and E.M. Adamo, Predict NO_x from gas-fired furnaces, *Hydrocarbon Processing*, **75**(11), 65–79, 1996.
24. Compilation of Air Pollutant Emission Factors, AP-42, Vol. I, 5th edn, Supplement D, Chapter 1: External Combustion Sources, Section 1.4, Natural Gas Combustion, U.S. EPA, Washington, DC, July 1998.
25. Compilation of Air Pollutant Emission Factors, AP-42, Vol. I, 5th edn, Supplement E, Chapter 1: External Combustion Sources, Section 1.3, Fuel Oil Combustion, U.S. EPA, Washington, DC, September, 1998.
26. General Rules: Definitions, 31 TAC, Chapter 101, Section 101.1, Texas Air Control Board (TACB), Austin, TX, Revised October 16, 1992.
27. Resolution of 'Natural Gas' and 'Pipeline Natural Gas' Definition Issues, U.S. EPA Clean Air Markets, available at http://www.epa.gov/airmarkets/emissions/gasdef.html, last updated August 14, 2007.
28. Inorganic Gaseous Pollutants: Nitrogen Oxides Emissions from Natural Gas-Fired Boilers and Water Heaters, Regulation 9, Rule 6, Section 9-6-208, Bay Area Air Quality Management District (BAAQMD), San Francisco, CA, November 7, 2007.
29. Rule 414, Natural Gas-Fired Water Heaters, Definition 205, Natural Gas, Sacramento Metropolitan AQMD, Sacramento, CA, adopted August 1, 1996.
30. Sulfur Content of Gaseous Fuels, Rule 431.1 (b) (10), South Coast Air Quality Management District (SCAQMD), Diamond Bar, CA, amended June 12, 1998.
31. Technical Guidance Package for Combustion Sources: Reciprocating Engines & Stationary Gas Turbines, Draft Report, p. 26 of 48, Texas Natural Resource Conservation Commission, Austin, TX (undated).
32. R. Hamilton and A. Talianchich, BACT Determinations for Boilers >40 MMBtu/hr, Interoffice Memorandum to New Source Review Permit Reviewers, Texas Commission on Environmental Quality (TCEQ), Austin, TX, March 3, 2003.
33. 30 TAC Part J, Chapter 17, Subchapter B, Division 3, Section 117.206 (c) (8) (B) (effective May 19, 2005).
34. Documentation for an Evaluation of Air Pollution Control Technologies for Small Wood-Fired Boilers, Technical Report, Resource Systems Group, Inc., White River Junction, VT, Revised September 2001; available at www.rsginc.com/pdf/R_Wood_Bact_Sept_2001.PDF, accessed on June 16, 2008.
35. Compilation of Air Pollutant Emission Factors, AP-42, Vol. I, 5th edn, Supplement B, Chapter 1: Liquefied Petroleum Gas Combustion, U.S. EPA, Washington, DC, October 1996.
36. U.S. Environmental Protection Agency, Standards of Performance for Petroleum Refineries, 40 CFR 60, Subpart J, U.S. Government Printing Office, Washington, DC, Electronic Code of Federal Regulations, http://ecfr.gpoaccess.gov/cgi/t/text/text current as of June 12, 2008.
37. D.R. Bartz, K.W. Arledge, J.E. Gabrielson, L.G. Hays, and S.C. Hunter (KVB Engineering, Inc.), *Control of Oxides of Nitrogen in the South Coast Air Basin (of California)*, Appendix C (Oil Refineries), PB-237 688, National Technical Information Service (NTIS), Springfield, VA, September 1974.
38. S.H. Tan, Find SO_2 concentration by nomograph, *Oil & Gas Journal*, **77**(22), 118, 1979.

39. A. Zanker, Estimating the dew points of stack gases, *Chemical Engineering*, **85**(9), 154–155, 1978.
40. R.R. Pierce, Estimating acid dewpoints in stack gases, *Chemical Engineering*, **84**(8), 125–128, 1977.
41. R.J. Jaworowski, Condensed sulfur trioxide: particulate or vapor? *Journal of Air Pollution Control Association*, **23**(9), 791–793, 1973.
42. A. Attar, Evaluate sulfur in coal, *Hydrocarbon Processing*, **58**(1), 175–179, 1979.
43. C.F. Prutton and S.H. Maron, *Fundamental Principles of Physical Chemistry*, Revised Edition, Sixth Printing, Macmillan, New York, 356–361, 1957.
44. C.F. Cullis, R.M. Henson, and D.L. Trimm, The kinetics of homogeneous gaseous oxidation of sulphur dioxide, *Proceedings of the Royal Society of London A*, **295**, 72–83, 1966.
45. R.R. Martin, F.S. Manning, and E.D. Reed, Watch for elevated dew points in SO_3-bearing stack gases, *Hydrocarbon Processing*, **53**(6), 143–144, 1974.
46. A. Levy, E.L. Merryman, and W.T. Reid, Mechanisms of formulation of sulfur oxides in combustion, *Environmental Science & Technology*, **4**(8), 653–662, 1970.
47. F.H. Verhoff and J.T. Banchero, Predicting dew points of flue gases, *Chemical Engineering Progress*, **70**(8), 71–72, 1974.
48. E.S. Lisle and J.D. Sensenbaugh, The determination of sulfur trioxide and acid dew point in flue gases, *Combustion*, **36**, 12–16, 1965.
49. R. J. Jaworowski and S. S. Mack, Evaluation methods for measurement of SO_3/H_2SO_4 in flue gas, *Journal of Air Pollution Control Association*, **29**(1), 43–46, 1979.
50. Y-H Kiang, Predicting dewpoints of acid gases, *Chemical Engineering*, **88**(3), 127, 1981.
51. *Sampling Procedures Manual*, Stack Facilities—Figure 2-2, Texas Air Control Board, Austin, TX, January 1983, p. 2–6 (revised July 1985).
52. U.S. Environmental Protection Agency, Test Methods, 40 CFR 60, Appendix A, U.S. Government Printing Office, Washington, DC, July 1, 1995.
53. W. S. Smith and D. J. Grove, Selecting a stack sampling consultant, *Pollution Engineering*, **9**(6), 36–39, 1977.
54. Model 3 Orsat Analyzer, Apex Instruments, Inc., available at http://www.apexinst.com/products/orsat.htm, accessed on May 28, 2008.
55. Instruction 11-9026 FYRITE Gas Analyzer CO and O Indicators, available at www.bacharach-inc.com/PDF/Instructions/11-9026.pdf, accessed on May 28, 2008.
56. Fyrite Gas Analyzers, available at www.midatlanticdiagnostics.com/documentation/pdf/bach/fyrite_gas_analyzers.pdf, accessed on May 28, 2008.
57. R.H.F. Pao, *Fluid Mechanics*, Wiley, New York, 1961, p.110.
58. R.L. Daugherty and J.B. Franzini, *Fluid Mechanics with Engineering Applications*, 6th edn, McGraw-Hill, New York, 1965, p. 332.
59. J.B. Franzini and E.J. Finnemore, *Fluid Mechanics with Engineering Applications*, 9th edn, WCB/McGraw-Hill, Boston, MA, 1997, pp. 515–516.
60. H.A. Haaland, editor, *Methods for Determination of Velocity, Volume, Dust, and Mist Content of Gases*, Bulletin WP-50 7th edn, Western Precipitation Division/Joy Manufacturing Company, Los Angeles, CA, 1970, 38 pp.
61. Emission Measurement Branch Technical Support Division, OAQPS, EPA, Method 202-Determination of Condensible Particulate Emissions from Stationary Sources, U.S. EPA Technology Transfer Network Emission Measurement Center, Text of Method Plus Frequently Asked Questions (FAQs), available at http://www.epa.gov/ttn/emc/promgate.html, posted September 25, 1996.
62. V.U. Garcia, P.A. Guris, and M. Hampton, Continuous emissions monitoring system in ammonia plant, *Ammonia Plant Safety & Related Facilities: A Technical Manual*, Vol. **34**, American Institute of Chemical Engineers, New York, NY, 1994, pp. 125–145.
63. T.C. Elliott, CEM system: lynchpin holding CAA compliance together, *Power*, **139**(5), 31–40, 1995.
64. J.A. Jahnke, *Continuous Emission Monitoring*, 2nd edn, John Wiley & Sons, Inc., New York, 2000, 404 pp.
65. A.M. Cheng and G.F. Hagen, An accurate predictive emissions monitoring system (PEMS) for an ethylene furnace, *Environmental Progress*, **15**(1), 19–27, 1996.

References

66. R.L. Croy, C.W. Simpson, D.B. Tipton, P. Sinha, and M.E. Lashier, NO_x PEMS experience on operating ethylene furnaces, *Environmental Progress*, **16**(4), 297–300, 1997.
67. U.S. Environmental Protection Agency, Continuous Emission Monitoring, 40 CFR 75, U.S. Government Printing Office, Washington, DC, current as of July 10, 2008, available at http://ecfr.gpoaccess.gov/cgi/t/text/text-idx, accessed on July 13, 2008.
68. U.S. Environmental Protection Agency, Specification and Test Procedures for SO_2 and NO_x Continuous Emission Monitoring Systems in Stationary Sources, 40 CFR 60, Appendix B, Performance Specification 2, U.S. Government Printing Office, Washington, DC, 1995.
69. W. E. Mitchell, Quality assurance support branch, *Quality Assurance Handbook*, Interim edn, Vol. **III**, Section 3.0.7, (November 5, 1985), U.S. Environmental Protection Agency, Cincinnati, OH, February 24, 1994, available as PDF document at www.eps.gov/ttnemc01/qahandbook3/ (accessed on July 11, 2008).
70. J.J. Morgester, D.L. Frisk, G.L. Zimmerman, R.C. Vincent, and G.H. Jordan, Control of emissions from refinery valves and flanges, *Chemical Engineering Progress*, **75**(8), 40–45, 1979.
71. R. Wetherold and L. Provost, *Emission Factors and Frequency of Leak Occurrence for Fittings in Refinery Process Units*, EPA-60/2-79-044, U.S. EPA, Washington, DC, 1979.
72. Control of Volatile Organic Compound Leaks from Petroleum Refinery Equipment, EPA-450/2-78-036 OAQPS No. 1.2-111, U.S. EPA, Research Triangle Park, NC, June 1978.
73. Protocol for Equipment Leak Emission Estimates, EPA 453/R-93-026, p. 2-11, U.S. EPA (June 1993); also available as PB 93-229212 from National Technical Information Service (NTIS), Springfield, VA (1993).
74. Protocol for Equipment Leak Emission Estimates, EPA 453/R-95-017, U.S. EPA, November 1995; also available as PB 96-175401 from National Technical Information Service, Springfield, VA, 1996, available online http://www.epa.gov/ttn/chief/efdocs/equiplks.pdf, as of July 6, 2008.
75. Compilation of Air Pollutant Emission Factors, AP-42, Vol. I, 5th edn, Chapter 5: Petroleum Industry, Section 5.1, Petroleum Refining, U.S. EPA, Washington, DC, January 1995.
76. T.W. Hughes, D.R. Tierney, and Z.S. Khan, Measuring fugitive emissions from petrochemical plants, *Chemical Engineering Progress*, **75**(8), 35–39, 1979.
77. Protocols for Generating Unit-Specific Emission Estimates for Equipment Leaks of VOC and VHAP, EPA-450/3-88-010, U.S. EPA, Research Triangle Park, NC, October 1988.
78. Protocol for Equipment Leak Emission Estimates, EPA 453/R-93-026, p. 2-10, U.S. EPA, June 1993; also available as PB 93-229212 from National Technical Information Service (NTIS), Springfield, VA, 1993.
79. Technical Guidance Package for Chemical Sources: Equipment Leak Fugitives, Texas Natural Resource Conservation Commission (TNRCC), Austin, TX, March 1995.
80. R.G., Wetherhold, L.P. Provost, and C.D. Smith,(Radian Corporation). Assessment of Atmospheric Emissions from Petroleum Refining: Volume 3. Appendix B. Prepared for U.S. Environmental Protection Agency. Research Triangle Park, NC. Publication No. EPA-600/2-80-075c, April 1980.
81. Compilation of Air Pollutant Emissions Factors, AP-42, Vol. I, 5th edn, Chapter 13: Miscellaneous Sources, Section 13.4, Wet Cooling Towers, U.S. EPA, Washington, DC, January 1995.
82. N. C. J. Chen and S. R. Hanna, Drift: modeling and monitoring comparisons, Paper presented at the Cooling Tower Institute, Houston, TX, January 31–February 2, 1977.
83. A.C. Hile III, L. Lai, K. Kolmetz, and J.S. Walker, Monitor cooling towers for environmental compliance, *Chemical Engineering Progress*, **97**(3), 37–42, 2001.
84. C.L. Yaws, L. Bu, and S. Nijhawan, New correlation accurately calculates water solubilities of aromatics, *Oil & Gas Journal*, **92**(35), 80–82, 1994.
85. Technical Supplement 2: Cooling Towers, TCEQ Publication RG-360 (Revised), Texas Council on Environmental Quality (TCEQ), Austin, TX, January 2006.
86. M. Fitzpatrick, User's Guide Emission Control Technologies and Emission Factors for Unpaved Roads Fugitive Emissions, EPA/625/5-87/022, U.S. Environmental Protection Agency, Cincinnati, OH, September 1987.
87. Compilation of Air Pollutant Emission Factors, AP-42, Vol. I, 5th edn, Chapter 13: Miscellaneous Sources, Section 13.2.2, Unpaved Roads, U.S. EPA, Washington, DC, November 2006.

88. Compilation of Air Pollutant Emission Factors, AP-42, Vol. I, 5th edn, Chapter 7: Liquid Storage Tanks, Section 7.1, Organic Liquid Storage Tanks, U.S. EPA, Washington, DC, November 2006.
89. J. S. Seitz, Calculating Potential to Emit (PTE) for Emergency Generators, U.S. EPA Guidance Memorandum to the Regional Air Directors, September 6, 1995, PDF document available at http://www.epa.gov/ttn/oarpg/15/memoranda/emgen.pdf, updated July 15, 2008.
90. Compilation of Air Pollutant Emission Factors, AP-42, Vol. I, 5th edn, Chapter 3: Stationary Internal Combustion Engines, Section 3.3, Gasoline and Diesel Industrial Engines, U.S. EPA, Washington, DC, October 1996.
91. U.S. Environmental Protection Agency, Standards of Performance for Stationary Compression Ignition Internal Combustion Engines, 40 CFR 60, Subpart IIII, promulgated in *Federal Register* (71 FR 39172 of July 11, 2006), U.S. Government Printing Office, Washington, DC; Electronic Code of Federal Regulations, http://ecfr.gpoaccess.gov/cgi/t/text/text, current as of July 25, 2008.
92. Compilation of Air Pollutant Emission Factors, AP-42, Vol. I, 5th edn, Chapter 5: Petroleum Industry, Section 5.2, Transportation and Marketing of Petroleum Liquids, U.S. EPA, Washington, DC, June 2008.
93. S. Strelzoff, Choosing the optimum CO_2-removal system, *Chemical Engineering*, **82**(19), 115–120, 1975.
94. M. S. DuPart, T. R. Bacon, and D. J. Edwards, Understanding corrosion in alkanolamine gas treating plants, *Hydrocarbon Processing*, International Edition, Part 1, **72**(4), 75–80, 1993, Part 2, 72(5), 89–94, 1993.
95. R. L. Pearce and M. S. DuPart, Amine inhibiting, *Hydrocarbon Processing*, **64**(5), 70–75, 1985.
96. S. Hopson, Amine inhibitor copes with corrosion, *Oil & Gas Journal*, **83**(26), 44–47, 1985.
97. R.N. Vaz, G.J. Mains, and R.N. Maddox, Ethanolamine process simulated by rigorous calculation, *Hydrocarbon Processing*, **60**(4), 139–142, 1981.
98. J. Gmehling, U. Onken, and J. R. Rarey-Nies, *Vapor–Liquid Equilibrium Data Collection Aqueous Systems*, Supplement 2, DECHEMA Chemistry Data Series, Vol. **I**, Part 1b, Deutsche Gesellschaft für Chemisches Apparatewesen, Chemisches Apparatewesen, Chemische Technik und Biotechnologie e.V. (DECHEMA), Frankfurt, Germany, 1988, pp. 129–133.
99. J. Gmehling and U. Onken, *Vapor–Liquid Equilibrium Data Collection Aqueous Systems*, Supplement 3, Br_2–$C_4H_{10}O$, DECHEMA Chemistry Data Series, Vol. I, Part 1c (in conjunction with Part 1d), Deutsche Gesellschaft für Chemisches Apparatewesen, Chemische Technik und Biotechnologie e.V. (DECHEMA), Frankfurt, Germany, 2003. pp. 290–296.
100. K. Tochigi, K. Akimoto, K. Ochi, F. Liu, and Y. Kawase, Isothermal vapor–liquid equilibria for water + 2-aminoethanol + dimethyl sulfoxide and its constituent three binary systems, *Journal of Chemical Engineering Data*, **44**(3), 588–590, 1999.
101. S. Cheng, A. Meisen, and A. Chakma, Predict Amine Solution Properties Accurately, *Hydrocarbon Processing*, **75**(2), 81–84, 1996.

Chapter 4

Air Control Devices

A pop-up behind third base is the shortstop's ball.

4.1 OVERVIEW

This chapter deals with various techniques for abatement of air pollution. Examples are chosen from among several pollutants. The explanation is at the level of process calculations, a conceptual design, or a screening study as an aid in completing the permit application. A properly chosen abatement strategy may make the difference between being able to construct a source at all or may reduce the emissions sufficiently to transfer the case into a much simpler permitting regime. It is expected that input from the technology vendor will be necessary for the final design.

With any abatement system, the ratio of outlet concentration to inlet concentration is known as the *penetration*. Multiply by 100, and it is the % penetration. The factor 1 − penetration or 100 − % penetration ((inlet − outlet)/inlet) is the removal, reduction, or collection *efficiency*. This efficiency is calculated in several example problems.

Finally, the last section deals with the stack as a sort of abatement device to provide dispersion of flue gas (FG) and its constituents to ground-level concentrations. A few simple situations are considered, but this is a specialized area in which calculations are carried out by EPA approved dispersion models requiring inputs of specific detailed meteorological data. Entire texts including Ref. [1], known to professionals in the field as "Turner's Workbook," have been written on this one topic, and any further consideration of the subject is beyond the scope of this book.

4.2 FLARES

Flaring is a high-temperature oxidation process employed to dispose of combustible components in waste gases from industrial processes. There are two types of flares—elevated flares and ground flares. Since elevated flares, with their larger capacities,

---

Environmental Calculations: A Multimedia Approach, by Robert G. Kunz
Copyright © 2009 John Wiley & Sons, Inc.

are the more common type, discussion here will be focused on them. A flow diagram of a typical steam-assisted smokeless elevated flare is shown in Chapter 13 of EPA AP-42 [2].

4.2.1 Flare Operation

Many waste gases burn in a flare without emitting smoke. However, waste gases containing heavy hydrocarbons such as paraffins (alkanes) above methane, olefins, diolefins, and aromatics will produce smoke upon combustion unless sufficient steam and/or air is injected at the flare tip. This promotes proper mixing of the waste gas and combustion air. Steam requirement is on the order of 0.1–1 lb of steam per lb of gas being flared, dictated by the constituents in the flare gas [3]. Typical refineries are said to use 0.25 lb/lb and chemical plants 0.5 lb/lb because of greater amounts of unsaturated hydrocarbons (olefins and aromatics) being flared [4]. Flare operation must be smokeless per 40 CF 60.18 [5].

In addition to atmospheric emissions, noise, light, and heat also accompany flare operation. Elevating the flare high enough seeks to minimize the "forbidden zone" encircling the flare where thermal radiation is too high for longer term personnel exposure [6]. An alternate design allows access to an area in which exposure is tolerable for the brief time interval until workers can retreat to a safe distance whenever the flare ignites unexpectedly [7]. Unless adversely affecting the plant's neighboring population, this constitutes more of an occupational health and safety issue than an environmental matter. Further discussion is contained in the references cited above.

The waste gas must have a heating value of 200–250 Btu/SCF or more for complete combustion without the addition of auxiliary fuel [2]. Heating values, compositions, and flare velocities for stable operation are specified in 40 CFR 60.18 [5]. These items should all be addressed in the flare design by the manufacturer of the flare tip.

Flares usually operate at high overall ratios of excess air. Approximately 60% excess combustion air was noted for tests of flaring natural gas [3,8].

A pilot gas must be provided to ignite the waste gas at the flare tip. Up to four pilot burners are required for flare tip diameters 1–60 in. and beyond [4,9], and 70 SCF/h (SCFH) of 1000 Btu/SCF fuel gas (natural gas or equivalent) is required for each pilot burner [4]. In general, more pilots are required for a waste stream containing a significant amount of ammonia or CO because ignition of these compounds is more difficult than for hydrocarbons [9].

4.2.2 Flare Emissions

Like other hydrocarbon combustion sources, flares produce environmental contaminants. EPA Manual AP-42 [2] provides emission factors given in Table 4.1 for total unburned hydrocarbons, carbon monoxide (CO), nitrogen oxides (thermal NO_x), and unburned carbon particles (soot). Emissions of sulfur oxides and fuel NO_x must be

Table 4.1 AP-42 Flare Emission Factors

Component	Emission factor lb/MMBtu (except as noted)
Total hydrocarbons (measured as methane equivalent)	0.14
CO	0.37
NO_x	0.068
Soot[a]	0–274[a]

[a] Soot is in concentration values—0 μg/L for nonsmoking flares, 40 μg/L for lightly smoking flares, 177 μg/L for average smoking flares, and 274 μg/L for heavily smoking flares.

determined separately by material balance if sulfur- or nitrogen-containing compounds are found in the waste gas.

The EPA's flare emission factors [2] are based on tests using a mixture of 80% propylene (MW = 42.08, HHV = 2332.8) and 20% propane (MW = 44.11, HHV 2516.1). The average molecular weight of this mixture calculates to 42.49 lb/lb mol, with a higher heating value of 2369.4 Btu/SCF (60°F, 1 atm). Other data sources, for example, Ref. [10], make use of a similar fuel.

One million Btu (1 MMBtu, 10^6 Btu) of this gas corresponds to

$$10^6 \, \text{Btu} \left(\frac{\text{SCF}}{2369.4 \, \text{Btu}} \right) \left(\frac{\text{lb mol}}{379.5 \, \text{SCF}} \right) = 1.112 \, \text{lb mol} \times 42.49$$

$$= 4725 \, \text{lb of gas flared}$$

Hydrocarbons : $(1 \, \text{MMBtu})(0.14 \, \text{lb}/\text{MMBtu})/(16.04 \, \text{lb}/\text{lb mol})$
$= 0.00873 \, \text{lb mol of HC escaping}$

CO: $(1 \, \text{MMBtu})(0.37 \, \text{lb}/\text{mm Btu})/(28.01 \, \text{lb}/\text{lb mol}) = 0.01321 \, \text{lb mol of CO formed}$

$0.00873 + 0.01321 = 0.02194 \, \text{lb mol of flare gas not destroyed}$

The emission factor for hydrocarbons is expressed as methane (MW = 16.04)

$100(0.02194/1.112) = 1.97\%$ of flare gas not destroyed

$100(1.112-0.02194)/1.112 = 98.03\%$ of flare gas destroyed on a molar or volume basis

This explains the statement in AP-42 that "properly operated flares achieve at least 98% combustion efficiency, meaning that hydrocarbon and CO emissions amount to less than 2% of hydrocarbons in the gas stream." Here, the molecular weight of unspecified hydrocarbons (HC) has been taken as 16.04, the molecular weight of methane, as noted previously. Using another convention, the molecular weight of carbon (12.01), or the molecular weight (25.98–26.14) of the average measured composition of hydrocarbons escaping (Table 4.2), makes little difference.

Table 4.2 Composition of Hydrocarbon Species[a] in Flare Emissions

Constituent	vol% Average	vol% Range
Methane	55	14–83
Ethane/ethylene	8	1–14
Acetylene	5	0.3–23
Propane	7	0–16
Propylene	25	1–65

[a] From EPA Publication AP-42, Table 13.5-2. Reference to the flare efficiency study (1983) from which this table was developed as well as a brief summary of test conditions is contained in Ref. [2]. Flared gas was a mixture of 80% propylene and 20% propane.

Destruction efficiency on a weight basis for the propylene–propane mixture of the EPA flare tests is close to, but not exactly, 98% compared to the AP-42 emission factors for hydrocarbons and CO. Table 4.2 supports the conclusion that the degradation products are different from the species originally contained in the flared gas.

Guidelines issued by the Texas Council on Environmental Quality (TCEQ), the agency responsible for environmental permitting of projects within the state of Texas, address the provisions of 40 CFR 60.18 in permitting of new or modified flares [9]. They require a flare destruction efficiency of at least 98% by weight and allow a 99% destruction efficiency to be used in calculations for C_1–C_3 hydrocarbons containing only carbon and hydrogen in the molecule, C_1–C_3 alcohols, ethylene oxide, and propylene oxide. A 99.5% destruction efficiency may be claimed when providing adequate documentation that the hydrocarbon compounds "are not difficult to combust." In addition, ethane and hydrogen, as well as methane, are not regulated as pollutants in the state of Texas. However, acceptable destruction efficiency for reporting hydrocarbon emissions on the annual emission inventory is limited by Texas guidelines [11] to 93% for existing flares not meeting the stability criteria of 40 CFR 60.18.

Furthermore, Texas guidelines consider the composition of the uncombusted flared gas to remain unchanged [11]. Although it is acknowledged that complex reactions in the flare may generate new compounds, it is stated that there exists no definitive method to identify them. Thus, the flare's combustion products as calculated would contain the same hydrocarbon species, reduced to a nominal 2% of what was present in the original waste gas. Nevertheless, the EPA flare test found that propane plus propylene broke down into the compounds shown in Table 4.2.

Finally, TCEQ lists emission factors different from EPA's for the other air contaminants. Similar supplementary rules certainly exist in other environmental jurisdictions, and one must be aware of such requirements in the permitting process. Discussion of TCEQ procedures here is simply to highlight one example of said differences.

PROBLEM 4.1

A waste hydrocarbon gas is sent to a smokeless flare at a rate of 30 MMBtu/h. This gas contains no sulfur or nitrogen compounds. Estimate the emissions of environmental contaminants to the atmosphere using the EPA AP-42 emission factors.

SOLUTION Using the factors in Table 4.1,

HC (expressed as methane) = $(30)(0.14) = 4.20$ lb/h $\times 4.38 = 18.40$ ton/year
CO = $(30)(0.37) = 11.1$ lb/h $\times 4.38 = 48.62$ ton/year
NO_x = $(30)(0.068) = 2.04$ lb/h $\times 4.38 = 8.94$ ton/year
Soot = $(30)(0) = 0$ lb/h $= 0$ ton/year ∎

In preparing the emission inventory, do not forget to include the emissions, however small, from combustion of the natural gas fed to the pilot tip(s).

4.3 SELECTIVE CATALYTIC REDUCTION

Selective catalytic reduction (SCR) is a process in which oxides of nitrogen (NO_x) are reacted in the presence of a suitable catalyst to nitrogen and water vapor [12,13]. It is called *selective* because other reagents such as methane react with NO_x, but they react with other constituents in the flue gas as well. For typical base metal catalysts composed of various percentages of titanium dioxide (TiO_2), vanadium pentoxide (V_2O_5), tungsten trioxide (WO_3), and molybdenum trioxide (MoO_3) [12], a typical operating temperature window is 600–750°F (316–399°C) [14,15].

SCR reactions take place within a flow-through catalyst composed of many parallel small passages normal to the flow. Flue gases containing soot, ash, or other high particulate matter (PM) require larger openings to preclude plugging by solids. The square openings for a honeycomb catalyst range from about 3 mm on a side for clean gas applications to about 7 mm for gases heavily laden with particulates [15–18]. Photographs of extruded SCR catalysts are contained in Refs [17,19–24].

The mechanism is thought to involve absorption of ammonia on active catalyst sites on the inside surfaces of the flow passages, where it lies in wait for unsuspecting NO_x molecules. Ammonia and NO_x then react on the surface, and the reaction products diffuse from the surface back into the main flow. This then frees the site for further absorption, reaction, diffusion, and so on.

4.3.1 SCR Equations

For a combustion effluent with a high NO to NO_2 ratio, reactions are as follows:

$$4NO + 4NH_3 + O_2 \rightarrow 4N_2 + 6H_2O \quad (4.1)$$
$$NO + NO_2 + 2NH_3 \rightarrow 2N_2 + 3H_2O \quad (4.2)$$

According to these reactions, the theoretical molar ratio of ammonia (NH_3) to NO_x (NO + NO_2) is 1:1.

Some competing side reactions can also occur. These include ammonia oxidation by flue gas oxygen to NO, conversion of NO_x to other nitrogen compounds, and reaction of ammonia with flue gas SO_3. If allowed to become significant, these would affect the theoretical ammonia to NO_x reaction ratio. CO was found to be unaffected across the SCR [12,13].

A design equation describing the SCR process is given in the literature [18]:

$$\eta = m[1 - \exp(-k/SV)] \qquad (4.3)$$

where η = fractional NO_x conversion efficiency

$$(NO_x\text{in} - NO_x\text{out})/NO_x\text{in} \qquad (4.4)$$

m is the molar ratio of NH_3 to NO_x used, exp is the exponential function, k is the effective kinetic constant (1/time), a lumped parameter reflecting the catalyst composition, temperature, composite rate of reaction and mass transfer, loss of catalyst activity with age, and the effects of flue gas moisture and oxygen on reaction rate. Mass transfer depends on catalyst geometry, surface properties, and flue gas velocity. Space velocity (SV) (1/time), the reciprocal of residence time, is the flue gas volumetric flow [32°F (0°C) and 1 atm] divided by the superficial catalyst volume. Area velocity (AV), volumetric flow divided by the planar catalyst surface area not including catalyst pore structure, is sometimes used in lieu of space velocity. AV and SV are related by the area to volume ratio of the catalyst. A space velocity of 10,000/h, for example, is reported for an SCR installed on a small oil-fired, fire-tube packaged boiler [24].

4.3.2 SCR Operation

A key to achieving SCR design performance is uniformity of distribution of flow, ammonia, and temperature across the face of and throughout the catalyst structure.

A typical plot of outlet data against normalized ammonia feed rate, (NH_3in/NO_xin), at a given space velocity is depicted in Figure 4.1. (In some literature, this ratio (NH_3in/NO_xin) is denoted as alpha (α).) An aqueous ammonia solution is often used as the feed, and NH_3in is calculated from solution flow rate multiplied by ammonia concentration. Concentration is determined from solution density.

Ideal NO_x conversion/removal is shown as a straight line at a 1:1 slope corresponding to the reactions of Equations (4.1) and (4.2). When reactions are complete at NH_3in/NO_xin = 1.0 and beyond, NO_x removal efficiency levels off at 1.0, or 100%.

Ideal slip in the absence of significant side reactions is calculated by material balance.

$$NH_3\text{slip} = NH_3\text{in} - (NO_x\text{in} - NO_x\text{out}) \qquad (4.5)$$

from which

$$NH_3\text{slip}/NO_x\text{in} = NH_3\text{in}/NO_x\text{in} - (NO_x\text{in} - NO_x\text{out})/NO_x\text{in} \qquad (4.6)$$

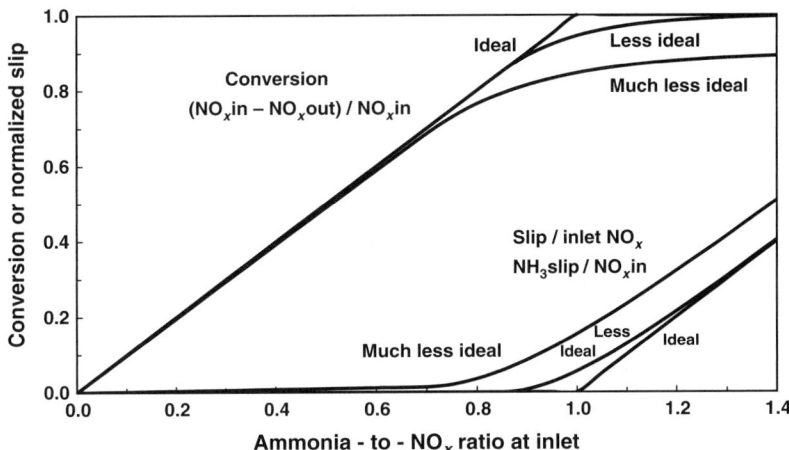

Figure 4.1 Typical NO_x conversion and NH_3 slip curves for selective catalytic reduction.

Actual SCR data, however, deviate somewhat from the ideal. For fresh catalyst, the "less ideal" NO_x removal curve in Figure 4.1 follows the 1:1 line more or less closely for low NO_x removals but approaches complete conversion on an asymptote at higher ammonia feed rates. Consistent with Figure 4.1, the design value for $NH_{3}in/NO_{x}in = \alpha$ is often in the range of 1.1–1.2. Ammonia slip also deviates from the ideal.

In use over time, the activity of SCR catalyst can be expected to degrade. There always seems to be something in the flue gas that will decrease, use up, poison, occupy, or mask active sites on the catalyst. If it is not an ongoing change in catalyst crystal structure, then it is trace components such as arsenic and phosphorous from coal combustion that affect catalyst activity. For steam methane reformer (SMR) hydrogen plant furnaces [22,25] and ethylene pyrolysis furnaces [20,26,27], it is chromium.

This results in a shift in conversion and slip toward the "much less ideal" curves of Figure 4.1 as time goes on. Enough catalyst must be installed to meet permit limits not only at start-up but also at end of run, which one would prefer to last on the order of years. Similar curves from a TVA pilot plant operation are shown elsewhere [28].

Without changing plant production (altering firing rate, flue gas flow, and space velocity) or to a lesser degree by modifying excess air, once an SCR is in place and operating, the only practical means of control is by adjusting the ammonia feed rate. As the catalyst ages and activity declines (for whatever reason), more ammonia must be injected to control NO_x even at the same outlet condition. At the higher ammonia feed rate, more unreacted ammonia (ammonia slip) appears at the outlet. When both the NO_x limit and the ammonia slip limit in the permit cannot be met simultaneously, as they say in sports, "the ballgame is over," and the catalyst must be replaced.

In the field, $NO_{x}in$ and $NO_{x}out$ of the SCR is commonly measured using the instrumental version of EPA Method 7 (Method 7E) [29]. Ammonia slip can be determined by Bay Area Air Quality Management District (California) Method ST-1B, a test method in existence since 1982 [30]. This procedure absorbs flue gas ammonia in 0.1 N HCl solution contained in the impingers of a Method 5-type sampling train. The impinger solutions are subsequently analyzed for ammonia by BAAQMD Analytical Procedure Lab-1A by means of a specific ion electrode technique [31].

AP-42 indicates NO_x reduction efficiencies of 75–85% achieved by SCR on oil-fired boilers in the United States [32] and efficiencies ranging from 80% to 90% for utility boilers [33]. More recent information points to a requirement in California for greater than 95% NO_x reduction to meet emission limits of 9 ppm or less [34].

Be also prepared to have to meet an ammonia slip limit in the single digit whole number ppm range. Some regulatory agencies require monitoring and reporting of ammonia feed rate as well.

A practical calculation [35,36] involving SCR NO_x removal and ammonia slip is discussed in Problem 4.2.

PROBLEM 4.2

A combustion source emits NO_x at a relatively steady 180 vppmd (corrected to 3% O_2 dry). The flue gas is treated using SCR to achieve an outlet NO_x concentration of 9 vppmd, also corrected to 3% O_2 dry. NO_xin and NO_xout are determined on a day-to-day basis by the continuous emission monitoring system (CEMS) of the plant. Ammonia is measured annually by stack testing. In the absence of significant side reactions in the SCR, estimate the ammonia slip on a daily basis between stack tests, assuming that NO_x removal and ammonia slip follow the "less ideal" curves of Figure 4.1.

SOLUTION From Equation (4.4), NO_x destruction efficiency

$$\eta = (180 - 9)/180 = 0.95$$

As read from Figure 4.1, $\alpha = NH_3\text{in}/NO_x\text{in} = 1.0$.
Then

$$NH_3\text{in} = (1.0)(180) = 180$$

(In an actual case, one would use the ammonia feed rate from plant measurements.)
From Equation (4.5),

$$NH_3\text{slip} = 180 - (180 - 9) = 9 \text{ vppmd at } 3\% \text{ } O_2 \text{ dry}$$

This value of ammonia slip may be higher than permitted in some jurisdictions. In that case, a different SCR design would be necessary for compliance. One solution might involve some combination of burners emitting a lower NO_x concentration and additional SCR catalyst. ■

4.4 SELECTIVE NONCATALYTIC REDUCTION

Selective noncatalytic reduction (SNCR) reacts ammonia with NO_x but without a catalyst. There are two variations of SNCR: one uses urea (NH_2CONH_2) as the source of ammonia, and the other uses ammonia directly. Both are patented [37, p. 1–13]. The ammonia-based system is known as Thermal DeNOx®, developed and patented by Exxon Research and Engineering Company in 1975 [38].

4.4 Selective Noncatalytic Reduction

The urea-based process was developed and patented by the Electric Power Research Institute (EPRI) in 1980 and is known under the trade name NOx OUT® [39,40]. The technology was licensed to (Nalco) Fuel Tech, which holds several additional patents claiming improvements and enhancements to the basic process. Fuel Tech has several sublicensees authorized to supply and install SNCR technology in several industrial sectors.

4.4.1 SNCR Design and Operation

Information on key aspects of the design of SNCR and SCR systems is considered proprietary by vendors; it is therefore necessary to develop correlations from available data using regression and curve fitting techniques in lieu of using proprietary design parameters [37, p. iii].

Reaction of NO, the predominant form of NO_x, is the same as for SCR [41]:

$$4NO + 4NH_3 + O_2 \rightarrow 4N_2 + 6H_2O \quad (4.1)$$

As with SCR, the ammonia can be used in either anhydrous or aqueous form. For dry solid forms, urea breaks down into ammonia [42]

$$NH_2CONH_2 \rightarrow NH_3 + HNCO \quad (4.7)$$

and reacts as above.

For urea–water solutions,

$$NH_2CONH_2 + H_2O \rightarrow 2NH_3 + CO_2 \quad (4.8)$$

Without a catalyst, the ammonia–NO_x reactions take place at a higher temperature than in an SCR. According to one source [43], the temperature window for a urea-based SNCR is 1400–2100°F (760–1149°C) and 1400–1780°F (760–972°C) for SNCR utilizing ammonia. The optimal temperature for a urea system is said to be about 900°C (1652°F) [24] and about 955°C (1750°F) for ammonia addition [44].

Besides temperature, other variables affecting the effectiveness of SNCR are mixing of the reagent in the flue gas, residence time, ratio of reagent to NO_x, and the sulfur content of the fuel leading to ammonium (bi)sulfate by-products [32]. Excess O_2 in the flue gas is also an important consideration [44].

The advantages of SNCR over SCR are that there is no catalyst to plug, foul, or blind, and the cost is less. The disadvantage is a lower NO_x removal efficiency.

4.4.2 SNCR Performance

AP-42 [33,34] cites a maximum SNCR performance of 25–40% in limited applications including natural gas-fired boilers. Some applications mentioned are even lower (13–24%) [33]. The ammonia-based system has demonstrated NO_x removal efficiencies of approximately 60%, up to 70%, as reported in an EPA document [44].

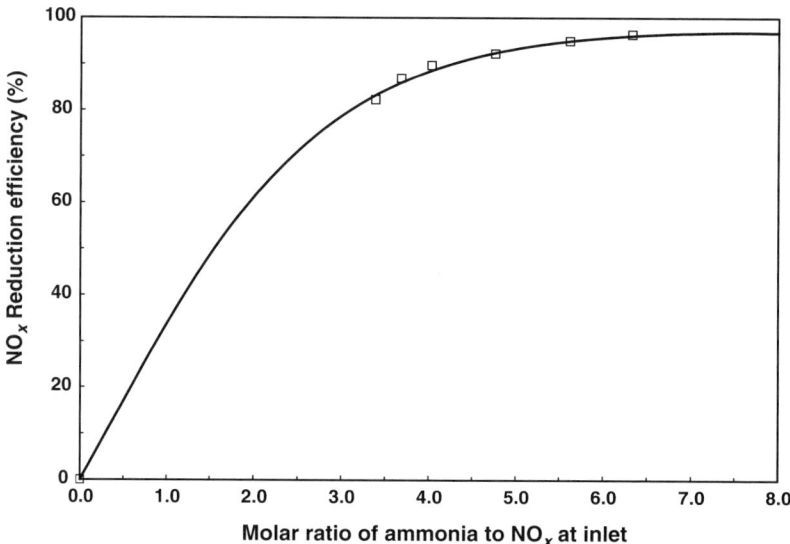

Figure 4.2 SCR performance on coal-fired circulating fluidized bed boiler [41].

In other literature, typical SNCR NO_x removal was 30–40% at 680°C (1256°F) (urea based on a small fire-tube packaged boiler) [24], 40–55%, up to as high as 66%, under performance test controlled conditions (urea-based on the CO boiler (1400–1850°F, 760–1010°C) of a fluid catalytic cracking (FCC) unit) [43], and 89% at about 1650°F (~900°C) (ammonia based on a 49.9 MW coal-fired fluidized bed boiler (CFB)) [41]. (See Appendix J for a discussion of the FCC process.)

One explanation for the variability in NO_x removal efficiency is the ammonia feed rate.[1] Similar to SCR, NO_x removal responds to the amount of ammonia injected, as in the curve of Figure 4.2. This curve was prepared from values read from a tiny graph presented in the CFB article [41] along with other information contained there. Data from the testing conducted after a maintenance outage and internal modifications to the boiler were used.

The curve of Figure 4.2 has the same general shape as that for the fire-tube boiler [24]. Increasing the inlet ammonia to NO_x ratio is maximally effective initially. However, the rate of change becomes less as ammonia feed is increased past a point of diminishing returns.

References [37, p. 1–11; 44] indicate that optimal process efficiency occurs at a (NH_3/NO_x)in molar ratio of 2.0. The CFB's NO_x control efficiency at approximately 89% corresponds to a molar ratio of about 4:1.

The ammonia slip is also an important consideration. Slips tend to be higher than for SCR. Allowable slip for the CO boiler application was 20 ppm in about 1988 [43],

---

[1] In the nomenclature of SNCR, the molar ammonia to NO_x ratio at the inlet, (NH_3/NO_x)in, is known as the normalized stoichiometric ratio (NSR) [24; 37, p. 1–11]. It is the same variable known as alpha (α) in SCR technology.

but undoubtedly would be much lower today. Measured ammonia slip concentration from 18 stack test runs for the CFB in the same time frame averaged 21 ppm with a 10 ppm standard deviation; the excess ammonia fed is said to be expected to form N_2 and H_2O [41].

PROBLEM 4.3

The curve of Figure 4.2 shows a NO_x removal efficiency of about 89% at an $(NH_3/NO_x)_{in}$ ratio of 4.0. Measured slip averaged 21 ppm (at 12% CO_2 dry). Estimate the NO_x removal efficiency at a ratio of 2.0.

SOLUTION From the curve at $(NH_3/NO_x)_{in} = 2.0$, removal efficiency $[(NO_{x\,in}-NO_{x\,out}/NO_{x\,in}]$ is approximately 61%. Ammonia slip is anticipated to be below, yet no worse than, 21 ppm. Not much more can be said with certainty since the investigator found little or no correlation between ammonia slip and NO_x [41]. However, strictly as an estimate, Figure 1.7 in an EPA publication [37] relating SNCR NO_x reduction efficiency and ammonia slip shows an ammonia slip of about 15 ppm for 61% efficiency. This graph is redrawn and reproduced as Figure 4.3.

Note: The values in this problem are specific to this particular case, and generalizations to other SNCR units should not be drawn without a proper basis. ∎

4.4.3 SNCR plus SCR

SNCR can also be used in conjunction with SCR to increase the overall efficiency. One example is a combined SNCR/SCR system in a technical paper referred to previously [24]. The SCR temperature was approximately 230°C (446°F) (233°C up

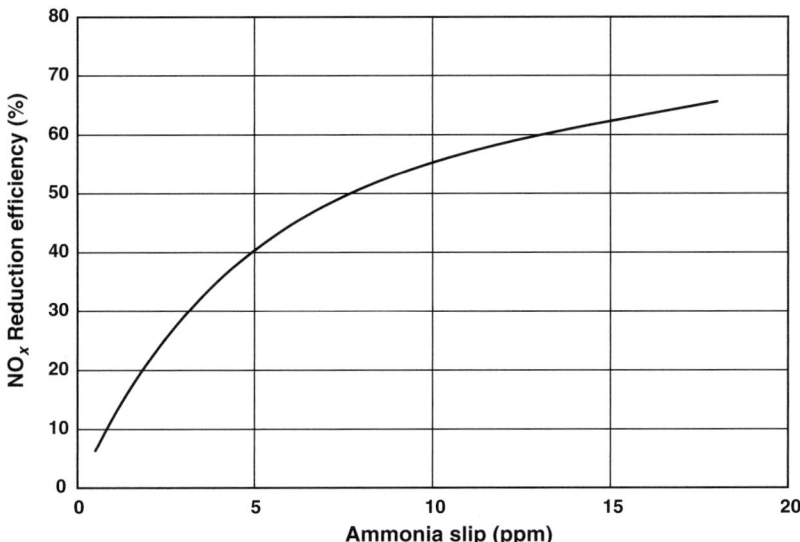

Figure 4.3 NO_x reduction for various ammonia slip levels [37].

to 300 °C, 451–572°F). The ammonia slip from urea fed to the upstream SNCR becomes ammonia feed to the SCR. Ammonia slip from the SCR was less than 6 ppm.

PROBLEM 4.4

If the overall NO_x removal efficiency of an SNCR/SCR combination in series is 85%, and the upstream SNCR removes 30–40% of the NO_x, what is the efficiency of the SCR alone?

SOLUTION This is a typical problem in which the overall collection/destruction efficiency is related to efficiencies of devices in series.

Basis 100 units of NO_x entering the SNCR

At 30–40% efficiency (E), the percent penetration (P) is 70%.

$$\%P = 100 - \%E$$
$$\%P = 100 - (30 - 40) = 60 - 70\%$$
(4.9)

$(100)[(60–70)/100] = 60–70$ NO_x units at the exit of the SNCR/entrance of the SCR.

Overall efficiency is 85%

Overall $\%P = 100 - 85 = 15\%$

At the exit of the SCR

$(100)(15/100) = 15$ units

Across the SCR

In: 60–70 units

Out: 15 units

$(100)(15/60) = 25\%$ $P = 75\%$ E

$(100)(15/70) = 21.4\%$ $P = 78.6\%$ E

Answer: Across the SCR, NO_x removal efficiency is approximately 75–79% when NO_x removal efficiency is 30–40% across the SNCR upstream and overall NO_x removal efficiency is 85%. Depending on the amount of SCR catalyst present and the space velocity, this SCR efficiency will most likely lie on the linear section of the SCR NO_x removal curve, exemplified by Figure 4.1. ∎

4.5 FLUE GAS RECIRCULATION

Flue gas recirculation (FGR) is the diversion of a portion of the flue gas from a combustion source back to the burners. The recirculated flue gas can be added to the fuel, to the air, or to both. Before its recycle to the burners, the flue gas is cooled from the very high temperature of the combustion zone. The presence of this cooler gas lowers the flame temperature and, depending on how it is controlled, may also reduce the excess O_2 in the firebox. It may also shorten the residence time of combustion products in the flame.

4.5.1 FGR Performance

AP-42 indicates that FGR can reduce NO_x by as much as 40–50% in some boilers when used alone [32] and by 40–85% in conjunction with low-NO_x burners [33], both with respect to uncontrolled levels. The emission factor in AP-42's Table 1.4-1 [33] for natural gas combustion reflects a NO_x reduction of 64% for large wall-fired boilers, a reduction of 55% for all sizes of tangential-fired boilers, and a 68% reduction in NO_x when combined with low-NO_x burners in wall-fired boilers firing less than 100 MMBtu/h. Many low-NO_x burners use the principle of flue gas recirculation internal to the burner to lower NO_x.

4.5.2 FGR Data Analysis

An example of FGR on a small experimental boiler took place in the 1930s long before today's concern with atmospheric pollution from contaminants such as nitrogen oxides [45,46]. This work was conducted not for environmental control, but to develop an inexpensive way of generating an inert gas containing a reduced concentration of nitric oxide (NO) and its reaction product, nitrogen dioxide (NO_2), found to be corrosive to equipment. (Recall that NO constitutes some 95% of the NO_x (NO + NO_2) in a combustion effluent [32].)

The process was developed by an engineer with Lone Star Gas Company, Dallas, Texas. Before adding FGR to existing larger scale commercial production, natural gas with a heating value of 1250 Btu/SCF was fired at an excess air of 9–16% in a boiler rated at 2.4 MMBtu/h. In many of the experimental runs, excess air was held at 12%. A portion of the flue gas was cooled from combustion temperatures to 140°F (60°C), mixed with the fuel, and recycled to the gas burner(s).

Furnace temperature, measured with an ordinary thermocouple, was seen to decrease from 1338°C (2440°F) to as low as 977°C (1791°F) with increasing flue gas recirculation. However, a high-velocity thermocouple, used on occasion as a check, revealed that the observed temperatures were 300–400°F (167–222°C) lower than actual firebox temperatures. Preliminary data (without recycle) showed a firebox temperature of ~2800°F (1538°C), but it was thought that maximum flame temperatures were likely to be as high as 3400–3500°F (1871–1927°C) from the data of Ref. [47].

Nitric oxide in the stack gas was reported along with the extent of flue gas recirculation, expressed as a volume basis ratio of flue gas recirculated to natural gas fired. At ratios of 6 or more, the NO concentration could be maintained at less than 3 ppm without putting out the flame. Nitric oxide concentration without FGR was measured at 65 ppm. A U.S. patent was issued for this process [45].

Results are tabulated and plotted in Ref. [46]. The graph is redrawn and reproduced here (Figure 4.4). A version of this graph has appeared before in one of author's previous papers [48]. The data and two correlating curves are shown—one from the original investigator and the other derived from a custom fit using calculated

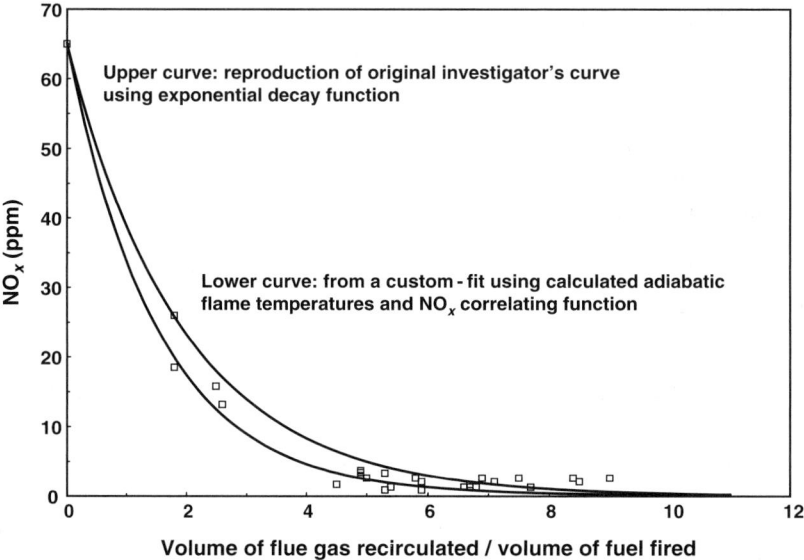

Figure 4.4 Flue gas recirculation decreases NO_x experimental furnace natural gas data NO_x versus volume of flue gas recirculated/volume of fuel fired.

adiabatic flame temperatures (AFTs) and the NO_x correlating functional form from Appendix C.[2] This function describes the data[3] better than the original investigator's line that resembles a simple exponential decay curve.

A more conventional presentation plots percent NO_x reduction versus the percent of flue gas recycled (Figure 4.5). The original investigator's curve, transformed from Figure 4.4, does not fit the data in this format very well at all and has been dropped. The only thing left from the original graph are the data points, recast as percent reduction and plotted against percent of flue gas recycled. Values from the curve of Figure 4.5 fit in nicely with a graph from a more recent technical publication [50]. That author claims that a 10–20% recirculation is typically used for NO_x control.

The uncontrolled NO_x from the Lone Star boiler and the percent reduction by flue gas recirculation vis-à-vis AP-42 emission factors are the subjects of the next example problem.

---

[2] A fuel composition is necessary to compute AFT. (See Chapter 3 for AFT calculation.) Because the natural gas analysis was not reported, a published composition for Dallas natural gas [49] was enriched with typical proportions of higher hydrocarbons to agree with the stated heating value of 1200 Btu/SCF used in the calculations.

[3] The reported NO was assumed to be on a wet basis since there is no mention of gas drying/moisture removal in the analytical methods cited. When flue gas composition remains constant or reasonably constant with recycle, this assumption is not critical for comparing results internal to a particular case.

4.5 Flue Gas Recirculation

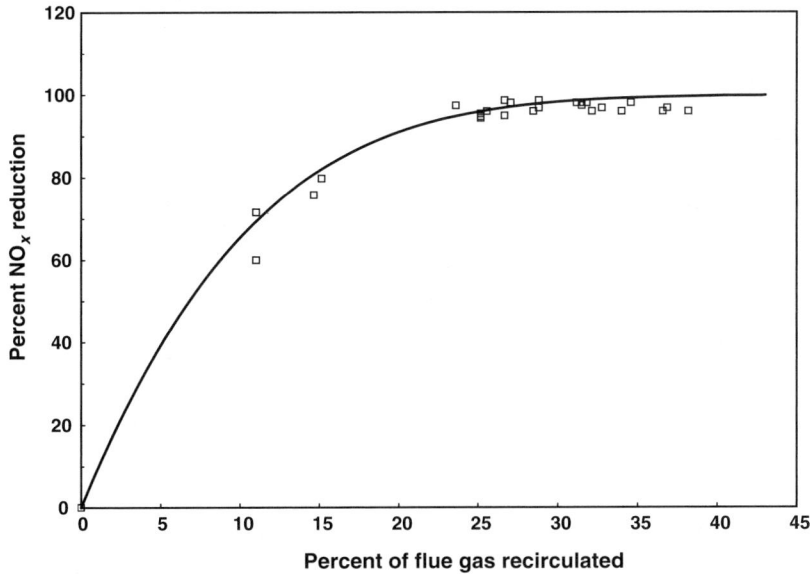

Figure 4.5 Flue gas recirculation decreases NO_x experimental furnace natural gas data [46] percent NO_x reduction versus percent of flue gas recirculated.

PROBLEM 4.5

Verify that NO_x results from the experimental furnace discussed in this section are consistent with AP-42 emission factors.

SOLUTION

For the Case Without Flue Gas Recirculation
Combustion of 100 lb mol of a natural gas with a higher heating value (HHV) of 1250 Btu/SCF results in the following flue gas characteristics. (Refer to Chapter 3 for details of combustion calculations.)

Characteristic	Air 100°F, 60% RH	Air 60°F bone dry
FG flow (wet) (lb mol)	1455	1441
% H_2O	16.85	16.04
% O_2 wet	2.04	2.06
% O_2 dry	2.46	2.46

Humidity in the ambient combustion air does not have a major influence here on flue gas properties.

Using the more realistic case including ambient moisture,

$$\frac{(1455 \text{ lb mol FG})}{h} \frac{(65 \text{ lb mol NO}_x)}{10^6 \text{ lb mol FG}} \frac{(46.01 \text{ lb NO}_x)}{\text{lb mol NO}_x} = 4.35 \text{ lb NO}_x/h$$

$$(100 \text{ lb mol fuel}) \frac{(1250 \text{ Btu})}{\text{SCF}} \frac{(379.5 \text{ SCF})}{\text{lb mol}} = 47.44 \text{ MMBtu/h}$$

$4.35/47.4 = 0.092$ lb NO_x/MMBtu for the case without flue gas recirculation

From AP-42 [33], Table 1.4.1, uncontrolled NO_x from a small boiler is

$$(100 \text{ lb}/10^6 \text{ SCF})/(1020 \text{ Btu/SCF}) = 0.098 \text{ lb NO}_x/\text{MMBtu}$$

The NO_x from the conventional burner(s) of the experimental furnace is certainly in the ballpark of the AP-42 factor (uncontrolled, no flue gas recirculation) $\cong 0.1$ lb/MMBtu.

Calculations for No FGR Using Method 19 [51]
Dry flue gas at stoichiometric conditions

$$(1455)[(100-16.85)/100][(20.9-2.46)/(29.9-0)] = 1067.43 \text{ lb mol/h}$$

or

$$(1441)[(100-16.04)/100][(29.9-2.46)/(20.9-0)] = 1067.46 \text{ lb mol/h}$$

$\times 385.3 = 411,300$ SCF/h (68°F, 1 atm) dry flue gas at stoichiometric conditions

$\div 47.44$ MMBtu/h (computed above)

$= 8670$ SCF (@68°F, 1 atm)/10^6 Btu (HHV)

This is the F_d factor for this particular natural gas versus $F_d = 8710$ listed for natural gas in Table of 40 CFR 60, Appendix A, Method 19 [51].
There is only a 0.5% difference between the two.
Using the EPA F_d factor of 8710 and Eq. (19.4) of Method 19 (see Table 3.4),

$$E = \frac{(1.194 \times 10^{-7})(70)(8710)(20.9)}{(1-0.1685)(20.9-2.46)} = 0.0992 \text{ lb NO}_x/\text{MMBtu}$$

Again, very close to the AP-42 emission factor.

The Case with FGR
AP-42 Table 1.4-1 [33] gives an emission factor for small boilers with FGR only in combination with low-NO_x burners. This is 32% of the NO_x from the uncontrolled case, or a 68% reduction. The text of the natural gas combustion section indicates a range of 60–90% reduction using FGR and low-NO_x burners together. The fuel oil combustion section of AP-42 [32] indicates a reduction by means of FGR alone of as much as 40–50% in some boilers.

As noted above, a range on the order of 10–20% of the combustion air FGR is said to be typical [50]. A graph in that reference shows a NO_x reduction of about 60% at 10% FGR and 80% at 20% FGR. For the natural gas combustion in the experimental furnace under discussion here, recirculation of 10% of the flue gas results in a 65% reduction and a 20% recirculation in a NO_x reduction of 91%. The combustion air amounts to about 92% of the flue gas that it generates.

In any event, especially for low to mid-range values, the FGR curve of Figure 4.5 is apparently in reasonable agreement with the graph presented in a technical journal. However, it predicts somewhat higher performance than AP-42.

The amount of NO_x reduction is a strong function of how the FGR process is conducted, for example, to what temperature the flue gas is cooled before being reintroduced into the combustion zone and how the excess O_2 is controlled. One should therefore always be sure of the percent NO_x reduction to be realized by FGR before making any commitments in a permit application in order to secure a permit.   ∎

4.6 WATER/STEAM INJECTION INTO COMBUSTION GAS TURBINE

Combustion gas turbines are used in electrical power generation and other applications. They are composed of three major components—a compressor, a combustor, and a power turbine. Air is drawn into the compressor, where it is compressed up to 30 times ambient pressure. The compressed air proceeds to the combustor, where fuel is introduced, ignited, and burned. Expansion of the hot exhaust gases causes a rotary motion in the shaft to drive the compressor and an external load, such as an electrical generator. Gas turbines are fired primarily on natural gas or No. 2 distillate fuel oil [52].

Operation of a gas turbine with exhaust to atmosphere of the hot flue gas typically in excess of 1000°F (+810°K) without heat recovery is called *simple cycle*. It is thermally inefficient. When a heat exchanger is added to heat the combustion air, combustion efficiency improves, and the process is called *regenerative cycle*. However, the added cost of the heat exchanger may not be economically justified.

Addition of a heat recovery steam generator (HRSG, pronounced *her-sig*) to a simple cycle gas turbine results in *combined cycle*, in which the flue gas temperature is reduced by generating steam. At least some of the steam can be directed to a steam turbine driving an electrical generator to produce additional electricity.

4.6.1 Emissions and Abatement

The primary atmospheric contaminants are NO_x and CO. Volatile organic compounds (VOCs), particulate matter, and hazardous air pollutants (HAPs) are also formed. Sulfur compounds, mainly sulfur dioxide (SO_2), also appear in the exhaust. Nitrogen oxides arise from the usual mechanisms of thermal NO_x and when distillate oil containing fuel-bound nitrogen is fired, fuel NO_x. Carbon monoxide, VOCs, PM, and HAPs arise from incomplete combustion; sulfur dioxide (SO_2) originates from the sulfur content of the fuel.

Atmospheric emissions from stationary gas turbines are regulated by 40 CFR 60, Subpart GG [53], as well as by other emission regulations in a particular environmental regulatory jurisdiction. Such regulation may require abatement of uncontrolled emissions. Thermal NO_x can be controlled by lowering the peak temperature in the combustion zone by injection of water or steam, and fuel NO_x by burning a fuel containing a lower concentration of fuel bound nitrogen. Both types of NO_x can be abated by an add-on NO_x control such as selective catalytic reduction.

With water or steam injection, the amount injected is typically less than 1:1 water to fuel on a weight basis. Steam could be provided by the turbine's HRSG itself; high-purity water or steam is required for injection into gas turbines [54].

Carbon monoxide, as well as other pollutants from incomplete combustion, can be controlled through CO oxidation catalysts. NO_x and CO (and possibly the other pollutants) can be controlled by means of other technology in various stages of commercial development.

The emission factors in AP-42 reflect both uncontrolled emissions and abatement of NO_x and CO by water/steam injection. The NO_x and CO factors carry an "A" (excellent) rating. However, EPA recognizes that the uncontrolled emission factor for CO is higher than the water–steam injection factor for CO, noted as "contrary to expectation." No explanation could be identified by EPA except that the data sets used for developing these factors are different.

4.6.2 Permitting

For a permit application, it is common practice to use AP-42 factors along with turbine manufacturer's guarantees/test data when available. One's own historical data may also be used. Bear in mind that the numbers appearing in the application become part of the permit and that compliance with permit limits will have to be verified by source testing during operation. This should prompt the use of the most realistic entries in the judgment of the applicant.

Use of the AP-42 factors with an eye to pollutant abatement by water/steam injection is exemplified in Problem 4.6.

PROBLEM 4.6

Estimate the emissions from an uncontrolled gas turbine rated at 40.5 MMBtu/h (LHV) heat input, and the degree of NO_x control to be achieved from water or steam injection. The gas turbine is fired on pipeline natural gas with a HHV of 1020 Btu/SCF, a lower heating value (LHV) of 918 Btu/SCF, and a sulfur content of 0.3 grains of H_2S per 100 SCF. The gas turbine is followed by a HRSG, but no additional firing takes place in the HRSG.

SOLUTION

Strategy

The major flue gas constituents Table 4.3 will be calculated using the F factor method described in Chapter 3. The contaminant emissions, uncontrolled and controlled, will be computed using the AP-42 factors for combustion turbines fired on natural gas [52], found in Tables 3.1-1, 3.1-2a, and 3.1-3 of the cited document. All sulfur in the fuel is assumed to be converted to SO_2, and the SO_2 will be calculated by material balance.

Combustion calculations will be performed at ISO standard dry conditions, as defined in 40 CFR 60, Subpart GG, New Source Performance Standards for Stationary Gas Turbines [53]. These are 288 °K, 60% relative humidity (RH), and 101.3 kPa total pressure (P_T). These were taken to be 15°C (59°F), 10% RH, and 14.696 psia.

4.6 Water/Steam Injection into Combustion Gas Turbine

Table 4.3 Flue Gas Composition (Major Constituents)

Constituent	% wet	% dry
CO_2	3.14	3.37
$N_2 + Ar$	76.14	81.63
O_2	13.99	15.00
H_2O	6.73	—
Total	100.00	100.00

At 59°F, vapor pressure (p_w) of water (H_2O) = 0.24713 psia [55, p. 88]. At a total pressure of 14.696 psia and % RH = 60, absolute humidity is given by

$$H = p_w(\% \text{RH}/100)/[P_T - p_w(\% \text{RH}/100)] \quad (3.5)$$

$$= 0.24713(60/100)/[14.696 - 0.24713(60/100)]$$

$$= 0.01019 \text{ mol } H_2O/\text{mol dry air}$$

$$\times (18.02/28.96) = 0.00633 \text{ lb } H_2O/\text{lb dry air}$$

The ambient air moisture factor ratio from Method 19 [51] is equivalent to H.

$$B_{wa}/(1 - B_{wa}) = H \quad (3.36)$$

Turbine Combustion Calculations
Natural gas: from Table 3.4

$$F_d = 8710, \quad F_w = 10,610, \quad F_c = 1040$$

From Table 3.7, per MMBtu (HHV) fired,

$$\text{Total dry flue gas} = F_d[20.9/(20.9-15)] = 30,854.1$$

A flue gas oxygen concentration of 15% (dry) is the standard for combustion gas turbines.

$$O_2 \text{ in FG} = (15/100)(30,854.1) = 4628.1$$

$$CO_2 \text{ in FG} = F_c = 1040$$

$$(N_2 + Ar) \text{ in FG} = \text{total dry FG} - O_2 - CO_2 = 25,186.0$$

$$(N_2 + Ar) \text{ in air} = (N_2 + Ar) \text{ in flue gas} = 25,186.0$$

$$\text{Total dry air} = (100/79.1)(25,186.0) = 31,840.7$$

$$O_2 \text{ in air} = (20.9/79.1)(25,186.0) = 6654.7$$

$$CO_2 \text{ in air} = 0$$

$$H_2O \text{ in air} = (31,840.7)(0.01019) = 324.5$$

$$H_2O \text{ in flue gas} = (F_w - F_d) + H_2O \text{ in air} = (10,610 - 8710) + 324.5 = 2224.5$$

$$\text{Total wet flue gas} = 30,854.1 + 2,224.5 = 33,078.6$$

Maximum Fuel Gas Fired

Heat input based on HHV $= 40.5(1020/918) = 45.0 \text{ MMBtu/h (HHV)}$

$$\left(45 \times \frac{10^6 \text{ Btu}}{\text{h}}\right) \frac{(\text{SCF})}{1020 \text{ Btu}} = 44,118 \text{ SCF/h of natural gas}$$

Sulfur Calculation

$$\frac{(44.118 \text{ SCF})}{\text{h}} \frac{(0.3 \text{ grains H}_2\text{S})}{100 \text{ SCF}} \frac{(\text{lb H}_2\text{S})}{7000 \text{ grains H}_2\text{S}} \frac{(\text{lb mol H}_2\text{S})}{34.08 \text{ lb H}_2\text{S}} \frac{(64.06 \text{ lb SO}_2)}{\text{lb mol SO}_2}$$

$$= 0.0355 \text{ lb SO}_2/\text{h} \cong 0.04 \text{ lb SO}_2/\text{h}$$

$$\times 4.38 = 0.16 \text{ ton/year} \cong 0.2 \text{ ton/year}$$

Sample Calculation for Other Contaminants

$$\text{NO}_x \text{ (uncontrolled)} = (0.32)(45.0) = 14.40 \text{ lb/h}$$

$$\times 4.38 = 63.1 \text{ ton/year}$$

Sample ppm Calculation

$$(0.32) \frac{(45 \text{ lb NO}_x)}{\text{h}} \frac{(\text{lb mol NO}_x)}{46.01 \text{ lb NO}_x} \frac{(385.3 \text{ SCF NO}_x)}{\text{lb mol NO}_x}$$

$$\times 10^6 = 86.8 \text{ ppm dry}$$

$$\left(\frac{30,854.1 \text{ SCF dry } FG}{\text{MMBtu}}\right) \left(\frac{45 \text{ MMBtu}}{\text{h}}\right)$$

Particulate Concentration

$$\frac{(0.0066 \text{ lb})}{\text{MMBtu}} \frac{(45 \text{ MMBtu})}{\text{h}} \frac{(7000 \text{ grains})}{\text{lb}}$$

$$= 3.3 \times 10^{-5} \text{ grams per dry standard cubic ft}$$

$$\frac{(30,854.1 \text{ SCF dry FG})}{\text{MMBtu}} \frac{(45 \text{ MMBtu})}{\text{h}}$$

NO_x concentration without abatement appears to be less than the new source performance standard (NSPS) for stationary gas turbines between 10 and 100 MMBtu/h (LHV). Abatement of NO_x may still be necessary to meet some other regulatory limit. According to Table 4.4, steam/water injection would reduce NO_x by slightly less than 60%.

If this degree of abatement is not sufficient to meet a specific regulatory requirement, NO_x could be reduced further by SCR. Reduction of NO_x from 42 to 8.4 ppm in an SCR on a gas turbine and HRSG combination, for a NO_x reduction efficiency of 80%, is noted in Refs [35,36]. A similar NO_x reduction efficiency for the controlled case in the table would result in the NO_x exit concentration of about 7 ppm. ∎

4.7 LOW-TEMPERATURE OXIDATION (OZONE REACTION/SCRUBBING)

4.7.1 Background

This process was developed by Cannon Boiler Works, Inc. of New Kensington, PA, starting in 1985 [56]. Marketing of the low-temperature oxidation (LTO) system for

Table 4.4 Gas Turbine Contaminant Emissions

Contaminants[a]	Emission factor lb/MMBtu (HHV)	lb/h	ton/year	ppm wet	ppm dry
NO_x					
Uncontrolled	0.32	14.40	63.1	81.0	86.9
Controlled	0.13	5.85	25.6	32.9	35.3
CO					
Uncontrolled	0.082	3.69	16.2	34.1	36.6
Controlled	0.030	1.35	5.9	12.5	13.4
N_2O	0.003	0.14	0.6	0.8	0.9
SO_2[b]	0.94S	0.04	0.2	0.1	0.2
CH_4[c]	0.0086	0.39	1.7	6.2	6.7
VOC	0.0021	0.09	0.4	1.5	1.6
TOC	0.011	0.50	2.2	8.0	8.6
PM[d]					
Condensable	0.0047	0.21	0.9	—	1.07×10^{-3}
Filterable	0.0019	0.09	0.4	—	0.43×10^{-3}
Total	0.0066	0.30	1.3	—	1.50×10^{-3}
HAP[e]					
Formaldehyde	7.1×10^{-4}	—	—	—	—
Toluene	1.3×10^{-4}	—	—	—	—
Sum of others	1.8733×10^{-4}	—	—	—	—
Total	10.2733×10^{-4}	0.05	0.2	—	—

[a] Uncontrolled, except as noted for NO_x and CO. Control is by water/steam injection.
[b] S is the percent sulfur in the fuel. Here, SO_2 is calculated by material balance.
[c] CH_4, VOC, and TOC concentrations are all expressed using the molecular weight (MW) of methane (CH_4, MW = 16.04).
[d] PM concentration is expressed as grains per dry standard cubic ft (grains/DSCF).
[e] HAP is estimated from the sum of formaldehyde, toluene, and the remaining entries in AP-42 HAP Table.

NO_x and SO_x reduction was begun by Cannon Technology, Inc., a wholly owned subsidiary. A licensing agreement with British Oxygen Corporation (BOC) resulted in the LoTOx™ technology offered by the BOC Group. Much later, Belco Technologies Corporation (BELCO®) acquired a license from the BOC Group to supply this technology to the petroleum refining industry; Belco is now affiliated with DuPont [57]. Patents have been issued, including Refs [58] and [59].

4.7.2 Process Description

The process utilizes ozone to oxidize the insoluble NO and NO_2 of NO_x to soluble higher oxides such as N_2O_5 and nitric acid (HNO_3). These are then removed from a flue gas in a wet gas scrubber (WGS). Ozone is manufactured on-site from oxygen or oxygen enriched air and electricity in a corona arc device.

Slipstream testing has demonstrated this technology on fossil fuel boilers and furnaces, a smelting plant, a steel pickling process, a chemical plant, and a refinery FCC unit [60,61]. There are several commercial applications [60], now including one on an FCC unit with four others scheduled for start-up later [61]. Optimal temperature for the FCC unit was observed to be 140–150°F (60–66°C), during prior slipstream testing [62], more or less typical of other applications. A NO_x outlet concentration of 10–20 ppm was achieved in the commercial unit [61]. Some additional detail is contained in Appendix J.

4.7.3 Chemical Reactions

Gas-phase chemical reactions in the process are exemplified by the equations contained in Appendix J [63,64]:

$$NO + O_3 \rightarrow NO_2 + O_2 \tag{J.1}$$

$$NO_2 + O_3 \rightarrow NO_3 + O_2 \tag{J.2}$$

$$2NO_2 + O_3 \rightarrow N_2O_5 + O_2 \tag{J.3}$$

$$NO_2 + NO_3 \rightarrow N_2O_5 \tag{J.4}$$

$$2N_2O_5 + H_2O \rightarrow 2HNO_3 \tag{J.5}$$

Ozone does not react with gaseous SO_2, nor with CO, to any appreciable extent [60,63,64] but does decompose into molecular oxygen (O_2) [56, p. 5].

$$O_3 \rightarrow O_2 + O \tag{4.10}$$

$$O + O \rightarrow O_2 \tag{4.11}$$

In the liquid phase, the nitric acid is neutralized by an alkali such as sodium hydroxide (NaOH) [56, p. 5]. A calcium- or magnesium-based reagent can also be used [59].

$$HNO_3 + NaOH \rightarrow NaNO_3 + H_2O \tag{J.6}$$

Reactions also occur with any CO_2 [56, p. 5]

$$CO_2 + 2NaOH \rightarrow Na_2CO_3 + H_2O \tag{4.12}$$

or SO_2 scrubbed from the gas

$$SO_2 + 2NaOH \rightarrow Na_2SO_3 + H_2O \tag{4.13}$$

As in any scrubbing process, the carbonate and sulfite species in solution equilibrate into bicarbonate (HCO_3^-) and carbonate (CO_3^{2-}) and into bisulfite (HSO_3^-) and sulfite (SO_3^{2-}), respectively. The sulfite and bisulfite further react with ozone in the water to sulfate (SO_4^{2-}) to satisfy their chemical oxygen demand (COD). This latter reaction serves to scavenge excess ozone (O_3) in solution and hinders its escape into the exhaust gas as ozone slip.

4.7.4 Ozone to NO_x Ratio

The ozone to NO_x ratio employed depends on such variables as temperature, residence time, the extent of NO_x reduction, and on the application. According to the patent literature, this ratio is 0.2–2.8 mol O_3 to mole of NO_x for boilers, furnaces, fired heaters, or a chemical process; a molar ratio of 0.75 to 1 of ozone to NO_x removes 95% of the NO_x from a steel strip pickling line using nitric acid as the pickling agent [58]. For the catalytic cracking application, the ozone to NO_x removal is indicated to range from 0.5 to 3.5, or preferably from 1.2 to 2.2; a greater extent of NO_x removal is obtained for a higher ozone to NO_x ratio [57]. (However, higher ozone application rates can lead to increased ozone slip.) In all these applications, residence times are on the order of seconds.

4.7.5 Possible Problem Areas

It was felt that ozone slip is a legitimate concern for an environmental regulatory agency [65, p. 9] and that there is a potential for sodium hydroxide emissions in the scrubber exhaust if it were used in the scrubbing solution [66, p. 8]. The nitrate reaction by-product ending up in the scrubbing solution may be of concern in the treatment/discharge of the wastewater generated [60].

4.7.6 Demonstration on Natural-Gas Fired Boiler

The first commercial demonstration of this process took place on a natural-gas fired package boiler at the Alta Dena Dairy in City of Industry, California. This location is in the Los Angeles area, within the jurisdiction of the South Coast Air Quality Management District (SCAQMD). That project's objectives included demonstration of successful scale-up; development of additional design, operating, and cost data; and minimization of NO_x emissions at the host site [56].

The Alta Dena Dairy processes dairy products around the clock, 24/7, 52 weeks a year. The dairy operates three boilers to produce approximately 20,000 lb/h of steam in the manufacture of dairy products, cleaning and sterilization of process units, and evaporation of milk. Only two of these boilers are needed to operate at any one time. The number 2 boiler, a 400 hp Model CB700 by Cleaver-Brooks [67], carries the load preferentially, and one of the others makes up the difference as necessary [65].

The ozone scrubbing demonstration took place on number 2 boiler. At the time of the commercial demonstration, a flue gas recirculation process capable of deactivation was used on this boiler. As an economy measure, liquid oxygen from a 1500 gal cryogenic storage tank [56,65] was used to produce the requisite ozone instead of the pressure swing adsorption (PSA) oxygen generator in the standard design.

On several occasions during the demonstration period, independent source testing was conducted on number 2 boiler, including once by SCAQMD [65]. One such test at high, medium, and low load was reported in Ref. [68]. As in the other tests, NO_x was reduced to well below 5 ppmd, corrected to 3% O_2 dry. In those same tests, an outlet CO of less than 5 ppmvd at 3% O_2 dry was also measured. Ozone was not reported. However, SCAQMD staff estimated a long-term ozone to NO_x ratio of 1.8 consistent with steam production logs and oxygen purchases for the ozone generator [65]. Elsewhere in the same SCAQMD document [65], an ozone to NO_x molar ratio of about 2 to 1 is mentioned as being required to achieve low outlet levels.

PROBLEM 4.7

From the excerpt of test results shown in Table 4.5 [68], estimate the size of ozone generator necessary.

SOLUTION Calculate the molar flow rate of NO_x at the inlet from the flue gas flow and the NO_x concentration noting that ppmd of $NO_x = DSCF\ NO_x/(10^6\ DSCF\ FG)$:

$$\text{High load}: \frac{(2711\ \text{DSCF}\ FG)}{\text{min}} \frac{(144\ \text{DSCF}\ NO_x)}{10^6\ \text{DSCF}\ FG} \frac{(\text{lb mol}\ NO_x)}{385.3\ \text{DSCF}\ NO_x} \frac{(1440\ \text{min})}{\text{day}}$$
$$= 1.459\ \text{lb mol}\ NO_x/\text{day}$$

NO_x concentration measured during source testing would be on a dry basis. Similarly,

$$\text{Medium load}: (1523) \frac{(103)}{10^6} \frac{(1440)}{385.3} = 0.586\ \text{lb mol}\ NO_x/\text{day}$$

$$\text{Low load}: (942) \frac{(41)}{10^6} \frac{(1440)}{385.3} = 0.144\ \text{lb mol}\ NO_x/\text{day}$$

The high load case controls. Using ozone to NO_x ratio at the upper end of the range stated in the Cannon patent [58], 0.2–2.8 to 1, one obtains

$$(2.459\ \text{lb mol}\ NO_x/\text{day})(2.8\ \text{lb mol}\ O_3/\text{lb mol}\ NO_x)(47.9982\ \text{lb}\ O_3/\text{lb mol}\ O_3)$$
$$= 196\ \text{lb/day}\ O_3$$

Estimates for the ozone to NO_x ratio, 1.8 and "about 2," contained in a SCAQMD Report [65] lie within the range indicated in the patent. The design for the Alta Dena Dairy demonstration called for an ozone generator with a capacity of 50 lb/day of ozone as a 6 wt % solution in oxygen [56]. The ozone to NO_x ratio would have to have been about 0.7 mol of ozone per mole of NO_x during the source test at high load to stay within an ozone usage rate of 50 lb/day. ■

Table 4.5 Excerpt of Alta Dena Dairy: Source Test Results

	High load	Medium load	Low load
Inlet NO_x (ppmv)	144	103	41
O_2 (% dry)	4.2	4.1	9.8
Flow rate (DSCFM) (Method 19)	2711	1523	942

PROBLEM 4.8

Estimate the heat input for each of the loads in Problem 4.7.

SOLUTION EPA Method 19 [51] is referenced as the source of the tabulated flow rates in the previous problem. Back-calculating the heat input from the tabulated information yields

High load: $(2711\,\text{DSCF FG}) \dfrac{(60\,\text{min})}{\text{h}} \dfrac{(\text{MMBtu (HHV)})}{8710\,\text{DSCF}} \dfrac{(20.9-4.2)}{20.9} = \dfrac{14.92\,\text{MMBtu (HHV)}}{\text{h}}$

Medium load: $(1523) \dfrac{(60)}{8710} \dfrac{(20.9-4.1)}{20.9} = \dfrac{8.43\,\text{MMBtu (HHV)}}{\text{h}}$

Low load: $(942) \dfrac{(60)}{8710} \dfrac{(20.9-9.8)}{20.9} = \dfrac{3.45\,\text{MMBtu (HHV)}}{\text{h}}$

The 8710 value is the F_d factor for natural gas listed in Method 19 (see Table 3.4). It is the ratio of the dry flue gas generated at stoichiometric conditions (0% O_2 in the flue gas) at 68°F and 1 atm pressure to the HHV of the heat input (Section 3.8). The dry flue gas is corrected to other excess O_2 conditions by multiplying F_d and the heat input by the factor $20.9/(20.9 - \%\,O_2\,\text{dry})$. ∎

PROBLEM 4.9

Estimate the maximum heat *input* for the boiler in Problems 4.7 and 4.8 and the percentage of maximum heat *input* corresponding to each of the source test conditions.

SOLUTION The rating of the boiler is 400 hp (horsepower) [67,68]. This is the useful heat *output*, and in this case, the capacity to make steam. By definition, a boiler horsepower is the energy needed to evaporate 34.5 lb of water at 212°F in 1 h

$$\text{Boiler hp} = (34.5\,\text{lb/h})(970.3\,\text{Btu/lb})\,[55,\ \text{p. 86}] = 33{,}475\,\text{Btu/h}$$

Then (400 boiler hp)(33475 Btu/h per boiler hp) = 13.39 MMBtu/h (LHV), as indicated in the manufacturer's literature [69]. This energy is supplied by the LHV of a fuel, such as natural gas, burned in the boiler. The rated steam capacity of this boiler is [69]

$$(13.39 \times 10^6\,\text{Btu/h})(970.3\,\text{Btu/lb steam}) = 13{,}800\,\text{lb/h steam production}$$

The natural gas supplier, Southern California Gas Company, delivers a product with a heating value between about 1000 and 1060 Btu/SCF (60°F, 30 in. Hg) in the major portion of its service area [70]. This is higher heating value. The ratio of HHV to LHV is not overly sensitive to the composition of a natural gas having a heating value in this typical range. It is about 1.11, similar to the same ratio for pure methane, and minimal error is introduced by using this value here for the ratio of HHV to LHV. Therefore, with the results for heat input in HHV from Problem 4.8, heat input in LHV is calculated as follows:

High load: $14.92/1.11 = 13.44\,\text{MMBtu/h (LHV)}$

Medium load: $18.43/1.11 = 7.59\,\text{MMBtu/h (LHV)}$

Low load: $3.45/1.11 = 3.11\,\text{MMBtu/h (LHV)}$

Only part of the heat *input* appears as useful heat *output*. The difference is radiation, conduction, and convection losses, including the hot flue gas that exits the stack. The ratio of useful heat output to heat input is the *thermal efficiency*. Efficiency for this boiler is estimated from the manufacturer's literature to be in the range of 82.2–84.4% [69] at 100% firing, depending on the unknown pressure of the steam produced. The manufacturer's data show a minimal effect of load on efficiency. Using an arithmetic average of 83.3%,

Estimated maximum heat input = maximum heat output/estimated efficiency

$$=13.39/(0.833) = 16.07 \, \text{MMBtu/h (LHV)}$$

Keep in mind that this calculation is only as good as the assumed efficiency. It is a substitute for not knowing the actual efficiency or the natural gas composition used to fire the boiler during the tests.

However, it can be used to get at least a rough idea of the percentage of maximum heat input (16.07 MMBtu/h (LHV)) corresponding to each of the three reported tests.

In terms of LHV values,

High load: $13.44/16.07 = 84\%$
Medium load: $7.59/16.07 = 47\%$
Low load: $3.11/16.07 = 19\%$

■

4.8 PARTICULATE REMOVAL

The petroleum refining industry provides an example of particulate removal from a gas stream. In the FCC process, discussed further in Appendix J, fine cracking catalyst particles are entrained in the flue gas leaving the fluidized bed in the catalyst regenerator vessel. The particle-laden gas passes through two stages of cyclone separators before exiting the regenerator vessel, and the captured solids are returned to the bed. A certain percentage of the solids escape collection in the cyclones, and further particulate removal is necessary to meet present day air quality standards. Current practice includes an additional stage of cyclones external to the regenerator, electrostatic precipitation (ESP), and wet scrubbing (in conjunction with the removal of sulfur dioxide).

Collection efficiency varies with particle size via the so-called *grade efficiency curve* for that particular device [71]. Although written many years ago, this information is still valid. In the cyclones and other devices, the larger particles are collected preferentially with higher efficiency, and the size distribution leaving the collector contains a greater proportion of finer particles.

The problem that follows starts out with an estimated particle size distribution at the inlet of the internal cyclones. Calculations using the grade efficiency curve allow both an overall collection efficiency and an exit particle size distribution to be derived. This exit size distribution is the inlet distribution to the external collection devices—third-stage cyclones, ESP, and wet scrubber applied separately.

PROBLEM 4.10

An estimated particle size distribution at the entrance of the cyclones in an FCC regenerator vessel [72] is listed in Table 4.6. (a) Estimate the collection efficiency and the particle size

Table 4.6 Estimated Size Distribution of Material Entering First-Stage Cyclones

As reported [72]		Interpolated to coincide with grade efficiency curves[a] [73,74]	
Size (μ)	Cum. % less than	Size (μ)	Cum. % less than
108	99.95	108	99.95
100	99.20	104	99.8
96	97.99	75	84.69
88	95.67	60	67
82	90.46	40	36
75	84.69	30	22
68	78.34	20	12
63	70.98	15	9.5
56	62.71	10	7.5
50	52.50	7.5	6.4
42	41.16	5	5
34	26.41	2.5	3
26	16.78	1	1.6
14	9.3		

[a] Sizes below 14 micrometers (microns, μ) are extrapolated from the as reported distribution, using the curve of Figure 4.6.

distribution at the outlet of the second-stage cyclones. (b) Using the result from (a), estimate the collection efficiency and the particle size distribution exiting external third-stage cyclones, an electrostatic precipitator, and a wet scrubber applied one at a time on the flue gas leaving the regenerator for discharge to the atmosphere. Compare the results.

SOLUTION The particle size distribution of Table 4.6 is plotted in Figure 4.6 and extrapolated somewhat for use in interpolating Table 4.6 entries to values coinciding with the ranges of the grade efficiency curves [73,74] listed in the subsequent tables.

Calculations proceed as in Refs [73,74], in which the percentage of particles in a given size range, or grade, is multiplied by the collection efficiency for that range. The amounts collected are summed, along with the amounts escaping collection. The percentage distribution in each range is determined for the escaping material, and the cumulative distribution is computed. Results are shown in Tables 4.7 and 4.8, respectively, for the first- and second-stage cyclones. Collection efficiency is calculated for each stage separately and in combination.

The second-stage exit then becomes the inlet for each of the external collection devices. These results are shown in Tables 4.9–4.11.

Resulting particle size distributions are plotted as point values in Figure 4.6. It is noticeable at the lower end that the sizes become increasingly smaller, proceeding from the third-stage cyclones through the ESP to the wet scrubber.

The purpose of this exercise is to illustrate the calculation procedure only, without regard to the different types of cyclones, with differing grade efficiency curves, which might be used in an actual design as primary, secondary, and tertiary cyclones. The grade efficiency curves chosen for the cyclones and the external devices are from public information. These particular curves were selected because they are either available in tabular form or could be easily interpolated

Table 4.7 First-Stage Cyclone Calculation (Basis 100 lb at Inlet)

Size range (μ)	Weight particulate w/in range	Efficiency at mean size	Overall collection (%)	Weight particulate escaping	% Particulate escaping	Cum. % less than
+104	0.2	100	0.20	0	0.00	100.00
75–104	15.11	99.1	14.97	0.14	2.34	97.66
60–75	17.69	98.5	17.42	0.27	4.51	93.15
40–60	31.0	97.3	30.16	0.84	14.02	79.13
30–40	14.0	96.0	13.44	0.56	9.35	69.78
20–30	10.0	94.3	9.43	0.57	9.51	60.27
15–20	2.5	92.0	2.30	0.20	3.34	56.93
10–15	2.0	89.3	1.79	0.21	3.51	53.42
7.5–10	1.1	84.2	0.93	0.17	2.84	50.58
5–7.5	1.4	76.7	1.07	0.33	5.51	45.07
2.5–5	2.0	64.5	1.29	0.71	11.85	33.22
0–2.5	3.0	33.5	1.01	1.99	33.22	0.00
Total	100.00	—	94.01	5.99	100.00	

$(100)(94.01/100) \cong 94\%$ collection efficiency.

Table 4.8 Second-Stage Cyclone Calculation

Size range (μ)	Weight particulate w/in range	Efficiency at mean size	Overall collection (%)	Weight particulate escaping	% Particulate escaping	Cum. % less than
+104	0	100	0	0	0	100.00
75–104	0.14	99.1	0.139	0.001	0.06	99.94
60–75	0.27	98.5	0.266	0.004	0.22	99.72
40–60	0.84	97.3	0.817	0.023	1.28	98.44
30–40	0.56	96.0	0.538	0.022	1.22	97.22
20–30	0.57	94.3	0.538	0.032	1.78	95.44
15–20	0.20	92.0	0.184	0.016	0.89	94.55
10–15	0.21	89.3	0.188	0.022	1.22	93.33
7.5–10	0.17	84.2	0.143	0.027	1.50	91.83
5–7.5	0.33	76.7	0.253	0.077	4.28	87.55
2.5–5	0.71	64.5	0.458	0.252	14.01	73.54
0–2.5	1.99	33.5	0.667	1.323	73.54	0.00
Total	5.99	—	4.191	1.799	100.00	

$(100)(4.191/5.99) = 69.97 \cong 70\%$ collection efficiency. Overall first and second stages: $(100 - 1.799)/100 = 0.9820 \cong 98.2\%$.

4.8 Particulate Removal

Table 4.9 Third-Stage Cyclone Calculation

Size range (μ)	Weight particulate w/in range	Efficiency at mean size	Overall collection (%)	Weight particulate escaping	% Particulate escaping	Cum. % less than
+104	0	100	0	0	0	100
75–104	0.001	99.1	0.0010	0	0	100
60–75	0.004	98.5	0.0039	0.0001	0.010	99.99
40–60	0.023	97.3	0.0224	0.0006	0.060	99.93
30–40	0.022	96.0	0.0211	0.0009	0.090	99.84
20–30	0.032	94.3	0.0302	0.0018	0.180	99.66
15–20	0.016	92.0	0.0147	0.0013	0.130	99.53
10–15	0.022	89.3	0.0196	0.0024	0.240	99.29
7.5–10	0.027	84.2	0.0227	0.0043	0.431	98.86
5–7.5	0.077	76.7	0.0591	0.0179	1.793	97.07
2.5–5	0.252	64.5	0.1625	0.0895	8.963	88.10
0–2.5	1.323	33.5	0.4432	0.8798	88.103	0.00
Total	1.799	—	0.8004	0.9986	100.000	

$(100)(0.8004/1.799) = 44.5\%$ collection efficiency. Overall three stages: $(100 - 0.9986)/100 = 99.0\%$ collection efficiency.

Table 4.10 ESP Calculation

Size range (μ)	Weight particulate w/in range	Efficiency at mean size	Overall collection (%)	Weight particulate escaping	% Particulate escaping	Cum. % less than
+104	0	100	0	0	0	100
75–104	0.001	99.2	0.0010	0	0	100
60–75	0.004	98.7	0.0039	0.0001	0.030	99.97
40–60	0.023	97.7	0.0225	0.0005	0.148	99.82
30–40	0.022	96.8	0.0213	0.0007	0.207	99.62
20–30	0.032	96.5	0.0309	0.0011	0.325	99.29
15–20	0.016	96.0	0.0154	0.0006	0.177	99.11
10–15	0.022	95.5	0.0210	0.0010	0.296	98.82
7.5–10	0.027	95.0	0.0257	0.0013	0.384	98.43
5–7.5	0.077	94.0	0.0724	0.0046	1.361	97.07
2.5–5	0.252	90.5	0.2281	0.0239	7.069	90.00
0–2.5	1.323	77.0	1.0187	0.3043	90.003	0.00
Total	1.799	—	1.4609	0.3381	100.000	

$(100)(1.4609/1.799) = 81.2\%$ collection efficiency. Overall two stages of cyclones plus ESP: $(100 - 0.3381)/100 = 99.7\%$ efficiency.

Table 4.11 High-Energy Venturi Scrubber Calculation

Size range (μ)	Weight particulate w/in range	Efficiency at mean size	Overall collection (%)	Weight particulate escaping	% Particulate escaping	Cum. % less than
+104	0	100	0	0	0	100
75–104	0.001	100	0.0010	0	0	100
60–75	0.004	100	0.0040	0	0	100
40–60	0.023	100	0.0230	0	0	100
30–40	0.022	99.87	0.0220	0	0	100
20–30	0.032	99.78	0.0319	0.0001	0.424	99.58
15–20	0.016	99.71	0.0160	0.0000	0.000	99.58
10–15	0.022	99.67	0.0219	0.0001	0.424	98.15
7.5–10	0.027	99.63	0.0269	0.0001	0.423	98.73
5–7.5	0.077	99.61	0.0767	0.0003	1.271	97.46
2.5–5	0.252	99.35	0.2504	0.0016	6.780	90.68
0–2.5	1.323	98.38	1.3016	0.0214	90.678	0.00
Total	1.799	—	1.7754	0.0236	100.000	

$(100)(1.7754/1.799) = 98.7\%$ collection efficiency. Overall two stages of cyclones plus venturi scrubber: $(100 - 0.0236)/100 = 99.98\%$. Calculated collection efficiencies are summarized in Table 4.12.

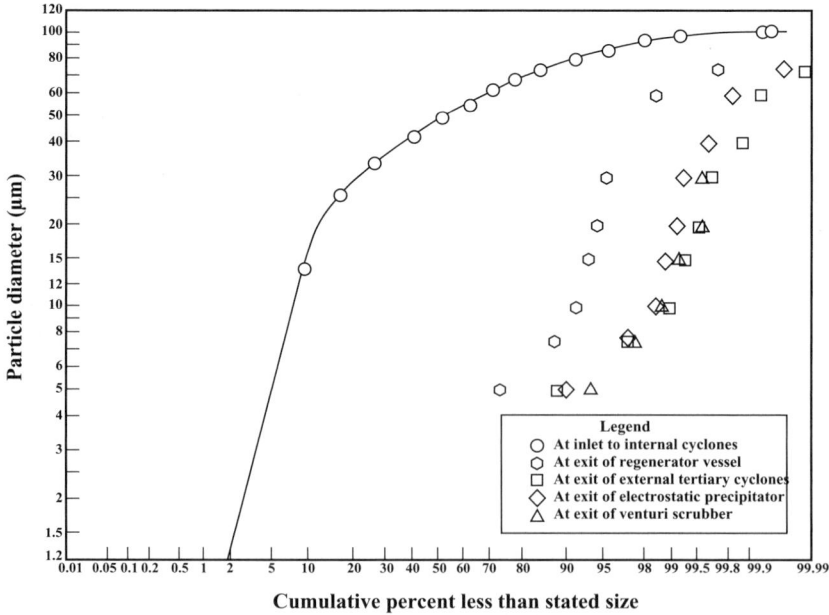

Figure 4.6 Calculated particle size distributions in/from FCC regenerator vessel.

Table 4.12 Summary of Calculated Collection Efficiencies

Collection device	Calculated efficiency (%)
First-stage (primary) cyclones	94
Second-stage (secondary) cyclones	70
First- and second-stage combination	**98.2**
Third-stage (tertiary) cyclones	44.5
Electrostatic precipitator	81.2
Wet scrubber (high-energy venturi)	98.7
Overall—two stages of cyclones plus	
Third-stage cyclones	99.0
ESP	99.7
High-energy venturi scrubber	99.98

from the tables or calculated from the small graphs in the Refs [73,74] with a minimal amount of direct reading of the graphs. The calculation procedure works for any standard size distribution from any industry, using the applicable grade efficiency curve.

Furthermore, the entire problem depends on the initial reported estimate for the size distribution at the inlet of the primary cyclones. Finally, the extrapolation here to particle sizes below 14μ in order to proceed with the problem is arbitrary.

Results for the case of an actual operating FCC unit will certainly be different. One company's experience is mentioned in Ref. [75]. Collection efficiency for two stages of cyclones was reported there at 99.973%. Collection efficiency noted here in Table J.2 is 99.995%. The bold entry in Table 4.12 is only 98.2%. ∎

4.9 FLUE GAS SCRUBBING

A WGS contacts a flue gas with an atomized spray of fine water droplets that encapsulate suspended solids and absorb sulfur oxides and other water-soluble contaminants. The droplets are then separated from the gas by cyclonic action in a second vessel. Particulate collection of 90–95% on cat cracker regenerator flue gas is regularly achieved along with sulfur oxide removals of 90–99% [76]. Particulate removal by wet scrubbing using a high-energy venturi scrubber is treated in Problem 4.10. Addition of ozone to the water can remove NO_x along with SO_2 and PM in a single step [57].

Untreated water used as a scrubbing medium to absorb SO_2 quickly assumes a pH of 2.5 or less. This allows significant SO_2 to exist in the vapor phase in equilibrium with dissolved molecular SO_2 in solution and thereby reduces its removal efficiency. This is overcome by addition of an alkaline material, such as caustic soda, sodium carbonate, lime, limestone, magnesium oxide, or some proprietary agent [77], to raise the solution pH to where the nonvolatile ionic species predominate in solution.

Sulfur dioxide dissolved in scrubbing water becomes sulfite ion (SO_3^{2-}) and bisulfite ion (HSO_3^-) when an alkaline pH control agent is added. These sulfite species exert a COD, which must be satisfied by oxidation to sulfate (SO_4^{2-}) before wastewater from a wet gas scrubber can be disposed of. In addition, when a calcium-based pH control agent is used, the resulting calcium sulfite precipitate is more difficult to settle out and separate than calcium sulfate after oxidation. Suspended particulate removed by scrubbing must also be dewatered before disposal of solids and liquid waste.

In addition to having to contend with such waste streams, there is also the disadvantage of a water saturated corrosive exhaust gas at relatively low temperature and a visible steam plume possibly containing traces of sulfuric acid. Calculation of the adiabatic saturation temperature assumed by the scrubbed gas and the scrubbing solution is the subject of Problem 2.37.

The disadvantages of wet scrubbing can be overcome by the use of so-called *dry scrubbing*. In a dry scrubber, the same types of alkaline materials employed in a wet scrubber are used in a dry condition to capture SO_2 and the sulfate materials that form are then removed via some sort of particulate collector [78].

The CFB of Section 4.4 is itself a form of dry scrubber. Pulverized limestone and crushed coal are fluidized together in the boiler vessel, and the entrained materials (in theory—coal ash, unburned coal, calcium sulfate, and unspent limestone) in the combustion gas are processed in internal cyclone separators and removed by an external baghouse [41].

4.10 ATMOSPHERIC DISPERSION

It will often be necessary to perform calculations to prove that an air emission conforms to some standard, not at the point of discharge but at ground level after dispersion of contaminants into the atmosphere. As mentioned in the introductory section of this chapter, this is a specialized area in which various EPA approved prepared computer models are employed. Extensive meteorological data, sometimes from a preferred time period, are used to determine long-term effects and identify the worst day or days for pollutant concentration at ground levels. Multiple sources can be modeled along with a variety of terrain features.

Such calculations are clearly beyond the scope of a general book such as this. Only a brief introduction to the necessary meteorology and mathematical functions is covered here along with some simple problems to reinforce those concepts. This treatment might provide an insight into how the more complicated models work and could possibly be used for screening purposes as well.

This section first explains some elementary meteorological concepts, examines stack height and plume rise, shows equations for Gaussian dispersion, and concludes with the aforementioned simple sample problems. The term *Gaussian* means that dispersed pollutant profiles follow a statistical normal, or Gaussian, distribution in their turbulence-caused departure from some central value. The normal distribution *per se* is discussed in Chapter 2.

4.10.1 Atmospheric Conditions—Stability [79]

The atmosphere cools with increasing altitude, a phenomenon known as the *environmental lapse rate*. For air containing water vapor at an unsaturated condition, the rate of cooling without gain or loss of heat with the surroundings is called the *dry adiabatic lapse rate*. It is approximately 10°C per 1000 m, 3°C per 1000 ft, or 5°F per 1000 ft. Once the air cools to its dew point, condensation takes place, cloud formation occurs, and the consequent lesser rate of cooling is known as the *moist adiabatic rate*.

Stability of atmosphere air is defined by comparing the temperature of a mass of rising air with that of its environment. When the rising air is colder, it is denser than its surroundings ($\rho = PM/RT$ by the ideal gas law). This packet of air will therefore sink and return to its original level. This situation is denoted as a *stable atmosphere*. *Absolute stability* prevails when the actual environmental lapse rate is less than the moist adiabatic lapse rate.

In contrast, a parcel of warmer air will continue to rise until it attains the same temperature as its surroundings. This air is said to be *unstable*. *Absolute instability* occurs when the environmental lapse rate exceeds the dry adiabatic lapse rate. *Conditional instability* means that the environmental lapse rate falls between the moist and the dry adiabatic rates.

The so-called *neutral stability* happens when the actual lapse rate for unsaturated air equals the dry adiabatic lapse rate, such that rising or falling air would cool or warm at the same rate as its surroundings. Neutral stability also applies to saturated air when the environmental lapse and the moist adiabatic lapse rates are equal.

Dispersion of flue gas from a stack depends upon the stability condition of the atmosphere. The stability class also affects the appearance of a stack plume and the type of atmospheric clouds that form.

4.10.2 Atmospheric Conditions—Wind Speed and Air Temperature

Dispersion of a stack gas into the atmosphere also depends on local conditions of air temperature and the velocity and direction of the wind at stack height. These factors, along with the indicators of atmospheric stability and the physical characteristics of the emission, are entered into a mathematical model to estimate the extent of dispersion into the atmosphere.

4.10.3 Characteristics of the Emission

These include the properties of the gas being discharged (temperature, velocity, and pollutant concentration) and the dimensions (height and diameter) of the stack.

The plume of gas emitted from a stack achieves a height fostered by its momentum and buoyancy, limited by meteorological conditions. Dispersion modeling calculates both and uses the higher value for plume rise. Plume rise for a hot buoyant plume is calculated from equations developed by G. A. Briggs in 1965 [80].

A buoyancy flux (F) is calculated from the acceleration of gravity (g), the exit velocity at the stack (v), the inside diameter (i.d.) at the top of the stack (d), and the difference in temperature (ΔT) between the temperature of the stack gas (T_s) and the ambient temperature (t) [81, p. 3-2].

$$F = gvd^2\Delta T/(4T_s) \qquad (4.14)$$

Units employed are meters, seconds, and Kelvin (°K). The flux has units of m^4/s^3. For buoyant rise under unstable and neutral conditions, buoyant plume rise (ΔH, meters) is given by one of two empirical formulations involving F and the wind speed (u, m/s) at the stack top (u_h).

$$\Delta H = 21.425 F^{3/4}/(u_h), \quad \text{for } F < 55 \qquad (4.15)$$

$$\Delta H = 38.71 F^{3/4}/(u_h), \quad \text{for } F \geq 55 \qquad (4.16)$$

Formulas to be used for other atmospheric conditions are given elsewhere [1]. Momentum rise under the same atmospheric conditions [82]

$$\Delta H = 3dv/(u_h) \qquad (4.17)$$

is compared with buoyant rise, and the greater of the two is used. The plume rise selected (ΔH) is then added to the lesser of actual physical actual height (h) or good engineering practice (GEP) stack height to provide the height used in modeling (H)

$$H = h + \Delta H \qquad (4.18)$$

4.10.4 Good Engineering Practice (GEP) Stack Height

The option to disperse environmental contaminants into the atmosphere simply by building a stack of unlimited height is addressed by the Clean Air Act Amendments of 1977 [83]. Implementing this Federal Law, EPA has defined GEP stack height (H_g) [84], measured from the ground-level elevation from the base of the stack as the lesser of

- 65 m
- H + 1.5 L
- The height demonstrated by a fluid model or field study.

The intent of GEP is to prevent *downwash* of concentrated stack gas to ground level in the immediate vicinity of the stack. This effect is accentuated by the presence of buildings, structures, terrain features, and other obstructions close to the stack.

The term H is the height of nearby structure(s) measured from the ground-level elevation from the base of the stack, and L is the lesser dimension, height or projected width, of nearby structures. For the purpose of the mathematical formula above, *nearby* means a distance up to five times the height or width dimension of a structure, but no greater than 0.8 km (1/2 mile).

The demonstration must be approved by the EPA, state, or local control agency. The demonstrated height ensures that the emissions from a stack do not result in excessive concentrations of any air pollutant as a result of atmospheric downwash, wakes, or eddy currents created by the source itself, nearby structures, or nearby terrain features. For conducting demonstrations, *nearby* means not greater than 0.8 km (1/2 mile), except that the portion of a terrain feature may be considered to be nearby if it falls within a distance of up to 10 times the maximum height (H_t) of the feature, not to exceed 2 miles if such feature achieves a height (H_t) 0.8 km from the stack that is at least 40% of the GEP stack height determined by the formula or 26 m, whichever is greater, as measured from the ground-level elevation at the base of the stack. The height of the structure or terrain feature is measured from the ground-level elevation of the base of the stack [84].

The GEP stack height does not limit the actual physical height of a stack that one may build, but rather limits the amount of stack height one can utilize when performing dispersion modeling calculations. If the actual stack height exceeds GEP, only that portion of the height up to GEP may be used in such situations.

4.10.5 Dispersion Modeling Calculations

The formula for dispersion of flue gas pollutants [1, p. 2–6] is

$$\chi = \left[\frac{Q}{2\pi u \sigma_y \sigma_z}\right] \frac{\exp[-y^2]}{2\sigma_y^2} \frac{\{\exp[-(H-Z)^2]}{2\sigma_z^2} + \frac{\exp[-(H+Z)^2]\}}{2\sigma_z^2} \quad (4.19)$$

where χ (the Greek letter chi) is downwind pollutant concentration (g/m³), Q is the emission rate (g/s), u is the wind speed at the point of release (m/s), σ_y (sigma y) is the Gaussian standard deviation in the crosswind direction at the downwind location, σ_z (sigma z) is the Gaussian standard deviation in the vertical direction at the downwind location, H is the effective height of the centerline of the plume (meters, m), y is the crosswind distance perpendicular to the centerline of the plume (m), z is the vertical distance measured from ground-level elevation (m), and π (pi) is the standard mathematical constant.

For changing conditions, this calculation is repeated over and over in a computer model and summed. Averaging time for the parameters σ_y and σ_z are a "few" minutes (about 3–10) [1, p. 2–11].

To calculate ground-level concentration below the centerline of the plume, Equation (4.19) reduces to

$$\chi = \frac{[Q]}{\pi u \sigma_y \sigma_z} \frac{\exp[-H^2]}{2\sigma_z^2} \quad (4.20)$$

242   Chapter 4   Air Control Devices

Table 4.13   Excerpt of Parameters a and b used in Equation (4.23)

Distance (km)	a	b
<0.3	34.459	0.86974
0.3–1.0	32.093	0.81066
1–3	32.093	0.64403
3–10	33.504	0.60486

Multiplying this equation by 10^6 expresses the downwind pollutant concentration (χ) in micrograms per cubic meter ($\mu g/m^3$).

For a neutral atmosphere (Stability Class D of dispersion modeling) [1, p. 2–8],

$$\sigma_y = (1000)(x)[\tan(T)/2.15] \qquad (4.21)$$

and T (in degrees) is given by

$$T = 8.3333 - 0.72382[\ln(x)] \qquad (4.22)$$

$$\sigma_z = ax^b \qquad (4.23)$$

where x is the downwind distance in kilometers (km) from the emission source, used in both the σ_y and σ_z equations [1, p. 2–4]

The parameters a and b are excerpted in the Table 4.13 for Stability Class D.

Equations for σ_y and σ_z for other stability classes are given elsewhere [1, pp. 2–8 and 2–11].

PROBLEM 4.11

Convert the following into the metric units and °K to be used for dispersion modeling. Firing 100 MMBtu/h (HHV) of natural gas at a flue gas oxygen concentration of 2.5% O_2 dry results in a NO_x concentration of 100 ppmd as measured in the stack gas. Stack diameter is 3 ft. Flue gas temperature of discharge is 350°F, and ambient air temperature is 68°F. Wind speed at the top of the stack is a steady 8 miles/h (mph). Stack height is 213 ft, 3 in. (equivalent to a GEP height of 65 m). No downwash occurs, the surrounding countryside is rural, and terrain is flat.

SOLUTION   Flue gas volumetric flow is

$$\frac{100 \text{ MMBtu}}{\text{h}} \frac{8710 \text{ dry SCF}}{\text{MMBtu}} \frac{[(20.9)/(20.0-2.5)]}{(3600 \text{ s/h})} = 274.8 \text{ dry SCF/s}$$

Pollutant (NO_x) emission flow rate (Q) is

$$\frac{(274.8 \text{ dry SCF FG})}{\text{s}} \frac{(100 \text{ SCF } NO_x)}{10^6 \text{ dry SCF FG}} \frac{(\text{lb mol})}{(385.3 \text{ SCF})} \frac{(46.01 \text{ lb } NO_x)}{\text{lb mol } NO_x} \frac{(453.59 \text{ g})}{\text{lb}} = 1.489 \text{ g/s}$$

Flue gas temperature (T_s) is

$$(350 + 459.67)/1.8 = 440.82 \,°K$$

Ambient air temperature (t) is
$$(68 + 459.67)/1.8 = 293.15\,°K$$

Flue gas volumetric flow at temperature is
$$274.8(449.82/293.15) = 421.7 \text{ actual cubic ft (ACF)/s}$$

Stack diameter (d)
$$(3 \text{ ft})(0.3048 \text{ m/ft}) = 0.9144 \text{ m}$$

Stack diameter to stack area
$$3^2(\pi/4) = 7.069 \text{ ft}^2, \quad 0.9144^2(\pi/4) = 0.6567 \text{ m}^2$$

Exit velocity (v)
$$421.7/7.069 = 59.65 \text{ ft/s}$$
$$\times\, 0.3048 \text{ m/ft} = 18.18 \text{ m/s}$$

Wind speed (u)
$$(8 \text{ mph})[(88 \text{ ft/s})/60 \text{ mph}] = 11.7 \text{ ft/s} \times 0.3048 = 3.58 \text{ m/s} \quad \blacksquare$$

PROBLEM 4.12

(a) Estimate the maximum ground-level concentration for the case outlined in Problem 4.11. Round the input values as follows:

$Q = 1.5$ g/s $\qquad\qquad u = u_h = 3.6$ m/s (wind speed)
$h = 65$ m $\qquad\qquad T_s = 450\,°K$ stack temperature
$d = 0.91$ m $\qquad\qquad t = 293\,°K$
$v = 18$ m/s (exit velocity) $\qquad \Delta t = 450 - 293 = 157\,°K$

(b) Keeping the other parameters constant, repeat (a) for a stack height of 55 m and a wind speed of 6.7 m/s ($\sim$15 mph).

SOLUTION
(a) From Equation (4.14),
$$F = (9.8)(18)(0.91)^2(157)/[(4)(450)] = 12.7$$

With $F < 55$, from Equation (4.15),
$$\Delta H = 21.425(12.7)^{3/4}/3.6 = 40.0 \text{ m}$$

and from Equation (4.17),
$$\Delta H = 3(0.91)(18)/3.6 = 13.65 \text{ m}$$

The buoyant plume rise controls ($40.0 > 13.65$).
From Equation (4.18),
$$H = 65 + 40 = 105 \text{ m}$$

Equation (4.20) is solved for χ (chi) as a function of distance (x) for case 1 in column 2 of Table 4.14. The variables Q, u, and H are entered directly. The Gaussian standard deviations (σ_y and σ_z) are calculated from Equations (4.21) and (4.23), with the intermediate variable T given by Equation (4.22).

244 Chapter 4 Air Control Devices

Table 4.14 Atmospheric Dispersion Modeling Calculations

	Case 1				Case 2		
	$Q = 1.5$ g/s				$Q = 1.5$ g/s		
	$u = 3.6$ m/s				$u = 6.7$ m/s		
	$H = 105$ m				$H = 86.5$ m		
x (km)	chi 1	chi 2	T degrees	T radians	sigma y	sigma z	sigma product
0.10	0.0000	0.0000	10.0000	0.1745	8.20	4.7	38.1
0.20	0.0000	0.0000	9.4982	0.1658	15.56	8.5	132
0.30	0.0000	0.0000	9.2048	0.1607	22.61	12.1	273
0.40	0.0000	0.0000	8.9965	0.1570	29.46	15.3	450
0.50	0.0000	0.0015	8.8350	0.1542	36.15	18.3	661
0.60	0.0007	0.0193	8.7030	0.1519	42.72	21.2	906
0.70	0.0080	0.0928	8.5915	0.1499	49.19	24.0	1.18E + 03
0.80	0.0410	0.2600	8.4948	0.1483	55.58	26.8	1.49E + 03
0.90	0.1272	0.5256	8.4096	0.1468	61.89	29.5	1.82E + 03
1.00	0.2874	0.8623	8.3333	0.1454	68.13	32.1	2.19E + 03
1.10	0.4599	1.1311	8.2643	0.1442	74.31	34.1	2.54E + 03
1.20	0.6635	1.3889	8.2013	0.1431	80.44	36.1	2.90E + 03
1.30	0.8869	1.6249	8.1434	0.1421	86.52	38.0	3.29E + 03
1.40	1.1189	1.8334	8.0898	0.1412	92.56	39.9	3.69E + 03
1.50	1.3502	2.0122	8.0398	0.1403	98.55	41.7	4.11E + 03
1.60	1.5735	2.1617	7.9931	0.1395	104	43.4	4.54E + 03
1.70	1.7836	2.2836	7.9492	0.1387	110	45.2	4.99E + 03
1.80	1.9774	2.3803	7.9078	0.1380	116	46.9	5.45E + 03
1.90	2.1528	2.4546	7.8687	0.1373	122	48.5	5.93E + 03
2.00	2.3093	2.5094	7.8316	0.1367	128	50.2	6.42E + 03
2.10	2.4468	2.5472	7.7963	0.1361	134	51.8	6.92E + 03
2.20	2.5661	2.5706	7.7626	0.1355	139	53.3	7.44E + 03
2.30	2.6682	2.58177	7.7304	0.1349	145	54.9	7.97E + 03
2.31	2.6775	2.58229	7.7273	0.1349	146	55.0	8.02E + 03
2.32	2.6866	2.58271	7.7242	0.1348	146	55.2	8.08E + 03
2.33	2.6956	2.58303	7.7210	0.1348	147	55.3	8.13E + 03
2.34	2.7044	2.58326	7.7179	0.1347	148	55.5	8.18E + 03
2.35	2.7131	2.58339	7.7149	0.1346	148	55.6	8.24E + 03
2.36	2.7216	2.58342	7.7118	0.1346	149	55.8	8.29E + 03
2.37	2.7300	2.58337	7.7087	0.1345	149	55.9	8.35E + 03
2.38	2.7382	2.58322	7.7057	0.1345	150	56.1	8.40E + 03
2.39	2.7463	2.58298	7.7026	0.1344	150	56.2	8.46E + 03
2.40	2.7542	2.58266	7.6996	0.1344	151	56.4	8.51E + 03

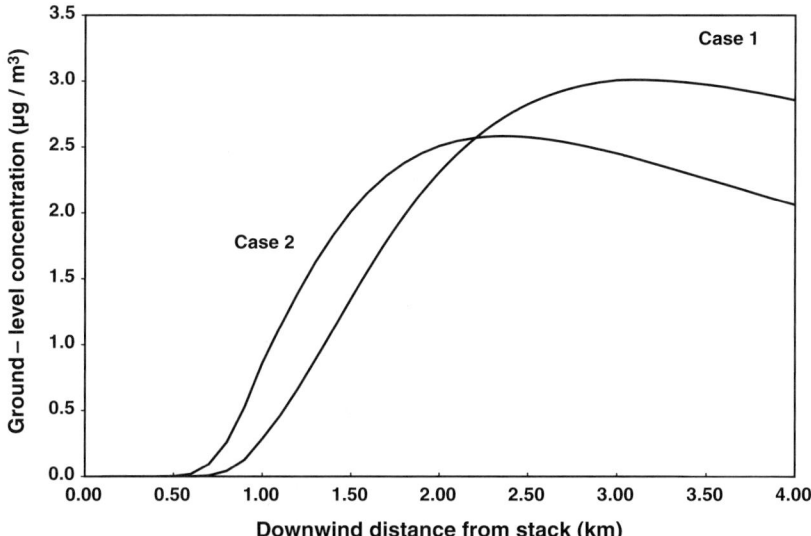

Figure 4.7 Atmospheric dispersion modeling two simple cases.

T is in degrees, and make sure that your calculator or spreadsheet accepts an angle in degrees as the argument of the tangent function in Equation (4.21). A column for T in radians is provided in the table for convenience.

Results for (a) are plotted as the upper curve in Figure 4.7. The maximum ground-level concentration is located at approximately 3.11 km downwind of the stack. Unfortunately, the function of Equation (4.20) is not easily differentiated to find the exact maximum point. It is located here within 0.01 km (10 m). This is within 1 s of latitude or longitude on a USGS map. At 3.11 km, the ground-level concentration is calculated as 3.0094 µg/m³ (approximately 3).

Changing only the wind speed or the pollutant emission rate while keeping all other variables constant in Equation (4.20) would change the height of the curve, including its maximum point, while leaving the location of the maximum point the same.

Results for a different pollutant emission rate can be obtained by direct proration of the concentration values in Table 4.14. Concentrations for a different wind speed are inversely proportional to concentration results in Table 4.14.

$$Q = 1.5 \text{ g/s}$$
$$u = 6.7 \text{ m/s}$$
$$h = 65 \text{ (actual physical stack height)}$$

All other variables are the same as in (a).

Nothing changes to affect Equation (4.14), and F remains at 12.7. From Equation (4.15), buoyant $\Delta H = 21.425(12.7)^{3/4}/6.7 = 21.5$ m. From Equation (4.17), momentum $\Delta H = (3)(0.91)(18)/6.7 = 7.33$. Buoyancy still controls, and $H = 65 + 21.5 = 86.5$ m.

Ground-level concentrations for this case are listed in column 3 of Table 4.14. The maximum point occurs at 2.36 km±. At this point, ground-level concentration is 2.5834 µg/m³ (approximately 2.6). A plot of column 3 for case 2 is shown as the lower curve in Figure 4.7. ∎

4.10.6 Stack Draft

A tall stack does more than allowing dispersion of contaminants into the atmosphere. It also creates a natural draft that moves the flue gas out of the stack. The theoretical stack draft (D_t) in inches of water column (in w.c.) for stacks of circular cross sections is given by the following equation (in English units) [85]

$$D_t = 0.52 P_T h[(1/t)-(1/T_s)] \qquad (4.24)$$

where P_T is the barometric pressure (psia), h is the stack height above flue gas inlet (ft), t and T_s are the temperature ambient and stack gas, respectively (°R). Frictional losses (F_s) and expansion losses (F_e) must be subtracted to provide a better estimate of the actual available draft [85].

$$F_s = (0.008 h v^2)/(d T_s) \qquad (4.25)$$

and

$$F_e = 0.012 v^2 / T_s \qquad (4.26)$$

where v is the stack gas velocity (ft/s), d is the stack diameter (ft), and the other symbols are as defined above.

The following equation from another source [86] gives the stack gas flow rate (Q, m/s) that can be induced by the natural draft in a stack. The equation assumes that the molecular weight of the ambient air and the flue gas are the same (not a bad assumption) and that frictional losses and heat losses can be neglected. *Caution*: metric units and degrees Kelvin are used in this equation:

$$Q = CA[2gh(T_s-t)/T_s]^{1/2} \qquad (4.27)$$

where C is the discharge coefficient (0.65–0.70) (dimensionless), A is the stack cross-sectional area (m²), g is the acceleration of gravity (~9.8 m/s²), H is the stack height (m), and t and T_s = temperature ambient and stack gas, respectively (°K).

PROBLEM 4.13

Using the data from Problem 4.11, determine whether the stack gas flow in Problems 4.11 and 4.12 can be induced by natural draft.

SOLUTION Theoretical draft in inches of water column is calculated from Equation (4.24) with English units and R.

$$D_t = (0.52)(14.696)(213.25)[(1/528)-(1/810)]$$
$$= 1.0745 \text{ in w.c.}$$

Induced flow is calculated from Equation (4.27) with metric units and °K.

$$Q = (0.70)(0.6567)[(2)(9.8)(65)(450-293)/450]^{1/2}$$
$$= 9.69 \text{ m}^3/\text{s} = 342.2 \text{ ft}^3/\text{s}$$
$$v = Q/A = 14.76 \text{ m/s} = 48.42 \text{ ft/s}$$

Without considering frictional losses, the stack gas flow and velocity are less than the values from Problem 4.11.

Furthermore, even at the lower flow rate and exit velocity, the frictional losses from Equations (4.25) and (4.26) would exceed the available stack draft, negating a basic assumption of Equation (4.27).

The conclusion is that natural draft is not sufficient to provide the flow rate and exit velocity for the case of Problem 4.11 and that forced and/or induced draft must be provided. ∎

REFERENCES

1. D.B. Turner, *Workbook of Atmospheric Dispersion Estimates*, 2nd edn, Lewis Publishers CRC, Boca Raton, FL, 1994.
2. Compilation of Air Pollutant Emission Factors, AP-42, Fifth Edition, Volume I, Chapter 13: Miscellaneous Sources, Section 13.5 Industrial Flares, US EPA, Washington, DC, September 1991, reformatted January 1995.
3. O.C. Leite, Smokeless efficient, nontoxic flaring, *Hydrocarbon Processing*, **70**(3), 77–80, 1991.
4. D.K. Stone, S.K. Lynch, R.F. Pandullo, L.B. Evans, and W.M. Vatavuk, Flares. Part I: flaring technologies for controlling VOC-containing waste streams, *Journal of the Air and Waste Management Association*, **42**(3), 333–340, 1992.
5. Code of Federal Regulations 40 CFR 60 Section 60.18, General Control Device Requirements, U.S. Government Printing Office, Washington, DC, July 1, 2006.
6. P.R. Oenbring and T.R. Sifferman, Flare design... are current methods too conservative? *Hydrocarbon Processing*, **59**(5), 124–129, 1980.
7. J.G. Seebold, Practical flare design, *Chemical Engineering*, **91**(25), 69–72, 1984.
8. J.F. Straitz III, Make the flare protect the environment, *Hydrocarbon Processing*, **56**(10), 131–135, 1977.
9. Anonymous, Air Permit Technical Guidance for Chemical Sources: Flares and Vapor Oxidizers, TCEQ Publication RG-109 (Draft), Air Permits Division, Texas Commission on Environmental Quality (TCEQ), Austin, TX, October 2000.
10. J.F. Straitz III, Flaring for gaseous control in the petroleum industry, Paper No. 78-58.8 presented at the Annual Meeting of the Air Pollution Control Association, Austin, TX, June 29, 1978.
11. Anonymous, Technical Supplement 4: Flares, TCEQ Publication RG-360A (Revised), Texas Commission on Environmental Quality (TCEQ), Austin, TX, January 2007.
12. R.G. Kunz, SCR performance on a hydrogen reformer furnace—a comparison of initial and first- and second-year anniversary emissions data, Paper No. 96-RA 120.01 presented at the 89th Annual Meeting Air & Waste Management Association, Nashville, TN, June 23–28, 1996.
13. R.G. Kunz, SCR performance on a hydrogen reformer furnace, *Journal of the Air and Waste Management Association*, **48**, 26–34, 1998.
14. J. Czarnecki, C.J. Pereira, M. Uberoi, and K.P. Zak, Put a lid on NO_x emissions, *Pollution Engineering*, **26**(12), 26–29, 1994.
15. D. Fusselman and D. Lipsher, Several technologies available to cut refinery NO_x, *Oil & Gas Journal*, **90**(44), 45–50, 1992.
16. T. Koyanagi and K. Suyama, Development of SCR system by MHI for clean gas, Paper 89-96B.4 presented at the 89th Annual A&WMA Meeting, Anaheim, CA, June 25–30, 1989.
17. Y. Seo, N. Iiyama, and T. Suzuki, Recent application trend of IHI selective catalytic reduction technologies, *IHI Engineering Review*, **25**(1), 7–12, 1992.
18. S.M. Cho, Properly apply selective catalytic reduction for NO_x removal, *Chemical Engineering Progress*, **90**(1), 39–45, 1994.
19. J.W. Beeckman and L.L. Hegedus, Design of monolith catalysts for power plant NO_x emission control, *Industrial and Engineering Chemistry Research*, **30**(5), 969–978, 1991.

20. K. Funahashi, T. Kobayakawa, K. Ishii, and H. Hata, SCR DeNO$_x$ in New Marazen ethylene plant, Proceedings of the 13th Ethylene Producers' Conference, Vol. 10, American Institute of Chemical Engineers, New York, NY, 2001, pp. 741–755.
21. R.G. Kunz, D.C. Hefele, R.L. Jordan, and F.W. Lash, Consider SCR to mitigate NO$_x$ emissions, *Hydrocarbon Processing*, **82**(11), 43–50, 2003.
22. J.R. O'Leary, R.G. Kunz, and T.R. von Alten, Selective catalytic reduction (SCR) performance in steam–methane reformer service: the chromium problem, *Environmental Progress*, **23**(3), 194–205, 2004.
23. R.G. Kunz and T.R. von Alten, SCR treatment of ethylene furnace flue gas (a steam–methane reformer in disguise), Paper presented at Institute of Clean Air Companies (ICAC) Forum '02, Houston, TX, February 2002.
24. P.W. Groff and B.K. Gullett, Industrial boiler retrofit for NO$_x$ control: combined selective noncatalytic reduction and selective catalytic reduction, *Environmental Progress*, **16**(2), 116–120, 1997.
25. J.R. O'Leary, R.G. Kunz, and T.R. von Alten, Selective catalytic reduction (SCR) performance in steam–methane reformer service: the chromium problem, Paper ENV-02-178 presented at the 2002 NPRA Environmental Conference, New Orleans, LA, September 9–10, 2002.
26. A. Suwa, Operating experiences of SCR De NO$_x$ unit in Idemitsu ethylene plant, Proceedings of the 13th Ethylene Producers' Conference, Vol. 10, American Institute of Chemical Engineers, New York, NY, 2001, pp. 766–773.
27. M. Karrs and P.S. Crepinsek, Cracking heater convection retrofit with integral SCR, Proceedings of the 20th Ethylene Producers' Conference, Vol. 17, American Institute of Chemical Engineers, New York, NY, 2008, pp. 276–287.
28. A. Kokkinos, J.E. Cichanowicz, D. Eskinazi, J. Stallings, and G. Offen, NO$_x$ controls for utility boilers: highlights of the EPRI July 1992 Workshop, *Journal of the Air and Waste Management Association*, **42**(11), 1498–1505, 1992.
29. U.S. Environmental Protection Agency, Test Methods, 40 CFR 60, Appendix A, U.S. Government Printing Office, Washington, DC, 1995.
30. Anonymous, Ammonia Integrated Sampling, Source Test Procedure ST-1B in Manual of Procedures (MOP) Volume 4 Source Test Policy and Procedures, Bay Area Air Quality Management District, San Francisco, CA, adopted January 20, 1982.
31. Anonymous, Determination of Ammonia In Effluents Collected in Acid Media Using the Specific Ion Electrode, Laboratory Method 1A in Manual of Procedures (MOP) Volume 3 Laboratory Policy and Procedures, Bay Area Air Quality Management District, San Francisco, CA, September 2, 1998.
32. Compilation of Air Pollutant Emission Factors, AP-42, Volume I, Fifth Edition, Supplement E, Chapter 1: External Combustion Sources, Section 1.3 Fuel Oil Combustion, U.S. EPA, Washington, DC, September 1998.
33. Compilation of Air Pollutant Emission Factors, AP-42, Volume I, Fifth Edition, Supplement D, Chapter 1: External Combustion Sources, Section 1.4 Natural Gas Combustion, U.S. EPA, Washington, DC, July 1998.
34. S. Blackenship, Sophisticated SCR now available for small power plants, *Power Engineering*, **107**(11), 126–127, 2003.
35. C.M. Anderson and J.A. Billings, Calculating NH$_3$ slip for SCR equipped cogeneration units, Paper 90-GT-105 presented at the Gas Turbine and Aeroengine Congress and Exposition, Brussels, Belgium, June 11-14, 1990, American Society of Mechanical Engineers (ASME), New York, NY, 1990.
36. C.M. Anderson, and J.A. Billings, Simple calculation measures NH$_3$ slip for cogeneration units, *Power Engineering*, **95**(4), 42–44, 1991.
37. D.C. Mussatti, R. Srivastava, P.M. Hemmer, and R. Strait, Selective noncatalytic reduction, in: P. Mussatti, editor, *The EPA Air Pollution Cost Manual*, EPA 452/B-02-001, U.S. EPA, Research Triangle Park, NC, January 2002.
38. R.K. Lyon, Method for the reduction of the concentration of NO in combustion effluents using ammonia, U.S. Patent 3,900,554, assigned to Exxon Research and Engineering Company, August 19, 1975.

39. J.K. Arand, L.J. Muzio, and J.G. Sotter, Urea reduction of NO_x in combustion effluents, U.S. Patent 4,208,386 assigned to Electric Power Research Institute, Inc., June 17, 1980.
40. J.K. Arand, L.J. Muzio, and D.P. Teixeira, Urea reduction of NO_x in fuel rich combustion effluents, U.S. Patent 4,325,924, assigned to Electric Power Research Institute, Inc., April 20, 1982.
41. T.C. Hess, Advanced NO_x and SO_2 emission control performance on a CFB, *Power Engineering*, **93**(7), 47–49, 1989.
42. A.M.G. Gentemann and J.A. Caton, Decomposition and oxidation of a urea-water solution as used in selective non-catalytic removal (SNCR) processes, Paper presented at the 2nd Joint Meeting of the United States Sections of the Combustion Institute 2001 Spring Technical Conference, Oakland, CA, March 25–28, 2001.
43. M.P. Younis, Implementation of NO_x control technologies in petroleum refining applications, Mobil Torrance Refinery, Conference Proceedings from Clean Air '94—First North American Conference & Exhibition, Emerging Clean Air Technologies and Business Opportunities: Meeting Global Air Challenges through Partnerships, Toronto, Ontario, Canada, September 26–30, 1994, pp. 301–314.
44. C. Castaldini, K.G. Salvensen, and H.B. Mason, Technical Assessment of Thermal $DeNO_x$ Process, EPA-600/7-79-17, U.S. EPA, Washington DC (May 1979), available as PB 297 949 from National Technical Information Source, Springfield, VA, 1979.
45. T.S. Bacon, Process for producing inert gas, U.S. Patent 2,051,125, August 18, 1936.
46. T.S. Bacon, Production of non-corrosive inert gas by the combustion of fuel gas, *Petroleum Engineer*, **11**(3), 51–56, 1940.
47. G.W. Jones, B. Lewis, J.B. Friauf, and G. St. J. Perrott, Flame temperatures of hydrocarbon gases, *Journal of the American Chemical Society*, **53**(3), 869–883, 1931.
48. R.G. Kunz, D.D. Smith, N.M. Patel, G.P. Thompson, and G.S. Patrick, Control NO_x from gas-fired hydrogen reformer furnaces, Paper AM-92-56 presented at the 1992 NPRA Annual Meeting, New Orleans, LA, March 22–24, 1992.
49. R.H. Perry and C.H. Chilton, *Chemical Engineers' Handbook*, 5th edn, McGraw-Hill, New York, 1973, p. 9–12.
50. S.C. Wood, Select the right NO_x control technology, *Chemical Engineering Progress*, **90**(1), 32–38, 1994.
51. U.S. Environmental Protection Agency, Determination of Sulfur Dioxide Removal Efficiency and Particulate Matter, Sulfur Dioxide, and Nitrogen Oxide Emission Rates, 40 CFR 60, Appendix A, Method 19, July 1, 2007.
52. Compilation of Air Pollutant Emission Factors, AP-42 Volume I, Fifth Edition, Chapter 3: Stationary Internal Combustion Sources, Section 3.1 Stationary Gas Turbines, U.S. EPA, Washington, DC, April 2000.
53. U.S. Environmental Protection Agency, Standards of Performance for Stationary Gas Turbines, 40 CFR 60, Subpart GG, U.S. Government Printing Office, Washington, DC, Electronic Code of Federal Regulations, http://ecfr.gpoaccess.gov/cgi/t/text/text, current as of August 7, 2008.
54. D.A. Kolp, S.R. Gagnon, and M.J. Rosenbluth, Water treatment and moisture separation in steam-injected gas turbines, Paper 90-GT-372 presented at the Gas Turbine and Aeroengine Congress and Exposition, Brussels, Belgium, June 11–14, 1990, American Society of Mechanical Engineers (ASME), New York, NY, 1990.
55. C.A. Meyer, R.B. McClintock, G.J. Silvestri, and R.C. Spencer Jr., *1967 ASME Steam Tables*, 2nd edn, American Society of Mechanical Engineers, New York, 1968, p. 88.
56. Cannon Technology, Inc., Closeout Final Report on a Demonstration Test and Evaluation of the Cannon Lo-NO_x Digester System, DOE/PC/92161-1, 81 pp., U.S. Dept. of Energy, Office of Scientific and Technical Information (OSTI), Pittsburgh, PA, April 23, 1995, available at www.osti.gov/bridge/product.biblio.jsp?osti_id=52736, accessed August 20, 2008.
57. DuPont™ Belco® Clean Air Technologies—LoTOx™ Technology, http://www.belcotech.com/products/NOx.html, accessed March 6, 2008.
58. A.P. Skelley, J.M. Koltick Jr., N.J. Suchak, and W.M. Rohrer Jr., Process for removing NO_x and SO_x from exhaust gas, U.S. Patent 6,162,409, December 19, 2000.

59. J. Hsieh, K. R. Gilman, D. Philibert, S. Eagleson, and A. Morin, Wet scrubbing apparatus and method for controlling NO_x emissions, U.S. Patent 7,214,356 issued to Belco Technologies Corporation, May 8, 2007.
60. Anonymous, Assessment of NO_x Emissions Reduction Strategies for Cement Kilns—Ellis County, Attachment C (ERG Report on LoTOx™ Application to Refineries), Final Report to Texas Commission on Environmental Quality, ERG Inc., July 14, 2006.
61. N. Confuorto and J. Sexton, Wet scrubbing based NO_x control LoTOx™ Technology—first commercial FCC start-up experience, Paper ENV-07-100 presented at the 2007 NPRA Environmental Conference, Austin, TX, September 24–25, 2007.
62. J. Sexton, N. Confuorto, M. Barasso, and N. Suchak, LoTOx™ Technology demonstration at Marathon Ashland Petroleum LLC's Refinery in Texas, Paper AM-04-10 presented at the 2004 NPRA Annual Meeting, San Antonio, TX, March 21–23, 2004.
63. E.H. Weaver, N. Confuorto, and M.J. Barrasso, FCCU NO_x control scrubbing system—state of the art technology, Paper ENV-03-127 presented at the 2003 NPRA National Environmental & Safety Conference, New Orleans, LA, April 23–24, 2003.
64. N. Confuorto, M. Barasso, and N. Suchak, Clean generation, Reprint *Hydrocarbon Engineering*, June 2003, 5 pp.
65. M. Kay, and K. Beruldsen, Technical Assessment of the Low Temperature Oxidation System Achieved-in-Practice Demonstration at Alta Dena Dairy (Facility ID # 005903), South Coast Air Quality Management District, Diamond Bar, CA, October 22, 1999, available at http://www.aqmd.gov.bact/Response toBWG.DOC, accessed on August 20, 2008.
66. Anonymous, Scientific Review Committee Meeting No. 10, South Coast Air Quality Management District, Diamond Bar, CA, Oct. 29, 1999, available at http://www.aqmd.gov.bact/Minutes10-29-99.html, accessed on August 20, 2008.
67. Anonymous, Basic Equipment or Process: Boiler Application No. 259724, South Coast Air Quality Management District, Diamond Bar, CA, issuance date: February 5, 1992, available at http://www.aqmd.gov.bact/259724.htm, accessed on August 20, 2008.
68. R. Ferrell, Controlling NO_x emissions: a cooler alternative, *Pollution Engineering*, **32**(4), 50–52, 2000.
69. Cleaver-Brooks, Prior to 2005 Boiler Book, Section A2 Model CB Boilers and Section J Glossary and Index, available at http://www.cleaver-brooks.com/contact.htm, accessed on August 25, 2008.
70. Southern California Gas Company, Description of Service, Filed August 28, 2002; effective October 7, 2002, available at www.socalgas.com/regulatory tariffs/tm2/pdf/02.pdf, accessed on August 25, 2008.
71. G.D. Sargent, Dust collection equipment, *Chemical Engineering*, **76**(2), 130–150, 1969.
72. A. Tesoriero, Predict particle size distribution from FCC beds, *Hydrocarbon Processing*, **63**(11), 139–141, 1984.
73. C.J. Stairmand, The Design and Performance of Modern Gas-Cleaning Equipment, *Journal of the Institute of Fuel*, **29**, 58–81, 1956.
74. C.J. Stairmand, Selection of gas cleaning equipment: a study of basic concepts, *Filtration & Separation*, Jan./Feb. 1970, pp. 53–66.
75. J.G. Wilson and D.W. Miller, The removal of particulate matter from fluid bed catalytic cracking unit stack gases, *Journal of the Air Pollution Control Association*, **17**(10), 682–685, 1967.
76. G.D. Bouziden, J.K. Gentile, and R.G. Kunz, Selective catalytic reduction of NO_x from fluid catalytic cracking—case study: BP Whiting Refinery, Paper ENV-03-128 presented at the 2003 NPRA National Environmental & Safety Conference, New Orleans, LA, April 23-24, 2003.
77. A.J. Buonicore and W.T. Davis, editors, *Air Pollution Engineering Manual*, Van Nostrand Reinhold, New York, 1992, p. 232, Table 8.
78. R.W. Boubel, D.L. Fox, D.B. Turner, and A.C. Stern, *Fundamentals of Air Pollution*, 3rd edn, Academic Press, San Diego, CA, 1994, pp. 472, 491,
79. C.D. Ahrens, *Meteorology Today: An Introduction to Weather, Climate, and the Environment*, 4th edn, West Publishing Company, St. Paul, MN, 1991, pp. 189–213.

80. G.A. Briggs, A Plume Rise Model Compared with Observations, *J. Air Pollution Control Association*, **15**, 433–438, 1965.
81. G.A. Briggs, Plume rise predictions, in: D.A. Haugeen,editor, *Lectures on Air Pollution and Environmental Impact Analysis*, American Meteorological Society, Boston, MA, 1975, pp. 59–111, as cited in Ref. [1, p. 3–2].
82. G.A. Briggs, Some recent analysis of plume rise observations, in: H.M. Englund and W.T. Berry editors, *Proceedings of the Second International Clean Air Congress*, Academic Press, New York, 1971, pp. 1029–1032, as cited in Ref. [1, p. 3–2].
83. Clean Air Act (42 U.S.C. 7401 *et seq.*, as amended by Public Law 91-604, 84 Stat. 1676; Public Law 95-95, 91 Stat. 682, and Public Law 95-190, 91 Stat. 1399.
84. U.S. Environmental Protection Agency, Requirements for Preparation, Adoption, and Submittal of Implementation Plans, 40 CFR 51, Subpart F—Procedural Requirements, Subpart G—Control Strategy, and Subpart I—Review of New Sources and Modifications, U.S. Government Printing Office, Washington, DC, Electronic Code of Federal Regulations, http://ecfr.gpoaccess.gov/cgi/t/text, current as of August 7, 2008.
85. R.E. George and J.E. Williamson, On-site incineration of commercial and industrial wastes with multiple-chamber incinerators, in: R.C. Corey, editor, *Principles and Practices of Incineration*, Wiley-Interscience, New York, 1969, p. 121.
86. Anonymous, Flue Gas Stack, p. 2 of 4, Wikipedia, the Free Encyclopedia, available at http://en.wikipedia.org/wiki/Flue_gas_stack, accessed on August 11 2008.

Chapter 5

Water/Wastewater Composition

Water and wastewater are separated by only a few parts per million of pollution.

5.1 INTRODUCTION

Important concentration-based measurements are discussed in this chapter. In addition to dissolved oxygen and other individually identified species, these include certain nonspecific indicators of pollution in wastewater and the relationships among various ions and dissolved gases in water, wastewater, and process streams. The explanation is supplemented by material contained in the Appendix.

5.2 CONCENTRATIONS EXPRESSED AS mg/L AS CaCO₃

As indicated in Appendix E, the term *mg/L as calcium carbonate* ($CaCO_3$) is employed in water calculations. It is understandable that *hardness* and the somewhat mysterious *alkalinity* should be expressed in terms of this unit. However, concentrations of other constituents such as sodium, chloride, and sulfate are also frequently expressed as calcium carbonate as well, even when there is no actual calcium carbonate within 100 miles (160.9 km) of the sample.

The concept of hardness came from the ability of a metal ion to precipitate soap. The principal contributors to hardness are calcium (Ca^{2+}), magnesium (Mg^{2+}), strontium (Sr^{2+}), ferrous (divalent iron, Fe^{2+}), manganous (Mn^{2+}), and aluminum (Al^{3+}) ions. Of these, only calcium and magnesium are of practical importance in natural waters, and hardness is commonly understood to include only these two ions, total hardness being the sum of calcium hardness and magnesium hardness. Calcium and magnesium can form scale deposits in cooling systems and steam boilers. One such scale happens to be $CaCO_3$. However, there are others.

Environmental Calculations: A Multimedia Approach, by Robert G. Kunz
Copyright © 2009 John Wiley & Sons, Inc.

5.2 Concentrations Expressed as mg/L as CaCO₃

Alkalinity is the ability to neutralize an acid, and its counterpart, *acidity*, is the ability to neutralize a base. These are also expressed as $CaCO_3$, rather than the amount of acid or base added to bring about a change in a solution's pH. The pH is a measure of the free hydrogen ion activity in solution. Nothing about these definitions should indicate that there is any actual $CaCO_3$ involved.

In fact, the use of $CaCO_3$ is a convenient way of expressing chemical equivalents as numbers of the same order of magnitude as the original concentrations of the contributing ions without having to multiply by 10^{-4} to obtain gram equivalents. Calcium carbonate has a molecular weight of ~100, and an equivalent (EW) [molecular weight (MW) divided by the valence (number of positive or negative charges) 100/2≅50], thereby simplifying the calculations. This works for any ion.

$$\text{mg/L as CaCO}_3 = \text{mg/L as X(EW of CaCO}_3/\text{EW as X)} \quad (E.12)$$

where X represents any species under consideration.

Equivalent weight ratios for several important ions in water chemistry are listed in Table E.2, where a further explanation of these topics is provided. A more complete listing is contained in Ref. 1.

This is particularly important when verifying the electroneutrality of a water analysis. In this calculation, the positive charges have to equal the negative charges, or the total cations expressed as $CaCO_3$ must add up to the total anions also expressed as $CaCO_3$.

$$\text{Total cations (as mg/L as CaCO}_3\text{)} = \text{total anions (as mg/L as CaCO}_3\text{)} \quad (E.14)$$

In the pH range 6–9, the hydrogen [H⁺] and the [OH⁻] ions are small compared to other ions and also tend to cancel each other out.

In practice, a typical water analysis is never complete and correct enough so that Equation (E.14) is satisfied exactly. The analysis is therefore adjusted, typically by adding sodium (Na⁺) to make up for a deficit of cations and an anion such as chloride (Cl⁻) to compensate for an anion deficit.

5.2.1 Balancing a Typical Water Analysis

In the 1970s, a water service company ran a series of advertisements in *Chemical Engineering Magazine* and perhaps elsewhere, listing the chemical analysis of such famous waters as Crater Lake in Oregon, United States; the Thames River in London, England; the Rhine River in Germany; and the Amazon in Brazil.

The published analysis for Crater Lake, shown in Table 5.1, is a mixture of mg/L as individual ions and mg/L as $CaCO_3$ for hardness and alkalinity [2]. Entries from the original analysis are shown in bold. Problem 5.1 checks the water analysis for electroneutrality and compares the analysis with a data set of 91 water quality samples analyzed by the USGS collected between June 1967 through September 1995 [3] as part of a study mandated by the U.S. Congress on the lake's water quality. (Sampling frequency averaged only three samples per year because the access road was closed in winter from mid-October to about mid-June.)

Table 5.1 Water Services Company Analysis Reported for Crater Lake, Oregon[a]

Constituent	mg/L as the ion	mg/L as CaCO$_3$
Conductivity[b]	117	—
pH	7.2[c]	—
Total dissolved solids	83[d]	—
Cations		
Calcium hardness	7.6	19
Magnesium hardness	2.2	9
Total hardness	—	28
Sodium (Na$^+$)	11	23.98
Iron (Fe^{2+})	0.02	0.04
Total cations	—	52.02
Anions		
M-alkalinity	36.0 as HCO$_3^-$	29.5
P-alkalinity	0.0	0.0
Chloride (Cl$^-$)	12	16.9
Sulfate (SO$_4^{2-}$)	12	12.5
Total Anions	—	58.9
Difference (cations − anions)	—	−6.88[e]
Silica (SiO$_2$)[f]	18	—

[a] Adapted from Typical Water Analysis Table in Ref. [2].
[b] Conductivity in units of μmho/cm (equivalent to μS/cm).
[c] pH in standard units.
[d] Total dissolved solids determined by gravimetric analysis.
[e] For electroneutrality, 6.88 mg/L of Na$^+$ as CaCO$_3$ or 3.2 mg/L of Na$^+$ as the ion must be added.
[f] Silica expressed as SiO$_2$.

The lake is located high in the Cascade Range of southwestern Oregon within the boundaries of Crater Lake National Park. A paved perimeter road on the crater rim encircles the lake, but access to the lakeshore is very difficult. The park contains a visitor center, a lodge and cafeteria, and a septic tank sewage disposal facility on the south side of the lake, several picnic areas along the perimeter road, and a research station on an island in the lake.

The lake's average surface elevation is 1882 m (6174.5 ft), its surface area 53.2 km^2 (13,146 acres), and its maximum depth 589 m (1932.4 ft). The lake was formed from the collapsed caldera of Mount Mazama, which erupted about 6850 years ago. It is fed by direct precipitation and streams and springs that emanate from the caldera walls; outflow is by evaporation and seepage from the lake bottom. The lake elevation varies annually by about 60 cm ($\sim$2 ft) in response to seasonal variations in precipitation and evaporation. The lake is affected at least indirectly by discharge from a domestic wastewater septic tank drainfield, a facility that accommodates more than 500,000 park visitors annually, especially during July and August. Studies indicate that the nitrate concentrations in nearby springs discharging into the lake have increased.

5.2 Concentrations Expressed as mg/L as CaCO$_3$

Table 5.2 Median Values of USGS Crater Lake, Oregon Water Analyses (1967–1995)^a

Parameter	Equiv Wt^b	Number of samples	Microequiv per liter^c	mg/L as the ion^b	mg/L as CaCO$_3$^{b}
Lake stage^d	—	88	1881.77^d	—	—
Spec. cond., field^e	—	81	120	—	—
pH, field^f	—	77	7.5^f	—	—
Cations					
Calcium (Ca^{2+})	20.0	91	350	7.0	17.5
Magnesium (Mg^{2+})	12.2	91	210	2.6	10.5
Sodium (Na$^+$)	23.0	90	480	11.0	24.0
Potassium (K$^+$)	39.1	90	46	1.8	2.3
Ammonium (NH$_4$$^+$)	18.0	44	1.4	0.03	0.1
Total cations	—	—	1087.4	—	54.4
Anions					
Alkalinity, laboratory	50.0	88	600	36.6^g	30.0
Sulfate (SO$_4$$^{2-}$)	48.0	90	210	10.1	10.5
Chloride (Cl$^-$)	35.5	91	280	9.9	14.0
Nitrite (NO$_2$$^-$) plus	46.0	65	7.1	0.3^h	0.4
Nitrate (NO$_3$$^-$)	62.0			0.4^h	
Total anions	—	—	1097.1	—	54.9
Difference (cations – anions)	—	—	−9.7	—	−0.5^i
Silica (SiO$_2$)^j	—	90	300^i	18	—

^{a}Excerpted from Table 34 of Ref. [3].
^{b}Equivalent weights, mg/L columns, and difference entries added by the author.
^{c}As reported.
^{d}Lake stage reported in ft (sic).
^{e}In μS/cm at 25°C (equivalent to μmho/cm).
^{f}pH in standard units.
^{g}As HCO$_3$$^-$ (equivalent weight = 61.0).
^{h}Assuming first all nitrite, then all nitrate.
^{i}For electroneutrality, 0.5 mg/L of Na$^+$ as CaCO3 or 0.2 mg/L as the ion must be added.
^{j}Silica in μmol/L.

The USGS median water analysis during 1967–1995 is shown in Table 5.2. The analytical results are indicated to be of high quality. USGS calculated ion balances plotted in Ref. [3] for 86 samples having a complete major ion analysis range from −10% to +9.6%, 85% had values within the ±5% range, the average ion balance for all samples was −10%, and 65% of the samples showed a slight excess of measured cations to measured anions. This was attributed by the USGS as possibly indicating that unmeasured constituents such as organic anions could have contributed a small amount to the ionic content of lake water at the measurement station. A check of the USGS median values for electroneutrality is performed in Problem 5.2.

PROBLEM 5.1

Review the water analysis of Table 5.1 and balance the water, if necessary.

SOLUTION The entries in Table 5.1 from the original published analysis are shown in bold type. Some concentrations are reported as the ionic species and some as $CaCO_3$. To balance the water, they must all be expressed as $CaCO_3$. For example, sodium as the sodium ion must be multiplied by the ratio of the equivalent weight of $CaCO_3$ to the equivalent weight of sodium in accordance with Equation (E.12).

$$(11)(50.05/23) = (11)(2.18) = 23.98$$

This ratio and its reciprocal (for going from $CaCO_3$ to Na^+) are listed in Table E.2, along with others for other common ions. A more complete list is contained in Ref. [1].

The cations and the anions expressed as $CaCO_3$ are summed separately in Table 5.1 and compared. Anions exceed cations by 6.88 mg/L as $CaCO_3$. Sodium is commonly added to make up the difference. This corresponds to

$$6.88(23/50.05) = (6.88)(0.46) = 3.2 \text{ mg/L as } Na^+$$

This number is shown in a footnote to Table 5.1. If cations had been in excess, an anion such as chloride would typically be added.

The analysis compares favorably with that of Table 5.2. ∎

PROBLEM 5.2

Check the ion balance for the median of the USGS samples from 1967 to 1995 listed in Table 5.2.

SOLUTION The original USGS entries in column 4 are given in microequivalents (μ equiv) per liter. These can be summed directly to yield 1087.4 cations versus 1097.1 anions. They can also be converted, using Na^+ and Ca^{2+} as examples, into mg/L as the ion or mg/L as $CaCO_3$ as follows:

$$Na^+ \text{ as ion}: \quad (480 \, \mu\text{equiv/L})(10^{-6} \, \text{equiv}/\mu\text{equiv})$$
$$\times (23.0 \, \text{g/equiv})(1000 \, \text{mg/g}) = 11.0$$
$$Na^+ \text{ as } CaCO_3: \quad (480)(10^{-6})(50.05)(1000) = 24.0$$
$$Ca^{2+} \text{ as ion}: \quad (350)(10^{-6})(20.0)(1000) = 7.0$$
$$Ca^{2+} \text{ as } CaCO_3: \quad (350)(10^{-6})(50.05)(1000) = 17.5$$

For the monovalent ion Na^+, the equivalent weight is the molecular weight; for the divalent ion Ca^{2+}, the equivalent weight is the molecular weight divided by 2.

For the median of the samples, anions exceed cations, and some sodium must be added, as shown in Table 5.2. The difference amounts to less than 1%. However, the sample median is not like the 65% of the individual samples showing an excess of cations over anions, as mentioned by USGS [3] and quoted here. ∎

Further examples of ion balance calculations for less famous waters are provided in Appendix E.

Table 5.3 Results of Conductivity Calculation for Crater Lake Water

Ion	Conductivity factor	Water services analysis (mg/L)	Conductivity product (A)	USGS median analysis (mg/L)	Conductivity product (B)
Ca^{2+}	2.6	7.6	19.76	7.0	18.20
Mg^{2+}	3.82	2.2	8.40	2.6	9.93
Na^+	2.13	11 + 3.2	30.25	11.0 + 0.2	23.86
Fe^{2+}	1.93	0.02	0.04	—	—
K^+	1.84	—	—	1.8	3.31
NH_4^+	4.08	—	—	0.03	0.12
HCO_3^-	0.715	36.0	25.74	36.6	26.17
Cl^-	2.14	12	25.68	9.9	21.19
SO_4^{2-}	1.54	12	18.48	10.1	15.55
NO_2^-	1.56	—	—	0.3	either 0.47
NO_3^-	1.15	—	—	0.4	or 0.46
Conductivity Calculated total			128.35		118.79–118.80
Calculated total (rounded)			128		119
Measured			117		120

5.2.2 Calculation of Conductivity

PROBLEM 5.3

Estimate the conductivity of Crater Lake water of Tables 5.1 and 5.2. Compare with measured values.

SOLUTION The method presented in Appendix E for cooling water and cooling water makeup will be used. Activity coefficients will *not* be calculated since the water is so dilute. The method employs a factor tabulated in Appendix E for each ion to be multiplied by its concentration in mg/L. These individual products of the conductivity factor and the concentration are then summed to arrive at the conductivity as shown in Table 5.3. Factors for the ammonium ion (NH_4^+), nitrite (NO_2^-), and nitrate (NO_3^-) (not contained in Appendix E) were determined from Table 6-7 of Ref. [4]. Calculated conductivity and the water services measurement agree within 10%. The difference is within 1% of the USGS median value. ∎

5.3 DISSOLVED OXYGEN

To say that dissolved oxygen (DO) in the chemistry and biology of natural waters is important would be an understatement. It supports life of aquatic creatures and fosters the assimilation of water-borne waste materials. On the other hand, it proves corrosive in a cooling tower or boiler environment. This is all the more remarkable in light of its limited solubility in an aqueous medium.

One may encounter effluent limits mandating a minimum level of DO. Fortunately, as oxygen in solution is used up, it becomes easier to replenish it by mass transfer of oxygen from the atmosphere, such as by aeration. Unfortunately, it is difficult to keep it out when an oxygen-free boiler feedwater is desired.

Dissolved oxygen is measured by wet chemistry and by instrumental means [5, pp. 4–147 to 4–162; 6, p. 419]. Oxygen electrodes are especially convenient and useful in field work to capture fragile on-the-spot concentrations susceptible to change in transit back to the laboratory [6, pp. 408–409]. Otherwise, the samples must be "fixed," or treated with the reagents used in the DO test in the field, but the final titration is performed in the laboratory [6, p. 409; 7, p. 386].

Additional comments concerning DO are contained in the next paragraphs. The subject is explored in greater detail in Appendix D, and in particular Table D.1.

PROBLEM 5.4

Calculate the solubility of oxygen in pure water at atmospheric pressure and 20°C (68°F) in contact with 68°F air at a relative humidity (RH) of 60%. Compare with values computed for the limiting cases of dry air and water-saturated air in Table D.1.

SOLUTION Pressure for this example is standard atmospheric = 14.696 psia. Temperature is 20°C (68°F) with 60% relative humidity.
Vapor pressure $p_w = 0.33889$ psia [8, p. 88].

Partial pressure (p') at less than saturated conditions

$$p' = p_w(\text{RH}/100) \\
= (0.33889)(60/100) = 0.2033 \text{ psia} \tag{5.1}$$

Pressure of dry atmospheric air is then

$$14.696 - 0.2033 = 14.4927 \text{ psia}$$

Atmospheric oxygen amounts to 20.947% of dry atmospheric air (Table 3.3).

$$\text{Partial pressure of } O_2 = (20.947/100)(14.4927) = 3.0358 \text{ psia}$$
$$\div 14.696 = 0.2066 \text{ atm}$$

Henry's law constant for O_2 at 20°C

$$H = 4.01 \times 10^4 \text{ atm/mol fraction [9, p. 675]}$$

From Henry's law

$$P = Hx \tag{2.30}$$

where p is the partial pressure and x is the mole fraction in solution. Mole fraction is moles of constituent per total moles.

By rearrangement

$$x = 0.2066/(4.01 \times 10^4) = 5.1514 \times 10^{-6} \text{ m.f. of } O_2 \text{ in solution}$$

Neglecting the moles of other materials in solution,

$$\text{m.f. of } O_2 = \text{mol } O_2/(\text{mol } O_2 + \text{mol } H_2O)$$

Solving for mol $O_2 = (\text{mol } H_2O)[\text{m.f. } O_2/(1 - \text{m.f. } O_2)]$
and mol O_2 per liter $= (\text{mol } H_2O \text{ per liter}) [\text{m.f. } O_2/(1 - \text{m.f. } O_2)]$.

From steam tables [8, p. 88], density of water (in equilibrium with its vapor at 68°F)

$$= 1/\text{specific volume} = (1/0.026046) = 62.3208 \text{ lb}_m/\text{ft}^3$$

This will be used for the density at atmospheric pressure since moderate pressure has a minimal effect on the density of liquid water.

$$\left(\frac{62.3208 \text{ lb}_m}{\text{ft}^3}\right)\left(\frac{\text{lb mol H}_2\text{O}}{18.02 \text{ lb}_m \text{ H}_2\text{O}}\right)\left(\frac{\text{ft}^3}{28.317 \text{ L}}\right) = 0.1221 \text{ lb mol H}_2\text{O/L}$$

$$\text{lb mol O}_2/\text{L} = (0.1221)[(5.1514 \times 10^{-6})/(1 - 5.1514 \times 10^{-6})] = 6.2899 \times 10^{-7}$$

$$(6.2899 \times 10^{-7})\left(\frac{32.0 \text{ lb}_m \text{ O}_2}{\text{lb mol O}_2}\right)\left(\frac{453.59 \times 10^3 \text{ mg}}{\text{lb}_m}\right) = 9.1 \text{ mg/L } O_2$$

This value falls within the range of values in columns 2 and 3 of Table D.1.

Values for dissolved oxygen at various temperatures are readily available in that table and elsewhere, and one would not ordinarily go through this calculation except perhaps to understand their origin. Dissolved oxygen depends on the temperature of the water, on the atmospheric pressure, on the relative humidity as seen in this example, and on the dissolved solids content of the water as well. In the extreme, oxygen solubility in seawater is considerably less than in fresh water. The chloride content of seawater is almost 19,000 mg/L [10, p. F-203], and the solubility of oxygen at 20,000 mg/L of chloride at 20°C is about 20% less [11, pp. 446–447]. ∎

5.4 NONSPECIFIC INDICATORS OF WATER POLLUTION

In many cases, it is not possible, or even necessary, to characterize the contributors to water pollution, molecule by molecule, element by element, compound by compound, and species by species. Instead, it is typical to describe the ingredients by some overall measurement. The following are in common use.

5.4.1 Biochemical Oxygen Demand

As discussed at greater length in Appendix D, the biochemical oxygen demand (BOD) is the amount of dissolved oxygen required by microorganisms in the oxidation of organic material in a waste under aerobic conditions. The amount of DO used depends on the incubation time and the temperature. Five days at 20°C (68°F) has been adopted as the standard for BOD and is denoted as BOD_5. This amounts to about 65–85% of the ultimate carbonaceous BOD (Figure D.1).

BOD_5 is determined experimentally. The test is a nonspecific bioassay test but includes only those chemical species that are biodegradable by the specific microorganisms with which the diluted wastewater has been inoculated. For an accurate assessment of BOD, it may be necessary to acclimate first the microorganisms to the waste. The microorganisms responsible for the biodegradation of the carbonaceous material do not assimilate ammonia (NH_3), and any NH_3 present does not show up as a part of the measured BOD_5.

Experimental determinations of BOD_5 from DO measurements on a sewage sample are illustrated in the example problem at the end of this section.

5.4.2 Chemical Oxygen Demand

Chemical oxygen demand (COD), as discussed in Appendix D, is the oxygen equivalent of the wastewater constituents, as determined by boiling/refluxing for 2 h with an acidic solution of potassium dichromate ($K_2Cr_2O_7$) (a strong oxidizing agent) and a silver sulfate (Ag_2SO_4) catalyst to facilitate the oxidation of certain types of chemicals (Appendix D). It is intended as an indication of organic matter, but inorganics such as ferrous ion (Fe^{2+}), the bisulfite ion (HSO_3^-), and others are affected as well. Since this is a chemical test, results for the same wastewater are highly reproducible. They are also obtained quickly, 2 h versus 5 days for BOD_5.

The COD test was originally intended to determine the ultimate BOD, including the nitrification step where the ammonia and organic nitrogen react to nitrate. However, many organic species are not completely oxidized under the COD test conditions, even with the addition of a catalyst. Some materials that show up as COD are not biodegradable even by acclimated microorganisms, and biodegradable substances that are picked up as BOD are not attacked by acidified potassium dichromate. Ammonia, originally present in a sample or formed by the decomposition of organic nitrogen, is not oxidized without significant free chloride ions present [5, p. 5–11].

Nonetheless, where a wastewater contains only readily biodegradable organic matter and nothing toxic, the experimental COD can be used to approximate the ultimate carbonaceous BOD, although it is likely that both this ultimate BOD and the COD will be less than the theoretical oxygen demand, as observed for wastewaters of known chemical composition. For a given source of wastewater, it may be possible to develop a working correlation between BOD and COD [5, p. 5–10; 12, p. 31; 13, p. 6].

5.4.3 Permanganate Oxygen Demand Test

This test is in use in various international locations, with the same purpose as the dichromate COD test employed in the United States. It determines the amount of chemical oxidation of a wastewater sample by reaction with 1/80 N potassium permanganate in dilute sulfuric acid at 80°F (26.7°C, standardized to 27°C); reaction time in a stoppered bottle is 4 h, or a quick 3 min [12, pp. 28–30]. It was noted that oxygen consumed by permanganate is highly variable from compound to compound tested and shows considerable variation with the strength of the potassium permanganate reagent [7, p. 414].

In a manner similar to the COD test, an excess of potassium iodide (KI) is added to react with the excess permanganate, and the resulting iodine (I_2^0) is back titrated with a standardized solution of sodium thiosulfate ($Na_2S_2O_3$). Interferences such as iron, chloride, nitrite, and hydrogen peroxide are acknowledged but can be compensated for by various techniques.

The 3 min. permanganate test shows the immediate oxygen demand of inorganic constituents plus easily oxidized organics. The ratio of the 4 h to the 3 min value for a wastewater diluted into the receiving stream is on the order of 3:1 for domestic sewage and 2:1 for certain industrial wastes. Higher ratios are indicative of less polluted streams.

Greater oxidation of organic compounds occurs with the potassium dichromate reagent of the COD test at its test conditions than with potassium permanganate even for 4 h, at its test conditions. With certain exceptions, the dichromate COD test oxidizes most inorganic compounds to 95–100% of the theoretical value [5, p. 5–10]. This makes the ratio of COD to BOD_5 for domestic sewage greater than 1.0. The analogous ratio for 4 h permanganate to BOD_5 is less than 1.0, also noted on p. 414 of Ref. [7].

5.4.4 Total Organic Carbon

A total organic carbon (TOC) determination is another attempt to estimate the organic matter in a wastewater sample. It is to be used in conjunction with and not as a replacement for BOD and COD. As with BOD and COD, a correlation may be possible for a given wastewater between TOC and BOD and/or COD.

It is basically an instrumental method in which the organic matter in an injected aqueous sample is oxidized to carbon dioxide (CO_2), and the CO_2 is determined chemically or instrumentally to obtain the total carbon in the sample. Suspended particulate must first be filtered out to remove its contribution to the total carbon determined. The dissolved inorganic carbon arising from the carbonate–bicarbonate–CO_2 system must be stripped out and determined separately to provide the contribution of inorganic carbon in solution. (In doing so, however, some of the volatile organic species in solution may be lost.) The TOC is determined by difference between total carbon and the inorganic carbon contribution. Additional comments on the TOC test are contained in Ref. [5, pp. 5–17 to 5–26].

Table 5.4 Test Results for Experimental BOD_5

Sample number	mL sample in 300 mL	% Dilution	DO (mg/L)
1	0	Blank	8.7
2	0	Blank	8.1
3	0	Blank	8.4
4	3	3/300 = 1.00%	7.3
5	5	5/300 = 1.66...%	6.6
6	6	6/300 = 2.00%	7.2
7	10	10/300 = 3.33...%	6.1
8	15	15/300 = 5.00%	4.9
9	20	20/300 = 6.66...%	3.9
10	30	30/300 = 10.00%	0.9

Table 5.5 BOD₅ Determination from DO Results

Sample number	$DO_{Blank} - DO_{Sample}$	Is sample valid?	BOD₅
4	1.1	No (<2 mg/L used)	—
5	1.8	No (<2 mg/L used)	—
6	1.2	No (<2 mg/L used)	—
7	2.3	Yes	69
8	3.5	Yes	70
9	4.5	Yes	68
10	7.5	No (<1 mg/L remaining)	—

PROBLEM 5.5

Using the data given in Table 5.4, determine the BOD of a wastewater after 5 days of incubation at 20°C in 300 mL BOD bottles.

The difference between the DO of the blank and the DO of the undiluted sample can be considered negligible.

SOLUTION BOD is calculated from the following equation [6, p. 426; 7, p. 404]:

$$BOD_5 \text{ (mg/L)} = [DO_{blank} - DO_{sample}](100\% \text{ dilution})] - [DO_{blank} - DO_{undiluted\ sample}] \quad (5.3)$$

DO of the blank is the average of the above three blank samples:

$$(8.7 + 8.1 + 8.4)/3 = 25.2/3 = 8.4 \text{ mg/L}$$

BOD₅ of sample no. 7 = [(8.4 − 6.1)(100/3.33)] − (0) = 69 mg/L

Results for this and the other samples are given in Table 5.5.
For a sample to be valid, at least 2 mg/L must be consumed (that is, a difference of 2 mg/L or more in ($DO_{Blank} - DO_{Sample}$)), and at least 1 mg/L DO must remain in the sample at the conclusion of the test.

$$\text{Average BOD}_5 = (69 + 70 + 68)/3 = 207/3 = 69 \text{ mg/L} \quad \blacksquare$$

Further details about the BOD test are given in Appendix D.

5.5 BOD, COD, AND TOC IN INDUSTRIAL WASTEWATER

For treatment or disposal of an industrial wastewater of known chemical composition, it will often be necessary to estimate its wastewater parameters—BOD, COD, and/or TOC. This can be accomplished through a table such as Table 5.6. This table was assembled from experimental data [13–21] for BOD₅ and COD. Entries for the theoretical oxygen demand (ThOD) and total organic carbon are calculated values. Molecular weights were taken from a standard handbook [10].

ThOD is the amount of oxygen necessary for reaction of a chemical compound completely to CO_2 and water. TOC is the fraction of an organic chemical that is

Table 5.6 BOD, COD, and TOC of Selected Chemical Compounds

Chemical	Formula	Molecular weight	BOD$_5$	COD	Calc ThOD	Calc TOC
Acids						
Formic	HCOOH	46.03	0.27	0.36	0.35	0.2609
Acetic	CH$_3$COOH	60.05	0.91	1.01	1.07	0.4000
Propionic	C$_2$H$_5$COOH	74.08	1.10	—	1.51	0.4864
Butyric	C$_3$H$_7$COOH	88.12	1.40	1.75	1.82	0.5452
Benzoic	C$_6$H$_5$COOH	122.13	1.45	1.95	1.97	0.6884
Alcohols						
Methanol	CH$_3$OH	32.04	1.14	1.5	1.50	0.3749
Ethanol	C$_2$H$_5$OH	46.07	1.67	2.11	2.08	0.5214
1-Propanol	n-C$_3$H$_7$OH	60.11	1.54	2.18	2.40	0.5995
2-Propanol	iso-C$_3$H$_7$OH	60.11	1.59	2.30	2.40	0.5995
1-Butanol	n-C$_4$H$_9$OH	74.12	2.0	2.45	2.59	0.6482
Isobutyl alcohol	(CH$_3$)$_2$CHCH$_2$OH	74.12	1.66	2.39	2.59	0.6482
Cyclohexanol	C$_6$H$_{11}$OH	100.16	0.08	2.5	2.72	0.7195
Aldehydes						
Acetaldehyde	CH$_3$CHO	44.05	1.3	—	1.82	0.5453
Formaldehyde	HCHO	30.03	1.07	1.06	1.07	0.4000
Amines						
Monoethanolamine (MEA)	H$_2$NC$_2$H$_4$OH	61.09	0.8–1.1	1.27	1.31[a], 2.36[b], 2.49[c]	0.3932
Aromatics						
Aniline	C$_6$H$_5$NH$_2$	93.13	1.42	2.34	2.41[a], 3.09[b], 3.18[c]	0.7738
Benzene	C$_6$H$_6$	78.12	1.80	1.41	3.07	0.9225
Toluene	C$_6$H$_5$CH$_3$	92.15	0.86–2.28	1.41	3.13	0.9124
Chlorobenzene	C$_6$H$_5$Cl	112.56	0.03	0.41	1.99[d], 2.06[e]	0.6402
Nitrobenzene	C$_6$H$_5$NO$_2$	123.11	0	—	1.43[a], 1.95[b], 2.01[c]	0.5854

(continued)

Table 5.6 (*Continued*)

Chemical	Formula	Molecular weight	BOD$_5$	COD	Calc ThOD	Calc TOC
Phenol	C$_6$H$_5$OH	94.11	1.81–2.16	2.30	2.38	0.7658
Pyridine	C$_5$H$_5$N	79.10	0.06	0.05	2.33^a, 3.03^b, 3.14^c	0.7586
Chlorinated aliphatics						
Carbon tetrachloride	CCl$_4$	153.82	0	—	—	0.0781
Chloroform	CHCl$_3$	119.38	0.008	—	0.21^e	0.1006
Ethylene (di)chloride	ClCH$_2$CH$_2$Cl	98.96	0.002	1.025	0.13^d, 0.34^e	0.2427
Ethylene chlorohydrin	ClCH$_2$CH$_2$OH	80.52	0, 0.50^f	—	0.81^d, 0.97^e	0.2983
					0.99^d, 1.09^e	
Esters						
Ethyl acetate	CH$_3$COOC$_2$H$_5$	88.12	1.24	1.69	1.82	0.5452
n-Propyl acetate	CH$_3$COOC$_3$H$_7$	102.13	1.26	2.04	2.04	0.5880
iso-Propyl acetate	CH$_3$COOCH(CH$_3$)$_2$	102.13	1.24	1.67	2.04	0.5880
n-Butyl acetate	CH$_3$COOC$_4$H$_9$	116.16	1.28	2.32	2.20	0.6204
Glycols						
Ethylene glycol	HOCH$_2$CH$_2$OH	62.07	0.95–1.08	1.29	1.29	0.3870
Ketones						
Acetone	CH$_3$COCH$_3$	58.08	1.54	2.07	2.20	0.6204
Methyl ethyl ketone (MEK)	CH$_3$COC$_2$H$_5$	72.12	2.14	2.23	2.44	0.6662
Diethyl ketone	C$_2$H$_5$COC$_2$H$_5$	86.14	1.00	—	2.60	0.6972

^{a}To NH$_3$.
^{b}To HNO$_3$.
^{c}To NO$_3^-$.
^{d}To HCl.
^{e}To Cl$_2$.
^{f}10 days.

carbon (C). All numerical entries for BOD_5, COD, ThOD, and TOC in the table are in units of weight per weight (wt/wt) of compound—lb/lb, g/g, mg/mg, and so on. For those few cases where the measured COD is shown as greater than the theoretical oxygen demand, the difference is attributed to experimental error.

In constructing the table, entries for ThOD and TOC are fairly straightforward. For example, for acetic acid (CH_3COOH), ThOD is given by

$$CH_3COOH + 2O_2 \rightarrow 2CO_2 + 2H_2O$$
$$O_2 = [(2)(32)]/60.05 = 1.0657 \cong 1.07 \text{ wt/wt} \quad (5.4)$$

where 32 and 60.05 are the molecular weights of O_2 and acetic acid, respectively. Since acetic acid contains two carbon atoms (MW = 12.011), TOC is given by $[(2)(12.011)]/60.05 = 0.4000$ wt/wt. The TOC in units of wt/wt will always be a fraction, less than 1.0.

An exception to the straightforward nature occurs in computing the ThOD of chemical compounds containing nitrogen or chlorine atoms in the molecule. End products could be ammonia, nitric acid (HNO_3), or nitrate ion for molecules containing organic nitrogen, and hydrogen chloride (HCl) or chlorine (Cl_2) for molecules containing Cl. The multiple entries in the table for these materials are so specified. It is important to understand the basis when ThOD is reported, but its basis is *not* specified.

PROBLEM 5.6

(a) Calculate the theoretical oxygen demand for monoethanolamine (MEA) going to CO_2, H_2O, and ammonia; CO_2, H_2O, and HNO_3; and CO_2, H_2O, and nitrate.

(b) Repeat for chlorobenzene (C_6H_5Cl) to HCl and to Cl_2.

SOLUTION

(a)

(1) Monoethanolamine (MEA) ThOD to ammonia
$$H_2NC_2H_4OH + (5/2)O_2 \rightarrow 2CO_2 + 2H_2O + NH_3 \quad (5.4)$$
$$O_2 = [(5/2)(32)]/(61.09) = 1.31 \text{ wt/wt}$$

(2) MEA ThOD to nitric acid
$$H_2NC_2H_4OH + O_2 \rightarrow CO_2 + 2H_2O + HNO_3 \quad (5.5)$$
$$O_2 = [(9/2)(32)]/(61.09) = 2.36 \text{ wt/wt}$$

The HNO_3 can be neutralized by alkalinity (e.g., bicarbonate (HCO_3^-))

$$HNO_3 + HCO_3^- \rightarrow NO_3^- + H_2O + CO_2 \quad (5.6)$$

(3) MEA ThOD to nitrate
$$2H_2NC_2H_4OH + (19/2)O_2 \rightarrow 4CO_2 + 7H_2O + 2NO_3 \quad (5.7)$$
$$O_2 = [(19/2)(32)]/[(2)(61.09)] = 2.49 \text{ wt/wt} \text{ [13, 16]}$$

The charge on the nitrate ion (NO_3^-) is ignored in this convention.

(b)

(1) For chlorobenzene to HCl

$$C_6H_5Cl + 7\,O_2 \rightarrow 6\,CO_2 + 2\,H_2O + HCl \tag{5.8}$$

$$O_2 = [(7)(32)]/(112.56) = 1.99 \text{ wt/wt}$$

(2) And to Cl_2

$$2\,C_6H_5Cl + (29/2)O_2 \rightarrow 12\,CO_2 + 5\,H_2O + Cl_2 \tag{5.9}$$

$$O_2 = [(29/2)(32)]/[(2)(112.56)] = 2.06 \text{ wt/wt}$$

∎

Entries in Table 5.6 for BOD_5 and COD are in general the highest values found. The BOD at 5 days depends on whether the seed culture of microorganisms is *acclimated* to the chemical food in question. Acclimation means that microorganisms having the proper enzymes for ready assimilation of a particular chemical have grown to sufficient numbers to consume a significant quantity of it as food. Otherwise, the microbial culture as constituted in unsuited to the task.

There is no such thing as "teaching" a particular strain of unsuited organisms to eat their "liver, spinach, and broccoli," or ethylene glycol, for example, to like it, or even to get used to it. Either they are up to the job or they are not. If there is no other food available, this first biological strain will die out. In any event, they will be supplemented or replaced by others with the ability to metabolize the subject chemical(s). While this acclimation process is taking place, a lag occurs in the BOD versus time curve of Figure D.1, making BOD_5 results low, as depicted in Figure D.2.

Ethylene glycol is a good example with regard to acclimated organisms. It is an important industrial chemical and one that is used as well in aircraft deicing [22,23]. Wastewater or storm water runoff containing a substantial amount of ethylene glycol must be treated biologically before discharge. Reported BODs at 5 days range from 12% to 84% of the theoretical oxygen demand.

The table entry chosen for its BOD_5 (0.95–1.08 wt/wt), obtained with a seed culture of acclimated organisms, is the highest range reported for a single investigation [19]; one figure is contained as a point value in a table, and the other is the 5-day value read from a curve of BOD versus incubation time. Measured COD for ethylene glycol in Table 5.6 (1.29 wt/wt) [20] is the same as its calculated theoretical oxygen demand.

For this and the other chemicals in Table 5.6, selection of the largest number found for BOD_5 assumes that the BOD is being stabilized by fully acclimated organisms. Nevertheless, BOD_5 could even be higher, depending on the results of some unknown experiment not yet performed or not yet found and not being considered here.

For some of the chemical constituents in Table 5.6, there may be no organisms that metabolize the waste. Hence, BOD_5 will be zero or close to it, and the waste is said to be *refractory* to biological oxidation. In fact, the chemical may be downright toxic to all microorganisms.

In addition, there are certain chemicals that are resistant to chemical oxidation. These are exemplified by most of the aromatic compounds in Table 5.6, whose BOD_5 is also negligible. There seems to be universal agreement that the BOD of chlorinated hydrocarbons is virtually nil as well. Interestingly enough, some materials that show up as COD are not biodegradable even by acclimated organisms, and some biodegradable substances with a BOD are not oxidized in the COD test, even when run with the catalyst as called for in the procedure.

PROBLEM 5.7

A wastewater from a chemical process contains 1000 mg/L of the following chemical constituents:

Acetaldehyde	10%	Ethanol	15%
Acetic acid	15%	Ethyl acetate	5%
Acetone	30%	Methanol	25%

Estimate its BOD, COD, and TOC. (Isn't it amazing that textbook problems start out with such even numbers?)

SOLUTION Using the entries of Table 5.6, wastewater parameters of acetaldehyde are calculated as follows:

$$\text{mg/L acetaldehyde in solution} = (1000\,\text{mg/L})(10\%/100) = 100\,\text{mg/L}$$

$$\text{BOD}_5 : (100\,\text{mg/L})(1.3\,\text{mg/mg}) = 130\,\text{mg/L}$$

$$\text{COD} : (100\,\text{mg/L})(1.82\,\text{mg/mg}) = 182\,\text{mg/L}$$

Here, the theoretical oxygen demand substitutes for the unlisted COD. For the other species, the COD is listed and will be used.

$$\text{TOC} : (100\,\text{mg/L})(0.5453\,\text{mg/mg}) = 54.5\,\text{mg/L}$$

In a similar manner, BOD_5, COD, and TOC are calculated for the other individual wastewater constituents. Calculated values are summarized in Table 5.7, and the results are totaled.

The estimated BOD_5, COD, and TOC are approximately 1330, 1730, and 500 mg/L, respectively. The BOD_5 and COD are derived from experimental data; TOC is calculated from molecular structure. A wastewater of this strength would have to be treated before discharge/disposal.

Note: This technique for BOD_5 assumes that there is no interaction or interference among the various chemical constituents when a composite BOD test is run. It also assumes a mixed

Table 5.7 Estimated BOD, COD, and TOC for a Chemical Process Wastewater

Constituent	mg/L in solution	mg/L BOD_5	mg/L COD	mg/L TOC
Acetaldehyde	100	130.0	182.0	54.5
Acetic acid	150	136.5	151.5	60.0
Acetone	300	462.0	621.0	186.1
Ethanol	150	250.5	316.5	78.2
Ethyl acetate	50	62.0	84.5	27.3
Methanol	250	285.0	375.0	93.7
Total	1000	1326.0	1730.5	499.8

population of "equal opportunity microorganisms," having no preference for one type of substrate over another in such a test. Since the waste materials are metabolized at different rates, this may not be completely true, but the method is probably the best we have. COD by chemical analysis and TOC by calculation should indeed be additive. ∎

5.6 DOMESTIC WASTEWATER (ALSO KNOWN AS SEWAGE)

In addition to the industrial wastewater from such a facility, one must be concerned with the sewage (euphemistically termed *domestic wastewater*) generated by personnel working in the plant.

PROBLEM 5.8

Your plant will be staffed by 30 people per shift on each of three shifts plus 10 additional administrative persons on the day shift. Estimate (a) the amount and (b) the composition (BOD_5 and total suspended solids (TSS)) of sewage from this facility.

SOLUTION

(a) *Amount*: Estimates in various state and local guidelines range upward from 20 gal/day per person. The highest number that the author has seen is 50 gal/day per person. This figure is based on the number of actual persons onsite, not the number of positions being filled. Thus, for a total of 100 employees present during each day

$$(100 \text{ persons})(50 \text{ gal/day per person}) = 5000 \text{ gal/day}$$

(b) *Composition*: A typical composition of domestic sewage is listed in Ref. [24]. BOD_5 determined at 20°C for strong sewage is given as 400 mg/L, revised upward from 300 mg/L in a previous edition [25]. TSS remains at 350 mg/L for strong sewage. COD in both editions is 1000, 500, and 250 mg/L for strong, medium, and weak sewage, respectively. Use of the entries for strong sewage, versus medium or weak sewage, provides a more conservative (higher) estimate for its concentration, whether the sewage is discharged to a publicly owned treatment works (POTW) or must be treated onsite in a packaged activated sludge treatment plant because a connection to a public sewer system is not available. For the latter case, typical allowable effluent concentrations are 30/30 or 20/20 mg/L of BOD_5 and TSS, respectively. In some cases, it is possible to handle such sewage in a properly designed septic system.

The original reference(s) should be consulted for these and other parameters in (strong, medium, and weak) sewage. ∎

5.6.1 Septic Tanks

Septic tanks are basins in which sewage solids are allowed to settle out and undergo some degree of digestion [26, p. 36–33]. The accumulated solids are removed periodically for disposal. The liquid effluent is allowed to seep into the ground via perforated pipes in a so-called *drain field*. Ideally, both the sludge and the seepage effluent will be odorless.

Typically, the domestic wastewater flows by gravity into a watertight rectangular box that functions as a type of primary treatment and a biological digestion system. Here, the heavier solids settle to the bottom, and the lighter solids and oil/grease accumulate as scum at the surface of the liquid. The clarified liquid effluent exits to the underground drain field, composed of multiple parallel lengths of perforated pipe placed on sand or gravel, and is allowed to seep into the ground [27,28].

Septic tank effluent quality depends on the characteristics of the incoming wastewater and the condition of the tank [27]. Effluent concentration data for single family homes and small communities vary from about 130–220 mg/L BOD, 50–160 mg/L TSS, and 230–450 mg/L COD. Other parameters are shown in the original source [27]. The organic constituents are worked on further by the soil bacteria.

The septic tank must provide enough residence time for suspended solids and oil/grease removal. Accumulated sludge and scum reduce residence time and must be pumped out periodically, on the order of once every 1–5 years [27,29]. A procedure to estimate pumping frequency is given in Ref. [29].

The entire contents of the septic tank, the scum, sludge, and partially clarified liquid lying between them, is defined as *septage*. It is an odiferous slurry made up of 3–10% solids composed of organic and inorganic materials typically containing high levels of grit, hair, nutrients, pathogenic microorganisms, and oil/grease, as well as a pH in the range of 1.5–12.6. Septage is controlled under federal regulations (40 CFR 503), and states and municipalities have established public health and environmental protection regulations for pumping, handling, transport, treatment, and reuse/disposal of septage. Accumulated sludge and scum material from a septic tank should be removed by a certified, licensed, or trained service provider and reused or disposed of in accordance with applicable federal, state, and local codes [27].

These units are typically designed according to local codes by a more or less "cookbook" method. The size of the tank depends on the estimated peak daily flow and the number of people using the system, among other considerations, and the drain field design depends in part on the percolation rate of the soil in question. The size of the building serviced and/or the number if bedrooms is often specified. A minimum tank volume of 1000 gallons or more may be required [27].

The author can remember seeking approval for one such design done "according to the rules." The approving official concurred but said that the size of the tank should be bumped up one or two standard sizes as he marked up the drawing. The approval process concluded with the words, "Pay the fee to the lady on your way out."

5.7 DISSOLVED OXYGEN CONCENTRATION IN A RECEIVING STREAM

When organic pollution is discharged into a flowing stream, it begins to consume the DO. As the oxygen is depleted, the driving force between the O_2 value at saturation and the actual O_2 value increases, and more oxygen from the atmosphere dissolves into the water.

The Streeter–Phelps equation describes the behavior between deoxygenation and reaeration [26, pp. 33–22 to 33–24; 30, pp. 150–153; 31, pp. 240–243].

$$D = \frac{kL_a}{(r-k)}[\exp(-kt) - \exp(-rt)] + D_a \exp(-rt) \qquad (5.10)$$

where D is the dissolved oxygen deficit (DO saturation − DO actual) at any downstream point in the stream (mg/L), D_a is the dissolved oxygen deficit at point of pollution (mg/L), L_a is the ultimate first-stage (carbonaceous) BOD at the point of pollution (mg/L), k is the deoxygenation constant (days^{-1}), r is the reaeration constant (days^{-1}), and t is the stream flow time from point of pollution (days).

The DO and the BOD in this equation are for the waste as diluted into the receiving stream. This function, the so-called *oxygen sag curve*, is depicted in Figure 5.1 as DO versus t, using input values from Problem 5.9.

This curve passes through a minimum point. By methods of differential calculus, one finds that this minimum in DO, the point of maximum oxygen deficit (D_{MAX}), occurs at a time t_{MIN}

$$t_{MIN} = \frac{1}{k[(r/k)-1]} \ln\{(r/k)[1-[(r/k)-1](D_a/L_a)]\} \qquad (5.11)$$

from which

$$D_{MAX} = L_a \frac{\exp(-kt_{MIN})}{(r/k)} \qquad (5.12)$$

When the oxygen sag exceeds a critical value or, worse yet, goes to zero, the stream becomes septic and a number of undesirable effects occur, including fish kills.

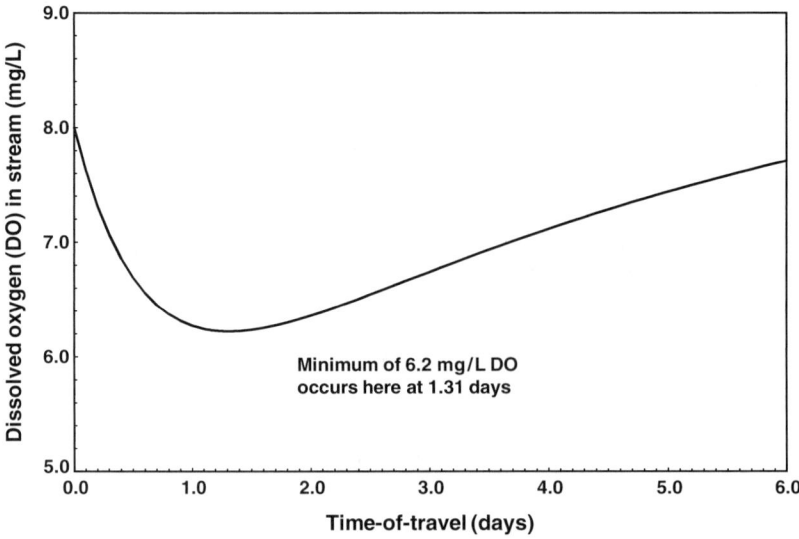

Figure 5.1 Streeter–Phelps oxygen sag curve, parameters as noted in the text.

A minimum dissolved oxygen concentration on the order of 2–5 mg/L [31, pp. 169, 266] is necessary to maintain the health of fish, the actual requirement depending on such factors as the type, age, and general condition of the fish; the temperature and composition of the water; and the presence of toxic substances [31, pp. 160–170, 264–268]. This does not leave much of a margin since the oxygen saturation value at 20°C (68°F) is only about 9 mg/L. (See Appendix D for further details on oxygen saturation values.)

PROBLEM 5.9

Assume that the saturated DO at 20°C in round numbers is 9.0 mg/L and that the wastewater mixed in flow proportion with the stream has a DO of 8.0 mg/L and an ultimate carbonaceous BOD of 30 mg/L.

SOLUTION Rate constants are $k = 0.19$ days^{-1}, $r = 1.6$ days^{-1}, and $r/k = 8.42$.

$$D_a = (9.0 - 8.0) = 1.0 \quad \text{and} \quad L_a = 30 \text{ mg/L}$$

Calculate the minimum value achieved by the dissolved oxygen.
From Equation (5.11),

$$t_{MIN} = \frac{1}{0.19(8.42-1)} \ln\{(8.42)[1-(8.42-1)(1.0/30)]\} = 1.31 \text{ days}$$

From Equation (5.12),

$$D_{MAX} = \frac{30}{8.42} \exp[(-0.19)(1.31)] = 2.8 \text{ mg/L}$$

Assuming the oxygen saturation value remains the same at the downstream location,

$$\text{minimum DO} = (9.0 - 2.8) = 6.2 \text{ mg/L}$$

For a stream moving at an average velocity of 0.7 ft/s, for example, this effect is felt at

$$(1.31 \text{ days})(0.7 \text{ ft/s})(86,400 \text{ s/day})(1/5280 \text{ ft/mile})$$

$$= 15.0 \text{ miles downstream from the point of discharge}$$

It does not take much for the stream to go septic. Increases in the ultimate BOD to 45 mg/L, the deoxygenation rate constant k to 0.25 days^{-1}, and a decrease in the reaeration coefficient to 0.7 days^{-1} make $t_{MIN} = 2.2$ days and produce a D_{MAX} of 9.3 and a minimum DO of $(9.0 - 9.3) = $ negative number, set equal to zero. (See p. 151 in Ref. [30] for a complete oxygen sag curve showing septic conditions.) ∎

5.8 ALKALINITY

As indicated in Section 5.2, as well as in Appendix E, alkalinity is the measure of resistance to change in solution pH upon addition of an acid. It is the capacity to neutralize the acid and is measured by titration with a strong acid. Phenolphthalein and methyl orange are commonly used as indicators during the titration. As acid is added, phenolphthalein changes from pink to colorless at approximately pH 8.3, and methyl orange goes from a yellow-orange to a salmon-pink at approximately pH 4.3.

Titration with a strong base such as sodium hydroxide (NaOH) would provide a measurement of acidity. The end points with methyl orange and phenolphthalein would be the same, but the color changes would be reversed.

5.8.1 Indicators Used in Alkalinity Titrations

As the author can attest to, the pH 4.3 end point is not an easy one to detect when using methyl orange indicator, and substitutes have been developed to overcome this difficulty [1]. The methyl orange indicator mentioned in the 13th and 14th editions of Standard Methods [11,32] has disappeared by the 17th edition, having been substituted for first by a mixture of bromcresol green-methyl red in the 13th and 14th editions and then replaced by bromcresol green alone in the 17th edition. The color change for the mixed indicator is a blue [1] or greenish-blue [32, p. 55] on the alkaline side to a red [1] or light pink [32] on the acid side, with shades of gray at the end point [1,32]. The individual indicator goes from blue to yellow and from yellow to red for bromcresol green and methyl red, respectively [6, p. 291].

In the meantime, metacresol purple with its sharper color change [5, p. 2–31] has been introduced to determine the pH 8.3 end point. Even though the color-indicating reagents may have changed, the alkalinity values so determined are still referred to as *methyl orange* (pH 4.3) and *phenolphthalein* (pH 8.3) *alkalinity*.

The author prefers the "superior electrometric technique" [6, p. 291] of titrating with a pH meter electrode present in the solution [6, p. 291; 5, p. 2–36; 33, p. 359] to determine numerical values for the pH of a sample versus the amount of reagent added. Color indicators can be used as a guide during the titration in addition to measuring pH.

5.8.2 Alkalinity of Natural Waters

The principal contributors to the alkalinity of a natural water are the hydroxyl ion (OH^-) and the carbonate (CO_3^{2-}) and bicarbonate ions of the CO_2–bicarbonate–carbonate system. Alkalinity measurements are often interpreted exclusively in terms of those ions, and the contribution of other species including phosphates, silicates, borates, sulfides, sulfites, and the ammonium ion, as well as organic acids that might also be present, is ignored. Simplified mathematical relationships among the various sources of alkalinity are shown in Table E.1.

In the absence of a strong base polluting the water, hydroxyl ion comes from the dissociation of water

$$H_2O \leftrightarrow H^+ + OH^- \tag{5.13}$$

$$[H^+][OH^-] = K_w = 10^{-14} \text{ at } 25°C \text{ [4, p. 5-7]} \tag{E.6}$$

and

$$pH = -\log_{10}[H^+] \tag{E.3}$$

Ionic equilibrium in the CO_2 system is governed by equations provided in Chapter 2 and reproduced here for convenience.

$$CO_2 + H_2O \xrightleftharpoons{K_1} HCO_3^- + H^+ \quad (2.32)$$

and

$$HCO_3^- \xrightleftharpoons{K_2} CO_3^{2-} + H^+ \quad (2.33)$$

$$\frac{[H^+][HCO_3^-]}{[CO_2]} = K_1 = 4.47 \times 10^{-7} \text{ at } 25°C \text{ [4, p. 5-40]} \quad (2.34)$$

$$\frac{[H^+][CO_3^{2-}]}{[HCO_3]} = K_2 = 4.68 \times 10^{-11} \text{ at } 25°C \text{ [4, p. 5-40]} \quad (2.35)$$

Mathematical manipulation of the above equations results in the following expressions for percentage composition of the individual species of the total $[T]$:

$$\%\,[CO_2] = 100/\{1 + (K_1/[H^+]) + (K_1 K_2/[H^+]^2)\} \quad (5.14)$$

$$\%\,[HCO_3^-] = 100/\{1 + ([H^+]/K_1) + (K_2/[H^+])\} \quad (5.15)$$

$$\%\,[CO_3^{2-}] = 100/\{1 + ([H^+]/K_2) + ([H^+]^2/(K_1 K_2))\} \quad (5.16)$$

These equations are consistent with Equations (2.37–2.39) but are written in a slightly different form.

The distribution of species in the CO_2–bicarbonate–carbonate system at 25°C has been shown in Figure 2.1 as a function of pH between pH 4 and 13. Ionization/dissociation constants are taken from Ref. [4, p. 5–40].

Dissolved CO_2 is at its highest at low pH; it is over 99% of the total at pH 4.3, the methyl orange end point in a manual titration. At this same point, bicarbonate ion amounts to less than 1%, and carbonate ion (CO_3^{2-}) is nil. Bicarbonate and CO_2 are equal at a mid-range pH of 6.35, and carbonate is still virtually absent.

As pH is increased, CO_2 decreases to about 1% at pH 8.3, the phenolphthalein end point, where bicarbonate passes through a maximum at approximately 98%, and carbonate makes up the remainder. The bicarbonate and carbonate curves cross at a pH of 10.3± for a 50/50 split since CO_2 is negligible at this point. At pH 13, carbonate is 99.8%, bicarbonate 0.2%, and CO_2 even more negligible. At every point, the three species add up to 100%, the horizontal line at the top of the chart. Use of the equations is illustrated in Problems 2.21 and 2.22.

This distribution of species and their reaction to titration with an acid give rise to the rules of thumb for alkalinity listed in Table E.1. The concepts of alkalinity and acidity are presented there in greater detail.

5.9 THE NITROGEN CYCLE

Nitrogen is all around us. As a gas, it comprises about 79% of the earth's atmosphere but is difficult to "fix" or convert into other forms. Nonetheless, organic matter contained in solids and liquids is an essential nutrient needed for cellular metabolism.

Decomposition of nitrogenous organic matter in nature or its excretion by humans and animals results in decomposition into ammonia and further degradation into nitrite and nitrate [6, p. 441; 7, pp. 421–422; 30, p. 135–136].

$$\text{Organic nitrogen} \xrightarrow[\text{bacteria}]{\text{saprophytic}} NH_3 \qquad (5.17)$$

Saprophytic bacteria are those that feed only on dead organic matter [31, p. 86].

$$2\,NH_3 + 3\,O_2 \xrightarrow[\text{Nitrosococcus bacteria}]{\text{Nitrosomonas and}} 2\,HNO_2 + 2\,H_2O \qquad (5.18)$$

$$2\,HNO_2 + O_2 \xrightarrow[\text{bacteria}]{\text{Nitrobacter}} 2\,HNO_3 \qquad (5.19)$$

The reaction of Equation (5.17) occurs under either aerobic (presence of oxygen) or anaerobic (absence of oxygen) conditions and is fostered by the action of a large variety of bacteria. For the specialized group of nitrifying bacteria to work their magic [Equations (5.18) and (5.19)], aerobic conditions are necessary, along with phosphate as a nutrient and an alkali to neutralize the nitrous (HNO_2) or nitric acid formed [30, p. 135]. These autotrophic bacteria derive their energy from the carbon contained in inorganic compounds (CO_2, bicarbonates, and carbonates), rather than from a food supply consisting of organic compounds [30, p. 128]. They are sensitive to certain organic molecules, which may be toxic to them, even in low concentrations. Under anaerobic conditions, other bacteria may reduce nitrates and nitrites to ammonia.

5.9.1 Kinetic Analysis of Organic Nitrogen Decomposition

Ammonia and its degradation products, nitrite and nitrate, in a stream indicate the presence of raw sewage or improperly treated sewage. Newly discharged pollution will contain organic nitrogen and ammonia primarily, the organic nitrogen degrading to ammonia. After some time, nitrite builds up before reacting to nitrate. At long times, nitrate becomes the predominant species. Before reliable tests for biological contamination from sewage were developed, the relative amounts of ammonia, nitrite, and nitrate from chemical testing were used as a surrogate [6, pp. 442–444; 7, p. 423]. Test methods for the various types of nitrogen are discussed elsewhere [5, p. 4–110 to 4–149; 6, pp. 446–452; 7, pp. 425–432]. This principle is illustrated in the curves of Figure 5.2.

The curves are conceptual only. They were derived from a kinetic model of sequential first-order reactions [34], which may or may not be the case. The system starts out with organic nitrogen along with some ammonia initially present. The decomposition of the organic nitrogen into ammonia-nitrogen is shown as the fastest *on a relative basis*. The ammonia degrades into nitrite via biological action but is constantly replenished from the decomposition of the organic nitrogen. With its rate of decomposition assumed to be less than that of the breakdown of the organics, the ammonia goes through a maximum. Meanwhile, nitrite builds up somewhat and then dies out as nitrate becomes predominant. At a long time, the other forms of nitrogen are essentially gone and only the nitrate remains.

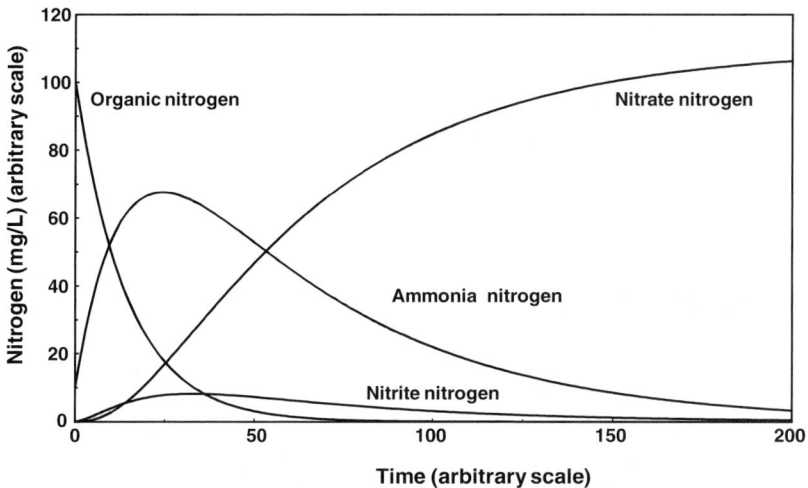

Figure 5.2 Diagram illustrative of nitrifying reactions in polluted water.

A similar set of curves has been published previously [6, p. 443; 7, p. 424]. The difference is that the model of Figure 5.2 does not show the lag in the formation of nitrite and nitrate depicted in the literature references cited above. However, the mathematical treatment of the consecutive decomposition of ammonia into nitrite and nitrate is similar to the kinetic analysis provided on other pages of these same references [6, pp. 79–81; 7, pp. 200–203] and is considered adequate for present purposes.

5.9.2 Ammonia-Nitrogen

The terms *ammonia-nitrogen*, *nitrite-nitrogen*, and *nitrate-nitrogen* used above frequently arise in an environmental context. They are simply the fraction of nitrogen contained within a given molecule, as computed in Problem 5.10. Putting these species on a common basis facilitates their use in material balance calculations.

PROBLEM 5.10

A hypothetical wastewater contains 10 mg/L each of ammonia, nitrite ion, and nitrate ion. Express each of these in terms of its nitrogen content.

SOLUTION   This problem requires the use of molecular/atomic/ionic weights.

Ammonia-nitrogen: (10 mg/L)(14.01 for N)/(17.03 for NH_3) = 8.23 mg/L
Nitrite-nitrogen: (10 mg/L)(14.01 for N)/(46.0 for NO_2^-) = 3.05 mg/L
Nitrate-nitrogen: (10 mg/L)(14.01 for N)/(62.0 for NO_3^-) = 2.26 mg/L

Although numerically the same when expressed as the species themselves, they are significantly different when expressed on a common basis of nitrogen content.

Each of these nitrogen species in water or wastewater may be regulated differently in a particular geographical area. Nitrate often receives special attention because it has been identified as the cause of methemoglobinemia (Blue Baby Syndrome in infants), limiting its content to 10 mg/L of nitrate-nitrogen or less in public water supplies [6, p. 444; 7, p. 424]. It may also cause problems in sewage treatment plants, although at much higher concentrations. ∎

5.9.3 Ammonia Ionization

Ammonia in water is toxic to fish. It is the nonionized portion [NH_3] of the total dissolved ammonia that is the culprit, rather than the accompanying ammonium ion [NH_4^+]. The ammonia molecule passes through the gills and gets into the bloodstream to exhibit its toxicity [30, pp. 168, 175, 372]. Since NH_3 and NH_4^+ are related by chemical equilibrium with [H^+], ammonia toxicity is a function of pH. A concentration of 25 mg/L of ammonia is considered harmful [30, pp. 111–112, 168] at a pH from 7.4 to 8.5 [35].

Ammonia ionizes as follows [4, p. 5-13; 26, pp. 28-15 to 28-16]:

$$NH_3 + H_2O \leftrightarrow NH_4^+ + OH^- \tag{5.20}$$

and acts as a weak base with a basic dissociation constant for the equilibrium

$$\frac{[NH_4^+][OH^-]}{[NH_3][H_2O]} = K_b = 1.758 \times 10^{-5} \text{ at } 25°C \tag{5.21}$$

Conversely, the ammonium ion behaves as a weak acid to release [H^+]

$$NH_4^+ \leftrightarrow NH_3 + H^+ \tag{2.41}$$

with an acidic ionization constant

$$\frac{[NH_3][H^+]}{[NH_4^-]} = K_a = 5.689 \times 10^{-10} \text{ at } 25°C \text{ [4, p. 5-40]} \tag{2.42}$$

The two reactions are related through the ionization of water

$$H_2O \leftrightarrow [H^+] + [OH^-] \tag{5.13}$$

and the equilibrium for the ion product of water.

$$K_w = [H^+][OH^-] = 10^{-14} \text{ at } 25°C \text{ [4, p. 5-7]} \tag{E.6}$$

Their equilibrium constants are related by

$$K_w = K_a/K_b \tag{5.22}$$

and

$$pK_w = pK_a - pK_b \tag{5.23}$$

For a known value of K_w, either of K_a or K_b can be obtained from the other, but one must be careful to interpret correctly what is being reported in a handbook.

The distribution of ammonia species has been plotted for a dilute solution in Figure 2.2. Since fish are affected at a very few mg/L or less, one recommendation for a discharge limit is 2.5 mg/L [30, pp. 111–112]. This corresponds to 1.47×10^{-4} g mol/L as NH_3.

Ammonium ion exists in high concentration at low pH; unionized ammonia predominates at high pH. The system pH and the $[H^+]$ concentration are related by

$$pH = -\log_{10}[H^+] \tag{E.5}$$

Ammonium ion constitutes at least 99.4% of the total up to a pH of 7; ammonia is at least 99.4% from pH 11.5–14. Both occur in significant proportions (5%/95% or 95%/5%) in the range of pH 8–10.5. Concentrations are about equal at approximately pH 9.25. The total at every point adds up to 100%.

PROBLEM 5.11

Calculate the proportion of nonionized ammonia $[NH_3]$ at 25°C (77°F) in the typical discharge range of pH 6–9. At this temperature $pK_b = 4.755$.

SOLUTION pK_a is needed for the ionization equilibrium equations. Rearranging Equation (5.23),

$$pK_a = pK_w - pK_b \tag{5.24}$$

Since $pK_w = 14$ at 25°C

$$pK_a = 14 - 4.755 = 9.245$$

Exponentiating

$$K_a = 10^{-pK_a} = 10^{-9.245} = 5.689 \times 10^{-10}$$

Then

$$\% [NH_3] = 100/(1 + [H^+]/K_a) = 100/\{1 + ([H^+]/5.689 \times 10^{-10})\}$$

Substituting $[H^+] = 10^{-6}, 10^{-7}, 10^{-8}$, and 10^{-9} at pH 6, 7, 8, and 9 in the equation, one obtains % values of 0.057, 0.57, 5.38, and 36.26 for pH 6, 7, 8, and 9, respectively. These values are confirmed by visual inspection of Figure 2.2.

If ammonia in one's effluent is a problem, it might be advisable to stay within the lower end of the permitted discharge range. That decision may well be made by the permit writer in setting ammonia and pH limits. ∎

5.10 CHLORINATION/DECHLORINATION

Chlorine is used as a disinfectant for biological control in water supplies, sewage, industrial wastewaters, cooling water, and swimming pools. It is important to understand the chemistry for such an important treatment process. Chlorine "works its magic" as a powerful oxidizing agent. Control is achieved by adding enough chlorine to leave an excess, or residual, in the treated water.

However, when a chlorinated effluent is discharged from a sewage treatment plant or an industrial facility, the chlorine residual in the wastewater may have a deleterious effect on the chemistry and biology of the receiving stream. This often requires that the excess chlorine be removed before discharge. This process goes under the name of *dechlorination*.

This section provides a background on chlorine chemistry, covering the special case when dissolved ammonia is present. It then addresses techniques for dechlorination, including the sulfur dioxide (SO_2)/sulfurous acid system.

5.10.1 Chlorine in Solution

When chlorine gas is dissolved in water, the following hydrolysis reaction takes place [26, pp. 31–16 to 31–18; 36, pp. 182–188]

$$Cl_2 + H_2O \leftrightarrow HOCl + H^+ + Cl^- \tag{5.25}$$

It is extremely rapid, essentially complete in a few seconds or less.

This reaction is described by an equilibrium constant for hydrolysis (K_h)

$$K_h = [HOCl][H^+][Cl^-]/[Cl_2] \tag{5.26}$$

with the bracketed terms expressed in concentrations of g mol/L. The value of K_h is 4.5×10^{-8} at 25°C [26, p. 31–16]. Values of K_h from 0 to 45°C are tabulated in Ref. [36, p. 183]. For total chloride [Cl^-] concentration of 1000 mg/L or less and a pH of about 3 or greater, only a minor amount of molecular chlorine [Cl_2] can be found in solution.

Instead, the hypochlorous acid [HOCl] ionizes into hypochlorite ion [OCl^-] according to the following ionization equilibrium

$$HOCl \leftrightarrow H^+ + OCl^- \tag{5.27}$$

with an ionization or dissociation constant (K_i)

$$K_i = [H^+][OCl^-]/[HOCl] \tag{5.28}$$

with the bracketed concentrations again in g mol/L.

The percentage contributions of HOCl and OCl^- to the total can be derived from this equation

$$\% \, HOCl = 100/\{1 + (K_i/[H^+])\} \tag{5.29}$$

$$\% \, OCl = 100/\{1 + ([H^+]/K_i)\} \tag{5.30}$$

These equations are plotted in Figure 5.3 as a function of $pH = -\log_{10}[H^+]$ between pH 4 and 11 using a dissociation constant of 3.7×10^{-8} at 25°C [36, p. 185].

In the higher pH range, OCl^- predominates, and HOCl is predominant at lower pH. If the pH gets low enough, some molecular chlorine will appear. Hypochlorous acid amounts to about 73% of the total at pH 7 and about 21% at pH 8. The ratio of

5.10 Chlorination/Dechlorination

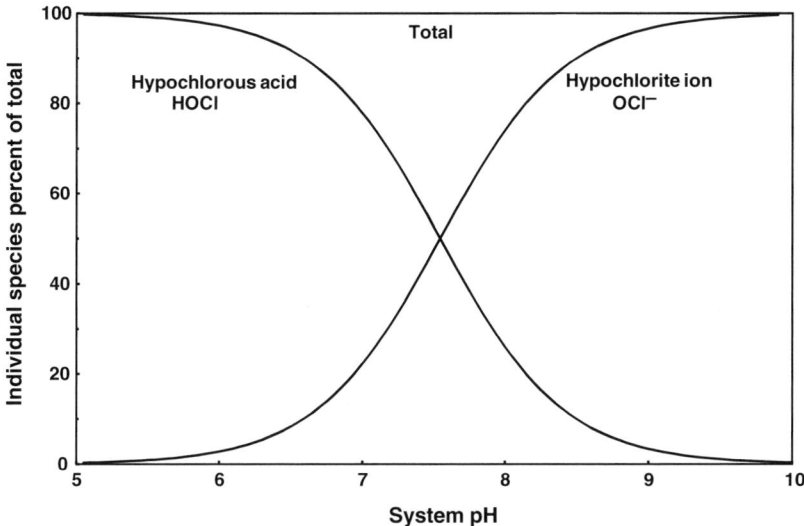

Figure 5.3 Distribution of species in the system HOCl–OCl⁻ at 25°C.

hypochlorous acid to hypochlorite ion increases with decreasing temperature (not shown in the figure, but addressed in Problem 5.12).

PROBLEM 5.12

Calculate the relative amounts of HOCl and OCl⁻ at pH 7 and 8 as in Figure 5.3 but at a temperature of 0°C. Compare with values for 25°C.

SOLUTION   At 0°C, K_i is reported as 2.0×10^{-8}. $[H^+] = [10^{-7}]$ at pH 7 and $[10^{-8}]$ at pH 8. From Equation (5.29),
  At pH 7 and 0°C

$$\%HOCl = \frac{100}{1 + 2.0 \times 10^{-8}/[10^{-7}]} = 83.3\%$$

compared to ~73% at 25°C.
  At pH 8 and 0°C

$$\%HOCl = \frac{100}{1 + 2.0 \times 10^{-8}/[10^{-8}]} = 33.3\%$$

compared to ~21% at 25°C.

Because of the relative magnitudes of the temperature-dependent dissociation constants, the HOCl species is a larger proportion of the total at the lower temperature.
Although both HOCl and OCl⁻ provide a *free chlorine residual* useful as a biological control agent, the unionized hypochlorous acid is the more effective, allowing a lower dosage and/or a shorter contact time for disinfection [26, pp. 31–16 to 31–18]. This is thought to occur because the uncharged molecule is better at penetrating the cell membrane of a targeted organism. Hence, application of a chlorine biocide is better at a lower pH. ∎

5.10.2 Oxidizing Power of Chlorine-Containing Chemicals

Chlorine in solution can be provided by chlorine-containing chemicals other than gaseous chlorine. Sodium hypochlorite [NaOCl] (liquid bleach) and calcium hypochlorite [Ca(OCl)$_2$] (used in swimming pools) are two examples. Their "equivalent chlorine" is calculated in Problem 5.13.

PROBLEM 5.13

(a) Calculate the chlorine equivalent for hypochlorous acid, sodium hypochlorite, and calcium hypochlorite. (b) Calculate the equivalent oxidizing power for each. This is known as *available chlorine*.

SOLUTION

(a) This gets back to how these chemicals were made.

$$Cl_2 + H_2O \rightarrow HOCl + H^+ + Cl^- \qquad (5.25)$$
$$HOCl + NaOH \rightarrow NaOCl + H_2O \qquad (5.31)$$
$$2\,HOCl + Ca(OH)_2 \rightarrow Ca(OCl)_2 + 2\,H_2O \qquad (5.32)$$

1 mol of HOCl corresponds to 1 mol of Cl$_2$.
1 mol of NaOCl corresponds to 1 mol of HOCl.
Therefore, 1 mol of NaOCl corresponds to 1 mol of Cl$_2$.

2 mol of HOCl correspond to 2 mol of Cl$_2$.
2 mol of HOCl correspond to 1 mol of Ca(OCl)$_2$.
Therefore, 1 mol of Ca(OCl)$_2$ corresponds to 2 mol of Cl$_2$.

(b) Oxidizing power on a weight basis

Chlorine Species	Molecular weights
Cl$_2$	70.906
Cl	35.453
HOCl	52.4608
NaOCl	74.4427
Ca(OCl)$_2$	142.9859
HOCl	$(35.453/52.4608)(100) = 67.6\% \times 2^a = 135.2\%$
NaOCl	$(35.453/74.4427)(100) = 47.7\% \times 2^a = 95.2\%$
Ca(OCl)$_2$	$(70.906/142.9859)(100) = 49.6\% = 2^a \times 99.2\%$

[a]The factor of 2 [36, p. 189] is used to account for the $+1$ valence of Cl in the (OCl) radical. When Cl^{1+} is reduced to Cl^{1-}, it is equivalent to the oxidizing power of Cl$_2^0$ (a molecule of twice its weight) going to 2Cl$^-$. This weight of molecular chlorine is the same required to liberate the same amount of elemental iodine from the potassium iodide used in the iodometric test method [5, pp. 4–48 to 4–51] for chlorine.

The chloramines (mono-, di-, and trichloramine) also possess germicidal properties, but their so-called *combined chlorine residuals* are less effective than the *free chlorine residual* provided by hydrolysis of chlorine. Their formation will be explored in the next section. ■

5.10.3 Breakpoint Chlorination

This is a phenomenon that occurs when chlorine is applied to a water/wastewater containing ammonia. With ammonia present, the system in theory goes through a series of consecutive chemical reactions before a free chlorine residual is produced [7, p. 368; 26, p. 31–21; 36, p. 377]. The same reactions occur whether it is desired deliberately to remove ammonia or whether the ammonia is simply a nuisance that must be overcome before a free chlorine residual can be attained during disinfection.

$$HOCl + NH_3 \rightarrow \underset{\text{monochloramine}}{NH_2Cl} + H_2O \qquad (5.33)$$

$$HOCl + NH_2Cl \rightarrow \underset{\text{dichloramine}}{NHCl_2} + H_2O \qquad (5.34)$$

$$HOCl + NHCl_2 \rightarrow \underset{\substack{\text{trichloramine} \\ \text{(a.k.a nitrogen trichloride)}}}{NCl_3} + H_2O \qquad (5.35)$$

Reactions are not instantaneous but could take many hours at certain unfavorable conditions of pH, concentration, and temperature. At favorable conditions, reaction times are on the order of minutes to hours [36, p. 193]. While reaction to monochloramine may be relatively rapid, the dichloramine reaction is much slower.

These reactions produce chlorinated compounds of ammonia known as *combined chlorine residuals*, having a disagreeable odor and a lesser disinfecting power than the free chlorine residual of HOCl, OCl$^-$, or Cl$_2$ [26, pp. 31–19 to 31–20]. A higher residual concentration and a longer contact time are necessary for combined chlorine residuals to achieve the same disinfection effect as a residual of free chlorine. In theory, a molar ratio of 3:1 hypochlorous acid to ammonia is required for completion. The phenomenon was discovered in 1939 and has been investigated in great detail subsequently to explain its mechanism [36, pp. 194–195].

Figure 5.4, manually digitized and redrawn from Ref. 7, p. 370, depicts a typical course of reaction for a solution containing 1 mg/L of ammonia-nitrogen (NH$_3$-N).

Chlorine residual is plotted against applied chlorine, both expressed in mg/L. Since the molar ratio of chlorine to ammonia-nitrogen is 70.906/14.0067 $\cong$ 5, the x-axis values of 5, 10, and 15 mg/L of chlorine applied correspond approximately to molar ratios of 1, 2, and 3, respectively.

In the first stage of reaction up to about 5 mg/L of applied Cl$_2$ (a 1:1 molar ratio), the reaction of Equation (5.33) is taking place, and the product is primarily monochloramine (NH$_2$Cl) up to the peak or *hump* in the curve [36, p. 195]. Dichloramine (NHCl$_2$) is also produced, with a greater relative proportion of NHCl$_2$ at lower pH [6, pp. 392–393].

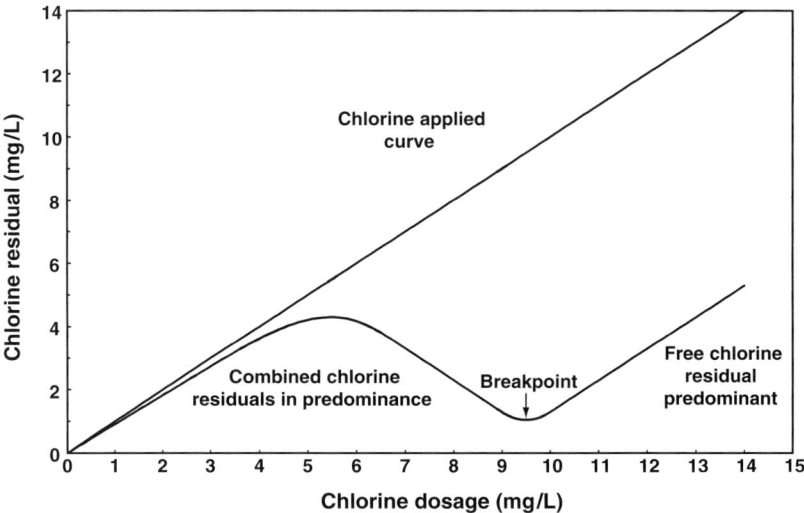

Figure 5.4 Typical breakpoint chlorination curve ammonia-nitrogen content: 1.0 mg/L.

As chlorination progresses along the downslope of the curve, some more dichloramine in equilibrium [6, p. 369] with monochloramine appears.

$$2\,NH_2Cl + H^+ \leftrightarrow NH_4 + NHCl_2 \tag{5.36}$$

It is formed by disproportionation of monochloramine [36, p. 195, 198] and the slower reaction step of Equation (5.34). The author of Ref. [36, p. 196] does not recommend stopping the chlorination reactions in this area because of the malodorous and disagreeably tasting dichloramine present there.

The dip in the curve at a 2:1 molar ratio (approximately 10 mg/L applied chlorine to 1 mg/L of ammonia-nitrogen) is known as the *breakpoint*. At this point, ammonia-nitrogen disappears, and an irreducible minimum chlorine residual is present, composed of essentially equal amounts of mono- and dichloramine along with a trace of HOCl.

Although in theory a 3:1 molar ratio of applied chlorine to ammonia-nitrogen originally present would be needed to complete the reactions of Equations (5.33), (5.34) and (5.35) to nitrogen trichloride (NCl_3) and water (H_2O), a ratio closer to 2:1 has been found in practice. Other species including nitrous oxide (N_2O), elemental nitrogen (N_2), and nitrate ion (NO_3^-) have been identified at the breakpoint [6, p. 394; 7, p. 370; 36, pp. 199–200]. This suggests that other reactions such as

$$NH_2Cl + NHCl_2 + HOCl \rightarrow N_2O + 4\,H^+ + 4\,Cl^- \tag{5.37}$$

$$NHCl_2 + NH_2Cl \rightarrow N_2 + 3\,H^+ + 3\,Cl^- \tag{5.38}$$

$$2\,NHCl_2 + H_2O \rightarrow N_2 + HOCl + 3\,H^+ + 3\,Cl^- \tag{5.39}$$

$$NHCl_2 + 2\,HOCl + H_2O \rightarrow 5\,H^+ + 4\,Cl^- + NO_3^- \tag{5.40}$$

$$NCl_3 + H_2O \rightarrow NHCl_2 + HOCl \tag{5.41}$$

$$\underset{\text{in solution}}{NCl_3} + \text{aeration} \rightarrow \underset{\text{gas}}{NCl_3} \uparrow \tag{5.42}$$

may also be taking place. The breakpoint curve in one text [6, pp. 393–394] has been revised to reflect a lower breakpoint value at 1.5:1 molar ratio of chlorine to ammonia-nitrogen, consistent with an assumed overall reaction of

$$2\,NH_3 + 3\,Cl_2 \rightarrow N_2 + 6\,H^+ + 6\,Cl^- \tag{5.43}$$

Regardless, a free chlorine residual appears along the upslope beyond the breakpoint. The free chlorine residual then follows the applied chlorine dosage at a 1 to 1 slope similar to the straight line at a 45° through the origin in Figure 5.4 for chlorination without ammonia present.

Free available chlorine (residual) and the various forms of combined chlorine can be determined by analytical procedures contained in "Standard Methods" [5, pp. 4-45 to 4-67]. Methods involving orthotolidine as a colorimetric indicator, as mentioned in some textbooks, are no longer listed along with others for various reasons. For actual measurements of chlorine residuals, the latest edition of this reference should be consulted.

PROBLEM 5.14

For a wastewater containing 5 mg/L of ammonia-nitrogen, calculate the amount of chlorine (as Cl_2) that must be added to reach the breakpoint and estimate the free chlorine residual when 55 mg/L of Cl_2 is added. There are no other chemicals in this wastewater that would react with chlorine.

SOLUTION Five mg/L of ammonia-nitrogen corresponds to

$$(5\,mg/L)(1\,g/1000\,mg)(g\,mol\,NH_3\text{-}N/14.01\,g)$$
$$= 3.57 \times 10^{-4}\,g\,mol/L\,of\,NH_3\text{-}N$$

To reach the breakpoint, we will assume that a more conservative 2 mol Cl_2 per mole of NH_3-N must be added.

$$(2)(3.57 \times 10^{-4}) = 7.14 \times 10^{-4}\,g\,mol/L\,of\,Cl_2$$
$$\times (70.906)(1000) = 50.6\,mg/L\,of\,Cl_2$$

The excess chlorine added (55 – 50.6 = 4.4 mg/L) then becomes the residual.
In a real-world situation, residual chlorine and its constituents should be verified by test.
Note: chlorination can generate unwanted reaction by-products in a treated effluent. Two reports on such by-products are listed in Refs [37,38]. ∎

5.10.4 Dechlorination

It may be necessary to remove the chlorine residual from a chlorinated effluent before discharge. Two methods will be used as examples—reaction with sulfur dioxide and related chemicals, and treatment with activated carbon.

The free chlorine residual HOCl and the combined chlorine residuals of monochloramine (NH_2Cl) and dichloramine ($NHCl_2$) react with SO_2 as follows

[36, p. 346; 39]:

$$Cl_2 + SO_2 + 2\,H_2O \rightarrow H_2SO_4 + 2\,HCl \tag{5.44}$$
$$HOCl + SO_2 + H_2O \rightarrow H_2SO_4 + HCl \tag{5.45}$$
$$NH_2Cl + SO_2 + 2\,H_2O \rightarrow NH_4HSO_4 + HCl \tag{5.46}$$
$$NHCl_2 + 2\,SO_2 + 4\,H_2O \rightarrow NH_4\,HSO_4 + H_2\,SO_4 + 2\,HCl \tag{5.47}$$

Sulfur dioxide is an attractive reagent because it can be applied from cylinders using the same type of equipment as the chlorinator that originally chlorinated the wastewater. For smaller installations, sodium sulfite (Na_2SO_3)/bisulfite ($NaHSO_3$) or sodium metabisulfite ($Na_2S_2O_5$) can be employed. The speed of all these reactions is said to be a matter of seconds [36, p. 346]. Sodium thiosulfate is not recommended because a longer time is required for completion, and it can impart an offensive dechlorination flavor or odor [36, p. 347].

Unfortunately, a competing reaction occurs in which the SO_2 and its cousins react with the DO in the water [39]:

$$2\,SO_2 + 2\,H_2O + O_2 \rightarrow 2\,H_2\,SO_4 \tag{5.48}$$

This oxygen scavenging reaction is slower at low concentrations without a catalyst; however, aeration of the effluent to an acceptable level would be necessary to the extent that it occurs.

The second dechlorination technique under discussion is the use of granular activated carbon. Presumed reactions are as follows, with the symbols C^* representing the activated carbon and CO^* a carbon–oxygen complex on the activated carbon [39]:

$$Cl_2 + C^* + H_2O \rightarrow 2\,HCl + CO^* \tag{5.49}$$
$$HOCl + C^* \rightarrow HCl + CO^* \tag{5.50}$$
$$NH_2Cl + C^* + H_2O \rightarrow NH_3 + HCl + CO^* \tag{5.51}$$
$$2\,NH_2\,Cl + \underset{\substack{\text{"aged"}\\ \text{carbon-oxygen}\\ \text{complex}}}{CO^*} \rightarrow N_2\uparrow + 2\,HCl + H_2O + C^* \tag{5.52}$$

$$2\,NHCl_2 + C^* + H_2O \rightarrow N_2\uparrow + 4\,HCl + CO^* \tag{5.53}$$

Dechlorination of drinking water using activated carbon can produce a finished water with a pleasant taste [36, p. 347].

PROBLEM 5.15

A treated effluent shows a free chlorine residual of 2.5 mg/L expressed as Cl_2. Calculate the minimum required dosage of SO_2, assuming that the effluent contains an extreme concentration of 9 mg/L DO.

SOLUTION From Equation (5.44),

$$\underset{70.906}{\overset{2}{Cl_2}} + \underset{64.06}{\overset{x}{SO_2}} + H_2O \rightarrow H_2SO_4 + 2HCl$$

$x = (2)(64.06)/70.906 = 1.8$ mg/L SO_2 for dechlorination

From Equation (5.48),

$$\underset{(2)64.06}{2\overset{y}{SO_2}} + 2H_2O + \underset{32.0}{\overset{9}{O_2}} \rightarrow 2H_2SO_4$$

$y = (2)(9)(64.06/32.0) =$ up to 36 mg/L SO_2 for reaction with DO

In this case, the potential participation of SO_2 in the oxygen scavenging reaction amounts to 20 times its requirement in the dechlorination reaction.

In reality, the SO_2 reaction with DO should be much less and not as much SO_2 would need to be provided to satisfy this competing reaction. First, the treated effluent would *not* likely be saturated in DO. Second, the uncatalyzed reaction of SO_2 with DO is relatively slow and would probably not go to completion within the residence time allowed [40]. On the other hand, the metal ions that catalyze the reaction at trace concentrations may be inadvertently present in the effluent. Additional comments on scavenging of DO with sulfite salts are provided in Section 5.13.2.

Whatever SO_2 residual survives in the effluent will impart a COD to the treatment process or receiving water. ∎

5.10.5 The Sulfurous Acid System

Another important system of ionic equilibrium from an environmental standpoint is comprised of dissolved SO_2 and the resulting ions, bisulfite (HSO_3^-) and sulfite (SO_3^{2-}). Dissolved SO_2 interacts with H_2O to form sulfurous acid (H_2SO_3)

$$SO_2 + H_2 \rightarrow H_2SO_3 \quad (5.54)$$

This in turn ionizes into bisulfite

$$H_2SO_3 \leftrightarrow HSO_3^- + H^+ \quad (5.55)$$

and sulfite

$$HSO_3^- \leftrightarrow SO_3^{2-} + H^+ \quad (5.56)$$

Ionic equilibria are as follows:

$$K_1 = [HSO_3^-][H^+]/[H_2SO_3] \quad (5.57)$$
$$K_2 = [SO_3^{2-}][H^+]/[HSO_3^-] \quad (5.58)$$

with $pK_1 = 1.76$ ($K_1 = 1.72 \times 10^{-2}$) and $pK_2 = 7.20$ ($K_2 = 6.3 \times 10^{-8}$) at 25°C [26, p. 28–16] and concentrations in the bracketed terms expressed as g mol/L.

Calculated distribution of species is shown in Figure 5.5.

Dissolved SO_2 predominates at very low pH but with a significant contribution of bisulfite ion. At a pH of 4.3 (the methyl orange end point), bisulfite [HSO_3^-] constitutes some 99.6% of the equilibrium mixture. In the mid-pH range, it is almost all bisulfite and sulfite (SO_3^{2-}), and 92.6% sulfite at a pH of 8.3 (the phenolphthalein end point). The proportion of sulfite is even greater at higher pH. The curves will

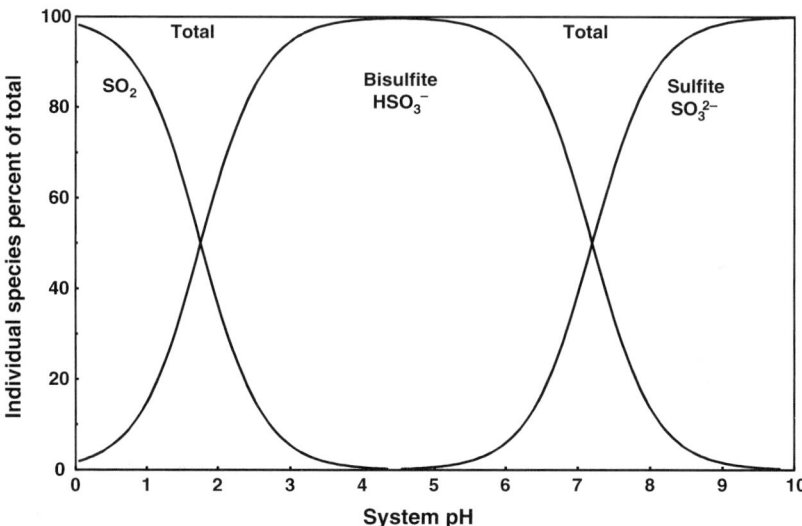

Figure 5.5 Distribution of species in the sulfurous acid (SO_2)–bisulfite–sulfite system at 25°C.

shift somewhat with temperature and with dissolved solids concentrations, but the low-concentration curves of Figure 5.5 at 25°C are instructive.

In addition to being important for dechlorination, this system arises in flue gas desulfurization (FGD) when scrubbing SO_2 from a flue gas with an aqueous solution containing a soluble alkali such as sodium hydroxide or sodium carbonate (Na_2CO_3). The purpose of the alkali is to shift the natural pH of the scrubbing liquor to a higher level where the captured SO_2 in solution does not exert an appreciable back-pressure.

A graph of FGD sulfur removal versus scrubber pH, with many, many data points in evidence and a correlating curve, is presented in Ref. 41. The curve shows about 60% SO_2 removal between a pH of 3 and 4, climbing steeply to about 90% at pH 6. The curve then tapers off between pH 6 and 8, where most of the data points are located, the scrubbing liquid is well buffered, and SO_2 removal is 90–95%. Ninety percent SO_2 removal appears to be a design value quoted by a number of sources. Sulfur dioxide removal is also a function of liquid to gas ratio (L/G); the type of scrubber, number of stages, and pressure drop across each stage; and the alkalinity of the scrubbing solution. We shall see the effect of buffering in the next problem.

PROBLEM 5.16

Verify that pH buffering exists in the SO_2 system between the bisulfite and sulfite ions in the pH range of 6–8.

SOLUTION The task will be to construct a titration curve for the system as NaOH is added to adjust pH. A gradual change in pH with added reagent indicates buffering; a steep precipitous change indicates little or no buffering.

Illustrative calculations are performed using an SO_2 solution containing 0.1 g mol/L, the limit of validity indicated for tabulated handbook values of ionization/dissociation constants

[10, p. D-151], and corresponding to about a 0.6 wt% solution (as SO_2). This temperature is somewhat low for flue gas scrubbing, where the actual gas and liquid temperatures should be closer to 150°F (65.6°C) than to 25°C (77°F), but instructive nonetheless.

An ion charge balance on this theoretical solution containing only H^+ and Na^+ positive ions (cations) plus OH^- and the HSO_3^- and SO_3^{2-} from SO_2 ionization is as follows:

$$[H^+] + [Na^+] = [HSO_3^-] + 2[SO_3^{2-}] + [OH^-] \tag{5.59}$$

with $[T]$ defined as the total SO_2 species and K_w the ion product of water at approximately 10^{-14}, and the equation for $[Na^+]$ becomes

$$[Na^+] = \frac{[T]}{1 + ([H^+]/K_1) + (K_2/[H^+])} + \frac{2[T]}{1 + ([H^+]/K_2) + ([H^+]^2/K_1 K_2)} + \frac{K_w}{[H^+]} - [H^+] \tag{5.60}$$

Solution for the initial pH where $[Na^+] = 0$ is by trial and error. The value of

$$pH = -\log_{10}[H^+] \tag{E.5}$$

for this condition is approximately 1.47.

Solution for $[Na^+]$ is straightforward by plugging values of $[H^+]$ into Equation (5.60). For example, at pH 7 ($[H^+] = 10^{-7}$),

$$[Na^+] = \frac{0.1}{1 + (10^{-7}/1.72 \times 10^{-2}) + (6.3 \times 10^{-8}/10^{-7})}$$
$$+ \frac{(2)(0.1)}{1 + (10^{-7}/6.3 \times 10^{-8}) + (10^{-14}/(1.72)(6.3) \times 10^{-10})} + \frac{10^{-14}}{10^{-7}} - 10^{-7}$$
$$= 0.06135 + 0.07730 + 10^{-7} - 10^{-7} = 0.1387$$

The complete titration curve is depicted in Figure 5.6 by reversing the roles of the x- and y-axes and plotting the results backwards as pH versus NaOH added. In the figure, the

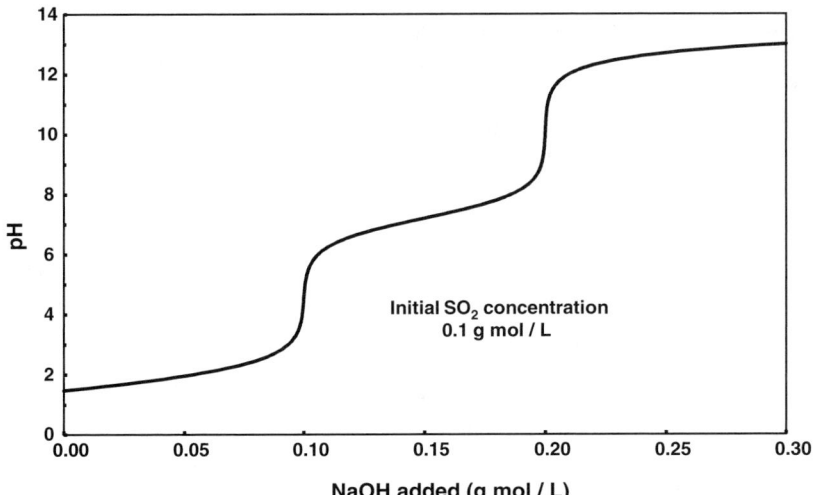

Figure 5.6 Calculated titration curve for dissolved SO_2 against NaOH at 25°C.

pH ranges from its initial value to 13, and the NaOH added varies from zero to 0.3 g mol/L.

Addition of 0.1 g mol/L of NaOH corresponds to conversion of all dissolved SO_2 to HSO^-, and 0.2 to its complete conversion in turn to SO_3^{2-}. These changes are sharp with little or no buffering. The conversion of bisulfite to sulfite is more gradual in the pH range of 6–8, where buffering is apparent. Sodium hydroxide addition beyond 0.2 g mol/L results in free hydroxide (OH^-) in solution. ∎

5.11 PETROLEUM OIL

Oil appears in a wastewater effluent from situations such as, for example, storm water runoff from outdoor areas containing oil-lubricated machinery or from the hose-down of a garage floor. This oil is commonly treated in an oil–water separator. In its simplest form, this is a large box receiving the wastewater in horizontal flow. It allows what is termed *free oil* in a separate liquid phase to be removed by gravity before discharge. Gravity separation does not work on emulsified oils soluble in the aqueous phase, such as cutting oils or degreasing oils, and it also does not work on other types of emulsified oil [30, pp. 24–25, 502].

For gravity separation to be effective on emulsified oil, the emulsion must first be broken [30, p. 502]. Various chemicals are often tried in a series of the so-called *jar tests* conducted using the ever popular six-paddle stirrer, whose paddles fit neatly inside 1000 mL beakers. Those unfamiliar with the jar test apparatus may wish to consult the manufacturer's literature or Refs [42,43], where examples are pictured.

Design of an oil–water separator is based on the buoyant rising velocity (v_r) of an oil droplet, which is less dense than the surrounding water phase. The velocity is described by Stokes' law for spherical particles in a Newtonian fluid [44]

$$v_r = g(\rho_w - \rho_o)D_o^2/(18\mu) \tag{5.61}$$

where g is the acceleration due to gravity, ρ_w and ρ_o are the density of water and the oil droplet, respectively, D_o is the diameter of an oil droplet, and μ is the viscosity of water.

This form of Stokes' law is valid for particle/droplet Reynolds numbers less than 1.0 [44], with droplet diameter denoted as d_p

$$Re = d_p \rho_w v_r / \mu \tag{5.62}$$

Both v_r in Equation (5.61) and Re in Equation (5.62) must be calculated in consistent units.

In the worst case, the droplet is allowed enough time to rise all the way to the surface, where it can be skimmed off in a quiescent zone before the average horizontal wastewater flow velocity causes it to exit the equipment uncollected.

The parallel-plate separator is a variation on the standard oil–water separator, otherwise denoted as an *API separator*. In this variation, a module of parallel undulating plates, spaced on the order of inches apart vertically, is inserted with the wavy shape in the flow direction. Now, rather than having to traverse the entire vertical depth to be collected, the droplet needs to travel only from the top surface of one plate to the bottom of the one above. The small oil droplets coalesce there and grow to much larger sizes at the high points of the corrugation. Weep holes, lined up one over another in the plates at the high points, allow the agglomerated oil droplets to move upward rapidly to the liquid surface for collection.

Oil–water separation equipment is available commercially in a series of sizes as a function of volumetric throughput. The assumed droplet size or size distribution plus whatever safety factor has been employed is the basis for the equipment manufacturer's guarantee. Guaranteed effluent concentrations also depend on the inlet concentration and range typically from 5 to 10 mg/L oil. Some equipment vendors even specify an effluent oil droplet diameter down to 20 μm.

The lowest possible effluent concentration of oil from an oil–water separator occurs when absolutely all of the free (floating) oil is removed and only the soluble portion remains at the limit of its solubility. This figure is quoted for new, unused oil at roughly 1 mg/L in fresh water and some 40% higher in seawater [45]. As the oil is used, or "weathered," its solubility is reduced significantly.

Regulatory effluent limits for oil, characterized as oil and grease or total petroleum hydrocarbons (TPH), can be in the range of 10–15 mg/L and possibly as low as 5 mg/L. The oil remaining in the aqueous liquid after oil separation is determined by EPA Method 1664 using *n*-hexane extraction in lieu of chlorofluorocarbon 113 formerly employed [46].

The uncollected oil imparts its own BOD_5 and COD. Estimates of the BOD_5 and COD contributed by the oil are 1 mg/L BOD_5 per mg/L of oil and 2.25 mg/L COD per mg/L of oil, based on the highest numbers obtained and reported for petroleum-based oils [47], rounded upward by your author.

PROBLEM 5.17

(a) A traditional rectangular oil–water separator is 5 ft wide and 5 ft deep. What must its length be to capture spherical oil droplets 100 μm (0.1 mm) in diameter at a 200 gpm throughout?

(b) Repeat for the same throughout, but using a corrugated plate separator 2 ft wide by 2 ft deep with plates spaced 1 in. apart. Assume that the specific gravity of the oil is 0.85 compared to 1.0 for water.

Note: The purpose of this example with these figures is to provide a feel for the relative dimensions involved and not to serve as a design basis for actual equipment.

SOLUTION The worst case for part (a) occurs for an oil droplet located at the bottom of the separator when the wastewater enters the oil–water separator chamber at the inlet end. This droplet must be able to rise to the surface for skimming before it is carried out of the equipment by the average wastewater flow velocity. In part (b), the vertical travel of the droplet is only 1 in.

- Density of water is taken as 62.4 $lb_m/ft^3 \cong 1000\, kg/m^3$.
- Then the density of the oil is $(0.85)(1000) = 850\, kg/m^3$.
- Part (a): cross-sectional area for flow = $(5)(5) = 25\, ft^2$; in part (b) = $(2)(2) = 4\, ft^2$.

- Droplet diameter $(D_o) = 100 \times 10^{-6} = 10^{-4}$ m (meters); $D_o^2 = 10^{-8}$ m^2.
- Viscosity of water $(\mu) = 1.13$ cp (centipoises) $\rightarrow 10^{-3}[44] = 1.13 \times 10^{-3}$ kg/m s.
- Volumetric flow: 200 gpm $(2.228 \times 10^{-3}$ ft^3/s per gpm$) = 0.4456$ ft^3/s.

Substituting in Equation (5.61), the rising velocity v_r is found to be

$$v_r = \frac{(9.81 \text{ m/s})(1000 \text{ kg/m}^3 - 850 \text{ kg/m}^3)(10^{-8} \text{ m}^2)}{(18)(1.13 \times 10^{-3} \text{ kg/m s})} = 7.235 \times 10^{-4} \text{ m/s}$$

$$\div 0.3048 \text{ m/ft} = 2.374 \times 10^{-3} \text{ ft/s}$$

Time needed for oil droplet to rise

(a) From the bottom to the surface (5 ft) is given by dividing distance by the velocity of rise (v_r)

$$t = 5 \text{ ft}/2.374 \times 10^{-3} = 2106.6 \text{s}$$

(b) From one plate to another (in. = 1/12 ft)

$$t = (1/12)/2.374 \times 10^{-3} = 35.11 \text{ s}$$

Horizontal velocity (v_h)

This is the average velocity in the horizontal (flow) direction obtained by dividing the volumetric flow (Q) by the cross-sectional area

$$v_h = Q/A \quad (5.63)$$

(a) $v_h = (0.4456 \text{ ft}^3/\text{s})/(25 \text{ ft}^2) = 0.01782$ ft/s
(b) $v_h = 0.4456/4 = 0.1114$ ft/s

Minimum required length of separator (L)

This is the horizontal length necessary to provide just enough residence time for droplet collection at flow conditions.

$$L = (v_h)(t) \quad (5.64)$$

(a) $L = (0.01782 \text{ ft/s})(2106.6 \text{ s}) = 37.5$ ft
(b) $L = (0.1114)(35.11) = 3.9$ ft

Check on the Reynolds Number

The dimensionless particle Reynolds Number [Equation (5.62)] is the same in both cases.

$$Re = [(10^{-4})(7.235 \times 10^{-4})(1000)]/(1.13 \times 10^{-3})$$
$$= 0.064 < 1.0$$

Therefore, Stokes' law in the form of Equation (5.61) applies.

Conclusion

Notice that the same separation can be accomplished in a much smaller box by using a corrugated parallel-plate separator. ∎

PROBLEM 5.18

For the 200 gpm throughput of Problem 5.17, estimate the mass flow rate of oil and its BOD_5 and COD at a manufacturer's guarantee of 10 mg/L. Use the BOD and COD factors mentioned in the text above.

SOLUTION

$$(200 \text{ gal/min})(3.785 \text{ L/gal})(10 \text{ mg/L})(1 \text{ g}/1000 \text{ mg})(1 \text{ lb}/453.59 \text{ g})(60 \text{ min/h})$$
$$= 1.0 \text{ lb/h of oil}$$

$$BOD_5 = (1)(1.0) = 1.0 \text{ lb/h}$$
$$COD = (2.25)(1.0) = 2.25 \text{ lb/h}$$

For a properly designed oil–water separator, these values are independent of the oil–water separator dimensions in Problem 5.17.

When the system is up and running, actual effluent concentrations and volumetric flow can be measured. ∎

5.12 COOLING WATER OPERATIONS

Appendix E outlines a methodology for performing calculations on an open recirculating cooling system. These procedures show how to calculate evaporation, makeup, and blowdown, and the chemical composition of the recirculating water. The principles are illustrated by worked examples and several case studies for operating plants.

Ion balance calculations are demonstrated in Section E 3.3 and Table E.4. An extended example problem is contained in Section E.8. In Tables E.7, E.8, and E.9, total dissolved solids (TDS) and conductivity are built from individual concentrations and conductivity factors. The case studies are addressed in Section E.10.

Such calculations are a useful tool to monitor both cooling tower operation and the composition of the blowdown for environmental control.

5.13 BOILER OPERATIONS

In a steam boiler, liquid water is evaporated to produce steam by burning fuel. Unless the produced steam is heated further, a process known as *superheating*, the (vapor) pressure of the steam is determined by the temperature alone, and *vice versa*.

The water fed to a steam boiler, also known as the *boiler feedwater*, must be conditioned in a series of external and internal treatment steps to mitigate corrosion and scaling and to ensure steam purity. The boiler feedwater is first pretreated as necessary to remove suspended solids via clarification and/or filtration, to remove or reduce certain dissolved species, and to adjust pH. Treatment chemicals are added to control the chemistry within the boiler, in the steam, and of any returned condensate.

5.13.1 Dissolved Solids Removal from Boiler Feedwater

Dissolved solids are removed by means of softening, dealkalization, or demineralization. Hardness cations can be removed by softening using lime or cation ion exchange. Removal of alkalinity and other anions in addition can be accomplished by using anion ion exchange. Demineralization can be accomplished by cation–anion exchange or reverse osmosis. Complete demineralization becomes necessary as boiler pressure is increased and the boiler water quality guidelines become more stringent. Block diagrams of typical demineralizer systems are depicted in Ref. [33, p. 65].

5.13.2 Deaeration

The final mechanical treatment step is deaeration, where steam is added to the boiler feedwater to strip out DO, which proves corrosive in a boiler. By definition, a *deaerating heater* is capable of removing DO to 0.03 cm^3/L (39 ppb); a *deaerator* is capable of achieving DO removal to the level of 0.005 cm^3/L (6.5 ppb) [33, pp. 76, 78; 48, pp. 131–133]. A small quantity of steam is vented, but the majority of steam is condensed and joins the deaerated boiler feedwater [33, p. 74].

To reduce the DO further, catalyzed sodium sulfite is added as an oxygen scavenger to react with and destroy the residual oxygen in low-pressure boilers. A noncarcinogenic substitute for hydrazine (N_2H_4) is added in high-pressure boilers; examples of hydrazine substitutes include hydroquinone ($C_6H_6O_2$), carbazine ($H_2NNHCONHNH_2$), and erythorbic acid (an isomer of ascorbic acid, $C_6H_8O_6$).

The sodium sulfite reacts rapidly to sodium sulfate (Na_2SO_4) in the presence of salts of cobalt, copper, nickel, iron, or manganese (or combinations thereof), which catalyze the reaction [33, p. 79; 48, p. 170].

$$Na_2SO_3 + 1/2 O_2 \rightarrow Na_2SO_4 \qquad (5.65)$$

Among the catalyst metals, cobalt is the most effective [49,50], 12 times faster than copper [50,51], and cobalt salts are the most widely used to catalyze the sulfite scavenging reaction [52]. Cobalt in the form of cobalt chloride ($CoCl_2$) or cobalt sulfate ($CoSO_4$) at concentrations ranging from 0.005 to 0.05 ppm [49,50,53,54], and as low as 0.001 ppm (1 ppb) [50], is sufficient for rapid reaction. A concentration of 0.1 ppm cobalt is reported for effective scavenging in a field application [50]. In contrast, reaction of sodium sulfite without a suitable catalyst takes much longer.

The sodium sulfite reaction increases the TDS somewhat. A small amount of additional TDS is easily tolerated in a low-pressure boiler, where this operating parameter is less critical than in a high-pressure boiler. When hydrazine was used, reaction with DO would produce no solid products, only nitrogen gas and water

$$N_2H_4O_2 \rightarrow N_2 + 2 H_2O \qquad (5.66)$$

and leave the very low boiler feedwater TDS at this point unchanged.

5.13.3 Boiler Blowdown

Even with pretreatment, a purge, or *blowdown*, is taken from the system to control the concentration of unwanted material inside the boiler. An overall material balance around the boiler is written for feedwater (F), steam production (S), and blowdown (B) as follows:

$$F = S + B \tag{5.67}$$

accompanied by a series of component material-balance equations for each nonvolatile constituent (X) numbering up to one less than the total number of constituents

$$F x_F = S(0) + B x_B \tag{5.68}$$

where x_F and x_B are the mass fractions of constituent X in the feedwater and blowdown, respectively. A truly nonvolatile species results in a zero for the concentration in the steam production term, and hence it drops out. Combining these two equations yields expressions for *blowdown on boiler feedwater*

$$\% \, (B/F) = 100(x_F/x_B) \tag{5.69}$$

and blowdown on total steam-make

$$\% \, (B/S) = 100(x_F/x_B)/[1-(x_F/x_B)] \tag{5.70}$$

It is important to state the difference unambiguously.

Rearrangement of these equations enables one to calculate the blowdown concentration of a given species when the percent blowdown and the feed concentration are known.

$$x_B = x_F/[\% \, (B/F)/100] \tag{5.71}$$
$$x_B = x_F/[1 + \% \, (B/S)/100]/[\% \, (B/S)/100] \tag{5.72}$$

Guidelines are available to select the maximum allowable concentration of various species in the blowdown and, in some cases, the boiler feedwater [55,56]. These are, of course, subject to change as new information is acquired; be sure to consult the latest version for an actual case.

Application of Equations (5.69) and (5.70) to different guideline parameters will result in a range of calculated blowdowns. The constituent causing the highest blowdown to occur is controlling and sets the blowdown rate.

5.13.4 Addition of Treatment Chemicals

The purpose of boiler treatment chemicals is to combat corrosion and scaling. Dissolved oxygen, highly corrosive in a boiler environment, is removed by deaeration and addition of an oxygen scavenger. A residual of the oxygen scavenger is maintained in the boiler. Corrosion is also controlled by operating at a pH clearly in the alkaline range. A corrosion inhibitor may also be added to lay down a protective film on the surface of the metal [33, pp. 78–81, 89–94, 147–149].

294 Chapter 5 Water/Wastewater Composition

Amines are used as control agents in condensate-return systems. The *neutralizing amines* neutralize the acid produced by CO_2 in the condensate. Examples are morpholine and cyclohexylamine. Filming amines such as octyldecylamine are long-chain molecules that lay down a protective film and protect against attack by CO_2 and DO.

Some boilers may be treated with phosphates to mitigate against hard calcium carbonate or calcium sulfate scale at high pH. Calcium salts exhibit a so-called *retrograde solubility*, which means that they are less soluble at high temperatures than at low temperatures. The phosphate is much less soluble with calcium, is relatively nonadherent, and forms a soft sludge that is more easily removed in the blowdown. Magnesium hardness can combine with silica from the makeup water to form magnesium silicate and with free hydroxide alkalinity at high pH to form magnesium hydroxide. Polymer materials are used as dispersing agents to keep calcium phosphate, magnesium silicate, and magnesium hydroxide in suspension.

A chelating agent is the sodium salt of a weak organic acid. These large molecules are used to surround metal ions and prevent them from forming deposits. Examples are ethylene diamine tetraacetic acid (EDTA) and nitrilo triacetic acid (NTA). They are particularly effective in combination with polymers in complexing and dispersing unwanted iron oxide deposits. All of these chemicals must be accounted for when boiler blowdown is to be discharged as wastewater.

5.13.5 Boiler Wastewater Composition

Wastewater consists of all streams generated and discharged necessary in order to enable the production of steam. Determination of the flow rate and composition of the boiler blowdown, including any added treatment chemicals, provide the particulars for one such wastewater stream. Clarifier underflow, filter backwash, ion exchange regenerant and rise flows, and reverse osmosis reject liquid are examples of others.

The concept of boiler wastewater streams will become clearer in the examples that follow.

PROBLEM 5.19

The water of composition given in Table 5.8 is to be used as boiler feedwater to a 300 psig, 40,000 lb/h steam generator. It will be treated for hardness removal in a sodium zeolite softener.

(a) What is its composition into and out of the boiler?
(b) Calculate flow rate and composition of the boiler blowdown needed to stay within the following guidelines [55,56]

Boiler feedwater	Boiler water
Total iron ≤ 0.1 mg/L	Silica ≤ 150 mg/L SiO_2
Total copper ≤ 0.05 mg/L	Total alkalinity 700 mg/L as $CaCO_3$[a]
Total hardness ≤ 0.3 mg/L as $CaCO_3$	Specific conductance 7000 μmho/cm or μS/cm

[a]Alkalinity not to exceed 10% of specific conductance.

(c) Comment on wastewater disposal of this boiler blowdown stream.

5.13 Boiler Operations

Table 5.8 Boiler Feedwater and Blowdown Composition, 40,000 lb/h Industrial Steam Generator

Parameter	Raw water		Sodium zeolite treated		Boiler blowdown[a]	
	mg/L as the ion	mg/L as $CaCO_3$	mg/L as the ion	mg/L as $CaCO_3$	mg/L as the ion	mg/L as $CaCO_3$
Ca^{2+}	10	25	0.10	0.25	1.7	4.2
Mg^{2+}	1.2	5	0.01	0.05	0.17	0.8
Na^+	8.94	19.43	22.6	49.13	376.7	818.8
K^+	1.7	2.17	1.7	2.17	28.3	36.2
Fe^{2+}	0.05	0.09	0.05	0.09	0.83	1.5
Ca^{2+}	0.02	0.03	0.02	0.03	0.33	0.5
Total cations	—	51.72	—	51.72	—	862.0
HCO_3^-	12.2	10	12.2	10	0.0	0.0[b]
Cl^-	15.0	21.15	15.0	21.15	250.0	352.5
SO_4^{2-}	199.0	19.76	19.0	19.76	316.7	329.3
NO_3^-	1.0	0.81	1.0	0.81	16.7	13.5
Total anions	—	51.72	—	51.72	—	862.0
SiO_2	9	—	9	—	150	—
pH std. units	6.5–7.5		6.5–7.5[c]		pH 10.5 +[d]	
Conductivity (μmho/cm)	124 (calculated)		123 (calculated)		2050 (estimated)	
Flow (gpm)	85.1		85.1		5.1	

[a]Boiler blowdown also contains mg/L quantities of boiler water treatment chemicals and sodium sulfate/excess sodium sulfite, as well as traces of cobalt or copper catalyst from the sodium sulfite treatment for DO removal following the deaeration step.

[b]Bicarbonate ion (HCO_3^-) at 166.7 mg/L as $CaCO_3$ is assumed to react completely to 56.7 mg/L as OH^-, in accordance with Equation (5.73). Heat-induced reactions are as follows [33, pp. 90, 146]:

$$HCO_3^- \xrightarrow[\text{goes to completion}]{} OH^- + CO_2 \quad (5.73)$$

$$CO_3^{2-} + H_2O \xrightarrow[\text{only about 80\% complete}]{} 2\,OH^- + CO_2 \quad (5.74)$$

The OH^- liberated by the reaction of Equation (5.73) is sufficient to raise the pH to 11.5. Bicarbonate ion also reacts to H^+ and CO_3^{2-} even at more moderate conditions to depress the pH [33, p. 146].

$$HCO_3^- \rightarrow H^+ + CO_3^{2-} \quad (5.75)$$

[c]pH 8.3–9.0 if some blowdown water is returned to the softened feedwater as a corrosion control measure [33, pp. 81, 99].

[d]Ref.[47, p. 168].

SOLUTION

(a) Composition of an ion balanced feed to a sodium zeolite ion exchange softener is given in columns 2 and 3 of Table 5.8. Sodium ion is determined by difference. Total hardness is 25 $[Ca^{2+}] + 5[Mg^{2+}] = 30$ mg/L as $CaCO_3$. Alkalinity (all HCO_3^- ion) is 10 mg/L as $CaCO_3$. Its conductivity is calculated as in Problem 5.3.

296 Chapter 5 Water/Wastewater Composition

After exchange with the sodium zeolite ion exchange resin, hardness is reduced to a residual of 0.3 mg/L as $CaCO_3$ in the same proportion as the incoming ratio of calcium to magnesium. Sodium from the resin replaces the hardness ions in the water as the hardness ions are adsorbed onto the resin surface. Calculated conductivity changes only slightly. The pH remains about the same.

(b) Percent blowdown is calculated from Equations (5.69) and (5.70), in turn, for silica, alkalinity, and conductivity.

SiO_2: $\quad 100(9/150) = 6.00\%$ on boiler feedwater
$\qquad\qquad 100(9/150)/[1 - (9/150)] = 6.38\%$ on total steam make
Alkalinity: $\quad 100(10/700) = 1.43\%$ on boiler feedwater
$\qquad\qquad 100(10/700)/[1 - (10/700)] = 1.45\%$ on total steam make
Conductivity: $\quad 100(123/7000) = 1.76\%$ on boiler feedwater
$\qquad\qquad 100(123/7000)/[1 - (123/7000)] = 1.79\%$ on total steam make

Calculated blowdown is highest based on silica. Therefore, silica is the limiting constituent.

The ratio of concentration in the blowdown to that in the feedwater at a blowdown of 6% on feedwater or 6.38% on steam water can be calculated from rearrangement of Equation (5.71) or Equation (5.72).

$$(x_F/x_B) = 1/[(6/100)] = 16.6667\ldots$$
$$(x_F/x_B) = [1 + (6.38/100)]/(6.38/100) = 16.6667\ldots$$

meaning that concentrations cycle up by a factor of more than 16 times in the boiler. For example, from Equation (5.71) or Equation (5.72)

SiO_2: $\quad x_B = 9\ (16.6667\ldots) = 150$ mg/L as SiO_2, as expected
Alkalinity: $\quad x_B = 10\ (16.6667\ldots) = 166.7$ mg/L as $CaCO_3$
Conductivity: $\quad x_B = 123\ (16.6667\ldots) = 2050\ \mu$mho/cm

At this condition, blowdown conductivity is less than 7000 μmho/cm, and the blowdown alkalinity is less than 10% of total alkalinity. Figures for silica, alkalinity, and conductivity are listed in the last two columns of Table 5.8. Blowdown concentrations calculated in a similar manner for the other feedwater species are also shown in Table 5.8.

By overall material balance

$$40,000\ \text{lb/h of steam}\ (1\ \text{gpm})/(500\ \text{lb/h per gpm}) = 80\ \text{gpm}$$
$$\text{Blowdown} = 6.38\% = 5.1\ \text{gpm}$$
$$\text{Boiler feedwater} = 80 + 5.1 = 85.1\ \text{gpm}$$

These flows are shown in the last row at the bottom of Table 5.8. Mass flow rate of each individual constituent in lb/h can be obtained by multiplying its concentration in mg/L as the ion by 5.0056×10^{-4} times blowdown in gpm. For the Na^+, this amounts to 2.0907, rounded to 2.1 lb/h.

The zeolite-softened feedwater proceeds through the deaeration step where DO is expelled from solution down to the ppb level, and catalyzed sodium sulfite is added to remove the last traces of oxygen. The sulfite reacts with the remaining DO to form sodium sulfate.

Treatment chemicals are added, and the boiler water and the blowdown (assumed to have the same concentration) will contain these chemicals and exhibit a higher pH. Boiler treatment chemicals and the reaction products from catalyzed sodium sulfite treatment in the deaeration step, specific to a given system, are not shown in Table 5.8.

(c) Such chemicals should be noted, however, when the blowdown is discharged. If the pH is too high for discharge, the blowdown must be mixed with other wastewater streams before disposal, or a pH adjustment step would have to be added. ∎

PROBLEM 5.20

Calculate the wastewater streams resulting from the ion exchange operation referred to in Problem 5.19. No condensate is returned to the system.

SOLUTION A sodium zeolite softener vessel containing a typical cation resin in the sodium form is 3 ft in diameter and 7 ft tall (tangent to tangent). The bed contains 30 ft^3 of resin. It has more than enough capacity to remove the hardness present in the untreated boiler feedwater

$$\frac{[(30-0.3) \text{ mg/L as } CaCO_3](85.1 \text{ gal/min})(1440 \text{ min/day})(3.785 \text{ L/gal})}{(50 \text{ g/equiv})(1000 \text{ mg/g})(30 \text{ ft}^3)(28.32 \text{ L/ft}^3)}$$

$$= 0.3243 \text{ equiv of hardness removed/day per liter of resin}$$

It can accommodate a service flow of 85.1 gpm at 3 gpm/ft^3 of resin and allow for a 75% expansion on backwashing. It is regenerated at 10 lb NaCl/ft^3 of resin. This conceptual design is based on guidance provided in Refs. [33, pp. 55–56; 57] supplemented by vendor information for a typical cation resin. No claim is made here that this is the most efficient design, but it serves the purpose for this example.

The calcium and magnesium hardness ions picked up during the service flow cycle must be removed by taking the softener out of service once per day and contacting the resin with a concentrated sodium chloride (NaCl) solution. A second identical softener on standby is activated and handles the service flow for the next 24 h, until *it* is replaced by the first unit after that unit's regeneration.

During each unit's regeneration cycle, it is backwashed, treated with the concentrated brine solution, and then rinsed. These flows contain the calcium and magnesium removed by the resin, the excess NaCl used in regeneration, as well as resin fines plus the accumulated debris filtered out by the resin bed. They are sent for ultimate treatment and disposal and are summarized in Table 5.9. For the purposes of this problem, all of these flows will first be collected in a holding tank, where they will be allowed to mix prior to disposal.

Assuming that all of the hardness coming off the resin bed and the excess sodium from regeneration remain in the regenerant stream, the backwash and rinses will contain only the constituents in the raw water at their original concentrations. (This is not true but convenient for the calculations here since all of the flows associated with regeneration will be mixed together.)

Regenerant Flow
For the 10% NaCl solution

$$(0.3243 \text{ equiv of hardness/liter of resin})(30 \text{ ft of } resin)(28.32 \text{ L/ft}^3) = 275.5 \text{ equiv/ft}^3$$

The sodium ion from the brine solution will replace 275.5 equiv of hardness (1/6 Mg^{2+} and 5/6 Ca^{2+}) on the resin. The sodium in the brine solution will be reduced by this amount. The

Table 5.9 Backwash, Regenerant, and Rinse Flows for 30 ft³ Ion Exchange Resin Bed[a]

Step	Time (min)	gpm	gal/ft³	gpm/ft³	gpm/ft²	Gallons to holding tank
Backwash ↑ (water)	10	56.5	22.6	1.9	8.0	565[b]
Regeneration (10% NaCl) ↓	22.4	15	11.2	0.5	2.1	336[c]
Slow rinse ↓ (water)	40	15	20	0.5	2.1	600[d]
Fast rinse ↓ (water)	20	45	30	1.5	6.4	900[e]
Total	92.4	—	—	—	—	2401

[a] Arrows indicate flow direction through bed.
[b] Contains suspended matter, resin fines, and soft water.
[c] Contains hardness ions in excess NaCl brine solution.
[d] For displacement of spent regenerant solution through the bed.
[e] Starts out with hardness ions and sodium and finishes when water concentrations reach inlet levels.

hardness ions will join the brine solution.

$$45.9 \text{ equiv of } Mg^{2+} \text{ as } CaCO_3$$
$$+ 229.6 \text{ equiv of } Ca^{2+} \text{ as } CaCO_3 \text{ to be added to the brine solution}$$
$$= 275.5 \text{ equiv of } Na^+ \text{ as } CaCO_3 \text{ to be removed from the brine solution}$$

Total amount of the 10% NaCl brine solution

$$336 \text{ gal } (1.0707)(8.34)(10/100) = 300 \text{ lb}$$
$$Na^+ \text{ is } [23/(23+35.5)](300) = 117.95 \text{ lb}$$
$$Cl^- \text{ is } [35.5/(23+35.5)](300) = 182.05 \text{ lb}$$

Na : $(275.5 \text{ eqiv}) \left(\dfrac{23 \text{ g}}{\text{eqiv}}\right) \dfrac{\text{lb}}{453.59 \text{ g}} = 13.97 \text{ lb of } Na^+$ to be removed from the brine solution

Ca : $(229.6 \text{ eqiv}) \left(\dfrac{20 \text{ g}}{\text{eqiv}}\right) \dfrac{\text{lb}}{453.59 \text{ g}} = 10.12 \text{ lb of } Ca^{2+}$ added to the brine solution

Mg : $(45.9 \text{ eqiv}) \left(\dfrac{12.2 \text{ g}}{\text{eqiv}}\right) \dfrac{\text{lb}}{453.59 \text{ g}} = 1.23 \text{ lb of } Mg^{2+}$ added to the brine solution

Na : $117.95 - 13.97 = 103.98$ lb Na^+ remaining in brine solution

Cl : $182.05 = 182.05$ lb Cl^- remaining in brine solution

Specific gravity of 10% NaCl solution is 1.0707 [4, p. 10–138]. This leads to

$$\left(\dfrac{8.34 \text{ lb water}}{\text{gal water}}\right)\left(\dfrac{\text{gal solution}}{8.34 \times 1.0707 \times 0.9 \text{ lb water in NaCl solution}}\right)$$
$$= \dfrac{1.04 \text{ gal NaCl solution}}{\text{gal of water}}$$

$336/1.04 = 323$ gal of water used to make up 10% NaCl solution

5.13 Boiler Operations

Table 5.10 Summary of Softener Regenerant Cycle Flows

Species	Brine solution (lb of constituent)	Backwash + rinses (lb of constituent)	Total (lb of constituent)	Blended stream (mg/L of species)
Ca^{2+}	0.0270 + 10.12 = 10.15	0.1723	10.32	515.1
Mg^{2+}	0.0032 + 1.23 = 1.23	0.0207	1.25	62.4
Na^+	0.0241 + 103.98 = 104.00	0.1540	104.16	5199 (0.52%)
K^+	0.0046	0.0293	0.034	1.7
Fe^{2+}	0.0001	0.0009	0.001	0.05
Cu^{2+}	0.0001	0.0003	0.0004	0.02
HCO_3^-	0.0329	0.2102	0.2431	12.1
Cl^-	0.0404 + 182.05 = 182.09	0.2585	182.35	9101 (0.90%)
SO_4^{2-}	0.0512	0.3274	0.38	19.0
NO_3^-	0.0027	0.0172	0.02	1.0
SiO_2	0.0243	0.1551	0.18	9.0
Volume	336 gal	2065 gal	2401 gal	—

The other species in the makeup water simply pass through as, for example, sulfate (SO_4^{2-})

$$(323 \text{ gal})(3.785 \text{ L/gal})(1 \text{ g}/1000 \text{ mg})(1 \text{ lb}/453.59 \text{ g})(19 \text{ mg/L}) = (2.695 \times 10^{-3})(19)$$
$$= 0.0512 \text{ lb}$$

These are summarized in Table 5.10 in the brine solution column.

Backwash and Rinse Flows
Total of backwash and rinse flows = 565 + 600 + 900 = 2065 gpm
For the backwash and rinses, the constituents from the makeup water will simply pass through, as for sulfate (SO_4^{2-})

$$(2065 \text{ gal})(3.785 \text{ L/gal})(1 \text{ g}/1000 \text{ mg})(1 \text{ lb}/453.59 \text{ g})(19 \text{ mg/L})$$
$$= (1.723 \times 10^{-2})(19) = 0.3274 \text{ lb } SO_4^{2-}$$

These are listed in Table 5.10 in the backwash plus rinses column.

Calculation of Final Volume

$$(2401 \text{ gal})(8.34 \times 1.0088) \text{ lb/gal} = 20{,}201 \text{ lb of solution}$$

This constitutes a 1.49 wt % NaCl solution based on the Cl^- ion. Specific gravity based on NaCl concentration is 1.0088 (by interpolation), confirming specific gravity used above.

$$\text{Final volume}: \frac{8.34}{[8.34(1.0088)(0.9851)]} = \frac{1.006 \text{ gal NaCl solution at } 1.49\% \text{ NaCl}}{\text{gal of water}}$$

Total water used in making up NaCl solution + backwash + rinses

$$323 + 565 + 600 + 900 = 2388 \text{ gal}$$
$$(2388)(1.006) = 2402.3 \text{ gal (total ``corrected'' volume)}$$

Use 2401 total volume from Table 5.9.

Concentration in mg/L for each constituent, for example, for SO_4^{2-}

$$\frac{1}{2401 \text{ gal}} \frac{\text{gal}}{3.785 \text{ L}} \frac{453.59 \text{ g}}{\text{lb}} \frac{1000 \text{ mg}}{\text{g}} (0.38 \text{ lb})$$
$$= (49.9121)(0.38) = 18.97 \rightarrow 19.0 \text{ mg/L}$$

Within roundoff, concentrations for all of the nonparticipating species are the same as in the makeup water. Calcium, magnesium, sodium, and chloride are changed.

Concentrations in wt% for Na^+ and Cl^- are as follows:

$$\frac{(100)(5199 \text{ mg/L as the ion})3.785 \text{ L/gal})(2401 \text{ gal solution})(1 \text{ g}/1000 \text{ mg})(\text{lb}/453.59 \text{ g})}{(2401 \text{ gal solution})(8.34 \times 1.0088 \text{ lb/gal})}$$

For Na^+ $(5199)(9.9182 \times 10^{-5}) = 0.516\% \rightarrow 0.52\%$
For Cl^- $(9101)(9.9182 \times 10^{-5}) = 0.903\% \rightarrow 0.90\%$

Calculations for a demineralizer system are much the same, except that there are more of them. At least one additional bed, for anion removal, must be considered. A degasifier or decarbonator for stripping out CO_2 is often placed after the cation bed. The cation bed is regenerated with acid, and the anion bed with caustic. A mixed bed of cation and anion resins may be added at the tail end of the train for polishing to obtain higher purity treated water. A mixed bed may also be used to treat any returned steam condensate. ∎

REFERENCES

1. F.N. Kemmer, editor, *Water: The Universal Solvent*, Nalco Chemical Company, Oak Brook, IL, 1977, pp. 50–51.
2. Olin Water Services, Olin Water Services explores in depth: Crater Lake, *Chemical Engineering*, 82(23),76–77, 1975.
3. M.A. Mast, and D.W. Clow, Environmental Characteristics and Water-Quality of Hydrologic Benchmark Network Stations in the Western United States, U.S. Geological Survey Circular 1173-D, 2000, 115 pp.
4. J.A. Dean, editor, *Lange's Handbook of Chemistry*, 11th edn, McGraw-Hill, New York, 1973.
5. L.S. Clesceri, A.E. Greenberg, R.R. Trussell, and M.A.H. Franson, editors, *Standard Methods for the Examination of Water and Wastewater*, 17th edn, American Public Health Association, Washington, D.C., 1989.
6. C.N. Sawyer and P.L. McCarty, *Chemistry for Environmental Engineering*, 3rd edn, McGraw-Hill, New York, 1978, 532 pp.
7. C.N. Sawyer and P.L. McCarty, *Chemistry for Sanitary Engineers*, 2nd edn, McGraw-Hill, New York, 1967, 518 pp.
8. C.A., Meyer, R.B. McClintock, G.J. Silvestri, and R.C. Spencer Jr., *1967 ASME Steam Tables: Thermodynamic and Transport Properties of Steam*, 2nd edn, American Society of Mechanical Engineers, New York, 1968, 328 pp.
9. J.H. Perry, editor, *Chemical Engineers' Handbook*, 3rd edn, McGraw-Hill, New York, 1950, 1942 pp.
10. R.C. Weast, editor, *CRC Handbook of Chemistry and Physics*, 58th edn, CRC Press, Inc., Cleveland, OH, 1977.
11. M.C., Rand, A.E. Greenberg, M.J. Taras, and M.A. Franson, *Standard Methods for the Examination of Water and Wastewater*, 14th edn, American Public Health Association, Washington, DC, 1976, 1193 pp.
12. L. Klein, *River Pollution I. Chemical Analysis*, Butterworths, London, 1959, 206 pp., reprinted 1967.

References 301

13. W.W. Eckenfelder, and D.L. Ford, *Water Pollution Control: Experimental Procedures for Process Design*, Pemberton Press, Austin, TX, 1970, 272 pp.
14. H. Heukelekian, and M.C. Rand, Biochemical oxygen demand of pure organic compounds, *Sewage Industrial Wastes*, **27**(9), 1040–1053, 1955.
15. P.E. Gaffney and H. Henkelekian, Oxygen demand of the lower fatty acids, *Sewage Industrial Wastes*, **30**(5), 673–679, 1958.
16. H.F. Lund, editor, *Industrial Pollution Control Handbook*, McGraw-Hill, New York, 1971, pp. 4-20 to 4-21.
17. C.B. Lamb and G.F. Jenkins, B.O.D. of synthetic organic chemicals, Proceedings of 7th Industrial Waste Conference, Purdue University, 1953, pp. 326–339.
18. R. Hatfield, Biological Oxidation of Some Organic Compounds, *Industrial and Engineering Chemistry*, **49**(2), 192–196, 1957.
19. E.J. Mills Jr., and V.T. Stack Jr., Biological Oxidation Parameter Applied to Industrial Wastes, *Industrial and Engineering Chemistry*, **48**(2), 260–262, 1956.
20. K.S. Price, G.T. Waggy, and R.R. Conway, Brine Shrimp Bioassay and Seawater BOD of Petrochemicals, *Journal of the Water Pollution Control Federation*, **46**(1), 63–77, 1974.
21. K. Verschueren, *Handbook of Environmental and Organic Chemicals*, Vol. 1, John Wiley & Sons, Inc. New York, 2001, pp. 1060–1062.
22. B.E., Jank, H.M. Guo, and V.W. Cairns, Activated sludge treatment of airport wastewater containing aircraft de-icing fluids, *Water Research*, **8**(11), 875–850, 1974.
23. W.H. Evans and E.J. David, Biodegradation of mono-, di- and triethylene glycols in river waters under controlled laboratory conditions, *Water Research*, **8**, 97–99, 1974.
24. Metcalf & Eddy, Inc., *Wastewater Engineering: Treatment Disposal Reuse*, 2nd edn, McGraw-Hill, New York, 1979, p. 64, revised by G. Tchobanoglous.
25. Metcalf & Eddy, Inc., *Wastewater Engineering: Collection Treatment Disposal*, McGraw-Hill, New York, 1972, p. 231.
26. G.M. Fair, J.C. Geyer, and D.A. Okun, Water and wastewater engineering. *Purification and Wastewater Treatment and Disposal*, Vol. **2**, Wiley, New York, 1968.
27. U.S. Environmental Protection Agency, *Onsite Wastewater Treatment Systems Manual*, EPA/625/R-00/008, U.S. EPA National Risk Management Research Laboratory, Cincinnati, OH, 2002, pp. 4-37 to 4-48.
28. A.W. Olivieri, R.J. Roche, and G.L. Johnston, Guidelines for control of septic tank systems, *Journal of the Environmental Engineering Division ASCE*, **107**(EE5), 1025–1034, 1981.
29. K. Mancl, Estimating septic tank pumping frequency, *Journal of Environmental Engineering*, **110**(1), 283–285, 1984.
30. L. Klein, *Aspects of River Pollution*, Butterworths, London, 1957, 621 pp.
31. L. Klein, *River Pollution II. Causes and Effects*, Butterworths, London, 1962, 456 pp.
32. M.J. Taras, A.E. Greenberg, R.D. Hoak, and M.C. Rand, editors, *Standard Methods for the Examination of Water and Wastewater*, 13th edn, American Public Health Association, Washington, DC, 1971, 874 pp.
33. *BETZ Handbook of Industrial Water Conditioning*, 8th edn, BETZ Laboratories, Inc., Trevose, PA, 1980, 438 pp.
34. O. Levenspiel, *Chemical Reaction Engineering: An Introduction to the Design of Chemical Reactors*, Wiley, New York, 1962, pp. 59–62.
35. J.E. McKee and H. W. Wolf, *Water Quality Criteria*, 2nd edn, California State Water Resources Control Board, Sacramento, CA, 1963, p. 134, Publication No. 3-A, reprinted Jan. 1973.
36. G.C. White, *Handbook of Chlorination*, Van Nostrand Reinhold, New York, 1972, pp. 374–381, 192–201.
37. Manufacturing Chemists Association, The Effect of Chlorination on Selected Organic Chemicals, U.S. EPA Report 12020 EXG 03/72, U.S. Government Printing Office, Washington, D.C., March 1972.
38. National Research Council, *The Chemistry of Disinfectants in Water: Reactions and Products*, PB-292 776, National Technical Information Service (NTIS), Springfield, VA, Mar. 1979.

39. V.L. Smoeyink and F.I. Markus, Chlorine residuals in treated effluents, *Water & Sewage Works*, **121** (4),35–38, 1974.
40. G.R. Helz and L. Kosak-Channing, Dechlorination of Wastewater and Cooling Water, *Environmental Science & Technology*, **18**(2), 48A–55A, 1984.
41. C.F. Cornell and D.A. Dahlstrom, Performance results on a 2500-ACT FT^3/MIN double-alkali plant for SO_2 removal, *AIChE Symposium Series*, No. 148, 71, 272–282, 1975.
42. W.A. Mracek and L. Greenberg, Control and automation of chromate waste reduction plants, Proceedings of the 30th International Water Conference, Engineers Society of Western Pennsylvania, Pittsburgh, PA 1969, p. 91, including prepared discussion.
43. J.G. Surchek and T.R. Tutein, Simplified method determines cost performance of polymeric flocculants, *Water & Sewage Works*, **123**(1), 37–39, 1976.
44. W.L., McCabe, J.C. Smith, and P. Harriott, *Unit Operations of Chemical Engineering*, 5th edn, McGraw-Hill, New York, 1993, pp. 160, 1083.
45. R.J., Irwin, M. Van Mouwerik, L. Stevens, M.D. Seese, and W. Basham, *Environmental Contaminants Encyclopedia*, National Park Service, Water Resources Division, Fort Collins, CO, 1997 (Distributed within the Federal Government as an Electronic Document, Projected public availability on the Internet or NTIS: 1998).
46. U.S. EPA, Analytical Methods—Method 1664, Revision A: *n*-Hexane Extractable Material (HEM; Oil and Grease) and Silica Gel Treated *n*-Hexane Extractable material (SGT-HEM; Non-polar Material) by Extraction and Gravimetry, http://www.epa.gov/waterscience/methods/method/oil/1664.html, accessed on October 5, 2008.
47. J.C., Groenewold, R.F. Pico, and K.S. Watson, Comparison of BOD relationships for typical edible and petroleum oils, *Journal WPCF*, **54**(4), 398–405, 1982.
48. *BETZ Handbook of Industrial Water Conditioning*, 6th edn, BETZ Laboratories, Inc., Trevose, PA, 1962, 427 pp.
49. E.S. Snavely and F.E. Blount, Rates of reaction of dissolved oxygen with scavengers in sweet and sour brines, *Corrosion*, **25**(10), 397–404, 1969.
50. R.L., Miron, Removal of aqueous oxygen by chemical means in oil production operations, *Materials Performance*, **20**(6), 45–50, 1981.
51. V. Linek and V. Vacek, Chemical engineering use of catalyzed sulfite oxidation kinetics for the determination of mass transfer characteristics of gas–liquid contactors, *Chemical Engineering Science*, **36** (11),1747–1768, 1981.
52. Drew Chemical Corporation, *Principles of Industrial Water Treatment*, 1st edn, Drew Chemical Corporation, Boonton, NJ, 1977, p. 169.
53. E.S., Snavely Jr., Chemical removal of oxygen from natural waters, *Journal of Petroleum Technology*, **23**, 443–446, 1971.
54. R.W., Edwards, M. Owens, and J.W. Gibbs, Estimates of surface aeration in two streams, *Journal of the Institution of Water Engineers*, **15**, 395–405, 1961.
55. Suggested Water Quality Limits, American Society of Mechanical Engineers (ASME) Research Committee on Steam and Water in the Thermal Power Systems. Boiler type: industrial water tube, high duty, primary fuel fired, drum type; makeup water percentage: up to 100% of feedwater; conditions: includes superheater, turbine drives, or process restriction on steam purity; saturated steam purity target. Quoted by such sources as Ref. [33, p. 99].
56. D.E., Simon II, Feedwater quality in modern industrial boilers—a consensus of proper current operating practices, Proceedings of the 36th International Water Conference, Engineers Society of Western Pennsylvania, Pittsburgh, PA, 1975, pp. 65–69, including prepared discussion.
57. R., Kunin, *Ion Exchange Resins*, Krieger, Huntington, NY, 1972, pp. 125–135, 388–389.

Chapter 6

Water/Wastewater Hydraulics

There is no such thing as spare time or extra money.

6.1 MEASUREMENT OF EFFLUENT FLOW

For an entity holding a Federal National Pollutant Discharge Elimination System (NPDES) wastewater discharge permit, a state wastewater discharge permit, or an industrial wastewater discharge permit issued by a local sewage authority, it is necessary to compile daily discharge data on what is commonly known as a *discharge monitoring report* (*DMR*) or equivalent. Typically, the DMR is filled out on a monthly basis, and the DMRs for every three calendar months are submitted quarterly to the governing regulatory agency.

Measured wastewater concentrations are multiplied by measured flows (and an appropriate conversion factor) to come up with mass flow rates of contaminant species. Very often, the wastewater is discharged by gravity into some sort of open channel, drainage ditch, or partially full pipe/conduit, where flow is measured. According to EPA's NPDES Compliance Inspection Manual [1, p. 6–1], open-channel flow is the most prevalent type of flow at NPDES-regulated discharge points. This flow can be measured by any of several techniques/devices, as discussed in this chapter. Techniques for measurement of liquid flow under pressure are treated elsewhere [2].

6.1.1 The Weir

Perhaps the simplest open-channel flow measuring device is the weir. A weir is an abrupt restriction to flow in an open channel. The liquid is forced to flow over the restriction, and the effluent flow rate is a function of the measured height of liquid above this sharp-edged obstruction, or weir plate. A steel weir plate can be mounted across a channel in a concrete structure, for example, or welded onto the end of an effluent pipe. The height of liquid can be measured manually with a calibrated

---

Environmental Calculations: A Multimedia Approach, by Robert G. Kunz
Copyright © 2009 John Wiley & Sons, Inc.

staff gage (a fancy name for a ruler) or a hook or point gage, or sensed using an electronic device. The weir itself uses no power and has no moving parts to service but is subject to a buildup of solids when suspended particulate matter settles out behind the weir.

The shape of the open area for flow may be either rectangular (simple or modified) or triangular. Rectangular weirs are used for measuring higher rates of flow, whereas triangular weirs are used for measuring lower flow rates ($\leq$90 gal/min, gpm), where increased sensitivity and greater accuracy are important [3]. Disadvantages of all types of weirs are high head loss and buildup of solids settling out behind the weir plate [4], as noted above. Textbooks on fluid flow or hydraulics may be consulted for typical weir configurations and construction details. Much useful information is provided in Ref. [5].

6.1.1.1 The Rectangular Weir

A general formula for flow over a sharp-edged rectangular weir is

$$Q = CWH^{3/2} \qquad (6.1)$$

where W is the horizontal width of the rectangular flow over the weir (ft), H is the height/head of water above the horizontal edge (crest of weir) (ft), and C is an empirical coefficient (ft$^{1/2}$/s), numerically in the range of $3\pm$ [3] when English units are used in the computations.

Caution: This coefficient is not dimensionless [6, p. 552], and it takes on values of order of magnitude 2 m$^{1/2}$/s in metric units. This constant accounts for the contraction of the liquid through the weir's open area, its viscosity, and the surface tension [7, p. 476]. The sheet of water flowing over the weir is termed the *nappe* [6, p. 550; 8, pp. 138–139; 9, p. 366; 10, p. 401].

One equation, the Francis formula in English units [8, p. 141; 10, p. 403]

$$Q = 3.33WH^{3/2} \qquad (6.2)$$

or in metric units [10, p. 403]

$$Q = 1.84WH^{3/2} \qquad (6.3)$$

is based on precise observations during a 4-year period in the mid-1800s over large-size weirs [8, p. 141]. Subsequently, accuracy has been found to be within 1–3% for heads above 0.3 ft but about 7% too low when head is 0.1 ft or less [8, p. 141].

Graphs, for example [4], show the effluent versus head relationship for rectangular weirs. However, for the most careful, accurate work, the weir should be calibrated in place and rechecked periodically using a "bucket and stopwatch" approach or other method [4,11]. Other methods include tracer dilution and pitot tube techniques.

PROBLEM 6.1

Using the Francis formula, calculate the flow rate of wastewater over a weir occupying the entire width of a channel 2 ft wide, if the head of liquid over the top edge of the weir plate is 4 in.

SOLUTION From Equation (6.2),

$$Q = (3.33)(2)(4/12)^{3/2} = 1.28 \text{ ft}^3/\text{s (cfs)}$$
$$(1.28 \text{ ft}^3/\text{s})(7.48052 \text{ gal/ft}^3)(60 \text{ s/min}) = (1.28)(448.83) = 575 \text{ gal/min (gpm)}$$

The product of 7.48052 times 60 = 448.83 can be used to convert cfs into gpm directly.

For a 2 ft wide rectangular weir of this type, the minimum head is 0.2 ft (2.4 in.) and the recommended flow rate is between 0.56 and 6.66 cfs [5, p. 33]. For this type and other weirs, the minimum flow recommendation is to prevent the nappe from clinging to the weir [5, pp. 29 and 32]. ∎

PROBLEM 6.2

Repeat Problem 6.1 if the channel is wider than the weir, and end contractions of the liquid flow over the weir must be considered.

SOLUTION When the width of the weir (W) is greater than three times the height (H) of liquid over the weir, but the weir does not extend all the way across the entire channel, the Francis formula can still be used. However, the effective width of the weir must be reduced by $H/10$ for each side of the weir not placed along a side wall of the channel; limiting distance from the end of the weir to the side wall of the channel is $2H$ [6, p. 552; 8, p. 142; 9, p. 370; 10, p. 403]. For these situations, the configuration is known as a *contracted weir*.

For a weir centered in the channel, W is decreased by $(2)(H/10)$.

$$W = 2 - (2)(4/12)(1/10) = 1.9333$$

For a weir located along one side wall, W is decreased by $(1)(H/10)$.

$$W = 2 - (1)(4/12)(1/10) = 1.9667$$
$$Q = (3.33)(1.9333)(4/12)^{3/2} = 1.24 \text{ cfs} = 556 \text{ gpm (for two end contractions)}$$
$$Q = (3.33)(1.9667)(4/12)^{3/2} = 1.26 \text{ cfs} = 566 \text{ gpm (for one end contraction)}$$

Alternatively, for two end contractions, the Gourley and Crimp formula [3; 8, p. 143]

$$Q = 3.10 W^{1.02} H^{1.47} \tag{6.4}$$

can be used without restriction regarding the minimum value of W/H.

$$Q = (3.10)(2)^{1.02} H^{1.47}$$
$$Q = (3.10)(2.0279)(0.1989) = 1.25 \text{ cfs} = 561 \text{ gpm (for two end contractions)}$$

The Schoder equation [3]

$$Q = 3.00 W H^{3/2} \tag{6.5}$$

is also available to describe the discharge over a weir with modified end contractions

$$Q = (3.00)(2)(4/12)^{3/2} = 1.15 \, \text{cfs} = 518 \, \text{gpm}$$

For a 2 ft wide rectangular weir with end contractions, minimum head is likewise 0.2 ft (2.4 in.); its minimum and maximum flow rates are 0.584 and 5.99 cfs, respectively [5, p. 32]. ∎

6.1.1.2 The Cipolletti Trapezoidal Weir

A variation of the rectangular weir is the Cipolletti trapezoidal weir with side slopes 4:1 vertical to horizontal [6, p. 553]. Its purpose is to offset the lateral contraction at the sides of a rectangular weir. The flow formula in ft³/s is

$$Q = 3.367 W H^{3/2} \tag{6.6}$$

with W the width of the weir crest (ft) and H (ft) the height of the liquid above the top of the weir plate.

PROBLEM 6.3

Compare the flow from the Cipolletti Weir formula with flows calculated in Problems 6.1 and 6.2.

SOLUTION In this problem, the width at the top of the weir and the head of liquid over the weir have not changed. From Equation (6.6),

$$Q = (3.367)(2)(4/12)^{3/2} = 1.30 \, \text{cfs} = 582 \, \text{gpm}$$

For a Cipolletti weir with a 2 ft crest length, minimum head should be 0.2 ft (2.4 in.). Minimum and maximum recommended flow rates are 0.602 and 6.73 cfs, respectively [5, p. 36]. ∎

6.1.1.3 Triangular or V-Notch Weirs

The triangular weir is constructed in a similar manner from a sharp-edged steel plate with a triangular opening of vertex angle theta (θ). Values of θ can be found between 10° and 90°. Vertex angles of 60° and 90° are commonly used [3]. The triangular weir is capable of measuring small to reasonably large flows [6, p. 553].

The general formula for all V-notch weirs is [8, p. 143]

$$Q = C_d (8/15)(2g)^{1/2} H^{5/2} \tan(\theta/2) \tag{6.7}$$

where C_d is a dimensionless discharge coefficient and g is the gravitational constant (32.1740 ft/s² in English units (~32.2) or 9.80665 m/s² in metric units (~9.81)).

The discharge coefficient (C_d) decreases with increasing head; for heads from 0.2 to 2.0 ft, experimental data suggest that Q varies with H to a power somewhat less than 2.5 [8, p. 144].

Formulas for the 60° and 90° V-notch weirs are given in Equations (6.8) and (6.9) for English units [8, p. 144].

$$60°: \quad Q\,(\text{cfs}) = 1.44[H\,(\text{ft})]^{2.48} \quad (6.8)$$

$$90°: \quad Q\,(\text{cfs}) = 2.48[H\,(\text{ft})]^{2.47} \quad (6.9)$$

Slightly different equations are given elsewhere for 60° and 90° sharp-crested V-notch weirs [5, Chapter 8]. The exponent is 2.5 and the coefficients are 1.443 and 2.500, respectively. In addition, an equation for the 30° weir is listed there with a coefficient of 0.6760 and an exponent of 2.5.

The horizontal length of the weir plate is not a factor in these equations.

At least as a first approximation, equations for triangular weirs with other vertex angles can be obtained by proration of the 90° case. Since tangent $(90°/2) = 1$, the factors for other values of θ equal $(2.500)\tan(\theta/2)$. For example, the factor for $\theta = 60° = (2.500)(0.5779) = 1.443$, as mentioned above. For $\theta = 30°$, the factor $= (2.500)(0.2679) = 0.6699$, close to the 0.6760 coefficient from the cited reference; the same is the case for V-notch weirs of 22.5°, 45°, and 120° [5, Chapter 8].

Graphs for 60°, 90°, and the rarely employed [3] 30° V-notch weirs are provided in the literature [4]. Again, caution should be exercised in the selection of metric versus English units when using any given formulation. The basic error in the head/flow rate relationship is quoted as ±3–6% for a properly installed and maintained V-notch weir; with errors of ±5–10% or more for a poorly constructed or silted weir. A calibration *in situ* of the actual weir in use (with periodic recalibrations) may also be in order for a V-notch weir, as noted above for rectangular weirs.

PROBLEM 6.4

For a flow rate of 90 gpm, calculate the corresponding liquid head for water in (a) a 60° weir and (b) a 90° weir.

SOLUTION

$$90\,\text{gpm}/(448.83\,\text{gpm/cfs}) = 0.2\,\text{cfs}$$

By rearrangement of Equations (6.8) and (6.9),

(a) 60°: $\quad H^{2.48} = Q\,(\text{cfs})/1.44 = 0.1393$

$\quad\quad\quad\quad H = 0.4516\,\text{ft} = 5.4\,\text{in}.$

(b) 90°: $\quad H^{2.47} = Q\,(\text{cfs})/2.48 = 0.0809$

$\quad\quad\quad\quad H = 0.3612\,\text{ft} = 4.3\,\text{in}.$

These values are well within the tables of Ref. [5, Chapter 8].

Minimum head for these weirs is 0.2 ft (2.4 in.) and maximum head is 2.0 ft; minimum and maximum recommended flow rates are 0.026 and 8.16 cfs, respectively, for the 60° weir and 0.045 and 14.1 cfs for the 90° weir [5, p. 29].

For the same rate of discharge, note that the 60° weir provides a higher head to measure than the 90° weir. ∎

6.1.1.4 Other Types of Weirs

Other sharp-crested weirs of a specialized nature are described in the literature [5, pp. 37–38; 11, p. 318]. Except for the compound weir, these are beyond the scope of this study.

The compound weir consists of a small 90° V-notch set in the center of a larger rectangular or Cipolletti weir. This design acts as a V-notch weir for low heads/flows and as a combined V-notch and rectangular weir for high heads/flows.

Calculations for this type of weir often use the separate equations for each part and sum the results. The analysis is correct at low heads when the device is acting solely as a V-notch weir. However, flow is not totally predictable for heads above the horizontal weir crest, especially in the intermediate transition zone where liquid just begins to overflow the weir. Results from this calculation technique have not seen extensive experimental verification and calibration in the laboratory or *in situ* is therefore recommended [5, p. 38].

6.1.2 Flumes

An alternative to the weirs discussed above for measurement of open-channel flow is a flume. There are several different types of flumes, including

- Parshall
- Palmer–Bowlus
- Cutthroat
- H-type
- Leopold–Lagco
- Montana
- Repolge–Bos–Clemmens (RBC)
- San Dimas
- Trapezoidal

Details are available on the Internet.

Many of these were originally developed to measure irrigation water or agricultural runoff. Some are totally different designs; some are closely related design derivatives. This list is not comprehensive, and only the commonly employed Parshall flume and Palmer–Bowlus flume are briefly discussed here.

6.1.2.1 The Parshall Flume

This measurement device, originally called the *Venturi flume*, was developed in the 1920s by Dr. Ralph L. Parshall of the U.S. Soil Conservation Service [5, p. 51; 12, p. 75; 13]. Unlike the weir, it is a low head-loss device and is not susceptible to the buildup of solids [4]. For properly fabricated and installed flumes, the error in flow measurement can be as little as $\pm 3\%$ [14, p. 23].

The flume is a device that constricts the flow in an open channel such that the volume of flow is proportional to the height or depth of liquid measured at one or more places in the flume [14]. It consists of a converging section, a throat, and a diverging section. The throat has a steep slope downward, followed by an incline in the diverging section. Specifying the throat size fixes all other dimensions of the flume, and several manufacturers offer prefabricated flumes in plastic/fiberglass. Schematic diagrams of the Parshall flume are provided in various references [5,11,14–17].

For the so-called *free-flow* condition, the critical depth occurs in the throat section, a hydraulic jump occurs in the diverging section, and the liquid downstream is not high enough to affect the flow. A hydraulic jump is an abrupt rise in water surface marked by violent turbulence [8, pp. 265–266; 12, pp. 45–46]. It occurs frequently, for example, downstream of the spillway of a dam 12, p. 45]. It is desirable that a flume be designed for free flow [5, p. 47].

The opposite of free flow is *submerged flow*, where the downstream flow depth partially submerges the overfall from the high point of liquid in the flume [14, p. 4]. This condition is handled by making an additional depth measurement in the diverging section and/or through the use of correction factors, but the free-flow condition is preferred for quantitative measurements.

For properly dimensioned Parshall flumes, the flow can be described by

$$Q = CH^n \tag{6.10}$$

Q is the volumetric flow, C is the flow coefficient, and n is an exponent on the liquid head (H). H is the liquid depth measured in the converging section, one-third of the way from the upstream entrance to the junction of the converging section and the throat.

Table 6.1 lists the values of C and n for flumes having throat widths in the range of 1 in. to 50 ft.

Different values of C are to be used for English and for metric units; values of n are the same in both units.

In the table, the maximum flow capacity for each flume is also listed [14, p. 10]. This is indicated in other references as the *maximum free-flow capacity* [5, p. 56; 12, p. 73]. In those references, the minimum flow capacity is also listed. These minimums are only a few percent of the maximum flows and are not reproduced here.

This table can be summarized in equation form as follows, with Q in cfs and H in ft [12, p. 75]:

$$Q = 0.992 H^{1.547} \quad \text{for } W = 3 \text{ in.} \tag{6.11}$$

$$Q = 2.06 H^{1.58} \quad \text{for } W = 6 \text{ in.} \tag{6.12}$$

$$Q = 3.07 H^{1.53} \quad \text{for } W = 9 \text{ in.} \tag{6.13}$$

$$Q = (4)[W \text{ (ft)}] H^{1.522 W \text{ (ft)}^{0.026}} \quad \text{for } W = 12 \text{ in. to 8 ft} \tag{6.14}$$

$$Q = (3.6875)[W \text{ (ft)} + 2.5] H^{1.6} \quad \text{for } W = 10\text{-}50 \text{ ft} \tag{6.15}$$

The exponent of H in Equation (6.11) is rounded to two decimal places in the table. The exponent of H calculated from Equation (6.14) must be rounded to three decimal

Table 6.1 Free-Flow Values of C and n and Maximum Q for Parshall Flumes[a]

Throat width (w)		Parshall parameters			Max. Q (free-flow capacity)			
in. or ft	cm or m	C English[b]	C metric[c]	n both[d]	cfs	gpm	MGD	m³/s
1 in.	2.5 cm	0.338	0.060	1.55	0.2	90	0.13	0.006
2 in.	5.1 cm	0.676	0.121	1.55	0.5	200	0.32	0.010
3 in.	7.6 cm	0.992	0.177	1.55	1.1	490	0.71	0.031
6 in.	0.152 m	2.06	0.381	1.58	3.9	1750	2.52	0.110
9 in.	0.229 m	3.07	0.535	1.53	8.9	4000	5.75	0.250
1 ft	0.305 m	4.00	0.691	1.522	16.1	7230	10.4	0.456
1.5 ft	0.457 m	6.00	1.056	1.538	24.6	11,000	15.9	0.697
2 ft	0.610 m	8.00	1.429	1.550	33.1	14,900	21.4	0.937
3 ft	0.914 m	12.00	2.184	1.566	50.4	22,600	32.6	1.43
4 ft	1.219 m	16.00	2.954	1.578	67.9	30,500	43.9	1.92
5 ft	1.524 m	20.00	3.732	1.587	85.6	38,400	55.4	2.42
6 ft	1.829 m	24.00	4.518	1.595	103.5	46,500	66.9	2.931
7 ft	2.134 m	28.00	5.313	1.601	121.4	54,500	78.5	3.438
8 ft	2.438 m	32.00	6.115	1.607	139.5	62,600	90.2	3.950
10 ft	3.048 m	39.38	7.463	1.6	—	—	—	—
12 ft	3.658 m	46.75	8.859	1.6	—	—	—	—
15 ft	4.572 m	57.81	10.955	1.6	—	—	—	—
20 ft	6.096 m	76.25	14.450	1.6	—	—	—	—
25 ft	7.620 m	94.69	17.944	1.6	—	—	—	—
30 ft	9.144 m	113.13	21.439	1.6	—	—	—	—
40 ft	12.19 m	150.00	28.426	1.6	—	—	—	—
50 ft	15.24 m	186.88	35.415	1.6	—	—	—	—

[a] Based on Table 1 and Table A.2 of Ref. [1].
[b] Use this value of C and H in ft to obtain Q in cfs.
[c] Use this value of C and H in meters to obtain Q in m³/s; for w of 10 ft and higher, C metric in the table is calculated by multiplying the tabulated C English by $(0.028316839 \text{ m}^3/\text{ft}^3)$ times $(1/0.4038)1.6$; w in meters is calculated from w in ft by multiplying by 0.3048 m/ft.
[d] The value of n is dimensionless and is the same in both systems of units.

places to coincide with corresponding entries of n in the table. The coefficient determined from Equation (6.15) is rounded, if necessary, to the hundredths place after the decimal point to obtain the tabulated values.

To these equations, the following can be added:

$$Q = 0.338 W \text{ (in.)} H^{1.547} \quad \text{for } W = 1 \text{ and } 2 \text{ in.} \quad (6.16)$$

It is likely that the effluent flow rate Q for a field installation would actually be computed automatically from continuous head measurements. The liquid depth would be measured by mechanical or electronic means in a stilling well outside the main body of the flume. Table 6.1 and its associated equations would be used in the design phase to select the proper size Parshall flume for the application.

PROBLEM 6.5

(a) Determine the proper size Parshall flume to measure an effluent flow in the range of 300–400 gpm.

(b) Calculate the variation in liquid head corresponding to this range of flow.

SOLUTION

(a) From Table 6.1, the given flow is greater than the maximum rating of a 2 in. Parshall flume but within the capacity of a 3 in. flume. This decision should be reconsidered if the flow is expected to exceed 490 gpm at any given time. For highly variable flows, an equalization tank upstream might perhaps be considered. For an actual project, obtain the flume manufacturer's input.

(b) Rearranging Equation (6.11) and taking logarithms, one obtains

$$\ln H = (1/1.547)\ln[Q \text{ (cfs)}/0.992]$$

For 300 gpm,

$$\ln H = (1/1.547)\ln[(300 \text{ gpm})/(448.83 \text{ gpm/cfs})/0.992]$$
$$= (0.6464)(-0.3948) = -0.2552$$

Exponentiating,

$$H = \exp(\ln H) = \exp(-0.2552) = 0.7747 \text{ ft} = 9.30 \text{ in.} \qquad \blacksquare$$

Similarly, results are shown in Table 6.2 for other flows between 300 and 400 gpm. Response here is linear with a maximum deviation of $10.27 - 10.25 = 0.02$ at the midpoint.

For a 3 in. Parshall flume, allowable head ranges from 0.10 ft (1.2 in.) to 1.5 ft (18 in.) [5, p. 56].

6.1.2.2 The Palmer–Bowlus Flume

The Palmer–Bowlus flume was developed by Harold W. Palmer and Fred D. Bowlus of the Los Angeles (CA, USA) County Sanitation Districts in the mid-1930s [5, p. 58; 13]. These flumes have been in use there for over 60 years [18] and are an important flow measuring device in other municipal and industrial sewer systems [13]. This flume is particularly suited for installation in circular pipes and

Table 6.2 Summary of Flows and Heads Calculated for 3 in. Parshall Flume

Flow (gpm)	Calculated H (in.)	Linear H (in.)
300	9.30	9.30
325	9.79	9.775
350	10.27	10.25
375	10.74	10.725
400	11.20	11.20

culverts or rectangular conduits, although it can also be employed in many other different types of open channels [15,19]. It is compact enough to be contained in a standard sampling manhole.

Like the Parshall flume, the Palmer–Bowlus flume is composed of three sections: a converging section upstream, a contracted throat section in the middle, and a diverging section downstream. It also should be operated in the free-flow condition.

Flow entering the flume must not be turbulent (no "white water"); however, the velocity increases in the throat and gives rise to a small hydraulic jump, which properly occurs in the diverging section. Acceleration of the flow through the throat serves to prevent the accumulation of solids. For proper operation, slope of the approach pipe from approximately 10 pipe diameters upstream of the flume should not be more than 2%, the flume itself flat, and the downstream liquid depth should not be more than 85% of the upstream depth [18]. When properly designed and operated, this device exhibits a relatively low head loss and a flow measurement accuracy comparable to that of the Parshall flume [18]. However, its sensitivity is somewhat less.

These flumes are commercially available in sizes from 4 in. to 6 ft, with an upper size of 42 in. often quoted. This dimension represents the diameter of the pipe into which it fits, rather than the throat width for a Parshall flume [5, pp. 60–61; 15,17]. To maximize the metering accuracy, the flume is specified, however, with regard to expected flow rate rather than the size of the pipe, as, for example, in the case of a large sewer regularly not flowing full. This means that the flume can be smaller than the pipe in which it is installed, but a larger flume should never be placed in a smaller size pipe [18].

The standard Palmer–Bowlus flume contains a throat of trapezoidal cross section [5,15,17]. This is the shape chosen by Palmer and Bowlus because of its greater accuracy at low flow and maximum flow at peak capacity [18]. A similar flume with a rectangular cross section is the Leopold–Lagco [5, p. 72].

Since the other dimensions of the flume are not standardized as they are for Parshall flumes, there is no standard flow equation, and one must rely on rating information and head-discharge tables/curves provided by the individual manufacturer [1, p. 6–12; 11]. Discharge characteristics supplied by another manufacturer are not to be used [1,11,20].

Table 6.3, based on an EPA publication [1, p. O-11] and manufacturer's data, summarizes maximum and minimum flow rates for flumes from one such manufacturer. This manufacturer has also made head-discharge curves marked "for estimating purposes" available on its website [18]. A decent fit for the 21 in. Palmer–Bowlus flume, for example, can be obtained from those tiny curves, using the form of Equation (6.10) with $C = 3$ and $n = 1.9$ (see Problem 6.6). This functional form is consistent with the expectations of at least one sewage authority for the information to be provided by the permittee during an inspection of an industrial wastewater monitoring station [20]. Flume discharge tables from the same manufacturer are provided in Ref. [5, pp. 64, 285–299]. It is to be noted that maximum and minimum flow data tabulated there are not completely consistent with Table 6.3 from Ref. [1], but agreement for the 21 in. flume between Equation (6.17) and those data is quite reasonable. Be careful!

Table 6.3 Minimum and Maximum Flow Rates for Free Flow Through Plasti-Fab® Palmer–Bowlus Flumes[a]

Flume size (in.)	Min. head (ft)	Minimum flow rate			Max. head (ft)	Maximum flow rate		
		cfs	gpm	MGD		cfs	gpm	MGD
6	0.11	0.035	16	0.023	0.36	0.315	141	0.203
8	0.15	0.074	33	0.048	0.49	0.670	301	0.433
10	0.18	0.122	55	0.079	0.61	1.16	521	0.752
12	0.22	0.198	89	0.128	0.73	1.83	821	1.18
15	0.27	0.334	150	0.216	0.91	3.18	1430	2.06
18	0.33	0.549	246	0.355	1.09	5.01	2250	3.24
21	0.38	0.780	350	0.504	1.28	7.44	3340	4.81
24	0.44	1.12	503	0.721	1.46	10.4	4670	6.70
27	0.49	1.46	655	0.945	1.64	13.8	6190	8.95
30	0.55	1.95	875	1.26	1.82	18.0	8080	11.6

[a] Abridged from Table O-6 of Ref. [1]; tabulated flows in cfs multiplied by 448.83 gpm/cfs and rounded to obtain the flows in gpm shown here.

A site-specific calibration can provide greater accuracy. An eight-page standard calibration procedure covers the measurement of volumetric flow rate of water and wastewater in sewers/open channels by means of a Palmer–Bowlus flume [19]. EPA expects a permitted facility to calibrate its flow measurement systems at least once a year [1, p. 6–3].

PROBLEM 6.6

Verify that the manufacturer's head-capacity curve [18] for a 21 in. Palmer–Bowlus flume can be described by the equation

$$Q \text{ (MGD)} = (3)[H \text{ (ft)}]^{1.9} \tag{6.17}$$

SOLUTION Several values of head and flow within the recommended minimum to maximum ranges noted in Table 6.3 [H (ft) between 0.38 and 1.28 and Q (MGD) between 0.504 and 4.81] are read from the manufacturer's curve. They are listed in Table 6.4 along with calculated values for flow. Greatest deviation found is within a maximum of about 3%. Although the (0, 0) point lies outside the recommended range, the calculated curve passes through the origin, as expected, since a zero-flow condition in the flume corresponds to a zero liquid head. ∎

6.2 FLOW IN RIVERS AND STREAMS

On a larger scale, flow in rivers and streams is determined by measuring or estimating the average velocity of flow and multiplying it by the cross-sectional area of the

Table 6.4 Summary of Flows and Heads Calculated for 21 in. Plasti-Fab Palmer–Bowlus Flumes

H (ft)	Q (MGD) from curve	Q (MGD) calculated	Deviation (%)
0	0	0.00	0
0.5	0.8	0.80	0
0.65	1.3	1.32	+1.5
0.8	2.0	1.96	−2.0
1.0	3.0	3.00	0.0
1.15	4.0	3.91	−2.3
1.26	4.8	4.65	−3.1

flowing body of water according to the continuity equation:

$$Q = AV \tag{6.18}$$

Methods include the pitot tube; current meters; observation of floating objects, dyes, and salt solutions; and calculation using such equations as the Chézy or Manning formula. As with any of these techniques, the contour of the stream normal to the flow must first be determined by manual measurements at multiple points close together from one stream bank to the other [3].

6.2.1 Pitot Tube

Point velocity measurements in the stream are made with a pitot tube, in a manner similar to measuring flue gas flow discussed in Chapter 3. Studies have shown that there is a vertical velocity distribution from the free liquid surface down to the stream bed [7, p. 457], and the average velocity in a vertical section is closely approximated by the point velocity at the 20% point measured from the liquid surface and even more closely by the arithmetic average of the 20% and 80% points. The U.S. Geological Survey (USGS) recommends that the stream be divided into a number of vertical strips and that the point velocity measured at the 20% point (or more accurately the average of the 20% and 80% points) for each strip be multiplied by the area of each vertical strip. The results are then summed over all such vertical strips to obtain the total discharge [7, p. 458]. Each vertical strip should contain less than 10% of the vertical flow [3].

6.2.2 Current Meter

A current meter is a streamlined device weighted and suspended in a stream. At one end, it contains a propeller, whose rate of rotation is calibrated to the velocity of flow. This is the standard method for river gauging in the United States [7, p. 472]; simultaneously, the depth (stage) of water in an adjacent stilling well connected to the stream is measured and recorded by means of a float arrangement connected to a depth indicator [7, p. 472]. The stilling well dampens out any instantaneous fluctuations in water depth in the main stream. Positioning of the current meter in the stream is the same as that of the pitot tube [3].

PROBLEM 6.7

The average velocity of a flowing stream has been found to be 2 ft/s. Its cross-sectional area approximates a trapezoid with side slopes of 1:2 vertical to horizontal, a depth of 2.5 ft, and a width of 15 ft at the free liquid surface. Estimate discharge (Q) in cubic feet per second (ft^3/s, cfs), gallons per minute (gpm), and cubic meters per hour (m^3/h).

SOLUTION Draw a sketch to follow along. From geometry, area of a rectangle 15 ft wide by 2.5 ft high is $15 \times 2.5 = 37.5$ ft^2. Subtract the area of two triangles 2.5 ft high by 5 ft wide $= (2)(2.5)(5)/2 = 12.5$ ft^2. Area of flow $= 37.5 - 12.5 = 25$ ft^2. (Width of the flat section of stream bed is 5 ft.)

From Equation (6.18)

$$Q = AV = (25)(2) = 50 \text{ ft}^3/\text{s (cfs)}$$

Conversion of units:

$$(50 \text{ cfs})(448.83 \text{ gpm/cfs}) = 22,442 \text{ gpm}$$
$$(50 \text{ cfs})(3600 \text{ s/h})(0.3048 \text{ m/ft})^3 = 5097 \text{ m}^3/\text{h}$$

∎

6.2.3 Tracer Techniques

This technique assumes that the average flow is uniform between two points along a stream. Therefore, the velocity (V) is given by

$$V = d/t \tag{6.19}$$

where d is the measured distance between two points in the flow direction and t is the elapsed time between the release at the upstream point and its appearance at the downstream location. These are known as time-of-travel methods and measurements, further discussed in Appendix G.

6.2.3.1 Chemical Tracers

Various fluorescent dyes, radioactive tracers, or salt solutions can be used. As in the use of tracers for groundwater monitoring, possible health implications must be considered and permits/approvals from governmental agencies may be required [21, p. 127].

Fluorescent dyes are detected by means of a fluorometer/fluorimeter [5, p. 7; 22, p. 12]. The greenish-yellow fluorescein can be detected in water at 1 part per 50 million parts and Rhodamine B at 1 part in 10^{11} [22, p. 12]. Rhodamine B is bluish-red with a strong yellow fluorescence [22, p. 12]. Radioactive tracers can be detected using a Geiger counter [5, p. 7; 22, p. 12].

Salt solutions, including sodium chloride, can also function as tracers. Lithium chloride [3] or lithium sulfate [22, p. 12] is often employed because lithium is not a common element of natural waters. The lithium is determined spectroscopically [22, p. 12] or by flame photometry [22, pp. 12 and 22].

Flow measurement with tracers involves two different techniques: the slug method, which makes use of Equation (6.19), and constant-rate injection [3,5, p. 7]. In the latter technique, the tracer is injected at a known constant rate, and the main flow is sampled at a location downstream after thorough mixing. This procedure can be used to measure flow in a stream or to calibrate a fixed flow measurement device such as a weir.

The following equation [3;22, p. 22] can be derived by material balance on the tracer and the water flow, assuming a constant density for all aqueous solutions:

$$Q = q(X - x_{out})/(x_{out} - x_{in}) \tag{6.20}$$

where Q is the flow in the main stream, q is the constant flow of the tracer solution in the same units as the main flow, X is the concentration of the tracer in solution, and x_{in} and x_{out} are, respectively, the upstream and downstream concentrations of the tracer material in the main stream. All concentrations must also be in the same units.

However, flow units and concentration units need not be consistent; for example, metric and English units may be mixed.

Typically, the upstream concentration of tracer will be zero, allowing simplification of Equation (6.20).

PROBLEM 6.8

A compound weir is to be calibrated at high flow rates. An excerpt of the test program is discussed below.

The flow and head are varied as a 1 wt% (~10,000 ppm) tracer solution of lithium sulfate is added at a constant rate of 1 gpm. No lithium is detected upstream of the weir. It is measured at some distance downstream after complete mixing has occurred. At steady-state conditions for four such runs, downstream concentration was measured at 40, 28, 25, and 20 ppm (mg/L), expressed as lithium sulfate. (a) Calculate the flow rate of water passing over the weir. (b) Compare the results when the density of the tracer solution is taken into account.

SOLUTION

(a) For zero upstream concentration, Equation (6.20) becomes

$$Q = q(X - x_{out})/x_{out} \tag{6.21}$$

$Q \text{ (gpm)} = (1 \text{ gpm})(10,000 \text{ ppm} - 40 \text{ ppm})/40 \text{ ppm} = 249 \text{ gpm}$

Similarly, $Q = 356.1$, 399, and 499 gpm at the other tracer concentrations measured.

(b) Specific gravity (sg) of a 1 wt% solution of Li_2SO_4 is 1.0068 (20°C/4°C) [23, p. 10–135]. Density of water at 4°C is 0.99973 [23, p. 10–127]. Therefore, density (ρ) of 1 wt%

$Li_2SO_4 = (1.0068)(0.99973) = 1.006772816$. In comparison, sg of LiCl is 1.0041 [23, p. 10–135].

$$\frac{1 \text{ g } Li_2SO_4}{100 \text{ g solution}} \cdot \frac{1.006772816 \text{ g solution}}{\text{mL solution}} \cdot \frac{1000 \text{ mL}}{L} \cdot \frac{1000 \text{ mg}}{g} = \frac{10,067.7282 \text{ mg}}{L}$$

Substituting in Equation (6.21) results in flows of 250.7, 358.6, 401.7, and 502.4 gpm. Deviation is 0.7% across the board. ∎

6.2.3.2 Floating Objects

With floating objects, Equation (6.19) is modified to account for the observed ratio of 1.2 for the surface velocity to average velocity in an open channel [3].

$$V = (d/t)/1.2 \tag{6.22}$$

Many years ago, an "old hand" at this sort of measurement told the author that oranges are a favorite floating object for use in open-channel flow studies. They are roughly spherical, maintaining the same orientation to the flow in any position, and are sufficiently neutrally buoyant to float with a majority of their bulk just below the surface. The buoyancy observation can be easily verified in a "kitchen experiment." Try it!

PROBLEM 6.9

Hypothetically, 10 numbered, nonpolluting floating objects are released 60 s apart into a uniform reach of a natural watercourse and are retrieved 1 mile downstream. (These represent no hazard to navigation.) The average cross-sectional area of the stream is 25 ft^2 (Problem 6.7). Estimate the average stream velocity and volumetric flow from Table 6.5.

Table 6.5 Elapsed Times for Hypothetical Floating Object Experiment

Object no.	Time of release	Time of retrieval	Elapsed time (t, s) (by difference)	Average V (ft/s), Equation (6.22)
1	1:05 p.m.	1:40:57 p.m.	2157	2.04
2	1:06 p.m.	1:43:25 p.m.	2245	1.96
3	1:07 p.m.	1:44:14 p.m.	2234	1.97
4	1:08 p.m.	1:44:40 p.m.	2200	2.00
5	1:09 p.m.	1:46:02 p.m.	2222	1.98
6	1:10 p.m.	1:45:36 p.m.	2136	2.06
7	1:11 p.m.	1:47:18 p.m.	2178	2.02
8	1:12 p.m.	1:49:36 p.m.	2256	1.95
9	1:13 p.m.	1:49:07 p.m.	2167	2.03
10	1:14 p.m.	1:50:51 p.m.	2211	1.99
	Total	—	22,006	20.00
	Mean	—	÷10 = 2200.6	÷10 = 2.00

Time-of-travel for the stream = 5280/2 = 2640 s ÷ 60 = 44.0 min (Equation (6.19)). Numbering is important because Object 6 showed up at the finish line before Object 5, and Object 9 before Object 8.

SOLUTION Sample calculation from Equation (6.22) for object no. 1,

$$\text{Average } V = [(1 \text{ mile} = 5280 \text{ ft})/(2157 \text{ s})]/1.2 = 2.04 \text{ ft/s}$$
$$1:40:57 \text{ p.m.} - 1:05 \text{ p.m.} = 60(35 \text{ min}) + 57 \text{ s} = 2157 \text{ s}$$

Overall average V = mean V above or $=[5280/2200.6]/1.2 = 1.9995 \cong 2$ ft/s. From Equation (6.18), volumetric flow or discharge $(Q) = (25)(2) = 50$ cfs (see Problem 6.7). ∎

6.2.4 Calculation of Open-Channel Flow

In 1775, Antoine Chézy, a French engineer, proposed an equation for the average velocity (V) of uniform flow in an open channel, based on theoretical considerations [7, p. 451]. The formula is the same in metric and English units.

$$V = C(RS)^{1/2} \tag{6.23}$$

where C is a parameter related to surface roughness of the channel bed and the gravitational constant, R is the hydraulic radius, the wetted area of flow divided by the wetted perimeter, and S is the slope of the channel in the flow direction.

It is difficult *a priori* to choose a correct numerical value representative of surface roughness in man-made channels and open streams, and work continued in the laboratory and in the field to resolve the issue.

In 1891, Robert Manning, an Irish engineer, came up with one of the several empirical formulations that were developed to describe open-channel uniform flow [5, p. 112; 7, p. 453]. This was neither the first nor the only such equation. However, it is widely employed because it is simple and easy to use with no trial and error, and accurate enough in practice [7, p. 453; 8, p. 237]. In metric units, with the same definition of V, R, and S as above,

$$V \text{ (m/s)} = (1/n)R^{2/3}S^{1/2} \tag{6.24}$$

and in English units

$$V \text{ (ft/s)} = (1.486/n)R^{2/3}S^{1/2} \tag{6.25}$$

The factor in the denominator is known as Manning's n. To four significant figures, 1.486 is the cube root of 3.28084, the number of feet in a meter. This conversion factor is included in the formula so that Manning's n retains the same numerical value in both systems of units.

Equations (6.23) and (6.25) are equivalent if the factor C in the Chézy equation is taken as the quantity $(1.486R^{1/6})/n$ in Manning's equation [7, p. 453; 12, p. 100].

Manning's n, the roughness coefficient, is a measure of channel roughness, tabulated in references describing open-channel hydraulics. It ranges from about 0.010 for a smooth surface to 0.025–0.150 for the roughest natural streams containing vegetation [6, p. 425; 7, p. 452; 12, pp. 104–113]. It is an empirical constant, varying also with such factors as the size of the stream, channel curvature, abrupt changes in the channel, obstructions, suspended material, time of year, stage and discharge (depth and flow), and other conditions [12, pp. 101–108]. This same reference contains photographs of typical channels exhibiting different values of n [12, pp. 115–123]. These are useful as a guide in selecting an appropriate value of n for use in calculations.

PROBLEM 6.10

Calculate the hydraulic radius (R) for the stream described in Problem 6.7.

SOLUTION Hydraulic radius is defined as the wetted area (A_w) divided by the wetted perimeter (P_w).

$$R = A_w/P_w \qquad (6.26)$$

Wetted area has already been calculated in Problem 6.7 as 25 ft^2.
From the geometry of the stream cross section,

$P_w = 5$ ft (bottom surface) $+ 2 \times b$ ft (two times the length of submerged bank)

The wetted perimeter does not include the free surface [8, p. 163].
From trigonometry, acute angle of bank (b) with horizontal (α) $= \tan^{-1}(1/2)$.

$$\alpha = 26.57°$$
$$\sin \alpha = 0.4472 = 2.5/b$$
$$b = 2.5/0.4472 = 5.59 \text{ ft}$$
$$P_w = 5 + (2)(5.59) = 16.18 \text{ ft}$$

From Equation (6.24), $R = 25/16.18 = 1.545$ ft. ∎

PROBLEM 6.11

Calculate the average velocity and the volumetric flow rate for the stream described in Problem 6.7 for a range of Manning's n from 0.010 to 0.150. Slope is 6.5 ft in 1 mile.

SOLUTION

$$R = 1.545 \text{ ft (Problem 6.10)}$$
$$S = 6.5 \text{ ft}/[(1 \text{ mile})(5280 \text{ ft/mile})] = 1.23 \times 10^{-3} \text{ ft/ft}$$
$$n = 0.010$$

From Equation (6.25),

$$V = (1.486/0.010)(1.545)^{2/3}(1.23 \times 10^{-3})^{1/2} = 6.97 \text{ ft/s}$$

From Equation (6.18),

$$Q = (25)(6.97) = 174.2 \text{ cfs}$$

Similarly, for $n = 0.150$,

$$V = (1.486/0.150)(1.545)^{2/3}(1.23 \times 10^{-3})^{1/2} = 0.46 \text{ ft/s} \quad \text{and} \quad Q = 11.6 \text{ cfs}$$

Percent deviation here around the average value is about 50%. "Careful" use of the Manning equation in the field is said to result in errors of 10–20% in flow rate; "less careful" use possibly results in errors of 20–50% or more [5, p. 106]. It is therefore important to be able to narrow down the value of Manning's n. ∎

PROBLEM 6.12

Assuming a relatively straight uniform channel for its entire reach, (a) back calculate Manning's n for the stream whose properties are given in Problems 6.7 and 6.9–6.11. The stream drops 6.5 ft in elevation over a mile in linear distance. (b) What is the corresponding description of this type of channel?

SOLUTION

Inputs: $V = 2$ ft/s (see Problems 6.7 and 6.9)

$R = 1.545$ ft (Problem 6.10)

$S = 6.5/[(1)(5280)] = 1.23 \times 10^{-3}$ ft/ft

Average time-of-travel $= 2200.6$ s (Problem 6.9)

Rearranging Equation (6.3) (English units) yields

$$n = (1.486/V)R^{2/3}S^{1/2}$$

$$n = (1.486/2)(1.545)^{2/3}(9.47 \times 10^{-4})^{1/2} \qquad (6.27)$$

$$n = (0.743)(1.3364)(0.0351) = 0.035$$

A Manning's n of 0.035 corresponds to a photograph showing a straight natural channel. As noted in its caption, it has a fairly even, clean, and regular bottom in silty clay to silt loam, with very little variation in cross section [12, p. 120].

Use of Manning's equation is discussed further in Section 6.6 and in the case study of Appendix G. ∎

PROBLEM 6.13

(a) The stream in the previous problem is swollen to a depth of 3.5 ft as a result of runoff from spring rains. Assuming that the stream remains within its banks and that Manning's n remains the same at 0.035, what are the average velocity, discharge, and time-of-travel?

(b) Repeat the calculation for a low-water drought depth of 1.5 ft.

This kind of approach, in which the Manning formula is first calibrated to a flow measured by some other method, is discussed at length in Ref. [5, pp. 106–108]. One must exercise caution in that Manning's n may vary with stage and discharge [5, pp. 102 and 106; 12, pp. 104–106].

6.2 Flow in Rivers and Streams

SOLUTION

(a) The values of slope and Manning's n remain the same. The values of A_w and P_w change. Refer to Problem 6.9 and prepare a sketch.
From similar triangles, $5(3.5/2.5) + 5(3.5/2.5) + 5 = 19$ ft length of free liquid surface.

$$\text{Length of one submerged bank} = (3.5)/0.4472 = 7.83 \text{ ft}$$
$$P_w = 5 + 7.83 + 7.83 = 20.66$$
$$A_w = (19)(3.5) - (2)(3.5)(7)/2 = 66.5 - 24.5 = 42 \text{ ft}^2 = A$$
$$R = A_w/P_w = 42/20.66 = 2.033 \text{ ft}$$

From Equation (6.23), $V = (1.486/0.035)(2.033)^{2/3}(1.23 \times 10^{-3})^{1/2} = 2.39$ ft/s.
From Equation (6.18) $Q = (42)(2.39) = 100.4$ cfs.
Time-of-travel for a 1 mile reach of the stream (Equation (6.19)) is

$$5280/2.39 = 2209.2 \text{ s} \div 60 \text{ s/min} = 36.8 \text{ min}$$

(b) Repeating the calculation for a depth of 1.5 ft,

width of stream at free surface = 11 ft
$$P_w = 11.71 \text{ ft}$$
$$A_w = A = 12 \text{ ft}^2$$
$$R = A_w/P_w = 1.025 \text{ ft}$$
$$V = (1.486/0.035)(1.025)^{2/3}(1.23 \times 10^{-3})^{1/2} = 1.51 \text{ ft/s}$$
$$Q = (12)(1.51) = 18.1 \text{ cfs}$$
$$t = 5280/1.51 = 3496.7 \text{ s} \div 60 \text{ s/min} = 58.3 \text{ min} \quad \blacksquare$$

Values were also calculated at depths of 2.0 and 3.0 ft. All results are summarized in Table 6.6.

A plot of calculated V versus calculated Q (not shown) resembles a similar graph for a real river as depicted in Figure G.14. A plot of depth versus flow rate (Q) (also not shown) looks like the generalized stage versus discharge rating curve of Ref. [21, p. 56].

For the particular example problem illustrated here, results for discharge (Q) versus depth (h) in the table below can be correlated very nicely using a second-degree polynomial, $Q = 8.30952h^2 - 0.397617h + 0$, with its constants derived from an exact fit at the end points ($h = 1.5$ and 3.5). A maximum deviation of 1.9% occurs at

Table 6.6 Stream Velocity and Flow Calculated from Manning's Equation

Depth (ft)	Flow Area (ft²)	V (ft/s)	Q (cfs)	t for 1 mile (min)
1.5	12	1.51	18.1	58.3
2.0	18	1.77	31.8	49.8
2.5	25	2.00	50.0	44.0
3.0	33	2.20	72.5	40.0
3.5	42	2.39	100.4	36.8

the midpoint ($h = 2.5$), with lesser percentage deviations found in between. The function passes through the origin (0, 0), as expected, but has not been investigated in detail below $h = 1.5$ ft, where the effects of differences between an actual stream bed and the assumptions made are more likely to become pronounced.

A similar mathematical or graphical approach with real data in lieu of the calculated values in the table above is another method discussed in Ref. [5, pp. 107–108].

6.3 MEETING WATER QUALITY LIMITS

There are basically three kinds of streams: ephemeral, intermittent, and perennial [21, p. 54]. The entire flow in an ephemeral stream is surface runoff, it may have no well-defined channel, and the bottom of the stream consistently remains above the water table.

Intermittent streams flow only part of the year, generally from spring to midsummer and during wet weather periods as well. Flow occurs during dry weather only for as long as the groundwater table lies above the base of the channel and the groundwater discharges into them; when the water table drops sufficiently, flow stops and the stream runs dry. Intermittent streams are often shown on topographical maps as (blue) dashed lines.

In contrast, perennial streams flow throughout the year. The water table lies above the bottom of the stream all year long. Flow is augmented during wet weather by surface runoff.

6.3.1 Low-Flow Conditions

Situations may arise when environmental regulations and/or one's wastewater discharge permit will require that water quality standards (concentrations) in the receiving stream not be exceeded, typically at the 7-day, 10-year low-flow condition ($7Q_{10}$). This statistic is defined as the lowest flow occurring in the stream for 7 consecutive days during a 12-month period, with a recurrence interval (RI) of 10 years. The recurrence interval in years is the decimal probability that the flow in the stream will be less than or equal to the stated flow. A 10-year recurrence interval corresponds to the 0.1 (10%) point of the statistical distribution; that is, only 10% of observed flow conditions lie at or below this value. Similarly, the $7Q_2$ is the 50% point.

In the United States, stream flow data, also known as *surface water records*, are measured and compiled by the U.S. Geological Survey and published in cooperation with the corresponding agency of the state in which the stream is located. In those cases where tabulated 7-day low-flow information is reported [24], one simply uses the figure in a mass balance calculation and compares the result to the applicable standard.

6.3.1.1 Calculation of the 7Q10

When the low-flow figure is not immediately available, it can be calculated from the daily mean flows, year by year, spanning a long period of record. The calculation is

Table 6.7 Low-Flow Data—Discharge (cfs)[a] Little Beaver Creek Near East Liverpool, Ohio (Tributary to The Ohio River)

		Recurrence interval (years)				
		2	5	10	20	50
	Number of	Probability (less than or equal to)				
Period	consecutive days	50%	20%	10%	5%	2%
April–March	1	29	20	17	15	12
	7	33	22	**19**	**16**	**14**
	30	46	32	27	23	20

Drainage area: 496 square miles. Average discharge: 517 cfs October 1915–September 1978 (63 years). Minimum daily discharge: 12 cfs on several days in 1918 and 1932.

[a] Excerpted and modified from Appendix 3, p. 65 of Ref. [24].

tedious and will only be outlined here. An example for the stream listed in Table 6.7 is given in Ref. [24, pp. 4–6] using observations from each year of over 60 years of flow records. A probability plot from which the recurrence interval for 7-day low-flow data can be read is presented in the cited reference.

That graph is prepared by first

- listing the lowest daily flow value for the 7-day time period from each year,
- arranging the values in increasing order from smallest to largest, and
- assigning to each value an "order number" (m), an integer starting at 1 for the smallest up to n, the total number of values.

The RI in years is then computed from the formula

$$\text{RI} = (n+1)/m \quad (6.28)$$

Each flow value is then plotted against its computed recurrence interval as a cumulative distribution on logarithmic probability paper (discussed in Chapter 2), and a smooth curve is drawn through the plotted points. (The curve in the referenced example resembles an exponential decay curve plotted on arithmetic coordinates.) An equation for a Log Pearson Type III frequency distribution is also given in the reference. To reiterate a previous statement here, probability expressed as a decimal is the reciprocal of the recurrence interval in years, and vice versa.

Results for 7-day flows ($7Q_2$, $7Q_5$, $7Q_{10}$, $7Q_{20}$, and $7Q_{50}$) as read from the log-probability curve are shown in bold type in Table 6.7, prepared from a more extensive tabulation in the original reference. The other figures in the table, 1-day and 30-day low flows, have been likewise computed by the USGS from flow data during each year (a) for the single day of lowest flow and (b) for the 30 consecutive days of lowest flow.

Obtaining a permit to discharge into a perennial stream where water quality standards are an issue makes use of the aforementioned $7Q_{10}$ and a material balance as in Equation (6.20). That equation can be rearranged to yield

$$x_c = [(q/Q)x_d + x_s]/[(q/Q) + 1] \quad (6.29)$$

where x_c is the resulting pollutant concentration in the receiving stream at the low-flow condition, x_d is the pollutant concentration in the discharge, x_s is the pollutant concentration already in the stream (may or may not be zero), and q and Q are the effluent flow and the stream flow (at low flow), respectively.

Use of this equation is explored in the following example problem.

PROBLEM 6.14

In this hypothetical example, let us assume that a water quality limit of 0.1 mg/L for a total phosphorous (P) concentration is of concern. The upstream P concentration is 0.08 mg/L and its concentration in the effluent is 1 mg/L. Calculate its concentration in the flowing stream downstream of the mixing zone if the effluent flow is (a) 1% and (b) 5% of the stream.

SOLUTION

(a) From Equation (6.29),

$$x_c = [(0.01)(1) + 0.08]/[(0.01) + 1] = 0.09 \text{ mg/L P} < 0.1 \text{ mg/L}$$

Discharge is feasible.

For a $7Q_{10}$ of 19 cfs (8528 gpm), effluent flow is

$$(0.01)(19)(448.83) = 85.3 \text{ gpm}$$

(b)

$$x_c = [(0.05)(1) + 0.08]/[(0.05) + 1] = 0.12 \text{ mg/L}$$

Water quality standard is exceeded at the effluent flow of

$$85.3(0.05/0.01) = 426 \text{ gpm}$$

When the concentration of P in the stream is already in excess of 0.1 mg/L, discharge limits for P will have to be negotiated. Good luck. ∎

6.3.2 Discharge Under No-Flow or Low-Flow Conditions

When wastewater is discharged into a public watercourse during a *no-flow* condition, the effluent becomes the stream. Contaminant concentrations in the discharge are then the same as the ambient water quality. When water quality standards are exceeded without benefit of dilution, discharge is prohibited. Alternatives include treatment prior to discharge or perhaps running a closed discharge pipe all the way to the nearest perennial stream.

6.4 GROUNDWATER FLOW

Assessment of groundwater quality requires knowledge of groundwater flow to follow the underground movement of contaminants. Groundwater movement is described by Darcy's law:

$$Q = KIA \tag{6.30}$$

where Q is the quantity of flow per unit of time (gallons/day, gpd); K is the hydraulic conductivity (gpd/ft^2), also expressed in units of ft/s and cm/s; I is the hydraulic gradient, differential head loss (Δh) measured between two points L ft apart (ft/ft); and A is the cross-sectional area through which the flow occurs (ft^2).

Darcy's law describes laminar flow and is a simplification of a more general expression [25]. Use of this equation is illustrated by the following example problem, taken from an EPA handbook [21, p. 76].

PROBLEM 6.15

A sand aquifer about 30 ft thick lies within the flood plain of a river that is about a mile wide. The aquifer is covered by a confining unit of glacial till, the bottom of which is about 45 ft below the land surface. The difference in water level in two wells a mile apart is 10 ft. The hydraulic conductivity of the sand is 500 gpd/ft^2. Find the quantity of underflow passing horizontally through the sand layer.

SOLUTION Use Equation (6.30) with the following inputs for the sand layer:

$$K = 500 \text{ gpd/ft}^2 \text{ (given)}$$
$$I = 10 \text{ ft}/[(1 \text{ mile})(5280 \text{ ft/mile})]$$
$$A = (30 \text{ ft})(1 \text{ mile})(5280 \text{ ft/mile})$$

Using Equation (6.30),

$$Q = (500 \text{ gpd/ft}^2)(10 \text{ ft}/5280 \text{ ft})[(30 \text{ ft})(5280 \text{ ft})]$$
$$= (500)(10)(30) = 150,000 \text{ gpd} \qquad \blacksquare$$

An additional step beyond the example in the EPA manual is equally instructive. Superficial (empty space without particles) velocity within the aquifer in the flow direction is

$$150,000 \text{ (gal/day)}/[(7.48052 \text{ gal/ft}^3)(5280 \text{ ft})(30 \text{ ft})(1440 \text{ s/day})] = 8.8 \times 10^{-5} \text{ ft/s}$$

Actual velocity in the interstitial voids would be about double this value, depending on the actual void fraction of the soil. Distance traveled is obtained by multiplying elapsed time by velocity. This type of analysis is used to trace an underground plume of contaminated groundwater resulting from an accidental release of an environmental pollutant.

The term *soil texture* characterizes soils by particle sizes [26, pp. 113–116]. The U.S. Department of Agriculture (USDA) classifies soils by the designations listed in Table 6.8.

The texture of a given soil sample is categorized on triangular graph paper by the percentage of "clay," "silt," and "sand" as defined in the table. Subareas on the triangular grid are denoted by such terms as *sand, clay, sandy loam, silty clay,* and *silt loam,* in addition to the terms *clay, sand,* and *silt,* whose areas of designation are located near the vertices of the triangle. For example, a soil analysis with 35% 0–2 μm "clay," 35% 2–50 μm "silt," and 30% 50–200 μm "sand" would fall within the area designated as *clay loam* on a USDA chart.

Table 6.8 Soil Texture Names/Designations

	Particle size diameter	
Designation	Range: millimeter (mm)	Range: micrometers (μm)
Clay	0–0.002	0–2
Silt	0.002–0.05	2–50
Sand	0.05–2	50–2000
Very fine	0.05–0.1	50–100
Fine	0.1–0.25	100–250
Medium	0.25–0.5	250–500
Coarse	0.5–1	500–1000
Very coarse	1–2	1000–2000
Gravel	2–70	2000–70,000

Hydraulic conductivity (K) varies over a wide range (orders of magnitude) for different materials and even for the same material. In general, K increases with increasing particle diameter. The value given in the problem above lies within the range quoted for sand and gravel in several other sources. Because of the broad range in K for natural materials, an experimental determination of K in the laboratory for the actual soil(s) under investigation is essential. A simultaneous particle size analysis would also be helpful in evaluating the results.

6.5 STORM WATER CALCULATIONS

This section is included to illustrate typical calculations to characterize the flow of water, as rainfall is transformed into runoff. Many jurisdictions are quite specific about how the calculations are to be done and also about the design features of storm water collection, conveyance, detention, treatment, and release. In these cases, one must follow prescribed procedures, including seeking waivers and variances for good cause.

6.5.1 The Rational Method

There are several methods of storm water assessment. One such summary is provided in Ref. [26]. The only one discussed here is the so-called rational method to estimate peak flow from small drainage areas, up to perhaps 100 acres [26, p. 205]. One jurisdiction limits its use to 10 acres for a single peak flow calculation [27, p. 3–11]. There may be others.

The basic equation is

$$Q = CIA \tag{6.31}$$

where Q is the peak flow (ft^3/s, cfs), C is the runoff coefficient (dimensionless), I is the rainfall intensity (in./h), and A is the drainage area (acres).

The units are consistent since the embedded conversion factor (1.008) is taken as being close enough to 1.0. The equation can also be used with an intensity converted to ft/s and dimensions in square feet to yield a peak flow in cfs.

The formula makes the following implicit assumptions:

- Uniform intensity over the "watershed."
- Duration of rainfall intensity equal to the time of concentration.
- Recurrence interval of peak flow equal to recurrence interval of rainfall intensity.

The time of concentration exemplified in Section 6.5.5 is the time required for runoff to travel from the farthest point in the drainage area to the point of collection [26, pp. 147–148]. This most remote point takes the distance, slope, and surface roughness into account. It evaluates runoff transport by sheet flow, concentrated flow, and channel flow. The time of concentration concept is to allow the creation of a peak flow from individual packets of water falling at different locations on the drainage path during a storm of time-varying intensity.

6.5.2 Coefficient

The coefficient C in Equation (6.31) is a measure of how much of the falling rain runs off versus how much infiltrates into the surface. A coefficient of 1.0 indicates complete runoff from a perfectly impervious surface; a coefficient of zero means that absolutely all of the rainfall is soaked up by the soil. In addition to the type of ground cover, factors such as soil composition/texture, degree of saturation from previous storms, slope of the land, rainfall intensity, and others are included in its value [26, p. 206]. Values of C by land use, soil type, surface slope, and storm recurrence/return period are tabulated in the literature, for example, Ref. [26, pp. 207–208]. A much shorter list of coefficients allowable for use in the cited jurisdiction [27, pp. 3–14] is summarized in Table 6.9. These values may not be

Table 6.9 Rational Method Runoff Coefficient $(C)^a$

Land cover	C
Forest	
Dense	0.10
Light	0.15
Pasture	0.20
Lawns	0.25
Playgrounds	0.30
Gravel areas	0.80
Pavement and roofs	0.90
Open water (ponds, lakes, wetlands)	1.00

^{a} Source: Ref. [27, p. 3–14].

completely consistent with those from other sources, but they are what they are. Be sure to use the approved values of C within your project's jurisdiction.

In addition, values of C ranging from 0.17 to 0.60 are tabulated in the cited reference for single-family residential units by type, assuming an average of 2500 ft² of impervious coverage per lot.

For acres of mixed cover/land use, an area-weighted runoff coefficient is calculated as follows:

$$C_{av} = (C_1 A_1 + C_2 A_2 + C_3 A_3 + \cdots)/A_{total} \tag{6.32}$$

6.5.3 Intensity

Sources of rainfall intensity data for U.S. locations include the following:

- National Weather Service Website (http://www.nws.noaa.gov/ohd/hdsc)
- Technical Paper 40 (1961)
- Technical Memorandum NWS Hydro 35 (1977)
- NOAA Atlas 2 (1973)
- NOAA Atlas 14, Volume 1 (2003) and Volume 2 (2004)
- U.S. Department of Agriculture Technical Release 55 (1986)
- Pennsylvania Department of Transportation Charts (1986)

Not all of these sources cover the entire country and not all are the most up to date. These sources are discussed further on pages 124–134 of Ref. [26].

6.5.4 Criteria for Storm Water Design

To mitigate the impacts of land development, the standard practice is to capture the surface runoff and detain the release at no higher than predevelopment runoff rates [26, p. 2]. However, the postdevelopment runoff must then flow for a longer duration, subjecting the receiving streams to erosion for a longer period of time. Because of this, the release rate and its inherent velocity may be limited by the authorities to something less than predevelopment rates. Runoff flowing across developed properties may pick up pollution from human activities and must somehow be addressed. Greater runoff also results in a lesser amount of groundwater recharge and increases the propensity for flooding. A case history pointing out the cumulative impacts of incremental development is discussed in Appendix F.

In addition, best management practices necessitate that materials and processes having the potential to contaminate storm water runoff be housed under roof out of the rain or that the storm water falling on these outdoor areas be segregated and treated prior to disposal. Examples include provisions to capture and neutralize any leaks from the storage of acids and oil–water separation facilities to treat possible oil leakage from mechanical equipment. While the overall philosophy may be the same, implementing procedures may vary somewhat from jurisdiction to jurisdiction.

PROBLEM 6.16

A 2 acre piece of farmland containing 75% tilled field, 10% pasture, and 15% light forest is to be converted into a small industrial plant containing 30% buildings, pavement, and driveways/roads; 60% gravel areas; and 10% lawn. Estimate the increase in runoff resulting from such development. Use the values for C from Table 6.9. Take C for tilled farmland equal to 0.08.

SOLUTION   Writing Equation (6.31) twice and canceling out the intensity term shows that

$$Q_{\text{After}}/Q_{\text{Before}} = (CA)_{\text{After}}/(CA)_{\text{Before}} \qquad (6.33)$$

The CA terms are evaluated using Equation (6.32).

$$(CA)_{\text{Before}} = (2 \text{ acres})[(0.75)(0.08) + (0.10)(0.20) + (0.15)(0.15)]$$
$$\text{Composite runoff factor} = 0.1025$$
$$(CA)_{\text{Before}} = (2)(0.1025) = 0.2050$$
$$(CA)_{\text{After}} = (2 \text{ acres})[(0.30)(0.90) + (0.60)(0.80) + (0.10)(0.25)]$$
$$\text{Composite runoff factor} = 0.7750$$
$$(CA)_{\text{After}} = (2)(0.7750) = 1.5500$$
$$(CA)_{\text{After}}/(CA)_{\text{Before}} = 1.5500/0.2050 = 7.5610$$

Postdevelopment runoff amounts to about 7.5 times predevelopment runoff, regardless of the intensity of the storm and regardless of the overall acreage. ■

6.5.5 Time of Concentration

PROBLEM 6.17

An industrial plant of a nominal 10 acre size occupies a plot of 235 ft by 1870 ft. The surface is moderately impervious to storm water infiltration. Runoff is collected at the corner of one of the short ends. Overall slope of the property is 1/8 in. in 1 ft. Calculate the time of concentration and peak runoff. Rainfall data [26, p. 130] are given in Table 6.10. (In an actual case, one would use rainfall data specific to the plant location.)

SOLUTION   Time-of-travel (t) for overland flow along a given path length (L) is calculated from

$$t \text{ (min)} = L \text{ (ft)}/[(60 \text{ s/min})V(\text{ft/s})] \qquad (6.34)$$

This is a rearrangement of Equation (6.19) with the addition of a conversion factor from seconds to minutes.

The flow path in this case is fairly simple to determine as the diagonal dimension of the property, calculated below. In other cases, it will be more involved as several paths are investigated to determine the one with longest time-of-travel and therefore the greatest time of concentration.

$$\text{tan of angle of diagonal} = 235/1870 = 0.1257$$
$$\text{angle} = 7.16°$$
$$\text{diagonal} = 1870/\cos(7.16°) = 1884.71 \text{ ft}$$

Table 6.10 Precipitation Frequency Estimates (in.) for Location in Virginia[a]

Duration	Recurrence interval				
	2 years	5 years	10 years	50 years	100 years
5 min	0.42	0.50	0.56	0.71	0.77
10 min	0.67	0.80	0.90	1.12	1.22
15 min	0.84	1.00	1.13	1.41	1.53
30 min	1.15	1.42	1.62	2.11	2.33
60 min	1.44	1.81	2.11	2.84	3.20
2 h	1.69	2.15	2.52	3.50	3.97
6 h	2.26	2.83	3.31	4.60	5.25
12 h	2.76	3.46	4.05	5.70	6.55
24 h	3.17	4.04	4.80	6.96	8.09
48 h	3.70	4.71	5.57	7.94	9.14

[a] Abridged from NOAA ATLAS 13 for Virginia 38.705N 78.046W, as quoted in Ref. [26, p. 130].

Average flow from the USDA Soil Conservation Service method as quoted in Ref. [26, p. 154] is

$$V = 16.1 S^{0.5} \text{ for unpaved surfaces} \quad (6.35)$$

$$V = 20.3 S^{0.5} \text{ for paved surfaces} \quad (6.36)$$

■

Other techniques for calculating runoff velocity and time-of-travel/time of concentration are summarized and compared in Ref. [26, pp. 147–175].

For a paved surface,

$$V = (20.3)(0.0104)^{0.5} = 2.0719 \text{ ft/s}$$
A slope of 1/8 in. to 1 ft = $(0.125)/(12) = 0.0104$

From Equation (6.34),

$$t = 1884.71/[(60)(2.0719)] = 15.2 \text{ min}$$

From Table 6.10, the 10-year (10% exceedance frequency) 15 min storm results in 1.13 in. of rain.

$$(1.13 \text{ in.}/15 \text{ min})(60 \text{ min/h}) = 4.52 \text{ in/h}$$

The "15.2 min" storm would result in a slightly lower average intensity; the use of 15 min storm is, therefore, conservative and safer than attempting to interpolate the table.

A more conservative design would employ the 100-year (1% exceedance frequency) 15 min storm, resulting in (1.53 in./15 min) and (6.12 in./h).

From Equation (6.31)

$$Q = CIA$$
$$Q = (0.85)(4.52)(235)(1870)/(43,560 \text{ ft}^2/\text{acre})$$
$$= 38.8 \text{ cfs}$$

For the 100-year storm, $Q = 52.5$ cfs. C is assumed to be 0.85 from a 50/50 mix of pavement/roofs (0.90) and gravel areas (0.80) (from Table 6.9).

Maximum storage capacity can perhaps be estimated from the 100-year 24 h storm (8.09 in.), the plant area (10.09 acres), and a coefficient of 0.85.

$$(0.85)(8.09 \text{ in.})(10.09 \text{ acres})(1 \text{ ft}/12 \text{ in.})(43,560 \text{ ft}^2/\text{acre})$$
$$= 252,000 \text{ ft}^3$$

As an alternative consideration, designing for the 100-year storm would handle two back-to-back 10-year storms.

This figure is conservative also from the standpoint that some storm water would be worked off before the next storm hits. Choice of such design conditions would likely be dictated by the storm water regulations of the governing jurisdiction.

Additional real estate would have to be acquired to accommodate a storm water detention pond.

6.6 BACK TO THE MANNING EQUATION

Ultimately, storm water runoff is conveyed to surface water bodies for disposal. The Manning equation, already discussed under open-channel flow, can also be used to describe the flow (q) in sewers and storm sewers not flowing full. A variety of cross sections are possible, including circular, rectangular, trapezoidal, triangular, or some combination.

6.6.1 Rectangular Channels

Let us begin with the rectangular cross section, the simplest shape to analyze. Even this shape results in a trial and error solution when solving for dimensions to accommodate a known flow.

PROBLEM 6.18 *(inspired by Example 11.1, p. 285 of Ref. [26]):*

A concrete channel of rectangular cross section of minimal width is to convey storm water via gravity at a flow of 20 cfs and a maximum depth of 1.5 ft. Slope of the channel is 1/4 in. per ft, and Manning's n is 0.013 for smooth concrete. What should be the width of the channel?

SOLUTION Use the Manning equation multiplied by the cross-sectional area to define the flow rate

$$Q = (1.486/n)AR^{2/3}S^{1/2} \tag{6.37}$$

with symbols as previously defined.

$A = wh$ (channel width times depth of water)
R (hydraulic radius) = wetted area/wetted perimeter
$\quad = (wh)/(w+2h)$
$S = (0.25 \text{ in.}/\text{ft})(1 \text{ ft}/12 \text{ in.}) = 0.0208 \text{ ft}/\text{ft}$

Substituting into Equation (6.37),

$$20 = (1.486/0.013)w(1.5)[(wh)/(w+2h)]^{2/3}(0.0208)^{1/2}$$

Rearranging,

$$w = 0.8081/[(1.5w)/(w+3)]^{2/3}$$

This equation is solved via trial and error, starting with an initial guess of $w = 1$ ft on the right-hand side to calculate w on the left-hand side $= 1.5540$. This result is then substituted into the right-hand side to generate another value for w.

The value of w from this procedure converges to 1.3468 in 13 iterations and to 1.34682590 in 26 iterations. Both of these figures are much more precise than necessary and than what is warranted by the precision of the inputs. A value of $w = 1.35$ ft (1 ft 4.2 in.) is close enough.

Substituting back into Equation (6.37),

$$Q = (1.486/0.013)(1.35)(1.5)\{[(1.35)(1.5)]/[1.35+2(1.5)]\}^{2/3}[0.0208]^{1/2}$$
$$Q = 20.05 \text{ cfs (again, close enough)}$$

The vertical walls of the channel should be increased in height to include a *freeboard*, the depth above the water surface elevation, to contain surges in the flow. Common default conditions for freeboard include 1 ft, half the maximum flow depth, and one velocity head ($V^2/2g$) plus 0.5 ft [26, p. 287].

Here,

$$V = 20 \text{ ft}^3/\text{s}/[(1.35)(1.5)] = 9.8765 \text{ ft/s}$$
$$V^2/2g = 97.5461/[(2)(32.2)] = 1.5147 \cong 1.5 \text{ ft}$$
$$1.5 + 0.5 = 2 \text{ ft}$$

Freeboard should perhaps be 1–2 ft. A freeboard of 1.5 ft would accommodate slightly more than twice the design flow ($2Q = 44.9$ cfs) for water flowing up to the top of the wall. ∎

6.6.2 Circular Channels/Pipe

Of particular interest in sewer/storm sewer design is the shape of a common round pipe with circular cross section.

When full, the wetted area (A_f) of a circle is $(\pi/4)D^2$, its wetted perimeter (P_f) is πD, and the hydraulic radius (R_f) is $D/4$, with D as the diameter. For a circular cross section not flowing full, the following ratios apply:

$$A/A_f = (1/\pi)\{\cos^{-1}(1-2h/D) - [1-2h/D][1-(1-2h/D)^2]^{1/2}\} \quad (6.38)$$
$$P/P_f = (1/\pi)\cos^{-1}(1-2h/D) \quad (6.39)$$
$$R/R_f = (A/A_f)/(P/P_f) \quad (6.40)$$
$$Q/Q_f = (A/A_f)(R/R_f)^{2/3} \quad (6.41)$$

The angle returned by the inverse cosine function must be expressed in radians and not in degrees (2π radians $= 360°$).

These relationships are plotted in Figure 6.1 as a function of the relative depth of fluid (h/D) measured from the inside bottom of the pipe. This plot reproduces the

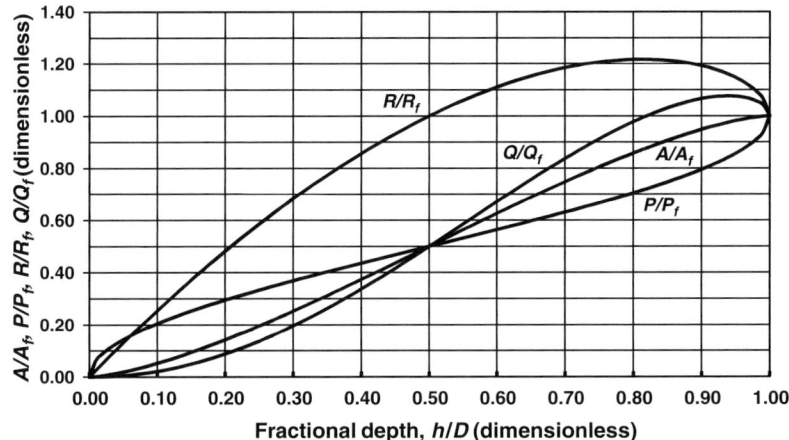

Figure 6.1 Hydraulic elements graph for pipe of circular cross section not flowing full.

values as shown in the literature and is known as a *hydraulic elements* graph [5, p. 109; 26, p. 89].

The functions (A/A_f) and (P/P_f), although not linear, show an increase with increasing relative depth (h/D). The area function (A/A_f), when multiplied by axial length, describes the partially filled volume of a horizontal cylindrical tank, discussed in Chapter 7. The relative hydraulic radius (R/R_f) shows a maximum of 1.217234± (1.2 as read from the graph) at (h/D) close to 0.81280 (0.81). Maximum flow, $(Q/Q_f) \cong 1.05706$ (1.1 from the graph), occurs when the relative liquid depth (h/D) is 0.93818± (approximately 94%).

PROBLEM 6.19 *(inspired by Ref. [26, pp. 89–90])*

The flow in Problem 6.18 is to be conveyed by gravity to a circular plastic pipe of minimal size at the same slope. Determine the size of the pipe and the depth of flow. This pipe comes in diameters of 15, 18, 21, 24, 27, and 30 in. Its Manning's n is 0.012.

SOLUTION The first step is to find the absolute minimum pipe size with the pipe flowing full. For a circular pipe at the limiting condition of full flow,

$$R_f = (\pi D^2/4)/(\pi D) = D/4 \tag{6.42}$$

Manning's equation for flow (Equation (6.37)) becomes

$$Q(\text{cfs}) = (1.486/n)(\pi D^2/4)(D/4)^{2/3} S^{1/2} \tag{6.43}$$

$$Q(\text{cfs}) = (0.4631/n) D^{8/3} S^{1/2} \tag{6.44}$$

Rearranging to solve for D,

$$D(\text{ft}) = (Qn/S^{1/2})^{3/8}/(0.4631)^{3/8} = 1.335(Qn/S^{1/2})^{3/8} \tag{6.45}$$

$$D(\text{ft}) \times 12 = D(\text{in.})$$

$$D(\text{in.}) \cong 16(Qn/S^{1/2})^{3/8} \tag{6.46}$$

Substituting numbers into this equation,

$$D(\text{in.}) = 16[(20)(0.012)/(0.0208)^{1/2}]^{3/8}$$
$$= 16[1.6628]^{3/8} = 16[1.2102] = 19.4 \text{ in.}$$

The second step is to try the next size (21 in. = 1.27 ft) pipe and compute the Q for a full pipe. From Equation (6.44),

$$Q \text{ (cfs)} = (0.4631/n)D^{8/3}S^{1/2}$$
$$= (0.4631/0.012)(21/12)^{8/3}(0.0208)^{1/2}$$
$$= (38.59)(4.4474)(0.1443) \qquad (6.47)$$
$$= 24.8 \text{ cfs}$$
$$Q/Q_f = 20/24.8 = 0.81$$

From Figure 6.1, h/D corresponding to $Q/Q_f = 0.81$ is approximately 0.68.

$$h = (0.68)(21) = 14.3 \text{ in.} \; (<21 \text{ in.})$$

The 21 in. pipe is not flowing full, flow is therefore by gravity, and Manning's equation applies. ∎

PROBLEM 6.20

A storm drainage channel made of somewhat rougher concrete is an estimated 20 ft deep and 20 ft wide at the bottom. Its cross section is trapezoidal with sides on a 1:1 slope. What is its carrying capacity when full if the slope of the channel in the flow direction is 1%, 2%, and 3%?

SOLUTION Draw a sketch, if necessary. Since sides are at a 1:1 slope, width at the top = 20 + 2(20) = 60 ft. Area for flow (A_w) = (60)(20) − (2)(1/2)(20)(20) = 1200 − 400 = 800 ft². Length of each inclined side = 20/sin 45 = 20/0.7071 = 28.3 ft or by the Pythagorean theorem $[20^2 + 20^2]^{1/2}$ = 28.3 ft. Wetted perimeter (P_w) = 28.3 ft + 20 + 28.3 = 76.6 ft. $R = A_w/P_w$ 800 ft²/76.6 ft = 10.4 ft, from Equation (6.26).

Use the Manning equation for flow to calculate Q (cfs):

$$Q = (1.486/n)AR^{2/3}S^{1/2} \qquad (6.48)$$

Take n for "somewhat rougher concrete" = 0.015.

Remember to put the slope into units of ft/ft and to take the square root.

$$\text{Slope}^{1/2} = (1/100)^{1/2}, (2/100)^{1/2}, (3/100)^{1/2} = 0.1, 0.1414, 0.1732, \text{ respectively}$$

Then,

$$Q = (1.486/0.015)(800)(10.4)^{2/3}S^{1/2}$$
$$= (377,600)(0.1; \; 0.1414; \text{ or } 0.1732)$$
$$= 37,800; \; 53,400; \text{ and } 65,400 \text{ cfs for slopes of 1\%, 2\%, and 3\%, respectively}$$

A large public storm drain of this type is shown in Figures 6.2 and 6.3. ∎

Figure 6.2 Photograph of large public storm drain of trapezoidal cross section.

PROBLEM 6.21

Suppose that the runoff coefficient in Problem 6.17 were 0.44 instead of 0.85. Then the total runoff would be 38.8 (0.44/0.85) $\cong$ 20 cfs. Calculate the time-of-travel for the 20 cfs flow in the circular pipe of Problem 6.19 over the same distance as in Problem 6.17 and compare with the overland time of concentration.

SOLUTION The flow (Q) in Problems 6.18 and 6.19 was given as 20 cfs, and a pipe diameter of 21 in. with a liquid level of 14.3 in. was calculated in Problem 6.19 to handle this flow.

$$h/D = 14.3/21 = 0.68$$

At $(h/D) = 0.68$, $(A/A_f) = 0.72$ from Figure 6.1. Therefore, the actual area of flow in the 21 in. pipe is

$$(\pi/4)(21/12)^2(0.72) = 1.73 \text{ ft}^2$$

From the continuity equation (Equation (6.18)),

$$V = Q/A = 20/1.73 = 11.6 \text{ ft/s}$$
$$11.6 \text{ ft/s} \times 60 \text{ s/min} = 694 \text{ ft/min}$$

336   Chapter 6   Water/Wastewater Hydraulics

Figure 6.3 Photograph of same open channel with water flowing after a storm.

Distance from Problem 6.17 is 1884.71 ft.

$$t = \text{distance}/V = 1884.71 \text{ ft}/(694 \text{ ft/min}) = 2.7 \text{ min}$$

Since this number is less than 15.2 min calculated in Problem 6.16, 15.2 min would govern as the time of concentration.

The purpose of this exercise is to compare in a gross manner the time-of-travel for pipe flow versus overland flow at the same Q. Not much more should be made of it. This simple problem would not represent a practical design for the 10 acre plant site of Problem 6.17 for the following reasons:

- A storm sewer would not be laid out in this manner to drain the plant site. Various subareas of the plant would drain to a single trunk line or to a pipeline network.
- Some storm water would run off via an overland route, and some would be collected in the storm sewer, requiring more detailed calculations from specific layout drawings.
- Time of concentration would actually increase if all of the runoff could be made to run overland from the near corner of the property back to the far corner

and then back to the near corner via the storm sewer. Again, this is not a practical design.

Note that the slope of the storm sewer is twice that of the slope of the land. Time-of-travel in a sewer at the lesser slope would increase by about 40%, without affecting the conclusion regarding time of concentration. ∎

REFERENCES

1. P. Bahlor, editor, *National Pollutant Discharge Elimination System (NPDES) Compliance Inspection Manual*, U.S. Environmental Protection Agency (EPA), Washington, DC, 2004.
2. N.P. Cheremisinoff and R.A. Niles, A survey of fluid flow measurement techniques and fundamentals, *Water & Sewage Works*, **122**(12), 74–78, 1975.
3. J.G. Rabosky and D.L. Koraido, Gauging and sampling industrial wastewaters, *Chemical Engineering*, **80**(1), 111–120, 1973.
4. T. Thorsen and R. Oen, How to measure industrial wastewater flow, *Chemical Engineering*, **82**(4), 95–100, 1975.
5. D.M. Grant, *Isco Open Channel Flow Measurement Handbook*, 3rd edn, Isco, Inc., Lincoln, NE, 1991, 356 pp.
6. J.B. Franzini and E.J. Finnemore, *Fluid Mechanics with Engineering Applications*, 9th edn, WCB McGraw-Hill, New York, 1997, 807 pp.
7. R.H.F. Pao, *Fluid Mechanics*, John Wiley & Sons, Inc., New York, 1961, 502 pp.
8. R.L. Daugherty and A.C. Ingersoll, *Fluid Mechanics with Engineering Applications*, 5th edn, McGraw-Hill, New York, 1954, 472 pp.
9. R.L. Daugherty and J.B. Franzini, *Fluid Mechanics with Engineering Applications*, 6th edn, McGraw-Hill, New York, 1965, 574 pp.
10. R.L. Daugherty and J.B. Franzini, *Fluid Mechanics with Engineering Applications*, 7th edn, McGraw-Hill, New York, 1977, 564 pp.
11. Anonymous, British Columbia Field Sampling Manual, Province of British Columbia, Canada, January 2003.
12. V.T. Chow, *Open-Channel Hydraulics*, McGraw-Hill, New York, 1988, 680 pp.
13. Anonymous, Tracom Fiberglass Products, TRACOM, Inc., Alpharetta, GA, http://www.tracomfrp.com, accessed on April 15, 2008.
14. G. Kulin, Recommended Practice for the Use of Parshall Flumes in Wastewater Treatment Plants, EPA-600/2-84-186, PB85-122745, National Technical Information Service, Springfield, VA, November 1984.
15. Anonymous, Guidance Document for Flow Measurement of Metal Mining Effluents, Appendix 1: Open Channel Flow Measurement, Environment Canada, available at http://www.ec.gc.ca, accessed on April 15, 2008.
16. Anonymous, Waste Water Flow Measurement, NALCO TECHNIFAX, Nalco Chemical Company, Oak Brook, IL, July 1974, 8 pp.
17. B. Dawson, Open channel flowmeters, Measurements & Control Magazine, reprint available at http://www.isco.com, accessed on April 15, 2008.
18. Anonymous, Plasti-Fab® Product Bulletin Palmer–Bowlus Flumes, Plasti-Fab® Inc., Tualatin, OH, http://www.plasti-fab.com, accessed on April 15, 2008.
19. Anonymous, ASTM D5390-93 (2007) Standard Test Method for Open-Channel Flow Measurement of Water with Palmer–Bowlus Flumes, American Society for Testing and Materials, West Conshohocken, PA, 2007, 8 pp.
20. Anonymous, Industrial Monitoring Station Requirements, Metropolitan Sewer District of Greater Cincinnati Division of Industrial Waste, August 1, 2002.
21. M. Barcelona, J. F. Kelly, W. A. Pettyjohn, and A. Wehrmann, *EPA Ground Water Handbook*, EPA/625/6-87/016, U.S. Environmental Protection Agency, Cincinnati, OH, 1987.

22. L. Klein, *River Pollution III. Control*, Butterworths, London, 1969, 484 pp.
23. J.A. Dean, editor, *Lange's Handbook of Chemistry*, 11th edn, McGraw-Hill, New York, 1973.
24. D.P. Johnson and K.D. Metzker, Low-Flow Characteristics of Ohio Streams, Open-File Report 81-1195, U.S. Geological Survey prepared in cooperation with the Ohio Environmental Protection Agency, Columbus, OH, 1981, 285 pp.
25. R. H. Perry and C. H. Chilton, *Chemical Engineers' Handbook*, 5th edn, McGraw-Hill, New York, 1973, p. 5–54.
26. T. A. Seybert, *Stormwater Management for Land Development: Methods and Calculations for Quality Control*, John Wiley & Sons, Inc., Hoboken, NJ, 2006, 372 pp.
27. Anonymous, King County, Washington Surface Water Design Manual, King County Department of Natural Resources, January 24, 2005.

Chapter 7

Water/Wastewater Draining of Tanks

What we have is not what we want.
What we want is not what we need.
What we need is not available.

7.1 INTRODUCTION

There may be occasions when one will be concerned with the draining of tanks, either under controlled circumstances or under sudden and catastrophic conditions. This chapter discusses tank volumes, time to drain, and the trajectory of a jet of liquid emanating from a leaking tank. The chapter consists essentially of two parts: the volume of a tank as a function of height plus time to drain the tank, and the trajectory of a leaking jet of liquid.

Tanks of commonly encountered shapes are considered, alone or in combination. These include vertical and horizontal tanks with various types of tank ends (heads). They are treated in an order such that each sample problem builds upon what has gone before. Some types of tanks are as follows:

- Vertical tank of constant cross section and flat bottom
- Cone-shaped tank
- Spherical tank
- Spheroidal or ellipsoidal tank
- Vertical cylindrical tank with cone bottom
- Vertical cylindrical tank with hemispherical bottom
- Vertical cylindrical tank with ellipsoidal bottom
- Vertical cylindrical tank with dished bottom

Environmental Calculations: A Multimedia Approach, by Robert G. Kunz
Copyright © 2009 John Wiley & Sons, Inc.

- Vertical tank of elliptical cross section with ellipsoidal bottom
- Horizontal cylindrical tank with flat ends
- Horizontal cylindrical tank with hemispherical ends
- Horizontal cylindrical tank with elliptical ends
- Horizontal cylindrical tank with dished ends (approximation)
- Horizontal tank of elliptical cross section with flat ends
- Horizontal tank of elliptical cross section with elliptical ends

Consideration of the leaking jet consists of a series of problems leading up to a determination of the height of a secondary-containment dike wall designed to intercept the leaking jet. Volume of the diked area is also set to contain the spilled liquid from the entire volume of a full tank.

7.2 TIME TO DRAIN TANKS

7.2.1 Vertical Tank of Constant Cross Section

Let us start with the simplest example, a vertical tank with a constant cross section. The cross section can be circular (cylindrical tank), elliptical, square, rectangular, or any shape whatsoever.

From Bernoulli's equation written in terms of head between two points (denoted as "1" and "2"), no mechanical work added (e.g., a pump), and neglecting friction for the moment,

$$p_1/w + v_1^2/(2g) + h_1 = p_2/w + v_2^2/(2g) + h_2 \tag{7.1}$$

where p is the pressure (lb_f/ft^2), w is the specific weight of fluid (lb_f/ft^3), v is the velocity of liquid (ft/s), h is the height above a specific datum (ft), and g is the gravitational constant ($\cong 32.2 \, ft/s^2$).

With both locations at the same atmospheric pressure, the p terms drop out. For the condition that $v_1 \gg v_2$, the v_2 term vanishes. By taking h_1 as the location of the datum, h_1 equals zero. What is left, with some algebraic manipulation and upon rearrangement, is

$$v = (2gh)^{1/2} \tag{7.2}$$

where v is the velocity of the jet leaving the enclosure (ft/s) and h is the liquid head height above the discharge (ft). This is known as *Torricelli's law* or *Torricelli's theorem, without friction*.[1]

The actual velocity of the jet leaving the vessel is less than that calculated from Equation (7.2) because of friction, accounted for by an empirical coefficient of

---

[1]Evangelista Torricelli (1608–1647), Italian mathematician, physicist, and inventor of the barometer, was a student of Galileo (1564–1642) at the Academy of Florence and succeeded Galileo as professor upon his death [1–3].

velocity (C_V). The area of the jet is less than the area of the opening, a phenomenon known as the *vena contracta* and accounted for by a contraction coefficient or coefficient of area (C_A). The two coefficients multiplied together give the coefficient of discharge (C_D).

$$C_D = C_A \times C_V \qquad (7.3)$$

Equation (7.2) then becomes

$$v = C_A \times C_V (2gh)^{1/2} = C_D (2gh)^{1/2} \qquad (7.4)$$

7.2.1.1 Typical Values of C_V and C_A

The velocity coefficient (C_V) for sharp-edged circular orifices may range from about 0.95 to 0.99, and from 0.97 to 0.99 for nozzles with smooth walls, and much lower for other devices, say 0.7–0.8 [2, p. 123; 3, p. 347]. The contraction coefficient (C_A) may range from about 0.5 to 1.0 [2, pp. 113 and 123; 3, p. 347], with a theoretical value of 0.611 for a sharp-edged rectangular orifice [4].

The composite product (C_D) for standard sharp-edged circular orifices may range from 0.59 to 0.68 [2, pp. 124–125]. In other cases, C_D can be estimated if C_V and C_A are known separately. If necessary, the configuration in question can be calibrated in place.

7.2.1.2 Velocity Coefficient for Valves

An alternate treatment makes use of a head-loss friction factor (h_f) for valves and fittings [5].

$$h_f = Kv^2/(2g) \qquad (7.5)$$

The discharge velocity from Bernoulli's equation then becomes

$$v = [1/(1+K)]^{1/2}(2gh)^{1/2} \qquad (7.6)$$

A comparison of Equations (7.4) and (7.6) shows that C_V for a valve is equivalent to

$$C_V = [1/(1+K)]^{1/2} \qquad (7.7)$$

when C_A can be taken as 1.0.

Over the range of K values (0.2–10.0) listed for wide open and partially open/closed valves [5; 6, p. 317], C_V for valves from Equation (7.7) ranges from 0.30 to 0.91. Values for C_A for a valve may be assumed to be similar to those already quoted (0.5–1.0). They will be closer to 1.0 when there is a uniform diameter for some distance prior to the exit of the jet [2, p. 123].

7.2.1.3 Draining of the Tank

Draining the tank is an unsteady-state rate problem:

Change in volume in the tank = Flow in (influent) − Flow out (effluent)

$$dV/dt = Q_{in} - Q_{out} \qquad (7.8)$$

342 Chapter 7 Water/Wastewater Draining of Tanks

In the case of zero flow in, the first term on the right-hand side of the equation drops out. For a vertical tank of constant cross-sectional area (A_T), the change in volume is linear with the instantaneous height (h) of liquid in the tank.

$$V = (A_T)(h) \qquad (7.9)$$

and

$$dV/dt = A_T(dh/dt) \qquad (7.10)$$

The effluent equals the flow out of the opening of physical cross-sectional area A_o.

$$Q_{out} = C_D A_o (2g)^{1/2} h^{1/2} = k h^{1/2} \qquad (7.11)$$

Then,

$$-(dh/dt) = C_D(A_o/A_T)(2g)^{1/2} h^{1/2} = (k/A_T) h^{1/2}$$

$$-h^{1/2} dh = (k/A_T) dt \qquad (7.12)$$

Integration between the limits $h = H$ at time $= 0$ (initial condition) and $h = h$ at time $= t$ yields

$$[h^{1/2}]_h^H = +(k/(2A_T))[t]_0^t \qquad (7.13)$$

$$H^{1/2} - h^{1/2} = [k/(2A_T)](t-0) = (1/2)(A_o/A_T)(2g)^{1/2} C_D t \qquad (7.14)$$

C_D can be found experimentally by plotting $H^{1/2} - h^{1/2}$ versus t. The slope is the coefficient of t in the plot of Equation (7.14), and

$$C_D = [(A_o/A_T)(2g)^{1/2}] \times \text{Slope} \qquad (7.15)$$

A typical experiment can be found in Appendix H.

Time (T) to drain the tank completely is given by rearranging Equation (7.14) with $h = 0$.

$$T = [(A_T/A_o)(2H/g)^{1/2}]/C_D \qquad (7.16)$$

Equation (7.16) works for a constant cross-sectional area of any shape and is consistent with the formula given in Ref. [7] for a vertical cylinder (circular cross section).

PROBLEM 7.1

A tank 5 ft in diameter and 10 ft high drains halfway in 30 min. (a) How much longer will it take to empty the tank? (b) Estimate the value of C_D.

SOLUTION

(a) From Equation (7.14), using seconds as the unit of time, find $k/(2A_T)$.

$$H^{1/2} - h^{1/2} = [k/(2A_T)]t$$

$$k/(2A_T) = (10^{1/2} - 5^{1/2})/(1800 \text{ s})$$

$$= (3.16228 - 2.23607)/1800 = 0.926212/1800 = 0.000514562$$

Then with the value of $[k/(2A_T)]$, find time to drain tank completely from Equation (7.16)

$$t = (10^{1/2} - 0)/[k/(2A_T)] = 3.16228/0.000514562 = 6145.57169$$

$$\div 3600 \text{ s/h} \cong 1.7071 \text{ h} = 1 \text{ h } 42 \text{ min } 26 \text{ s total time}$$

To find how much longer will it take to empty the tank, subtract 30 min to obtain 1 h 12 min 26 s.

(b) Now for $k = C_D A_o (2g)^{1/2}$ and a 1 in. diameter drain valve, where

$$A_o = 0.00545 \text{ ft}^2, \text{ and } A_T = (\pi/4)5^2 = 19.635 \text{ ft}^2$$

$$C_D = \{[k/(2A_T)](2)(19.635)\}/[A_o(2g)^{1/2}]$$

$$= \frac{(0.000514562)(2)(19.635)}{(0.00545)(2 \cdot 32.2)^{1/2}} = 0.46 \qquad \blacksquare$$

7.2.2 (Vertical) Cone-Shaped Tank [8]

This has the shape of a funnel, with the small end down.
As before,

$$-(dV/dt) = C_D A_o (2g)^{1/2} h^{1/2} \qquad (7.17)$$

$$\text{Volume } (V) \text{ of cone} = (\pi/3)R^2 H$$

$$\text{and area at top of inverted cone} = A_c = \pi R^2$$

where R is the largest radius of the cone at maximum height (H).
At any lesser height and radius

$$\text{Volume} = (\pi/3)r^2 h$$

From similar triangles (cross section of right cone),

$$r/h = R/H, \quad r = (Rh)/H$$

$$\text{Volume } (V) = (\pi/3)[(Rh)/H]^2 h$$

$$\text{Volume } (V) = (\pi/3)\{(R/H)^2 h^3\}$$

$$dV = (\pi)(R/H)^2 h^2 dh$$

$$-(\pi)(R/H)^2 h^2 (dh/dt) = C_D A_o (2g)^{1/2} h^{1/2} = kh^{1/2}$$

$$(\pi)(R/H)^2 h^{3/2} dh = -k dt$$

$$A_c h^{3/2} dh = -kH^2 dt$$
$$h^{3/2} dh = -[kH^2/A_c] dt$$

When integrated between the limits of $h = h$ and $h = H$, this yields

$$[h^{5/2}]_h^H = (5/2)[kH^2/A_c][t]_0^t$$

Upon rearrangement,

$$H^{5/2} - h^{5/2} = (5/2) C_D (A_o/A_c)(2g)^{1/2} H^2 t \qquad (7.18)$$

Time (T) to drain a full conical tank occurs with initial height H, and $h = 0$.

$$H^{5/2} = (5/2) C_D (A_o/A_c)(2g)^{1/2} H^2 T$$

Solving for T,

$$T = (1/5)(2H/g)^{1/2}[A_c/(C_D A_o)] \qquad (7.19)$$

Area of discharge must be small compared to area of tank, and there must be no vortex formation at the drain. When liquid level is low, neglect of the $v^2/(2g)$ term in the cone becomes less and less valid as liquid level drops lower and lower, approaching the drain.

Equation (7.19) for draining a cone is also consistent with the formula given in Ref. [7].

PROBLEM 7.2

Estimate the time to drain a vertical cylindrical tank and a conical tank of (a) the same height and (b) the same volume. Compare the two cases.

SOLUTION

(a) Dimensions: 5 ft in diameter (D) (constant for cylinder, at top for cone), 1 in. diameter opening at bottom, 10 ft in height (H). Assume $C_D = 0.5$.

Volume of cylinder = $(\pi/4) D^2 H = (3.141592.../4)(25)(10) = 196.35 \text{ ft}^3$

Volume of cone = $(1/3)(\pi/4) D^2 H = [3.141592.../(3 \cdot 4)](25)(10) = 65.4 \text{ ft}^3$

For cylinder,

$$T = [(A_T/A_o)(2H/g)^{1/2}]/C_D \qquad (7.16)$$

$$A_T = (\pi/4)(D^2) = (3.141592.../4)(5)^2 = 19.635 \text{ ft}^2$$

$$A_o = (\pi/4)(1\text{in.})^2/(144 \text{in}^2/\text{ft}^2) = (3.141592.../4)(1)^2/(144) = 0.00545 \text{ ft}^2$$

$$T = [(19.635/0.00545)(2 \cdot 10/32.2)^{1/2}]/0.5 = 5679 \text{ s}$$

$$\div 60 \text{ min/h} = 94.7 \text{ min}$$

$$\div 3600 \text{ s/h} = 1.58 \text{ h} \cong 1 \text{ h } 35 \text{ min}$$

For cone,
$$T = (1/5)(2H/g)^{1/2}[A_c/(C_D A_o)] \tag{7.19}$$
$$A_c = (\pi/4)(D)^2 = (3.141592\ldots/4)(5)^2 = 19.635 \text{ ft}^2$$
$$T = (1/5)(2 \cdot 10/32.2)^{1/2}(19.635)/[(0.5)(0.00545)] = 1136 \text{ s}$$
$$\div 60 \text{ min/h} = 18.9 \text{ min} \cong 19 \text{ min}$$
$$\div 3600 \text{ s/h} = 0.316 \text{ h}$$

Although the volume of the cylinder is three times that of the cone, it takes more than three times longer duration to drain the cylinder than the cone (94.7 min/18.9 min) = 5 (within roundoff of the times). This is the result of the factor (1/5) in Equation (7.19).

(b) For cylinder of the same volume as cone, keeping the same diameter as the cone,
$$(\pi/4)(5)^2(H) = 65.4 \text{ ft}^3$$
$$H = 65.4/[(\pi/4)(5)^2] = 3.33 \text{ ft}$$
$$T = [(A_T/A_o)(2H/g)^{1/2}]/C_D \tag{7.16}$$
$$T = \{[(19.635)/(0.00545)][(2 \cdot 3.33/32.2)^{1/2}]\}/(0.5) = 3277 \text{ s}$$
$$\div 60 \text{ s/min} \cong 55 \text{ min}$$

It takes just about three times longer duration to drain a cone of the same volume and diameter as the cylinder.

Keeping the same height and changing the diameter to achieve the same volume,
$$(\pi/4)(D)^2(10) = 65.4 \text{ ft}^3$$
$$D = \{(65.4)/[(10)(\pi/4)]\}^{1/2} = 2.89 \text{ ft}$$
$$A_T = (\pi/4)(D^2) = (3.141592\ldots/4)(2.89)^2 = 6.56 \text{ ft}^2$$
$$T = \{[(6.56)/(0.00545)][(2 \cdot 10/32.2)^{1/2}]\}/(0.5) = 1897 \text{ s}$$
$$\div 60 \text{ s/min} \cong 32 \text{ min (only about 1.6 times longer)}$$

Keeping the height (H) of the cone the same does a better job of compensating for the factor of 1/5 in Equation (7.19). ∎

7.2.3 Spherical Tank

The volume (V) of a partially filled spherical tank can be calculated from
$$V = \pi h^2[R-(h/3)]$$
$$V = (\pi/3)h^2[(3/2)D-h] \tag{7.20}$$

where h equals the depth of liquid measured from the bottom of the tank and D is the tank diameter. It is obtained by integrating the incremental volume of the sphere.
$$dV = \pi r^2 dy \tag{7.21}$$

where dy is the incremental distance from the center of the sphere and translation of coordinates.

From trigonometry, the square of the radius of the circle at any vertical distance y is given by $r^2 = R^2 - y^2$, with R being the radius (D/2) of the spherical tank.

Equation (7.20) reduces to $(4/3)\pi R^3$ or $(\pi/6)D^3$ for a full sphere and $(2/3)\pi R^3$ or $(\pi/12)D^3$ for a half-full spherical tank.

In terms of volume relative to a full sphere,

$$\frac{V}{(\pi/6)D^3} = (h/D)^2[3-2(h/D)] \quad (7.22)$$

Equation (7.22) takes on a value of 1.0 at $h = D$, 0.5 at $h = 0.5 D$, and 0 at $h = 0$. A plot is shown in Figure 7.1

7.2.3.1 Draining of Spherical Tank

$$dV/dt = Q_{in} - Q_{out} \quad (7.8)$$

For zero flow in and Q_{out} given by Torricelli's law with friction,

$$Q_{out} = C_D A_o (2g)^{1/2} h^{1/2} \quad (7.11)$$

Then,

$$(dV/dt) = -C_D A_o (2g)^{1/2} h^{1/2} = -kh^{1/2}$$

with

$$k = C_D A_o (2g)^{1/2}$$

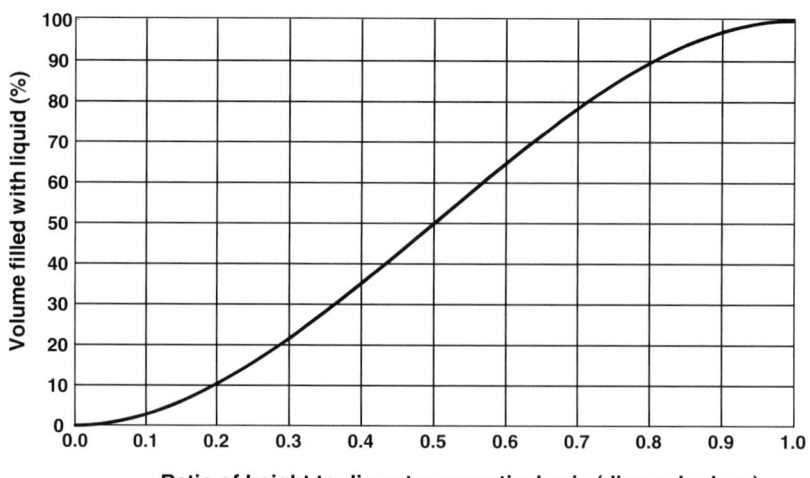

Figure 7.1 Spherical or ellipsoidal tank volume versus height of liquid.

Differentiation of Equation (7.22) with respect to h yields

$$dV = \pi[Dh-h^2]dh$$
$$\pi[Dh-h^2](dh/dt) = -kh^{1/2}$$
$$-(k/\pi)dt = \{[Dh-h^2](dh)\}/h^{1/2} = [Dh^{1/2}-h^{3/2}]dh$$

Integrating to find time (T) to drain the tank from height (h_o) to the bottom ($h=0$),

$$[t]_0^T = (\pi/k)[(2/3)Dh^{3/2}-(2/5)h^{5/2}]_0^{h_o}$$
$$T = (2\pi/k)[(1/3)Dh_o^{3/2}-(1/5)h_o^{5/2}] \qquad (7.23)$$

Equation (7.23) for draining a spherical tank from an arbitrary depth is also consistent with the formula given in Ref. [7].

For a full tank ($h_o = D$),

$$T = (2\pi/k)[(1/3)D^{5/2}-(1/5)D^{5/2}]$$
$$T = (4\pi D^{5/2})/(15k) = (4\pi D^{5/2})/[15C_D A_o(2g)^{1/2}] \qquad (7.24)$$

PROBLEM 7.3

Calculate the time to drain a full spherical tank 7.21 ft in diameter (having the same volume as a cylindrical tank 5 ft in diameter and 10 ft long) with a 1 in. diameter opening, as in Problem 7.2. Again use $C_D = 0.5$.

SOLUTION From Equation (7.4),

$$T = (4\pi D^{5/2})/(15k)$$

where

$$k = C_D A_o(2g)^{1/2} \text{ and } (2g)^{1/2} = (2 \cdot 32.2)^{1/2} = 8.0250$$
$$k = (0.5)(0.00545)(8.0259) = 0.021868 \text{ ft}^{5/2}/\text{s, and } 7.21^{5/2} = 139.5848$$
$$T = \{(4)((3.141592\ldots)(139.5848)/[(15)(0.021868)]\} = 5347 \text{ s}$$
$$\div 3600 \text{ s/h} = 1.49 \text{ h}, \div 60 \text{ s/min} \cong 89 \text{ min} \qquad \blacksquare$$

PROBLEM 7.4

Repeat calculation for a 5 ft diameter sphere with the same C_D and A_o.

SOLUTION

$$5^{5/2} = 55.9017$$
$$T = \{(4)((3.141592\ldots)(55.9017)/[(15)(0.021868)]\} = 2142 \text{ s}$$
$$\div 3600 \text{ s/h} = 0.59 \text{ h}, \div 60 \text{ s/min} \cong 36 \text{ min} \qquad \blacksquare$$

7.2.4 Spheroidal or Ellipsoidal Tank

This shape is of the type exhibited by some elevated storage tanks. It is *not* a "squashed" cylindrical tank with an end in the shape of an ellipse. That tank configuration is dealt with in another section.

The volume (V) of a full spheroidal tank is given by

$$V = (4/3)\pi abc \tag{7.25}$$

Equation (7.25) reduces to $V = (4/3)\pi R^3$ for the volume of a sphere when a, b, and c (each analogous to a radius in a sphere) are equal and equal to R, the radius of a sphere. The factors a, b, and c are the semiaxes in three dimensions. They are not necessarily equal.

A *prolate* spheroid is generated by rotating an ellipse about its major (longer) axis. An *oblate* spheroid is created by rotation about the ellipse's minor (shorter) axis. The formula in Equation (7.25) applies to the general case where all three axes are of different lengths. In the equations for spheroidal geometric figures of rotation, two of the axes are of the same length and the product abc is replaced by ab^2 or a^2b, as appropriate.

For the general case, the liquid volume in a partially filled spheroidal tank can be calculated from

$$V = \pi[(abc)(h/c)^2][1-(1/3)(h/c)] \tag{7.26}$$

The dimension h is measured from the bottom of the tank along the vertical (c) semiaxis. This reduces to the partial volume formula for a sphere.

$$V = \pi h^2[R-(h/3)] \tag{7.27}$$

by replacing a, b, and c in Equation (7.26) by R.

The fractional volume occupied by the liquid is obtained by dividing Equation (7.26) by Equation (7.25).

$$V_{\text{fractional}} = (3/4)(h/c)^2[1-(1/3)(h/c)] \tag{7.28}$$

For $h = 0$, $V_{\text{fractional}} = 0$. For $h = c$ (half-full), $V_{\text{fractional}} = 0.5$. For $h = 2c$, $V_{\text{fractional}} = 1.0$. For the complementary quarter points $h = 0.5c$ and $1.5c$, $V_{\text{fractional}} = 5/32$ (0.15625) and 7/32 (0.84375), respectively. The relative volume of a spheroidal tank follows the same curve as for a spherical tank (Figure 7.1).

7.2.4.1 Draining of the Spheroidal Tank

Time (t) to drain from initial height (h_o) to final height (h),

$$t = (\pi abc^{1/2}/k)\{(4/3)[(h_o/c)^{3/2}-(h/c)^{3/2}]-(2/5)[(h_o/c)^{5/2}-(h/c)^{5/2}]\} \tag{7.29}$$

Time (T) to empty a full tank ($h=2c$) completely to final height ($h=0$),

$$T = (\pi abc^{1/2}/k)\{(4/3)(2c/c)^{3/2}-(2/5)(2c/c)^{5/2}\}$$
$$T = (\pi abc^{1/2}/k)\{(8/3)(2)^{1/2}-(8/5)(2)^{1/2}\}$$
$$T = (16/15)(2)^{1/2}(\pi abc^{1/2}/k)$$
$$T = (4.7391)(abc^{1/2}/k) \qquad (7.30)$$

PROBLEM 7.5

(a) Estimate the time to empty a full tank of spheroidal shape having semiaxes (in ft) of $a=5$, $b=3$, and $c=2$, with $k=0.021868$ ft$^{5/2}$/s as in previous problems.

(b) Recalculate for $a=2$, $b=3$, $c=5$, and explain the difference.

SOLUTION Applying Equation (7.30) for T,

(a) $T = (4.7391)(abc^{1/2}/k)$

$$T = (4.7391)(5)(3)(2^{1/2})/(0.021868) = 4597 \text{ s}$$
$$\div 3600 \text{ s/h} = 1.28 \text{ h} \cong 1 \text{ h } 17 \text{ min}$$

(b) By proration, for $(2)(3)(5^{1/2}) = 13.42$ versus $(5)(3)(2^{1/2}) = 21.21$, time is shortened to $(13.42)/(21.21)(4597 \text{ s}) = 2908.7 \text{ s} \cong 0.81 \text{ h} \cong 48.5$ min because of the higher overall head of liquid above the drain when the longest dimension is in the vertical direction. ∎

7.2.5 Vertical Cylindrical Tank with Cone Bottom

The next step is to build a cone-bottom tank by placing the cylinder on top of the cone.

$$-(dh/dt) = \underbrace{[C_D(A_o/A_T)(2g)^{1/2}]}_{k'}h^{1/2}\text{ (Cylinder)} + \underbrace{[C_D(A_o/A_T)(2g)^{1/2}]}_{k'}H^2 h^{-3/2}\text{ (Cone)}$$

$$-k'dt = h^{-1/2}dh + H^2 h^{+3/2}dh$$

$$k'[t]_0^T = [2h^{1/2}]_{H_\text{Cone}}^{H_\text{Total}} + (2/5)(H^2)[h^{5/2}]_0^{H_\text{Cone}}$$

The cylinder drains from $h = H_\text{Total}$ to $h = H_\text{Cone}$.
The cone drains from $h = H_\text{Cone}$ to $h = 0$.
For cylinder,

$$(H_\text{Total})^{1/2} - (H_\text{Cone})^{1/2} = (1/2)(A_o/A_T)(2g)^{1/2}C_D(T_\text{Cylinder})$$
$$T_\text{Cylinder} = [(H_\text{Total})^{1/2} - (H_\text{Cone})^{1/2}]/[C_D(A_o/A_T)(g/2)^{1/2}]$$

For cone,

$$T_\text{Cone} = (1/5)(2H/g)^{1/2}[A_c/(C_D A_o)]$$

PROBLEM 7.6

Compute the drainage times for the cylindrical section, the conical section, and the entire cone bottom assembly cylinder. For tank 5 ft in diameter, 10 ft cylindrical section, 10 ft conical section, total height = 20 ft. Assume $C_D = 0.5$.

SOLUTION For cylinder,

$$T_{Cylinder} = [(20)^{1/2} - (10)^{1/2}]/\{(0.5)[(0.00545)/(19.635)](32.2/2)^{1/2}\}$$
$$= [4.472 - 3.162]/\{(0.5)[0002776](4.0125)\}$$
$$= [1.310]/\{0.0005569\} = 2352 \text{ s} = 39.2 \text{ min}$$

For conical section, from Problem 7.2a,

$$T_{Cone} = 18.9 \text{ min}$$

$$\text{Total} = 39.2 + 18.9 = 58.1 \text{ min} \cong 1 \text{ h}$$

Total drainage time is not the sum of the separate drainage times for a cylinder with $H = 10$ ft and a cone with $H = 10$ ft. That would be 94.7 min (also from Problem 7.2a) plus 18.9 min 113.6 min, or 1.89 h (almost 2 h). The difference is that the cylindrical section benefits from the increased head above the outlet for a decreased drainage time. ∎

7.2.6 Vertical Cylindrical Tank with Hemispherical Bottom

Analysis of this configuration follows the same technique as the development of the vertical cylinder on top of the cone in Section 2.5.

$$\overset{\text{Cylinder}}{dt = -(A_T/k)h^{-1/2}dh} \overset{\text{Hemispherical bottom}}{- (\pi/k)[Dh^{1/2} - h^{3/2}]dh}$$

where

$$k = (C_D)(A_o)(2g)^{1/2}$$

For the cylinder, h goes from H_{Total} to $h = D/2$, and for the hemispherical bottom from $h = D/2$ to $h = 0$.

$$T = (2A_T/k)[h^{1/2}]_{H=D/2}^{H_{Total}} + (2\pi/k)[(D/3)h^{3/2} - (1/5)h^{5/2}]_{0}^{H=D/2}$$

$$T = (2A_T/k)[(H_{Total})^{1/2} - (D/2)^{1/2}] + (14/15)(\pi/k)[(D/2)^{5/2}] \quad (7.31)$$

PROBLEM 7.7

Find the time to drain a full tank consisting of a vertical cylinder with a hemispherical bottom and a flat top. Diameter = 5 ft. Cylinder height = 10 ft. Total height = 12.5 ft (cylinder + hemisphere). $C_D = 0.5$. Value of $k = 0.021868$ ft$^{5/2}$/s as in Problem 7.5.

SOLUTION

$$A_T = \pi D^2/4 = 19.635 \text{ ft}^2$$
$$T = [(2)(19.635)/(0.021868)][(12.5)^{1/2} - (5/2)^{1/2}]$$
$$+ (14/15)(3.141592\ldots)/(0.021868)[(5/2)^{5/2}]$$
$$T = [1795.7746][3.5355 - 1.5811] + (134.0842)[9.8821]$$
$$T = 3510 \text{ s for the cylinder} + 1325 \text{ s for the hemispherical bottom}$$
$$T = 58.5 \text{ min} + 22.1 \text{ min} = 80.6 \text{ min} \quad \text{total} \cong 1.34 \text{ h} \cong 1 \text{ h } 20 \text{ min} \quad \blacksquare$$

7.2.7 Vertical Cylindrical Tank with Ellipsoidal Bottom

This arrangement is a cylinder on an elliptical head.

$$\overset{\text{Cylinder}}{dt = -(A_T/k)h^{-1/2}dh} - \overset{\text{Elliptical bottom}}{(\pi/k)(ab)[2(h/c) - (h/c)^2]h^{-1/2}dh}$$

where again

$$k = (C_D)(A_o)(2g)^{1/2}$$

For the cylinder, h goes from H_{Total} to $h = c$, and for the elliptical head (bottom) from $h = c$ to $h = 0$.

$$\int_0^T dt = +(2A_T/k)[h^{1/2}dh]_{H=c}^{H_{\text{Total}}} + (\pi/k)(abc^{1/2})[(4/3)(h/c)^{3/2} - (2/5)(h/c)^{5/2}]_0^{H=c}$$

Total drainage time (T),

$$T = (2A_T/k)[H_{\text{Total}}^{1/2} - c^{1/2}] + (\pi/k)(D/2)^2(c^{1/2})[(4/3) - (2/5) = (14/15)] \quad (7.32)$$

since $a = b = R = (D/2)$ and $(4/3 - 2/5) = 14/15$.

For the specific case where $c = (R/2) = (D/4)$, a typical type of elliptical head,

$$T = (2A_T/k)[H_{\text{Total}}^{1/2} - (D/4)^{1/2}] + (14/15)(1/2)^{1/2}(\pi/k)(D/2)^{5/2} \quad (7.33)$$

PROBLEM 7.8

Find the time to drain a full tank consisting of a vertical cylinder with an ellipsoidal bottom and a flat top. Depth of the elliptical bottom is half the radius of the tank. Cylinder diameter (D) = diameter of elliptical head = 5 ft. Cylinder height (L) = 10 ft. $C_D = 0.5$. Value of k 0.021868 ft$^{5/2}$/s as in previous problems.

SOLUTION

$$\text{Radius of tank } (R) = D/2 = 5/2 = 2.5 \text{ ft}$$
$$\text{Total height } (H_{\text{Total}}) = L + D/4 = 10 + 1.25 = 11.25$$

Cross-sectional area $(A_T) = \pi D^2/4 = (3.141592\ldots)(5^2)/(4) = 19.635$ ft^2

Substituting into Equation (7.33), in which the lead constant $[(14/15)(1/2)^{1/2}(\pi)]$ in the second term equals 2.0733.

$$T = (2)(19.635)/(0.021868)[(11.25)^{1/2}-(1.25)^{1/2}] + [(2.0733)/(0.021868)](2.5)^{5/2}$$
$$T = (1795.7746)[3.3541-1.1180] + [94.8118](9.8821)$$
$$T = 4016 \text{ s for the cylinder} + 937 \text{ s for the elliptical head}$$
$$T = 66.9 \text{ min} + 15.6 \text{ min} = 82.5 \text{ min} \quad \text{total} \cong 1.38 \text{ h} \cong 1 \text{ h } 23 \text{ min}$$

The smaller volume compared to the tank with the hemispherical bottom is offset by the decreased head of liquid, and the overall drainage time is almost the same. ∎

7.2.8 Vertical Cylindrical Tank with Dished Bottom

A dished head is a head having a shape other than flat, hemispherical, or elliptical and is welded onto the end(s) of a cylindrical-type tank, be it horizontal or vertical. Various types of heads are depicted in the cited references [9,10]. Most of them are described by somewhat complicated mathematical formulas. These and other references cited [9–14] serve as useful sources of information in this area.

7.2.8.1 Volume of Horizontal Dished Head

The simplest kind of dished head is fabricated from a section of a sphere whose radius (ρ) is larger than the radius (R) of the cylinder to which it is attached. This larger sphere's center is located along the axis of the cylinder. Distance between the center of the sphere to any point on the head, including points on the end circumference of the cylinder is ρ. Distance between the center of the sphere and the tangent line of the cylinder is ℓ. The depth (d) of the dished head is then

$$d = \rho - \ell \qquad (7.34)$$

or by rearrangement

$$\ell = \rho - d \qquad (7.35)$$

Within the cylinder, for any plane through the central axis,

$$R^2 + \ell^2 = d^2 \qquad (7.36)$$

by the Pythagorean theorem.

Then, substituting for ℓ,

$$R^2 + (\rho-d)^2 = \rho^2 \qquad (7.37)$$
$$d^2 - 2\rho d + R^2 = 0 \qquad (7.38)$$

and

$$d = \rho - (\rho^2 - R^2)^{1/2} \qquad (7.39)$$

by the quadratic formula.

This is the maximum height (h) achieved by a liquid contained within a dished bottom head. Volume for lesser depths of liquid within the head is given by the formula for the volume of a hemispherical head with ρ replacing R.

$$V = \pi h^2[\rho-(h/3)] \tag{7.40}$$

However, this equation applies only for values of h up to and including $h = d$, and not to the point $h = \rho$ as would apply if the radius were the same size as the radius of the cylinder.

Maximum volume contained in the head at $h = d$.

$$V_{\max} = \pi[\rho-(\rho^2-R^2)^{1/2}]^2\{\rho-(1/3)[\rho-(\rho^2-R^2)^{1/2}]\} \tag{7.41}$$

7.2.8.2 Time to Drain Cylindrical Tank with Dished Head

As before, time (T) to drain such a tank is comprised of the time to drain the cylindrical section from $h = H_{\text{Total}}$ to $h = d$ plus the time to drain the dished bottom head from $h = d$ to $h = 0$, where Total = height of the cylindrical portion plus the height (d) of the bottom dish given by Equation (7.39)

$$d = \rho-(\rho^2-R^2)^{1/2} \tag{7.39}$$

$$\overset{\text{Cylinder}}{T = (2A_T/k)[H_{\text{Total}}^{1/2}-d^{1/2}]} + \overset{\text{Dished head}}{(2\pi/k)[(2/3)\rho d^{3/2}-(1/5)d^{5/2}]} \tag{7.42}$$

PROBLEM 7.9

Find the time (T) to drain a vertical cylindrical tank 5 ft in diameter and 10 ft tangent-to-tangent, having a flat top and a bottom comprised of a dished head with a 10 ft dish radius. $A_T = (\pi/4) D^2 = 19.635$ and k (with a 1 in. diameter drain) $= 0.021868$ ft$^{5/2}$/s, as before.

SOLUTION From Equation (7.39), d for a dished head $= \rho - (\rho^2 - R^2)^{1/2}$

$$d = (10)-[10^2-(5/2)^2]^{1/2} = 10-[100-6.25]^{1/2} = 10-[9.6825] = 0.3175 \text{ ft}$$

$$\text{Total height of tank} = 10+d = 10+0.3175 = 10.3175$$

From Equation (7.42),

$$T = [(2)(19.635)/(0.021868)][10.3175^{1/2}-0.3175^{1/2}]$$

$$+ [(2)(3.141592\ldots)/(0.021868)][(2/3)(10)(0.3175)^{3/2}-(1/5)(0.3175)^{5/2}]$$

$$T = [1795.7746][3.2121-0.5635] + [287.3233][1.1927-0.0114]$$

$$T = 4756 \text{ s for the cylinder} + 339 \text{ s for the dished head}$$

$$T = 79.3 \text{ min} + 5.7 \text{ min} = 85.0 \text{ min} \quad \text{total} \cong 1.42 \text{ h} \cong 1 \text{ h } 25 \text{ min} \qquad \blacksquare$$

7.2.9 Vertical Tank of Elliptical Cross Section with Ellipsoidal Bottom

This type of tank can be handled in the same way as a cylindrical tank with an ellipsoidal bottom. In this case, the cross-sectional area of the tank (A_T) is πab rather than πR^2, where a and b are the semiaxes of the elliptical cross section, and R is the radius of the circular cross section of a cylindrical tank. In the first part of the equation for the drainage time, corresponding to the cylindrical section, A_T must be changed from πR^2 or $\pi D^2/4$ to πab. The semiaxes a and b also appear in the formula for the elliptical head forming the bottom of the tank, and the semiaxis c is the depth of the bottom head. The general equation for drainage time is then modified as follows:

$$T = (2\pi ab/k)[H_{\text{Total}}^{1/2} - c^{1/2}] + (14/15)(\pi/k)(abc^{1/2}) \tag{7.43}$$

7.2.10 Horizontal Cylindrical Tank with Flat Ends

7.2.10.1 Volume of Tank

Let us now turn to a series of cylindrical tanks in the horizontal position. The simplest of these is a true cylinder, equipped with flat ends, which are known as *heads*. The end of the cylinder is a circle. The horizontal liquid surface in a partially filled tank is a chord of that circle, parallel to the horizontal diameter, which passes through the center of the circle. The end area of the liquid is a segment of the circle bounded by the chord at the liquid surface and the circular arc of the tank body. It is found by subtraction: the total area of the sector contained between the two radii connecting the center of the circle with the ends of the chord minus the triangular area formed by the center of the circle and the points at both ends of the chord.

$$\text{Area of sector} = R^2\theta \tag{7.44}$$

where R is the radius and 2θ is the central angle (expressed in radians), and $D = (2R)$ is the length of the diameter.

$$\text{Triangular area} = R^2 \sin\theta \cos\theta \tag{7.45}$$

By subtraction, area of liquid (segment of circle),

$$\text{Area of segment} = R^2[\theta - \sin\theta \cos\theta] \tag{7.46}$$

End area of liquid equals area of segment. In Cartesian coordinates [13],

$$\text{End area of liquid} = \{R^2 \cos^{-1}[(R-h)/R]\} - [R-h][2Rh-h^2]^{1/2} \tag{7.47}$$

where h is the vertical height of liquid measured from the bottom of the horizontal tank to the liquid surface and $\cos^{-1}$ denotes the angle in radians whose cosine is the stated argument. A single formula can be used for liquid heights both above and below the centerline because here the cosine of an angle θ greater than 90° is negative. This is computed automatically on a multifunction calculator. (*Caution:* Some calculators/computers/spreadsheet programs are programmed for angles in radians and some in degrees.)

Multiplying by the length of the cylinder (*L*) plus a slight rearrangement gives the volume of the liquid in a partially filled tank.

$$V = R^2 L \cos^{-1}[(1-(h/R)] - [1-(h/R)][1-(1-h/R)^2]^{1/2} \qquad (7.47)$$

A further rearrangement and the introduction of π put the equation on a relative basis with respect to a full tank, whose volume is $\pi R^2 L = \pi (D^2/4) L$.

$$V/(\pi R^2 L) = (1/\pi)\{\{\cos^{-1}[(1-(h/R)]\} - [1-(h/R)][1-(1-h/R)^2]^{1/2}\} \qquad (7.48)$$

This relationship is plotted against $D = (D = 2 \cdot R)$ in Figure 7.2. It runs from a relative volume of zero at $(h/R) = 0$ to 1.0 at $(h/R) = 2.0$ $(h/D = 1.0)$. When the liquid level reaches the center of the circle/cylinder ($h = R$ or $D/2$), the cylinder is exactly half-full.

Graphs of the liquid level–volume relationship for the flat-head horizontal tank and with several other types of heads are depicted in the cited reference [13], although in a slightly different form.

PROBLEM 7.10

Using Equation (7.48) or Figure 7.2, calculate the volume at the 1/4 and 3/4 points in a horizontal cylindrical tank with flat ends.

SOLUTION At the 1/4 point, $(h/R) = 0.5$, $(1 - h/R) = 0.5$, $(1 - h/R)^2 = 0.25$, and $[1 - (1 - h/R)^2] = 0.75$. Computation of the various intermediate functions yields a $(V/(\pi R^2 L)$ of 0.1955 (19.55%). The $\cos^{-1}$ of 0.5 is 1.0472 radians (60°). At the 3/4 point, $h/R = 1.5$,

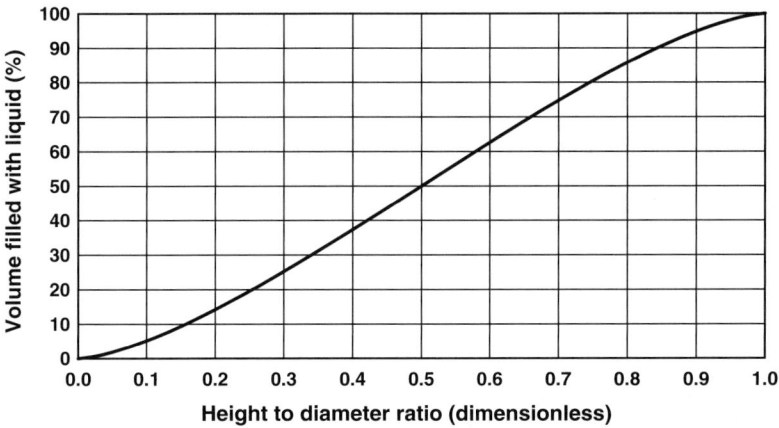

Figure 7.2 Horizontal cylindrical tank with flat ends volume versus height of liquid.

$(1 - h/R) = -0.5$, $(1 - h/R)^2 = 0.25$, and $[1 - (1 - h/R)^2] = 0.75$. $(V/\pi R^2 L)$ is 0.8045 (80.45%). The $\cos^{-1}$ of (-0.5) is 2.0944 radians (120°). The heights at these related points are complementary (19.55% + 80.45% = 100%). Values read from the graph are the same, but *not* with the same degree of precision. ■

7.2.10.2 Draining of the Tank

Derivation of a relationship for the time to drain a horizontal cylindrical tank, where the tank's cross-sectional area in the horizontal plane varies with the height/depth of liquid, requires use of the chain rule of calculus: $(dV/dt) = (dV/dh)(dh/dt)$.

The governing equation,

$$dV/dt = Q_{in} - Q_{out} \qquad (7.8)$$

with $Q_{in} = 0$ reduces to

$$(dV/dt) = -C_D A_o (2gh)^{1/2} \qquad (7.11)$$

as before. Next, the volume relationship of Equation (7.47) is differentiated with respect to t. The result is

$$(dV/dt) = 2L[2Rh - h^2]^{1/2}(dh/dt) \qquad (7.49)$$

This relationship can be verified by integration and rearrangement back to the original equation.

The two expressions Equations (7.11) and (7.49) for (dV/dt) are then equated. Subsequent integration of the limits of $t = 0$ to t and $h = h_o$ to h with $k = C_D A_o (2g)^{1/2}$ followed by substitution of D for $2R$ results in

$$t = [(4/3)(L/k)]\{(D - h)^{3/2} - (D - h_o)^{3/2}\} \qquad (7.50)$$

For $h = 0$, time to *empty* the tank from some arbitrary initial height/depth of liquid (h_o) becomes

$$t = [(4/3)(L/k)]\{D^{3/2} - (D - h_o)^{3/2}\} \qquad (7.51)$$

This equation is consistent with the formula given in Ref. [7] for emptying a horizontal cylinder.

Then, when $h_o = D$, the time to drain a full tank completely to an empty tank is

$$T = (4/3)(L/k)(D)^{3/2} = (4/3)(L)(D)^{3/2}/[C_D A_o (2g)^{1/2}] \qquad (7.52)$$

PROBLEM 7.11

For the previous problem involving a vertical cylindrical tank 5 ft in diameter and 10 ft high, estimate the time to drain a full tank if the tank is turned on its side. Use $C_D = 0.5$, and $A_o = 0.00545 \text{ ft}^2$ for a 1 in. diameter drain. Compare with draining the full vertical tank.

SOLUTION

$$\text{Evaluate } k = C_D A_o (2g)^{1/2}$$

$$k = (0.5)(0.00545)(2 \cdot 32.2)^{1/2} = 0.021868 \text{ ft}^{5/2}/\text{s}$$

Then, from Equation (7.52),

$$T = (4/3)(10)(5)^{3/2}/[0.021868] = (4/3)(10)(11.1803)/[0.021868] = 6817 \text{ s}$$

$$\div 3600 \text{ s/h} = 1.89 \text{ h} \cong 1 \text{ h } 54 \text{ min}$$

Time is somewhat longer than for the same tank in the vertical position ($\sim$1 h 42 min). ∎

7.2.11 Horizontal Cylindrical Tank with Hemispherical Ends

This configuration can be verified alternatively as a horizontal cylinder with an added volume at both ends, or as a sphere cut into half with a cylinder sandwiched in between the two halves. The total volume of the cylinder plus the two hemispheres is

$$V_{\text{Total}} = \text{Cylinder} + 2 \text{ Hemispheres} \tag{7.53}$$
$$\pi R^2 L \qquad (4/3)\pi R^3$$

The volume of a partially filled cylinder [rearranged from Equation (7.48)] is

$$V = R^2 L \{\cos^{-1}[(1-(h/R)] - [1-(h/R)][1-(1-h/R)^2]^{1/2}\}$$

and the volume of a partially filled sphere (or two hemispheres side by side) is rearranged from Equation (7.22)

$$V = \pi h^2 [R - (h/3)]$$
$$V = \pi R^3 (h/R)^2 [1 - (1/3)(h/R)]$$

Relative volume for this configuration is given by

$$V_{\text{Rel}} = \frac{(1/\pi)\{\cos^{-1}[(1-(h/R))]\} - [1-(h/R)][1-(1-h/R)^2]^{1/2}\} + (R/L)(h/R)^2[1-(1/3)(h/R)]}{1+(4/3)(R/L)}$$
(7.54)

An important parameter (R/L) pops out of the mathematics. In tank terminology, it is more commonly expressed as the length–diameter (L/D) ratio.

Plots of the fractional volume of a horizontal cylinder with hemispherical ends are depicted in Figure 7.3. Closely spaced curves are shown for L/Ds of 0 (same as spherical tank), 2 (intermediate tank), and ∞ (cylinder with flat ends). The curves agree exactly at the endpoint and the midpoint ($H/D = 0$, 0.5, and 1.0); slight differences occur at other points.

7.2.11.1 Draining of the Horizontal Tank with Hemispherical Endcaps

Time to drain this type of tank can be found by summing the separate times to drain a cylinder with flat ends plus a sphere (two parallel hemispheres).

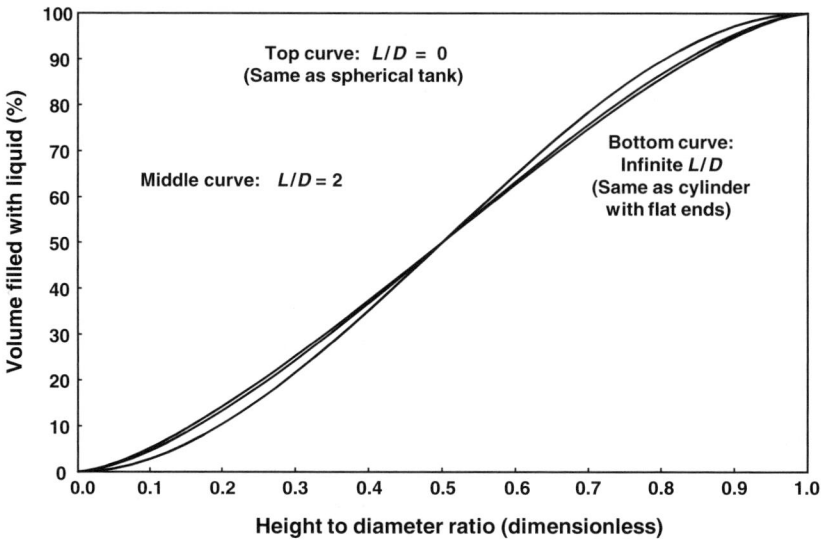

Figure 7.3 Horizontal cylinder with hemispherical ends volume versus height of liquid.

Time to drain completely (T), a totally full tank is given by

$$T = (4/3)(L)D^{3/2}/[C_D A_o (2g)^{1/2}] + 4\pi D^{5/2}/[15 C_D A_o (2g)^{1/2}] \quad (7.55)$$

Time (t) to drain a tank of diameter (D) initially filled to some intermediate point (h_o) all the way to the bottom ($h = 0$) is

$$t = (4/3)(L)[D^{3/2} - (D - h_o)^{3/2}]/[C_D A_o (2g)^{1/2}] + 2\pi[(D h_o^{3/2}/3) - (h_o^{5/2}/5)]/[C_D A_o (2g)^{1/2}] \quad (7.56)$$

Time to drain a tank partially from one intermediate point to another can be found by adding the results of the governing equations for the individual geometric figures.

PROBLEM 7.12

For a horizontal cylinder ($L/D = 2$) with spherical endcaps, calculate the percentage of a full tank occupied when the liquid level is located at three-fourth of the diameter.

SOLUTION Use Equation (7.54) with $h/D = 0.75$, $h/(2R) = 0.75$, $h/R = 1.5$, and $L/D = 2$, $L/(2R) = 2$, $L/R = 4$, $R/L = 0.25$.

$$V_{Rel} = \frac{(1/\pi)\{\cos^{-1}([(1-1.5])-[1-1.5]\cdot[1-(1-1.5)^2]^{1/2}\} + (0.25)(1.5)^2[1-(1/3)(1.5)]}{1+(4/3)(0.25)}$$

$$V_{Rel} = \frac{(1/\pi)\{(2/3)\pi + (0.5)(0.8660)\} + (0.25)(2.25)(0.15)}{1+(1/3)} = \frac{0.8045 + 0.2813}{(4/3)} = 0.8144$$

■

PROBLEM 7.13

Estimate the time to drain completely a 5 ft diameter cylindrical tank with 10 ft of straight side, spherical endcaps, and a 1 in. diameter opening at the bottom. Take $C_D = 0.5$ as in previous problems.

SOLUTION As before $A_o = (\pi/4)(1 \text{ in.})^2/(144 \text{ in.}^2/\text{ft}^2) = 0.00545 \text{ ft}^2$

$$k = C_D A_o (2g)^{1/2} = (0.5)(0.00545)(2 \cdot 32.2)^{1/2} = 0.021868 \text{ ft}^{5/2}/\text{s}$$

From Equation (7.55),

$$T = (4/3)(10)5^{3/2}/[0.021868] + 4\pi 5^{5/2}/[(15)(0.021868)]$$
$$= 6817 + 2142 = 8959 \text{ s for this configuration} = 2.49 \text{ h} \cong 2 \text{ h } 29 \text{ min} \quad \blacksquare$$

7.2.12 Horizontal Cylindrical Tank with Elliptical Ends

Take a horizontal cylindrical tank and attach hemispherical endcaps. Then squeeze it along its length such that the axial dimension (depth) of the endcap (head) is less than its diameter. The head is basically one half an oblate spheroid (i.e., an ellipse rotated around its minor axis). Its major axis corresponds to the diameter of the cylinder.

Table 7.1 lists the relative volume of a single elliptical head at selected points. The total volume of the cylinder plus the two elliptical ends is

$$V_{\text{Total}} = \text{Cylinder} + 2 \text{ Elliptical heads} \quad (7.57)$$
$$\phantom{V_{\text{Total}} = }\pi R^2 L \quad\quad\quad (4/3)\pi f R^3$$

where f is the ratio of the depth of the endcap to the radius of the cylinder. It is also the ratio of the minor axis of the elliptical cross section to its major axis.

The volume of the partially filled cylinder is given by Equation (7.48), and the volume of the two elliptical ends is given by Equation (7.26). The latter is rewritten

Table 7.1 Relative Volume of Vertical Elliptical Head Versus Liquid Depth

Relative depth of liquid (h/c)	Volume relative to full head (v/V)	
0	Empty	0
0.2	10% Full	0.028
0.5	Quarter-full	0.15625
0.7	35% Full	0.28175
1	Half-full	0.5
1.3	65% Full	0.71825
1.5	3/4 Full	0.84375
1.8	90% Full	0.972
2	Full	1.0

below in terms of the cylinder radius and the ratio f. Relative volume for this configuration is given by

$$V_{\text{Rel}} = \frac{(1/\pi)\{\cos^{-1}[(1-(h/R)]\} - [1-(h/R)][1-(1-h/R)^2]^{1/2}\} + f(R/L)(h/R)^2[1-(1/3)(h/R)]}{1+(4/3)(f)(R/L)}$$

(7.58)

The same parameter (R/L) appears in this formula also, as in the case of the cylinder with hemispherical heads. In addition, another parameter (f) appears. When $f = 1.0$, the heads are hemispherical, and the equations for the cylindrical tank with hemispherical ends and for the cylindrical tank with elliptical ends become identical.

Relative volumes for this tank configuration are given in Figure 7.4.

7.2.12.1 Draining of the Horizontal Tank with Elliptical Ends

Volumes for the cylinder and ends are additive, as are their derivatives with respect to height of liquid. The time to empty such a tank completely can, therefore, be found from the sum of Equation (7.52) for the cylinder with flat ends and Equation (7.30) for the spheroidal tank, with the substitution of half the tank diameter ($D/2$) for both a and b and of the product of (f) and ($D/2$) for c.

$$T = (4/3)(L/k)D^{3/2} + (f)(4/15)(\pi/k)D^{5/2}$$

(7.59)

where $k = C_D A_o (2g)^{1/2}$ and (f) is the ratio (between 0 and 1) of the depth of head to ($D/2$).

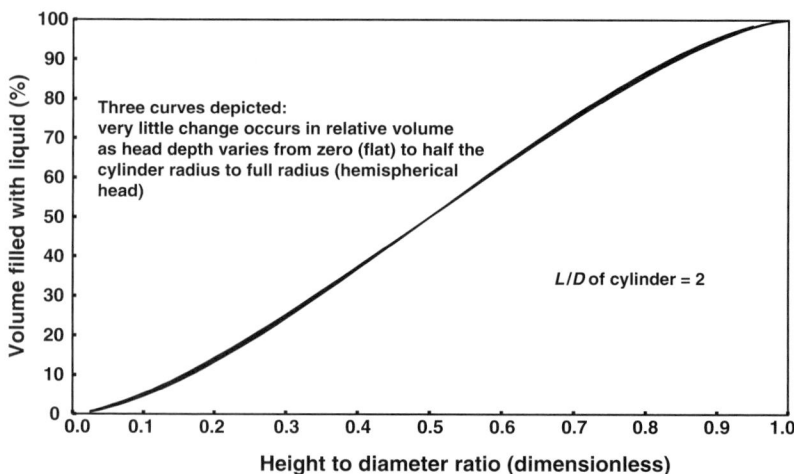

Figure 7.4 Horizontal cylinder with elliptical ends, volume versus height of liquid.

PROBLEM 7.14

Find the time to empty a full horizontal cylindrical tank with elliptical ends. Minor axis of ellipse is one half the tank diameter (a common configuration). Tank is 5 ft in diameter with 10 ft of straight side. Data on the outlet structure is the same as before, a 1 in. diameter opening and a C_D of 0.5.

SOLUTION As before, $A_o = 0.00545$ ft^2 and $k = C_D A_o (2g)^{1/2} = 0.021868$ ft$^{5/2}$/s. Inserting tank information into Equation (7.59),

$$T = (4/3)(10/0.021868)5^{3/2} + (0.5)(4/15)(3.141592\ldots/0.021868)5^{5/2}$$
$$T = [(1.3333)(457.2892)(11.1803)] + [(0.5)(0.2667)(143.6616)(55.9017)]$$
$$T = 6817 + 1071 = 7888 \text{ s for this configuration} = 2.19 \text{ h} \cong 2 \text{ h } 11.5 \text{ min} \quad \blacksquare$$

7.2.13 Horizontal Cylindrical Tank with Dished Ends

Consideration of the partial volume of vertical dished heads on a horizontal cylinder leads to complicated mathematical formulas [10–13]. One of these sources [13] may even contain an error [14]. Differentiating such expressions in which inverse trigonometric functions will survive and then dividing by $h^{1/2}$ will likely lead to terms requiring numerical integration. This situation is beyond the scope of the present study.

Depth (d) of a dished head of the type described in Section 7.2.8 for a vertical tank is

$$d = \rho - (\rho^2 - R^2)^{1/2} \qquad (7.39)$$

in which ρ and R are the radii of the dished head and the cylinder, respectively. When ρ becomes very large, d approaches the zero depth of a flat head. When $\rho = R$, d becomes R and the head is hemispherical. The volume of the dished head lies somewhere between these two extremes.

Time (T) to empty a full horizontal cylindrical tank with flat ends is given by Equation (7.52).

$$T = (4/3)(L/k)(D)^{3/2} = (4/3)(L)(D)^{3/2}/[C_D A_o (2g)^{1/2}] \qquad (7.52)$$

When hemispherical ends are present, drainage time is increased by an additional term from Equation (7.55)

$$(4/15)(\pi/k)D^{5/2} = 4\pi D^{5/2}/[15 C_D A_o (2g)^{1/2}]$$

Volume of the cylinder is $\pi R^2 L$ or $(\pi/4)D^2 L$ and the volume of two hemispherical heads is $(4/3)\pi R^3 = (\pi/6)D^3$. Both the added volume and the time to drain caused by the presence of the heads compared to the total tank including heads are a function of the L/D ratio, as summarized in Table 7.2.

Therefore, by substituting hemispherical ends for dished heads in calculation of volume and drainage time, one would not make more than the positive error shown above. Less discrepancy would occur between a tank with properly chosen elliptical heads and a tank with flat ends.

Table 7.2 Added Volume and Time to Drain for Hemispherical Heads on Horizontal Cylinder

(L/D)	Added volume (%)	Added time (%)
1	40.0	38.6
2	25.0	23.9
3	18.2	17.3
4	14.3	13.6
5	11.8	11.2
6	10.0	9.5
7	8.7	8.2
8	7.7	7.3
9	7.4	6.5
10	6.3	5.9

It is possible to do better for volume. One can approximate the volume of a dished head whose head depth (d) is given by Equation (7.39) by using an elliptical head with a depth of 0.78 times h for calculations. The error in volume is no more than $\pm 4\%$ for (ρ/R) of 1.25 or greater. The error increases from $+4$ to $+12\%$ for (ρ/R) in the range of 1.25–1.05. These figures are lower than the entries in Table 7.2 for (L/D) of 4 or less. For (ρ/R) = 1.0, one would perform the calculations exactly for hemispherical heads.

Furthermore, since the drainage times in Table 7.2 show the same trend as the volume changes, one might assume that drainage times using the elliptical head approximation for dished heads would also be reasonable. Example problems illustrating these points follow.

PROBLEM 7.15

For a horizontal cylindrical tank 12 ft in diameter and having dished heads with $\rho = 10$ ft, calculate the total volume of the *two* heads (a) using the exact formula for the dished heads and (b) using the elliptical approximation.

SOLUTION

(a) *Exact*

$$V = \pi d^2 [\rho - (1/3)d] \tag{7.39}$$

with d given by $d = \rho - (\rho^2 - R^2)^{1/2}$

$$R = (D/2) = 12/2 = 6 \text{ ft}$$
$$d = 10 - [10^2 - 6^2]^{1/2} = 10 - [64]^{1/2} = 10 - 8 = 2 \text{ ft}$$
$$V = (3.141592\ldots)(2^2)[10 - (2/3)] = 117.2861 \text{ ft}^3 \text{ for one head}$$
$$\text{For two heads} = (2)(117.2861) = 234.5723 \text{ ft}^3$$

(b) *Approximation*

$$\rho/R = 10/6 = 1.6667 > 1.25 \text{ approximation good}$$

$$a = (0.78)(d) = 1.56 \text{ ft (semiaxis of ellipse in the head–depth direction)}$$

$$f = a/R = (1.56)/(6) = 0.26$$

V for one head $= (2/3)pfR^3$; for two heads $= (4/3)\pi fR^3$

Calculating volumes

One head : $(2/3)(3.141592\ldots)(0.26)6^3 = 117.6212 \text{ ft}^3$

Two heads : $(4/3)(3.141592\ldots)(0.26)6^3 = 235.2425 \text{ ft}^3$

% Error compared to exact volume,

$$\frac{100 \times (235.2425 - 234.5723)}{234.5723} \cong -0.3\%$$

Volumes are carried out to four decimal places to show the small difference between the exact value and the approximation in this case. ∎

PROBLEM 7.16

Estimate volumes and drainage times for another tank 6 ft in diameter and 15 ft long tangent to tangent. For this other tank, $R = 3$ and $\rho = 30$.

SOLUTION

$$d = \rho - (\rho^2 - R^2)^{1/2} = 30 - (900 - 9)^{1/2} = 0.1504$$

Volume of head $= \pi d^2[\rho - (1/3)d] = (3.141592\ldots)(0.1504)^2[30 - (0.1504)/3] = 2.1283 \text{ ft}^3$

Approximation
Volume of head ellipse $= (2/3)\pi R^2 a = (2/3)(3.141592\ldots)(3^2)(0.1504)(0.78) = 2.2113$

(1) Ignore the heads in calculation of drainage time (d is only ~5% of R).
$T = [(4/3)(15)(6)3/2]/(0.021868) = 13,441.5 \text{ s} = 3.73 \text{ h} \cong 3 \text{ h } 44 \text{ min}$

(2) Assume hemispherical ends.
This adds $(4)(3.141592\ldots)(6)^{5/2}/[(15)(0.021868)] = 3378.2 \text{ s} \cong 0.94 \text{ h} \cong 56 \text{ min}$ and would add about 25% to the drainage time for a cylinder with flat ends.

(3) Use ellipse approximation: $a = (0.78)(d) = (0.78)(0.1504) = 0.1173 \text{ ft}$, where a is the semiaxis of ellipse in the head-depth direction. $f = (a)/(R) = (0.1173)/(3) = 0.0371$
This adds $(4/15)(\pi/k)D^{5/2}(f) = [(4/15)(3.141592\ldots/(0.021868)](6)^{5/2}(0.0371) = 125 \text{ s}$ and adds only about 1% to the time to drain a cylinder with flat ends. ∎

7.2.14 Horizontal Tank with Elliptical Cross Section and Flat Ends

By analogy with the horizontal cylindrical tank with circular cross section and flat ends, the volume of liquid partially filling the subject tank is given by

$$V = L\{(ab)\{\cos^{-1}[(b-h)/b]\} - (a/b)(b-h)[2bh-h^2]^{1/2}\} \quad (7.60)$$

where a is the semiaxis of the ellipse in the horizontal direction and b is the semiaxis of the ellipse in the vertical direction.

The factor (a/b) occurs because of the stretching or compression that occurs in an ellipse compared to a circle, in which all of the circle's radii are the same. The volume is once again the end area times the length (L) of the tank. Equation (7.60) can also be written in terms of the relative depth of liquid (h/b) as

$$V = abL\{\{\cos^{-1}[(1-(h/b)]\} - [1-(h/b)][1-(1-(h/b))^2]^{1/2}\} \quad (7.61)$$

The geometric formula for the total area of an ellipse is πab, and the volume of the full tank then becomes πabL. The fractional volume of liquid in a partially filled tank relative to the volume of a full tank is found by dividing Equation (7.61) by πabL.

$$V_{\text{Rel}} = V/(\pi abL) = (1/\pi)\{\{\cos^{-1}[(1-(h/b)]\} - [1-(h/b)][1-(1-(h/b))^2]^{1/2}\} \quad (7.62)$$

This is the same as the equation for a horizontal cylindrical tank having a circular cross section, with the radius (R) of the circle replaced by the ellipse's vertical semiaxis (b). It follows the same relative volume curve in Figure 7.2 for the tank with a circular cross section with R replaced by b and R^2 replaced by a times b.

7.2.14.1 Draining of the Tank

For substitution into the governing equation for draining the tank

$$(dV/dt) = -C_D A_o (2gh)^{1/2} \quad (7.63)$$

and using of the chain rule of calculus for derivatives: $(dV/dt) = (dV/dh)(dh/dt)$, the derivative (dV/dh) can be found. The result is

$$(dV/dt) = 2(a/b)(L)[2bh-h^2]^{1/2}(dh/dt) \quad (7.64)$$

This is similar to the horizontal cylindrical tank of circular cross section with the substitution of b for R and an additional factor containing the ratio (a/b).

Upon integration and rearrangement,

$$t = (4/3)(a/b)(L/k)\{[(2b-h)^3]^{1/2} - [(2b-h_o)^3]^{1/2}\} \quad (7.65)$$

when $h_o = 2b$ (full tank);

$$t = (4/3)(a/b)(L/k)[(2b-h)^3]^{1/2} \quad (7.66)$$

then when $h=0$, time (T) to empty completely an initially full tank is

$$T = (4/3)(a/b)(L/k)(2b)^{3/2} = (4/3)(a/b)(2b)^{3/2}/[C_D A_o (2g)^{1/2}] \quad (7.67)$$

PROBLEM 7.17

Find the time to drain a horizontal tank, 10 ft long tangent-to-tangent, with elliptical cross section having the same cross-sectional area as a circle 5 ft in diameter and the ratio of major to minor axes 3:2. Perform calculations (a) with the major axis in the horizontal position, and (b) with the major axis in the vertical position. Assume the same values of $A_o = 0.00545$ ft², $C_D = 0.5$, and $k = 0.021868$ ft$^{5/2}$/s as in previous problems. Comment on the results.

SOLUTION Volume of full cylindrical tank with circular cross section 5 ft in diameter and 10 ft long.
For circular cross section,

$$V = \pi L D^2/4 = (3.141592\ldots)(10)(5)(5)/(4) = 196.35 \text{ ft}^2$$

For elliptical cross section,

$$V = \pi a b L$$

Given

$$(b/a) = (3/2), \, b = (3/2)(a)$$
$$V = \pi L (3/2) b^2 = 196.35 \text{ ft}^2$$
$$a = \{196.35/[(\pi)(10)(3/2)]\}^{1/2} = 2.0412 \text{ ft}$$
$$b = (3/2)(a) = 3.0619 \text{ ft}$$

(a) Orient major axis in vertical direction with h measured perpendicular to it and parallel to the minor axis. Therefore, $b = 3.0619$ ft and $a = 2.0412$ ft.
  Plugging values into Equation (7.67) in which b is the *vertical* axis,

$$T = (4/3)(2/3)(10)[(2)(3.0619)]^{1/2}/(0.021868) = 6160 \text{ s} = 1.71 \text{ h} \cong 1 \text{ h } 43 \text{ min}$$

(b) Orient minor axis in vertical direction. Switch a and b and plug into Equation (7.67)

$$T = (4/3)(3/2)(10)[(2)(2.0412)]^{1/2}/(0.021868) = 7544 \text{ s} = 2.10 \text{ h} \cong 2 \text{ h } 6 \text{ min}$$

The same volume of liquid with the same tank discharge characteristics drains from the tank as follows:

(A) Elliptical cross section major axis vertical (elongated in the vertical direction)	Time: 1 h 43 min; $\Delta = -11$ min
(B) Circular cross section	Time: 1 h 54 min; Base case
(C) Elliptical cross section minor axis vertical (elongated in horizontal direction)	Time: 2 h 6 min; $\Delta = 12$ min

366 Chapter 7 Water/Wastewater Draining of Tanks

Tank A with the highest liquid head is the quickest to drain (approximately 6 ft from top of tank to bottom drain), while tank C with the lowest liquid head (approximately 4 ft from top of the tank to bottom drain) is the slowest. Tank B, the tank with the circular cross section and head (5 ft from top of tank to bottom drain) and designated as the base case in the summary above, shows an intermediate time to drain. The cylindrical tank (Tank B) drains even faster in the vertical position (1 h 42 min 26 s from Problem 7.1) because of its still higher head (10 ft from top of cylinder to bottom of tank). ∎

7.2.15 Horizontal Tank of Elliptical Cross Section with Elliptical Ends

This configuration can be evaluated by combining the flat-end case and elliptical ends.

$$V = (L)\{ab\{\cos^{-1}[(b-h)/b]\} - (a/b)(b-h)[2bh-h^2]^{1/2}\} + \pi[abc(h/b)^2][1-(1/3)(h/b)] \quad (7.68)$$

(Tank part; Two elliptical ends)

where a and b are the horizontal and vertical semiaxes, respectively, of the tank's elliptical cross section and c is the depth of the heads. Divide the above expression by $\pi abL + (4/3)\pi abc$ to obtain the relative volume (V_{Rel}) in the tank as a function of liquid height.

7.2.15.1 Draining of the Tank

Equations for time to empty the tank can be obtained by the usual technique—use of Torricelli's theorem in the drainage equation

$$(dV/dt) = -C_D A_o (2gh)^{1/2} = -kh^{1/2} \quad (7.63)$$

use of the chain rule for derivatives, and use of the volume relationships for the separate components to derive an expression for (dV/dh)

$$(dV/dh) = (2)(a/b)(L)(2bh-h^2)^{1/2} + \pi(ac/b^2)(2bh-h^2) \quad (7.69)$$

(Tank part; Two elliptical ends)

and integration of (dV/dh) divided by $-kh^{1/2}$. The results are as follows:
In general, from any level to another level,

$$t = [4aL/(3bk)][(2b-h)^{3/2} - (2b-h_o)^{3/2}]$$
$$+ [\pi ac/(kb^2)][(4/3)b(h_o^{3/2} - h^{3/2}) - (2/5)(h_o^{5/2} - h^{5/2})] \quad (7.70)$$

(Tank part; Two elliptical ends)

For a full tank $h_o = 2b$ to any level,

$$t = [4aL/(3bk)][(2b-h)^{3/2} - (0)] + [\pi ac/(kb^2)][(4/3)b[(2b)^{3/2} - h^{3/2}] - (2/5)[(2b)^{5/2} - h^{5/2}]] \quad (7.71)$$

(Tank part; Two elliptical ends)

For a full tank to completely empty, when $h = 0$ and $t = T$,

$$T = [(4/3)(a/b)(L/k)](2b)^{3/2} + [\pi ac/(kb^2)][(4/3)(b)(2b)^{3/2} - (2/5)(2b)^{5/2}]$$
$$T = [(4/3)(a/b)(L/k)](2b)^{3/2} + [(\pi ac/k)b^{1/2}][(4/3)(2)^{3/2} - (2/5)(2)^{5/2}]$$
$$T = (2^{3/2})(4/3)(a/b)(L/k)b^{3/2} + (2^{1/2})(16/15)(ac)(\pi/k)b^{1/2} \qquad (7.72)$$

PROBLEM 7.18

Replace the flat ends of the tank in Problem 7.17 and drain the tank completely from full to empty. The major semiaxis is 3.0619 ft oriented horizontally and the minor axis is 2.0412 ft, with the ratio 3:2. Values of $A_o = 0.00545$ ft^2, $C_D = 0.5$, and $k = 0.021868$ ft$^{5/2}$/s remain the same. The depth of the elliptical head is 2.0412 ft, the length of the minor semiaxis.

SOLUTION Applying Equation (7.72) with the major semiaxis oriented horizontally; here $a = 3.0619$, $b = 2.0412$ ft, and $c = 2.0412$ ft. The ratio $(a/b) = (3/2)$.

$$T = (2^{3/2})(4/3)(3/2)(10/0.021868)(2.0412^{3/2})$$
$$+ (2^{1/2})(16/15)(3.0619)(2.0412)(3.141592\ldots/0.021868)(2.0412^{1/2})$$
$$T = 7544 \text{ (as before)} + 1935 \text{ (for the heads)} = 9479 \text{ s (total)} = 158 \text{ min} \cong 2\text{ h } 38 \text{ min}$$

Adding heads to the cylindrical tank adds another 32 min to the drainage time ∎

7.3 TRAJECTORY OF THE JET FROM A LEAKING TANK

The jet's instantaneous trajectory from a spontaneous leak of any size can be found by using its source velocity calculated from Torricelli's theorem plus the equations of motion to evaluate downstream conditions.

For a constant acceleration (a),

$$v = at + v_o \qquad (7.73)$$

and

$$s = \frac{1}{2}at^2 + v_o t + s_o \qquad (7.74)$$

where t is time, v_o and v are the initial velocity and the velocity at any time, respectively, and s_o and s are the initial displacement and the total distance traveled, respectively [x (horizontal), y (lateral), and z (vertical)].

The equations are applied in three dimensions, as necessary.

PROBLEM 7.19

As an example, consider an elevated vertical tank that has sprung a leak 25 ft off the ground under atmospheric pressure and 20 ft of liquid head. Tank is cylindrical, 5 ft in diameter. Where will the first packet of water strike, at what velocity, and at what angle?

SOLUTION Note that the tank diameter and the shape of its cross section, if constant, are irrelevant to this problem, as is the size of the leak.

First determine the initial velocity of the jet issuing from the tank. Assume the worst case of $C_D = 1.0$. Disregard air friction, turbulence, and wind effects. Carry extra decimal places and round off at end.

$$v = (2gh)^{1/2} \tag{7.2}$$

$$v = [(2)(32.2)(20)]^{1/2} = 35.8887 \text{ ft/s}$$

Next, apply the equations of motion, first in the vertical direction (z) and then in the horizontal direction (x). Problems of this type can be more easily visualized by sketching the velocity (*not* the distance) versus time. The constant slope of such a curve (not shown) is the acceleration, and the area under the curve is the total distance traveled.

In the vertical direction, the acceleration is due to gravity with $a = -g$ (the gravitational constant, ~32.2 ft/s^2), but there is zero initial velocity (v_0) here. Total vertical distance traveled is the triangular area, equal to $s - s_0 = 25$ ft. Base of the triangle is time (t) and its elevation is $v_z = 0 - gt_{\text{final}}$.

$$\text{Triangular area} = 25 = \frac{1}{2}(t)(v_z) = \frac{1}{2}(t)(-gt) = -\frac{1}{2}gt^2 = -\left(\frac{1}{2}\right)(32.2)t^2$$

From this,

$$t = [(2)(25)/32.2]^{1/2} = 1.246112 \text{ s}$$

and

$$v_z = -gt = -(32.2)(1.246112) = 40.1248 \text{ ft/s}$$

In the horizontal (x) direction, the velocity is constant at its initial value, which is 35.8887 ft/s from above, and the time-of-travel (t) is the same at 1.246112 s. The area to be sketched is a rectangle, with its vertical dimension equal to v_0, its horizontal dimension equal to the time-of-travel, and the area equal to the total distance.

$$s - s_0 = v_0 t = (35.8887)(1.246112) = 44.7213 \text{ ft}$$

Vector components of velocity

$$v_x = 35.8887 \text{ ft/s (to the right)}$$

$$v_z = -40.1248 \text{ ft/s (downward)}$$

$$\text{Angle}(\theta) = \tan^{-1}[v_z/v_x] = \tan^{-1}[-40.1248/35.8887] = \tan^{-1}[-1.1180]$$

$$\theta = -48.1897° \text{ (with respect to the horizontal)}$$

By the Pythagorean theorem,

$$v = [v_x^2 + v_y^2]^{1/2} = [(35.8887)^2 + (-40.1248)^2]^{1/2}$$

$$= 2898^{1/2} = 53.8331 \text{ ft/s}$$

Check

$$v_x = v\cos(\theta) = (53.8331)(+0.6666665) = 35.8887 \text{ (to the right)}$$

$$v_x = v\sin(\theta) = (53.8331)(-0.7453562) = -40.1248 \text{ (downward)}$$

Summary (with rounded numbers)

Time of packet of water to reach the ground	1.25 s
Vertical distance traveled	25 ft (given)
Horizontal distance traveled	44.7 ft
Horizontal component of velocity (from Torricelli's theorem)	35.9 ft/s
Vertical component of velocity	40.1 ft/s
Velocity (at an angle of $-48.19°$ from the horizontal)	53.8 ft/s

Additional reading on the subject of jet trajectory can be found in Refs. [2,3,6,15]. ∎

PROBLEM 7.20

What is the complete trajectory of the jet in the previous problem from its origin in the side of the tank all the way to the ground?

SOLUTION With calculations as in the example above, the following tabulation can be developed:

A general equation describing a jet trajectory is given in the cited references [2, pp. 115–116; 3, p. 105–107; 6, pp. 169–170; 15, pp. 104–106].

$$z = (v_z/v_x)(x) - gx^2/(2v_x^2) \tag{7.75}$$

This relationship is useful in checking the results above. It is obtained by eliminating the time (t) between the x and z equations.

For horizontal flow from a vertical orifice, as in this case, $v_z = 0$ and the equation reduces to

$$z = -gx^2/(2v_x^2) \tag{7.76}$$

with z becoming increasingly negative in a traditional coordinate system.

The negative sign can be eliminated from Equation (7.76) by denoting z as positive downward. Then,

$$z = +gx^2/(2v_x^2) \tag{7.77}$$

If desired, elevation of the jet above grade from the entries in Table 7.3 can be plotted against its horizontal distance from the source of the leak, producing the curve shown generally to the right in Figure 7.5. Such a plot can also be produced by application of Equation (7.77) at selected values of x. The value of z calculated from these equations is measured with respect to the source of the leak. It must be adjusted by the elevation of the leak above grade to obtain its elevation with respect to the grade level.

The trajectory takes a familiar form as the liquid drops gradually over the horizontal distance and the vertical velocity increases from zero to a positive downward value under the influence of gravity. If the jet could hold together indefinitely and the exit velocity were constant over such a long period, it would become vertical (downward) at infinite distance from the source. For example, at a z of 10 million feet below the source, the x-coordinate would be over 5.33 miles. Theta (θ) for the jet velocity at this point would be greater in absolute value than 89.9° (downward), approaching the vertical.

Problem 7.21 introduces a vertical component of the jet arising from a leaking tank. The jet is assumed to leave the tank normal (perpendicular) to the tank surface at the source of the leak. ∎

370 Chapter 7 Water/Wastewater Draining of Tanks

Table 7.3 Items Related to Jet Trajectory

Vertical distance from leak	Elevation above grade (ft)	Time (t) in s	Horizontal displacement (x) in jet (ft)	v_x (ft/s)	v_z (ft/s)	v (ft/s)	$\theta°$
0	25	0	0	35.9	0	35.9	0
0.5	24.50	0.18	6.32	35.9	5.7	36.3	8.98
1.25	23.75	0.28	10.00	35.9	9.0	37.0	14.04
2.5	22.5	0.39	14.14	35.9	12.7	38.1	19.47
5	20	0.56	20.00	35.9	17.9	40.1	26.57
7.5	17.5	0.68	24.49	35.9	22.0	42.1	31.84
10	15	0.79	28.28	35.9	25.4	44.0	35.26
12.5	12.5	0.88	31.62	35.9	28.4	46.0	38.33
15	10	0.97	34.64	35.9	31.1	47.5	40.89
17.5	7.5	1.04	37.42	35.9	33.6	49.1	43.09
20	5	1.11	40.00	35.9	35.9	50.8	45.00
22.5	2.5	1.18	42.43	35.9	38.1	52.3	46.69
25	0	1.25	44.72	35.9	40.1	53.8	48.19

PROBLEM 7.21

Now turn the 5 ft diameter cylindrical tank on its side (25 ft above grade) and calculate the trajectory of a jet issuing from a hole in the top side of the tank at an angle of 30° from the horizontal.

SOLUTION
(a) Calculate the coordinates of the jet.
   x-coordinate (with respect to the center of the tank)
   $$= R\cos(\theta) = (D/2)\cos(30°) = (5/2)(0.8660) = 2.1651 \text{ ft}$$

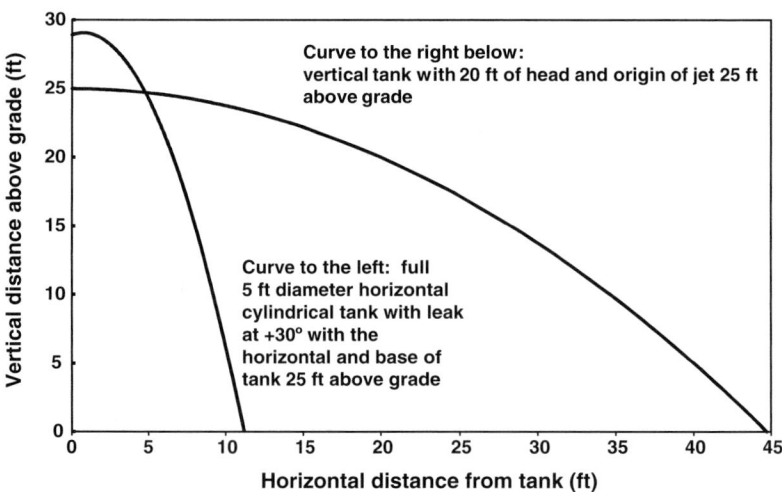

Figure 7.5 Jet trajectories from leaking tanks.

7.3 Trajectory of the Jet from a Leaking Tank

z-coordinate (with respect to the center of the tank)

$$= R\sin(\theta) = (D/2)\sin(30°) = (5/2)(0.5000) = 1.25 \text{ ft}$$

(b) Calculate jet velocity.

Head of the liquid above the point of leak (full tank)

$$= (D/2) - 1.25 \text{ (from above)} = 2.5 - 1.25 = 1.25 \text{ ft}$$

Velocity of the jet from Equation (7.2) (worst case with $C_D = 1.0$):

$$v = (2gh)^{1/2} = [(2)(32.2)(1.25)]^{1/2} = 8.9722 \text{ ft/s}$$

Resolve velocity into its vector components:

$$v_x = v\cos(\theta) = 8.9722(0.8660) = 7.7699 \text{ ft/s}$$
$$v_z = v\sin(\theta) = 8.9722(0.5000) = 4.4861 \text{ ft/s}$$

(c) Apply Equation (7.75)

$$z = (v_z/v_x)(x) - gx^2/(2v_x^2) \tag{7.75}$$

with inputs of v_x and v_z to calculate z versus x.

For example, at $x = 1$ with respect to the source of the leak,

$$z \text{ (with respect to the leak)} = (4.4861)/(7.7699)(1) - (32.2)(1^2)/(7.7699^2)$$
$$= 0.5774 - 0.2667 = +0.3107 \text{ ft}$$

The coordinates of x and z are determined with respect to the source of the leaking jet. The x-coordinate with respect to the tank surface at the horizontal diameter

$$= (x\text{-coordinate at leak}) - R[1 - \cos(30°)]$$
$$= (x\text{-coordinate at leak}) - 0.3349 = 1 - 0.3349 = 0.6651 \text{ ft}$$

The vertical z-coordinate (vertical) of the source of the leak with respect to grade

$$= 25 + (D/2) + R\sin(30°) = 25 + 2.5 + 1.25 = 28.75 \text{ ft}$$

This number is added to the value of z from Equation (7.75) to determine the elevation of the jet above grade.

$$z \text{ with respect to grade} = 28.75 + 0.3107 = 29.0607$$

The calculation is repeated to find values of z corresponding to other values of x. Calculations of z versus x carried out in a spreadsheet are summarized in the curve shown generally to the left in Figure 7.5.

With a nonzero vertical source jet velocity at the source (v_z), the curve of Equation (7.75) describing the trajectory is a parabola parametric in v_x and v_z and passing through the origin of coordinates (0, 0) situated at the leak [2, pp. 115–116; 3, p. 105–107; 6, pp. 169–170; 15, pp. 104–106]. The maximum point of the trajectory occurs at

$$x_{max} = v_x v_z / g \tag{7.78}$$

and

$$z_{max} = v_z^2/(2g) \tag{7.79}$$

found by setting the derivative $(dz/dx) = 0$ and solving for x_{max} and z_{max}.

Here $x_{max} = (7.7699)(4.4861)/32.2 = 1.0825$ ft, and $z_{max} = (4.4861^2)/[(2)(32.2)] = 0.3125$, both with respect to the leak. This puts the coordinates of the maximum point of the

372 Chapter 7 Water/Wastewater Draining of Tanks

jet trajectory for the subject curve in Figure 7.5 at

$$x: \quad 1.0825 - 0.3349 = 0.7475 \text{ ft from the tank}$$
$$z: \quad +0.3125 + 28.75 = 29.0625 \text{ ft above grade}$$

The curve around the maximum point is fairly steep. ∎

PROBLEM 7.22

A nominal 600 gal horizontal tank with hemispherical endcaps (heads), 3 ft in diameter and 10 ft long tangent-to-tangent, is mounted on 4 ft high saddles standing on grade. The tank is to be enclosed in a diked area for containment of its contents in the event of a leak. If the walls of the containment area are located 3 ft from the sides and ends of the tank, how tall should the walls be constructed to intercept a liquid jet leaking from the tank?

SOLUTION

(a) First, calculate the actual volume of the tank and the volume of the containment area per ft of wall height. Assume the stated measurements for the tank are inside dimensions.

$$\text{Volume of the tank} = \text{volume of cylinder} + \text{volume of heads}$$

$$(\text{two heads} = \text{one sphere})$$

$$= (\pi/4)D^2 L \text{ (cylinder)} + (\pi/6)D^3 \text{ (sphere)}$$

$$= (\pi/4)(3^2)10 + (\pi/6)(3^3) = 70.6858 \text{ ft}^3 + 14.1372 \text{ ft}^3 = 84.8230 \text{ ft}^3$$

$$\times 7.48052 \text{ gal/ft}^3 = 635 \text{ gal}$$

Volume of containment area:

$$\text{Length of tank} = (3/2) + 10 + (3/2) = 13 \text{ ft}$$
$$\text{Add } 3+3 \text{ to obtain length of containment area} : 13 + 3 + 3 = 19 \text{ ft}$$
$$\text{Width of tank} = 3$$
$$\text{Add } 3+3 \text{ to obtain width of containment area} : 3 + 3 + 3 = 9 \text{ ft}$$

Volume per feet of wall height

$$= 19 \times 9 = 171 \text{ ft}^3 \times 7.48052 = 1279 \text{ gal}$$

Therefore, only about 6 in. of diked area [(635/1297) × 12 = 5.96 in.] would be required to hold the contents of the tank when it is leaking. However, the jet must be considered.

(b) Calculate the elevation of the jet at a horizontal distance (x) 3 ft away from the tank. Start with a leak along the horizontal diameter of the tank, where the horizontal distance of the tank is closest to the wall. For this case, the leak is horizontal, as in previous problems. Use the governing equations and some geometry.

$$\text{Liquid head above the leak} = (D/2) = 1.5 \text{ ft}$$
$$\text{Velocity } (v), \text{ of the leak} = (2gh)^{1/2} = 9.8285 \text{ ft/s (horizontal)}$$
$$z = (0) - (32.2)(3^2)/[(2)(9.8285^2)] = -1.5000 \text{ ft, from Equation (7.75)}$$

This is the vertical distance with respect to the leak. For elevation above grade, subtract this number from the vertical elevation of the leak.

7.3 Trajectory of the Jet from a Leaking Tank

Elevation of the leak (above grade) $= 4 + (D/2) = 4 + 1.5 = 5.5$ ft

Then, $5.5 - 1.5 = 4.0$ ft.

Therefore, the enclosing wall must be a minimum of 4 ft above grade to intercept a leak halfway up the tank (5.5 ft, or 5 ft 6 in. above grade). In reality, some safety factor would be added to ensure that the jet and its splash are contained by the wall of the dike.

(c) However, one cannot assume that the leak is located only at the horizontal diameter. Investigate the jet from a leak at other locations along the tank's circumference from $+90°$ (upward) to $-90°$ (downward). In this notation, the horizontal is denoted as $0°$.

As a sample calculation, consider an angle of $20°$ upward from the horizontal. Find the elevation of the jet at the wall location and the coordinates of the maximum point of the jet.

Again, from geometry

Elevation of the leak with respect to centerline of tank
$= 5.5$ (above grade) $+ R \sin(20°) = 5.5 + 1.5(0.3420) = 6.0130$ ft

Elevation of top of tank above grade $= 4 + D = 4 + 3 = 7$ ft

Liquid head above leak $= 7 - 6.0130 = 0.9870$ ft

Horizontal distance from leak to wall
$= 3 + [(D/2) - R\cos(20°)] = 3 + 1.5 - 1.5(0.9397) = 3.0905$ ft

Velocity (v) of the leak $= (2gh)^{1/2} = [(2)(32.2)(0.9870)]^{1/2} = 7.9726$ ft/s

Vector components

v_x of leak $= v\cos(20°) = 7.4918$ ft/s

v_z of leak $= v\sin(20°) = 2.7268$ ft/s

Applying Equation (7.75)

$$z = (v_z/v_x)(x) - gx^2/(2v_x^2)$$

$$z = [(2.7268)/(7.4918)](3.0905) - (32.2)(3.09052)^2/[(2)(7.49182)^2]$$

$$= 1.1249 - 2.7398 = -1.6149 \text{ ft}$$

This is the elevation with respect to the leak.

Elevation with respect to grade $= 5.5 + 1.5\sin(20°) - 1.6149 = 4.3981$ ft, or ~ 4 ft 4.75 in.

This is already higher than the 4 ft calculated in part (b) for the $0°$ location. Orientation of the leak at other angles must be investigated in part (d) that follows.

(d) Repeat the calculation for z at the wall for other angles. This can be conveniently accomplished by setting up a spreadsheet to go through the calculations in part (c) as the angle is varied from $+90°$ to $-90°$. Results from $+25°$ to $-25°$, where a fairly shallow maximum point can be seen, are shown as the upper curve in Figure 7.6. The maximum height of wall required is 4.46 feet (approximately 4 ft 5.5 in.) when the leak occurs at $+15.41°$.

(e) This high wall would necessitate ladders, either built-in or portable, to climb over and enter the diked area. If, for example, it is possible to lower the tank elevation to 3 ft and move the wall location to 4 ft away, required wall height would be only 2.1044 ft (~ 2 ft 1.25 in.). In this case, critical angle changes to $10.10°$ (Figure 7.6 lower curve). This would allow one to traverse the dike more easily without mechanical assistance.

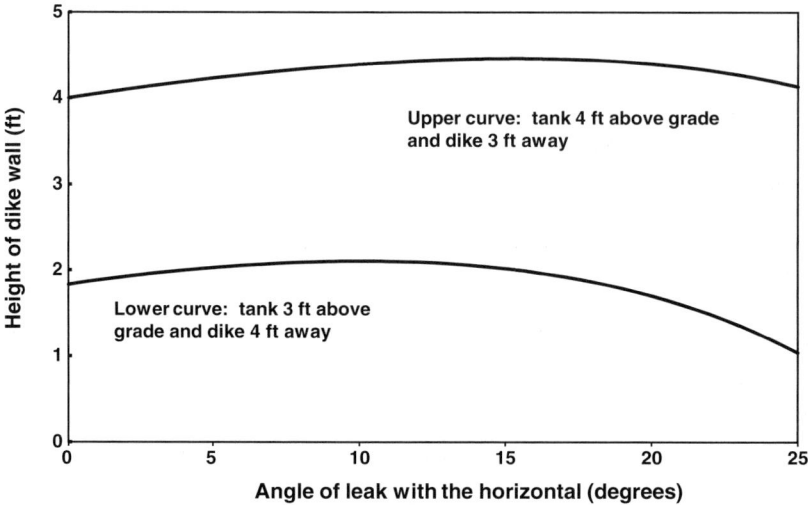

Figure 7.6 Height of dike required to intercept leaking jet from 3 ft diameter horizontal cylindrical tank.

There are, of course, an infinite number of combinations of height and distance that could be investigated to allow easier access within operational and cost constraints. Such an optimization is left to the reader. ∎

REFERENCES

1. *Funk & Wagnalls New Encyclopedia*, Vol. 10, Funk & Wagnalls, New York, 1971, Vol. 10, p. 452 and Vol. 23, p. 198.
2. R.L. Daugherty and A.C. Ingersoll, *Fluid Mechanics with Engineering Applications*, 5th edn, McGraw-Hill, New York, 1954, 472 pp.
3. R.L. Daugherty and J.B. Franzini, *Fluid Mechanics with Engineering Applications*, 6th edn, McGraw-Hill, New York, 1965, 574 pp.
4. V.L. Streeter, *Fluid Dynamics*, McGraw-Hill, New York, 1948, pp. 174–177.
5. W.L. McCabe, J.C. Smith, and P. Harriott, *Unit Operations of Chemical Engineering*, 5th edn, McGraw-Hill, New York, 1993, pp. 107–108.
6. J.B. Franzini and E.J. Finnemore, *Fluid Mechanics with Engineering Applications*, 9th edn, WCB/McGraw-Hill, Boston, MA, 1997, 807 pp.
7. T.C. Foster, Time required to empty a vessel, *Chemical Engineering*, **88**(9), 105, 1981.
8. R.B. Bird, W.E. Stewart, and E.N. Lightfoot, *Transport Phenomena*, Wiley, New York, 1960, pp. 226–229.
9. A. Tankha, Selecting formed heads for cylindrical vessels, *Chemical Engineering*, **88**(11), 89, 1981.
10. A.K. Escoe, *Mechanical Design of Process Systems*, 2nd edn, Vol. 2, Appendix A: Partial Volumes and Pressure Vessel Calculations, Gulf Publishing Co., Houston, TX, 1995, pp. 179–187. This same material is also contained in Vol. 1.
11. D. Jones, Computing fluid tank volumes, *Chemical Processing*, **65**(11), 46–50, 2002.
12. D. Jones, Compute fluid volumes in vertical tanks, *Chemical Processing*, **66**(12), 29–31, 2003.
13. G. Kowal, Quick calculation for holdups in horizontal tanks, *Chemical Engineering*, **80**(13), 130, and 132, 1973.
14. W.J. Uhlarik, Programmable calculation: letter to the editor, *Chemical Engineering*, **84**(16), 5, 1997.
15. R.L. Daugherty and J.B. Franzini, *Fluid Mechanics with Engineering Applications*, 7th edn, McGraw-Hill, New York, 1977, 564 pp.

Chapter 8

Solid Waste

Never buy a PIC in a POHC, and NEVER EVER wear a red shirt in a STAR TREK episode.

8.1 INTRODUCTION

Before beginning a chapter on solid waste calculations, it is instructive to present some regulatory background. This discussion is perforce general in nature because some specifics may have evolved between the writing and the publication of this book, and its main purpose is *not* to provide a detailed exposition of regulations. However, the general tenor of those regulations is not likely to have changed.

This chapter discusses getting rid of unwanted "stuff," either by having it hauled away or by burning it with or without energy recovery. Although this book is not about regulations *per se*, one must understand enough about regulations to know whom to call to have your stuff hauled away and what to do while it accumulates awaiting disposal. Then there is the question of which permits/registrations must be obtained by the owner of the stuff, by the hauler, and by the facility where the stuff will ultimately reside. Another question is whether a manifest must be prepared to allow transport of the stuff. Also to be understood from a liability standpoint is who is responsible for what, should something go wrong along the line.

Since regulations are subject to change, it is always a good idea (meaning "just do it!") to look up current regulations for oneself in the applicable subject area and governmental jurisdiction. Be sure to consult all levels of government since regulations are often built upon one another but can sometimes seem contradictory and conflicting.

The stuff to be disposed of is normally called *waste*. Various categories of waste material have been defined. Examples include the designations listed below in alphabetical order. Some of these categories are overlapping, and their exact meaning may vary from jurisdiction to jurisdiction. This list is in no way claimed to be complete or all inclusive.

Environmental Calculations: A Multimedia Approach, by Robert G. Kunz
Copyright © 2009 John Wiley & Sons, Inc.

8.2 SELECTED WASTE DESIGNATIONS/DEFINITIONS

8.2.1 Coal Wastes

Mining of coal produces waste solids, sludges, slurries, and liquids when the desired coal product is separated from contaminating earthen debris by a series of physical and/or chemical processing steps, including water washing. Coal preparation means chemical or physical processing and the cleaning, concentrating or other processing, or preparation of coal; *coal processing waste* means earth materials that are separated and wasted from the product coal during cleaning, concentrating, or other processing or preparation of coal (30 CFR 701.5).

For the most part, the separation is not perfect, and some coal material is left in the waste. Low-level coal refuse solids from anthracite (hard coal) are called *culm*, and from bituminous (soft coal), they are termed *gob*. Disposal of coal processing waste is regulated by the federal government and/or the states.

Burning of coal produces several types of leftover materials, including fly ash and bottom ash and by-products from flue gas desulfurization, if employed. These latter include the residues from alkaline scrubbing processes and fluidized bed combustion, in which coal is burned in a fluidized bed along with limestone material (see Problem 4.3).

All these wastes contain heavy metals concentrated from the coal plus compounds of the removed sulfur. Care must be exercised in their reuse or disposal, especially under acidic conditions. The various governing state and local regulations are in place to ensure that such care is taken.

8.2.2 Construction/Demolition Waste/Debris

This is waste material resulting from construction, alteration, rehabilitation, repair, or demolition of buildings or other man-made structures, including but not limited to

- concrete
- steel
- rock/stone
- brick
- lumber/wood
- plaster
- drywall/gypsum wallboard
- roofing materials
- paper
- cardboard
- rubber
- plastics

- plumbing/heating equipment/fixtures
- electrical wiring/components.

Definition may vary by state, as will the type of landfill authorized for disposal. In general, construction debris is not a hazardous waste unless a specific component exhibits some hazardous characteristic or contains such constituents as hazardous liquids or asbestos. Asbestos waste material, defined in 40 CFR 61 Subpart M, is subject to specialized removal and disposal practices by trained personnel.

8.2.3 Garbage (See Refuse/Rubbish/Trash)

According to a definition in use in the State of Texas (30 TAC 330.3), garbage is solid waste composed of putrescible animal and vegetable wastes that result from the handling, preparation, cooking, and consumption of food. These include wastes from facilities that store, handle, and sell produce and other food products. The dictionary definition of *putrescible* relates to the presence of organic matter capable of being decomposed typically by anaerobic bacteria and fungi to foul-smelling, only partially oxidized decay products.

8.2.4 Hazardous Waste [1]

Hazardous waste (a subcategory of solid waste) is defined in a section of the Code of Federal Regulations (40 CFR 261.3). It must first be a solid waste as defined in 40 CFR 261.2; those characteristics of a solid waste are summarized in another entry herein. A waste is considered hazardous if it is specifically listed as a hazardous waste and is not specifically excluded, or if it exhibits any of the four defined characteristics of hazardous waste—ignitability, corrosively, reactivity, toxicity. EPA also provides specific examples of hazardous wastes to assist small businesses/small quantity generators to determine the proper EPA hazardous waste code numbers for categories of commonly generated hazardous wastes [2, pp. 31–36]. However, tabulations and examples of what is and what is not a hazardous waste are too voluminous to reproduce here.

Hazardous waste regulations are administered by EPA and/or at the state level [1, Chapter 10]. A state program may be broader in scope and more stringent than the EPA's; it may not be narrower or less stringent. *Broader in scope* means, for example, that additional wastes may be included by the state. *More stringent* relates to a state's administrative requirements such as reporting frequency being stricter than its federal counterpart [1, p. 286]. The state's program must be equivalent to and consistent with the EPA's, it must be consistent with other state programs, and it must be capable of being adequately enforced [1, p. 273].

In summary, the topic of hazardous waste is complex and its definition somewhat convoluted. EPA acknowledges that the regulations are complex [2, p. 1]. It is therefore again strongly recommended that the most up-to-date version of the applicable laws and regulations (both federal and state) be consulted for details. Additional commentary can also be found elsewhere [1].

8.2.5 Hazardous Household Waste

There are a number of items that are exempted from the federal definition of hazardous waste simply because they are generated by individual households or similar sources. (See Section 8.2.9 for the definition of such sources.) Nevertheless, the source of these waste materials does not make them any less dangerous to human health or the environment.

These things include chemical agents commonly used as or a part of common household items, including their containers when "empty." A representative listing is shown below. Some of these items correspond on a household scale with items classified as universal wastes, discussed in another section.

- paints and solvents
- chemical cleaners and polishes
- automotive fluids and car care products
- garden chemicals (pesticides, herbicides, etc.)
- mothballs
- photographic chemicals
- dry cell batteries
- mercury-containing products (fluorescent and other lamps, thermometers, thermostats).

(Recall that the list of materials defined as hazardous can vary from state to state.)

Although the amount of hazardous household waste (HHW) per household is relatively small, the amount of waste in the aggregate becomes quite significant. EPA estimates that Americans generate 1.6 million tons of HHW per year and that the average home can accumulate as much as 100 pounds of HHW in the basement, garage, and storage closets [3]. Based on U.S. Census figures [4], the average per capita generation for the United States as a whole amounts to about 11 lb/year, compared to 4 lb/person/year published for residents of the Commonwealth of Pennsylvania as of 1999 [5]. With 2.59 persons/household in the year 2000 [4], the U.S. figure translates into about 30 lb of HHW/household/year. The analogous figure for Europe is given as 1–5 kg/household/year (2.2–11 lb/household/year) [6, p. 99].

In order to keep household hazardous waste out of municipal solid waste (MSW), as well as to prevent its being stored/stockpiled indefinitely in the home, numerous collection programs have sprung up at the state and local levels, more than 3000 as of 1997 [3]. These range from annual special collection days to periodic events and year-round programs (either open all the time or available by appointment). Waste collection may take place at a fixed site or at a transient special location; it may be serviced by a mobile facility. The collected material is then disposed of as hazardous waste by the governmental entity or recycled/reused when possible.

Other nasty materials such as new and used antifreeze, used motor oil and oil filters, and lead acid batteries may also be collected along with HHW at some of these centers, as well as latex paint and medical waste, including sharps. However, explosives, fireworks, ammunition, compressed gas cylinders, used tires, and radioactive materials are not normally accepted.

There may be a limit on the maximum size of container and/or the total amount of waste or amount of waste per visit to be dropped off by an individual resident. Items of hazardous waste from businesses are usually excluded altogether.

8.2.6 Industrial (Solid) Waste

This is waste material created as an undesired, unusable leftover by-product from an industrial or manufacturing process of some sort, including mining and electric power generation. A specific industrial waste may or may not be hazardous.

8.2.7 Infectious/Medical Waste

EPA defines *infectious waste* as a waste that contains pathogens with sufficient virulence and quantity so that exposure to the waste by a susceptible host could result in an infectious disease [7]. EPA estimates that only 10–15% of the medical waste generated at health care facilities is infectious. The more general term *medical waste* is defined as any solid waste that is generated in the diagnosis, treatment, or immunization of human beings or animals, in research pertaining thereto, or in the production or testing of biologicals. Sources of such waste include hospitals and clinics, veterinary hospitals and clinics, physicians' and dental offices, nursing homes, schools, blood banks, mortuaries, medical research facilities/laboratories, and the like. Infectious waste may also be termed *biohazardous waste*, *biomedical waste*, *regulated medical waste*, *biological and chemotherapeutic waste*, or *pathological waste* in the literature of other regulatory jurisdictions. It may also be considered to be a type of special waste or hazardous material.

Examples of at least potentially infectious wastes include

- removed human tissue, organs, body parts
- human blood and blood products (serum, plasma, other blood components)
- human waste and body fluids (e.g., urine, feces)
- blood soaked bandages/dressings
- disposable items contaminated with human blood or body fluids (surgical gloves, clothing, towels, sponges, sorbent liners, surgical instruments)
- contaminated disposable glass and plastic laboratory ware, culture dishes
- cultures, stocks, and swabs of/containing infectious agents
- dialysis waste, laboratory wastes, waste from postmortem examinations, waste from operating rooms and other high hazard areas in a medical facility
- contaminated animal tissue, body parts, carcasses; infectious animal bedding; animal urine, feces, blood; disposable animal cages
- sharps used in patient treatment/care

Sharps are syringes/needles; intravenous (IV) tubing with attached needles; scalpels and scalpel blades; lancets; glass vials, tubes, and slides; pipettes; and broken glass. Sharps are disposed of in special sealed containers (sturdy, leak-proof, puncture resistant, and regulated by the Food and Drug Administration (FDA)) to prevent waste-handling workers and others of the public from being exposed to potential stick injuries and potential exposure to infection from serious diseases such as HIV/AIDS and hepatitis [8].

Medical wastes generated in the United States and not containing mercury or other toxic metals or radioactive isotopes/materials are regulated at the state and local levels. Those toxic and radioactive wastes may be governed by Resource Conservation and Recovery Act (RCRA) hazardous regulations or by Nuclear Regulatory Commission (NRC) regulations. As of this writing, more than 90% of potentially infectious medical waste is incinerated and is subject to EPA regulations for medical waste incinerators. EPA also has jurisdiction over chemical treatment of medical waste, and the Department of Transportation (DOT) regulates medical waste transportation under 49 CFR, Sections 172 and 173.

8.2.8 Liquid Waste

A definition of liquid waste is provided by the State of Texas as any waste material containing "free liquids," determined in accordance with the EPA test method specified (30 TAC 330.3). The presence of free liquids in a representative sample of waste can be determined by EPA SW-846 [9] Method 9095, "Paint Filter Liquids Test."

8.2.9 Municipal Solid Waste (MSW)

Municipal solid waste, also known as *household waste*, *trash*, or *garbage*, means any material derived from single and multiple residences, hotels/motels, bunkhouses, ranger stations, crew quarters, campgrounds, picnic grounds, and day use recreation areas [40 CFR 261.4(b)(1)]. The definition(s) may vary by locality; Texas, for instance, excludes "brush" (the cuttings or trimmings from trees, shrubs, and lawns and similar materials) from its definition of household waste but is silent concerning "brush" with regard to municipal solid waste.

MSW is made up of things commonly thrown away from households. It includes containers and packaging, food scraps, grass clippings/yard trimmings, durable and nondurable goods, and other wastes [10]. *Durable goods* include appliances and furniture. *Nondurable goods* consist of paper and paperboard products (including office papers, newspapers and magazines, telephone directories, unwanted mail, etc.); rubber, plastics, and leather; clothing and textiles; and similar materials.

Municipal waste may also include ashes, street cleanings, dead animals, and abandoned automobiles (30 TAC 330.3), as well as nonhazardous sludge from a water supply treatment facility, wastewater treatment plant, or air pollution control facility (PA Code Title 25, Article VII).

In some areas, nonhazardous waste from nonmanufacturing commercial establishments is included in the definition of municipal waste; these would encompass businesses such as office buildings, stores, supermarkets, shopping centers, restaurants, and theaters. Industrial lunchroom or office waste might also be included.

EPA estimates residential waste (including waste from apartment houses) to be 55–65% of total MSW generation; waste from schools and commercial locations, such as hospitals and businesses, is estimated at 35–45% [10].

As of 2005, the EPA reports that about 32% of the MSW generated in the United States is recovered by recycling and composting, about 14% is combusted with energy recovery, and about 54% is landfilled or otherwise disposed of [10]. Total volume of waste has followed population growth with per capita MSW generation nearly constant at 4.5 lb per day since the 1990 s. Percentages as above have also changed only slightly in recent years (see previous reports). Recovery is on the increase accompanied by a decrease in disposal by incineration and landfilling. EPA plans to update this report every 2 years [11].

Ideally, the most desirable order of options for the disposition of MSW as well as for other types of waste is (1) waste reduction at the source, (2) reuse of the same materials, (3) recycling and composting to other usable products, (4) combustion/incineration with energy recovery, and finally (5) landfilling as a last resort [6, pp. 9–11].

Composting is the controlled microbiological decomposition process of an organic material such as an organic solid waste to yield a humus-like product [25 PA Code 287.1, 30 TAC 330.3]. Yard trimmings, grass clippings, food scraps, and other organic matter in MSW are amenable to composting [10]. *Compost* is the stabilized product of such a decomposition and is used or sold for use as a soil or growing medium, soil amendment, artificial top soil, or the like [30 TAC 330.3].

8.2.10 Nuclear/Radioactive Waste

This subject and the special licensing required are discussed in Chapter 10. Analytical methods for radioactive waste characterization are contained in EPA Manual SW-846 [9] as follows:

- Method 9310 Gross Alpha and Gross Beta (in Surface and Ground Waters): This method covers the measurement of gross alpha and gross beta particle activities.
- Method 9315 Alpha-Emitting Radium Isotopes (in Surface and Ground Waters): This method covers the measurement of the total alpha-emitting radioisotopes of radium (radium-223, radium-224, and radium-226).
- Method 9320 Radium-228 (plus Radium-226): This method measures beta activity from actinium-228, which is produced by decay of radium-228.

8.2.11 Refuse/Rubbish/Trash (See Garbage)

Texas defines these terms as synonymous and as nonputrescible solid waste (excluding ashes), consisting of both combustible and noncombustible waste materials

(30 TAC 330.3). The dictionary definition of *nonputrescible* relates to the absence of organic matter capable of being decomposed typically by anaerobic bacteria and fungi to foul-smelling only partially oxidized decay products. The combustible portion contains paper and wood products, rubber, plastics, brush (cuttings or trimmings from trees, shrubs, lawns, and similar materials), and so on. The remaining portion (including glass, ceramics, food and beverage cans, and similar materials) is noncombustible at 1600–1800°F.

8.2.12 Residual Waste

In general, this can be thought of as a nonhazardous waste derived from an industrial or manufacturing process as opposed to a domestic-type waste. The Commonwealth of Pennsylvania defines residual waste as a type of solid waste, including solid, liquid, semisolid, or contained gaseous materials. This term means *nonhazardous wastes* (garbage, refuse, other discarded materials, or other wastes) that result from industrial, mining, and agricultural operations, as well as the nonhazardous sludges from their water supply and wastewater treatment plants or air pollution control facilities (25 PA Code 287.1).

Residual waste does not include treatment sludges from coal mine drainage treatment plants (acid mine drainage), whose disposal is being carried out in compliance with a valid permit issued under the Pennsylvania Clean Streams Law (35 P.S. Section 691.1–691.1001). *Acid drainage* means water with a pH of less than 6.0 in which total acidity exceeds total alkalinity, discharged from an active, inactive, or abandoned surface coal mine and reclamation operation or from an area affected by surface coal mining and reclamation operations (40 CFR 701.5).

Acid mine drainage occurs when iron pyrites (primarily FeS_2) contained in layers of rock adjacent to coal seams are exposed to air and water as the coal is removed to form sulfuric acid and ferrous sulfate [12, pp. 100–106]. (FeS_2 is a yellowish mineral that along with the yellow chalcopyrite $FeCuS_2$ is known as *fool's gold*.) An additional hydrolysis reaction occurs at lower pH together with oxidation to form the less soluble ferric hydroxide (($Fe(OH)_3$), a yellow-orange gelatinous adherent precipitate known in coal regions as *yellow boy*. These noxious acidic materials then find their way into local streams and beyond.

The term *industrial establishment* means a facility that is engaged in manufacturing or processing, including factories, foundries, mills, processing plants, refineries, mines, and slaughterhouses. The term *mining* means the extraction of minerals from the earth, from waste or stockpiles, or from pits or banks. A *mine* is defined as a deep or surface mine—whether active, inactive, or abandoned. The term *agricultural* means related to the production, harvesting, and marketing of agronomic, horticultural, aquacultural, and silvicultural crops or commodities grown on or in farms, forests, fish hatcheries, or other agricultural lands.

The residual waste regulatory category is thought to be unique to Pennsylvania [13]. However, Ohio, in fact, has a similar designation, denoted also as *residual waste* or *residual solid waste* (OAC Chapter 3745-30: Residual Solid Waste Disposal Rule 3745-30-01 (Definitions) (Effective 8/15/03)). This is defined as a type of solid waste that arises from the following operations:

- coal burning
- foundries
- pulp and papermaking
- steel making
- gypsum processing
- lime processing
- Portland cement

These wastes include wastes from air and water pollution control; dusts; wastewater treatment plant sludges; other sludges; wood, paper, and gypsum wallboard wastes; and other wastes with similar characteristics. Details are contained in the cited regulations. These wastes may be disposed of in a sanitary landfill in which residual solid wastes are disposed of exclusively. Nontoxic fly ash, nontoxic bottom ash, and nontoxic spent foundry sand may also be disposed of in this type of landfill.

8.2.13 Solid Waste

Solid waste is defined in 40 CFR 261.2 as any discarded material that is not excluded by 40 CFR 261.4(a) or by variance granted under 40 CFR 260.30 and 260.31. Those exceptions and exclusions are too numerous to list here.

Solid wastes need not be solids in the physical sense but may be liquids, semisolids, or contained gases. For example, household sanitary waste in septic tanks is considered a solid (but nonhazardous) waste (40 CFR 261.4(b)(1)), and domestic sewage is specifically defined *not* to be a solid waste (40 CFR 261.4(a)(1)(i)).

A *discarded material* is, in turn, defined as any material that is abandoned, recycled, considered "inherently waste-like," or a military munition. These categories include materials that are accumulated, stored, treated, burned, or incinerated prior to or in lieu of being disposed of or recycled.

Some listed materials being recycled are solid wastes when they or their products or mixtures are "used in a manner constituting disposal" by being applied to or placed on the land in a manner "that constitutes disposal", when they or their products or mixtures are burned for energy recovery, when they are accumulated speculatively, or when they are any of a number of listed "inherently waste-like materials." There are strict requirements for what constitutes *recycling*.

Now that is all perfectly clear. (Or is it?) It is therefore strongly recommended that the most up-to-date version of the original wording of the definition in the Code of Federal Regulations be consulted for oneself, rather than relying on the summary presented above.

8.2.14 Special Waste

As defined by the State of Texas Administrative Code (30 TAC 330.3), special waste is any solid waste requiring special handling and disposal to protect human health or the environment because of the waste's concentration, its physical or chemical characteristics, or its biological properties. Conditions for disposal of some special wastes are stated explicitly in the governing regulations, while others require written approval on a waste-specific and site-specific basis. The type, or class, or landfill allowed for disposal varies, depending on the category of special waste. Only a summary of listed special wastes is provided here. Any conditions regarding proper disposal are contained in the code (30 TAC 330.3 Regulatory Guidance; 330.171 (c), (d); and 300.173).

A representative sample of Texas Special Wastes, as noted in the citations given above, follows:

- Small quantities of household hazardous waste
- Untreated medical waste from a natural or man-made disaster
- Dead animals and/or slaughterhouse waste
- Sludges from municipal wastewater treatment plants, domestic sewage treatment plants, and water supply treatment plants
- Wastes from commercial or industrial wastewater treatment plants and air pollution control facilities
- Incinerator ash
- Grease-trap and grit-trap wastes, sludge, domestic septage, or liquid waste from municipal sources treated to contain no free liquids in accordance with SW-846, Test Method 9095 (Paint Filter Test)
- Nonhazardous industrial solid waste not routinely collected with municipal solid waste
- Certain soils contaminated by petroleum products, crude oils, or chemicals
- Certain wastes from oil, gas, and geothermal activities
- Drugs and contaminated foods or beverages, exclusive of those contained in normal household waste
- Empty containers formerly used to contain pesticides, herbicides, fungicides, or rodenticides (triple rinsed and rendered unusable)
- Tanks, drums, or containers used for shipping hazardous components
- Used oil and properly treated used oil filters
- Lead acid storage batteries
- Properly packed asbestos waste
- Certain wastes originating outside the State of Texas

8.2.15 Toxic Waste

Toxic waste would be a waste that can be classified as hazardous simply by virtue of its toxicity, the ability to do harm to human health and the environment. *Toxic waste* is said to be a terminology used primarily by the media and the public [1, p. 2].

8.2.16 Universal Waste

According to 40 CFR 273.9, the term *universal waste* means any of the following hazardous wastes:

- batteries (40 CFR 273.2)
- pesticides (40 CFR 273.3)
- mercury thermostats (40 CFR 273.4)
- lamps (e.g., fluorescent bulbs, high intensity discharge (HID), neon, mercury vapor, high pressure sodium, and metal halide lamps) (40 CFR 273.5)

These wastes may be managed under 40 CFR 273 in lieu of regulation under 40 CFR parts 260 through 272. Specific differences in requirements can be found in the regulations. In addition, states can have more stringent requirements [1, p. 286].

This rule affects small and large businesses that generate hazardous waste in the above categories and eases the regulatory burden on businesses that generate these wastes. For example, the rule extends the amount of time that a business can accumulate these materials on-site. It also allows companies to transport the materials with a common carrier, instead of a hazardous waste transporter, and no longer requires companies to obtain a manifest [14]. The rule also promotes proper recycling, treatment, or disposal and provides for easy collection.

8.2.17 Waste Oil

Waste oil means used products primarily derived from petroleum, which include, but are not limited to, fuel oils, motor oils, gear oils, cutting oils, transmission fluids, hydraulic fluids, and dielectric fluids. *Used oil* means any oil that has been refined from crude oil or any synthetic oil that has been used and as a result of such use is contaminated by physical or chemical impurities (40 CFR 260.10).

Typical ways of handling used/waste oil are to reuse it for another purpose, to re-refine it oneself, to send it off-site for recycling, to burn it for energy recovery in a boiler or industrial furnace (BIF), to burn it in space heaters, to incinerate it, to store it indefinitely, and possibly to blend it with virgin material for reuse. In general, used oil is not considered as a full-fledged EPA hazardous waste. However, its storage, treatment, transportation, and disposal are subject to various degrees of regulation similar to the federal RCRA requirements for hazardous waste (40 CFR Part 279 and elsewhere). In addition, it may be defined as a hazardous waste under a state hazardous waste management program [1, pp. 287–292].

Household "do-it-yourselfer" used oil means oil that is derived from households, such as used oil generated by individuals through the maintenance of their personal vehicles (40 CFR 279.1). It is excluded from regulation as a hazardous waste by virtue of being a "household solid waste" (40 CFR 261.4(b)(1)) (see also Section 8.2.5).

Nonterne-plated used oil filters are excluded from the definition of a RCRA hazardous waste provided that certain specified methods or any equivalent method of hot draining is used that will remove the oil (40 CFR 261.4 (b) (13)). (Terne is an alloy of lead and tin, typically in the ratio of 4:1.)

8.3 WASTE ANALYSIS

There are times when it is necessary to analyze a waste material. This may occur, for example, as a result of a permit application, permit modification, or enforcement action. Accepted analytical methods are contained in "EPA Test Methods for Evaluating Solid Waste, Physical/Chemical Methods" (SW-846), EPA Methods for Chemical Analysis of Water and Wastes, (EPA 600/4-79-020), and "Standard Methods for the Examination of Water and Wastewater" prepared jointly by the American Public Health Association, American Waterworks Association, and the Water Pollution Control Federation.

The Manual SW-846, authored by the U.S. EPA Office of Solid Waste, Washington, DC, 20460 [9] is a continually updated multivolume document of approximately 3500 pages. It is primarily a guidance document presenting acceptable, though not required, methods for waste sampling and analysis. Paper copies are available from the National Technical Information Service (NTIS) in Springfield, VA, and an electronic version is available online from the EPA website.

Other test methods and procedures are referenced in the Code of Federal Regulations at 40 CFR 260.11, 264.1063, and 265.1034. It is also possible to petition for an alternate analytical method in accordance with 40 CFR 260.21.

Once again, consult any governing regulatory language for case-by-case requirements.

8.4 CALCULATIONS FOR SOLID/HAZARDOUS WASTE PERMITTING

Many of the calculations necessary for waste permitting have already been discussed in previous chapters on air and water/wastewater. For example, calculations for source testing of a waste incinerator or of a boiler burning hazardous waste are much the same as those presented in Chapter 3 for combustion sources in general. Similarly, calculations for the characterization of wastewater leaching from a landfill differ little, if at all, from those employed to describe a water supply or point source discharge (Chapter 5). Therefore, examples of only those calculations with some feature unique to solid waste and not available in previous chapters will be considered here.

8.4.1 Conversion of Units

We start with a simple example of changing mg/kg of waste into parts per million (ppm).

8.4 Calculations for Solid/Hazardous Waste Permitting

PROBLEM 8.1

Concentrations of contaminants in solid samples are often given in mg/kg (mg of contaminant per kilogram of sample). Change this unit into parts per million (by weight), expressed as ppm, wtppm, wppm, ppmw, or ppm by wt; into weight fraction; and into weight percent (wt%).

SOLUTION For example,

$$1000 \text{ mg/kg} = 1000 \text{ mg}/(1 \text{ kg} \times 1000 \text{ g/kg} \times 1000 \text{ mg/g}) = 10^3/10^6 = 1000 \text{ ppm}$$
$$10^3/10^6 = 10^{-3} = 0.001 \text{ (wt fraction)}$$
$$10^3/10^6 \times 100 = 0.1 \text{ wt\%}$$

■

8.4.2 More Unit Conversions

The RCRA hazardous waste permit application consists of the forms of Part A and the required narrative (but no forms) for Part B. The application is filed with the EPA and/or the state regulatory agency.

In the application of Part A, EPA Form 8700-23 [15] must be filled out. In this form, one must select units of measurement of rates, weight, area, volume, and heat input from a prescribed list, with letter codes denoted as A through Y. This list is reproduced in Table 8.1.

Factors are provided here to convert those units into mass or mass flow rate in kilograms or kilograms per unit time. In selecting these conversion factors, an attempt was made to use exact, defined factors and to carry sufficient significant figures when two or more of them are multiplied together. Use of these factors is illustrated in the next problem.

PROBLEM 8.2

A 20,000-gal storage tank is filled with a hazardous waste liquid having a density of 50 lb/ft^3. Compute the amount of hazardous waste in kilograms.

SOLUTION This is another exercise in the conversion of units.

$$20,000 \text{ gal } (1/7.480519481 \text{ gal/ft}^3)(50 \text{ lb/ft}^3)(0.45359237 \text{ kg/lb}) = 60,636 \text{ kg}$$

The mass in kilograms can also be found directly from the information provided in Table 8.1.
  Density in g/cc (footnote to Table 8.1)

$$(50)(0.016018463) = 0.800923169 \text{ g/cm}^3 = \text{g/cc} = \text{kg/L}$$

Using a formula under code G in the table,

$$(20,000)(3.785411784)(0.800923169) = 60,636 \text{ kg}$$

■

The numbers come out exactly the same when the conversion factors are carried out to a sufficient number of decimal places.

Table 8.1 To Convert Multiply Unit of Measure by Factor to Obtain Result

Unit of measure code	Unit of measure	Factor	To obtain result
G	Gallons (U.S.)	3.785411784	L
		$3.785411784\rho^a$	kg
E	Gallons (U.S.)/h	3.785411784	L/h
		$3.785411784\rho^a$	kg/h
		$90.8498\rho^a$	kg/day
U	Gallons (U.S.)/day	3.785411784	L/day
		3.785411784	kg/day
		$113.5623\rho^a$	kg/month
L	L	ρ^a	kg
H	L/h	ρ^a	kg/h
		$24\rho^a$	kg/day
		$720\rho^a$	kg/month
V	L/day	ρ^a	kg/day
		$30\rho^a$	kg/month
D	Short ton/h	907.18474	kg/h
		21,772.43376	kg/day
		653,173.0128	kg/month
W	Metric ton/h	1000	kg/h
		24,000	kg/day
		720,000	kg/month
N	Short ton/day	907.18474	kg/day
		27,215.5422	kg/month
S	Metric ton/day	1000	kg/day
		30,000	kg/month
J	Pounds/h	0.45359237	kg/h
		10.88621688	kg/day
		326.5865064	kg/month
R	Kilograms/h	24	kg/day
X	Million Btu/h	10^6	Btu/h

Table 8.1 (*Continued*)

Unit of measure code	Unit of measure	Factor	To obtain result
Y	Cubic yards	0.7645548	m^3
		$12.24699399\rho_{BULK}{}^b$	kg
C	m^3	1	m^3
		$16.01846337\rho_{BULK}{}^b$	kg
B	Acres	43,560	ft^2
		4840	$yard^2$
A	Acre-feet	54,560	ft^3
		1,613.3333	$yard^3$
Q	Hectares	10,000	m^2
F	Hectare-meter	10,000	m^3
I	Btu/h	0.45359237/heating value (Btu/lb)	kg/h
		10.88621688/heating value (Btu/lb)	kg/day
		326.5865064/heating value (Btu/lb)	kg/month

From EPA Form 8700-23 (Revised 3/2005) plus conversion factors derived here.

[a] ρ, liquid density in equivalent units of g/cc or kg/L. Multiply density in lb/gal (U.S.) by 0.119826 or density in lb/ft^3 by 0.016018463.

[b] ρ_{BULK}, solids bulk density in lb/ft^3 for the bulk material.

8.4.3 Conversion of Volume to Weight

A third problem related to permitting follows.

PROBLEM 8.3

A uniform water-insoluble granular material known to be a hazardous waste will be generated in a new process being considered for your plant. A small sample of this material is available from another plant. Density of an individual particle is 165 lb/ft^3. (*Note*: Change the number, as necessary, for one's own actual waste.) Estimate how many 55 gal drums of this material would qualify the plant as a large quantity generator (>1000 kg/month) under 40 CFR 260.10.

SOLUTION   The strategy is to find the void fraction (ε) and the solids fraction ($1 - \varepsilon$), and from there the solids bulk density, to determine the volume corresponding to a given weight of material in bulk. To do this

- Measure the tare weight of a laboratory pycnometer or other piece of precision glassware.
- Weigh this device full of water.
- Weight it again full of the sample plus water.
- Subtract the tare weight from each of these weights.
- Divide the net weight of sample plus water by the net weight of the water alone.

In this case, assume that this ratio comes out to be 1.97. (For an actual waste, substitute the experimentally determined value.)

$$\text{Net wt of water plus solids} = [\rho_{water}\varepsilon + \rho_{solids}(1-\varepsilon)] \times \text{Vol}$$
$$\text{Net wt of water alone} = \rho_{water} \times \text{Vol}$$

Divide one equation by the other.

$$\text{(net wt of water plus solids)}/\text{(net wt of water)} = \rho_{water}\varepsilon + \rho_{solids}(1-\varepsilon)]/\rho_{water}$$

Solving for ε and $(1 - \varepsilon)$ using $\rho_{water} = 62.4$ lb/ft, $\rho_{solids} = 165$ lb/ft (given), and the weight ratio (WR) = 1.97 (also given)

$$\varepsilon = [(165/62.4) - \text{WR}]/[(165/62.4) - 1)] = (2.6442 - 1.97)/1.6442 = 0.41$$
$$(1-\varepsilon) = 0.59$$
$$\text{Bulk density} = 165(1-\varepsilon) = 97.35 \text{ lb/ft}^3$$

Volume corresponding to 1000 kg/month $\times$ 2.20462 = 2204.62 lb/month

$$(2204.62 \text{ lb/month})/97.35 \text{ lb/ft}^3 = 22.6463 \text{ ft}^3/\text{month}$$
$$1 \text{ gal} = 231 \text{ in.}^3 = 231 \text{ in.}/(1728 \text{ in.}^3/\text{ft}^3) = 0.13368 \text{ ft}^3$$
$$(22.6463 \text{ ft}^3/\text{month})/0.13368 \text{ ft}^3 = 169.4068 \text{ gal/month}$$
$$169.4068/55 = 3.08 \text{ drums}$$ ■

Therefore, anything more than three drums of this particular material will trigger the provisions of 40 CFR 260.10 (not a whole lot!). Packing the drums more densely would decrease the number of drums to reach the trigger point. *Note*: The values in this problem are typical for a quartz sand particle density [16,17], void fraction [18], and bulk density [17].

8.4.4 Sampling of Solid Waste

According to SW-846, [9, Chapter 9] the initial and perhaps most critical element in a program designed to evaluate the physical and chemical properties of a solid waste is the plan for sampling the waste. This section discusses the basis of a sampling plan to compare waste properties with regulatory threshold levels.

There are two general categories of sampling—*authoritative* sampling versus *probablilistic* sampling. In authoritative sampling, an individual with expertise

8.4 Calculations for Solid/Hazardous Waste Permitting 391

concerning the solid waste to be sampled and/or the site under investigation selects a sample without regard to randomization. Despite its ability to obtain valid data in certain circumstances, authoritative sampling is not recommended for the chemical composition of most wastes [9, p. NINE 10].

Probabilistic, or statistical, sampling relies on mathematical and statistical principles to obtain an appropriately chosen number of random samples that allow one to test hypotheses regarding the hazardous nature of the waste. The appropriately chosen number of samples is the least number of samples required to generate a sufficiently precise estimate of the true mean concentration of a chemical contaminant in the waste.

In random sampling, every unit in the statistical population, for example, every location in a waste pile, has a theoretically equal chance of being sampled and measured. Once the appropriately chosen number of samples has been determined, their locations in the pile can be selected by constructing a grid of sampling locations and choosing among them by the use of a random number generator.

If little or no information is available concerning the distribution of chemical contaminants in the waste, simple random sampling is the most appropriate strategy. Its implementation will be illustrated in the Problem 8.4, reworked from the parameters of an example given in Ref. [19, Appendix C]. Other types of random sampling are stratified random sampling and systematic random sampling, both explained elsewhere [9, Chapter 9; 19].

PROBLEM 8.4

A waste pile from a metals recovery process consists of granular material suspected of being a hazardous waste because of its lead content. The purpose of the sampling plan is to determine if the material must be disposed of in a hazardous waste landfill, or whether a lesser type of landfill would be adequate. The waste pile is to be considered as a single unit; either the pile as a whole is hazardous, or it is not.

As part of the sampling program, a pilot study on the pile was conducted at five locations, selected at random. Results of the toxicity characteristic leaching procedure (TCLP) [9, Method 1311] and analysis for lead are shown in Table 8.2, compared to a regulatory limit of 5.0 mg/L.

Determine the minimum number of random samples that must be taken across the pile to characterize the waste pile as hazardous or nonhazardous.

Table 8.2 Results of Pilot Sampling Program

Sample no.	Lead concentration (mg/L)
1	5.8
2	10.5
3	4.9
4	2.1
5	5.4

SOLUTION

(1) The appropriate number of samples (n) is calculated from Equation (8) of SW-846 [9, p. NINE 3]

$$n = t^2 S^2 / \Delta^2 \tag{8.1}$$

where t is the statistic from the tabulated Student-t distribution for the probability of making a Type I error (i.e., that the waste pile would be declared nonhazardous when it really is). It is for a one-tailed confidence interval with a probability of $\alpha = 0.10$ and degrees of freedom (df) of the number of samples minus 1. S^2 is the variance of the full number of samples, estimated from the variance of the pilot samples. The standard deviation $(S) = \sqrt{S^2}$ or $(S^2)^{1/2}$. (For a normally distributed population, 95% of the values lie within the range of the mean ± 2 standard deviations.) Δ is the difference between the regulatory threshold (RT) and the mean of the sample ($\bar{x}$).

Here, RT = 5.0 mg/L, and the mean is estimated from the mean of the samples from the pilot testing.

This formula has been simplified by setting the probability of making a Type II (β) error at 0.5 (that is, the waste pile is not hazardous but would be declared hazardous). At $\beta = 0.5$, its Student-t statistic drops out to leave the equation above.

(2) The mean ($\bar{X}$) and variance (S^2) are estimated from the pilot study values. From Equation (2a) of SW-846,

$$\bar{X} = (1/n) \sum_{i=1}^{n} X_i \tag{8.2}$$

$$\bar{X} = (5.8 + 10.5 + 4.9 + 2.1 + 5.4)/5 = 5.74 \text{ mg/L}$$

(Note that the mean from the pilot study is greater than the regulatory level of 5.0 mg/L.) Equation (3a) of SW-846

$$S^2 = \frac{\sum_{i=1}^{n} X_i^2 - \left[\sum_{i=1}^{n} X_i\right]^2 / n}{(n-1)} \tag{8.3}$$

or the equivalent form

$$S^2 = \left[\sum_{i=1}^{n} X_i^2 - n\bar{X}^2\right] / (n-1) \quad i = 1 \tag{8.4}$$

is used to compute the variance

$$S^2 = [5.8^2 + 10.5^2 + 4.9^2 + 2.1^2 + 5.4^2 - 5(5.74)^2]/4 = 9.1830$$

From which

$$S = 3.0303$$

(3) Substituting into Equation (8.1) for the appropriate number of samples (n),

$$n = (1.533)^2 (9.1830)/(5 - 5.74)^2 = 39.4 \rightarrow \text{call } 39$$

The value of the Student-t statistic (1.533) is read from a table [9, p. NINE 9,19, p. 73] for $5 - 1 = 4$ degrees of freedom for the number of samples from the pilot study.

The number n is now recalculated using df = (39 − 1) = 38 corresponding to the number of samples just calculated.

$$n = (1.3044)^2(9.1830)/(5-5.74)^2 = 28.5 \rightarrow \text{call } 29$$

The Student-t statistic (2.3044) is interpolated by proportional parts from the cited tables.

A third iteration of df = (29−1) = 28 yields

$$n = (1.313)^2(9.1830)/(5-5.74)^2 = 28.9 \rightarrow \text{call } 29$$

The calculation stabilizes after the third iteration, and the appropriate number of samples to be taken at random positions across the pile is found to be 29.

This computation solves the problem as stated.

(4) Comments on the full sampling program

The cited example [19, Appendix C] goes on to provide two sets of "data" from 30 sampling locations (one more than the minimum calculated). In one case, the sample values obtained are concluded to be normally distributed with a 90% one-tailed upper confidence level (UCL) of 4.3 mg/L. In the second case, the values exhibit the characteristics of a lognormal distribution with a 90% UCL of 3.1 mg/L. The UCLs for both of these situations are less than the regulatory level of 5.0 mg/L. According to the decision rule suggested by SW-846 (mean of full sampling program less than UCL), the waste pile was determined *not* to be hazardous for lead. The interested reader is directed to the cited reference for details [19, Appendix C]. ■

8.5 WASTE INCINERATION

One alternative to direct landfilling, which is containment (hopefully) *ad infinitum*, is incineration. With a waste that is not hazardous, a primary consideration is reduction in volume that still must be landfilled. With hazardous wastes, it is the destruction removal efficiency (DRE) of its hazardous components more often than not organic compounds, designated as principal organic hazardous constituents (POHCs).

8.5.1 Principal Organic Hazardous Constituents

EPA regulations require owners or operators of hazardous waste incinerators to perform specific testing prior to issuance of a final permit [9, Chapter 13]. The regulations mandate that incineration result in the so-called *destruction and removal efficiency* of 99.99% or higher for hazardous wastes. By definition

$$\text{DRE} = 100(\text{mass in} - \text{mass out})/(\text{mass in}) \tag{8.5}$$

Demonstration will most often involve a "trial burn" or "test burn" prior to which the POHCs in the waste will have been determined. These are usually selected in consultation with the regulatory agency on the basis of high concentration in the waste, toxicity, and difficulty of combustion. POHCs are then monitored during the trial burn to ascertain whether the incinerator is meeting the required DREs.

Some or all of the POHCs may in essence be acting as "surrogates" for other chemicals that are more easily destroyed within the residence time and combustion temperature framework of the incinerator. Otherwise, all hazardous chemical species present would be designated as POHCs.

In some circumstances, there may be a difficulty in obtaining measurements for a particular chemical. This occurs, for example, with dioxins (polychlorinated dibenzo-*p*-dioxins, PCCDs) and benzofurans (polychlorinated dibenzofurans, PCDFs), products of incomplete combustion (PICs) occurring at trace concentrations in the incinerator exhaust when chlorine components are contained in the waste feed [20,21]. (Other PICs, in general, include carbon monoxide (CO), unburned carbon, hydrocarbons, aldehydes, ketones, organic acids, and polycyclic organic compounds.)

In those cases and others, a true surrogate compound, not originally in the waste stream but substituting for something that is, may be designated as a POHC. The surrogate must be more difficult to destroy than any waste constituent. It is injected at a known rate into the inlet along with the waste feed and measured at the outlet. The surrogate should be measured easily enough at extremely low outlet concentrations and must not interact with or cause an interference with the normal waste constituents. A surrogate might also be designated because the governing environmental agency may wish to prescribe a standard, difficult-to-incinerate material whose behavior during incineration is familiar in that jurisdiction.

Organic compounds containing bromine, chlorine, or fluorine atoms and refrigerant-type materials have been used as surrogates during test burns. If, for example, there are no bromine compounds present in the original waste, the DRE of a bromine surrogate compound can also be related to the hydrogen bromide (HBr) released and measured in the stack. Another example is sulfur hexafluoride (SF_6) [22]. It is nontoxic, resists breakdown, and is detectable at parts per trillion (10^{-12}) or better [22]. Its destruction mechanism may be pyrolysis rather than combustion. It is *not* recommended as the only POHC for a test burn in a cement kiln since its destruction is independent of O_2 concentration [23]. There are various lists ranking the incinerability of chemical compounds using various criteria [24; 25, p. 91], and the situation may continue to evolve, as necessary.

PROBLEM 8.5

This problem was inspired by a combination of two examples given in Refs [26, pp. 81–85; 27, pp. 533–536], as later modified by an update of regulations for one of them [25, pp. 123–235].

One gallon per minute of a hazardous waste of the composition given in Table 8.3 and density of 7.5 lb/gal are to be incinerated with heat recovery at 50% excess air and 2 s residence time. Combustion with auxiliary fuel, as necessary, in a test burn yields 12,000 SCFM of wet flue gas of the composition given in Table 8.4 determined by an average of stack gas measurements from three replicate test burn runs, as recommended in Ref. [28, Vol. III, p. 3]

Calculate the following and compare to applicable RCRA regulations/standard guidelines.

(1) Total hourly mass flow rate of waste stream being processed.

(2) The amount of waste processed in a month.

8.5 Waste Incineration

Table 8.3 Composition of Hazardous Waste to be Incinerated

Constituent	wt%	Designation
Chlorobenzene	2.5	POHC
Methanol	5.0	POHC
Toluene	10.0	POHC
Xylene	20.0	POHC
Other organics	62.5	—

(3) The destruction and removal efficiency (DRE) for each of the POHCs.

(4) The maximum amount of HCl produced.

(5) The concentration of particulate matter emitted in grains/dry SCF.

Comment on the results.

SOLUTION (1) and (2) Mass flow rates

$$(1\,\text{gal/min})(7.5\,\text{lb/gal})(60\,\text{min/h}) = 450\,\text{lb/h}$$
$$\times (24\,\text{hr/day})(30\,\text{days/month}) = 324{,}000\,\text{lb/month}$$

Constituent flow rates entering are obtained by multiplying the total flow rate by the *fraction* of each constituent in Table 8.5.

For example, chlorobenzene $(450\,\text{lb/h})(0.025) = 11.25$.

These entries are shown in column 3 of Table 8.5. As shown below, outlet mass flows in the table are calculated using Table 8.4, and DREs are calculated from Equation (8.5).

(3) Destruction and removal efficiency (DRE)
   Outlet flow

$$(12{,}000\,\text{SCF/min})(60\,\text{min/h})/(385.3\,\text{SCF/lb mol})^{*}$$
$$= 1868.6738\,\text{lb mol/h (wet)}/(100-16)/100 = 1569.6860\,\text{lb mol/h (dry)}$$

*The EPA uses 20°C (68°F) and 760 mmHg as the basis for the standard cubic ft (SCF) (40 CFR 60, Appendix A, Method 19)

Table 8.4 Hazardous Waste Incinerator Effluent

Constituent	Concentration
Chlorobenzene	4.0 vppb[a] (dry)
Methanol	774.0 vppb (dry)
Toluene	27.0 vppb (dry)
Xylene	84.0 vppb (dry)
H_2O	16% vol (wet)
O_2	7.5% vol (dry)
Particulate	6.0 lb/h

[a] Parts per billion by volume.

Chapter 8 Solid Waste

Table 8.5 Summary of Hazardous Waste Incinerator Calculations

Constituent	MW	Mass flow in (lb/h)	Mass flow out (lb/h)	DRE (%)
Chlorobenzene	112.56	11.25	7.067×10^{-4}	99.994
Methanol	32.04	22.5	3.893×10^{-2}	99.827
Toluene	92.15	45.0	3.905×10^{-3}	99.991
Xylene	106.17	90.0	1.400×10^{-2}	99.984
Others	—	281.25	—	—
Total		450.00		
HCl	36.46	—	3.4	—
Particulates	—	—	6.0	—
			0.0720 (grains/dry SCF corrected to 7% O2)	

Outlet mass flow rates (where FG = flue gas)

$$\text{Chlorobenzene} : \frac{4.0 \times 10^{-9} \text{ lb mol}}{\text{lb mol of FG (dry)}} \frac{1569.6860 \text{ lb mol of FG}}{\text{h}} \frac{112.56 \text{ lb}}{\text{lb mol}}$$
$$= 7.06 \times 10^{-4} \text{ lb/h}$$

and DRE

$$\text{Chlorobenzene DRE} = 100(\text{in}-\text{out})/\text{in} = (100)(11.25 - 7.067 \times 10^{-4})/11.25$$
$$= 99.994\%$$

Outlet mass flow rates and DREs for the other POHCs are calculated in a similar manner. Results are shown in columns 4 and 5 of Table 8.5.

(4) Maximum HCl produced

HCl arises from the combustion of chlorobenzene. Assuming that all the chlorobenzene is converted and that all the chlorine in that molecule goes to produce HCl, the stoichiometric reaction is

$$C_6H_5Cl + 7O_2 \rightarrow 6CO_2 + 2H_2O + HCl \tag{8.6}$$

1 mol of C_6H_5Cl forms 1 mol of HCl

$$(11.25/112.56)(34.46) = 3.44 \text{ lb of HCl/h}$$
$$\cong 3.4 \text{ lb of HCl/h}$$

There are no other chlorinated organics in the waste.

(5) Particulate matter concentration

$$(6.0 \text{ lb/h}) (7000 \text{ grains/lb})/[(12,000)(0.84 \text{ SCF (dry)/min})(60 \text{ min/h})]$$
$$= 0.0694 \text{ grains/dry SCF at stack } O_2 \text{ conditions}$$

Correction to 7% O_2 (dry)

$$[(20.9-7.0)/(20.9-7.5)](0.0694) = 0.0720 \text{ grains/dry SCF}$$

8.5 Waste Incineration

(6) Comments on the results

The incinerator is processing more than 1000 kg/month ($\cong$ 2200 lb/month) and is subject to the full extent of RCRA regulation for a large quantity generator. For less than 500 lb/h of this total waste processed, the standard is 99.99% DRE or greater for each component identified as a POHC. This requirement is being met for chlorobenzene and toluene but not for methanol and xylene.

For a total incineration capacity of less than 500 lb/h, no HCl scrubber would have been required under previous regulations since the outlet mass flow rate of HCl is less than the 4 lb/h trigger to install an outlet scrubber with 99% HCl scrubbing efficiency. However, the maximum achievable control technology (MACT) standards of September 20, 1999 require an outlet HCl emission not to exceed 21 ppm by volume (corrected to 7% O_2 (dry)) for this case [28, p. 124].

Stack gas flow (dry at 7% O_2)

$$12,000 \frac{\text{SCF (wet)}}{\text{min}} \frac{(100-16) \text{ dry}}{100 \text{ wet}} \frac{(20.9-7.5)}{(20.9-7.0)} \text{ (correction of flow to 7\% } O_2 \text{ (dry))}$$

$$= 9717.4101 \frac{\text{SCF}}{\text{min}} \text{ dry at 7\% } O_2 \text{(dry)}$$

$$(9717.4101 \text{ SCF/min})(60 \text{ min/h})/(385.3 \text{ SCF/lb mol})$$

$$= 1513.2224 \text{ lb mol/h of dry flue gas, corrected to 7\% } O_2 \text{ (dry)}$$

For 21 ppm by volume HCl

$$21 = \frac{(x \text{ lb mol/h HCl})(10^6)}{1513.2224 \text{ lb mol/h flue gas}}$$

$$x = (1513.2224)(21)/10^6 = 0.03178 \text{ lb mol/h of HCl out}$$

$$(0.03178 \text{ lb mol HCl})(36.46 \text{ lb HCl/lb mol HCl})$$

$$= 1.1586 \text{ lb/h HCl out}$$

Required scrubbing removal efficiency for HCl to achieve 21 ppm

$$(3.44-1.1586)/(3.44) = 0.6632 \cong 66.3\%$$

An HCl scrubber with 99% removal efficiency would more than achieve an HCl emission concentration of 21 ppm. In fact, at 99% removal efficiency, HCl at the outlet would be 0.0344 lb/h.

$$\frac{0.0344 \text{ lb/h}}{(36.46 \text{ lb/lb mol})} = 9.435 \times 10^{-4} \text{ lb mol/h}$$

$$9.435 \times 10^{-4} \times 10^6/(1513.2224) \cong 0.6 \text{ ppm}$$

The particulate loading of 0.0720 grains per dry SCF corrected to 7% O_2 (dry) in the flue gas is less than the previous standard of 0.08 at this flue gas condition. The more recent MACT limit of 0.015 grains/(dry SCF) would require a particulate removal device to be added with a removal efficiency of at least $(0.0720 - 0.015)/0.0720 = 0.7917$, or approximately 79%. Perhaps HCl and particulate could be removed in a single scrubber.

It is also back to the drawing board for methanol and toluene destruction. The destruction efficiency increases with higher incinerator temperature and longer residence time. Both increase

398 Chapter 8 Solid Waste

with decreased excess air. Perhaps the excess air can be lowered enough to increase the DREs without increasing the PICs (products of incomplete combustion) such as CO and other organics.

This problem demonstrates how changes in governmental regulations can alter conclusions and the basis for design. Additional regulations promulgated in the meantime would also have to be complied with above and beyond those noted in this exercise. The reader is cautioned to ferret out *all* applicable regulations of federal, state, and local authorities for an actual case of interest in real time. ∎

8.5.2 Significant Figures for DRE

The topic of significant figures for DRE is discussed in the EPA Handbook, *Hazardous Waste Incineration Measurement Guide and Manual*, Vol. III [28]. This document provides general guidance to permit writers in reviewing the measurement aspects of incineration permit applications and trial burn plans. It is oriented to how measurements are made, *not* what measurements to make.

The reported DREs, showing up to five digits in the problem above, in actuality contain only one or two significant figures. This depends on the least accurately measured value used in the calculations. The reason is that it is the percent penetration, rather than the DRE itself that is being measured [28, pp. 7–8]. The percent penetration is defined as $(100 - \% \text{ DRE})$. For a DRE of 99.99%, $(100 - \% \text{ DRE}) = 0.01\%$ (one significant figure); for a DRE of 99.9916%, $(100 - \% \text{ DRE}) = 0.0084\%$ (two significant figures).

The penetration must be rounded off first to the proper number of significant digits before calculating the DRE from it. For this reason, a value calculated for DRE such as 99.988% in the EPA example [28, p. 8] cannot be acceptably rounded off to 99.99%.

As indicated in the EPA Guidance Manual, the controlling measurement determining the number of significant figures is usually the stack concentration [28, p. 7]. GC/MS analytical methods can normally report concentrations with only one or two significant figures. This will result in a derived DRE with the same number of significant figures as the reported concentrations unless another measurement, such as waste feed concentration, waste feed flow rate, or stack gas flow rate has fewer significant figures.

In addition, one must also verify that the sample size for the stack gas outlet sample is large enough to ensure that the POHC being measured is above the lower detection limit for the analytical method being employed. An example adapted from the EPA *Hazardous Waste Incinerator Measurement Guidance Manual* [28, p. 6] is provided below to illustrate the methodology. The sampling train referred to is some form of Method 5, described in Chapter 3.

PROBLEM 8.6

Determine whether the volatile organic sampling train (VOST) sample size is sufficient to measure 99.9% DRE for carbon tetrachloride (CCl_4).

Waste feed flow rate is 2000 lb/h (15.2 kg/min) with 500 ppm (0.50 g/kg of feed) of CCl_4. Stack gas flow rate is 4500 SCFM (127.4 m^3/min). Lower detection limit for CCl_4 is 2 ng (ng = 10^{-9} grams) per trap.

Sampling is done using three pairs of traps in the stack gas sampling/collection train (one pair for each of the three replicate test burn runs). Sampling rate is 500 mL/min with 20 L of sample/pair of traps.

SOLUTION Computation is more straightforward in metric units.

(1) Input of CCl_4 POHC

$$(15.2 \text{ kg/min})(0.5 \text{ g/kg}) = 7.6 \text{ g/min}$$

(2) CCl_4 POHC output rate at 99.99% DRE

$$7.6 \text{ g/min } (1-0.9999) = 0.00076 \text{ g/min}$$

(3) CCl_4 POHC concentration in stack gas at 99.99% DRE

$$(0.00076 \text{ g/min})/(127.4 \text{ m}^3/\text{min}) = 6.0 \times 10^{-6} \text{ g/m}^3$$
$$\times (10^{-3} \text{ m}^3/\text{L})(10^9 \text{ ng/g}) = 6.0 \text{ ng/L}$$

(4) Sample amount collected in one pair of traps

$$(20 \text{ L})(6.0 \text{ ng/L}) = 120 \text{ ng (i.e., } > 2 \text{ ng)}$$

Since the VOST lower detection limit for CCl_4 is 2 ng, the sample is sufficient to detect CCl_4 to determine a DRE of 99.99%. A margin of safety for the detection limit is desirable. The cited guidance document [28, p. 6] goes on to note that this calculation assumes that both traps in a pair are combined for analysis. If they are analyzed separately, the distribution of mass between each trap must be considered. ∎

8.5.3 DRE Determination for POHC via a Surrogate Compound

PROBLEM 8.7

The higher molecular weight organic compound 2, 3-dimethyl chicken wire is extremely toxic. Its isolation and analysis as a waste constituent takes on the trappings of a major research project every time when samples are sent to the lab. Repetitive determination under field conditions on a routine basis, especially at low outlet concentrations, would be subject to major error because of the unique handling requirements necessary to ensure the integrity of a sample.

Toluene was tried as a surrogate, but its DRE could not be quantified aside from >99.99% because its outlet concentration, although more easily measured, turned out to be below the detection limit. Some members of the staff have suggested experimenting with the toluene injection rate to increase its inlet concentration and possibly bringing the outlet concentration up within the range of measurement. (Toluene may actually not be a good choice in the first place, since it might be a PIC from the decomposition of the original toxic material to be incinerated.)

Instead, someone from R&D has proposed to substitute SF_6 as the surrogate during the test burn. That person claims that an extensive survey has shown that SF_6 is more difficult to decompose at the same mass concentration, and its degree of destruction can be calculated because of its extremely low detection limit.

Is the lab person correct? Determine how to go about calculating the DRE of SF_6 and from it the DRE of the organic compound by inference.

SOLUTION At the residence time and combustion temperature of the test burn, very careful measurements with R&D analytical chemists present and a field laboratory set up on-site showed an inlet feed rate for the hazardous organic compound of 376 lb/h and 0.0015 lb/h at the outlet, based on the average of three replicate runs. Simultaneously, SF_6 was injected at 5 lb/h. Dry flue gas flow averaged 8000 SCF/min (68°F, 760 mmHg = 1 atm = 14.696 psia) corrected to 7% O_2 (dry).

Outlet concentration of SF_6 corrected to 7% O_2 (dry) was measured at 2 ppb (10^{-9}).

(1) For the hazardous organic

$$DRE = (100)(lb/h_{in} - lb/h_{out})/(lb/h_{in}) = 100(376 - 0.0015)/376 = 99.9996$$

$$DRE > 99.99\% \text{ at these conditions}$$

(2) For the SF_6 surrogate

$$\text{Inlet flow rate} = 5 \text{ lb/h}$$

Outlet flow rate :

$$\frac{(2 \times 10^{-9} \text{ mol } SF_6)}{\text{mol FG}} \frac{(146.05 \text{ lb } SF_6)}{\text{mol } SF_6} \frac{(8000 \text{ SCF FG})}{\text{min}} \frac{(60 \text{ min})}{\text{h}} \frac{(\text{mol FG})}{385.3 \text{ SCF FG}}$$

$$= 3.64 \times 10^{-4} \text{ lb/h}$$

$$DRE = (100)(5 - 3.64 \times 10^{-4})/5$$

$$= 99.9927 > 99.99\% \text{ but } < DRE \text{ of hazardous organic}$$

Score one this time for R&D. The trial burn worked out well. In addition, for routine monitoring during normal operation, the plant will purchase and install a continuous emission monitor for SF_6. ■

8.5.4 MSW Incineration with AP-42 Estimated Emissions

Finally, to wrap up the chapter, we will consider an incineration case with combustion calculations at several levels of excess air, computation of combustion temperature, and estimates of uncontrolled emissions.

This next calculation involves incineration of municipal solid waste, first in a mass burn refractory wall (MB/REF) unit without heat recovery and then in a mass burn waterwall (MB/WW) unit with heat recovery and in addition extraction of heat from the flue gas.

In a mass burn unit, no preprocessing or separation occurs except to remove items too large to fit through the feed system. The inside of the MB/REF is lined with fire brick and/or refractory material without provision for heat recovery. In contrast, the

8.5 Waste Incineration

Table 8.6 Composition of MSW to Be Incinerated

Constituent	wt% (dry)	wt% (wet)
Carbon	35.31	28.0
Hydrogen	4.41	3.5
Oxygen	28.25	22.4
Nitrogen	0.43	0.34
Sulfur	0.20	0.16
Noncombustibles	31.40	24.9
Moisture	—	20.7
Total	100.00	100.00

Heating Value ~ 6200 Btu/h.

MB/WW combustor contains tubes in the furnace wall; these tubes are filled with circulating pressurized water whose purpose is to recover heat, which is used to produce steam and/or electricity [29].

Nonetheless, the purpose of the present exercise is not to design the incinerator but rather to determine enough of the process conditions to understand the operation and estimate the emissions.

We begin with the following ultimate analysis on an elemental basis, as derived from an extensive table in Ref. [30, p. 7] (Table 8.6).

In round numbers, this particular MSW consists of about 44% paper and wood; 12% grass, leaves, and brush, and so on; 1.5% leather, rubber, and plastics; 1% oil-based materials; 0.5% clothing/rags/textiles; 4% street sweepings/dirt, and so on; 12% food wastes; 24% noncombustibles; and the remainder unclassified/miscellaneous. The calorific value of the composite refuse, as received, is about 6200 Btu/lb. This waste comes from a time before the extensive recycling of the present day, and therefore its content of yard waste, paper products, and noncombustible beverage and food containers will be somewhat different. However, it is felt to be typical enough to illustrate the underlying principles.

PROBLEM 8.8

Incinerate the given MSW, estimate the atmospheric emissions without air pollution controls, and determine roughly how much solid residue leftover after incineration must still be disposed of by other means.

SOLUTION Begin with detailed combustion calculations similar to those shown in Chapter 3. Calculations are carried out to a sufficient number of decimal places to allow the reader to follow along and are then rounded at the end.

Flow rate of fuel
Flow rate of fuel constituents is shown in Table 8.7 based on 100 lb total of the MSW to be combusted.

402 Chapter 8 Solid Waste

Table 8.7 Combustion of MSW Fuel—Fuel Composition

Constituent	lb in Fuel	Atomic/molecular weight (MW)	Moles in fuel (lb/MW)
Carbon (C)	28.0	12.010	2.3314
Hydrogen (H)	3.5	1.008	3.4722
Oxygen (O)	22.4	16.000	1.4000
Nitrogen (N)	0.34	14.007	0.0243
Sulfur (S)	0.16	32.060	0.0050
Moisture (H_2O)	20.7	18.015	1.1490
Noncombustibles	24.9	—	—

Basis 100 lb of MSW waste fuel.

Combustion stoichiometry
The following composition is used for dry combustion air

O_2	20.947% dry
N_2	78.086% dry
Ar	0.934% dry
CO_2	0.33% dry

Using the EPA default humidity = 0.27 mol H_2O/mol wet air = 0.27749 mol H_2O/mol dry air = 0.017266 lb H_2O/lb dry air from 40 CFR 60, Appendix A, Method 19.
 The following stoichiometry is assumed:

(1) $$C + O_2 \rightarrow CO_2$$

2.3314 mol of carbon yields 2.3314 mol of CO_2

and consumes 2.3314 mol of oxygen (as O_2)

 The incinerator will operate at sufficient excess oxygen, determined in the next step, to prevent the formation of carbon monoxide of the same order of magnitude as the other flue gas constituents and thereby influence the combustion calculations to any significant degree. The CO will be estimated, along with other flue gas contaminants, in a subsequent step.

(2) $$H + (1/4)O_2 \rightarrow (1/2)H_2O$$

3.4722 mol of H yields 1.7361 mol of H_2O

and consumes 0.8681 mol of oxygen (as O_2)

(3) Nitrogen in the waste
 This is assumed to go to NO_x in the flue gas. This NO_x is fuel NO_x because of the relatively low operating temperature [29], rather than thermal NO_x by reaction of N_2 and O_2 in the combustion air. Furthermore, the NO_x formed is assumed to be 95% NO and 5% NO_2 from a combustion device. The values computed here will be compared to the emission estimates

determined later. The quantities involved have a negligible influence on the nitrogen balance compared to the incoming N_2 in the combustion air.

(4) $$N + (1/2)O_2 \rightarrow NO$$

0.95×0.0243 mol of N yields 0.0231 mol of NO

$$N + O_2 \rightarrow NO_2$$

0.05×0.0243 mol of N yields 0.0012 moles of NO_2

The N in the waste fuel consumes

$$0.0231/2 + 0.0012 = 0.0128 \text{ mol of } O_2$$

(5) Sulfur (S) in the waste fuel is assumed to react to 98% SO_2 and 2% SO_3.

$$S + O_2 \rightarrow SO_2$$

0.98×0.0050 mol of S yields 0.0049 mol of SO_2 and consumes 0.0049 mol of O_2

$$S + (3/2)O_2 \rightarrow SO_3$$

0.02×0.0050 mol of S yields 0.0001 mol of SO_3 and consumes 0.0002 mol of O_2

Total O_2 consumed $= 0.0049 + 0.0002 = 0.0051$ mol

(6) Moisture (H_2O) in the fuel is evaporated directly into the flue gas as 1.1490 mol of H_2O. The necessary latent heat of evaporation is assumed to be accounted for in the calorific value determined for the waste.

Total oxygen (O_2) consumed

2.3314	from C combustion
0.8681	from H combustion
0.0128	from N combustion
0.0051	from S combustion
3.2174	Total

Less oxygen in the fuel

1.4000 mol of O $= 0.7000$ mol of O_2

$3.2174 - 0.7000 = 2.5174$ net O_2 from combustion air

This brings along

$N_2\ (78.086/20.947) \times 2.5174 = 9.3843$ mol
$Ar\ (0.934/20.947) \times 2.5174\ \ = 0.1122$ mol
$CO_2\ (0.033/20.947) \times 2.5174 = \underline{0.0040\text{ mol}}$
9.5005 mol

Dry moles of combustion air (including the oxygen that is consumed in the combustion)

$$9.3843(100/78.086) = 12.0178 \text{ mol of dry air}$$

Humidity in ambient air

$(12.0178$ mol of dry air$)(0.027749$ mol $H_2O)$/mol of dry air
$= 0.3335$ mol H_2O in combustion air

The results for the stoichiometric case, in which all of the oxygen is consumed exactly, are summarized in Table 8.8. The concentration of any CO present is ignored in this intermediate table.

Actual operation at the stoichiometric condition would result in CO emissions and other PICs and give rise to excessive incinerator temperatures, harmful to materials of construction and mechanical components. Instead, an excess of combustion air is employed. This both drives the combustion reactions toward completion and provides cooling to limit the combustion temperature.

Results of the combustion calculations are shown in Tables 8.9–8.11 for conditions of 50, 100, and 250% excess air. The definition of excess (XS) *air* is given by the following ratio:

$$\% \text{ XS air} = 100(\text{actual air} - \text{stoichiometric air})/(\text{stoichiometric air}) \quad (8.7)$$

Therefore, 50, 100, and 250% XS air correspond to 1.5, 2, and 3.5 times the required stoichiometric air. *Excess oxygen* is the percent O_2 present in the flue gas on a wet or dry basis; it is excess over the 0% O_2 of the stoichiometric flue gas.

The 50% excess air case shown in Table 8.9 results in an excess O_2 value of about 7% (dry) and a CO_2 concentration of about 13% (dry). Seven percent O_2 is the standard reference condition for contaminants in incinerator flue gas, similar to the 3% O_2 standard condition for boilers and furnaces. Correction to 13% CO_2 is also encountered in the literature [30, p. 26].

The 100% excess air case of Table 8.10 is a typical operating condition (80–100%) for a MB/WW incinerator with heat recovery. It will be explored further for a MB/WW unit in the second part of this example.

In this example, the 250% excess air of Table 8.11 is necessary to control the combustion temperature of an MB/REF incinerator within the desired range. At 250% excess air, temperature in the incinerator's combustion chamber is approximately 1,700 °F. Incinerators of this type operate typically between 150% and 300% excess air [29]. Ordinary incinerator temperatures are quoted as 1600–1800°F [30 TAC 330.3 (130)].

Flue gas temperature

Flue gas temperature is computed by heat balance. The energy released from burning the MSW heats up the combustion products, is carried away in the noncombustible "ash," and is lost by radiation and leakage into and through the walls of the incinerator [30, p. 198].

For a unit heat release of 6200 Btu/lb and a calculation basis of 100 lb of fuel total

$$(100 \text{ lb of fuel basis})(6200 \text{ Btu/lb}) = 620,000 \text{ Btu}$$

Assume 10% heat losses from radiation, leakage, and exit with noncombustibles [30, pp. 23, 199]. This leaves 558,000 Btu in combustion gas. This will serve to heat up the gas to combustion chamber temperature and flue gas exit temperature without heat recovery.

To illustrate this calculation without undue complication, take the heat capacity of water vapor equal to 0.5 Btu/lb/°F and the heat capacity of everything else in the flue gas as 0.25 Btu/lb/°F. (These numbers are easily remembered as a rule of thumb.) The calculation can be refined by using more precise individual heat capacity data as a function of temperature.

Take the base temperature as 77°F (25°C) for combustion air and MSW fuel, the standard temperature at which calorific values are normally determined. Assume that the calorific value represents the net or lower heating value and that the evaporation of the free moisture content in the waste fuel is included in this figure.

$$\text{Net heat released} = (\text{lb of gas})(C_p \text{ of gas})(\text{Btu/lb °F})(\Delta T = T_f - 77°F)$$

Table 8.8 Municipal Solid Waste Combustion at 0% Excess Air

Constituent		Molecular weight	Moles of combustion products at 0% excess air				% wet	% dry
			From fuel	From combustion	From comb. air	Total		
Carbon dioxide	(CO$_2$)	44.01	—	2.3314	0.0040	2.3354	15.49	19.69
Nitrogen	(N$_2$)	28.0134	—	—	9.3843	9.3843	62.23	79.12
Oxygen	(O$_2$)	32.00	—	—	0.0000	0.0000	0.00	0.00
Argon	(Ar)	39.948	—	—	0.1122	0.1122	0.74	0.95
Sulfur dioxide	(SO$_2$)	64.06	—	0.0049	—	0.0049	0.03 (325 ppm)	0.04 (413 ppm)
Sulfur trioxide	(SO$_3$)	80.06	—	0.0001	—	0.0001	0.00 (7 ppm)	0.00 (8 ppm)
Nitric oxide	(NO)	30.01	—	0.0231	—	0.0231	0.15 (1532 ppm)	0.19 (1948 ppm)
Nitrogen dioxide	(NO$_2$)	46.01	—	0.0012	—	0.0012	0.01 (80 ppm)	0.01 (101 ppm)
Subtotal		—	—	—	—	11.8613	78.66	100.00
Moisture (water)	(H$_2$O)	18.015	1.1490	1.7361	0.3335	3.2186	21.34	—
Total		—	—	—	—	15.0798	100.00	—

Basis: 100 lb of MSW fuel

Average molecular weight (AMW): dry flue gas 31.30; wet flue gas 28.46. Temperature of flue gas with 10% heat losses, without heat recovery, and fuel heating value of 6200 Btu/lb. [Moles H$_2$O × MW H$_2$O × 0.5 + moles dry flue gas × AMW DFG × 0.25] × (T_f − 77) = 0.9 × 6200 × 100. Solve for T_f: T_f = 4658°F.

Table 8.9 Municipal Solid Waste Combustion at 50% Excess Air

Constituent		Molecular weight	Moles of combustion products at 50% excess air				% wet	% dry
			From fuel	From combustion	From comb. air	Total		
Carbon Dioxide	(CO$_2$)	44.01	—	2.3314	0.0059	2.3373	11.00	13.08
Nitrogen	(N$_2$)	28.0134	—	—	14.0765	14.0765	66.23	78.77
Oxygen	(O$_2$)	32.00	—	—	1.2587	1.2587	5.92	7.04
Argon	(Ar)	39.948	—	—	0.1684	0.1684	0.79	0.94
Sulfur dioxide	(SO$_2$)	64.06	—	0.0049	—	0.0049	0.02 (231 ppm)	0.03 (274 ppm)
Sulfur trioxide	(SO$_3$)	80.06	—	0.0001	—	0.0001	0.00 (5 ppm)	0.00 (6 ppm)
Nitric oxide	(NO)	30.01	—	0.0231	—	0.0231	0.11 (1087 ppm)	0.13 (1293 ppm)
Nitrogen dioxide	(NO$_2$)	46.01	—	0.0012	—	0.0012	0.01 (56 ppm)	0.01 (67 ppm)
Subtotal		—	—	—	—	17.8702	84.07	100.00
Moisture (water)	(H$_2$O)	18.015	1.1490	1.7361	0.5002	3.3853	15.93	—
Total		—	—	—	—	21.2556	100.00	—

Basis: 100 lb of MSW fuel

Average molecular weight: dry flue gas 30.51; wet flue gas 28.52. Temperature of flue gas with 10% heat losses, without heat recovery, and fuel heating value of 6200 Btu/lb. [Moles H$_2$O × MW H$_2$O × 0.5 + moles dry flue gas × AMW DFG × 0.25] × (T_f − 77) = 0.9 × 6200 × 100. Solve for T_f: T_f = 3422°F.

Table 8.10 Municipal Solid Waste Combustion at 100% Excess Air

Constituent		Molecular weight	Moles of combustion products at 100 % excess air					
			From fuel	From combustion	From comb. air	Total	% wet	% dry
Carbon Dioxide	(CO$_2$)	44.01	—	2.3314	0.0079	2.3393	8.53	9.80
Nitrogen	(N$_2$)	28.0134	—	—	18.7687	18.7687	68.42	78.60
Oxygen	(O$_2$)	32.00	—	—	2.5174	2.5174	9.18	10.54
Argon	(Ar)	39.948	—	—	0.2245	0.2245	0.82	0.94
Sulfur Dioxide	(SO$_2$)	64.06	—	0.0049	—	0.0049	0.02 (179 ppm)	0.02 (205 ppm)
Sulfur Trioxide	(SO$_3$)	80.06	—	0.0001	—	0.0001	0.00 (4 ppm)	0.00 (4 ppm)
Nitric Oxide	(NO)	30.01	—	0.0231	—	0.0231	0.08 (842 ppm)	0.10 (967 ppm)
Nitrogen Dioxide	(NO$_2$)	46.01	—	0.0012	—	0.0012	0.00 (44 ppm)	0.01 (50 ppm)
Subtotal		—	—	—	—	23.8792	87.05	100.00
Moisture (water)	(H$_2$O)	18.015	1.1490	1.7361	0.6670	3.5521	12.95	—
Total		—	—	—	—	27.4313	100.00	—

Basis: 100 lb of MSW fuel

Average molecular weight: dry flue gas 30.12; wet flue gas 28.56. Temperature of flue gas with 10% heat losses, without heat recovery, and fuel heating value of 6200 Btu/lb. [Moles H$_2$O × MW H$_2$O × 0.5 + moles dry flue gas × AMW DFG × 0.25] × (T_f − 77) = 0.9 × 6200 × 100. Solve for T_f: T_f = 2711 °F.

Table 8.11 Municipal Solid Waste Combustion at 250% Excess Air

Constituent		Molecular weight	Moles of combustion products at 250% excess air					
			From fuel	From combustion	From comb. air	Total	% wet	% dry
Carbon Dioxide	(CO$_2$)	44.01	—	2.3314	0.0139	2.3453	5.10	5.60
Nitrogen	(N$_2$)	28.0134	—	—	32.8452	32.8452	71.47	78.38
Oxygen	(O$_2$)	32.00	—	—	6.2935	6.2935	13.69	15.02
Argon	(Ar)	39.948	—	—	0.3929	0.3929	0.85	0.94
Sulfur Dioxide	(SO$_2$)	64.06	—	0.0049	—	0.0049	0.01 (107 ppm)	0.01 (117 ppm)
Sulfur Trioxide	(SO$_3$)	80.06	—	0.0001	—	0.0001	0.00 (2 ppm)	0.00 (2 ppm)
Nitric Oxide	(NO)	30.01	—	0.0231	—	0.0231	0.05 (503 ppm)	0.06 (551 ppm)
Nitrogen Dioxide	(NO$_2$)	46.01	—	0.0012	—	0.0012	0.00 (26 ppm)	0.00 (29 ppm)
Subtotal		—	—	—	—	41.9061	91.18	100.00
Moisture (Water)	(H$_2$O)	18.015	1.1490	1.7361	1.1672	4.0523	8.82	—
Total		—	—	—	—	45.9584	100.00	—

Basis: 100 lb of MSW fuel

Average molecular weight: dry flue gas 29.63; wet flue gas 28.60. Temperature of flue gas with 10% heat losses, without heat recovery, and fuel heating value of 6200 Btu/lb [Moles H$_2$O × MW H$_2$O × 0.5 + moles dry flue gas × AMW DFG × 0.25] × (T_f − 77) = 0.9 × 6200 × 100. Solve for T_f; T_f = 1686°F.

For the 250% excess air case, lb of steam in flue gas/100 lb of fuel

$$(4.0523 \text{ mol } H_2O)(18.015) = 73.022 \text{ lb } H_2O/100 \text{ lb of fuel}$$

lb of everything else

$$(41.9061 \text{ mol of dry flue gas})(AMW = 29.63) = 1241.6779$$

$$\Sigma m\, C_p\, \Delta T = \text{heat released to the flue gas} \qquad (8.8)$$

lb of water vapor $\times C_{p\text{ water vapor}}(T_f - 77)$
+ lb of other flue gas constituents $\times C_{p\text{ flue gases}}(T_f - 77)$
= $(73.0022)(0.5)(T_f - 77) + (1241.6779)(0.25)$
= $558,000$

$$(36.5011 + 310.4195)(T_f - 77) = 558,000$$

$$346.9206\, T_f = 558,000 + 26,712.883$$

$$T_f = 1685°F \text{ (approximately } 1700°F)$$

Results for other conditions of excess air are computed in a similar manner. Flue gas temperatures without heat recovery for this particular incinerator and MSW fuel are summarized in Table 8.12. Dry molecular weight of the flue gas is about 30, as expected, and between 28.5 and 28.6 on a wet basis.

Now estimate the operation of a waterwall incinerator with heat recovery and an excess air rate of 100% (Table 8.10). Operate at a 1700°F firebox temperature. Remove sufficient heat to make this happen at 100% excess air.

$$\underset{\text{Moisture}}{[(3.5521)(18.015)(0.5)} + \underset{\text{Dry gas constituents}}{(23.8792)(30.12)(0.25)]}[1700 - 77] = \text{heat in flue gas}$$

$$[31.9955 + 179.8104][1623] = 343,761 \text{ heat content for } 1700°F$$

$$558,000 - 343,857 = 214,239$$

(Approximately 214,000 Btu/100 lb of waste must be removed within the incinerator)

Table 8.12 Flue Gas Temperature of Incinerator under Study

% XS air	Temperature (°F)
0.0	4658
50.0	3422
100.0	2711
150.0	2250
200.0	1926
224.5	1800
246.5	1700
250.0	1685
271.6	1600
300.0	1501

Now remove enough heat from the flue gas to lower the stack temperature to 350°F. This temperature is reasonable but arbitrary. In practice, the design stack temperature would be selected to allow enough buoyancy in the flue gas and to keep the flue gas temperature above the acid dew point in order to minimize corrosion of stack components.

$$\Delta H = mC_p \Delta t$$
$$(211.8059)(1700 - 350) = 285,938 \text{ (approximately 286,000)} \tag{8.9}$$

Another 286,000 Btu/100 lb of waste charged to incinerator must be removed prior to discharge from the stack. The heat recovered can be used to make steam and/or electricity. The steam produced at lower pressure/temperature can be used to heat the boiler feed water.

Those readers interested in the design of a waste heat boiler are referred to the pertinent sections in Refs [25,26].

Estimation of air emissions

Contaminants emitted from the incinerator without air pollution controls can be estimated using the factors listed in EPA AP-42, reproduced Table 8.13. The majority of the factors selected are rated "A", the highest quality. An explanation of their basis is given in the accompanying text material of AP-42. The emission factors must be adjusted for the heat

Table 8.13 AP-42 Emission Factors for Refuse Combustion without Air Pollution Controls[a]: Uncontrolled Mass Burn Air Combustion

| Contaminant | Refractory wall | | Waterwall | |
	lb/ton of refuse combusted	Rating	lb/ton of refuse combusted	Rating
PM	2.51×10^{-1}	A	2.51×10^{-1}	A
As	4.73×10^{-3}	A	4.73×10^{-3}	A
Cd	1.09×10^{-2}	A	1.09×10^{-2}	A
Cr	8.97×10^{-3}	A	8.97×10^{-3}	A
Hg	5.6×10^{-3}	A	5.6×10^{-3}	A
Ni	7.85×10^{-3}	A	7.85×10^{-3}	A
Pb	2.13×10^{-1}	A	2.13×10^{-1}	A
SO_2	3.46×10^{0}	A	3.46×10^{0}	A
HCl	6.40×10^{0}	A	6.40×10^{0}	A
CDD/CDF[b]	1.50×10^{-5}	D	1.67×10^{-6}	A
NO_x	2.46×10^{0}	A	3.56×10^{0}	A
CO	1.37×10^{0}	C	4.63×10^{-1}	A
CO_2	$1.97 \times 10^{+3}$	D	$1.97 \times 10^{+3}$	D

[a] From EPA AP-42, Fifth Edition, Volume I, Chapter 2: Solid Waste Disposal, Tables 2.1.2, 2.1.4, and 2.1.6 for waste with a heating value of 4500 Btu/lb. Instructions for other wastes are to multiply by the new heating value and divide by 4500. (Emission factors employing metric units are contained in other tables in this section of AP-42.)

[b] CDD/CDF = total tetra- through octa-chlorinated dibenzo-*p*-dioxin/chlorinated dibenzofurans; 2,3,7, 8-tetrachlorodibenzo-*p*-dioxin; and dibenzofurans.

8.5 Waste Incineration

Table 8.14 Calculated Uncontrolled Emissions lb/100 lb of 6200 Btu/lb MSW

Contaminant	Refractory wall	Waterwall
PM	1.7291×10^{-2}	1.7291×10^{-2}
As	3.258×10^{-4}	3.258×10^{-4}
Cd	1.12258×10^{-2}	1.12258×10^{-2}
Cr	6.179×10^{-4}	6.179×10^{-4}
Hg	3.857×10^{-4}	3.857×10^{-4}
Ni	5.407×10^{-4}	5.407×10^{-4}
Pb	1.46733×10^{-2}	1.46733×10^{-2}
SO_2	2.383552×10^{-1}	2.383552×10^{-1}
HCl	4.408883×10^{0}	4.408883×10^{0}
CDD/CDF	1×10^{-6}	1×10^{-7}
NO_x	1.694664×10^{-1}	2.452441×10^{-1}
CO	9.43776×10^{-2}	3.18955×10^{-2}
CO_2	1.3571093×10^{2}	1.3571093×10^{2}

content of the waste as indicated in the footnote to Table 8.13. AP-42 estimated emissions are listed in Table 8.14.

It is left for the reader to estimate incinerator emissions for other cases with various types of pollution control equipment using the companion tables in AP-42 or vendor design information.

Comparison of emission calculations from different methods

- $SO_2 + SO_3$ from combustion calculations

$$0.0049(64.06) + 0.0001(80.46) = 0.313894 + 0.008006 = 0.3219 \text{ lb}/100 \text{ lb}$$

as calculated from sulfur content of the refuse versus 0.2384 from AP-42.

Same order of magnitude: When sulfur content of the specific waste is available, use it in preference to AP-42, which represents an average for many different wastes plus seasonal variations.

- $NO + NO_2$ from combustion

$$0.0231(30.01) + 0.0012(46.01) = 0.693231 + 0.055212$$
$$= 0.748443 \text{ lb}/100 \text{ lb of}$$

NO_x from AP-42 (lb/100 lb of refuse)

Refractory wall	Waterwall	Ratings
0.1694664	0.2452441	A

The calculated $NO + NO_2$ may not be valid. Not all of the nitrogen in the fuel goes to form NO_x; some of the fuel nitrogen shows up as N_2 in the flue gas rather than into the products assumed. Use AP-42 value.

- CO_2 from AP-42 approximately 136 versus CO_2 from combustion calculation

$$(2.3454)(44.01) \cong 103 \text{ lb}/100 \text{ lb of refuse}$$

Values are the same order of magnitude. Use carbon content of the waste fuel rather than AP-42 value to calculate CO_2. CO_2 emission factors are rated "D".

Volume of incinerated MSW
With or without heat recovery, the mass and hence the volume of the MSW that remains to be landfilled or disposed of by other means is substantially reduced (here, to about 25% by weight, Table 8.6, assuming no unburned combustible material). Volume reduction is estimated at 10:1 or better [30, p. 200]. ∎

8.5.5 Wrap-Up of MSW Problem

Years ago, the author had the opportunity to visit one of New York City's municipal incinerator locations, on the East Side of Manhattan. The scale of the operation was simply awesome, with garbage trucks, one after another, dumping their waste loads into huge pits from which the MSW was being fed to the incinerators. If memory still serves correctly, three combustion units were operating in parallel, and at the time, air pollution equipment was being pilot tested at full scale on one or two of them.

All in all, there were 32 New York City MSW incinerators at 24 locations [31]. Twenty-one of these had closed before 1969. In the 1970s, high particulate emission rates became unacceptable as more stringent environmental laws were enacted. The high cost of advanced air pollution controls curtailed construction of new MSW incinerators in New York City, followed by the closure between 1969 and 1981 of 8 of the 11 remaining incinerators. The particular incinerator visited ran from 1957 to 1972.

The last three to operate were equipped with high-efficiency electrostatic precipitators between 1980 and 1984. As of 1994, all of the New York City MSW incinerators have been shut down. Further details on the air pollution from them can be found in Ref. [31].

REFERENCES

1. T.P. Wagner, *The Complete Guide to the Hazardous Waste Regulations: RCRA, TSCA, HMTA, OSHA, and Superfund*, 3rd edn, Wiley, New York, 1999, 536, pp.
2. U.S. EPA, Instructions for EPA Form 8700-12, Notification of Regulated Waste Activity (OMB # 2050-0028; Expires 6/30/2009).
3. U.S. EPA, Municipal Solid Waste—Household Hazardous Waste, available at http://www.epa.gov/garbage/hhw.htm, last updated on Thursday, August 9, 2007.
4. U.S. Census Bureau, State & County QuickFacts: USA, available at http://quickfacts.census.gov/qfd/states/00000.html, last revised on Monday, May 7, 2007 13:42:46 EDT.
5. Commonwealth of Pennsylvania Department of Environmental Protection, What is HHW? Brochure 2520-PA-DEP2315, 2 pp. (3/99).
6. P.T. Williams, *Waste Treatment and Disposal*, 2nd edn, John Wiley & Sons, Ltd, West Sussex, England, 2005, 380 pp.
7. U.S. EPA, Wastes—Medical Waste, available at http://www.epa.gov/epaoswer/other/medical/, last updated on Wednesday, January 31, 2007.

8. U.S. EPA, Wastes—Disposal of Medical Sharps, available at http://www.epa.gov/epaoswer/other/medical/disposal.htm, last updated on Wednesday, January 31, 2007.
9. U.S. EPA, Test Methods for Evaluating Solid Waste, Physical/Chemical Methods, (SW-846), EPA Office of Solid Waste, Washington, DC, paper copies available from National Technical Information Service (NTIS), Springfield, VA; also available on the Internet at EPA website (First published 1980, continually updated, containing approximately 3500 pages), 1980.
10. U.S. EPA Office of Solid Waste, Municipal and Industrial Solid Waste Division, Municipal Solid Waste in the United States: 2005 Facts and Figures Executive Summary, EPA530-S-06-001 (October 18, 2006), available on the Internet at www.epa.gov/osw, 2006.
11. U.S. EPA, Municipal Solid Waste in the United States: 2005 Facts and Figures, available at http://www.epa.gov/garbage/msw99.htm, last updated on Friday, June 1, 2007.
12. E.B. Besselievre, *The Treatment of Industrial Wastes*, McGraw-Hill, New York, 1969, pp. 100–106.
13. Commonwealth of Pennsylvania Department of Environmental Protection, DEP Secretary Addresses Residual Waste Issues, NewsRelease (2/12/2004), available at www.ahs.dep.state.pa.us/newsreleases/default.asp?ID=2766&varQueryType=Detail, accessed on July 18, 2007.
14. U.S. EPA, Universal Waste—Basic Information, available at http://www.epa.gov/epaoswer/hazwaste/id/univerwast/basic.htm, last updated on February 22, 2006.
15. U.S. EPA, RCRA Hazardous Waste Part A Permit Application—Instructions and Forms, EPA Form 8700-23 (OMB # 2050-0034; Expires 11/30/2005).
16. R.C. Weast, editor, *CRC Handbook of Chemistry and Physics*, CRC Press, Inc., Cleveland, OH, 1977, p. F-1.
17. J.H. Perry, editor, *Chemical Engineers' Handbook*, 3rd edn, McGraw-Hill, New York, 1950, pp. 194–195.
18. F.A. Zenz and D.F. Othmer, *Fluidization and Fluid-Particle Systems*, Reinhold, New York, 1960, p. 234.
19. W.M. Cosgrove, M.P. Neill and K.H. Hastie, editors, *RCRA Waste Management: Planning, Implementation, and Assessment of Sampling Activities*, American Society for Testing and Materials, West Conshohocken, PA, 2000, 76 pp.
20. P.M. Lemieux, The Use of Surrogate Compounds as Indicators of PCCD/F Concentrations in Combustor Stack Gases, EPA-600/R-04/024, U.S. EPA Office of Research and Development, Washington, DC, 2004.
21. W.H. Shaub and W. Tsang, Dioxin Formation in Incinerators, *Environmental Science & Technology*, **17**(12), 721–730, 1983.
22. C.L. Proctor, II, M.C. Berger, D.L. Fournier Jr., and S. Roychoudhury, Sulfur hexafluoride as a tracer for the verification of waste destruction levels in an incineration process, Final Report ESL-TR-86-47, Engineering & Services Laboratory, Air Force Engineering & Services Center, Tyndall AFB, Florida, April 1987, 94 pp.
23. M. Von Seebach, F.M. Miller, C. Lamb, R.S. Southdown, and D. Gossman, The use of monochlorobenzene as a principal organic hazardous constituent for destruction efficiency determinations in cement kilns, Paper submitted for presentation at Air & Waste Management Association Specialty Conference on New RCRA Regulations, Orlando, FL, March 17–19, 1992.
24. M. Williams, Request for Guidance in Designating POHC's, Memorandum to C. Simon, U.S. EPA Office of Solid Waste and Emergency Response, Washington, DC, 1988.
25. J.J. Santoleri, J. Reynolds, and L. Theodore, *Introduction to Hazardous Waste Incineration*, 2nd edn, Wiley-Interscience, New York, 2000, 635 pp.
26. L. Theodore and J. Reynolds, *Introduction to Hazardous Waste Incineration*, Wiley-Interscience, New York, 1987, 463 pp.
27. J.P. Reynolds, J.S. Jeris, and L. Theodore, *Handbook of Chemical and Environmental Engineering Calculations*, Wiley-Interscience, New York, 2002, 948 pp.
28. U.S. EPA, Handbook Hazardous Waste Incineration Measurement Guidance Manual: Volume III of the Hazardous Waste Incineration Guidance Series, EPA/625/6-89/021, Washington, DC, Research Triangle Park, NC, and Cincinnati, OH, June 1989.

29. U.S. EPA, AP 42, Fifth Edition, Volume I Chapter 2: Solid Waste Disposal, available at http://www.epa.gov/ttn/chief/ap42/ch02/index.html, last updated on Thursday, June 21, 2007.
30. R.C. Corey, *Principles and Practices of Incineration*, Wiley-Interscience, New York, 1969, 297 pp.
31. D.C. Walsh, S.N. Chillrud, H.J. Simpson, and R.F. Bopp, Refuse incinerator particulate emissions and combustion residues for New York City during the 20th century, *Environmental Science & Technology*, **35**(12), 2441–2447, 2001.

Chapter 9

Noise

Noise annoys.

9.1 GENERAL

Along with the topics discussed in previous chapters, noise is a pollutant producing adverse physiological effects on humans and other living beings. Noise in excess of applicable norms must be abated both for the health and welfare of those affected and to contribute to the success of a new project or a continuing operation. This chapter presents basic concepts, definitions, equations, and general mitigation measures for sound and its evil twin, noise. Its intent is to make the subject less mysterious for those having to perform or evaluate calculations and measurements to ensure compliance in this area and to secure or issue whatever regulatory approvals may be required.

Agencies of the U.S. federal government are involved in research and technical assistance to state and local governments and in setting standards for building materials, motor vehicles, aircraft, and other noise-producing equipment [1, pp. 630–632]. They are also responsible for enforcement of standards for safety in the workplace, including mines. Responsibility for controlling noise in the external community is in the hands of state and local governments. Further comments here on typical provisions of state and local noise regulations must wait until after an explanation of terms necessary to understand them.

9.2 SOUND VERSUS NOISE

To control the generation and propagation of noise (unwanted sound), one must first understand the physics of sound. The attributes of sound and noise are contrasted in Table 9.1.

Sound is perceived by the sense of hearing by a physical process in the ear. Noise is unwanted sound at the wrong time or in the wrong place. It is any sound that annoys, disturbs, or adversely affects living beings or interferes with their ability to

---

Environmental Calculations: A Multimedia Approach, by Robert G. Kunz
Copyright © 2009 John Wiley & Sons, Inc.

Table 9.1 Sound Versus Noise

Sound	Noise
Neutral	Annoying, unwanted sound at wrong time or place
Follows the laws of physics	Response highly subjective
Subjects a medium to repeated vibrations	Adversely affects living beings
Perceived by sense of hearing	Interferes with communication

communicate. Adverse auditory effects on humans include loss of sleep; muscle and nervous tension; irritability; fatigue; increased production of adrenaline and other hormones; effects on the heart, blood pressure, and circulatory system; gastrointestinal problems; and hoarseness/laryngitis from having to shout to be heard as well as overall behavioral changes and a general interference with communication. Sound adversely affects animals as well.

Whatever the reason, people do not like noise, especially when it is beyond their control. Reaction to noise is highly subjective and varies from person to person and with changing circumstances. Depending on the situation, public response may range from no reaction through various levels of complaints and ultimately community action [2, pp. 137–138]. It does not matter whether a noise that is deemed offensive is within the allowable range of any existing code or if a specific noise code even exists. Furthermore, once sensitized to a new noise, the public may continue to complain and clamor for action even after the offensive noise has been quieted below its original intensity.

9.2.1 Sound Characteristics

Sound is produced by vibration, a series of compressions and rarefactions in a medium such as atmospheric air. Its effect is perceived by the sense of hearing. A simple sound can be described mathematically by a sine wave (Figure 9.1) [3]. This pressure wave can be plotted versus time at a given location or versus distance at a given time. When it is plotted against time, the interval between any two successive recurring points on the curve is known as the *period* (τ). Against distance, this interval is called the *wavelength* (λ).

The greater the displacement of the wave from the *x*-axis (its *y*-value), the larger the compression, and the more energy is contained. This increase in amplitude translates into the higher *intensity or loudness* that occurs.

The frequency is determined by how often the displacement takes place. The shorter the period or wavelength (Figure 9.1), the higher the frequency, and conversely. This defines the *pitch* of the sound. The unit of frequency is the hertz (Hz); its more descriptive old name was cycles per second (cps). Other items of interest are indicated in the figure. Underlying mathematical relationships are presented in a subsequent section.

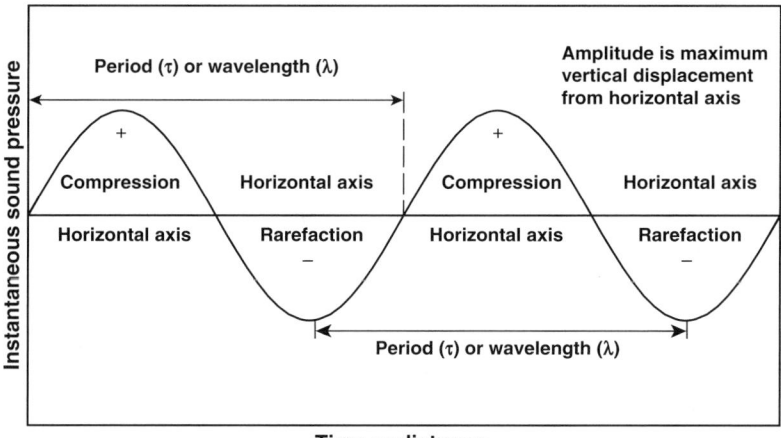

Figure 9.1 Representation of sound as a simple sine wave.

A third important characteristic of a sound is its *duration*, the length of time or the number of cycles to which a listener is exposed. Sounds are classified as continuous, intermittent, or impulse. An impulse sound is one of extremely short duration but usually of high intensity; it may be repeated on a regular or irregular basis. Examples include a sudden clap of thunder, a gunshot, a sonic boom, a slamming door, and operation of equipment such as a punch press, drop forge, or riveting machine [1, p. 105; 4, p. 108; 5, p. 279].

9.3 SOUND PROPERTIES

9.3.1 Loudness

To reiterate, the intensity, or loudness, of a simple sound is related to the amplitude of the pressure wave (refer again to Figure 9.1). Sound pressure intensities vary from a barely detectable 2×10^{-10} atm (atmosphere) to 2×10^{-3} atm for sounds so loud as to be painful. To put this range into workable units, the decibel (dB) is defined as

$$\mathrm{dB} = 20 \log_{10}[P_{\mathrm{actual}}/P_{\mathrm{reference}}] \tag{9.1}$$

where the reference pressure is taken more precisely as 2×10^{-4} µbar (1 bar $\cong$ 1 atm). The corresponding range in dB extends from 0 to 140 dB (Table 9.2). Further comments on the decibel scale are provided later in this chapter.

9.3.2 Pitch

Frequencies range upward without limit from zero (no oscillations = no sound). The human ear is sensitive to frequencies between 20 and 20,000 Hz. It is most sensitive in the range of 1000–4000 Hz. A frequency of 20 Hz is just below the lowest

Table 9.2 Typical Sound Pressure Levels[a]

Example	dBA	Effect
Aircraft carrier deck operations	140	Painfully loud
Warning siren	130	
Rock band, jet takeoff at 200 ft	120	Maximum vocal effort
Pile driver, loud motorcycle at 20 ft	110	Intolerable
Garbage truck, jackhammer	100	
Subway train at 20 ft, boiler room	90	Very annoying, hearing damage
Hair dryer, average street traffic	80	Annoying
Noisy restaurant	70	Telephone use difficult
Air conditioner at 20 ft	60	Intrusive
Classroom, private business office	50	Moderately quiet
Very quiet room	40	
Library, quiet bedroom at night	30	Very quiet
Broadcasting studio	20	
Rustling leaves	10	Just audible
Threshold of hearing (youths)	0	

[a]Information in table compiled and abbreviated from numerous sources.

note on a standard 88-key piano [1, p. 292; 4, p. 31]; 20,000 Hz is representative of jet engines and high-speed drills. In between, the chirping of crickets (2000 Hz) and the highest note on that same piano keyboard (about 4000 Hz) provide further reference points.

9.3.3 Timbre

Finally, there is timbre [1, pp. 125–227]. This is quality or richness. It is the quality of the sound that would allow a listener to differentiate between it and another of the same intensity and pitch. An example would be the same note produced by musical instruments of different types.

9.4 FREQUENCY SPECTRUM

The complex sounds of human speech and operating equipment are made up of a band of frequencies rather than the pure tone of a simple sine wave [1, p. 319; 6, p. 8; 7, pp. 94–95]. This frequency spectrum is monitored in a sound meter, an electronic instrument capable of changing tiny variations in pressure into an analogue or digital readout in decibels. When the meter admits sounds of all frequencies, a composite reading is obtained.

However, it is sometimes important to determine the component frequencies of the broad spectrum sound, but commercial sound meters used for field measurements do not discriminate by individual frequency. Rather, groups of adjacent frequencies are processed in an ancillary device known as an *octave band analyzer*.

This is accomplished by feeding the signal through an electronic filter network contained in the circuitry. The filter allows the selected band of frequencies to pass, to the exclusion of all others.

9.5 THE OCTAVE

9.5.1 Octave Definition—the Musical Scale

An octave is the interval between two tones having a frequency ratio of 2:1. To illustrate this concept, consider the musical scale. Western music divides the octave into 12 steps called *semitones*, as marked, for example, by the keys on a piano [1, pp. 100, 171–185]. The frequency ratio of two such adjacent keys (both black and white) is the 12th root of 2 (1.059463). The ratio between the center of the octave and either end (six keys) is the square root of 2 (1.414213), and the frequency ratio dividing the octave into thirds (four keys) is the cube root of 2 (1.25992).

The frequencies of selected musical notes for a piano tuned to equal temperament are listed in Table 9.3. Tuning to equal temperament means that all the semitones are the same 1/12 of the octave, having a frequency ratio of $2^{1/12}$, as opposed to other styles of tuning [1, pp. 177–178].

The notes depicted range from A below middle C (also known as A_3) to one octave above middle C. The A above middle C (A_4) is tuned to a frequency of 440 Hz, as set by international agreement in 1939 [1, p. 125]. A reference tone at 440 Hz has been broadcast since 1936 by the National Institute of Standards and Technology (NIST), formerly the National Bureau of Standards (NBS), once an hour to serve as a calibration aid [1, p. 125; 8]. Then A_3 at the lower end of this octave becomes $440/2 = 220$ Hz. Middle C (C_4) has a frequency of 261.63 HZ, and its octave (C_5) 523.25 Hz. The center frequency for the C octave shown is 369.99 (F sharp). The center frequencies for the 1/3 octaves C–E, E–A flat, and A flat to C are 293.66 (D), 369.99 (F sharp), and 466.16 (B flat), respectively.

Table 9.3 Frequencies of Selected Musical Notes

	Note		Frequency (Hz)	Remarks
	A------------------|		220.00	Lower end of A octave
	B flat		233.08	
	B		246.94	Center for lower 1/3 A octave
|--------- Middle C			261.63	Lower end of C octave
|	C sharp	|	277.18	1/3 Octave boundary for A octave
|	D	|	293.66	Center for lower 1/3 C octave
|	E flat	A octave	311.13	Center for A octave and for middle 1/3 A octave
|	E	|	329.63	1/3 Octave boundary for C octave
|	F	|	349.23	1/3 Octave boundary for A octave
C octave	F sharp	|	369.99	Center for C octave and for middle 1/3 C octave
|	G	|	392.00	Center for upper 1/3 A octave
|	A flat	|	415.30	1/3 Octave boundary for C octave
|	A------------------|		440.00	Upper end of A octave
|	B flat		466.16	Center for upper 1/3 C octave
|	B		493.88	
|------------------C			523.25	Upper end of C octave

Table 9.4 Preferred Octave Band and 1/3 Octave Band Frequencies[a]

Frequency band number	Octave center frequency	1/3 Octave center frequency	Calculated 1/3 octave end frequency	Calculated octave end frequency
			11	11
11		12.5		
			14	
12	16	16		
			18	
13		20		
			22	22
14		25		
			28	
15	31.5	31.5		
			35	
16		40		
			45	45
17		50		
			56	
18	63	63		
			71	
19		80		
			90	90
20		100		
			112	
21	125	125		
			141	
22		160		
			178	178
23		200		
			225	
24	250	250		
			282	
25		315		
			355	355
26		400		
			450	
27	500	500		
			560	
28		630		
			710	710
29		800		
			890	
30	1,000	1,000		
			1,120	
31		1,250		
			1,400	1,400
32		1,600		
			1,780	
33	2,000	2,000		
			2,240	
34		2,500		
			2,800	2,800
35		3,150		
			3,550	
36	4,000	4,000		
			4,470	
37		5,000		
			5,600	5,600
38		6,300		
			7,080	
39	8,000	8,000		
			8,900	
40		10,000		
			11,200	11,200
41		12,500		
			14,130	
42	16,000	16,000		
			17,780	
43		20,000		
			22,400	22,400

[a] End values rounded slightly.

9.5.2 Octave Band Analysis—Sound Meter

Sound meters with octave band analyzers separate the pressure generated by a complex sound into discrete bands of frequencies. The width of the octave band varies from band to band just as the octaves on a piano keyboard. Sound pressure within the band is allowed to pass through and be measured. Sound at frequencies outside the given bandwidth is shut out by the electronic filter networks. This function is controlled by moving a selector switch or similar control on the meter/analyzer.

The standard sound meter octaves do not coincide exactly with the musical notes of Table 9.3, but rather have been standardized as listed in Table 9.4. The preferred series of octave bands shown in Table 9.4 covers the audible range in 10 bands, with center frequencies ranging from 31.5 to 16,000 Hz. The lowest band shown, centered at 16 Hz, is found on some sound meters in conjunction with a 1/3 octave band analyzer.

For a more detailed frequency analysis, the octave is split into three parts, with the ratio between adjacent 1/3 octave center frequencies equal to the cube root of 2, as noted earlier for the musical scale. The center frequency for the middle 1/3 octave is the same as the center frequency for the corresponding octave. Upper and lower end frequencies of the bandwidth represented by each center frequency are calculated for the octave bands using the square root of 2 and for the 1/3 octave bands using the sixth root of 2 (1.22462). Some sound meters cover the audible range and slightly beyond in 33 such 1/3 octave bands (Table 9.4). Narrower bands of 1/10 octave and below have been used for specialized purposes.

Another set of octave bands was used before 1966 but is no longer the series preferred by the American National Standards Institute (ANSI). This older series may still be encountered in some test codes and published data, and one should be aware of its existence. Nevertheless, conversion of octave band levels between the two series is beyond the scope of the present treatment, and the reader is referred to ANSI standards for this purpose, should the need arise (Table 9.5).

Table 9.5 Older Series of Octave Bands

Octave band frequency range (Hz)	Geometric mean frequency (Hz)
18.75–37.5	26.5
37.5–75	53
75–150	106
150–300	212
300–600	424
600–1200	849
1200–2400	1700
2400–4800	3390
4800–9600	6790
9600–19,200	13,580

Note: Information from Ref. [9, p. 14].

9.6 COMBINING DECIBELS

There are numerous occasions in sound/noise work to add readings expressed in units of decibels. For an industrial situation, it is reasonable to assume that the various sources are incoherent, meaning that they radiate sound independently of one another and do *not* produce waveforms with a definite phase relationship to one another. In this case, phase relationships need not be considered, and it is appropriate to combine sound pressure decibel levels on a root-mean-square (RMS), or energy, basis. *Caution*: This procedure for combining dB levels is not valid for coherent noise, for example, pure tones, and may not be valid for equipment where a pure tone or tones may dominate, such as transformer hum or machinery running at synchronous speeds.

9.6.1 Decibel Addition

This basis is logarithmic, and decibels do *not* sum directly in arithmetic terms. The formula for decibel addition is

$$\text{Total dB} = 10 \log_{10}[10^{dB_1/10} + 10^{dB_2/10} + \cdots + 10^{dB_n/10}] \quad (9.2)$$

To add two sound levels of 60 dB each would *not* equal 120 dB, but rather would total to

$$10 \log_{10}[10^{6.0} + 10^{6.0}] = 10 \log_{10}[2,000,000] = 10[6.3010] \cong 63$$

This technique is equally valid for combining any number of decibel readings from separate sources, or for adding up the decibel readings from different frequencies, weighted or unweighted, making up the sound spectrum from the same source.

A special case arises when a number (n) of sources of equal decibel value are being added. Equation (9.2) reduces to

$$\text{Total dB} = 10 \log_{10}[n \times 10^{dB_1/10}] \quad (9.3)$$

and further to

$$\text{Total dB} = 10 \log_{10}[n] + dB_1 \quad (9.4)$$

The first term is the amount to be added to the original common dB value [1, pp. 89–90]. It is linear in the logarithm of the number of identical sources and is listed in Table 9.6 and plotted in Refs. [1, p. 89; 10, p. 17].

As before, two sources of equal intensity combine by adding 3 dB to either value. For four such sources, 6 dB would be added, and 9 dB would be added for eight.

Once again, the equation and table for addition of equal dB levels are applicable only for incoherent noise sources comprised of a broad spectrum of frequencies. In that case, there will be no interference between the waveforms to enhance or cancel one another according to phase.

Table 9.6 Adding dB for a Number (n) of Equal Noise Sources

n	$10\log_{10}(n)$	n	$10\log_{10}(n)$
1	0.00	6	7.78
2	3.01	7	8.45
3	4.77	8	9.03
4	6.02	9	9.54
5	6.99	10	10.00

9.6.2 Averaging Decibels

To obtain a single value for use in further calculations, it may be necessary to average decibel readings at various locations around a source or at one position at different times [7, pp. 15–16]. The averaging formula [7, p. 16] is derived from Equation (9.2) for decibel addition

$$\text{dB avg} = 10\log_{10}[(1/N)\sum_{i=1}^{N} 10^{\text{dB}_i}] \qquad (9.5)$$

PROBLEM 9.1

Compute the decibel average for dB values of 50, 60, 70, and 80.

SOLUTION

$$\begin{aligned}
\text{dB}_{\text{avg}} &= 10\log_{10}[(1/4)(10^{50/10} + 10^{60/10} + 10^{70/10} + 10^{80/10}] \\
&= 10\log_{10}[(1/4)(111,100,000) = 10\log_{10}(27,775,000)] \\
&= (10)(7.4437) \cong 74.4 \text{ dB}
\end{aligned}$$

$$\text{Arithmetic average dB} = \bar{x} = (1/N)\sum_{i=1}^{N}[\text{dB}_i] \qquad (9.6)$$

$$\text{Arithmetic average} = \bar{x} = (50+60+70+80)/4 = 65$$

There are methods to approximate the logarithmic average for dB readings within 0–10 dB from one another [7, pp. 16–17]. Equation (9.5) is simple enough to implement using an electronic calculator and provides an exact result; however, the following approximations can be accomplished rapidly in the absence of a calculator and may be equally as valid.

When the difference between the maximum and minimum values is 5 dB or less, a straight arithmetic average can be used to approximate the logarithmic average of Equation (9.5). When the difference is between 5 and 10 dB, add 1 dB to the arithmetic average. ■

PROBLEM 9.2

Find the average of (a) 50, 51, 52, and 55 by Equation (9.5) and (b) 50, 54, 56, and 60 and compare with the arithmetic average.

SOLUTION (a) From Equation (9.5),

$$\mathrm{dB}_{\mathrm{avg}} = 10\log_{10}[(1/4)(10^{5.0} + 10^{5.1} + 10^{5.2} + 10^{5.5})]$$
$$= 10\log_{10}[175{,}152.41] = (10)(5.2434) \cong 52.4\,\mathrm{dB}$$

From Equation (9.6),

$$\bar{x} = (50 + 51 + 52 + 55)/4 = 208/4 = 52\,\mathrm{dB}\,(\text{approximation to } 52.4\,\mathrm{dB}_{\mathrm{avg}})$$

For (b),

$$\mathrm{dB}_{\mathrm{avg}} = 10\log_{10}[(1/4)(10^{5.0} + 10^{5.4} + 10^{5.6} + 10^{6.0})]$$
$$= 10\log_{10}[437{,}323.95] = (10)(5.6408) \cong 56.4\,\mathrm{dB}$$

From Equation (9.6),

$$\bar{x} = (50 + 54 + 56 + 60)/4 = 220/4 = 55\,\mathrm{dB}$$
$$55\,\mathrm{dB} + 1\,\mathrm{dB} = 56\,\mathrm{dB}\,(\text{approximation to } 56.4\,\mathrm{dB}_{\mathrm{avg}})\qquad\blacksquare$$

9.6.3 Subtraction of Decibels

The equation for adding decibels can be rearranged to subtract 1 dB reading from another. This situation arises when it is necessary to separate the noise attributable to a given source from the general background level, the so-called neighborhood residual (NR). The frequency spectra of source and background are usually different [4, p. 204].

A study of outdoor noise levels was conducted by the City of Chicago, IL, as part of a program to establish "baseline noise levels" there [11]. Frequency distribution of the noise background was not reported. The study concluded in part that

- Transportation noise dominates. During the day, local traffic such as cars and buses determine the noise levels; at night, the noise comes from expressway traffic in the distance.
- Maximum urban noise levels occur during morning and evening traffic rush hours.
- Minimum background noise occurs between 2 a.m. and 3 a.m. in some areas removed from main arterial streets and expressway traffic.
- Residential district background noise can decrease to 35–40 dBA in a large metropolitan area during nighttime hours.

The Chicago findings would seem to be representative of most highly populated urban areas. Background traffic noise is expected to be of low frequency.

Using measurement of the total noise ($\mathrm{dB_T}$, for the source of concern plus all other contributing sources) and a background without the contribution of the noise in question, one can estimate the noise from the source alone.

$$\mathrm{dB_{source}} = 10\log_{10}[10^{\mathrm{dB_T}/10} - 10^{\mathrm{dB_{NR}}/10}] \qquad (9.7)$$

where dB_T is the total measured for the source and background together, dB_{NR} is for the background without the source, and dB_{source} is therefore for the source alone. It is termed the *corrected source strength*.

The ambient background/neighborhood residual can be estimated by any of the following methods [3]:

- An actual measurement of background noise with the particular noise source turned off (preferred method). (However, coming back on a different day to obtain a background when the source happens to be off may not be valid and is not recommended.)
- Walking away until the influence of the source no longer contributes significantly to the measured total and the total noise no longer changes with distance.
- Obtaining a reading behind a barrier where the contribution of the noise source of interest is not present.
- Measuring the background in a similar neighborhood far removed from the noise source.

A simple numerical example will serve to illustrate the point. A reading of 77 dBA is obtained with a particular noise source operating, and 72 dBA is measured in this same environment when the source is shut down. Its source strength can be calculated by plugging values into Equation (9.7) as follows:

$$dB_{source} = 10 \log_{10}[10^{77/10} - 10^{72/10}]$$
$$= 10 \log_{10}[50,118,723 - 15,848,931] = (10)(7.5349) \cong 75.3 \text{ dBA}$$

An additional solved problem showing what happens as the measured level of dB_T changes relative to the measured ambient background is contained in the next section.

9.6.4 "Shortcut" Graphical Procedure

There are tables and graphs in the literature to "simplify" dB addition. These were undoubtedly developed before the widespread availability of electronic calculators with functions for logarithms and powers of 10. One such graph is redrawn as the lower curve in Figure 9.2. The correction function is exact ($10 \log_{10} [1 + 10^{-x/10}]$), where x is the arithmetic difference between the original dB readings and independent of those dB levels. Use it if you wish, but Equation (9.2) always works, and for multiple sources in a single calculation.

To use the graphical technique when adding more than two sources, one must combine them two at a time and then the combined sources must be addressed, and so on. A careful choice of the pairs can often reduce computations. In the simple example given, a value of 3 dB would be added to either of the identical 60 dB readings to yield a composite value of 63 dB. For two readings 10 dB or more apart, the contribution of the lower reading is considered to be negligible. Results of an example problem done both ways are shown in Table 9.7.

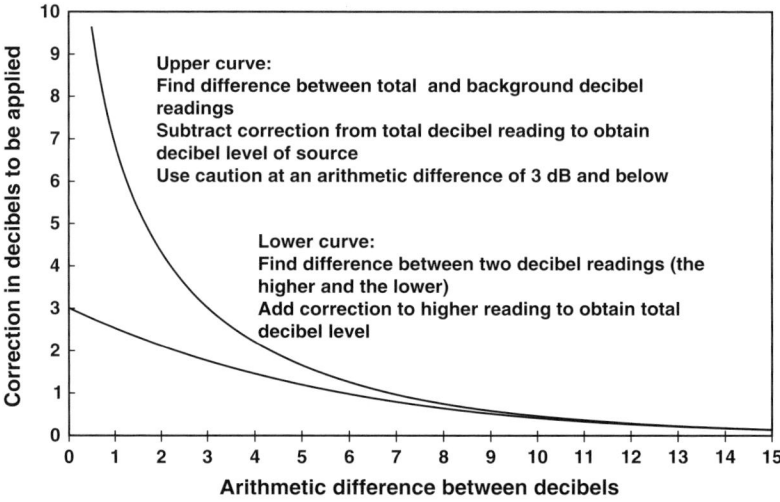

Figure 9.2 Adding and subtracting decibels.

There is also a "shortcut" method for subtracting decibels. This uses the arithmetic difference (x) between dB_T and dB_{NR} to make a correction to dB_T, as read from the upper curve in Figure 9.2. The equation of that curve is $-10\log_{10}[1 - 10^{-x/10}]$. This produces positive numbers that are to be subtracted from the total of source and background (dB_T) to find source noise alone (also known as the *corrected source strength*). The curve becomes steep at low values of the independent variable and blows up at $x = 0$ (i.e., becomes indeterminate). At $x = 3$ dB and below, the noise source is less than the background [6, pp. 36–37], and the value for the noise source alone is not reliable. This condition is generally regarded as unenforceable (u.e.) in the State of New Jersey [3].

PROBLEM 9.3

Six sources separately contribute to the dBA readings listed in Table 9.7. Find the combined dBA using both Equation (9.2) and the shortcut method based on Figure 9.2.

Table 9.7 Addition of dBA Contributions from Multiple Sources

Source no.	dBA	$10^{dBA/10}$
1	35	3162.3
2	60	1,000,000
3	63	1,995,262
4	49	79,433
5	35	3162.3
6	60	1,000,000
Total	*66.1*	4,081,019.6

9.6 Combining Decibels

Table 9.8 Addition of dBA Contributions from Multiple Sources Shortcut Method

Pair of Sources	dBA	dBA Combination
1 and 5	35 + 35	38
2 and 6	60 + 60	63
(1 and 5) and 4	38 + 49	49
(1, 4, and 5) and 3	49 + 63	63
(2 and 6) and (1, 3, 4, and 5)	63 + 63	**66**

SOLUTION Results for $10^{dBA/10}$ are calculated for each of the sources in the last column of Table 9.7, and the logarithm of the sum is taken according to Equation (9.2) to give 66.1 dBA. In Table 9.8, they are combined two at a time in conjunction with Figure 9.2, resulting in a dBA of 66. The shortcut calculations are simplified by choosing pairs of equal value to add 3 dBA, or pairs more than 10 dBA apart in order to be able to drop the lower dBA value. ∎

PROBLEM 9.4

For the set of total source plus background readings in Table 9.9, find the corrected source level using both Equation (9.7) and Figure 9.2. The background in all cases is 65 dBA.

SOLUTION A sample calculation using Equation (9.7) for the first point

$$dB_2 = 10 \log_{10}[10^{75/10} - 10^{65/10}] = (10)(7.5414) \cong 74.5 \text{ dBA}$$

Using graph,

$$\Delta = 75 \text{ (total)} - 65 \text{ (NR)} = 10 \text{ dBA}$$

For $x = 10$ dBA, read 0.5 dBA from Figure 9.2 to be subtracted from total dBA to give corrected source level of $75 - 0.5 = 74.5$. Results for this and the other points are listed in Table 9.9. ∎

Table 9.9 Subtracting dB to Obtain Corrected Source Level

Total dBA	Arithmetic Difference dBA	Factor from Graph dBA	Corrected Source dBA	
			Equation	Graph
75	10	0.5	74.5	74.5
72	7	1.0	71.0	71.0
69	4	2.2	66.8	66.8
67	2	4.3	62.7 u.e.	62.7 u.e.
65	0	∞	Impossible	Impossible

9.7 COMPOSITE SOUND LEVEL

9.7.1 Flat Weighting

The flat or unweighted setting on the sound meter/analyzer switch will allow sound at all frequencies to pass. This composite reading is known as dB_{flat}, $dB_{unweighted}$, dB_{linear}, or $dB_{all\ pass}$. Since the contribution from all frequencies is allowed to add, the $dB_{all\ pass}$ for a sound spanning multiple frequency bands must be greater than the contribution from any individual band. This summation is accomplished automatically in the meter, but it is instructive to calculate the total manually to see just how it is done. An example problem demonstrating how to combine dB levels is solved at the end of Section 9.7.3.

9.7.2 Weighting Networks

The human ear does not respond uniformly to sounds of all frequencies, being less efficient at low and high frequencies than at medium or speech frequencies. To the human ear, sounds of low frequency do not seem as loud as those of equal intensity but of higher frequency. The so-called *A-weighting network* in a sound meter simulates the human ear by electronically attenuating the low frequencies but allowing the high frequencies to pass through virtually unchanged. The A-weighting circuitry is accessible through still another setting on the meter.

Frequency response curves are shown in the literature for A-weighting along with two others, B and C [1, p. 93; 4, p. 8; 5, p. 280; 7, p. 48; 10, p. 14]. The C-weighting network is essentially flat except for some attenuation below 100 Hz. It is contained on some simple sound meters along with the A scale. B-weighting is used less often, with attenuation up to 500 Hz. There is also a D-weighting network, which is employed when measuring aircraft noise. Table 9.10 in the example problem below lists the numerical values to be added to the dB level for each of the preferred octave bands to obtain a sound level with A-, B-, or C-weighting [10, p. 14; 12, p. 77].

9.7.3 A-Weighting—The dBA Scale

The composite of the A-weighted values summed across all frequency bands is known as L_A. It is expressed in logarithmic decibel units on the A scale (dBA), also written as dB A or dB(A). The dBA reading is also called the *noise level*. This composite dBA reading is read directly by a sound meter with A-weighting, but it is also possible to calculate it step by step from the results of an octave band analysis and the entries in Table 9.10. This example problem goes on to show how to apply the attenuation factors to obtain an overall dBA (L_A) from measured octave band readings in dB. (Similar remarks apply to the calculation of B- or C-weighted values in the table to compute L_B or L_C.) Calculation of L_A to within 1 or 2 dBA of the meter's dBA reading provides a good check on the consistency of the separate recorded measurements. One source of error occurs when there is a varying traffic noise in the background and the individual frequency readings, and L_A are not all taken at the same time.

9.7 Composite Sound Level

Table 9.10 Calculating Composite (Broadband) Noise Levels: All Readings and Weighting Factors in dB

Center freq (Hz)	Octave band reading	A-weight factor	A-corr reading	B-weight factor	B-corr reading	C-weigh factor	C-corr reading
31.5	58	−39.5	18.5	−17.1	40.9	−3.0	55.0
63	62	−26	36.0	−9.3	52.7	−1.0	61.0
125	57	−16	41.0	−4.2	52.8	0	57.0
250	44	−8.5	35.5	−1.3	42.7	0	44.0
500	38	−3	35.0	−0.3	37.7	0	38.0
1000	33	0	33.0	0	33.0	0	33.0
2000	30	1	31.0	−0.1	29.9	0	30.0
4000	28	1	29.0	−0.7	27.3	−1.0	27.0
8000	26	−1	25.0	−2.9	23.1	−3.0	23.0
16,000	25	−6.5	18.5	−8.4	16.6	−8.5	16.5
Center freq (Hz)	$10^{(dB/10)}$		$10^{(dB/10)}$		$10^{(dB/10)}$		$10^{(dB/10)}$
31.5	630957.34		70.79		12302.69		316227.77
63	1584893.19		3981.07		186208.71		1258925.4
125	501187.23		12589.25		190546.07		501187.23
250	25118.86		3548.13		18620.87		25118.86
500	6309.57		3162.28		5888.44		6309.57
1000	1995.26		1995.26		1995.26		1995.26
2000	1000.00		1258.93		977.24		1000.00
4000	630.96		794.33		537.03		501.19
8000	398.11		316.23		204.17		199.53
16,000	316.23		70.79		45.71		44.67
Sum	2752806.76		27787.07		417326.20		2111509.5
Log sum	6.439775727		4.4438428		5.6204756		6.324593
dB$_{flat}$	**64.4**	dBA	**44.4**	dBB	**56.2**	dBC	**63.2**

Octave band readings in column 2 are combined according to the formula:

$$\text{Total dB}_{flat} = \text{dB}_{all\,pass} = 10 \times \log_{10}\left[10^{(dB1/10)} + 10^{(dB2/10)} + \cdots + 10^{(dB10/10)}\right]$$

Likewise, the weighted values of columns 4, 6, and 8 are processed to obtain dBA, dBB, and dBC. One can tell without octave band data that this is low-frequency noise since dBC ≫ dBA. Speech interference level (PSIL) = arithmetic average of dB at 500, 1000, and 2000 Hz. PSIL = 33.7 dB. It can be approximated by dBA −9 = 35.4 dB. Communication possible in a normal voice at normal distance. PSIL approximation decent.

With sound meters having no octave band analyzer but at least the A and C scales, it is still possible to get an indication of frequency distribution: if L_C is greater than L_A by several decibels, much of the sound is of low frequency, probably below 600 Hz; if L_C and L_A are more or less equal, middle frequencies around the 1000–2000 Hz region predominate; if L_A is greater than L_C by at least 2 dB, the sound is composed of high frequencies, 4000 Hz and higher. In the example problem worked in Table 9.10, the noise is of low frequency. Meters reading only dBA will not determine frequency information.

430 Chapter 9 Noise

PROBLEM 9.5

Calculate dB_{flat} ($dB_{all\ pass}$) and the A-, B-, and C-weighted composite noise levels (dBA, dBB, and dBC) from the frequency spectrum data listed in Table 9.10. Identify the type of noise under consideration.

SOLUTION Octave band data are contained in columns 1 and 2 of the spreadsheet of Table 9.10. In columns 3, 5, and 7 can be found the weighting factors to be applied to the raw dB data of column 2. These factors are added to the readings of column 2 to yield the corrected values of columns 4, 6, and 8. Decibel addition of the individual readings, according to Equation (9.2) and the explanatory note, is also shown in the table, resulting in the figures for dB_{flat}, dBA, dBB, and dBC in bold italics. As indicated in the note, this particular example represents low-frequency noise since dBC is much greater than dBA. The note on the speech interference level is discussed in the next section. ∎

9.8 SPEECH INTERFERENCE

The speech interference level (SIL) is defined as the arithmetic average in dB of the sound pressure levels of ambient noise in the three octave bands of center frequency—500, 1000, and 2000 Hz [1,2,4,5,7]. It is also called the preferred speech interference level (PSIL) to distinguish it from a similar average of sound pressure levels using the older series of octave bands—600–1200, 1200–2400, and 2400–4800 Hz (Table 9.5).

$$\text{PSIL (dB)} = (dB_{500\ Hz} + dB_{1000\ Hz} + dB_{2000\ Hz})/3 \qquad (9.8)$$

An approximate value for PSIL can be obtained by subtracting 9 from the dBA value [5, p. 285].

$$\text{PSIL (dB) (approximate)} = dBA - 9 \qquad (9.9)$$

Ability to communicate correlates well with this empirical parameter. For two people standing at a normal communication distance, communication is satisfactory when PSIL is 65 dB or less. For higher values of PSIL, speaker and listener must move closer, perhaps with shouting. Above a 90 dB PSIL, face-to-face communication is impossible even with maximum vocal effort. Various curves correlating ease of communication with PSIL are available in the literature, but they are not completely consistent with one another. For telephone communication, the value of PSIL must be slightly lower for comparable conversation quality [4, p. 62].

Communication indoors versus outdoors might be somewhat easier at a given PSIL because reflected sound will reinforce direct sound [1, p. 622]. Observations already mentioned are based on a male speaker. For a female speaker, the threshold PSIL values should be reduced by 5 dB [5, p. 51].

PROBLEM 9.6

Calculate the PSIL for dB readings of 53, 50, and 47 for the 1000, 2000, and 2000 Hz octave bands, respectively.

SOLUTION Calculate PSIL from Equation (9.8)

$$\text{PSIL} = (53 + 50 + 47)/3 = 50$$

This PSIL ought to be fine for easy communication.

An additional calculation of PSIL with different dB readings is contained in a footnote to Table 9.10, where that result is compared with the approximation of Equation (9.9). ■

9.9 A-WEIGHTING STATISTICS

9.9.1 Statistical Definitions

Sound is different in different locations, and even at the same location, sound is not constant at different times. Data may be presented as a graph of time-varying noise levels versus time or as a histogram. The variation in the data over time is treated statistically, using the definitions for the descriptors in Table 9.11 [1, p. 603; 5, p. 289].

- L_p: This is another name for a single, snapshot A-weighted sound pressure level, also known as L_A.
- L_x: Sound level in dBA exceeded x percent of the time during the measurement period; for example, L_{90} is the sound level exceeded 90% of the time. The L_{90} is indicative of background noise in the absence of local noise events. The L_{50} is the median or the arithmetic average sound level exceeded 50% of the time. The L_{10} is usually indicative of maximum noise from recurring events such as traffic during peak volumes. The L_{01} can arise from sudden loud short-term noise events (single peaks).
- L_{eq}: Equivalent sound level that provides an equal amount of acoustical energy as the time-varying sound, as defined in greater detail below. It is the level for a steady sound of uniform intensity that matches the acoustical energy of the time-varying sound. For steady, nontraffic-related noise sources, L_{eq} and L_{90} should agree to within 3 dBA or less.
- L_d: Daytime L_{eq} sound level for the period 7 a.m. to 10 p.m., or 7 a.m. to 7 p.m., depending on the specific calculation in which it is employed.
- L_n: Nighttime L_{eq} for the period from 10 p.m. to 7 a.m. the next calendar day.
- L_e: L_{eq} determined for evening hours, defined as 7–10 p.m.

Table 9.11 Some Typical A-Weighed Statistical Noise Descriptors

L_p	L_{01}	L_{eq}	L_{dn}
	L_{10}	L_d	CNEL
	L_{50}	L_n	
	L_{90}	L_e	

L_{dn}: The 24 h day–night average L_{eq}, with a 10 dBA penalty added to nighttime levels from 10 p.m. to 7 a.m. L_{dn} is a measure of noise over a 24 h period, with a penalty on operations generating noise between the hours of 10 p.m. and 7 a.m. (9 h) penalized by adding 10 dBA to the nighttime noise level. This is to account for the greater sensitivity of the neighboring public to nighttime noise, accompanied by the lower background levels of noise at night. Daytime L_{eq} from 7 a.m. to 10 p.m. (15 h) is not adjusted. L_{dn} values range from an average of 50 dBA for a quiet suburban residential neighborhood to 70 dBA for a very noisy urban residential setting [2, p. 137].

CNEL (community noise equivalent level) is a metric used in California. It is a 24 h L_{eq} with adjustments similar to L_{dn}. CNEL breaks up the day into three periods: a shorter daytime (12 h) from 7 a.m. to 7 p.m., for which no dBA adjustment is made; evening from 7 p.m. to 10 p.m. (3 h), which carry a 5 dBA penalty to be added; and nighttime still from 10 p.m. to 7 a.m. with a 10 dBA penalty addition as for L_{dn}. The motivation for the 5 dBA penalty is to compensate for speech interference during the evening hours. All or part of the evening period can represent a peak time for traffic for various reasons, even though other noises in the community are beginning to shut down.

9.9.2 Determination of L_{eq}

L_{eq} is normally measured using an integrating sound-level meter, which automatically samples many, many times at an extremely small interval, performs the requisite calculations, and produces an output. Some such meters are also capable of producing the other statistics, for example, L_{10}, L_{50}, and L_{90}.

L_{eq} is a single number characterizing the constantly varying noise level over the measurement period on an equivalent acoustical energy basis. It is defined as the integrated average over time of the sound pressure level (SPL) on an energy basis relative to the reference pressure chosen [10, p. 15]

$$L_{eq}(\text{dBA}) = 10 \log_{10}[(1/T) \int_0^t (P/P_{ref})^2 dt] \qquad (9.10)$$

By substitution of SPL and the definition of the decibel and replacement of the integral with a summation of discrete terms, it can be determined from the following relationship:

$$L_{eq}(\text{dBA}) = 10 \log_{10}[(1/T) \sum_{i=1}^{n} (t_i \times 10^{(L_{A_i})/10})] \qquad (9.11)$$

where T is the total of the individual time periods t_i.

For n equal time periods, where $\sum n = N$,

$$L_{eq}(\text{dBA}) = 10 \log_{10}\left[\frac{1}{N} \sum_{i=1}^{n} 10^{(L_{A_i}/10)}\right] \qquad (9.12)$$

The repeated dBA measurements needed as inputs are tedious to perform manually, but are theoretically possible. Sampling times are of necessity much less frequent (and the sampling interval much longer) with a human operator than when done automatically by the integrating sound-level meter. If, as expected, the sound level is changing more rapidly than the manual sampling period, the data will be less representative than when compiled automatically.

With today's instrumentation, it is doubtful that anyone would ever want to go through this exercise to compile data by hand with only a simple sound-level meter. However, a calculation of L_{eq} and the other statistics from hand collected data is instructive.

PROBLEM 9.7

Determine statistical descriptors from repeated manual dBA measurements at a single location. The noise data of Table 9.12 have been recast into a histogram [13] from entries in a data sheet used for sampling community noise levels by hand with a sound-level meter [4, p. 222]. Estimate L_{01}, L_{10}, L_{50}, and L_{90} and compute the L_{eq}. Data consist of 180 values sampled at 10 s each, for a total of 1800 s, or 30 min. Assume that the noise level was constant during each 10 s recording period. Note that the distribution is somewhat skewed from the normal distribution of statistics, and therefore the median and the arithmetic average or mean are not exactly identical.

SOLUTION The various L descriptors can be estimated from percentages of the data. Ten percent of 180 measurements or 1800 s represents 18 measurements or 180 s, respectively. These values are useful for estimating L_{10} (50 dBA) and L_{90} (42 dBA) as percentages of the dBA noise values above or below a certain value in the tails of the histogram in Table 9.12. L_{01} here is best estimated on a time basis. Noise was measured for more than 18 s [$1800 \times 0.01 = 18$] at 58 dBA and above.

L_{50} can be determined from the median of the distribution (45 dBA); the mean value of the 180 measurements is 45.6 dBA. L_{eq} calculated from Equation (9.12) for equal periods is summarized in the spreadsheet of Table 9.13; the L_{eq} value so determined is 47.5 dBA. Note that the L_{eq} and L_{50} here and in the general case are not the same. L_{eq} is greater here than L_{50} because it picks up the greater noise energy of the higher dBA values in the right-hand tail area of the distribution, which are more heavily weighted in the calculation of L_{eq}. The quarter-points of the distribution are shown for information.

Finally, for the discrete noise measurements of this distribution, it is possible to identify by inspection the absolute maximum value (L_{max}) and the absolute minimum value (L_{min}) as 60 and 41 dBA, respectively. ■

9.9.3 Time Periods for L_{eq}

L_{eq} can be measured for any length of time, but periods of 15 min, 1 h, and 24 h are common. Among other things, it depends on the type of noise and the purpose for which the data are to be used. In one jurisdiction, it was permissible to substitute a 20 min L_{eq} for the 1 h L_{eq} contained in the ordinance. During testing, the cumulative L_{eq} value as measured and reported automatically by the integrating sound-level meter remained virtually unchanged after 10–15 min into a run, like baseball batting averages at the end of the season. Primary noise sources consisted of an industrial plant, nearby traffic on a busy highway, and general background including other plants in the far distance.

434 Chapter 9 Noise

Table 9.12 STEM-AND-LEAF Histogram of Manual Noise Data Obtained for 30 min

				44														
			43	44														
			43	44														
			43	44														
			43	44	45													
			43	44	45	46												
			43	44	45	46												
			43	44	45	46												
			43	44	45	46												
			43	44	45	46												
			43	44	45	46	47											
			43	44	45	46	47											
			43	44	45	46	47											
			43	44	45	46	47											
Individual		42	43	44	45	46	47											
readings		42	43	44	45	46	47											
in dBA		42	43	44	45	46	47											
10 sec. each		42	43	44	45	46	47											
		42	43	44	45	46	47											
		42	43	44	45	46	47											
		42	43	44	45	46	47											
		42	43	44	45	46	47	48										
	41	42	43	44	45	46	47	48	49									
	41	42	43	44	45	46	47	48	49			52						
	41	42	43	44	45	46	47	48	49	50		52	53					
	41	42	43	44	45	46	47	48	49	50	51	52	53					
	41	42	43	44	45	46	47	48	49	50	51	52	53	54				
	41	42	43	44	45	46	47	48	49	50	51	52	53	54	56		58	60
Total no. of readings at each level	0	6	15	28	29	25	24	19	7	6	4	3	5	4	2	0	1	0	1	0	1

Grand Total Number of Readings = 180

Noise statistics: (dBA)

L_{max}	60		L_{min}	41
L_{01}	58			
L_{10}	50		L_{90}	42
L_{50}	45		L_{eq}	47.5 (See Table 9.1

In addition, 25% of the data lie at or below 43 dBA; 75% at or below 47 dBA.

9.9 A-Weighting Statistics

The L_{eq}'s for contiguous periods can be combined on an equivalent energy basis into a composite L_{eq} for a longer period. For example, a 24 h L_{eq} can be calculated from twenty-four 1 h L_{eq}'s determined over that same 24 h period. Equations (9.11) and (9.12) are used, with the individual shorter time L_{eq}'s substituted for the L_A's in the exponential terms. A worked example is given in the next section.

9.9.4 Adjustments to L_{eq}—L_{dn} and CNEL

PROBLEM 9.8

Values for 1 h L_{eq}'s are listed in column 3 of the spreadsheet of Table 9.14, along with start and stop times in columns 1 and 2. From these figures, calculate the corresponding 24 h L_{eq}, the L_{dn}, and the CNEL.

Table 9.13 Calculation of L_{eq} from a Series of Short-Term dBA Readings Obtained Manually

Number of readings (n)	10-s dBA reading	dBA/10	$10^{(dBA/10)}$	$n \times 10^{(dBA/10)}$
0	40	4.0	10,000.0000	0.0000
6	41	4.1	12,589.2541	75,535.5247
15	42	4.2	15,848.9319	237,733.9789
28	43	4.3	19,952.6231	558,673.4482
29	44	4.4	25,118.8643	728,447.0651
25	45	4.5	31,622.7766	790,569.4150
24	46	4.6	39,810.7171	955,457.2093
19	47	4.7	50,118.7234	952,255.7439
7	48	4.8	63,095.7344	441,670.1411
6	49	4.9	79,432.8235	476,596.9408
4	50	5.0	100,000.0000	400,000.0000
3	51	5.1	125,892.5412	377,677.6235
5	52	5.2	158,489.3192	792,446.5962
4	53	5.3	199,526.2315	798,104.9260
2	54	5.4	251,188.6432	502,377.2863
0	55	5.5	316,227.7660	0.0000
1	56	5.6	398,107.1706	398,107.1706
0	57	5.7	501,187.2336	0.0000
1	58	5.8	630,957.3445	630,957.3445
0	59	5.9	794,328.2347	0.0000
1	60	6.0	1,000,000.0000	1,000,000.0000
Total 180	—	—		10,116,610.4142

$$
\begin{aligned}
L_{eq} &= 10 \times \log_{10}[\{\text{total } n \times 10^{(dBA/10)}\}/\text{total } n] \\
&= 10 \times \log_{10}[\text{total of column 5/total of column 1}] \\
&= 10 \times \log_{10}[10,116,610.4142/180] \\
&= 10 \times \log_{10}[56,203.3912] \\
&= 47.49762521 \\
&= 47.5
\end{aligned}
$$

Table 9.14 Calculation of L_{eq}, L_{dn}, and CNEL (dBA) for Other Averaging Periods

Time		1 h L_{eq} (dBA)	$L_{eq}/10$	$10^{(L_{eq}/10)}$	$10^{((L_{eq}+A)/10)}$	$10^{((L_{eq}+B)/10)}$
Start	End					
7 a.m.	8 a.m.	49	4.9	79432.82	79432.82	79432.82
8 a.m.	9 a.m.	52	5.2	158489.32	158489.32	158489.32
9 a.m.	10 a.m.	53	5.3	199526.23	199526.23	199526.23
10 a.m.	11 a.m.	55	5.5	316227.77	316227.77	316227.77
11 a.m.	12 noon	59	5.9	794328.23	794328.23	794328.23
12 noon	1 p.m.	61	6.1	1258925.41	1258925.41	1258925.41
1 p.m.	2 p.m.	57	5.7	501187.23	501187.23	501187.23
2 p.m.	3 p.m.	56	5.6	398107.17	398107.17	398107.17
3 p.m.	4 p.m.	60	6.0	1000000.00	1000000.00	1000000.00
4 p.m.	5 p.m.	62	6.2	1584893.19	1584893.19	1584893.19
5 p.m.	6 p.m.	65	6.5	3162277.66	3162277.66	3162277.66
6 p.m.	7 p.m.	64	6.4	2511886.43	2511886.43	2511886.43
7 p.m.	8 p.m.	58	5.8	630957.34	630957.34	1995262.31
8 p.m.	9 p.m.	54	5.4	251188.64	251188.64	794328.23
9 p.m.	10 p.m.	50	5.0	100000.00	100000.00	316227.77
10 p.m.	11 p.m.	47	4.7	50118.72	501187.23	501187.23
11 p.m.	12 mid	42	4.2	15848.93	158489.32	158489.32
12 mid	1 a.m.	44	4.4	25118.86	251188.64	251188.64
1 a.m.	2 a.m.	40	4.0	10000.00	100000.00	100000.00
2 a.m.	3 a.m.	38	3.8	6309.57	63095.73	63095.73
3 a.m.	4 a.m.	35	3.5	3162.28	31622.78	31622.78
4 a.m.	5 a.m.	36	3.6	3981.07	39810.72	39810.72
5 a.m.	6 a.m.	37	3.7	5011.87	50118.72	50118.72
6 a.m.	7 a.m.	41	4.1	12589.25	125892.54	125892.54
Sum				13079568.03	14268833.15	16392505.48
Sum/24				544982.00	594534.71	683021.06
$10 \times \log_{10}[\text{sum}/24]$				**57.4**	**57.7**	**58.3**
				24 h L_{eq}	L_{dn}	CNEL

436

Last entry in column 5 is average 24 h L_{eq} (unadjusted) with units of dBA
Last entry in column 6 is L_{dn} calculated by adding 10dBA to each of the nine 10 p.m. to 7 a.m. nighttime readings; units of L_{dn} = dBA
Last entry in column 7 is CNEL calculated by adding 5 dBA to each of the 3 7 p.m. to 10 p.m. evening hourly readings and 10dBA to each of the nine 10 p.m. to 7 a.m.; hourly readings; units of CNEL = dBA

Notes

The value of A in the heading of column 6 is 0 from 7 a.m. to 10 p.m. (15 readings) and 10 from 10 p.m. to 7 a.m. (9 readings)
The value of B in the heading of column 7 is 0 from 7 a.m. to 7 p.m. (12 readings), 5 from 7 p.m. to 10 p.m. (3 readings), and 10 from 10 p.m. to 7 a.m. (9 readings)

Alternate method by first calculating unadjusted L_{eq}'s for day, evening, and nightime periods and then adjusting these longer-term L_{eq}'s by the appropriate penalty factor

	dBA	
Avg 24-h L_{eq} = $10 \times \log_{10}$[SUMCOL5/24] =	57.4	for 24 hs unadjusted
Avg 15-h L_{eq} from 7 a.m. to 10 p.m. =	59.4	L_d (daytime unadjusted)
Avg 9-h L_{eq} from 10 p.m. to 7 a.m. =	41.7	L_n (night unadjusted)
+ 10 dBA =	51.7	L_n with 10 dBA penalty
Avg 12-h L_{eq} from 7 a.m. to 7 p.m. =	60.0	L(7 a.m. to 7 p.m.) (unadjusted)
Avg 3-h L_{eq} from 7 p.m. to 10 p.m. =	55.2	L_e (7 p.m. to 10 p.m.)(unadjusted)
+ 5 dBA =	60.2	L_e with 5 dBA penalty

$L_{dn} = 10 \times \log_{10}[(15 \times 10^{(L_d/10)} + 9 \times 10^{(L_n+10)/10})]/24]$
$= 10 \times \log_{10}[\{15 \times 10^{(59.4/10)} + 9 \times 10^{(51.7/10)}\}/24]$

L_{dn} =   **57.7**   dBA

CNEL = $10 \times \log_{10}[\{(12 \times 10^{(L/10)} + 3 \times 10^{(L_e+5)/10)}$
$+ 9 \times 10^{(L_n+10)/10)})/24]$
$= 10 \times \log_{10}[\{12 \times 10^{(60.0/10)} + 3 \times 10^{(60.2/10)}$
$+ 9 \times 10^{(51.7/10)}\}/24]$

CNEL =   **58.3**   dBA

SOLUTION Calculations are performed using the spreadsheet of Table 9.14.

In Table 9.14, calculation of the 24 h L_{eq} is straightforward. First, the individual 1 h L_{eq}'s are divided by a factor of 10 to be used as the exponents in Equation (9.12). Then the individual hourly figures are summed in the spreadsheet to give a total. When multiplied by a factor of 10, the base ten logarithm of the total provides the 24 h L_{eq}.

A similar summation is performed in the calculation of L_{dn} and CNEL, but with adjustments to night and evening values. The L_{dn} and CNEL are calculated by two different but equivalent techniques. The first, in the upper part of the tabulation, adds the appropriate dBA penalty to each 1 h L_{eq} prior to exponentiation. It then sums these terms, takes the logarithm of the sum, and multiplies by 10 to find L_{dn} and CNEL, again employing Equation (9.12).

The second procedure calculates an unadjusted L_{eq} for the 9 h nighttime period from 10 p.m. to 7 a.m., for the 3 h evening period from 7 p.m. to 10 p.m., and for two different daytime periods (12 h from 7 a.m. to 7 p.m. and 15 h from 7 a.m. to 10 p.m.). These are then used in the appropriate combinations to substitute in the equations below. The 5- and 10-dBA penalties are applied to the calculated 3 h evening L_{eq} and the 9 h nighttime L_{eq}. (See the second part of the spreadsheet of Table 9.14.)

$$L_{dn} = 10 \log_{10}[(15 \times 10^{(L_d/10)} + 9 \times 10^{(L_n + 10)/10})/24] \quad (9.13)$$

$$\text{CNEL} = 10 \log_{10}[(12 \times 10^{(L_d/10)} + 3 \times 10^{(L_e + 5)/10} + 9 \times 10^{(L_n + 10)/10})/24] \quad (9.14)$$

Results are identical for this case and others used to test the method. ∎

9.10 NOISE REGULATIONS

9.10.1 General

Community noise from a stationary source is regulated on the state and/or county/city level. However, noise must be addressed as part of any state or federal environmental impact statement (EIS) that may be required in specific circumstances. Local noise ordinances are found typically in one or more of the following local departments:

- Planning and Zoning
- Health
- Police/Law Enforcement
- Air Quality Enforcement

In many cases, no formal approval, permit, or registration is required; rather one must simply design and operate so as to comply with the noise code. In many cases, noise is an agenda item that comes up during a public hearing/public comment process. Since the present discussion is only general in nature, it behooves one to seek out and find *all* of the applicable noise regulations pertaining to one's own specific source and location.

9.10.2 Two Approaches—Common Law Nuisance/Qualitative Versus Quantitative

There are two governmental approaches to control noise. One is through general common law nuisance provisions under which it must be proven that one's quality of life or peaceful enjoyment of one's property has been abridged by an allegedly offensive noise; noise may or may not be listed as a nuisance. The other contrasting approach spells out in numerical terms exactly who, what, when, where, and how and enumerates in detail just what is specifically allowed and/or prohibited in terms of noise generation and its effect on the community [14,15]. One purpose of a numerical noise ordinance is to preclude such outrageous behavior as, for example, the true story of the crushing of rocks under bright lights during construction at 10 p.m. some 50–100 ft (15–30 m) from an occupied residence.

Typical regulations range from the common law nuisance type with or without qualitative prescriptions and prohibitions for noise to specific quantitative performance requirements. The quantitative regulations usually refer to time of day, duration, and allowable noise levels at various locations. Most regulate by means of snapshot dBA readings (L_A); some specify longer term L_{eq} values. A few describe compliance in terms of octave band readings; and still fewer may mention 1/3 octave band measurements. The nuisance and common law type of regulation is more encompassing, and it is easier to assert a violation but more difficult to prove a case. A quantitative code is easier to enforce once the evidence of a violation has been gathered. The State of New Jersey regularly conducts a community noise control course (open to the public) for certification and periodic recertification of governmental officials responsible for enforcing the state regulations and any local regulations in effect concerning noise, in order that evidence collected during investigation of a suspected violation will stand up in court [16].

Very often, compliance is determined at the property line of a source, typically an industrial plant. However, some governmental regulations list criteria based on points of measurement other than the source boundary line. These include

- Place of nearest zoning change (e.g., industrial to commercial or residential)
- Property line of nearest dwelling unit
- A specified distance from the residence itself

Noise contained inside an industrial plant's boundaries falls under occupational health and safety regulations and is outside the scope of this discussion.

9.10.3 Adjustments to Permissible Noise Levels

Noise regulations are written for typical ambient steady-state noise of an incoherent nature and its effect on the community. Many noise codes contain provisions to subtract 5–10 dBA from the published allowances to account for impulse noise or a pure tone. There is usually a lower noise limit, expressed or implied, for nighttime hours compared to daytime. These items are regulated more strictly because of their greater annoyance potential, especially at night when people are trying to sleep. New Jersey's 50 dBA nighttime limit, for example, is predicated on not disturbing sleep; the 65 dBA daytime limit is to avoid speech interference.

9.10.4 Impulse Noises

Impulse noises are those of extremely short duration, usually less than 1 s, with an abrupt onset and rapid decay, and usually a large difference in intensity over the background. Examples of impulse noise such as gunshots/explosions and typical machine shop activities are enumerated in a previous section of this chapter. Impulse-type noises are most properly measured using the impulse setting of a sound meter so equipped.

9.10.5 Pure Tones

In an academic sense, a pure tone is one having a sinusoidal waveform of a single frequency (Figure 9.1). Examples are the 1000 Hz, 250 Hz, and other tones emitted by a field sound-level calibrator and the 440 Hz (A above middle C) tone and the others broadcast by NIST for tuning musical instruments and calibrating equipment [1, p. 125; 8].

For regulatory purposes, however, a *pure tone, prominent discrete tone*, or *audible discrete tone* is defined in terms of the human ear and field quality sound-measuring instrumentation. This is any sound that can be distinctly heard as a single pitch or a set of single pitches. (The transformer hum of an example to follow is likely to be picked up under this provision or the quantitative definition below involving octave bands or 1/3 octave bands.)

Defined in terms of octave bands, a pure tone condition is said to exist when the sound pressure level of any octave band center frequency exceeds the sound pressure level of each of the two adjacent octave band center frequencies by 3 dB or more. All dB values are unweighted in this definition. A more recent definition states that a pure tone shall exist if the unweighted sound pressure level in any octave band or 1/3 octave band is higher than the arithmetic average of the sound pressure levels of the two adjacent octave bands or 1/3 octave bands by more than the following amounts when averaged over a specified measurement period:

- 5 dB for center frequencies 500 Hz and above (bands 27 and above, Table 9.4)
- 8 dB for center frequencies between 160 and 400 Hz (bands 22–26)
- 15 dB for center frequencies less than or equal to 125 Hz (bands 21 and below)

A stricter version of this latter definition contains such a requirement, but *with no averaging period*. In one state, the rule does not apply to audible discrete tones having a 1/3 octave band sound pressure level 10 dB or more below the allowable sound pressure levels specified in the state's code for the octave band that contains the 1/3 octave band of interest.

9.10.6 Night Versus Day

Aside from having separate standards for different times of day, regulations written in terms of L_{dn}, CNEL (both defined earlier), and the like also penalize sounds emitted at night.

9.11 INDUSTRIAL NOISE

Some sources of industrial noise are summarized in Table 9.15.

Techniques for estimating noise impacts include published tables, equations, and graphs [2, pp. 120–136; 7, pp. 114–135]; manufacturers' information; and estimation from one's own similar equipment in operation [17,18]. Pay particular attention to the details for the required inputs and to any special conditions pertaining to the results.

The estimated noise at various locations in the plant can then be projected to off-site receptor points. Compliance with governmental codes can be determined and, if necessary, methods of noise abatement [1, pp. 635–646; 7, pp. 237–266] can be considered. Several example problems follow the discussion of noise propagation models.

9.12 SOUND PROPAGATION FROM POINT SOURCE TO RECEPTOR

9.12.1 Sound Power Versus Sound Pressure

Sound power is the inherent strength of a sound-emitting source. For given operating conditions, it is constant, independent of the location of the source with respect to the surrounding sound field and the distance to any receptor. It is also known as the power watt level (PWL).

Table 9.15 Some Sources of Industrial Noise

Alarms/warning sirens	Gears/drive belts or chains
Combustion (boilers/furnaces)	Impact noise
Comfort heating/air conditioning	In-plant transportation related
Construction equipment/activity	Machine shop noise
Cooling towers	Materials handling
Diesel generators	Pneumatic conveying
Electrical motors/generators	Pumps
Fans/blowers	Rotating and reciprocating machinery
Flares	Sandblasting
Flowing fluids in pipes/ducts	Transformers
Gas/steam turbines	Valves

It is defined in decibels as

$$\text{PWL} = 10\log_{10}[W_{\text{actual}}/W_{\text{reference}}] \quad (9.15)$$

where W_{actual} is the power in watts and $W_{\text{reference}}$ is the reference sound power, 10^{-12} by international agreement [12, pp. 27–30].

Whereas SPL, previously defined, is a measured quantity at some distance from the source under specific circumstances, there is no instrument for measuring the sound power. Instead, it is calculated from sound pressure using a mathematical model for sound propagation. In these models, the sound waves spread out more and more the farther they travel from the source, thereby decreasing in strength the farther they travel from the source, and diluting or attenuating the sound level at greater distances from the source. This is known as *geometric spreading*. For outdoor measurements, in an open area, the most popular models are spherical radiation and hemispherical radiation.

9.12.2 Sound Radiation Models

A point source in *spherical radiation* is assumed to radiate from a single point equally in all directions [1, pp. 87–89; 10, pp. 23–26]. The propagating sound waves diverge through an enclosing surface whose area increases with the square of the distance from the source. This model applies to sound emitted from a distant airplane to the ground, where reflection from the ground is negligible.

The equation relating sound power and sound pressure for this case is as follows:

$$\text{PWL (dB)} = \text{SPL (dB)} + 20\log_{10}[r] + 11 \quad (9.16)$$

where r is the straight line distance in meters from source to receiver, at which location the SPL is measured. The constant 11 in Equation (9.16) is a very close approximation of $10\log_{10}(4\pi)$, which arises in the model. The equation can be rearranged to solve for SPL when PWL is known

$$\text{SPL (dB)} = \text{PWL (dB)} - 20\log_{10}[r] - 11 \quad (9.17)$$

Hemispherical radiation occurs when the same point source radiates omnidirectionally except for a single flat reflecting planar surface, such as concrete pavement or hard ground. A simple example would be the noise from an isolated piece of machinery mounted on its foundation, sufficiently far away so that noise possibly generated from multiple components behaves as if originating from a single point.

The governing equation here is [1, pp. 87–89; 10, pp. 23–26]

$$\text{PWL (dB)} = \text{SPL (dB)} + 20\log_{10}[r] + 8 \quad (9.18)$$

and rearranged

$$\text{SPL (dB)} = \text{PWL (dB)} - 20\log_{10}[r] - 8 \quad (9.19)$$

The constant 11 from spherical radiation is reduced by 3 dB to account for the presence of a reflecting plane (planar surface, *not* an airplane).

9.12 Sound Propagation from Point Source to Receptor

Confusion arises because SPL and PWL are both expressed in dB. However, the definition of the decibel ratio is different in each case. The reference for the SPL decibel is a pressure, $2 \times 10^{-4}\,\mu\mathrm{bar}$; for PWL, the reference is acoustic power, 10^{-12} W. Sound power of a source is a stand-alone number computed from a mathematical model; for sound pressure (measurable), the location must always be stated. For example, in an open area (the so-called *far field*) [10, p. 24, 12, p. 141], the SPL 1 m from a hemispherical source of 100 dB PWL would be 92 dB.

9.12.3 Attenuation of Sound over Distance

Either model can be used to derive an expression for SPL at some distance from a point source when the SPL is known at some other distance [2, p. 141].

$$\mathrm{SPL}_2\,(\mathrm{dB}) = \mathrm{SPL}_1\,(\mathrm{dB}) - 20\log_{10}[r_2/r_1] \qquad (9.20)$$

Here, r_2 and r_1 are the respective distances (in any consistent units) from the source, corresponding to the locations of SPL_2 and SPL_1, expressed in dB. For simple geometric spreading, SPL_2 and SPL_1 can also both be expressed in dBA. Equation (9.20) leads to the familiar rule for a point source that the SPL decreases by 6 dB (or dBA) for every doubling of distance. For instance, 66 dB at 100 ft translates into 60 dB at 200 ft, 54 dB at 400 ft, 48 dB at 800 ft, and so on, as shown below:

$$\mathrm{SPL}_2 = 66 - 20\log_{10}[(200)/(100) = 2]$$
$$66 - 20(0.3010) \cong 60$$
$$66 - 20\log_{10}[(400)/(100) = (2)(2)]$$
$$66 - 20(\log[2] + \log[2]) = 66 - (20)(2)(0.3010) \cong 54$$
$$66 - 20\log_{10}[(800)/(100) = (2)(2)(2)]$$
$$66 - (20)(3)(0.3010) \cong 66 - 18 \cong 48$$

Other examples follow.

PROBLEM 9.9

The 5000 kVA, 480-V transformer mentioned in Ref. [19] produces a SPL of 80 dB outdoors at a distance of 6 ft from the transformer. Estimate its PWL and the SPL at 1000 ft. What is the corresponding SPL (dB and dBA)?

SOLUTION From Equation (9.18) for a hemispherically radiating source (with reflection from the floor/ground) with distance (r) in meters

$$\mathrm{PWL}\,(\mathrm{dB}) = \mathrm{SPL}\,(\mathrm{dB}) + 20\log_{10}[r] + 8\,\mathrm{dB}$$
$$= 80 + 20\log_{10}[(6\,ft)(0.3048\,\mathrm{m}/ft)] + 8$$
$$= 80 + 20(0.2622) + 8$$
$$= 80 + 5.2440 + 8 \cong 93.2\,\mathrm{dB}$$

From Equation (9.19),

$$\begin{aligned} \text{SPL at 100 ft} &= \text{PWL (dB)} - 20\log_{10}[r] - 8 \text{ dB} \\ &= 93.2(\text{from above}) - 20\log_{10}[(100)(0.3048)] - 8 \\ &= 93.2 - 20(1.4840) - 8 \\ &= 93.2 - 29.7 - 8 \cong 56 \text{ dB} \end{aligned}$$ ■

This noise produces the familiar audible transformer hum, a pure tone at twice the power line frequency (with a possible contribution from the even harmonics) [4, p. 268; 20,21]. The sound is produced by the expansion and contraction of the transformer's iron core plus the mechanical forces between primary and secondary windings, causing them to vibrate [20,21].

At a line frequency of 50–60 Hz, the resulting hum at 100–120 Hz will be picked up by the 125 Hz octave band of a sound meter. If all of the energy is contained in the 125-Hz octave band, dBA at 125 Hz = 56 − 16 = 40 dBA (from Table 9.10). This pure tone will undoubtedly be penalized by the applicable noise code, as discussed in a later section.

Also, because of the coherent nature of transformer noise, the decibel levels from multiple (two or more) transformers will not sum according to Equation (9.2). Instead, the level of combined noise can vary noticeably according to the position of the receptor [4, p. 10].

A word of caution is in order. The PWL calculated here should be regarded as only approximate. Although not strictly a large power distribution transformer discussed in Ref. [4, p. 197], the sound-level contours around it may not be circular. Unless the starting-point SPL is some sort of average (preferably on an energy basis), the calculated PWL will depend on from which direction of the transformer the SPL is determined. At a far enough distance away, however, the sound level should decrease by the expected 6 dB for each doubling of distance.

PROBLEM 9.10

Noise from a mechanical draft cooling tower consists primarily of fan noise, pump noise, flow noise, the sound of falling water, and possibly some structural vibration. If the SPL 50 ft on a perpendicular from the center of the open face of a cooling tower is 70 dBA (by rule of thumb), calculate its SPL and the total SPL (including background) on this same line 150 ft away. The background SPL is consistently measured as less than 60 dBA or less at all times when the cooling tower is not operating. (*Note*: measured dBAs reported for both mechanical draft and natural draft cooling towers at a total of 15 power stations are in the range of 74–79.5 dB at 50 ft, but background sound levels are not mentioned [22].)

SOLUTION Since background noise is specified at 60 dBA or less, the 70 dBA SPL is from the cooling tower alone. No correction for background must be made at 50 ft because the background is 10 dBA (or more) below that of the cooling tower.

From Equation (9.20),

$$\text{SPL}_2 \text{ (dB)} = \text{SPL}_1 \text{ (dB)} - 20\log_{10}[r_2/r_1]$$

9.12 Sound Propagation from Point Source to Receptor

Any consistent set of decibel units and distance units may be used since the actual units cancel. Substituting for the cooling tower noise and distance,

$$\text{SPL at 150 ft} = \text{SPL at 50 ft} - 20 \log_{10}[(150)/(50) = 3]$$
$$= 70 - 20(0.4771) = 60.4580 \cong 60.5 \text{ dBA}$$

However, to find the total SPL including background at 150 ft, the background noise must be added. The worst case for total SPL occurs when the background is 60 dBA.

Adding decibels (Equation (9.2)),

$$\text{SPL}_{\text{Total}} = 10 \log_{10}[(10^{60.4580/10} + 10^{60/10}]$$
$$10 \log_{10}[(1,111,220 + 1,000,000)]$$
$$= 10(6.3245) = 63.245 \cong 63 \text{ dBA}$$

Approximating the cooling tower's SPL at 60 dBA plus 60 dBA for background would have led immediately to 60 dBA + 60 dBA = 63 dBA$_{\text{Total}}$ by the shortcut method (Figure 9.2). ∎

PROBLEM 9.11

Estimate the source intensity of a 20 hp reciprocating air compressor whose PWL is given by the equation here [7, pp. 127–129]:

$$\text{PWL} = 10 \log_{10}[\text{hp}] + 86 \tag{9.21}$$

for compressors rated at 1–100 hp.

The PWL of this compressor is defined as evenly distributed among the 500, 1000, 2000, and 4000 Hz octave bands.

SOLUTION Substitution of the hp provided into Equation (9.21) results in

$$\text{PWL} = 10 \log_{10}[20] + 86$$
$$(10)(1.3) + 86 = 13 + 86 = 99 \text{ dB}$$

concentrated in the 500–4000 Hz octave bands.

Working backward by using Figure 9.2 for two pairs of octave bands would each exhibit 3 dB less than the total, or 96 dB for the pair. Applying this logic to each octave band for each pair would result in another 3 dB drop to a PWL of 93 in each band, such that

$$93 \text{ dB} + 93 \text{ dB} + 93 \text{ dB} + 93 \text{ dB} = 99 \text{ dB}_{\text{Total}}$$
$$500 \text{ Hz} \quad 1000 \text{ Hz} \quad 2000 \text{ Hz} \quad 4000 \text{ Hz}$$

as found above. ∎

PROBLEM 9.12

Convert the PWL found for the compressor of Problem 9.11 to an SPL outdoors at 60 ft (=30.48 m) expressed as both dB$_{\text{all pass}}$ and dBA.

SOLUTION From Equation (9.19) for each octave band

$$\text{SPL} = 93 - 20 \log_{10}[30.48] - 8 = 55.3 \text{ dB}$$

Table 9.16 Calculation of dBA for Compressor Problem

Hz	dB	Factor	Corrected dB
500	55.3	−3	52.3
1000	55.3	0	55.3
2000	55.3	1	56.3
4000	55.3	1	56.3
Total	61.3	—	61.3 dBA

Using Figure 9.2 in the forward direction, two at a time produces 58.3 dB for each of two pairs of octave bands and 61.3 dB$_{\text{allpass}}$ for the total. Then, applying the A-weighting factor adjustment for each octave band and summing yields the entries in Table 9.16.

Corrected dBs are added using Equation (9.2) to obtain dBA. Here, the SPL in dB$_{\text{all pass}}$ and dBA is the same, or more precisely dBA is slightly greater by 0.02 dB. Results are valid for the compressor alone in the absence of background, which would add frequency components in all octave bands and would cause a *measured* dB$_{\text{all pass}}$ and dBA to differ. ∎

PROBLEM 9.13

Given a single hemispherically radiating point source producing an SPL of 90, 84, and 78 dB at 1, 2, and 4 m, respectively [6, p. 12], calculate its source intensity (PWL) in dB.

SOLUTION The SPL–PWL relationship of Equation (9.19) can be rewritten as

$$\text{SPL} = \text{PWL} - [20 \log_{10}(\text{distance in meters}) + 8] \qquad (9.22)$$

By choosing the bracketed term as the independent variable (x) and SPL as the dependent variable (y), the relationship is seen to be linear with a slope of -1.0 and a y-intercept corresponding to the PWL.

The given input data can be recast as follows. For $y =$ SPL of 90, 84, and 78,

$$x = -[20 \log_{10}(r, \text{ meters}) + 8] = 8, 14.0206, \text{ and } 20.0402, \text{ respectively.}$$

Plotting these x and y values yields the linear graph of Figure 9.3, which extrapolates to a PWL of 98 dB at $x = 0$.

In this textbook example, one could have used Equation (9.18) to solve for PWL directly with any of the SPL–distance pairs given to obtain the same value for PWL. However, seldom are real data so kind, and the best estimate of PWL may have to be obtained by linear regression of imperfect data points in the field.

With a PWL of 98 dB and Equation (9.19), one can estimate the SPL at any distance from this point source of noise, for example,

$$\text{At } 100 \text{ m}: \text{SPL} = 98 - 20 \log_{10}[100] - 8 = 98 - 20(2) - 8 = 50 \text{ dB}$$
$$\text{At } 1000 \text{ m}: \text{SPL} = 98 - 20 \log_{10}[1000] - 8 = 98 - 20(3) - 8 = 30 \text{ dB}$$

To these values, corrections must be applied for ground absorption and reflection, plus the neighborhood residual, which usually depend on the time of day. These calculations form the basis of computerized models for predicting off-site noise effects from multiple plant noise sources. Such computer models are beyond the scope of this book, but a later simple manual example will illustrate the principle. ∎

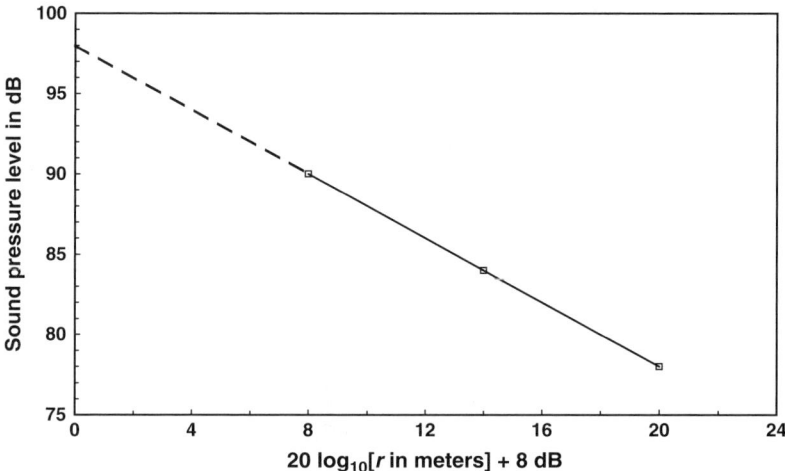

Figure 9.3 SPL–distance plot for ideal point source.

PROBLEM 9.14

A small plant is located on a plot 100 m × 100 m oriented exactly north-south and east-west. A single external receptor point of interest lies 1/10 of a mile (160.9 m) due north on the perpendicular through the midpoint of the northern property line. This is taken as the origin of coordinates.

Three sources of interest in the plant contribute to the SPL felt at the receptor. Source 1 with a PWL of 110 is located at −40 m, −30 m. Source 2 with a PWL of 98 dB is located at 0 m, −80 m. Source 3 with a PWL of 90 dB is located at +45, −50 m.

Find the combined SPL in dB at the receptor point.

SOLUTION This problem requires a sketch plus a little trigonometry, projection of sound pressure over distance, and addition of decibels from the individual sources.

For source 1, horizontal distance to the *y*-axis, the perpendicular from the plant to the receptor, is 40 m. Vertical distance from this source to the receptor is 160.9 + 30 = 190.9 m and straight line distance from source 1 to the receptor is given by the Pythagorean theorem.

$$[190.9^2 + 40^2]^{1/2} = 195.0 \text{ m}$$

Its SPL at the receptor point is given by Equation (9.19)

$$\text{SPL at } 195.0 \text{ m} = 110 - 20 \log_{10}[195.0] - 8$$
$$= 110 - 20(2.29) - 8 = 56.2 \text{ dB}$$

Straight line distance to the receptor point and the SPL there are calculated for the other sources, and results are summarized in Table 9.17. The total combined SPL (dB) is calculated from Equation (9.2):

$$\text{dB}_{\text{combined}} = 10 \log_{10}[10^{56.2/10} + 10^{42.4/10} + 10^{35.3/10}] = 56.4 \text{ dB}$$

In an actual case, these calculations would be performed for each octave band, and a combined SPL at the receptor would be determined octave band by octave band. From this, a $\text{dB}_{\text{all pass}}$ and a dBA can be computed and added to the ambient background (also known as the neighborhood residual) to obtain a total decibel level.

Table 9.17 Summary of SPL–Distance Results for Multiple Source Problem

Point no.	PWL dB	Distance to receptor (m)			SPL (dB)
		Vertical	Horizontal	Direct	
1	110	190.9	40	195.0	56.2
2	98	240.9	0	240.9	42.4
3	90	210.9	45	215.6	35.3
Total	—	—	—	—	56.4

For typical daytime background levels, one might not even notice the change in decibel level when this plant starts up unless the new sources are distinctly different in character from the pre-existing background. However, the plant noise would add significantly to typical nighttime background levels and may likely call for some sort of noise abatement. General methods of noise abatement are discussed in a subsequent section. ■

9.13 EXCESS ATTENUATION OF NOISE OVER DISTANCE

The equations in Section 9.12 provide the means to calculate attenuation over distance from distance alone (geometric spreading). In addition, propagation of sound waves is affected by atmospheric conditions, wind, ground cover, natural and man-made barriers, other obstacles, and even time of the day [1, pp. 604–605; 12, pp. 169–189]. Attenuation by atmospheric air with its temperature and humidity affects the higher frequencies more so than the lower, and the pitch of a high-frequency sound (perhaps a siren or fan noise) will appear lower to a distant observer [6, pp. 62–63].

Wind and time of day can enhance or diminish a distant sound. Reflection, reverberation, and absorption by structures over the intervening distance also modify the sound field. Absorption depends on the type of ground cover, thick growths of trees and other vegetation, and diffraction.

Most of these effects diminish the noise level from that calculated from geometric spreading alone. Ignoring them provides a more conservative (higher) estimate of the noise level at a receptor location. It may not be wise to take advantage of excess attenuation only to cause a violation when conditions change. If important to a specific situation, these effects can all be estimated one by one with appropriate corrections. A viable alternative is measurement under actual field conditions (Section 9.16).

9.13.1 Attenuation by Atmospheric Air

Besides a decrease in noise with distance alone, noise is attenuated to a lesser degree by passage through atmospheric air. This absorption depends on the frequency of the sound, the length of the path, and the temperature (T) and moisture content of the air. An equation is given for $68°F$ ($20°C$) as a function of the geometric mean frequency of an octave band, the distance (r) between source and receptor, and the percent relative humidity (RH) [7, p. 166; 12, p. 170].

$$\text{Atmospheric attenuation (AA) in dB} = 7.4 \times 10^{-8} \cdot (f^2 r/\text{RH}) \tag{9.23}$$

At 50% RH, the following approximation may be used to correct for other temperatures within a ΔT of $\pm 18°F$ ($\pm 10°C$) [12, p. 170]

$$\text{AA [at } T \text{ and 50\% RH]} = \text{AA [at } 68°F (20°C) \text{ and 50\% RH]}/(1 + \beta \Delta T f) \tag{9.24}$$

where $\beta = 2 \times 10^{-6}$ for ΔT in °F or 4×10^{-6} for ΔT in °C.

The absorption coefficient in decibel per meter at 68°F (20°C) is shown graphically as a function of frequency with % RH as a parameter from 0 to 100% [5, p. 154]. Other charts exist showing the coefficient of absorption in decibel per meter versus % RH, parametric in frequency [5, pp. 155–156].

There are tables and graphs for atmospheric attenuation for frequencies in the range of 125–12,500 Hz, for temperatures in the range of 14–86°F (from $-10°C$ to 30°C), and for RH in the range of 10–90% [23]. Additional charts have been published for aircraft to ground atmospheric attenuation [1, p. 605; 12, pp. 171–172] although the aircraft to ground stipulation may not be clearly stated [1, p. 605].

Additional factors can impact outdoor sound propagation [1, pp. 604–605; 12, pp. 169–189]. These include

- wind gradients,
- atmospheric turbulence,
- thermal gradients at different times of day,
- reflection from the ground surface,
- attenuation by ground cover (grass, shrubbery, trees, pavement),
- attenuation by precipitation or fog,
- attenuation by barriers, buildings, hilly terrain.

Excess attenuation by the other factors above will not be explored here in any detail. Except for reflection,[1] which would enhance noise levels when the receptor is impacted both by transmitted sound waves and by sound waves bouncing off nearby objects, these other factors decrease noise levels beyond what would be expected from distance alone.

A worked example involving atmospheric attenuation from an industrial point source is presented below. As demonstrated in the problem, atmospheric attenuation becomes important at long distances and affects the higher frequencies to a greater extent.

---

[1] For a single perfect reflection from some object in the sound field, sound pressure would double, that is, increase by 3 dB (Fig. 9.2); less than perfect reflection would add something less than 3 dB. Hence, providing an allowance for perfect reflection would provide a more conservative (higher) estimate of noise projected over distance. Unless it is critical to the analysis, this may be the best course of action.

Table 9.18 Decrease of Sound Levels from Distance and Atmospheric Attenuation

Temperature 68°F (20°C) and 50% RH

Center freq (Hz)	Delta (dB)	At 500 ft (152.4 m) Octave band reading	A-weight factor	A-Corr Reading	At 5000 ft (1524 m) dB Dist alone	Corr for Atm Atten	dB Dist + Atm Atten
31.5	0.0	60	−39.5	20.5	40	0.00	40.0
63	−2.0	58	−26	32.0	38	0.01	38.0
125	1.0	61	−16	45.0	41	0.04	41.0
250	−1.0	59	−8.5	50.5	39	0.14	38.9
500	−3.0	57	−3	54.0	37	0.56	36.4
1000	−5.0	55	0	55.0	35	2.26	32.7
2000	−8.0	52	1	53.0	32	9.02	23.0
4000	−12.0	48	1	49.0	28	36.09	0.0
8000	−23.0	37	−1	36.0	17	144.35	0.0
16000	−26.0	34	−6.5	27.5	14	577.41	0.0

	At 500 ft	At 1500 ft from distance alone	At 1500 ft from Dist + Atm Atten
dB$_{flat}$	66.7	46.7	46.3
dBA	60.0 (given)	40.0	37.6

Freq (Hz)	At 500 ft (starting basis) $10^{(dB/10)}$	5000 ft (from distance alone) $10^{(dB/10)}$	(distance + Atm Atten) 5000 ft $10^{(dB/10)}$
31.5	1.00	1.00	1.00
63	1000000.00	10000.00	9994.85
125	630957.34	6309.57	6296.58

450

250	1258925.41	12589.25		12487.51
500	794328.23	7943.28		7689.59
1000	501187.23	5011.87		4401.61
2000	316227.77	3162.28		1881.26
4000	158489.32	1584.89		198.51
8000	63095.73	630.96		1.00
16000	5011.87	50.12		1.00
Sum	472823.92	47283.23		42952.91
	66.7 dB$_{flat}$	46.7 dB$_{flat}$		46.3 dB$_{flat}$
31.5	1.00	1.00		1.00
63	112.20	1.12		1.12
125	1584.89	15.85		15.82
250	31622.78	316.23		313.67
500	112201.85	1122.02		1086.18
1000	251188.64	2511.89		2206.03
2000	316227.77	3162.28		1881.26
4000	199526.23	1995.26		249.91
8000	79432.82	794.33		1.26
16000	3981.07	39.81		0.79
Sum	995879.25	9959.78		5757.05
	60.0 dBA	40.0 dBA		37.6 dBA

Note: All readings and weighting factors in decibels.

PROBLEM 9.15

When far enough away from a noise-producing facility, all of the individual noise sources there can be lumped into a single virtual point source for the purpose of computation. Consider such a virtual source that produces the so-called *typical source spectrum* [24] said to be representative of many industrial sources of outdoor sound, and an overall noise level of 60 dBA at 500 ft (152.4 m). The typical source spectrum is defined in terms of differences between the individual octave band levels and the overall dBA (Table 9.18, column 2). Its octave band analysis at 500 ft is tabulated in column 3, and the corresponding 66.7 dB$_{flat}$ (dB$_{all\ pass}$) and 60.0 dBA are shown underneath.

Assume that an industrial plant with these noise properties is located in Central Nowhere, USA, surrounded by flat terrain in every direction with no wind. Calculate the octave band readings and the overall noise level to be expected at 5000 ft (1524 m) with and without atmospheric attenuation in the absence of any other effects. Estimate the PWL of the virtual source.

SOLUTION The necessary numbers are calculated in the spreadsheet of Table 9.18. Equation (9.20) is applied to each individual frequency and to dB$_{flat}$ (dB$_{all\ pass}$) to account for attenuation by distance alone at 5000 ft (1524 m) (column 6). This results in calculated values of 46.70 and 40.0 for dB$_{flat}$ (dB$_{all\ pass}$) and dBA, respectively, as listed in the table. The table also shows the details in computing dB$_{all\ pass}$ and dBA from the frequency spectrum.

Then Equation (9.23) is used to estimate atmospheric attenuation for each frequency (column 7). The lowest frequencies are affected very little. The frequency spectrum resulting from distance and air attenuation is tabulated in column 8. Where the equation produces an atmospheric attenuation greater than the sound level for the upper frequencies, the octave band reading is set to zero (no negative numbers allowed). Composite values of 46.3 dB$_{flat}$ (dB$_{all\ pass}$) and 37.6 dBA are calculated for this new frequency spectrum.

Equation (9.18) with an SPL of 66.7 dB at 152.4 m produces a PWL of 118.4 dB. ∎

PROBLEM 9.16

Recalculate the dB$_{all\ pass}$ and dBA entries of Table 9.18 at increments of 500 ft (152.4 m) up to 6000 ft (1828.8 m) with and without atmospheric attenuation. Add in a constant daytime background of 42 dBA and a nighttime background of 36 dBA. Plot the results.

SOLUTION In a manner similar to the method of the previous problem, dB$_{all\ pass}$ and dBA are calculated for various conditions as a function of distance and are shown in Table 9.19. Results are plotted against $\log_{10}$ [distance (ft/500)] in Figure 9.4. The resulting straight lines for dB$_{all\ pass}$ (square symbols) with and without atmospheric attenuation are virtually coincident. The line for dBA without atmospheric attenuation (triangles) comes out nearly parallel. Allowing for atmospheric attenuation steepens somewhat the slope of the dBA line (circles) because of the loss of the upper frequencies. Adding in the strong influence of the background causes both of these lines to begin to curve upward, for daytime (diamonds) more so than for nighttime (plus signs). Ideally, different conditions for temperature and relative humidity should be chosen for the nighttime conditions. Nevertheless, the background levels play an important role here regardless of the exact values of temperature and humidity chosen.

9.13 Excess Attenuation of Noise over Distance   453

Table 9.19 Tabulated Values for Attenuation of Typical Source Spectrum

Distance		Atten by Dist		Air Atten + Dist		Total Atten + Background	
ft	m	dB	dBA	dB	dBA	dBA + 42	dBA + 36
500	152.4	66.7	60.0	66.7	60.0	60.1	60.0
1000	304.8	60.7	54.0	60.6	53.1	53.4	53.2
1500	457.2	57.2	50.4	57.0	49.3	50.0	49.5
2000	609.6	54.7	47.9	54.5	46.6	47.9	47.0
2500	762.0	52.8	46.0	52.5	44.4	46.4	45.0
3000	914.4	51.2	44.4	50.9	42.7	45.4	43.5
3500	1066.8	49.8	43.1	49.5	41.2	44.6	42.3
4000	1219.2	48.7	41.9	48.3	39.8	44.0	41.3
4500	1371.6	47.7	40.9	47.3	38.7	43.7	40.6
5000	1524.0	46.7	40.0	46.3	37.6	43.3	39.9
5500	1676.4	45.9	39.2	45.5	36.6	43.1	39.3
6000	1828.8	45.2	38.4	44.7	35.8	42.9	38.9

If the values from these latter curves had been real data, measured in the field (with scatter), and were fitted with a straight line for the longer distances, it might lead to the erroneous conclusion that dBA is related to the logarithm of distance by a factor less in absolute value than the distance attenuation factor of 20 from Equation (9.20).

Caution: Be careful when using dBA over long distances in Equation (9.20) or Equation (9.28) (to follow), especially when the background noise level is significant. ∎

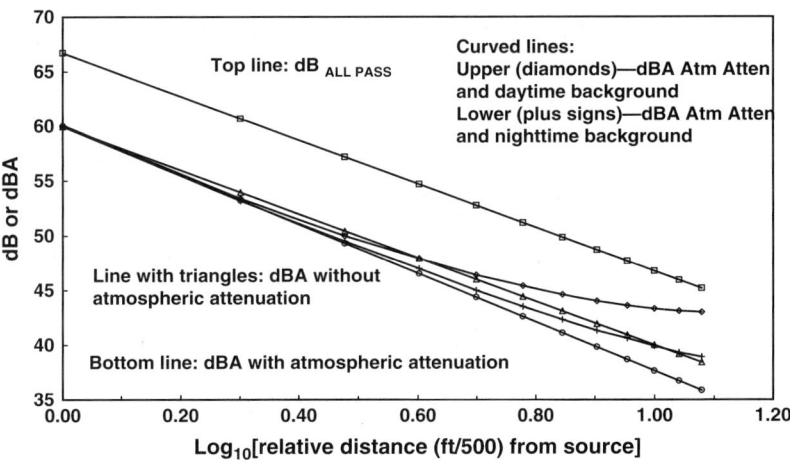

Figure 9.4 Attenuation of noise levels from typical source spectrum.

9.14 HIGHWAY NOISE (A LINE SOURCE)

This section will not make one an expert on the specialized area of highway noise so as to enable the design of highways. It is intended to provide a basic understanding and techniques to estimate noise as part of the background affecting one's project or operation. Nor will it instruct on *how to design* highway noise attenuation barriers, although comments on the performance of specific noise barriers can be found in the case studies of Appendix I.

The noise generated by a motor vehicle is made up mainly of engine/transmission/drive train/muffler noise and noise of tires on the pavement [1, pp. 609–611; 5, p. 295]. It is primarily of low frequency, said to be 550 Hz [25]. Tire noise predominates except for the slowest speeds [1, pp. 609–610; 5, p. 295]. There is also some aerodynamic noise from the vehicle body [1, p. 607].

Highway noise propagates as a line source [2, pp. 141–142] and is a function of vehicle speed, vehicle type, and traffic volume/density (vehicles per hour), as well as road geometry (including steep inclines), surrounding terrain, ground cover, and natural and man-made obstacles [26, pp. 2–3 of 8]. One truck at 55 mph (~88 kph) sounds as loud as 28 automobiles at the same speed. Traffic at 65 mph (~105 kph) sounds twice as loud as traffic at 30 mph (~50 kph). A traffic volume of 2000 vehicles per hour sounds twice as loud as 200 vehicles per hour [26, pp. 2–3 of 8].

Correlating equations are available to estimate the contribution of highway noise to the overall noise level at a standard distance of 50 ft (~15 m) from the centerline of the nearest traffic lane at a speed (v) in kilometers per hour. The *instantaneous* noise level in dBA for single vehicles is given by [5, p. 295]:

$$\text{For cars:} \quad L_A \text{ (dBA)} = 71 + 32 \log_{10}[v/88] \quad (9.25)$$

$$\text{For trucks:} \quad L_A \text{ (dBA)} = 84v \qquad\qquad \leq 48 \text{ kph} \quad (9.26)$$
$$= 88 + 20 \log_{10}[v/88] \; > 48 \text{ kph}$$

$$\text{For motorcycles:} \quad L_A \text{ (dBA)} = 78 + 25 \log_{10}[v/88] \quad (9.27)$$

As a point of reference, these equations reduce to 71 dBA for cars, 88 dBA for trucks, and 78 dBA for motorcycles at 88 kph (55 mph). The predominating tire noise for a single-chassis truck can range from 75 to 95 dBA [1, p. 630]. Noise from a *single* vehicle would propagate to a receptor on the ground as a moving point source [2, p. 142].

As the traffic density increases, the peak excursions decrease and the *average* noise level of the flow of traffic (as measured in L_{eq}) increases [5, pp. 293–294]. Different relationships that include the traffic volume apply. One such equation, for automobiles, is contained in the problem that follows.

The continuous noise from a busy highway in the distance contributes to the overall noise level and is included in the background. For a line source such as this busy highway containing a continuous flow of noise-producing vehicles, propagation is no longer characterized by a spherical or hemispherical spreading of sound from a single point. Instead, the reinforcement by one another of the line of point sources

9.14 Highway Noise (A Line Source)   455

makes the propagation surface a cylinder or half cylinder, and the drop-off with distance becomes [2, pp. 141–142]:

$$SPL_2 \text{ (dB)} = SPL_1 \text{ (dB)} - 10 \log_{10}[r_2/r_1] \quad (9.28)$$

where r_2 an r_1 are the corresponding perpendicular distances, in any consistent units, from the line source of the highway, to two different receptor points. This leads to the familiar rule for a line source that the SPL decreases by 3 dB for every doubling of distance (half of the 6 dB for a point source).

PROBLEM 9.17

(suggested by and expanded upon from Ref. [5, p. 295]:

Predict highway noise at 200 ft (~ 61 m) by

(a) simple cylindrical spreading, Equation (9.28)
(b) including the effects over the ground, as in the referenced problem
(c) using the more elaborate procedures of the Canadian Province of Ontario Ministry of the Environment (MOE) [27]

For negligible truck traffic, the L_{eq} at the standard receptor distance of 50 ft (~15 m) is given by

$$L_{eq} \text{ (dBA)} = 39 + 10 \log_{10}(Q) + 22 \log_{10}(v/88) \quad (9.29)$$

where Q is the number of vehicles per hour, and v is the average vehicle speed in kilometers per hour.

The road's surface and the surrounding terrain are assumed to be flat and of infinite extent. For such a short distance and low-frequency traffic noise, atmospheric air attenuation and its effect on the higher frequencies is not an important consideration, and L_{eq} (dBA) can be used. In the cited problem, Q is stated as 6000 vehicles per hour, and $v = 88$ kph (~ 55 mph).

SOLUTION

(a) By Equation (9.29),

$$L_{eq} \text{ (dBA)}_{\text{at 50 ft}} = 39 + 10 \log_{10}(6000) + 22 \log_{10}(88/88)$$
$$= 39 + 37.8 + 0 = 76.8 \text{ dBA}$$

By Equation (9.28),

$$SPL \text{ (dBA)}_{\text{at 200 ft}} = SPL_{\text{at 50 ft}} - 10 \log_{10}(200/50)$$
$$= 76.8 - 6.0 = 70.8 \text{ dBA} \cong 71 \text{ dBA } L_{eq}$$

(b) Adjustment for cylindrical spreading and ground absorption together is given as [5, p. 295]:

$$\Delta SPL = -a \log_{10}\{(r_2/15) - [(r_2-15)/(75)]^2\} \quad (9.30)$$
$$= 39 + 37.8 + 0 = 76.8 \text{ dBA}$$

where $a = 13.3$ over ground (including ground absorption) and 10.0 if the line of sight is 10° or more above the slope of the terrain, and r is the distance in meters.

$$\Delta SPL = -13.3 \log_{10}\{(61/15) - [(61-15)/(75)]^2\} = 7.6$$

Therefore,

$$\text{SPL}_{\text{at 200 ft (61 m)}} = 76.8_{\text{at 50 ft}} - 7.6 = 69.3 \cong 69 \text{ dBA}$$

With $a = 10.0$ (excluding ground absorption),

$$\Delta\text{SPL} = (10/13.3)(7.6) = 5.7 \text{ (by proration)}$$
$$\text{SPL}_{\text{at 200 ft (61 m)}} = 76.8 - 5.7 = 71.1 \cong 71 \text{ dBA}$$

(This is essentially the same as the answer to (a)).

(c) The MOE procedure [27] consists of tables, graphs, and equations too numerous to reproduce here. The procedure is straightforward and is based on a model developed by the U.S. Federal Highway Administration (FHWA) [28]. Factors affecting highway traffic noise and noise abatement techniques are described on FHWA's Web site [26].

The concept is to evaluate a reference sound level and adjust it using a series of factors to account for a finite length of roadway, its gradient, pavement surface, distance between the roadway and receptor, topography of the area, and whatever obstructions are in place. The prediction method is valid for distances from 15 m (~50 ft) up to about 500 m (~1640 ft, over 3/10 of a mile), with decreased accuracy from 200 to 500 m and for highly irregular topography. It is not applicable when traffic volume is less than 40 vehicles per hour, nor when speed of the traffic is less than 50 kph (~31 mph).

From Table 6 in the MOE procedure, the nearest reference sound level at 15 m, 40 vehicles per hour, and 88 kph is 56.1 dBA. (This figure actually includes 0.875% medium trucks and 0.125% heavy trucks, a total of 1% trucks, the closest figure in the tables to 100% automobiles.) From another table (MOE Table 2), the sound-level adjustment to a traffic volume nearest to 6000 vehicles per hour is read as an additive 22 dBA. Volume adjustment is also plotted in a semilogarithmic graph (MOE Figure 2). At 15 m, 88 kph, and 6000 vehicles per hour

$$L_{\text{eq}} = 56.1 + 22 = 78.1 \cong 78 \text{ dBA (at the roadway)} \tag{9.31}$$

Adjustment of sound level for distance to the receptor of -10.0 dBA over soft ground, or -6.0 dBA over a reflective surface, is read from still another table (MOE Table 7) or figure (MOE Figure 3). If more than half of the ground surface between the road and the receptor is a plowed field or is covered with grass, shrubs, or other vegetation, the ground is considered sound reflective. Corrections contained in the MOE procedure for heavy trucks on an uphill road grade; finite length and gradient of the roadway segment; other than flat topography; pavement surface other than asphalt or concrete; and any barriers, buildings, or other obstructions are not applicable for this simple example.

This makes the L_{eq} at the distant receptor equal to 68 dBA with ground absorption and 72 dBA with a reflective surface and without absorption. Compare these answers with those of (a) and (b). ■

9.15 NOISE CONTROL

There are three general methods to achieve noise control [1, pp. 629–630; [3,6], pp. 29–30]:

- Change the source.
- Change the path.
- Change the receptor.

Although any noise control technique can be classified into one of these categories, the methods can be employed in any combination. *Specific* designs for noise abatement are best left to experienced noise control specialists.

Noise control at the source is accomplished by choosing quieter equipment and processes; modifying flow velocities, pressures, and rotational speeds; reducing friction; balancing rotating parts; and isolating sources of vibration [1, pp. 629–648; 6, pp. 14–30,51–149; 7, pp. 237–266]. This list is illustrative and conceptual rather than exhaustive and design oriented. It suggests ways of keeping noise from ever being heard off the property and is often the best course of action in the long run.

Changing the path includes adding sound barriers and insulation, enclosures, and silencers and further decoupling sources of vibration. The type of abatement device that will prove successful depends on the frequency of the noise to be abated. Low-frequency noise from a slowly repeating series of events easily passes through small openings and bends around obstructions in its path; in contrast, high-frequency noise from rapid repetitions is easily reflected by a surface and does not diffuse around its edges [6, pp. 51–59]. The key to understanding this behavior is to consider the wavelength of the noise and the relationship of the wavelength to the physical dimensions of the noise abatement equipment, enclosures, and barriers. Several problems at the end of this section explore the subject of wavelength.

Changing the path also means making the most of distance to allow natural attenuation to occur. Noise sources should be located as far away as possible from sensitive noise receptors. This includes the overall plant siting issue and arrangement/orientation of noise sources on a given plot plan.

The last resort is to make changes to the receptor or at the receptor location to reduce exposure and/or to require personal protective equipment. Although these methods may work within a plant by posting a building for hearing protection, limiting access, and providing ear plugs/ear muffs for operating personnel, one can expect that the off-site public would not take kindly to this approach to "peaceful enjoyment of one's property." Nevertheless, if mutually agreeable and not otherwise prohibited by law, one possibility might be offering to install storm windows on a potential complainant's house and "just might make the problem go away." Reportedly, this was suggested at one time by a noise control official and was more recently echoed in another part of the country. There are drawbacks to this approach, however, and this course of action should not be taken lightly.

PROBLEM 9.18

(Frequency, wavelength, and the speed of sound)

Explore the relationship between frequency and wavelength for a sound wave in air.

SOLUTION By definition, the wavelength (λ) is the ratio of the speed of sound (c) in a given medium to the frequency

$$\lambda = c/f \tag{9.32}$$

or $$c = f \times \lambda \tag{9.33}$$

The wavelength has units of length. Units of the speed of sound are length per time. Units of frequency are hertz (cycles per second, or simply s^{-1}). These concepts are illustrated in Figure 9.1 for a simple sound wave of sinusoidal shape.

In a gas, the velocity of sound waves, or the speed of sound, is given by

$$c = [\gamma P/\rho]^{1/2} \text{ m/s or ft/s} \tag{9.34}$$

where γ is the dimensionless ratio of heat capacities, P is the absolute pressure, and ρ is the density of the gas.

For an ideal gas like air at ordinary ambient temperature and pressure

$$\rho = (PM)/(RT) \tag{2.15}$$

and therefore

$$c = [\gamma RT/M]^{1/2} \tag{9.35}$$

where R is the universal gas constant (Table 2.3), M is the molecular weight of the gas, and T is the absolute temperature (°K) or (°R).

For air at atmospheric pressure, the relationship simplifies to

$$c = 20.05 \cdot [T \, (°K)]^{1/2} = 14.94 \cdot [T \, (°R)]^{1/2} \text{ m/s} \tag{9.36}$$

Over the range of temperatures normally encountered, the speed of sound (c) is approximately 340 m/s, 1106 ft/s, or 750 mile/h. It changes by about 0.6 m/s per °C or °K [1, p. 41]. Humidity affects c indirectly through the physical properties γ and M in Equation (9.35). Since both γ and M decrease with increasing moisture, the effects are compensating, with less than a 1% increase in c over an extreme increase in humidity compared to dry air. The speed of sound in other materials is different from that in air [7, p. 4; 10, p. 7].

For further information on frequency, wavelength, and the speed of sound, see Refs [1, pp. 40–41; 4, pp. 197–198; 5, pp. 105–106; 7, pp. 2–5; 12, pp. 2–6]. ∎

PROBLEM 9.19

Show that Equation (9.35) for the speed of sound in a gas reduces to Equation (9.36) for air at atmospheric pressure.

SOLUTION From Equation (9.35),

$$c = [\gamma RT/M]^{1/2}$$

for air, $\gamma = 1.4$ (dimensionless)

$$R = 10.731 \, (\text{lb}_f/\text{in.}^2)(\text{ft}^3)/[(\text{lb mol})(°R)]$$

$$T \text{ in } °R = (T \text{ in } °F) + 459.67$$

$$M \text{ for dry air} = 28.96 \, \text{lb}_m/\text{lb mol}$$

g_c to convert from lb_f (pounds force) to lb_m (pounds mass) = 32.17 $(\text{lb}_m/\text{lb}_f)(\text{ft/s}^2)$

$$\text{Then } c = [(1.4)(10.731)(32.17)(144 \text{ in.}^2/\text{ft}^2)(T \, (°R)/28.96)]^{1/2}$$

$$c = 49.02 \cdot [T \, (°R)]^{1/2} \text{ ft/s}$$

$$\times 0.3048 \text{ m/ft} = 14.94 \cdot [T \, (°R)]^{1/2} \text{ m/s}$$

$$c = 14.94 \cdot [T \, (°K)(1.8 \, R/K)]^{1/2} = 20.05 \cdot [T \, (°K)]^{1/2} \text{ m/s}$$

$$= 65.77 \cdot [T \, (°K)]^{1/2} \text{ ft/s} \quad \blacksquare$$

Table 9.20 Frequency and Wavelength

Frequency (Hz)	λ (m)	λ (cm)	λ (ft)	λ (in.)
10	34	—	111.55	—
20	17	—	55.77	—
50	6.8	—	22.31	—
100	3.4	—	11.15	—
200	1.7	170	5.58	66.9
500	0.68	68	2.23	26.8
1000	0.34	34	1.12	13.4
2000	0.17	17	0.56	6.7
5000	0.068	6.8	0.22	2.7
10,000	0.034	3.4	0.11	1.3
20,000	0.017	1.7	0.056	0.7

PROBLEM 9.20

With $c = 340$ m/s, calculate the wavelength (λ) in both metric and English units for the frequencies in Table 9.20.

SOLUTION Sample calculation for 1000 Hz (from Equation (9.32)):

$$\lambda = c/f = (340 \text{ m/s})/(1000 \text{ s}^{-1}) = 0.34 \text{ m}$$
$$0.34 \text{ m}/0.3048 \text{ (m/ft)} = 1.12 \text{ ft}$$
$$(0.34 \text{ m})(100 \text{ cm/m}) = 34 \text{ cm}$$
$$\div 2.54 \text{ cm/in.} = 13.4 \text{ in.}$$

See also calculation and table on pp. 197–198 of Ref. [4].

Wavelength is a meaningful concept not only for a simple sine wave (Figure 9.1) but also for more complex sounds. Wavelength is important in determining where changes in the sound field occur and the behavior of noise abatement structures [6, p. 108]. In theory, the maximum absorption for acoustical insulation occurs when the thickness of the absorption material is approximately one-quarter of the wavelength of the lowest frequency of interest [10, p. 45]. ■

9.16 THE COMMUNITY NOISE SURVEY

Despite the usefulness of calculation methods, there will be times when field measurements will be necessary to verify those calculations or determine a baseline. These include

- Addition of new noise-producing elements
- Documentation of existing operations
- Response to noise complaints from neighbors
- When specifically required by regulations/regulatory agencies

Tables 9.21–9.25 list the where, what, how, and documentation of the community noise survey.

Table 9.21 Typical Locations for Community Noise Measurements

---
Facility boundaries
Place of nearest zoning change
Property line or specific distance from
    Nearest residence in each direction
    Nearest commercial/industrial properties
    Critical receptors

Table 9.22 Community Noise Measurements—What is Measured

Sound pressure levels (octave band data, L_A (dBA), L_{eq} (dBA), others)
Weather data
    Temperature
    Relative humidity
    Barometric pressure
    Wind
Ancillary measurements
    Time
    Distance
    Directions
Photographs

Table 9.23 Equipment for Community Noise Measurements

Calibrated sound-level meter	Weather radio (optional)
Microphone	Wind gauge
Windscreen	Timepiece
Headphones (desirable)	Tape measure
Tripod (desirable)	Automobile odometer
Field calibrator	Maps
Thermometer[a]	Compass
Sling psychrometer[a]	Camera
Barometer[a]	Flashlight/miner's hat with light
Clipboard with data sheets	Multipocket coat or vest ("safari jacket" or "press correspondent's coat")—(helpful!)

[a] Data from National Oceanographic and Atmospheric Administration (NOAA) weather broadcasts may be used for items of weather data except for wind velocity; this must be measured locally [29].

9.16 The Community Noise Survey

Table 9.24 Community Noise Measurement Procedures

Take both daytime and nighttime readings
Normally use slow meter response
Calibrate, calibrate, calibrate!
Follow standards
Employ sound measuring equipment properly
Follow manufacturers' instructions
Avoid proximity to reflecting surfaces (such as the observer's automobile)
Have any accompanying assistant stand behind you
Use prepared data sheets
Take pictures during daylight; use flashlights at night
Identify audibly evident tones

Note: Field calibration and a battery check at least once per hour are typical requirements; however, calibrate and verify the battery condition in the field more often and as often as possible to avoid loss of valid data should the calibration drift more than the allowable amount, typically 0.5 dB, or the battery check fail. Calibrate at the same ambient conditions as the field measurements. Do not waste your time using a meter whose manufacturer's certified annual calibration has lapsed.

Sometimes calculations and/or a noise survey will point out the need to mitigate the noise that would otherwise be generated at a plant. The final case of Appendix I describes the classic methodology:

- Consultation with regulatory authorities
- Conducting a background noise survey before the start of new construction
- Determining the noise generated by new sources to be added
- Projecting the noise to locations of critical receptors in the neighborhood
- Comparing the projected noise with code values
- Installing noise abatement measures, if necessary, to mitigate the noise impacts
- Conducting a follow-up noise survey to verify compliance

Table 9.25 Community Noise Measurements—Suggested Report Content

Description of/reason for test
Nomenclature and serial numbers of test equipment
Original completed data sheets containing information noted here
Field calibration data including time of calibration and battery checks
Manufacturers' calibration data certificates
Meteorological data
Description of the physical setting of the surrounding area (noting also wet pavement or fallen snow)
Map/sketch and, if possible, captioned photographs
Location and description of barriers and other structures
The applicable noise regulations
Summarized results with tables, graphs, histograms, annotated maps, and so on

Besides the noise sources themselves, there are several other factors that affect the measured sound pressure levels. These include distance(s) from the source(s); terrain and conditions on the surface; wet pavement; snow cover; reflecting and interfering objects; wind speed and direction; atmospheric temperature, pressure, and humidity; and position and orientation of the microphone and operator/observer. These factors affect the propagation of the sound wave, its attenuation/absorption in air, its reception, the character of the sound field, and the operation of the sound measuring device. The measured sound level at some remote distance from a source under investigation can change appreciably from time to time, even when the source remains constant. Although many of the conditions cannot be controlled, it is important to note those conditions and any differences that occur.

Procedures for conducting a community noise survey are described in Refs. [3,4; pp. 217–224; 12, pp. 74–99; 29], including specific do's and don'ts. A sample report for a simple sound survey is presented in the first case study of Appendix I. Some other helpful hints are provided here.

9.16.1 Data Sheets

Standard data sheets allow one to record data and observations covering both the sound readings and ancillary measurements. Two different sample data sheets have been published [7, p. 85; 12, p. 91]. The format is not nearly as important as the information itself. The data sheet should provide for entries to identify the operator; time, location, and layout; measured noise levels; frequency (if performing an octave band analysis); dBA; overall dB or dBC; noise sources and conditions being evaluated; meteorological conditions; and recommendations. Remarks such as "can hear birds or crickets chirping, continuous traffic in background, source cannot be heard at this location, and so on." can also be entered in the blank spaces of such a data sheet. A vocabulary of 116 words such as "bang, boom, rumble, crash, and roar" has been published to facilitate the description of the general nature of a sound being encountered in the field [4, p. 300].

Sample data sheets [4, p. 222; 29] are also available to serve as a model for repeated manual sampling of dBA over time at a given location to construct a histogram if an integrating sound-level meter is not available. From this histogram, various statistics such as L_{10}, L_{50}, L_{90}, and L_{eq} can be estimated, as exemplified in Problem 9.7.

9.16.2 Effect of Wind

Since sound/noise measurements are sensitive measurements of pressure, wind affects pressure readings and therefore the sound readings [12, pp. 87–88]. A windscreen covering the microphone is to be used for outdoor data acquisition.

A windscreen is a ball or other shape of polyvinyl foam material supplied by the sound meter manufacturer to fit over the microphone and mitigate wind effects by moving the turbulent airflow away from the microphone [12, p. 87].

Data are not to be taken at a steady wind speed in excess of 12 mph (20 km/h) or when gusts exceed 12 mph. A graph of wind noise as measured using a windscreen shows a rapid increase above 12 mph. Without a windscreen, the effect is much greater [12, p. 88 (Fig. 4.7)].

Even at lower wind speeds within the acceptable range and using a windscreen, wind may still influence the readings, especially octave band readings at the lower end of the frequency spectrum and consequently the $dB_{\text{all pass}}$ reading. The dBA reading, heavily attenuated at those frequencies, is not so affected. Use of headphones or careful observation of the meter reading is necessary to determine whether the fluctuations/excursions observed are caused by actual noise or the wind. Headphones are also useful to detect the effects of high humidity, high-voltage transmission lines, or other electromagnetic emissions on certain microphones [29].

9.16.3 Precipitation

Precipitation affects the sound field, along with its aftereffects—wet pavement or fallen snow. For example, traffic tire noise is higher with wet pavement [29]. Measurements are not to be taken during rain or snow events, and the presence of previously caused wet pavement or snow cover must be noted in the noise report. Even ignoring the effect of precipitation on the sound field, getting the sound meter wet from rain, snow, or condensation is *not* recommended for the health of the meter.

9.16.4 Temperature

Temperature affects both the battery and the microphone. Follow the manufacturer's recommendations regarding maximum and minimum temperatures. An allowable temperature range of 14°F ($-10°C$) to 122°F (50°C) is noted in Ref. [3]. The meter should be calibrated during the measurement period at the same ambient temperature as the measurements [3]. Do not, for example, calibrate the meter inside a heated building when the operator takes a "warming-up" break during cold weather measurements.

9.16.5 Sound Meter Setting

Most codes specify the slow response reading for community noise measurements. Slow response has a high dampening of the otherwise rapidly moving meter display with a time constant of about 1 s; it is suitable for capturing moderately time-varying sounds like an automobile pass-by [4, p. 108; 6, p. 33].

Impulse-type noises are most properly measured using the impulse setting of a sound meter so equipped. Impulse noises are those of extremely short duration, usually less than 1 s, with an abrupt onset and rapid decay, and usually a large difference in intensity over the background. The impulse setting of a sound meter has a very fast rising time constant and a very slow falling time constant to follow rapid changes in sound level, even more so than the fast response setting [4, p. 108; 6, p. 33]. Examples such as gunshots, explosions, and machine shop noises are noted in a previous section.

9.16.6 Other Phenomena Encountered during Noise Surveys

9.16.6.1 Reflection

Reflection will reinforce the sound pressure level when the receptor is impacted both by sound waves coming directly from the source and by the same sound waves reflected off nearby objects. For a single perfect reflection, the measured sound pressure will double, that is, increase by 3 dB. Less than perfect reflection will add less than 3 dB. Strategically placed objects can also reflect sound waves away from a receptor. When taking sound measurements, it is therefore important not to introduce the effects of reflecting (or absorbing) surfaces not ordinarily present, for example, by erroneously locating the sound meter too close to the operator's body or to the operator's vehicle.

9.16.6.2 Barriers

A barrier decreases the sound level at a receptor in the acoustic shadow behind the barrier through a combination of reflection away from the receptor and absorption of the sound by the barrier material. One situation that comes to mind from experience is a single-story building at some distance from an industrial plant. Not only did the sound meter readings decrease in the acoustical shadow of the building, but the sound of the plant also disappeared completely.

Like light waves at the boundary of different media, sound waves are also able to be diffracted around acoustic barriers [1, pp. 47–48, 637–638]. Diffraction occurs when sound waves pass through and spread out behind a narrow opening. This diffraction is more pronounced for the lower frequencies [6]. A highway noise barrier will block the upper frequencies of traffic noise from appearing behind it until distance and some air attenuation dissipate their intensity; however, the less offensive low-frequency noise appears to "crawl over" much closer to the wall. Highway noise barriers are addressed in the first two case studies of Appendix I.

9.16.6.3 Trains, Planes, and Automobiles

Unless one is trying to characterize noise from these sources, they are excluded or noted specifically as excursions or extraneous noise when measuring ambient noise or noise from a stationary source under investigation. It is customary to wait for individual vehicles and their accompanying noise to pass before resuming an ongoing manual sound survey. However, the continuous noise from a busy highway, often in the distance, contributes to the overall ambient noise level and is measured in with the background. If traffic noise cannot be separated from other ambient noise, it becomes part of the neighborhood residual. Noise from these transportation sources contribute to a measurement of L_{eq}. Highway noise is discussed further in Section 9.14 and Appendix I.

9.16.6.4 Highway Noise

As explained in Section 9.14, the noise generated by a motor vehicle is primarily of low frequency, with tire noise predominant at all but the slowest speeds. Highway noise propagates as a line source. In general, single-vehicle noise at 50 ft (~15 m) from the centerline of the nearest traffic lane for light cars and trucks ranges from about 65–75 dBA, heavy trucks and highway buses 75–95 dBA, and motorcycles 75–80 dBA. Correlating equations are available to estimate highway noise when it is necessary to separate its contribution from that of plant noise in a value measured for overall ambient noise (Section 9.14).

9.16.6.5 Barking Dogs and Crickets

Sporadic barking of dogs is likewise excluded from spot dBA readings although it may take some time for the barking to stop. The barking would show up as part of L_{eq} measurements. It is virtually impossible, however, to exclude steady cricket noise once it begins in the evening. (This shows up typically in the 2000 Hz octave band.) Typical dBA readings varying continuously over time at a fairly steady noise level but showing spikes from animal noises, local and distant cars, and an aircraft flyby are depicted in Ref. [5, pp. 287–289].

9.16.6.6 Pure Tones

Transformer hum, a pure tone at twice the power line frequency (and its harmonics), is often encountered during a community noise survey, and other equipment producing a pure tone in the regulatory sense may be encountered as well. The ramifications of pure tones are explored elsewhere in this chapter.

9.16.6.7 Beats

The beat phenomenon sometimes arises in community noise work. Beats are periodic variations in amplitude that occur when the frequencies of two slightly different sounds reach one's ears simultaneously. Audible beats can be detected for pure tones, notes played by musical instruments, and complex sounds as well. This throbbing pulsation can occur when two pieces of rotating machinery running at slightly different frequencies come into and out of phase.

The distinctive slow wobbling beat sound is easily perceived when the frequency difference is 10–15 Hz. The ear perceives the overall sound at the average frequency, along with a striking variation in intensity at the difference frequency [1, p. 147; 5, p. 267].

Musicians use the beat phenomenon to tune their instruments against a standard reference. If an instrument is sounded together with the reference and tuned until the beat disappears, the instrument is in tune with the standard.

9.16.6.8 Personal Safety

This final item is of utmost importance. Sound surveys in the community involve at least some readings at night, when the extraneous noise sources in the area are at a minimum. One must exercise extreme caution and be concerned for one's personal safety while present in strange neighborhoods in the dead of night. Some commonly encountered difficulties experienced include dogs and other predators, human, or otherwise; physical hazards; and citizens or police protecting their own lives and property against someone out of place in their community. Use of the "buddy system," staying off private property and out of active traffic lanes, and prior notification to law enforcement agencies are all common-sense practices, but one can never be too careful out there even when preoccupied with data collection. Imagine the author's startle reflex when a curious local resident and his dog suddenly showed up on foot out of nowhere during late night readings in a really, really rural agricultural area, wanting to know what was going on and how it was going. Lesson learned!

Further comments on noise survey measurements are contained in Refs. [3,4,12,29]. The last case study of Appending I describes the role of the noise survey in the successful completion of a grassroots project abutting an established residential neighborhood.

REFERENCES

1. T.D. Rossing, *The Science of Sound*, 2nd edn, Addison-Wesley, Reading, MA, 1990, 686 pp.
2. L.W. Canter, *Environmental Impact Assessment*, McGraw-Hill, New York, 1977, 331 pp.
3. E.M. Zwerling, editor, *Community Noise Enforcement*, various editions, Rutgers Noise Technical Assistance Center, Rutgers-The State University of New Jersey, New Brunswick, NJ, 1994–2001.
4. A.P.G. Peterson, *Handbook of Noise Measurement*, 9th edn, GenRad, Inc., Concord, MA, 1980, 394 pp.
5. L.E. Kinsler, A.R. Frey, A.B. Coppens, and J.V. Sanders, *Fundamentals of Acoustics*, 3rd edn, Wiley, New York, 1982, 480 pp.
6. Anonymous, *Noise Control Principles and Practice*, 1st edn, Brüel & Kjaer, Naerum, Denmark, 1982, 156 pp.
7. J.D. Irwin and E.R. Graf, *Industrial Noise and Vibration Control*, Prentice-Hall, Englewood Cliffs, NJ, 1979, 436 pp.
8. National Institute of Standards and Technology (NIST), Department of Commerce, WWV-WWVH-WWVB Broadcasts, 15 CFR 200.107 U.S. Government Printing Office, Washington, DC, Revised January 1, 2003, pp. 364–366.
9. A.P.G. Peterson and E.E. Gross Jr., *Handbook of Noise Measurement*, 6th edn, General Radio Company, West Concord, MA, 1967, pp. 14.
10. A.J. Rosing, technical editor, Biometrics Corporation, *Handbook for Industrial Noise Control*, Report NASA SP-5108, available as Publication N82-11 858 from, National Technical Information Service (NTIS), Springfield, VA, 1981, 137 pp.

11. C. Caccavari and H. Schechter, Background noise study in Chicago, *Journal of Air Pollution Control Association*, **24**(3), 240–244, 1974.
12. L.L. Beranek, *Noise and Vibration Control*, McGraw-Hill, New York, 1971, 623 pp.
13. A.F. Siegel, *Statistics and Data Analysis: An Introduction*, Wiley, New York, 1988, pp. 23–24.
14. New Jersey Noise Control Act, N.J.S.A. 13:1G, 1971.
15. New Jersey Noise Control Regulations, N.J.A.C., 7:29 (as amended), effective March 18, 1985.
16. New Jersey Noise Control Procedures for Determination of Noise from Stationary Sources, N.J.A.C., 7:29B, Department of Environmental Protection, Trenton, NJ, adopted March 5, 1982.
17. R.D. O'Neal, Noise barrier insertion loss: a case study in an urban area, Paper 90-156.3 presented at the 83rd Annual Air & Waste Management Association (A&WMA) Annual Meeting & Exhibition, Pittsburgh, PA, June 24–29, 1990.
18. R.D. O'Neal, Predicting sound levels: a case study in an urban area, *Journal of Air & Waste Management Association*, **41**(10), 1355–1359, 1991.
19. J.P. Reynolds, J.S. Jeris, and L. Theodore, *Handbook of Chemical and Environmental Engineering Calculations*, Wiley-Interscience, New York, 2002, pp. 864–865.
20. Anonymous, All about Electric Circuits, Heat and Noise—Chapter 9: Transformers—Volume II—AC, 4 pp., 2003, available at http://www.allaboutcircuits.com, accessed on November 2, 2006.
21. R.S. Masti, W. Desmet, and W. Heylen, On the influence of core laminations upon power transformer noise, Vibro-Acoustic Modelling and Prediction, Proceedings of ISMA2004, 2004, pp. 3851–3862.
22. G.A. Capano and W.E. Bradley, Noise prediction techniques for siting large natural draft and mechanical draft cooling towers, Paper presented at the 38th Annual Meeting of the American Power Conference, Chicago, IL, April 20–22, 1976, 9 pp.
23. C.M. Harris, Absorption of sound in air versus humidity and temperature, *Journal of Acoustic Society of America*, **40**(1), 148–159, 1966.
24. L.L. Beranek and I.L. Ver, *Noise and Vibration Control Engineering: Principles and Applications*, Wiley-Interscience, New York, 1992, p. 115.
25. Steven P., The Effects of Sound Barrier Design on Highway Noise Attenuation, available at http://www.selah.k12.wa.us/soar/sciproj2000/StevenP.html, accessed on November 5, 2006.
26. United States (U.S.) Department of Transportation Federal Highway Administration (FHWA), Highway Traffic Noise. Available at http://www.fhwa.dot.gov/environmental/htnoise.htm, accessed on November 5, 2006, 8 pp.
27. V. Schroter and C. Chiu, ORNAMENT Ontario Road Noise Analysis Method for Environment and Transportation, Technical Document (October 1989, reprinted February 1993, February 1997, September 1999), Queen's Printer for Ontario, 1990.
28. T.M. Barry and J.A. Reagan, FHWA Highway Traffic Noise Prediction Model, Report FHWA-RD-77-108, U.S. Federal Highway Administration, Washington, DC, December 1978.
29. ASTM Committee on Standards, Standard Guide for Measurement of Outdoor A-Weighted Sound Levels, Designation: E 1014-84, Annual Book of ASTM Standards, American Society for Testing and Materials, Philadelphia, PA, 1984.

Chapter 10

Radioactive Decay

Unless you are a medical practitioner, it is not good to have a disease named after you.

10.1 DEFINITIONS AND UNITS

10.1.1 Definitions

As discussed in an earlier chapter, the nuclei of atoms are made up of protons and neutrons. The number of protons (the atomic number) determines the identity of the element; protons and neutrons together, collectively known as *nucleons*, determine the atomic mass number. Atoms containing the same number of protons but differing in the number of neutrons are known as isotopes [1–4].

Not all isotopes are stable. Some will spontaneously decompose (over a period measured in fractions of a second to many, many years) by emitting material and ionizing radiation in an attempt to achieve stability. The emitting, or parent, nucleus undergoes a transmutation into the nucleus of a different element. The element is called the *daughter species* or *decay product*. This process is termed *radioactivity* or *radioactive decay* and the unstable isotopes *radionuclides* or *radioisotopes*. Radioactivity arises from both natural and man-made sources.

10.1.2 Units of Radioactive Decay and Exposure

The activity of a radioactive source is measured in curies (Ci), named after Madame Marie Curie and her husband Pierre Curie, Nobel Prize winners for their early work in radioactivity. The curie corresponds to 3.7×10^{10} radioactive disintegrations per second. It is the decay rate of 1 g of radium. Fractions of this unit, such as milli (10^{-3}), pico (10^{-12}), and others are also used. Other units of activity are the becquerel (Bq), one disintegration per second, and the TBq (10^{12} Bq). Becquerel, another pioneer in

---

Environmental Calculations: A Multimedia Approach, by Robert G. Kunz
Copyright © 2009 John Wiley & Sons, Inc.

this field, shared the Nobel Prize for Physics in 1903 with the Curies [3, pp. 246–247; 4, pp. 1089–1090].

The roentgen (R) (from Wilhelm Roentgen, the discoverer of X-rays in 1895) is a measure of the ability of a beam of X-rays or gamma rays to *deliver* energy to a given material. For example, a dental X-ray can provide an exposure of 300 mR/s, with 1 R corresponding to the *deposition* of 93.3×10^{-7} J of energy per gram of tissue or 87.8×10^{-7} J/kg of dry air.

The *rad* (radiation *a*bsorbed *d*ose) is a measure of the dose actually absorbed by the subject, whether it be the whole body or some specific part. An absorbed dose of 1 rad corresponds to 10^{-3} J/kg of material. Death will occur in 50% of the population exposed to a whole-body short-term gamma ray dose of 300 rads. Typical radiation exposure from both natural and man-made sources is about 200 mrad per year. One joule per kilogram is called the gray (Gy), named after a British scientist, making 1 mGy (milligray) equal to 1 rad.

The *rem* (*r*oentgen *e*quivalent in *m*an) is a measure of the dose equivalent to account for the differing biological effect of different types of radiation at the same energy per unit body mass. It is obtained by multiplying the absorbed dose in rad by a tabulated RBE (*r*elative *b*iological *e*ffectiveness) factor.

Film badges are calibrated to register a dose equivalent in rems, limited to 500 millirems or less per year from all types of radiation; background radiation is estimated at 360 millirems per year. A dose of approximately 600 rad or 600 rems (presumably all at one time) is fatal for humans. Exposure from a typical dental X-ray amounts to about 0.5 millirems and a typical chest X-ray is 25 millirems [4, p. 1099]. The RBE for electrons (β-radiation), X-rays, and γ-radiation is 1.0; for neutrons, it is from 5 (slow moving) up to 20 for those of higher energy. It is 20 for α-radiation. The equivalent dose is also given in sieverts (Sv), named after a Swedish scientist (1 Sv = 1 J/kg = 100 rem).

10.1.3 Other Measures of Dosage

The effects of certain types of radiation on particular body organs and the persistence of that radiation over a person's lifetime following ingestion or inhalation from the source are beyond the scope of this treatment.

10.2 SOME SOURCES OF RADIOACTIVITY

Exposure to radioactivity occurs from extraterrestrial cosmic radiation and from long-lived naturally occurring materials such as $_{92}U^{235}$, $_{90}Th^{232}$, and $_{19}K^{40}$ contained in soils, phosphate rock, and coal [1, p. 473]. These are released into the environment during mining, processing, or combustion. Radioactivity is leached into ground or surface water while in contact with geologic materials containing natural radioactivity [2, p. 229]. Radon ($_{86}Rn^{222}$) is contained in natural gas and, as a decay product from certain underground rocks, is responsible for chronic exposure for humans from seepage into basements and underground mines [3, p. 272; 1, pp. 10, 93, 175–176, 281–282].

Naturally occurring radionuclides are used to date various antiquities such as the Shroud of Turin and the Dead Sea Scrolls [3, pp. 261–263] and even the age of the earth itself [4, pp. 1083–1084]. Carbon-14 and potassium-40 are often employed for this purpose. Radioactive isotopes are used in medicine for the detection, diagnosis, and treatment of illness. Examples include $_6C^{14}$, $_{43}Tc^{99}$, and $_{53}I^{128}$ and $_{53}I^{131}$, $_{27}Co^{60}$, and $_{55}Cs^{137}$. Isotopes such as $_{55}Cs^{137}$ and $_{36}Kr^{85}$ are put to use in industry for level and leak detection, thickness gauging, weld testing, and determination of wear. Tracers in chemical and biochemical reactions include carbon-14 and tritium ($_1H^3$) and radioactive isotopes of oxygen.

Consumer products contain $_{95}Am^{241}$ in smoke detectors and radionuclides such as tritium and $_{61}Pm^{247}$ to produce luminous clock, watch, and telephone dials. These latter materials have largely replaced radium ($_{88}Ra^{226}$) for this purpose. Radioactive isotopes have been used as pesticides and to irradiate foodstuffs, strawberries, for example, to preserve freshness [3, p. 271].

Radionuclides, of course, arise in the production of nuclear power and the production and use of nuclear weapons.

10.3 TYPES OF RADIOACTIVE DECAY

There are several types of radioactive decay [1, pp. 20–24, 29–32; 2, pp. 227–228; 3, pp. 213–251; 4, pp. 1083–1087; 5, pp. 465–466]:

- Alpha decay
- Beta decay
- Gamma radiation
- Other types

10.3.1 Alpha Particle Decay

This type of radioactivity occurs when an unstable radionuclide emits at high speed the so-called alpha (α) particle, in essence the nucleus of the helium atom consisting of two protons and two neutrons. An example is the breakdown of an isotope of uranium (U-238) to yield thorium-244 and the helium nucleus (He minus electrons). The reaction is commonly written as follows:

$$_{92}U^{238} \rightarrow {_{90}Th^{234}} + {_2He^4}(\alpha) \tag{10.1}$$

Another typical alpha decay reaction involves polonium-210 to produce an isotope of lead.

$$_{84}Po^{210} \rightarrow {_{82}Pb^{206}} + {_2He^4}(\alpha) \tag{10.2}$$

With alpha decay, the resulting daughter product exhibits an atomic number (protons) two less than the parent and an atomic mass number (protons and neutrons) four less than the parent.

In the reaction shown in Equation (10.1), the thorium product is itself unstable and goes on to decay to another element in the so-called *radioactive series*. The isotope of lead in Equation (10.2) is a stable end product.

Alpha particles travel only short distances and are easily stopped by paper, clothing, skin, or one or more sheets of aluminum foil. They are extremely hazardous, however, if ingested, inhaled, or possibly taken in through open wounds.

10.3.2 Beta Particle Decay

Some radioactive elements spontaneously eject an electron of negative charge when a neutron in the nucleus is transformed into a proton.

$$_0n^1 \rightarrow {_1}H^1 + {_{-1}}e^0 (\beta^-) \tag{10.3}$$

The neutron (n) and proton possess virtually the same unit mass; the mass of the electron is negligible by comparison. The proton is represented in Equation (10.3) by the symbol for the hydrogen atom. During beta (β) decay, a *neutrino* or *antineutrino*, a neutral particle of negligible mass and difficult to detect, is also emitted.

An example of a beta transformation is shown below for another isotope of uranium (U-235) to form neptunium-235.

$$_{92}U^{235} \rightarrow {_{93}}Np^{235} + {_{-1}}e^0 (\beta^-) \tag{10.4}$$

In a beta transformation, the atomic mass number remains unchanged but the atomic number (determining the element) increases by one.

Beta particles are emitted from the nucleus at close to the speed of light and can penetrate several millimeters of living body tissue or bone. Depending on their maximum energy, they can be stopped by a moderately thick pane of glass or at least a 1/8 in. sheet of aluminum.

10.3.3 Gamma Radiation

This is a high-energy electromagnetic radiation with no change in mass, resulting from a change in the state of excitation of the product nucleus as initially formed to its final, or ground, state. Emission of gamma radiation usually accompanies alpha and beta decay. Although of a different origin, their effect is similar to but more energetic than that of X-rays. They are the most penetrating form of nuclear radiation and cannot be completely shielded against. Their penetration is attenuated only by certain high-density materials such as thick layers of lead or concrete.

10.3.4 Other Types

Another example of radioactive decay is the emission of a *positively* charged electron, or *positron* [3, p. 251],

$$_{29}Cu^{64} \rightarrow {_{28}Ni^{64}} + {_{+1}e^0}(\beta^+) \tag{10.5}$$

$$_{84}Po^{207} \rightarrow {_{83}Bi^{204}} + {_{+1}e^0}(\beta^+) \tag{10.6}$$

In this case, the atomic number steps down by one unit, while the atomic number remains constant.

Still another type is called *electron capture*, exemplified by

$$_4Be^7 \text{ (beryllium)} + {_{-1}e^0}(\beta^-) \rightarrow {_3Li^7} \text{ (lithium)} \tag{10.7}$$

involving capture by the nucleus of an electron belonging to that atom. It too results in the same atomic mass number but a decrease by one in the atomic number.

10.4 PATHWAYS OF RADIOACTIVE DECAY

Radioactive disintegration reactions can occur singly, in parallel, in combination with other unrelated decay reactions, and in series in what is termed a *radioactive decay chain*.

10.4.1 Radioactive Decay—Single Reaction

The factor k is known in the literature of chemical kinetics as the rate constant. This rate constant is often written as the Greek letter lambda (λ) in the literature of radioactivity.

$$A \xrightarrow{k} B \tag{10.8}$$

Regardless of the types of particles emitted, the governing relationship for radioactive decay is that the rate of change of the number of unstable atoms in the nucleus is proportional to the number (A) of such atoms remaining there at any given time. Furthermore, this number decreases with time, and therefore the rate of change is negative. Mathematically,

$$dA/dt = -kA \tag{10.9}$$

(In the radioactive decay literature, the number of atoms is often referred to as N_1, N_2, N_3, etc. for different species.)

This first-order rate equation is a homogenous differential equation with initial condition $A = A_0$ at time $(t) = 0$. Its solution is

$$A = A_0 \exp[-kt] \text{ or } A/A_0 = \exp[-kt] \tag{10.10}$$

where exp is the exponential function e^x. The solution can be verified by differentiation and substitution into Equation (10.9). At $t = 0$, A becomes identically equal to A_0.

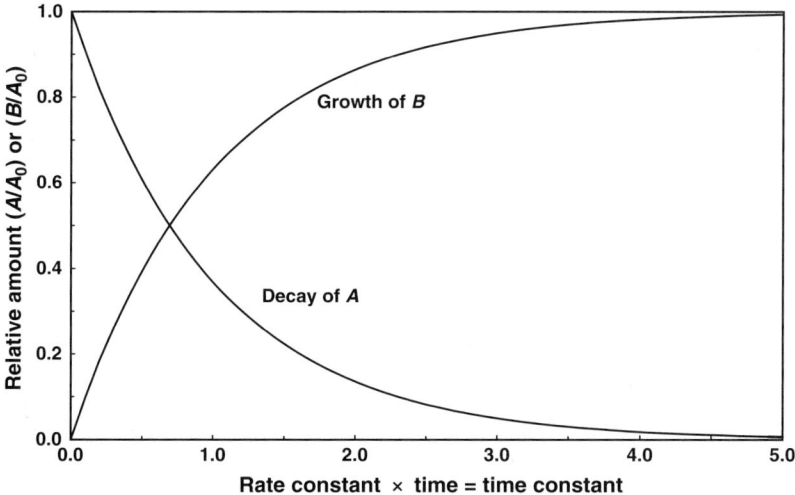

Figure 10.1 Radioactive decay of species A into species B (linear coordinates).

Equation (10.9) describes an exponential decay curve on linear coordinates and plots as a straight line on semilogarithmic coordinates (Figures 10.1 and 10.2).

$$\ln A = \ln A_0 - kt \text{ or } \ln(A/A_0) = -kt \qquad (10.11)$$

The slope of the normalized semilogarithmic plot is the rate constant $-k$, and its intercept is $\ln(A/A_0)$. The ln function is the natural, or Napierian, logarithm to the base e (2.71828...).

The starting material never disappears completely, except perhaps at infinite time. Therefore, other measures are used to describe the state of completion of the reaction, the mean life (τ) [1, 34–35], and the more familiar half-life ($t_{1/2}$).

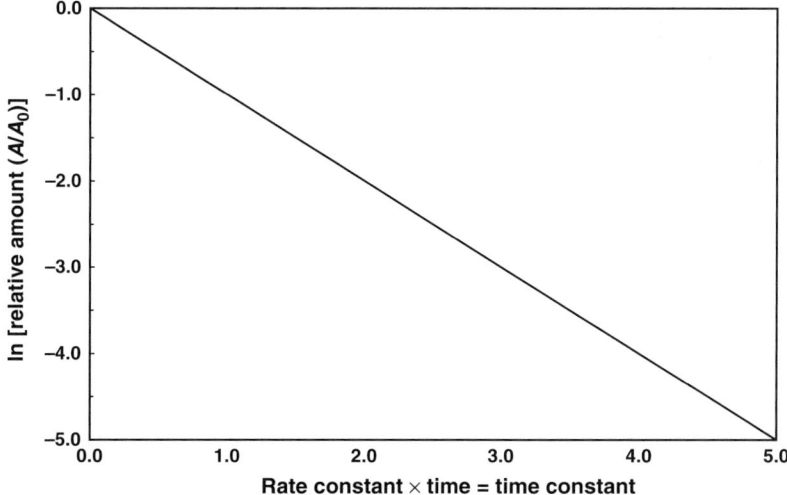

Figure 10.2 Radioactive decay of species A (semilogarithmic coordinates).

474 Chapter 10 Radioactive Decay

The *mean life* is defined as the average life expectancy of an individual radioactive atom, at which time the unstable atoms remaining in the nucleus have decayed to 1/e of their initial value. This mean life is the sum of the lifetimes of all such atoms divided by their initial number and can be represented by the following integral for a larger number of atoms:

$$\tau = (1/A_0) \int_0^\infty t \, dA \tag{10.12}$$

Substitution and integration yields

$$\tau = 1/k \tag{10.13}$$

Substitution back into Equation (10.10) shows that τ corresponds to the condition that the unstable atoms remaining in the nucleus have decayed to 1/e of their initial value.

Half-life is defined as the reaction time to achieve half of the initial value.

$$\ln(1/2) = -kt_{1/2}$$
$$t_{1/2} = [\ln(2)]/k \cong 0.69315/k \tag{10.14}$$

This concept can be extended to define multiple half-lives, during which the remaining half of the starting material disintegrates repeatedly by half, as shown in Table 10.1.

At 10 half-lives, less than 0.1% of the original remains.

The half-lives of some common radionuclides listed in Table 10.2 range from tiny fractions of a second to billions of years. Generally speaking, the ones most hazardous to health and the environment are those with relatively short half-lives; those with long half-lives exhibit low levels of radiation with limited penetrating ability [1, p. 34].

Table 10.1 Definition of Multiple Half-Lives

Number of half-lives	Fraction of initial amount remaining
0	$(1/2)^0 = 1/1$
1	$(1/2)^1 = 1/2$
2	$(1/2)^2 = 1/4$
3	$(1/2)^3 = 1/8$
4	$(1/2)^4 = 1/16$
5	$(1/2)^5 = 1/32$
6	$(1/2)^6 = 1/64$
7	$(1/2)^7 = 1/128$
8	$(1/2)^8 = 1/256$
9	$(1/2)^9 = 1/512$
10	$(1/2)^{10} = 1/1024$

Table 10.2 Half-Lives of Selected Radionuclides[a]

Nuclide/isotope	Symbol	Half-life
Aluminum-26	Al	7.4×10^5 y
Americium-241	Am	248 y
Argon-37	Ar	35.04 d
Argon-39	Ar	269 y
Bismuth-210	Bi	5.013 d
Bismuth-212	Bi	60.55 m
Bismuth-214	Bi	19.9 m
Carbon-14	C	5.73×10^3 y
Cesium-137	Ce	30.17 y
Chlorine-36	Cl	3.01×10^5 y
Cobalt-137	Co	5.27 y
Copper-64	Cu	12.7 h
Hydrogen-3 (tritium)	H	12.26 y
Iodine-131	I	8.04 d
Krypton-85	Kr	10.76 y
Lead-208	Pb	Stable
Lead-210	Pb	22 y
Lead-214	Pb	26.8 m
Manganese-54	Mn	312.5 d
Oxygen-15	O	2.0 m
Plutonium-239	Pu	2.41×10^4 y
Polonium-210	Po	138.38 d
Polonium-212	Po	3.0×10^{-7} s
Polonium-216	Po	0.15 s
Polonium-218	Po	3.10 m
Potassium-40	K	1.28×10^9 y
Radium-224	Ra	3.66 d
Radium-226	Ra	1.6×10^3 y
Radon-210	Rn	3.82 d
Radon-220	Rn	56 s
Sodium-22	Na	950.8 d
Strontium-90	Sr	28.8 y
Thallium-208	Tl	3.05 m
Thallium-210	Tl	1.3 m
Thorium-232	Th	1.045×10^{10} y
Thorium-234	Th	24.1 d
Uranium-235	U	7.038×10^8 y
Uranium-238	U	4.1×10^9 y

y = years, d = days, h = hours, m = minutes, s = seconds.
[a]Compiled largely from Refs [1,4].

PROBLEM 10.1 (Based on a similar problem in Ref. 5, p. 471)

A 10 g sample of radium-224 has a half-life of 3.66 days (Table 10.2). The decay product radon-220 gas escapes. How much of the original sample remains after 11 days?

SOLUTION

$$_{88}Ra^{224} \rightarrow {}_{86}Rn^{220} + {}_2He^4(\alpha) \tag{10.15}$$

(a) Half-life method

$$11 \text{ days elapsed time}/3.66 \text{ days half-life} = 3.005 \text{ half-lives}$$

From Table 10.1, 1/8 of sample remains after 3 half-lives.

$$(1/8)(10) = 1.25 \text{ g}$$

Because $3.005 > 3.0$, the amount will be slightly less.

(b) Rate constant method
From Equation (10.14), rate constant

$$(k) = \ln 2/t_{1/2} = 0.6931/3.66 = 0.1894 \text{ days}^{-1}$$

From Equation (10.10),

$$A = A_0 \exp[-kt] = 10 \text{ g} \cdot \exp[(-0.1894 \text{ days}^{-1})(11 \text{ days})]$$
$$= (10)(0.1245) = 1.245 \text{ g}$$

The half-life method works exactly (within calculation roundoff) when an exact number of half-lives is specified. The rate constant method works all the time to within the precision of computing the exponential function. ■

PROBLEM 10.2

By chemical analysis, a certain radioactive isotope in your laboratory now contains three times as much stable end product (B) as radioactive element (A). Thirty days ago, the amounts were equal. What is the half-life of this material, and how long will it take to deplete 90% of the active species?

SOLUTION There may be several ways to work this problem, but let us project the radioactive decay process back to the point where there was only pure $A = A_0$ and no B_0 in the reaction of Equation (10.8).

(a)
$$A = A_0 \exp[-kt] \tag{10.10}$$
$$B = (A_0 - A) = A_0\{1 - \exp[-kt]\} \tag{10.16}$$

Dividing Equation (10.16) by Equation (10.10)

$$B/A = \{1 - \exp[-kt]\}/\exp[-kt]$$

Buildup of the decay product B is also shown in Figure 10.1.

Then, by algebraic manipulation

$$\exp[-kt] = 1 + (B/A)$$

and exponentiation

$$kt = \ln[1 + (B/A)]$$

30 days ago at time t

$$kt = \ln[1 + (1/1)] = \ln(2)$$

Currently, at time $(t + 30)$

$$k(t+30) = \ln[1 + (3/1)] = \ln(4) = 2\ln(2)$$

Dividing one equation by the other and rearranging

$$k(t+30) = 2kt$$

Therefore, $t = 30$ (time from initial condition to 30 days ago) and $(t + 30) = 60$ (present time as measured from initial condition).

At 30 days from initial condition,

$$k = [\ln(2)]/t = 0.69315/30 = 0.023105 \text{ day}^{-1}$$

$$t_{1/2} = [\ln(2)]/k = 0.69315/0.023105 = 30 \text{ days}$$

(b) For 90% depletion of A

$$A = 0.1 A_0$$

$$A_0 \exp[-kt] = 0.1 A_0$$

$$\exp[+kt] = 10$$

$$kt = \ln(10)$$

$$t = [\ln(10)]/k = 2.302585/0.023105 = 99.66 \cong 100 \text{ days}$$

Check: At present time (60 days from initial condition)
(a) $(B/A) = \{1 - \exp[-0.023105(30)]\}/\exp[-0.023105(30)] = (1 - 0.25)/0.225 = 3$.
   Check: At 99.65 days
(b) $(A/A_0) = \{1 - \exp[-0.023105(99.66)]\} = 0.1 = 10\%$, that is, 90% depletion. ∎

10.4.2 Single-Reaction Rate of Decay

Often, what is desired is not the amount of material (A), but rather it is the rate of decay (R_A). Mathematically, the rate of decay can be obtained by differentiating Equation (10.10) and defining the rate to be positive [4, p. 1083]. This is measured by some sort of device that converts radioactive emissions into ionization, excitation, or chemical/physical effects in material(s) contained in the detector [1, pp. 199–230 and ff]. Possibly the most familiar of these is the Geiger–Müller counter [1, pp. 202–204]

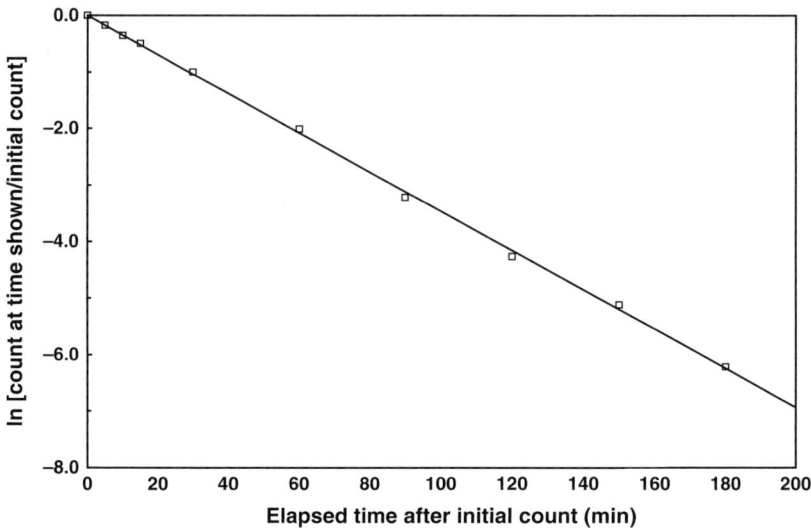

Figure 10.3 Radioactive rate of decay of laboratory source.

invented by Hans Geiger and Ernest Rutherford in 1908 [3, p. 259].

$$R_A = -dA/dt = +kA_0 \exp[-kt] = kA \tag{10.17}$$

At initial conditions

$$R_{A_0} = kA_0(1) \tag{10.18}$$

Dividing R_A by R_{A_0}, one obtains

$$R_A/R_{A_0} = \exp[-kt]$$

Also, by comparison with Equation (10.18)

$$(R_A/R_{A_0}) = (A/A_0) \tag{10.19}$$

Since the exponential function is its own derivative differing only by a constant factor, the number of atoms of A remaining and its rate of decay will both plot as semilogarithmic straight lines. For disintegration of a single radioactive species directly to product(s), Figure 10.3 works equally well with the absolute amount of radioactive material or its rate of decay.

PROBLEM 10.3   (Based on Ref. 3, p. 260)

(a) How long will it take a stored radioactive waste with a half-life of 200 years and an activity of 2.9 Ci to degrade to a relatively harmless 50 µBq (3×10^{-3} disintegrations per minute)?
(b) Repeat for a 50 µBq weak waste with a half-life of 10 years.

SOLUTION

(a) Since the answer is likely to be in years, convert disintegration rates to yearly basis.

$$(2.9 \text{ Ci})(3.7 \times 10^{10} \text{ disintegrations/s per Ci})(86,400 \text{ s/day})(365 \text{ days/year})$$
$$= 3.384 \times 10^{18} \text{ disintegrations per year}$$

$$(50 \times 10^{-6} \text{ Bq})(1 \text{ disintegration/s per Bq})(86,400(\text{s/day}))(365 \text{ days/year})$$
$$= 1576.8 \text{ disintegrations/year}$$

or

$$(3 \times 10^{-3} \text{ disintegrations/min})(1440 \text{ min/day})(365 \text{ days/year})$$
$$= 1576.8 \text{ disintegrations/year}$$

Find $k = [\ln(2)]/t_{1/2} = 0.69315/200 = 0.003466 \text{ year}^{-1}$ \hfill (10.20)

Then $t = [\ln(R_{A_0}/R_A)]/k$

Holding time $= \ln[3.384 \times 10^{18})/1.5768 \times 10^{3})]/0.003466$
$= 1.02 \times 10^4$ years (over 10,000 years)

(b) $k = 0.69315/10 = 0.06931$

Holding time $= 510$ years

In both cases, this is more than 50 half-lives, making the half-life method impractical to implement here. ■

PROBLEM 10.4   (Based on Ref. 3, p. 258, A-33)

Strontium-90 is a radionuclide associated with atomic bomb fallout and nuclear power plant accidents. It mimics calcium and enters bones and bone marrow and poses a long-term danger to human and animal life. How long will it take an initial sample emitting 2000 beta particles per minute to decay to 125 per minute? The half-life of this isotope is 28.8 years (Table 10.2).

SOLUTION

$$_{38}\text{Sr}^{90} \rightarrow {}_{39}\text{Y}^{90} + {}_{-1}\text{e}^{0}(\beta^{-}) \tag{10.21}$$

Half-life method

$$(2000)(x) = 125$$
$$x = (125/2000) = (1/16) = (1/2)^4$$

From Table 10.1, 4 half-lives are necessary

$$(4)(28.8) = 115.2 \text{ years}$$

Rate constant method (rearranging Equation (10.14))

Rate constant$(k) = \ln 2/t_{1/2} = (0.69315)/(28.8) = 0.02407 \text{ year}^{-1}$
$t = [\ln(R_A/R_{A_0})]/k = [\ln(2000/125)]/0.02407 \text{ years}^{-1} = 115.2 \text{ years}$ \hfill (10.20)

■

480 Chapter 10 Radioactive Decay

Table 10.3 Disintegration of Unknown Radionuclide

Time (min) = t	Radioactive counts, R_A (Bq)	R_A/R_{A_0}	$\ln(R_A/R_{A_0}) = y$
0	500	1.0000	0.0000
5	421	0.8420	−0.1733
10	350	0.7000	−0.3466
15	304	0.6080	−0.5199
30	184	0.3680	−1.0398
60	67	0.1340	−2.0796
90	20	0.0400	−3.1194
120	7	0.0140	−4.1592
150	3	0.0060	−5.1990
180	1	0.0020	−6.2388

PROBLEM 10.5 (Inspired by Ref. 4, p. 1083)

Some disintegration "data" for a radionuclide are listed in the first two columns of Table 10.3. Measurements are provided for the initial decay rate; every 5 min for the first 15 min, at 30 min; and every 30 min thereafter until the count starts to disappear. Find the rate constant and the half-life?

SOLUTION The ratio of the decay rate to the initial decay rate and the natural log of that ratio are shown in columns 3 and 4 of the table. The decay constant will be found from a least-squares fit of $y = [\ln(R_A/R_{A_0})]$ versus t (time in minutes), whose points are shown in the semilogarithmic plot of Figure 10.3. Data regression via least squares is being used to prevent errors that might arise from using any two individual points in the determination of the slope (and intercept) of the correlating line.

For a fit of $y = bt$, choosing to believe the initial rate, and forcing a straight line through the point (0, 1), data regression yields $b = (\Sigma t_i y_i)/(\Sigma t_i^2)$ [6, pp. 344, 625]. For this problem, the slope $(b) = (-2850.44)/(82,250) = -0.03466 = -k$

$$t_{1/2} = [\ln 2]/k = -0.69315/0.03466 = 20.00 \text{ min}$$

For a least-squares fit of the form $y = a + bt$ [6, p. 320; 7], allowing the intercept to be determined from the totality of the measured points,

$$\text{the slope } (b) = \{n(\Sigma t_i y_i) - (\Sigma t_i)(\Sigma y_i)\}/\{n(\Sigma t_i^2) - [(\Sigma t_i)]^2\}$$
$$\text{and the intercept } (a) = \{(\Sigma y_i)(\Sigma t_i^2) - (\Sigma t_i)(\Sigma t_i y_i)\}/\{n(\Sigma t_i^2) - [(\Sigma t_i)]^2\}$$

where n is the number of points in the data regression.
Here,

$$\text{the slope } (b) = \{10(-2850.44) - (660)(-22.8539)\}/\{10(82,250) - (660)(660)\}$$
$$= -0.03469 = -k$$
$$t_{1/2} = -0.69315/0.03469 = 19.98 \text{ min}$$

The intercept $(a) = \{(-22.8539)(82,250)-(660)(-2850.44)\}/\{10(82,250)-(660)(660)\}$

$= 0.00402 = \ln[R_A/R_{A_0}]$

Exponentiation of $\ln[R_A/R_{A_0}]$ at $t=0$ gives an intercept in terms of relative counts of 1.00402 and an absolute calculated intercept $=(1.00402)(500) = 502$.

In actual practice, there are many commercial computer programs that will calculate a least-squares regression fit without the user's having to perform the necessary manual summations, multiplication, additions, and divisions noted here. These programs will also do statistical tests to determine whether the intercept is significantly different from zero, and hence which formulation is appropriate. In this particular case, it makes little difference in the rate constant and the half-life.

One final note on this problem:

If the initial count is missing from the data, the next point could be defined as the initial point (and the times adjusted accordingly) to proceed with the data fit. Natural logarithms of the measurements could also be fitted directly without taking ratios. In that case, the governing relationship is Equation (10.11), with a slope of $-k$ and an intercept of $\ln(R_{A_0})$. The least-squares linear function to be fitted would then be $y = a + bt$, or specifically $\ln R_A = \ln R_{A_0} - kt$. ∎

10.4.3 Parallel Decay

A few radioactive isotopes decay in parallel to more than one product. Examples are $_{13}Al^{26}$, $_{17}Cl^{36}$, and $_{19}K^{40}$ [1, pp. 32, 430]. These reactions can be represented as follows:

$$A \xrightarrow{k_1} B$$
$$A \xrightarrow{k_2} C \tag{10.22}$$

$$dA/dt = -(k_1 + k_2)A \tag{10.23}$$

$$dB/dt = +k_1 A \tag{10.24}$$

$$dC/dt = +k_2 A \tag{10.25}$$

A, B, and C are the number of atoms of the parent (A) and the decay products $(B$ and $C)$. Solutions for the initial conditions of A_0, B_0, and C_0 are given below:

$$A = A_0 \exp[-(k_1 + k_2)t] \tag{10.26}$$

$$B - B_0 = A_0[k_1/(k_1 + k_2)]\{1 - \exp[-(k_1 + k_2)t]\} \tag{10.27}$$

$$C - C_0 = A_0[k_2/(k_1 + k_2)]\{1 - \exp[-(k_1 + k_2)t]\} \tag{10.28}$$

When starting from A_0 alone with no initial B and C, B_0 and C_0 are set equal to zero in the above equations.

The constants k_1 and k_2 are known as partial decay constants and the related half-lives $[\ln(2)/k]$ as partial half-lives [8].

PROBLEM 10.6

(a) Calculate the distribution of products and unreacted parent species for the reactions [1, p. 32; 4 pp. 1083, 1084, 1088, 1089].

$$_{19}K^{40} \rightarrow {_{20}Ca^{40}} + {_{-1}e^0}(\beta^-), \text{ beta decay } 89.3\% \quad (10.29)$$

$$_{19}K^{40} \rightarrow {_{18}Ar^{40}} + {_{+1}e^0}(\beta^+), \text{ positron emission } 10.7\% \quad (10.30)$$

The half-life for the combined reactions is 1.28×10^9 years.

(b) Estimate the initial number of disintegrations per second (Bq) from a 10 g sample of potassium chloride. The $_{19}K^{40}$ potassium isotope constitutes 1.17% of normal potassium. Convert the radioactive count into curies.

SOLUTION

(a) Calculate the decay constants.
   From Equation (10.14),

$$k = (\ln 2)/1.28 \times 10^9 \text{ years} = 5.415 \times 10^{-10}$$

$$5.415 \times \frac{10^{-10}}{\text{year}} \frac{\text{day}}{864000 \text{ s}} \frac{\text{year}}{365 \text{ d}} = 1.717 \times 10^{-17} \text{ s}^{-1}$$

This is $(k_1 + k_2)$.

(b) Calculate the other rate constants.
   Then,

$$k_1/(k_1 + k_2) = 89.3\% = 0.893 \text{ (given)}$$

and

$$k_2/(k_1 + k_2) = 10.7\% = 0.107 \text{ (also given)}$$

$$k_1 = 0.893(k_1 + k_2) = (0.893)(5.415 \times 10^{-10}) = 4.836 \times 10^{-10}$$

and

$$k_2 = 0.107(k_1 + k_2) = (0.107)(5.415 \times 10^{-10}) = 0.579 \times 10^{-10})$$

Partial half-lives are $[\ln(2)]/4.836 \times 10^{-10} = 1.43 \times 10^9$ years for k_1 and $[\ln(2)]/0.579 \times 10^{-10} = 1.20 \times 10^{10}$ years for k_2.
From Equations (10.27) and (10.28) with $B_0 = C_0 = 0$, the distribution remains constant at all times.

(c) The distribution of reactant and products is depicted in Figure 10.4. Note the time scale of the plot in this case.

(d) Calculate A_0

10 g KCL / 74.56 g /g mol KCL = 0.1342 g mol KCL

$$0.1341 \text{ g mol KCL} \frac{1 \text{ g atom K}}{\text{g mol KCL}} \frac{6.022 \times 10^{23} \text{ atoms}}{\text{g atom K}} K = 8.076 \times 10^{22} \text{ atoms K}$$

The figure 6.022×10^{23} is Avogadro's number, the number of atoms or molecules in a gram atom or gram molecular weight.

$$8.076 \times 10^{22}(0.0.0117) = 9.448 \times 10^{20} \text{ atoms of } _{19}K^{40} = A_0$$

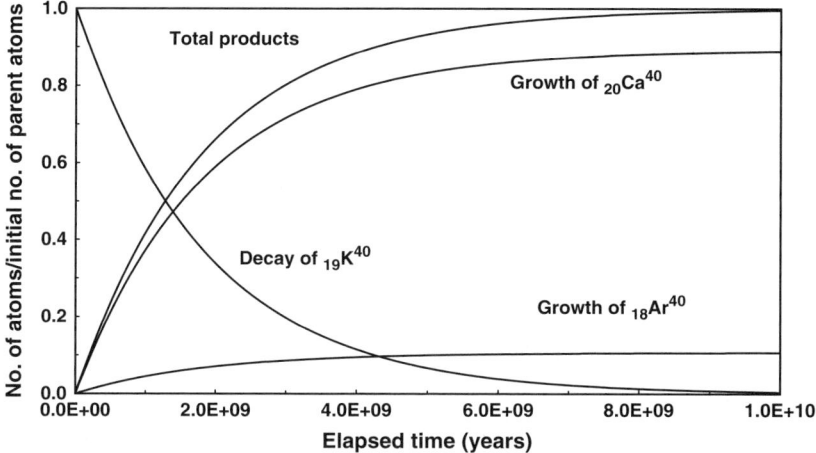

Figure 10.4 Example of parallel radioactive decay.

From Equation (10.18)

$$R_{A_0} = (k_1 + k_2)A_0 \tag{10.31}$$

$R_A = (1.717 \times 10^{-17}\ \text{s}^{-1})(9.448 \times 10^{20}\ \text{atoms radioactive K})$

$\times (1\ \text{disintegration/atom})$

$= 1.622 \times 10^4\ \text{disintegrations/s (Bq)}$

$1.622 \times 10^4\ \text{disintegrations/s}\ (1\ \text{Ci}/3.7 \times 10^{10}\ \text{disintegrations/s}) = 0.43\ \mu\text{Ci}$

Because of the size of the rate constants and the resulting value of $\exp[-(k_1 + k_2)]$, the disintegration rate will remain constant for many years to come.

In this case, the products, being stable [4, p. 1088; 3, p. 253], will not contribute to the radioactivity of the sample. ∎

10.4.4 Independent, Unrelated Simultaneous Decay Reactions

Up to this point, we have seen radioactive decay producing a semilogarithmic straight line plot of decay rate or number of atoms remaining. However, the independent decay of two (or more) radionuclides will produce noticeable curvature in the rate of decay, as in the solid line of Figure 10.5. In this figure, the solid curved line is the logarithm of the sum of two independent species, one with a relatively short half-live (A_2) and the other whose half-life is somewhat longer (A_1). Decay rates for these species are shown as the broken lines in the figure. In some cases, it is possible to separate the composite curve into its components and determine their decay constants and half-lives.

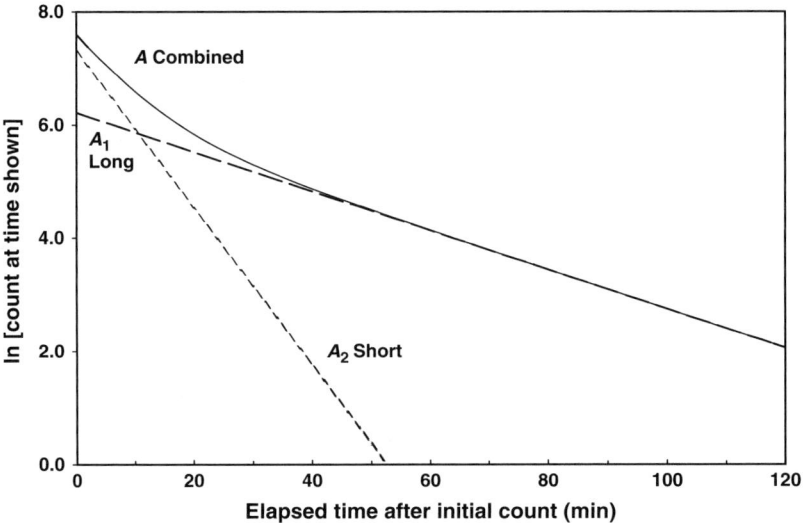

Figure 10.5 Simultaneous radioactive decay of two independent species.

PROBLEM 10.7

Given only the "measured" composite curve of radioactive decay of a laboratory material from Figure 10.5, and an excerpt of the points read from that curve (Table 10.4), estimate the decay constants, half-lives, and initial radioactive counts for each of its components. It is assumed that the dotted lines are unknown at the outset.

SOLUTION From the logarithmic values at 90 and 120 min obtained after the curve has reached a straight line asymptote, the slope is calculated to be

$$\text{Slope} = \frac{3.0956 - 2.0541}{90 - 120} = -0.034717 = -k$$

$$t_{1/2} = \ln(2)/(k) = 0.69315/0.034717 = 19.97 \text{ min}$$

The slope could also be computed via least squares using more points along the asymptote as the slope becomes constant. The asymptote occurs because the decay rate of the component

Table 10.4 Excerpt of Radioactive Decay Counts for Unknown Laboratory Sample

Time (min)	Counts (Bq)[a]	ln (Counts)
0	2000	7.6009
2	1603.3	7.3798
4	1296.8	7.1677
90	22.1	3.0956
120	7.8	2.0541

[a] Decay count figures are averages, over several seconds.

10.4 Pathways of Radioactive Decay

with the short half-life has become vanishingly small during the still active lifetime of the longer lived component.

Extrapolation back to time $= 0$ to find ln (intercept)

$$\text{Slope} = 0.034717 = \frac{\ln(\text{at } t = 4) - \ln(\text{at } t = 90)}{0-90}$$

$$\text{or} -0.034717 = \frac{\ln(\text{at } t = 0) - \ln(\text{at } t = 120)}{0-120}$$

$$\ln(\text{at } t = 0) = (-0.034717)(-90) + 3.0956 = 6.2201$$

$$\ln(\text{at } t = 0) = (-0.034717)(-120) + 2.0541 = 6.2201$$

$$R_{A_0} = \exp[\ln(\text{intercept})] = 502.8$$

These values can be used to reconstruct the function A_1, as follows: Subtracting A_1 from the composite function given provides A_2 by difference. Reconstructing the line for decay of A_1 alone at 0, 2, and 4 min,

$$A_1 = A_{1_0} \exp[-k_1 t] \quad A_{1_0} = 502.8, \quad k = -0.034717$$

t (min)	A_1	$\ln A_1$
0	502.8	6.2202
2	469.1	6.1508
4	437.6	6.0813

A_2 is then obtained by difference:

t (min)	$A_2 = A - A_1$	A_2	$\ln(A_2)$	(A_2/A_{2_0})	$\ln(A_2/A_{2_0})$
0	2000–502.8 =	1497.2	7.3114	1.0000	0
2	1603.3–469.1 =	1134.2	7.0337	0.7575	−0.2777
4	1296.8–437.6 =	859.2	6.7560	0.5739	−0.5554

A_2 at $t=0$ and its natural log are the intercepts of the linear and semilogarithmic curves. Find the slope in semilogarithmic coordinates from these three points using the least-squares formula for $y = bt$.

$$\text{Slope } (b) = (\Sigma t_y)/(\Sigma t^2) = -0.138850$$

where t is the time and y is $\ln(A_2/A_{2_0})$.

Results are summarized in Table 10.5.

Table 10.5 compares the estimated parameters with those used to construct the curves of Figure 10.5 for this problem. Agreement is quite good for this made-up example. Had this been a case involving real data, one would not know for sure the extent of agreement.

In this instance, the half-lives of the two components are related by a factor of four. It is said that the limit for resolution of such two-component curves is that this ratio not be less than about a factor of two [9]; additional techniques are mentioned in Ref. 9.

For the general case, the difference curve A_2 represents decay of all unstable constituents except for the one with the longest half-life. In the event that this residual shows significant

Table 10.5 Comparison of Estimated Parameters with Those Used to Construct the Curves in Figure 10.5

Parameter	A_1 (Longer half-life)	A_2 (Shorter half-life)
A_0 calculated	502.8	1497.2
A_0 original	500	1500
k calculated	0.034717	0.138850
k original	0.034657	0.138629
$T_{1/2}$ calculated	19.97	4.99
$T_{1/2}$ original	20.00	5.00

curvature, the procedure is repeated to separate that curve further into its components. Because of uncertainties in the experimental observations, the analysis of systems of more than three components may prove difficult [9]. ∎

10.5 DECAY SERIES

A sequence of radioactive decay steps involving unstable intermediates is known as a *radioactive series* or *radioactive decay chain*. After transmutation of the parent nucleus into the first daughter species, this in turn decays further, and each successive daughter species continues the decay process until a stable end product is achieved.

There are three such series found in nature, the thorium-232, the uranium-238 series, and the actinium series starting with uranium-235. Each ends up with a different stable isotope of lead [1, pp. 41–42]. Half-lives are longer than the age of the earth, making those series of the so-called *primordial* origin [1, p. 42]. The composition of certain rocks containing members of these species can be used to date the age of the earth.

There is a fourth man-made series, the neptunium series, starting with plutonium-241 and ending with bismuth-209. Half-lives in this series are considerably shorter, and the bismuth end product is found in nature, suggesting that the series may have been active during primordial times but may have played itself out long ago.

This process can be represented in general by the following reaction scheme:

$$A \xrightarrow{k_1} B \xrightarrow{k_2} C \xrightarrow{k_3} D \xrightarrow{k_4} E \cdots \quad (10.32)$$

Mathematically, the rate of decay is described by a set of irreversible first-order reactions

$$dA/dt = -k_1 A \quad (10.8)$$
$$dB/dt = +k_1 A - k_2 B \quad (10.33)$$
$$dC/dt = +k_2 B - k_3 C \quad (10.34)$$
$$dD/dt = +k_3 C - k_4 D \quad (10.35)$$
$$dE/dt = +k_4 D \quad (10.36)$$

where A, B, C, and so on indicate the *number* of radioactive atoms of each species.

10.5 Decay Series

Let us begin with the simple case of $A \to B \to C$, consisting of one parent material, one daughter species, one stable end product, and rate constants k_1 and k_2. For an initial number of parent atoms A_0, the solution (by traditional methods) is shown below [10, pp. 59–61, 169–173; 11]:

$$A = A_0 \exp(-k_1 t) \tag{10.9}$$

$$B = [(k_1 A_0)/(k_2 - k_1)][\exp(-k_1 t) - \exp(-k_2 t)] + B_0 \exp(-k_2 t) \tag{10.37}$$

$$C = C_0 + [A_0/(k_2 - k_1)]\{k_2[1 - \exp(-k_1 t)] - k_1[1 - \exp(-k_1 t)]\} + B_0[1 - \exp(-k_2 t)] \tag{10.38}$$

With no initial B and C present at time zero, the B_0 and C_0 terms drop out and simplify the B and C equations. The solution is valid for $k_1 \neq k_2$.

The competing buildup and decay of the intermediate B causes the curve for the *number* of B atoms to go through a maximum, while A constantly decays and C increases at a rate that depends on the availability of B. Typical curves, normalized on A_0, with no B_0 present initially may resemble Figure 10.6.

The shape of the curves will vary depending on the relative size of k_1 and k_2. An example of $k_1 > k_2$ comes from the simplified version of decay of an unstable isotope of bismuth to the stable lead-206 during the tin smelting process [1, pp. 179–180, 432–435].

$$_{82}Pb^{210} \xrightarrow[\beta]{22 \text{ years}} {}_{83}Bi^{210} \xrightarrow[\beta]{k_1} {}_{84}Po^{210} \xrightarrow[\alpha]{k_2} {}_{82}Pb^{206}_{\text{stable}} \tag{10.39}$$

Half-life for bismuth-210 is 5.013 days ($k = 0.13869934 \text{ day}^{-1}$) and for polonium-210 is 138.38 days ($k = 0.005009013 \text{ day}^{-1}$) (Table 10.2). The bismuth had been formed from lead-210 with a half-life of 22 years, orders of magnitude greater. In Figure 10.7, the rate-limiting step is the decay of polonium to lead-206, since the

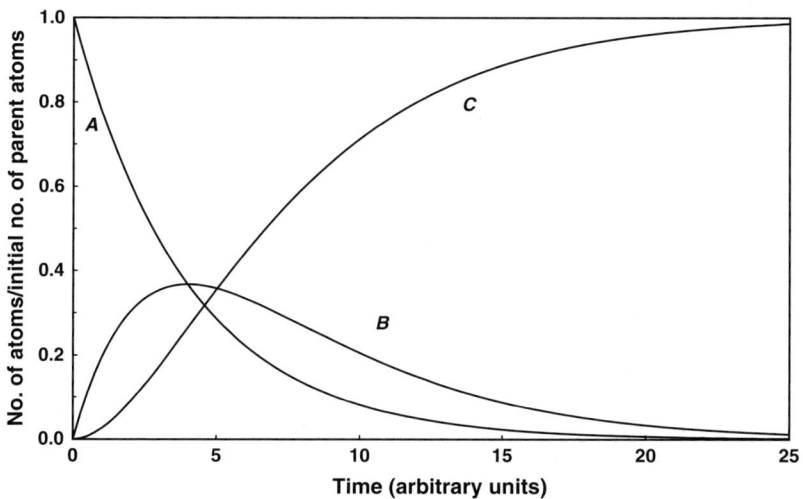

Figure 10.6 Product distribution $A \to B \to C$ typical curves.

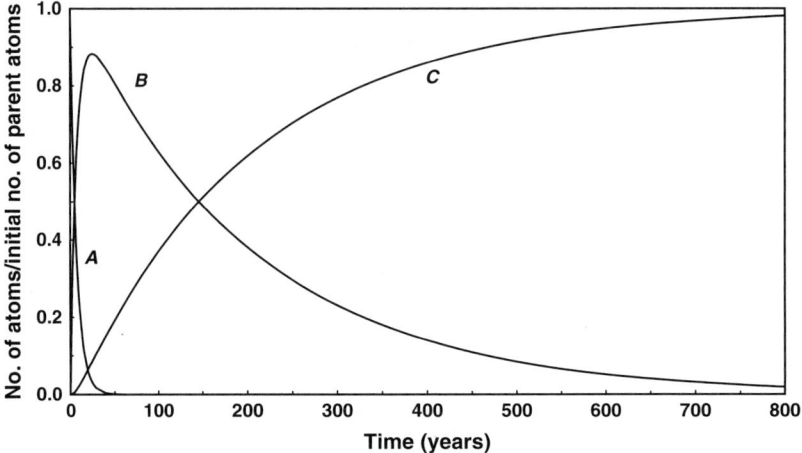

Figure 10.7 Product distribution $A \to B \to C$, $k_1 > k_2$.

polonium formed relatively rapidly from the bismuth "bides its time" while bunching up waiting to react.

An example of $k_2 > k_1$ is a simplified sequence taken from the thorium-232 decay chain:

$$_{83}\text{Bi}^{212} \xrightarrow[\alpha]{k_1} {_{81}\text{Tl}^{208}} \xrightarrow[\beta]{k_2} {_{82}\text{Pb}^{208}_{\text{stable}}} \quad (10.40)$$

Here, half-lives are 60.55 min for bismuth-212 and 3.05 min for thallium-208; k_1 and k_2 are 0.011447 and 0.227261 min^{-1}, respectively (Table 10.2). The intermediate curve in Figure 10.8 builds up only slightly as the rate-limiting step, and then dies off.

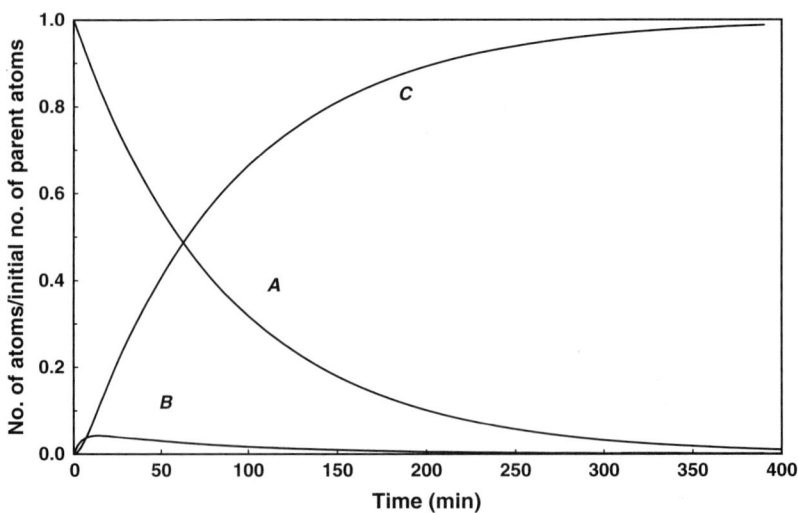

Figure 10.8 Product distribution $A \to B \to C$, $k_2 > k_1$.

In the unlikely event that k_1 equals k_2 exactly, the solution given in Equations (10.37) and (10.38) is not valid, and blows up at $k_1 - k_2 = 0$ in the denominator. For the case of $A \rightarrow B \rightarrow C$ with $k_1 = k_2 = k$.

$$dA/dt = -kt \qquad (10.8)$$
$$dB/dt = kA - kB \qquad (10.41)$$
$$dC/dt = -kB \qquad (10.42)$$

The solution of these differential equations becomes

$$A = A_0 \exp[-kt] \qquad (10.9)$$

as before and

$$B = kA_0 t \exp[-kt] + B_0 \exp[-kt] \qquad (10.43)$$
$$C - C_0 = A_0\{1 - \exp[-kt]\} - kA_0 t \exp[-kt] + B_0\{1 - \exp[-kt]\} \qquad (10.44)$$

The solution process is greatly simplified by means of the integrating factor technique and employing definite integrals. Consult any good book on differential equations for details. Again, the equations simplify for $B_0 = C_0 = 0$.

Although the equations might appear to be only of academic interest since a situation in which $k_1 = k_2$ exactly is not likely to be encountered in the real world, this solution defines the mathematical limit as $(k_1 - k_2)$ approaches zero. This can be proven by expansion of the exponentials from the $k_1 \neq k_2$ solution in an infinite power series, algebraic manipulation (with $k_1 = k_2$), and regrouping the powers of t back into the exponential of $(-kt)$. In fact, it is this limiting case of $k_1 = k_2$ that is shown as the typical example in Figure 10.6.

10.5.1 Maxima and Inflection Points

The maximum point in the amount of the daughter B atoms can be easily found using differential calculus by setting its derivative equal to zero and solving for time [10, pp. 60, 173]. This curve must also pass through an inflection point, where its concave downward slope becomes concave upward in preparation for achieving its asymptotic value of long reaction time. This time is obtained by setting the second derivative equal to zero and solving the resulting equation.

Substitution into the equations for A, B, and C yields the values of the functions corresponding to these times. Values of t_{MAX}, t_{INF}, and the functions A, B, and C at t_{MAX} and t_{INF} are summarized in Table 10.6 for both $k_1 \neq k_2$ and $k_1 = k_2$ cases.

Use of the definition of e as the limit as x approaches zero of $[1 + x]^{1/x}$ [12] or the limit as x approaches infinity of $(1 + 1/x)^x$ [12,14] is necessary to obtain some of the entries in the table. For both cases, the point of inflection in the B curve occurs at a time twice that of the maximum. For the limiting case of $k_1 = k_2$, A at $t_{MAX} = B_{MAX}$, that is, the decay curve for A intersects the B curve at its maximum point, and A at $t_{INF} = 1/2$ of B_{INF}. The values for $k_1 \neq k_2$ reduce to those for $k_1 = k_2$ in the limit as $k_2 - k_1$ approaches zero. Aside from the location of t_{MAX} and t_{INF}, entries normalized on A_0

Table 10.6 Values of A, B, and C at t_{MAX} and t_{INF}

	$k_1 \neq k_2$		$k_1 = k_2$	
	$t_{MAX} = \dfrac{\ln(k_2/k_1)}{(k_2-k_1)}$	$t_{INF} = \dfrac{2\ln(k_2/k_1)}{(k_2-k_1)} = 2t_{MAX}$	$t_{MAX} = (1/k_1)$	$t_{INF} = (2/k_1) = 2t_{MAX}$
A	$A_0(k_1/k_2)^{[k_1/(k_2-k_1)]}$	$A_0(k_1/k_2)^{[2k_1/(k_2-k_1)]}$	$A_0/e = 0.3679 A_0$	$A_0/e^2 = 0.1353 A_0 = (1/2)B_{INF}$
B	$A_0(k_1/k_2)^{[k_2/(k_2-k_1)]}$	$A_0 \dfrac{(k_2-k_1)}{k_1}(k_1/k_2)^{[2k_1/(k_2-k_1)]}$ $= A_0(B_{MAX}/A_0)^2(k_2+k_1)/k_1$	$A_0/e = 0.3679 A_0 = B_{MAX}$	$2 A_0/e^2 = 0.2707 A_0 = B_{INF}$
C	$A_0 - A$ at $t_{MAX} - B_{MAX}$	$A_0 - A$ at $t_{INF} - B_{INF}$	$A_0(1 - 2/e) = 0.2642 A_0$	$A_0(1 - 3/e^2) = 0.5940 A_0$

are functions only of $e = 2.71828\ldots$, the basis of the natural logarithm, and are independent of the numerical values for $k_1 = k_2$.

10.6 LONGER DECAY CHAINS

For longer decay chains, the defining differential equations require more effort to solve, and the resulting solutions become increasingly complicated. With the stipulation that the initial quantities are all zero except for that of the original parent ($A_0 \neq 0$, $B_0 = C_0 = D_0 = E_0, \ldots = 0$), the general solution [15] is given below (with a slight change in nomenclature from what we have been using).

$$N_i \text{ (with } i = 1 \text{ to } n) = C_1 \exp(-k_1 t) + C_2 \exp(-k_2 t) + \cdots + C_n \exp(-k_n t)$$

where N_i is successively A, B, C, D, E, and so on in the present nomenclature ($N_1 = A$ and $N_{1_0} = A_0$).

The coefficients (C_1 through C_n) of the exponentials in $-k_1 t$ through $-k_n t$ are as follows:

$$C_1 = \frac{k_1, k_2, \ldots, k_n}{(k_2 - k_1)(k_3 - k_1) \cdots (k_{n-1} - k_1)} (N_{1_0} = A_0) \quad (10.45)$$

$$C_2 = \frac{k_1, k_2, \ldots, k_n}{(k_1 - k_2)(k_3 - k_2) \cdots (k_{n-1} - k_2)} (A_0) \quad (10.46)$$

$$C_n = \frac{k_1, k_2, \ldots, k_n}{(k_1 - k_n)(k_2 - k_n) \cdots (k_{n-1} - k_n)} (A_0) \quad (10.47)$$

Each member of the series requires coefficients and exponentials up to and including its own subscript. The first member of the series ($N_1 = A$) is given by

$$N_1 = A = A_0 \exp[-k_1 t] \quad (10.9)$$

Succeeding members, for example,

$$N_2 = B = \frac{k_1 k_2 A_0}{(k_2 - k_1)} \exp[-k_1 t] + \frac{k_1 k_2 A_0}{(k_1 - k_2)} \exp[-k_2 t]$$

$$N_3 = C = \frac{k_1 k_2 k_3 A_0 \exp[-k_1 t]}{(k_2 - k_1)(k_3 - k_1)} + \frac{k_1 k_2 k_3 A_0 \exp[-k_2 t]}{(k_1 - k_2)(k_3 - k_2)} + \frac{k_1 k_2 k_3 A_0 \exp[-k_3 t]}{(k_1 - k_3)(k_2 - k_3)}$$

reduce algebraically to the forms previously employed (Equations (10.37) and (10.38)).

10.6.1 Example—A Five-Member Chain

The product distribution of a five-member chain, a simplified version excerpted from the uranium-238 series [1, pp. 432, 434–435], is depicted in Figure 10.9.

$$P_0^{218} \xrightarrow[\alpha,\ 3.10\ \text{min}]{k_1} Pb^{214} \xrightarrow[\beta,\ 26.8\ \text{min}]{k_2} Bi^{214} \xrightarrow[\alpha,\ 19.9\ \text{min}]{k_3} Tl^{210} \xrightarrow[\beta,\ 1.3\ \text{min}]{k_4} Pb^{210} \quad (10.48)$$

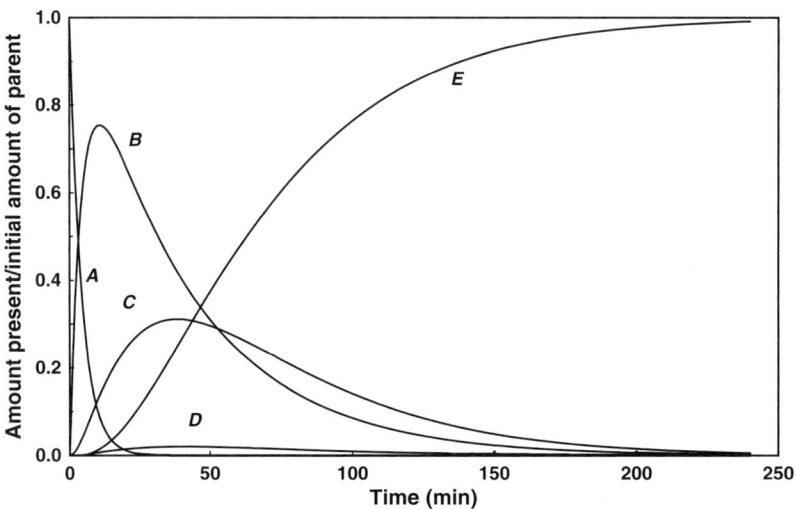

Figure 10.9 Product distribution five-member chain $A \rightarrow B \rightarrow C \rightarrow D \rightarrow E$.

The total radioactivity at any point in time is caused by the decay of the unreacted parent augmented by the decay of the daughter species that are formed. This will be illustrated in the next section for a simpler system.

While not a stable isotope, lead-210 decays to bismuth-210 (half-life of 22 years, compared to the time frame minutes of the decay chain above. Once again, these curves assume that the decay sequence begins with no daughter species present at the onset of polonium-218 disintegration. Two separate peaks are apparent in the figure, shown by lead-214 and bismuth-214, as the polonium-218 parent decays quickly relative to the decay rates of the lead and bismuth. The buildup curve for thallium-210 is long and low since its decay proceeds more rapidly than its formation. The rate-limiting step (longest half-life, lowest decay constant) is the decay of lead-214 to bismuth-214.

10.7 RATE OF DECAY IN A RADIOACTIVE SERIES

In a radioactive decay series, the decay rate is made up of the decay of the parent species plus the simultaneous decay of the daughters as they are formed. In terms of numbers of atoms, the decay of a parent species and the formation of daughter products have been illustrated for several three-member series $A \rightarrow B \rightarrow C$ in Figures 10.6–10.8.

In each of these figures, the intermediate material B goes through a maximum, where the rate of formation and the rate of decay are in balance. Both buildup and decay occur at the same time, even to the left at the maximum. Decay still takes place there, although the decay rate is less than the rate of buildup. To the right of the maximum, decay exceeds buildup.

10.7 Rate of Decay in a Radioactive Series

The decay rate of the parent is the disintegration constant times the number of parent (A) atoms present at any given time.

$$R_A = -(dA/dt) = k_1 A = k_1 A_0 \exp[-k_1 t] \quad (10.17)$$

By definition, the rate of disintegration of the daughter (intermediate) species is also the product of its decay constant times the number of atoms (B) present.

$$R_B = k_2 B = [k_1 k_2 A_0/(k_2 - k_1)]\{\exp[-k_1 t] - \exp[-k_2 t]\} \quad (10.49)$$

This equation assumes that the decay sequence started with only A present with no initial B (B_0). A maximum occurs also in the decay rate of B, at the same value of t_{MAX} that applies for the curve describing the number of B atoms. Total radioactivity is the sum of the two. The end product C is stable, and consequently contributes no radioactivity.

$$R_A + R_B = k_1 A + k_2 B \quad (10.50)$$

$$R_A + R_B = k_1 A_0 \exp[-k_1 t] + [k_1 k_2 A_0/(k_2 - k_1)]\{\exp[-k_1 t] - \exp[-k_2 t]\} \quad (10.51)$$

For the ($k_1 = k_2$) case (also with no initial B), Equation (10.51) is slightly different.

$$R_A + R_B = k_1 A_0 (1 + k_1 t) \exp[-k_1 t] \quad (10.52)$$

PROBLEM 10.8

Calculate the decay rates for the product distribution of Figure 10.6 ($k_1 = k_2 = k$)

SOLUTION Decay rates, listed in Table 10.7, are calculated from Equation (10.52). Decay rates for this case are shown in Figure 10.10. ∎

Table 10.7 Decay Rates for the Product Distribution of Figure 10.6

Time (1/k)	$R_A/kA_0 = \exp[-kt]$	$R_B/R_1 A_0 = kt \exp[-kt]$	Total
0	1.000000	0.000000	1.00000
0.25	0.778800	0.194700	0.973500
0.50	0.606531	0.303265	0.909796
0.75	0.472467	0.354275	0.826642
$1 = t_{MAX}$	0.367880	0.367880	0.735760
$2 = t_{INF}$	0.135335	0.270671	0.406006
5	0.006738	0.003689	0.040427
10	0.000045	0.000454	0.000499

The ratio of decay rates of parent and daughter are given below:

$$R_B/R_A = [k_2/(k_2-k_1)]\{1 - \exp[-(k_2-k_1)t]\} \quad (10.53)$$

$$(R_A + R_B)/R_A = [1 + k_2/(k_2-k_1)]\{1 - \exp[-(k_2-k_1)t]\}, \quad k_2 \neq k_1 \quad (10.54)$$

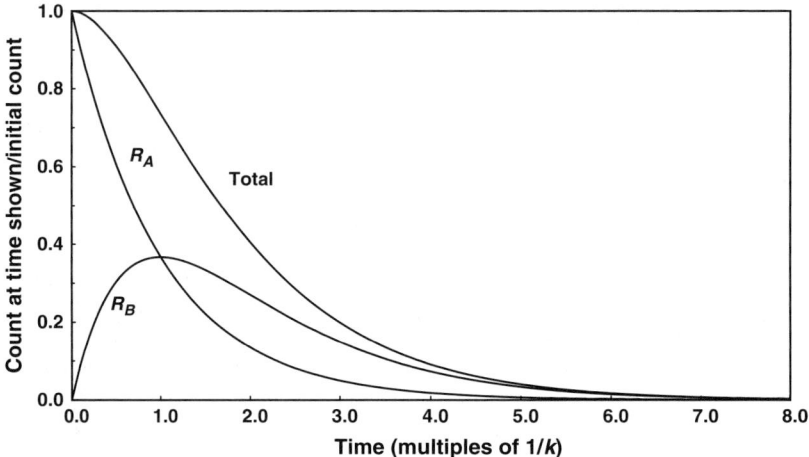

Figure 10.10 Radioactive decay rates for the typical product distribution shown previously.

In general, several situations arise, depending on the relative size of k_1 and k_2.

- Secular equilibrium where $k_2 \gg k_1$
- Transient equilibrium where k_1 is equal to or is just barely less than k_2
- No equilibrium where $k_1 \gg k_2$

10.7.1 Secular Equilibrium

At long times, the exponential drops out and k_2 overwhelms k_1. Equation (10.53) reduces to 1 and Equation (10.54) equals 2; that is, the daughter activity reaches that of the parent, and the total activity exhibited is twice that of the parent alone.

10.7.2 Transient Equilibrium

The activity of the daughter increases to its maximum at t_{MAX}, where R_A and R_B are equal when $k_1 = k_2$ exactly and almost equal when k_1 and k_2 are almost equal. At t_{MAX}, R_A and R_B are each less than R_{A_0} and therefore together less than $2R_{A_0}$, a mark of secular equilibrium. Activity of the daughter is then greater than that of the slower decaying parent. This situation is illustrated in Figure 10.10.

10.7.3 No Equilibrium

The half-life of the daughter greatly exceeds that of the parent, that is, $k_1 \gg k_2$. The exponential term $\exp[-k_1 t]$ then drops out more quickly than $\exp[-k_2 t]$, and

R_B becomes

$$R_B = \frac{(k_1 k_2 A_0)}{(k_1 - k_2)}(-\exp[-k_2 t]) \qquad (10.55)$$

$$= k_2 A_0 \exp[-k_2 t] \qquad (10.56)$$

$R_A = k_1 A_0(\exp[-k_1 t]$ approaches zero, and the total of $R_A + R_B$ becomes R_B, the decay rate of the daughter. In this case, the condition of no equilibrium ensues.

A more complete explanation of the various types of equilibrium in a radioactive decay series along with graphs depicting each situation is presented elsewhere [1].

10.8 A TRANSITION FROM SCIENCE TO THE REALM OF REGULATORY CONTROL

We now turn from the chemistry, physics, and mathematics of radioactive decay and the definitions of exposure to ionizing radiation to a brief discussion of regulatory practices. First, let us consider several horrible examples of radiation run amok, all of which provide a motivation for regulatory control. The chapter concludes with a numerical example for discharge of a wastewater containing radioactive material.

10.9 SOME NOTABLE ACCIDENTS INVOLVING NUCLEAR MATERIALS

The three most well known of these accidents occurred in 1957 at the Windscale nuclear power plant in England, in 1979 at the Three Mile Island reactor in Pennsylvania, USA, and in 1986 at the Chernobyl power station in the Ukraine in the former Soviet Union. Failures of various types led to uncontrolled releases of radioactivity to the atmosphere. The effects from the fire at Windscale and the explosions and fire at Chernobyl were widespread, crossing international borders. Fortunately, the off-site impacts of the partial core-meltdown at Three Mile Island, although significant, were more localized. All three of these catastrophes were a cause for concern to human health, Chernobyl being the worst. Details can be found elsewhere [1, pp. 149–157; 3, pp. 268–270].

Perhaps less well known are the incident that took place in 1987 in Brazil and the earlier one in Mexico in 1983. The Brazilian incident occurred in Guiana, Brazil, a city with a population of about 1 million, some 170 km southeast of the capital city of Brasilia [1, pp. 195–196; 2 p. 226]. An unregistered radiotherapy source containing a powdery soluble chloride salt of $_{55}Cs^{137}$ (50.9 TBq of Cs (1376 Ci), half-life 30.07 years) was left behind in an abandoned building at the time a medical facility moved from the premises. The building was partially demolished, and scavengers looking for scrap to sell removed the source and compromised its surrounding protective stainless steel jacket and platinum capsule, allowing the escape and dispersion of some of the contents into the environment.

What remained was sold to a scrap dealer, who noticed its glow-in-the-dark properties. The material was distributed among friends as "carnival glitter" and was spread onto their hands, faces, and clothing. Authorities were notified days later after several people became seriously ill. Only then did the extent of the contamination become apparent.

About 110,000 people were monitored for contamination, more than 10% of the city's population. Four people had died by the spring of 1988, with an additional five fatalities projected over the next 5 years. Some people showed localized skin burns, bone marrow suppression, or acute radiation syndrome. Others, including health care personnel and workers participating in the subsequent cleanup, also received substantial doses of radiation.

One square kilometer of the city had become contaminated. Many houses were evacuated. Decontamination consisted of washing the buildings; removing contaminated surfaces inside; taking out contaminated items; and removing walls, roofs, pavement, and topsoil. Several structures were completely demolished. Over 40 tons of radioactive contaminated material including clothing, shoes, and structural components were contaminated (at an estimated 86% of the activity of the original source), all from the release of less than 1 g of radioactive cesium. Decontamination required about 6 months to accomplish.

A similar case of tampering and its dire consequences happened in Mexico [1, p. 196]. A used radiotherapy unit containing $_{27}Co^{60}$ (37 TBq (1000 Ci), half-life 5.27 years) was imported into Mexico on behalf of a medical center in the city of Juarez.

The unit, never placed into service, was instead stored in a warehouse, and the source was not registered with the authorities. Subsequently, a medical technician employed at the center attempted to dismantle the source to sell it as scrap. It wound up as scrap metal, went through a smelting process, and became a constituent of steel reinforcing bars (rebars) and table legs. The situation was discovered when a truckload of contaminated rebars passed close to the Los Alamos, New Mexico, USA, nuclear laboratory and set off the radiation detection alarms there.

Following an intensive investigation to trace the source of contamination, many hundreds of houses containing the contaminated rebars were torn down in Mexico, and several thousand contaminated items were returned from the United States to Mexico for disposal. Areas in and around the scrap yard were also contaminated. Total radioactive waste was 4500 tons of metal and 16,000 m^3 soil.

Unlike the incident in Brazil, however, no deaths occurred, although several people were exposed to radiation doses of 7 Sv (700 rem). As discussed earlier in this chapter, a dose of 600 rem is fatal to humans and a typical annual limit for individuals is $\leq$500 millirems, more than 1000 times less.

Unfortunately, these cases are not unique. A multitude of radioactive sources have been found with scrap metal or have been assimilated into the smelting process, where the radioactivity contaminates the steel ($_{27}Co^{60}$) or dust in the furnace ($_{55}Cs^{137}$). A similar occurrence of rebar contamination took place in Taiwan (circa 1982–1983), where hundreds of buildings were contaminated and thousands of people were exposed; furnace dusts were contaminated by a $_{55}Cs^{137}$ source in scrap metal in 1985.

10.10 GOVERNMENTAL REGULATIONS AND LICENSING PROCEDURES

The lesson to be learned from nuclear disasters, accidents, and incidents is that sources of radioactivity are not to be trifled with. To control the handling of radioactive materials and to track their movement and disposition, registration and licensing procedures are in place in the United States and other countries as well.

In the United States, control of radioactive sources is under the jurisdiction of the Nuclear Regulatory Commission (NRC) [16,17]. For activities such as operation of a nuclear power plant or import/export of nuclear material, a specific license is required.

10.10.1 Nuclear Regulatory Commission

The Energy Reorganization Act in 1974 [2, p. 225] divided the Atomic Energy Commission (AEC) into the NRC and the Energy Research and Development Commission (ERDA), which was later to become the U.S. Department of Energy (DOE). Its headquarters is located in Rockville, MD, with four regional offices, one each in Pennsylvania, Georgia, Illinois, and Texas [16].

The NRC regulates commercial nuclear power reactors; use of nuclear materials; decommissioning of nuclear facilities; and storage, transportation, and disposal of nuclear materials and waste [18].

10.10.2 Regulations Governing Nuclear Material

Two important issues that arise in this area are the safe handling of radioactive sources and the disposal of waste products. The U.S. EPA does not have a large role in the regulation of radiation safety; however, there are a large number of federal, state, tribal, and local standards that pertain to radiation safety and the possession of sources of ionizing radiation [17, p. 57].

10.10.3 Waste Management Consists of Licensing

In the United States, the NRC is the primary agency involved in overseeing the handling of radioactive material. The NRC has regulations that govern the possession and use of radioactive material in three categories:

- Special nuclear material (SNM)
- Source material
- By-product material

definitions of which are provided below [17, pp. 57–58]

SNM consists of uranium-238, uranium-235, enriched uranium or plutonium, and any other material that the NRC determines to be SNM but does not include source material, as defined below. It means as well any material artificially enriched by any of the foregoing, but also does not include source material.

Source material means natural uranium or thorium (or any combination thereof in any physical or chemical form); depleted uranium that is not suitable for use as reactor fuel; and ores that contain 0.05% by weight or more of uranium, thorium, or any combination thereof. Source material does not include special nuclear material. Source material, if placed in a breeder reactor, can be turned into SNM, and this is why the source material is placed in a special class.

By-product material is defined in two separate ways. It is any radioactive material (except SNM) that is produced or made radioactive by a nuclear reactor. This includes fission products such as $_{38}Sr^{90}$, $_{53}I^{131}$, $_{55}Cs^{137}$, and numerous others. It also means material made radioactive by its exposure to neutron radiation emitted during the fission process. Some examples are $_{27}Co^{60}$, $_{25}Mn^{54}$, $_{26}Fe^{59}$, $_{30}Zn^{65}$, and many others. In addition, by-product material consists of the tailings and waste produced by extraction or concentration of uranium or thorium from an ore processed primarily for its source material content. Underground ore bodies depleted by these solution extraction operations do not constitute by-product material within this definition.

10.10.4 Licensing

Keeping track of nuclear materials and their safe handling is accomplished by means of a licensing procedure administered by the NRC or one of the so-called *agreement states* [17, p. 58; 19]. Approximately two-thirds of the states have entered into an agreement with NRC to take over most of the responsibility of licensing and regulating the use of by-product, source, and special nuclear material within their borders. NRC, however, retains the authority to license federal facilities located in agreement states.

Several types of radioactive material and sources of ionizing radiation fall outside the definitions above and are therefore not regulated by NRC, for example:

- Naturally occurring radioactive material (NORM),
- Accelerator-produced radioactive material,
- Radiation-producing machines, such as diagnostic and therapeutic X-ray machines, accelerators, industrial X-ray machines, scanning electron microscopes, ion implanters, and the like.

These items may need to be licensed or registered with the state in which they are located and to follow the state regulations that apply. Some states also have laser regulations.

10.10.5 Requirements

The rules of NRC for licensing, inspection, and radiation protection practices are contained in Title 10 of the Code of Federal Regulations. Standards for radioactive material transfer and disposal are also contained therein [17, p. 59, 19]. At facilities required to have a specific license from NRC, NRC rules supersede Occupational Safety and Health Administration (OSHA) rules on radiation protection. At facilities not required to have a specific license from NRC, Subpart Z of 29 CFR 1910.1096, the OSHA standard on ionizing radiation, and Subpart G of 29 CFR 1910.97, for nonionizing radiation, apply. Nevertheless, specific details concerning worker protection and occupational health and safety are beyond the scope of the present discussion of environmental issues.

10.10.6 General Licenses and Generally Licensed Devices

Organizations seriously involved in such activities as operation of a nuclear power plant or import, export, or sale of nuclear power plant fuel will require a specific license for those activities. Others, the so-called *general licensees*, whose use of nuclear materials in plant or laboratory is more incidental, may be covered by a *general license*.

It is believed that most readers of this text in industry will be more likely to be involved with a general license as part of their duties than with a specific license. Therefore, general licenses, general licensees, and generally licensed materials will be discussed here briefly. As with the details of radiation safety, discussion of specific licenses is beyond the scope of this text.

According to 10 CFR 31.5, a general licensee is a person or organization that acquires, uses, or possesses a generally licensed device (GLD) and has received the device through an authorized transfer by the device manufacturer/distributor, or by change in company ownership where the device remains in use at a particular location [20].

GLDs are devices containing radioactive material and are typically used to detect, measure, gauge, or control the thickness, density, level, or chemical composition of various items. Examples of such devices are gas chromatographs (detector cells), density gauges, fill-level gauges, and static elimination devices. Nickel-63, used in gas chromatographs, and polonium-210, used in static eliminators, are employed as sealed radioactive sources in these devices [16, p. 57]. NRC decided to register and track generally licensed devices because it wanted to increase control of and accountability for GLDs and to prevent them from becoming "orphaned" [21,22], that is, lost, abandoned, or whose disposition is unable to be supported financially. Refer again to the sad tale of the small sealed radioactive source in Brazil (Section 10.9).

10.10.7 Further Comments on General Licenses

A general license is a *de facto*, automatic authorization to acquire, use, and possess a generally licensed device [19]. General licensees do not initiate the registration of their devices. Instead, the registration process is initiated by the manufacturer upon transfer of the device to the licensee.

Requirements on the licensee are summarized below. The exact wordings in the applicable sections of Title 10 of the Code of Federal Regulations should be consulted to ensure compliance in an actual situation.

- Conduct routine maintenance (performing tests every 6 months, or as otherwise indicated on the label, and maintaining records of tests for 3 years).
- If the device becomes damaged or fails a test, suspend operation at once, have the device repaired or properly disposed of (see below), and report to the NRC within 30 days a brief description of the event, remedial action taken, and a plan for ensuring that the premises and environs are acceptable for unrestricted use if contamination is greater than 0.005 μCi (185 Bq).
- Report transfer or disposal of the device. A GLD can *only* be transferred (for replacement or disposal) to a person holding a specific license, such as the device manufacturer or licensed waste broker. In the case of the change of ownership where a GLD remains in use at a particular location, the new owner will be the new licensee.
- Maintain accountability and control and *not* abandon the device.
- If devices are lost or stolen, provide a report describing
  -The radioactive material
  -The circumstances under which the loss or theft occurred
  -The disposition of the radioactive material
  -Radiation exposure to individuals
  -Actions taken to recover the material
  -Actions taken to prevent recurrence
- Pay the annual fee to the NRC or to the agreement state, as appropriate.

General licensees are subject to registration if they possess at least one device containing one of the isotopes shown in Table 10.8 at or above the activity shown.

Table 10.8 Radioisotopes and Activity Levels Triggering Registration Under a General License

Isotope	Activity	
	mCi	MBq
Americium-241	1	37
Cesium-137	10	370
Cobalt-60	1	37
Curium-244	1	37
Strontium-90	0.1	3.7
Plutonium-238	1	37
Plutonium-239	1	37
Californium-252	1	37

10.10.8 Periodic Testing for Contamination—Wipe Tests

A standard method for determining radioactive contamination in a laboratory where nuclear materials and/or devices are in use is the so-called *wipe test*. This test consists in rubbing filter paper, cloth, or a swab with moderate finger pressure over work surfaces, floor areas, fume hoods, equipment, storage compartments, handles, door knobs, sinks, waste containers, and so on to assess contamination. A layout of the area should be prepared. The wipe test material may be dry or wetted by water or another solvent, as appropriate to the particular test procedure called for. The test must be performed by an individual deemed qualified to do so. A variation of the wipe-test procedure has been used to detect chemical contamination/fallout from environmental pollutants.

In the procedure, a square area $10\,cm \times 10\,cm$ is wiped where possible. The wiping material is then placed in a vial to be assessed with a radiation counter capable of determining the presence of $0.005\,\mu Ci$ ($185\,Bq$). The background level and the radiation emitted by one or more standards (typical of what is used in that laboratory), which function as (a) control(s), are also evaluated. The net radioactive contamination count is the difference between that of the wipe sample and the background. Contamination is considered to be present if the net radiation is greater than or equal to $0.005\,\mu Ci$ ($185\,Bq$), and leaking sources must be removed from service immediately and decontaminated, repaired, replaced, or disposed of properly.

Wipe tests are often conducted as a matter of routine on a weekly, monthly, quarterly, or semiannual basis. Tests should also be performed before resuming operations after maintenance or any change in operation or procedure or modification of radiation-producing equipment. For a sealed source, testing is required every 6 months, with records to be maintained for inspection for a period of 3 years, in accordance with 10 CFR 39.35. A *sealed source* is one in which encapsulated radioactive material is contained, protected from damage, and prevented from dispersal. It is typically part of a larger device such as a gas chromatograph or a radiotherapy unit.

10.11 RADIOACTIVE WASTE DISPOSAL

In general, radioactive (nuclear) waste is a material that is no longer useful and contains radioactivity of various types [2, p. 224]. The waste generated in the application of nuclear materials must be handled safely and given sufficient time to allow its activity to decay to an acceptable level. There are basically two options [1, pp. 100–112].

- Dilute and disperse
- Concentrate and contain

The first of these has been used to address slightly contaminated liquids and gases arising from spent nuclear reactor fuel, as at spent fuel processing facilities and at the storage ponds of nuclear power plants. This method also pertains to small radioactive sources used in industrial plants, laboratories of various kinds, and medical facilities. This technique affords protection by dilution and dispersion of the higher activity levels at the source but may result in exposure to individuals via contamination of air, drinking, water, crops, foodstuffs, fish, and so on by the dispersed radioactive materials.

The second option provides protection by isolation in a sealed repository to separate the source of radioactivity from man, other species, and the environment at least until its activity reaches a "safe" level (see Problem 10.3). Containment requires secure facilities with adequate shielding (plus cooling for removal of generated heat, if necessary) and safeguards against adverse geological events and human intrusion, accidental or otherwise. Concentration of a radioactive waste by such methods as incineration, compaction, or dewatering may be necessary to minimize the amount of material requiring secure disposal in a nuclear waste containment facility.

Nuclear waste can be classified into NORM, low-level radioactive waste (LLRW), and high-level radioactive waste (HLRW). An example of a NORM is phosphate rock, which is processed into phosphates for fertilizers and detergents [1, pp. 176–179]. Deposits of phosphate rock occur in Florida in the United States and elsewhere in the world [1, p. 176]. (See also NORM on p. 498.)

The phosphate rock contains uranium, and thorium to a lesser extent, and daughter products including radium-226 and its decay product radon-222. When the phosphate rock is processed, a waste by-product known as *phosphogypsum* is produced, containing most of the radium from the original ore. The radioactive phosphogypsum is discharged as a sludge to the aquatic environment, reprocessed in some way, or simply piled up near the plant or in mines [1, pp. 176–179]. The fertilizers, themselves containing some radioactivity, are spread over agricultural land [2, p. 230].

There are currently no federal regulations pertaining directly to wastes from NORM, and the volume generated is so large that disposal in an LLRW disposal facility is generally not feasible [2, p. 230].

LLRW is the broadest category of radioactive waste; it is defined as radioactive waste not classified as HLRW, transuranic waste (TRU, wastes containing elements heavier than uranium), spent nuclear fuel, or by-product material [2, p. 230]. This type of waste can be generated by any of a number of institutions—nuclear power plants, industry, government, academia, or medical in any process where radionuclides are employed.

Current regulations call for each individual state or interstate compact to dispose of all the LLW generated within its boundaries; an *interstate compact* is a group of states banding together, as approved by the U.S. Congress, to dispose of LLRW [2, p. 232]. There are also governmental regulations pertaining to collection, storage, packaging, handling, posting of warning signs limiting access, and transportation on the way to disposal.

HLRW encompasses spent nuclear fuel, liquid wastes from reprocessing irradiated reactor fuel, solid waste from the solidification of high-level liquid waste, and nuclear waste generated by the government in weapons facilities. Plans are being made in the United States to store HLRW in one central location [2, p. 231].

10.11.1 NRC Limits

Annual limits on intake (ALIs) and derived air concentrations (DACs) of radionuclides for occupational exposure, effluent concentrations, and concentrations for release to sewerage are contained in Tables 1, 2, and 3 of 10 CFR Appendix B to Part 20 [23]. These tables are too extensive to reproduce here but are available on the Internet. The rationale for the limits, listed individually for each radioisotope, is explained in the cited reference.

Tables 2 and 3 denoting maximum allowable effluent concentrations to air, water, and sewer are of particular interest. These values are set to limit the exposure to the public. The emission concentrations to air and water (Tables 2) are equivalent to the radionuclide concentrations that, if inhaled or ingested over the course of a year, would produce a total effective dose equivalent of 0.05 rem (50 millirems or 0.5 mSv). The values in Tables 3 are all maximum monthly average concentrations allowable for release to sanitary sewers, based on a committed effective dose equivalent of 0.5 rem assuming that the sewage released by the licensee were the only source ingested by the so-called *reference man*.

For a practical application of the wastewater/sewage concentrations, let us revisit the parameters of the radioactive source of the Brazilian incident (Section 10.9).

PROBLEM 10.9

Assuming that 1 g of radioactive $_{55}Cs^{137}$ with an activity level of 50.9 TBq was dissolved in wastewater, what is the maximum allowable concentration as cesium chloride (CsCl) allowable for discharge directly and via sanitary sewer? The applicable limits for cesium-137 are given in Table 10.9:

SOLUTION

(a) Convert activity in TBq to microcuries (Ci)

$$(50.9 \text{ TBq})(10^{12} \text{ Bq/TBq})(1 \text{ Ci}/3.7 \times 10^{10} \text{ Bq})(10^6 \text{ }\mu\text{Ci/Ci})$$
$$= (50.9/3.7) \times 10^8 = 13.76 \times 10^8 = 1.38 \times 10^9 \text{ }\mu\text{Ci}$$

(b) Calculate how much water is necessary to dissolve this amount of cesium-137 radioactivity to comply with the allowable basis for wastewater discharge.

Table 10.9 Applicable Limits for Cesium-137 from Tables 2 and 3 of 10 CFR Part 20, Appendix B

Effluent concentrations (Table 2)		Releases to sewers (Table 3)
Air	Water	Monthly average calculation
µCi/mL	µCi/mL	µCi/mL
2×10^{-10}	1×10^{-6}	1×10^{-5}

$(1.38 \times 10^9 /? \text{ mL}) = 1 \times 10^{-6} \text{ μCi/mL}$

$? = 1.38 \times 10^9 / 10^{-6} = (1.38 \times 10^{15} \text{ mL})(1 \text{ L}/1000 \text{ mL})$

$= 1.38 \times 10^{12} \text{ L}$

In English units $(1.38 \times 10^{12} \text{ L})(1 \text{ gal}/3.785 \text{ L}) = 3.65 \times 10^{11}$ gal $= 365$ billion gallons. To put this number in perspective, this is close to 5% of the volume of Lake Champlain in Upstate, New York, between New York State and Vermont and about 10% of Lake Seneca, the largest of New York State's Finger Lakes. It is also comparable to the entire volume of some smaller lakes.

(c) Calculate molecular weight of cesium-137 chloride

$_{55}\text{Cs}^{137}$ 137

Cl   35.453

Total 172.453 molecular weight of cesium-137 chloride assuming that the Cesium-137 is 100% pure

$1 \text{ g} = (1000 \text{ mg of }_{55}\text{Cs}^{137}) / (1.38 \times 10^{12} \text{ L})(172.453 \text{ CsCl}/137 \text{ Cs})$

$$\frac{(10 \times 10^2)}{138 \times 10^{12}} \frac{(172.453)}{137} = 9.12 \times 10^{-10} \text{ mg/L} \cong 10^{-9} \text{ ppm}$$

or 10^{-6} ppb of cesium-137 chloride corresponding to 1×10^{-6} μCi/mL.

Allowable limit for wastewater discharge as the chloride $= 9.12 \times 10^{-4} \cong 0.001$ ppb.

(d) For discharge into a sanitary sewer, volume of water necessary would be 10 times less since allowable concentration is 10 times higher than that for direct discharge. This volume is still a gigantic number and the chemical concentration still miniscule.

Bear in mind that not all sewage treatment plants will allow discharge of radioactive materials. Those that do will require that the federal standards be complied with. ∎

10.11.2 Radioactive Waste Disposal—Conclusion

Further details of the regulatory aspects of radioactive waste disposal constitute a specialized area and not considered here.

REFERENCES

1. J.R. Cooper, K. Randle, and R.S. Sokhi, *Radioactive Releases in the Environment: Impact and Assessment*, John Wiley & Sons, Ltd., Chichester, 2003, 473 pp.
2. L. Theodore and R.G. Kunz, *Nanotechnology: Environmental Implications and Solutions*, Wiley–Interscience, Hoboken, NJ, 2005, 378 pp.
3. J.C. Kotz and K.F. Purcell, *Chemistry and Chemical Reactivity*, 2nd edn, Saunders, Philadelphia, PA, 1990, 1260 pp.
4. D. Halliday, R. Resnick, and J. Merrill, *Fundamentals of Physics*, 3rd Extended, Vol. 2, John Wiley & Sons, Inc., New York, 1988, pp. 1149 and other pages.
5. J.P. Reynolds, J.S. Jeris, and L. Theodore, *Handbook of Chemical and Environmental Engineering Calculations*, Wiley–Interscience, New York, 2002, 948 pp.

6. R.E. Walpole and R.H. Myers, *Probability and Statistics for Engineers and Scientists*, 3rd edn, Macmillan Publishing Company, New York, 1985, 639 pp.
7. J.E. Freund and R.M. Smith, *Statistics: A First Course*, 4th edn, Prentice-Hall, Englewood Cliffs, NJ, 1986, 557 pp.
8. G. Choppin, J.-O. Liljenzin, and J. Rydberg, *Radiochemistry and Nuclear Chemistry*, 3rd edn, Chapter 4, Butterworth-Heinemann, 2002, p. 84. Can be viewed but not printed on the Internet at http://book.nc.chalmers.se/, last accessed on April 20, 2007.
9. Equations of Radioactive Decay and Growth, IUSS 2005 Lesson 3, www.unipv.it/oddone/oddone/IUSS%202005%203.doc, accessed on January 22, 2007.
10. O. Levenspiel, *Chemical Reaction Engineering: An Introduction to the Design of Chemical Reactors*, John Wiley & Sons, Inc., New York, 1962, 501 pp.
11. S.M. Walas, *Reaction Kinetics for Chemical Engineers*, McGraw-Hill, New York, 1959, 338 pp.
12. W.A. Granville, P.F. Smith, and W.R. Longley, *Elements of the Differential and Integral Calculus*, revised edition, Ginn and Company, Boston, 1941, pp. 87–88.
13. C.E. Love and E.D. Rainville, *Differential and Integral Calculus*, 5th edn, Macmillan, New York, 1954, pp. 159–160.
14. G.B. Thomas, Jr. and R.L. Finney, *Calculus and Analytic Geometry*, Addison-Wesley, Reading, MA, 1992, 424 pp.
15. H. Bateman, Solution of a system of differential equations occurring in the theory of radio-active transformations, *Proceedings of the Cambridge Philosophical Society*, **IS**, 423, 1910.
16. U.S. NRC, Our Locations, http://www.nrc.gov/who-we-are/locations.html.
17. United States Environmental Protection Agency, Environmental Management Guide for Small Laboratories, Section 3.6, Radioactive Materials, pp. 57–64, EPA 233-B-00-001, Washington, DC, May, 2000.
18. Wikipedia Article: Nuclear Regulatory Commission, http://en.wikipedia.org/wiki/Nuclear_Regulatory_Commission, accessed on February 18, 2007.
19. U.S. NRC General License Registration and Tracking, mhtml:file://A:\NRC%20General%20License%20Registration%20and%20Tracking.mht, accessed on February 4, 2007.
20. U.S. NRC Frequently Asked Questions About General Licenses, mhtml: file//A:\NRC%20Frequently%20Asked%20Questions%20(FAQ)%20About%20Ge..., accessed on January 5, 2007.
21. U.S. NRC, Orphan Sources, http://www.nrc.gov/materials/miau/miau-reg-initiatives/orphan.html, accessed on February 10, 2007.
22. U.S. EPA, Orphan' Radioactive Sources in Scrap Metal, Radtown USA, http://www.epa.gov/radtown/orphan-sources.htm, accessed on February 10, 2007.
23. U.S. NRC, Tables 1, 2, and 3 of 10CFR Appendix B to Part 20—Annual Limits on Intake (ALIs) and Derived Air Concentrations (DACs) of Radionuclides for Occupational Exposure; Effluent Concentrations; Concentrations for Release to Sewage, mhtml:file//A:\10%20CFR%20 Appendix%20B%20to%20Part%2020--Annual%20Limits..., accessed on February 4, 2007.

Appendix A

Suggested Undergraduate Environmental Curriculum

No good deed goes unpunished.

On the basis of extensive experience in industry in the environmental area, the author presents here a suggested list of courses in an undergraduate program (Table A.1) to prepare the student for environmental permitting work. This listing is based on mathematics, fundamental science, and chemical engineering process principles, in accordance with the author's own admitted prejudice for the need for such an approach. The author strongly believes that solutions based on firm science and engineering rather than on rhetoric and warm feelings are necessary for industry and others to "save the planet" from whatever ills are waiting to befall us. Such an approach is necessary to question the existing manufacturing processes, rather than relying strictly on end-of-pipe treatment of business as usual.

Filling out an environmental permit application form when one has access to data from an existing process is relatively easy. Putting credible numbers in the boxes is much more difficult for a new plant when one has to rely entirely on fundamental process principles to prepare such estimates. Training of "all-around players" requires a balance among various environmental media—air, (waste)water, solid/hazardous waste, and noise.

For solving problems in the chemical process industries, a background is needed in organic and physical chemistry as well as in the chemistry and biology of the earth, air, and water. Individual building blocks of mathematics, chemical engineering science and stoichiometry, plus specific pollution control courses are capped by an environmental unit operations course with a two-semester laboratory and a comprehensive design course.

Environmental Calculations: A Multimedia Approach, by Robert G. Kunz
Copyright © 2009 John Wiley & Sons, Inc.

Appendix A Suggested Undergraduate Environmental Curriculum

Table A.1 Suggested 4-Year Environmental Curriculum

Semester I	CR	Semester II	CR
Freshman Year			
Calculus I	3	Calculus II	3
Physics I	4	Physics II	4
Chemistry I	4	Chemistry II	4
Technical Writing[a]	3	Literature Elective[a]	3
Nontechnical Elective[b]	3	Computer Science	3
Orientation[c]	0	Orientation[c]	0
	17		17
Sophomore Year			
Calculus III	3	Differential Equations	3
Physics III	3	Physical Chemistry I	3
Chem Eng Process Calcs (Stoichiometry)	3	Statistics[d]	3
Intro Thermodynamics	3	Statics and Strength of Materials[e]	3
Electrical Systems	3	Fluid Mechanics[f]	3
Nontechnical Elective[b]	3	Environmental Law	3
	18		18
Junior Year			
Chem Eng Heat Transfer	3	Chem Eng Mass Transfer	3
Chem Eng Thermodynamics	3	Reaction Kinetics	3
Physical Chemistry II	3	Air Quality[g]	3
Physical Chemistry Lab	1	Organic Chemistry II	3
Organic Chem I	3	Organic Chemistry Lab	1
Environmental Chemistry and Lab[h]	3	Applied Microbiology[i]	3
Nontechnical Elective[b]	3	Nontechnical Elective[b]	3
	19		19
Senior Year			
Unit Operations in Environmental Engineering[j]	3	Environmental Process Design[k]	3
Chem Eng Lab I[l]	2	Chem Eng Lab II[l,m]	2
Water and Wastewater Treatment Processes[n]	3	Solid/Hazardous Waste Engineering	3
Surface Water Quality	3	Hydrology[o]	3
Industrial Air Pollution Control[p]	3	Noise Generation, Effects, Control, and Measurement[q]	3
Nontechnical Elective[b]	3	Nontechnical Elective[b]	3
	17		17

[a] Writing and literature courses carry English Department credit.

[b] Electives in Economics, Psychology, Sociology, Philosophy, Religious Studies, and Business Courses such as Accounting, Marketing, or Management may also be possible.

[c] An overview of environmental problems and the environmental engineering profession. Guest speakers and student presentations.

(continued)

Appendix A Suggested Undergraduate Environmental Curriculum 509

[d] A Mathematics Department course or an equivalent course in any other department. This course should come early enough in the program to allow use of the concepts in subsequent laboratory courses.

[e] Intended for students not in Civil/Structural Engineering.

[f] Includes pressurized laminar and turbulent flow through pipes, conduits, and process vessels; pump and compressor performance; and open-channel flow.

[g] This course must include a thorough discussion of meteorology, atmospheric (photo)chemistry, and an introduction to dispersion modeling, to be followed by a more in-depth dispersion modeling analysis in the air pollution course in the senior year.

[h] Otherwise known as Chemistry for Environmental Engineers.

[i] A study of bacteria, fungi/yeasts/molds, and viruses, as related to environmental control.

[j] This course should tie together all of the individual elements of fluid flow, heat transfer, mass transfer, reaction kinetics, and so on, as applied to pollution abatement processes for all media. Lecture to accompany laboratory courses, listed separately.

[k] Design and economic evaluation of environmental control processes using the knowledge and skills gained in previous courses.

[l] These courses should be set up with a separate section for environmental engineering students to emphasize experiments different from the classical chemical engineering process variety; for example, activated sludge wastewater treatment, adsorption, absorption/scrubbing, flue gas source testing, percolation/leaching, noise measurements. Includes planning and performance of experiments and preparation of laboratory reports.

[m] Substitution of Independent Project could be permitted.

[n] To include industrial wastewater in addition to sanitary wastewater.

[o] Topics to include quantitative treatment of hydrologic cycle and hydrograph, surface and subsurface hydrology, precipitation and atmospheric transport, and storm water collection.

[p] This course should include air pollution abatement processes plus a discussion of stack sampling and a completion of the dispersion modeling started in the Air Quality course of Junior Year, Second Semester.

[q] This course should include the fundamentals of sound, noise generation sources, control techniques, sound/noise level meters, and measurement procedures. It could possibly be taught by the Acoustics, Physics, and/or Mechanical Engineering Departments.

A basic course providing a background on legal principles is included as an introduction to the complex maze of environmental laws, regulations, and enforcement. This should bring out a fundamental difference between scientists and engineers vis-à-vis the legal profession: technical people attempt to solve a problem by assembling the parts to form the whole, a possibly unique solution; lawyers take the specific facts and fit them into the framework of legal documents, rules/regulations, and precedents—with all possible logical conclusions being potentially correct [1]. Their object is to win the case.

The proposed 4-year curriculum, detailed in Table A.1, is a rigorous one. There are only a few nontechnical electives, and the junior year is especially heavy. Graduates should be well rounded and well qualified to face the challenges of industry or the rigors of regulatory agency work. The full program should qualify one for at least a minor in chemical engineering. Institutions of higher education choosing to offer this program may have to make some case-specific modifications to qualify for approvals from various accreditation bodies. It is also possible to augment this curriculum and offer it at the MS level after 5 years of study. Individuals not pursuing the curriculum *in toto* can simply use it as a convenient menu to select courses to fill in topics not otherwise covered in related programs.

Further comments are provided as footnotes to the table.

REFERENCE

1. M.A. Tumeo, Editorial: need for law courses in environmental engineering curricula, *Journal of Environmental Engineering*, **123**(11), 1067–1068, 1997.

Appendix B

Relationship among Expressions for Atmospheric Contaminants as Concentrations (ppm), Mass Flow Rates (lb/h), and Emission Factors (lb/MMBtu)

We never make the same mistake three (IIII) times.

B.1 SUMMARY

Expressions for pollutant criteria were developed during the analysis, treatment, and evaluation of plant data, using oxides of nitrogen (NO_x) as an example [1]. It was observed that curves for dry ppm adjusted to 3% O_2 on a dry basis (ppmd @ 3% O_2, dry), lb/h, and lb/MMBtu of fuel fired differ by only fixed constants across the board when plotted against excess combustion oxygen (O_2). This has practical implications when environmental permit conditions specify emission limits in terms of two or more of these units. Such limits are at best redundant if calculated correctly; if not, the more/most restrictive limit applies.

Environmental Calculations: A Multimedia Approach, by Robert G. Kunz
Copyright © 2009 John Wiley & Sons, Inc.

B.2 CONCENTRATION LIMITS

Pollutant concentrations are measured in a flue gas at the operating conditions of an emission source. They can be expressed on a wet basis (including combustion moisture) or on a dry (water-free) basis. They are also expressed on a dry basis adjusted to some standard oxygen concentration in the flue gas. For combustion units—boilers, heaters, and furnaces—the standard is combustion with the amount of excess air necessary to provide a flue gas concentration of 3% O_2 on a dry basis. For gas turbines, it is 15% O_2 (dry).

The standard oxygen content was adopted by regulatory agencies to prevent a source from satisfying permit concentration limits simply by dilution with a convenient amount of excess air. The convention for combustion units traces back to coal- and oil-fired boilers, for which the typical 15% excess air gave rise to 3% O_2 on a dry basis in the flue gas. For turbines, the typical target for operation is the 15% O_2 (dry) figure.

$$\text{ppm X, dry (ppmd)} = \text{ppm X (wet)}[100/(100 - \% H_2O)] \tag{B.1}$$

where X represents the pollutant species.
Similarly,

$$\text{volume or mole\% } O_2 \text{ (dry)} = \text{volume or mole\% } O_2 \text{ (wet)}[100/(100 - \% H_2O)] \tag{B.2}$$

The change in basis is calculated for a given operating condition using the point value of % H_2O measured (or calculated) for that condition.

This value of % H_2O is a function of the excess O_2 for that condition, the moisture from combustion at 0% excess O_2 (the stoichiometric case), and the humidity (H) in the combustion air.

$$\frac{\% H_2O}{100}\bigg|_{\text{at conditions}} = \frac{\% H_2O @ 0\% O_2}{100} - \frac{\% O_2 \text{ (wet)}}{20.9}\left[(1+H)\frac{\% H_2O @ 0\% O_2}{100} - H\right] \tag{B.3}$$

or

$$\frac{\% H_2O}{100}\bigg|_{\text{at conditions}} = \frac{\dfrac{\% H_2O @ 0\% O_2}{100} - \dfrac{\% O_2 \text{ (wet)}}{20.9}\left[(1+H)\dfrac{\% H_2O @ 0\% O_2}{100} - H\right]}{\left\{1 - \dfrac{\% O_2 \text{ (dry)}}{20.9}\left[(1+H)\dfrac{\% H_2O @ 0\% O_2}{100} - H\right]\right\}} \tag{B.4}$$

where H is expressed in moles of water per mole of dry air. Flue gas moisture overall decreases with increasing excess O_2 because of dilution of water from combustion counterbalanced by the additional atmospheric moisture entering with the excess air.

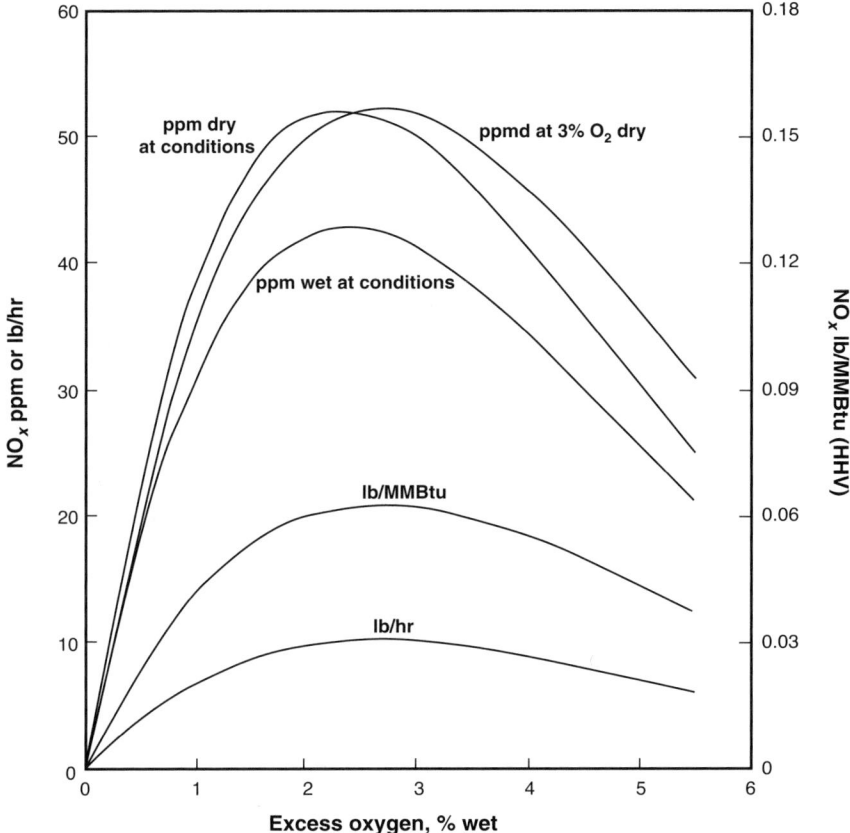

Figure B.1 NO_x concentrations, mass flow, and emission factor.

Here, combustion air is assumed to contain 20.9% O_2 on a dry basis, as in EPA method 19 [2]. In other cases, the dry % O_2 may be taken as 20.95 [3] or even as 20.947 (Table 3.3) when differing amounts of argon plus trace atmospheric constituents are considered.

Curves calculated for NO_x as a function of excess O_2 are shown in Figure B.1.

The central curve reproduces a fit of data in ppm wet at conditions for a natural gas fired furnace using author's the mathematical model based on kinetic theory.[1]

---

[1] The functional form of the NO_x model predicts NO_x to be zero at zero O_2 [4]. However, the function is based on the reaction kinetics for NO formation in an oxygen-rich environment. Experimental data show that NO_x begins to deviate from the "theoretical" curve below about 1.5% O_2 (wet) to intercept the y-axis at some finite positive value. The NO_x correlation is therefore not recommended for use below 1.5% O_2 although the curves in Figure B.1 are extended to the origin. (Even so, the predicted NO_x at 1.5% O_2 ought to provide a conservative (high) estimate for NO_x at lower values of flue gas O_2.) Further details and additional references can be found in Ref. [4].

The top left curve in this fit is ppm dry at conditions. Conversion from wet to dry is achieved by multiplying every point on the ppm wet curve by the factor from Equation (B.1), with % H_2O given by Equation (B.3) or (B.4).

Concentration in ppmd @ 3% O_2 is obtained by multiplying ppmd at conditions by the factor from Equation (B.5):

$$\text{ppm X @ 3\% } O_2 \text{ (dry)} = \text{ppmd X at conditions } [(20.9-3)/(20.9-\% O_2 \text{ (dry)})] \quad \text{(B.5)}$$

with O_2 (dry), as necessary, from Equation (B.2).

The expression for lb/h (for the bottom curve in the figure) is the concentration multiplied by the flow rate (in consistent units) times the molecular weight.

$$\text{lb/h X} = (\text{ppmd X @ 3\% } O_2 \text{ (dry)})(\text{mol/h of dry flue gas @ 3\% } O_2 \text{ (dry)})(\text{MW}) \quad \text{(B.6)}$$

$$\text{lb/h X} = (\text{ppm X @ 3\% } O_2 \text{ (dry)}) \left[(\text{mol/h of dry flue gas @ 0\% } O_2) \frac{(20.9-0)}{(20.9-3)} (\text{MW}) \right] \quad \text{(B.7)}$$

The functional relationship of flue gas flow rate (Q) with excess O_2 is

$$Q_{dry} = Q_{dry} \text{ @ 0\% } O_2 \text{ (dry)} \{(20.9-0)/(20.9-O_2 \text{ (dry)})\} \quad \text{(B.8)}$$

$$Q_{wet} = Q_{wet} \text{ @ 0\% } O_2 \text{ (wet)} \{(20.9-0)/(20.9-O_2 \text{ (wet)}(1+H))\} \quad \text{(B.9)}$$

Equations (B.8) and (B.9) are plotted in Chapter 3. A plot of Equation (B.9) results in a series of parametric curves, similar in shape, with parameter $(1 + H)$.

B.3 MASS FLOW RATE LIMITS

The lb/h curve assumes a constant firing rate, independent of excess combustion air and the resulting flue gas O_2. It ignores the need to burn slightly more (or less) fuel to maintain the same process temperature and throughput as excess air is increased (decreased) around a normal operating point. For the typical case, an increased (decreased) flue gas flow is expected to compensate for the dilution (concentration) of contaminant per unit volume of flue gas, and vice versa, keeping the product of concentration and flow more or less constant. A discussion specific to NO_x is contained in Ref. [1].

B.4 EMISSION FACTOR

The emission factor in lb/MMBtu is plotted second from the bottom in Figure B.1. It is obtained by dividing the lb/h curve by the firing rate in MMBtu/h based on the higher heating value (HHV).

$$(\text{lb/MMBtu}) = (\text{lb/h})/(\text{MMBtu/h}) \tag{B.10}$$

$(\text{lb/MMBtu}) = \text{ppmd @ 3\% O}_2 \text{ (dry)}$

$$\times \left[\frac{(\text{mol/h of dry flue gas @ 0\% O}_2)(20.9-0)(\text{MW})}{(\text{MMBtu/h})(20.9-3)} \right] \tag{B.11}$$

As discussed in Chapter 2, the gross or higher heating value is a convention adopted by many regulatory agencies, rather than the net or lower heating value (LHV) more representative of the actual heat realized; the LHV is used by burner manufacturers. When the HHV for a generic natural gas is taken as 1000 Btu/SCF, conversion between a natural gas emission factor in lb/MMSCF of fuel flow and the one based on firing rate in lb/MMBtu (HHV) of fuel fired is accomplished simply by moving the decimal point. This may explain the origin of the convention.

The calculated emission factor is identical whether a constant firing rate is assumed or the heat input is allowed to vary to maintain constant process conditions. This is consistent with the observation that flue gas volume divided by gross heat input is constant for a given fuel and % excess air [5].

Three curves in Figure B.1—ppmd @ 3% O_2 (dry), lb/h, and lb/MMBtu (HHV)—differ only by fixed constants. The bracketed terms [ ] in Equations (B.7) and (B.11) are the fixed constants representing the ratio of lb/h and lb/MMBtu (HHV) to ppm @ 3% O_2 (dry). Experimentally determined emission factors (lb/MMBtu), for example, plotted against ppmd @ 3% O_2 (dry) should scatter around a straight line passing through the origin.

Mol/h of dry flue gas at 0% O_2 is the stoichiometric flue gas, a constant dependent on the composition of a given fuel. The constancy of the MMBtu/h firing rate factor is addressed above.

The curves of Figure B.1 could just as easily be plotted against % excess O_2 dry or % excess O_2 wet, with these same constant ratios throughout.

B.5 CONCLUSION

As stated in Section B.1, the environmental permit emission limits specifying more than one of the units of ppmd @ 3% O_2 (dry), lb/h, and lb/MMBTU are redundant if correct; otherwise, the most restrictive permit limit applies. This is something to be considered either when issuing or when accepting a permit.

REFERENCES

1. R.G. Kunz, D.D. Smith, and E.M. Adamo, Predict NO_x from gas-fired furnaces, *Hydrocarbon Processing*, **75**(11) 65–79, 1996 (Appendix A).

2. U.S. Environmental Protection Agency, Standards of Performance for Non-Stationary Sources—Test Methods, 40 CFR 60, Appendix A, U.S. Government Printing Office, Washington, DC (various editions).
3. R. Kneile, Anatomy of combustion calculations, *Hydrocarbon Processing*, **74**(5), 87–96, 1995.
4. R.G. Kunz, B.R. Keck, and J.M. Repasky, Mitigate NO_x by steam injection, *Hydrocarbon Processing*, **77**(2), 79–84, 1998.
5. I. Frankel, Shortcut calculations for flue gas volume, *Chemical Engineering*, **90**(11), 88–89, 1981.

Appendix C

Burner NO_x from Ethylene Cracking Furnaces[1]

Serendipity is the art of making important discoveries by accident.

C.1 GENERAL

This appendix traces the development of and discusses the results from a correlation developed by the author for emissions of nitrogen oxides (NO_x) to be expected from ethylene cracking furnaces. The previous correlation for NO_x generated in steam-methane reformer (SMR) hydrogen (H_2) production on which it is based is already contained in published sources [1–3]. The ethylene correlation, presented at technical conferences and on record to date only in their proceedings [4,5], is reproduced here to make it more widely available to interested parties.

C.2 SUMMARY

Allowable emission limits for NO_x have become more stringent as the environmental regulatory agencies strive to improve ambient air quality for ground-level ozone. The ozone is formed in the atmosphere by the sunlight-induced reaction of NO_x with certain hydrocarbons. It is understandable that operators of affected combustion sources would prefer to achieve compliance with any such NO_x reductions by means of burner modifications, rather than through more costly and complex postcombustion controls, such as selective catalytic reduction (SCR).

---

[1] Adapted from material presented at the 2005 National Petrochemical & Refiners Association (NPRA) Environmental Conference, Dallas, TX, September 19–20, 2005 and the American Institute of Chemical Engineers (AIChE) 2007 Ethylene Producers' Conference, Houston, TX, April 22–26, 2007.

Environmental Calculations: A Multimedia Approach, by Robert G. Kunz
Copyright © 2009 John Wiley & Sons, Inc.

For smooth project execution, accurate prediction of the extent of burner NO_x reduction is critical. However, computational fluid dynamics (CFD), used to model flame patterns, furnace temperatures, and the like, has done a poor job in predicting burner NO_x. Likewise, burner testing in a manufacturer's pilot facility often produces low estimates for NO_x when compared to a full-scale furnace. A viable alternative is an empirical approach based on kinetic theory and validated by numerous field data.

This appendix discusses the results from a new correlation of NO_x emissions for ethylene cracking furnaces. It is derived from an established NO_x correlation for commercial SMR furnaces, while recognizing the differences in fuels and furnace conditions between the two processes. It uses adiabatic flame temperature (AFT), excess furnace oxygen (O_2), and furnace temperatures. Calculations can be accomplished rapidly and allow one to compute absolute values of NO_x, explore changes in NO_x from a base case, and explain experimental observations. Calculated values compare favorably with the available NO_x data reported for commercial ethylene furnaces spanning a wide range of conditions. The correlation can also be tailored to fit individual furnace data for even better agreement.

C.3 INTRODUCTION

It would indeed be nice to be able to estimate quickly the NO_x emissions from properly operating furnace burners under a variety of conditions. It would then be possible to evaluate changes in emission levels brought about by regulatory-driven improvements in burner technology.

Accurate prediction of NO_x emissions before implementation of burner modifications to comply with changing regulatory mandates will avoid unpleasant surprises in the field after start-up. The correlation whose development and use are charted step by step here provides just that capability, a tool to estimate NO_x and/or changes in NO_x in a wide variety of conditions. But first, the following caveat (the fine print) applies...

C.4 DISCLAIMER

("Some restrictions apply; batteries not included; your mileage may vary")

The information contained herein is offered in good faith but without guarantee, warranty, or representation of any kind (expressed or implied) as to its usefulness, correctness, completeness, or fitness for any particular purpose. The user assumes all risks for its implementation and should seek independent professional verification of its accuracy. The author assumes no responsibility and shall not be liable for any loss of profit or any special, incidental, consequential, or other damages that may result from the use of any of the information contained in this presentation, be it oral or written. Any statements concerning design, construction, operation, what constitutes regulatory compliance, and/or how to achieve such compliance should not be construed as recommendations on the part of the author and/or his organization.

C.5 REGULATORY CONSIDERATIONS

The majority of ethylene production in the United States is concentrated along a system of interconnected pipelines [6] on the Texas–Louisiana Gulf Coast [7] Canadian production is contained in the provinces of Alberta, Ontario, and Quebec [7].

A number of the U.S. ethylene plants are located in ozone nonattainment areas. In the Houston–Galveston–Brazoria (HGB) Nonattainment Area of Texas, especially, plants are faced with having to comply with increasingly more stringent emission control limits for NO_x. Along with certain hydrocarbons, NO_x is a critical ingredient in the formation of ground-level ozone. Hydrocarbons specifically identified as "bad actors" in ozone formation include ethylene, propylene, butenes, and butadiene [8], emissions of which are also being closely regulated.

Ethylene pyrolysis furnaces typically fire what are essentially the mixtures of hydrogen (H_2) in methane (CH_4) [9], the primary constituent of natural gas. Test results for low-NO_x burners range from 55 to 60 parts per million by volume (ppm) for natural gas and from 70 to 100 ppm for 20–70% H_2 in the fuel gas, presumably using ambient air for combustion; a typical range for ultralow-NO_x burners was quoted in 1993 as 25–40 ppm [10]. Historical NO_x emissions from conventional burners in ethylene plant furnaces are stated as 100–120 ppm [11] and 220–250 ppm for a "hydrogen-rich" fuel gas [10]. The 100–200 ppm range corresponds to 0.082–0.118 pounds of NO_x per million British thermal units (lb/MMBtu) for natural gas/methane, based on the gross or higher heating value (HHV) of the fuel fired. A more recent emission inventory for the HGB area indicates a range of 0.06–0.25 lb/MMBtu [12]. With certain exceptions, the current standard for pyrolysis reactors in the HGB area is 0.036 lb/MMBtu (HHV) [13].

The relationship between ppm NO_x (dry), corrected to 3% O_2 on a dry basis in the flue gas (ppmd @ 3% O_2, dry),[2] and lb/MMBtu (HHV) varies slightly with the composition of the fuel being burned. The NO_x standard of 0.036 lb/MMBtu (HHV) corresponds to a NO_x concentration of 30 ppmd @ 3% O_2 (dry) for natural gas/methane and roughly 44 ppmd @ 3% O_2 (dry) for pure hydrogen. Values for the standard are shown in Figure C.1 for the full range of methane–hydrogen mixtures. Other sets of corresponding values between ppmd @ 3% O_2 (dry) and lb/MMBtu (HHV) for a given gas composition can be obtained by proration. The general relationship between ppm, pounds per hour (lb/h), and lb/MMBtu has been discussed previously [2]. (See also Appendix B.)

Additional NO_x emission standards and the ground-level ozone standards in effect in other regulatory jurisdictions at the time of the first conference presentation are discussed there in greater detail [4]. It is still fair to say that the situation will continue to evolve [5].

---

[2]The term ppmd @ 3% O_2 (dry) = ppmd NO_x (dry basis) adjusted from its concentration as measured at a given flue gas % O_2 to a standard reference value of 3% O_2 (dry). Mathematically, ppmd @ % O_2 (dry) = ppmd × [(20.9 − 3)/(20.9 − % O_2 (dry))]; ppm on a dry basis is related to ppm on a wet basis as follows: ppmd = ppm (wet)([100/(100 − % H_2O)].

520 Appendix C Burner NO$_x$ from Ethylene Cracking Furnaces

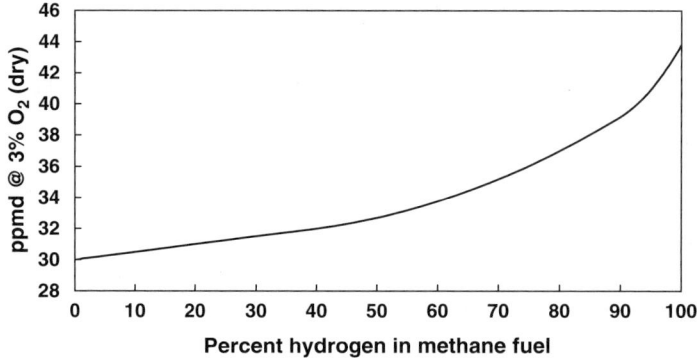

Figure C.1 NO$_x$ Concentration equivalent to 0.036 lb/MMBtu (HHV).

C.6 TECHNICAL CONSIDERATIONS

Driven by the plant operator's desire to achieve NO$_x$ compliance in the most cost-effective manner, burner developments have focused on making the latest regulatory NO$_x$ targets within reach using burners. However, lower NO$_x$ is often accompanied by flame instability—rollover and impingement on the furnace tubes [14–16]—unless and until corrected by redesign. Part of the (re)design effort for ethylene furnace burners has involved CFD to study burner flame patterns and temperature profiles [14–17]. CFD has been employed for SMR furnaces as well [18].

Although CFD has proven to be a useful tool in the cited studies, it is unable to predict burner and furnace NO$_x$ emissions [14–17]. This has prompted the present investigation to develop a NO$_x$ prediction technique for ethylene furnaces as a complement to CFD. The resulting model is an extension of a successful correlation for SMR furnace burner NO$_x$, whose functional form is derived from theory, with empirical constants based on the experimental data. This type of approach is a proven technique for the estimation of physical property data [19]. The new correlation also turns out to be effective in its own right, independent of CFD, for a variety of situations.

C.7 NO$_X$ CORRELATION FOR SMR FURNACE BURNERS

The original correlation was first presented at the 1992 NPRA Annual Meeting [20] and has been elaborated upon in a series of subsequent presentations and articles as more data have become available [1–3,21–24]. Its generalized form [$\ln(NO_x/O_2) = A - B \times 10{,}000/\text{AFT}$ (°R)] is shown in Figure C.2, with constants specified in the figure for conventional and low-NO$_x$ burners. NO$_x$ at the furnace is in parts per million by volume; furnace O$_2$ is in vol%. Concentration units for NO$_x$ and O$_2$ at furnace flue gas conditions must be consistent, either both wet or both dry.

The existing generalized correlation contains lines only for conventional burners and low-NO$_x$ burners. A patient and careful count shows that these relationships are

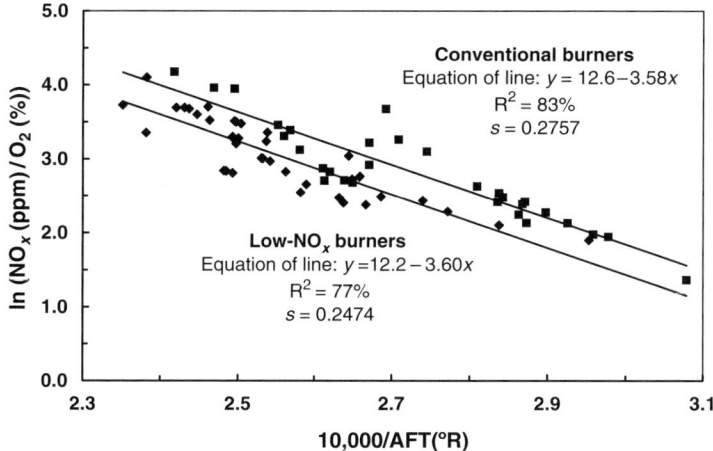

Figure C.2 SMR NO_x correlation.

derived from regression of seventy data points. The two lines have come out virtually parallel, as one would hope, to avoid the anomaly of their crossing at some point along the x-axis.

It is largely empirical but with a functional form derived from Zeldovich kinetics [25] for the formation of nitric oxide (NO) from oxygen (O_2) and nitrogen (N_2) in the combustion air (thermal NO_x). NO is the primary constituent (95%) of NO_x along with nitrogen dioxide (NO_2) (5%) in furnace combustion. It does not apply to prompt NO_x, formed by a more rapid free-radical mechanism, nor to fuel NO_x, arising from chemically bound nitrogen compounds in the fuel.

The variables identified were the temperature in the flame, where the thermal NO_x reaction takes place, and the excess O_2 concentration. The predicted dependence on the N_2 concentration turns out not to be statistically significant for combustion with atmospheric air. The correlation employs the AFT for combustion of the given fuel under the specified conditions plus the furnace firebox excess O_2, normally measured on a wet basis, as the independent variables. Firebox O_2 and dry basis O_2 measured at the stack during source testing are not the same if significant infiltration of tramp air occurs in between.

The theoretical AFT assuming complete combustion is the temperature attained when a fuel is burned, without mechanical work or gain or loss of heat, to the theoretical end products such as carbon dioxide (CO_2) and water vapor (H_2O), regardless of any equilibrium condition that might apply [26]; it is a function of the heating value of the fuel, the combustion products generated, and the inlet conditions [20]. Another research has shown the AFT to agree reasonably well with the actual peak temperature in a combustion flame [27–31]. AFT is a useful surrogate for that temperature and therefore serves as a good correlating variable. Similarly, O_2 concentration at the furnace exit is a surrogate for oxygen in the flame.

Use of an additional parameter, the so-called elusive third variable, such as the furnace firebox temperature at the bridgewall, to reduce the scatter has been discussed [20]. The rationale is that the calculated AFT more closely approaches the actual flame temperature, the higher the overall temperature being maintained in the surrounding furnace into which the flame radiates. For an individual furnace, one can also improve on the prediction from the generalized correlation [3].

C.8 EXTENSION OF THE CORRELATION TO ETHYLENE CRACKING FURNACES

The correlation of Figure C.2 can be modified to apply to ethylene furnace burners by considering the similarities and differences between those burners and their fuels compared to the burners and fuels in a steam-methane reformer.

A brief review of the two processes is in order.

C.8.1 Processes Are Different—A Review

C.8.1.1 SMR Process and Fuels

Steam-methane reforming, also known as steam reforming [32–37], catalytic steam reforming [38], and methane reforming [39], is a commercial process for the simultaneous production of hydrogen and carbon monoxide (CO). In this continuous process, steam and desulfurized hydrocarbon feed react at elevated temperatures over a solid nickel-based catalyst [33,34,39,40], which is contained inside tubes suspended in a furnace [41–42]. To maximize the H_2 yield for a hydrogen plant, additional steam is provided in one or more shift-converter vessels downstream outside the furnace. Feed is usually natural gas but can also be refinery gas, propane, liquefied petroleum gas (LPG), butane, or straight-run naphtha [33,35,36].

The H_2 product is separated from the resulting synthesis gas (syngas), a generic term for mixtures of H_2, CO, and CO_2. In a hydrogen–carbon monoxide plant, where CO is a desired product (along with a hydrogen coproduct), the shift conversion step is omitted. The product CO is recovered by low-pressure distillation at cryogenic temperatures following CO_2 removal by regenerative amine absorption and a drying step [2].

In either case, the hydrogen is separated from the syngas in a pressure-swing-adsorption (PSA) unit capable of producing a hydrogen purity of 99.9–99.999%. Other components in the syngas end up in the PSA purge gas, resulting from the periodic regeneration of the PSA unit [23,24]. Hydrogen plant PSA purge gas contains unrecovered H_2 plus the nonhydrogen constituents in the syngas—unreacted methane (CH_4) excess steam (H_2O), CO, CO_2, and impurities such as nitrogen (N_2) from the feed [20,23,24]. Combustible components include H_2, CO, and unreacted CH_4, the so-called methane slip [41,42].

The internally generated PSA purge gas is recycled as fuel to the reformer furnace burners. This purge gas can provide 50% [20] up to 90% [20,33] of the fuel requirement for the furnace. The low-Btu hydrogen plant PSA purge gas containing some 40% CO_2 is supplemented by an auxiliary, or trim, fuel to make up the firing requirement, typically natural gas or refinery fuel gas [23,24]. A typical purge gas/ natural gas composition can be found in previous documentation [1,2,20]. In contrast, the PSA purge gas from a hydrogen–carbon monoxide plant contains about 90% H_2 and 10% CO without the high CO_2 concentration usually found in hydrogen plant PSA purge gas [2]. Older hydrogen plants, built before the mid-1970s, typically used amine absorption, carbon dioxide removal, and methanation to make a final product of lower hydrogen purity [23,24]. This design also produces high purity CO_2, which can be liquefied for sale or used in further processing [20,43,44]. These older SMR plants without a PSA are fired solely on external fuels.

C.8.1.2 Ethylene Cracking Process and Fuels

Thermal cracking (without a catalyst) of hydrocarbon feedstocks in the presence of steam is the primary commercial route to ethylene and its coproducts [11]. In the continuous thermal cracking process for ethylene production, hydrocarbon feed is mixed with steam and reacted inside tubes known as radiant coils suspended in a furnace. The ethylene cracking unit is also referred to as a steam cracker, ethylene cracker, thermal cracker, or pyrolysis furnace [6,11,45,46]. Hydrocarbons ranging from ethane, propane, butane, and LPG through naphtha, kerosene, and gas oil can be used as feed [9].

Following thermal cracking to the desired conversion, the cracked gas exiting the furnace coils is rapidly quenched and separated downstream into its constituents. Many by-products or coproducts are generated in addition to ethylene, with greater amounts formed from the heavier feeds, and the distribution of products is strongly influenced by operating conditions. Components in the cracked gas include ethylene, propylene, butadienes, butanes, butenes, higher olefins, and aromatics as useful products; hydrogen recovered as a product or used as fuel; methane used as fuel; and coke, which lies down on the inside surface of the radiant coils and interferes with the operation.

To remove the unwanted by-product coke from inside the radiant coils, elements of the process train must be taken out of service periodically and "decoked" using a mixture of steam and air. The decoking off gas, containing CO, CO_2, and carbon particles [11], is either diverted to quench water and a knockout pot [6,11,47] and then to atmosphere or is combusted in a firebox [11]. This might be the ethylene furnace itself [14,47] or some external heater with its own separate stack.

The steam-cracking process generates an impure hydrogen-rich gas, which may be purified or upgraded for chemical uses or used as fuel [9], plus an impure methane stream known as methane-rich gas [9], or the pyrolysis methane fraction [6] that is used as fuel [6]. The hydrogen-rich fuel is only 85% to a maximum of 95% pure [9,11]. The methane fuel gas stream consists of 95% methane with some minor impurities of hydrogen, carbon monoxide, and traces of ethylene [11]. By-product ethane and propane may also be contained in the fuel [6].

In addition, a so-called pyrolysis fuel oil product is obtained from cracking heavier feedstocks [9,45]. A lighter cut known as pyrolysis gasoline [9,45], or pygas [45], a gasoline-like liquid high in unsaturated compounds and rich in aromatics

Table C.1 Fuels Used in SMR Furnaces and Ethylene Plant Cracking Furnaces

SMR	Ethylene plant
Generated in the process:	
PSA purge gas	Hydrogen-rich fuel gas
A portion of the feed gas	Methane-rich gas (pyrolysis methane fraction)
	By-product ethane
	By-product propane
	Pyrolysis fuel oil
From external sources:	
Natural gas	Natural gas
Refinery fuel gas	Refinery fuel gas
Distillate fuel oil	Distillate fuel oil

Source: Compiled from Refs [6,9,11,35,39].

(benzene, toluene, xylenes (BTX)) [48], is used as a chemical feedstock. An ethylene cracking furnace can also be supplied or supplemented with other fuels available in a petroleum refinery or petrochemical complex.

Fuels for both processes are summarized in Table C.1.

C.8.2 But Combustion Is Similar

Although the processes are different, the combustion sides are quite similar. Both processes

- Employ a furnace to supply the heat of reaction.
- Use many small burners to deliver the required heat as uniformly as possible [49].
- Must be capable of burning significant concentrations of hydrogen in the fuel mixture.
- Must be designed to prevent flame rollover and impingement on the tubes/coils.

C.8.3 Firebox Temperatures Are Different

It is no secret that ethylene furnaces tend to run hotter than SMR furnaces, and this difference must be accounted for in predicting NO_x. In a reformer furnace, flue gas temperature exiting the radiant firebox, referred to as the bridgewall temperature [39], is 1800–1900°F (~980–1040°C) [33,35]; firebox temperature in an ethylene furnace

C.8 Extension of the Correlation to Ethylene Cracking Furnaces 525

Table C.2 Approximate Temperatures in Process Furnaces

Process	Furnace	Tube metal
SMR	1800–1900°F	1600–1925°F
Ethylene	1830–2200°F	1750–2100°F

is typically 1000–1200°C (~1830–2200°F) [6] (Table C.2) [50]. These figures average about 1850°F (~1010°C) and 2000°F (~1100°C), respectively. Tube metal temperatures, drawn from multiple sources [50] and shown for information only, may not be completely consistent with the furnace temperatures.

C.8.4 Adiabatic Flame Temperatures Are Both Different and Similar

Adiabatic flame temperatures for typical fuel gases used in SMR and ethylene cracking furnaces are examined below.

C.8.4.1 Adiabatic Flame Temperatures for Typical SMR Furnace Fuels

Adiabatic flame temperatures for fuels containing both hydrogen plant PSA purge gas and hydrogen–carbon monoxide plant PSA purge gas can be found in Refs [1,2,20]. The AFT for a natural gas at various combustion air temperatures is also addressed [20].

In summary, with hydrogen plant PSA purge gas (containing about 40% CO_2 in the mixture) supplemented by natural gas, the AFT is some ±400°F (222°C) below that for firing natural gas or pure methane alone. Because of this lower AFT, hydrogen plant furnaces have a natural tendency to produce lower NO_x. However, hydrogen–carbon monoxide plant purge gas (H_2 and CO without the high concentration of CO_2) mixed with a natural gas trim fuel produces AFTs 100–200°F (56–111°C) above that of the natural gas/methane reference.

C.8.4.2 Adiabatic Flame Temperatures for Typical Ethylene Furnace Fuels

Typical fuel gases combusted in an ethylene furnace are impure waste gases recovered in the process. These consist mainly of methane and hydrogen with some minor constituents. However, to keep things simple in the calculations that follow, the fuels are assumed to be various combinations of pure methane and pure hydrogen. Furthermore, "natural gas," also a common fuel brought in to fire the furnace, is assumed to be 100% methane.

AFTs for combustion of methane, hydrogen, and various mixtures thereof are shown in Figure C.3. Its basis is fuel and air temperatures of 60°F (15.6°C) and 60%

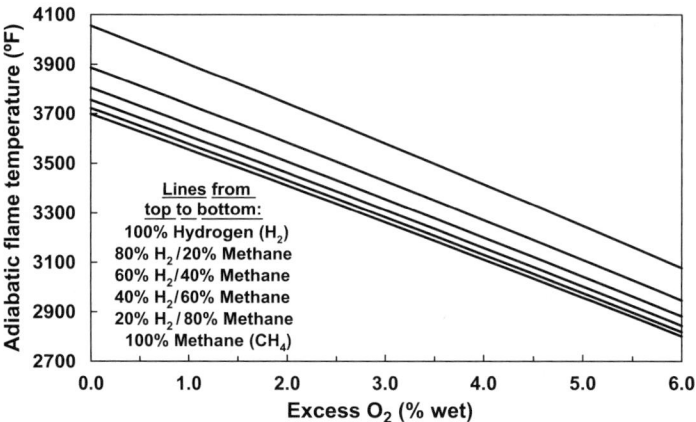

Figure C.3 Adiabatic flame temperatures for CH_4/H_2 mixtures, ambient combustion air.

relative humidity for the ambient air. (Unless otherwise stated, that same basis is assumed here throughout.)

As depicted in Figure C.3, AFTs are higher as more hydrogen is present in the fuel. The curves are linear in excess % O_2 (wet) up to about 5–6%, although a slightly better fit can be obtained through the addition of a quadratic term. Slope of the line is on the order of 150°F per % O_2 (83°C per % O_2) but increases slightly with increasing hydrogen. The line for methane is essentially the same as that calculated for a typical natural gas in a previous presentation [20]. AFTs regressed against % O_2 (dry) (not pictured here) also show a decent linear fit.

The AFTs calculated for the methane–hydrogen blends fired in an ethylene cracker (Figure C.3) are higher than that for methane alone. Their AFTs are comparable to those for SMR hydrogen–carbon monoxide plant PSA purge gas (mixtures of H_2 and CO) supplemented by natural gas [2] and fall within the range of applicability of the SMR NO_x correlation (Figure C.2).

AFT increases as fuel and air temperatures are raised. The effect is more dramatic for combustion air temperature (air preheat) because of the relative amounts of air and fuel taking part in the combustion process. AFTs for combustion of the fuel gases of Figure C.3 recalculated for the same starting ambient air and a combustion air temperature of 350°F (177°C) are shown elsewhere [4]. In round numbers, corresponding values between the cases depicted in Figure C.3 and the air preheat cases are about 200°F (a delta of 111°C) higher for preheated air, but the slopes of AFT versus % O_2 remain nearly the same.

C.9 NO_x CORRELATING EQUATIONS FOR ETHYLENE FURNACES

The proposed correlating equations for NO_x from ethylene cracking furnaces derived from the SMR correlation of Figure C.2 are shown in Figure C.4, with an additional

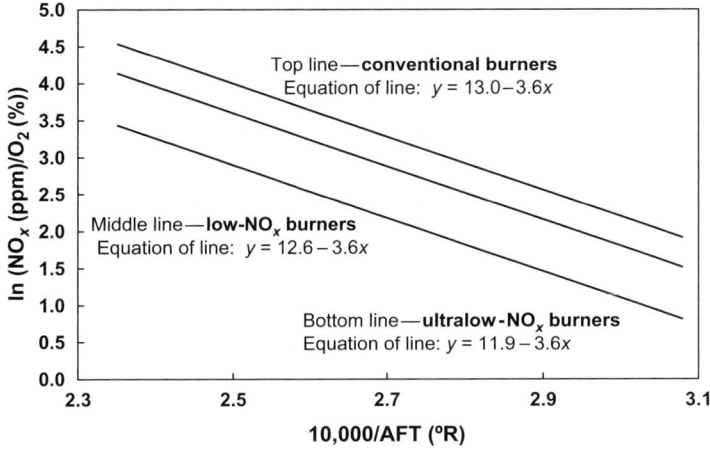

Figure C.4 Proposed NO_x correlation for ethylene furnaces.

line estimated for ultralow-NO_x burners [4].[3] The relationship to AFT is assumed to be the same. (The coefficient B, slope of NO_x versus 10,000/AFT (R), equals 3.6 to two significant figures.) The constant A is taken as 13.0, 12.6, and 11.9, respectively, for conventional, low-NO_x, and ultralow-NO_x burners to account for a difference in furnace temperatures between the two processes (Table C.2). It turns out that about 1.5 times as much NO_x would be generated from the "average" or "typical" ethylene furnace compared to the "average" or "typical" SMR furnace, caused by a difference in furnace temperature alone, everything else being equal. It is interesting to note that the low-NO_x burner equation for ethylene furnaces (Figure C.4) is nearly the same as the conventional burner equation for SMRs (Figure C.2). The decoking step in the ethylene process is not modeled in this correlation.

C.10 INFLUENCE OF THE VARIABLES

C.10.1 Oxygen Dependence

Oxygen dependence in the NO_x correlation is both direct through the $\ln[NO_x/O_2]$ term and indirect through the AFT (Figure C.3). The $\ln[NO_x/O_2]$ relationships from Figure C.2 or C.4

$$[\ln(NO_x/O_2) = A - B \times 10,000/AFT\ (^\circ R)] \qquad (C.1)$$

can be plotted on linear coordinates as NO_x versus O_2 for combustion of a given fuel at a specified set of fuel, air, and furnace conditions. There then results a curve passing

---

[3] *A cautionary note:* The corresponding SMR ultralow-NO_x equation is $[\ln(NO_x/O_2) = 11.5 - 3.6 \times 10,000/AFT\ (^\circ R)]$. The user is strongly cautioned, however, that this proposed correlation for SMR ultralow-NO_x burners is relatively untested compared to the other two lines in Figure C.2. The low values that it predicts, especially at the upper end of the reciprocal AFT range, may not be attainable in practice.

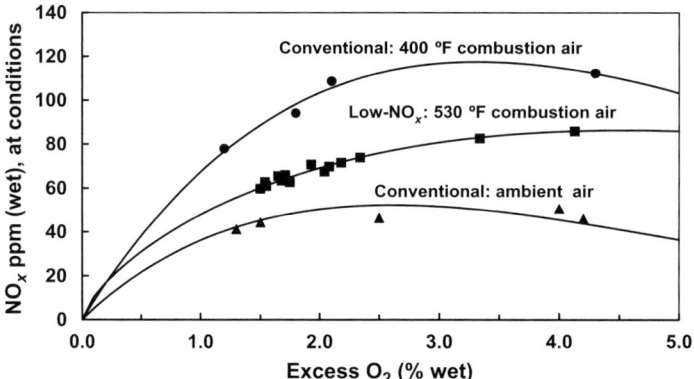

Figure C.5 NO_x from natural gas fired SMR furnaces, burners and air preheat as noted.

through a maximum point, where the competing influences of excess O_2 and AFT are in balance [20].

Some previously published NO_x data points [1,3,20,22] and customized correlating curves for three commercial SMR furnaces firing natural gas are plotted against furnace excess O_2 in Figure C.5.

NO_x concentrations in the figure are expressed as ppm wet at conditions, as reported in the original publications. Concentrations in ppm at conditions on a dry basis would be about 20–25% higher [1,2], and adjustment of ppm (dry) to a flue gas concentration of 3% O_2 (dry) (a common regulatory standard) would shift the curves somewhat [2]. As discussed previously [3,20], NO_x values below about 1.5% O_2 (wet) may not be reliable since the actual intercept is expected to be positive, rather than zero [51], and NO_x in this region may be somewhat higher than that predicted from the curve as drawn. Use of the NO_x correlation below this value of furnace O_2 is therefore not recommended [3].

C.10.1.1 Location of the Maximum Point

Through differential calculus, it is possible to calculate the value of furnace oxygen corresponding to the peak of the ppm NO_x wet at conditions versus % O_2 (wet) curve, such as those depicted in Figure C.5. For the more general case of

$$\ln[NO_x] = A - B \times 10,000/\text{AFT}(^\circ R)] + C \ln[O_2] \quad (C.2)$$

with an additional coefficient of the $\ln[O_2]$ term, the O_2 at the maximum point is given as

$$O_{2 \text{ at the peak}} = \{[(B \times 10,000/C) + 2b] - \text{SQRT}[[(B \times 10,000/C) + 2b]^2 - 4b^2]\}/(2a) \quad (C.3)$$

where a is the absolute value of the linear slope of AFT versus O_2 (°F or °R per % O_2), b is the AFT (°R) at 0% O_2 (linear intercept) (*Note:* temperature in °R), C is an additional

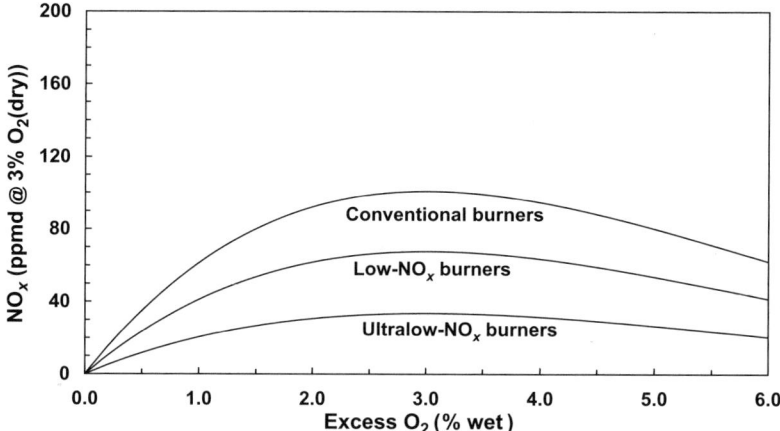

Figure C.6 Predicted ethylene furnace NO_x: 100% methane fuel and 60°F air.

constant that reduces to 1.0 for the standard NO_x equation, and SQRT is the square-root operator.

Both the slope *a* and intercept *b* are obtained by regression of AFT against % O_2 (wet), as plotted in Figure C.3, but in °R. Flue gas moisture (not shown) is also linear in % O_2 (wet). Equation (C.3) also works to find the peak for ppm NO_x dry at conditions versus % O_2 (dry) for the standard NO_x equation ($C = 1.0$). AFT must first be regressed against % O_2 (dry) to obtain a different slope and intercept.

NO_x in ppmd @ 3% O_2 (dry) versus % O_2 (wet) curves predicted for ethylene cracking furnaces from the equations of Figure C.4 are shown in Figures C.6–C.9 for several representative cases.

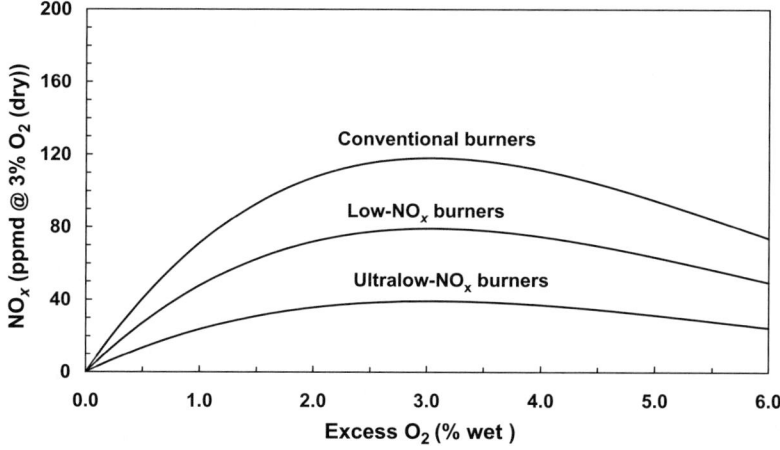

Figure C.7 Predicted ethylene furnace NO_x: 40% H_2/60% methane and 60°F air.

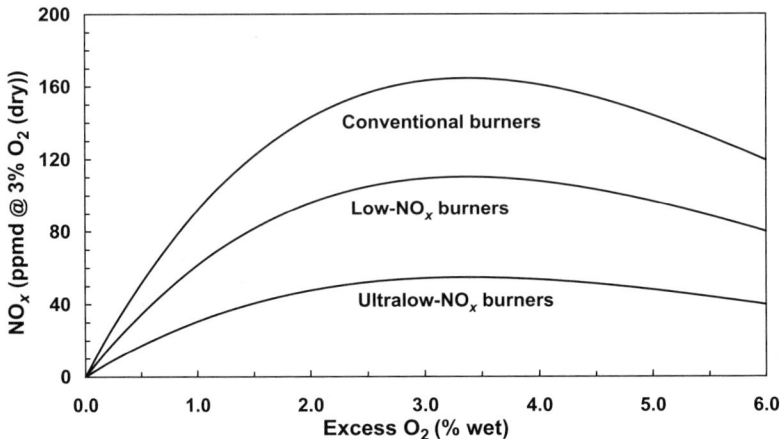

Figure C.8 Predicted ethylene furnace NO_x: 100% methane and 350°F air preheat.

Peak values of NO_x for these and other fuel compositions are listed in Table C.3. The peak occurs at 3.0–3.2% O_2 (wet) for ambient air and at 3.3–3.6% O_2 (wet) for air preheat. Oxygen at the peak increases with higher AFT as H_2 in the fuel increases (Figure C.3). The curves become steeper and the maximum point moves to the right with added air preheat [1].

Finding the peak value of the ppmd @ 3% O_2 curve by calculus requires differentiating a more complicated function than Equation (C.2) followed by a trial and error solution for O_2. It is more straightforward then to locate the maximum point from the function itself.

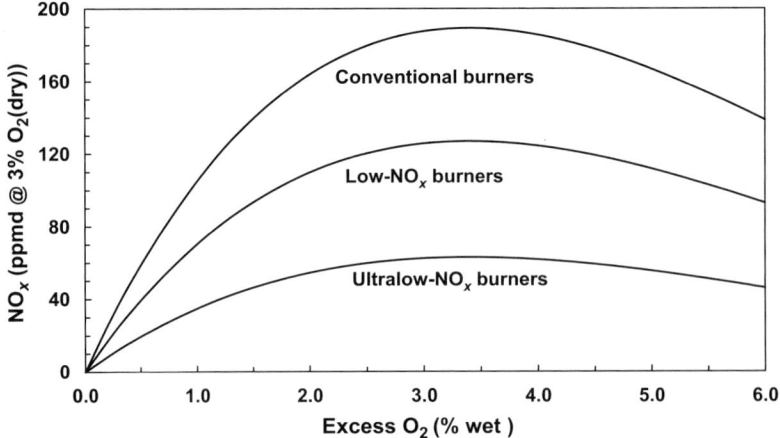

Figure C.9 Predicted ethylene furnace NO_x: 40% H_2/60% methane and 350°F air preheat.

Table C.3 Predicted Ethylene Furnace NO_x Peak Values, Burners and Air Preheat as Shown

	ppmd @ 3% O_2 (dry)			lb/MMBtu (HHV)		
% H_2 in fuel	Conventional	Low NO_x	Ultralow NO_x	Conventional	Low NO_x	Ultralow NO_x
Ambient						
0	101	68	34	0.121	0.081	0.040
10	104	70	35	0.123	0.083	0.041
20	108	72	36	0.126	0.084	0.042
30	112	75	37	0.129	0.087	0.043
40	118	79	39	0.134	0.090	0.044
50	126	84	42	0.139	0.093	0.046
60	136	91	45	0.146	0.098	0.049
70	149	100	50	0.155	0.104	0.052
80	170	114	57	0.168	0.113	0.056
90	205	137	68	0.188	0.126	0.063
100	270	181	90	0.222	0.149	0.074
350°F air						
0	165	110	55	0.197	0.132	0.066
10	169	113	56	0.200	0.134	0.067
20	174	117	58	0.204	0.137	0.068
30	181	121	60	0.209	0.140	0.069
40	189	127	63	0.214	0.144	0.071
50	200	134	67	0.221	0.148	0.074
60	214	144	71	0.230	0.154	0.077
70	234	157	78	0.243	0.163	0.081
80	263	176	88	0.260	0.174	0.086
90	310	208	103	0.286	0.191	0.095
100	397	267	133	0.328	0.220	0.109

C.10.2 Combustion Air Temperature

Another way to depict the effect of combustion air temperature on AFT and NO_x (Figures (C.6–C.9) and Table C.3) is to plot the relative concentration of NO_x formed for a given combustion air temperature to NO_x at some reference ambient air temperature, say 60°F (15.6°C). At otherwise constant conditions, the relative NO_x can be derived from the functional form of the NO_x correlation (Figures C.2 and C.4) to yield

$$NO_{x \text{ with air preheat}}/NO_{x \text{ ambient}} = \exp[B \times (10{,}000/\text{AFT}\,(°R)_{\text{ambient}} - 10{,}000/\text{AFT}\,(°R)_{\text{air preheat}})] \quad \text{(C.4)}$$

This relationship comes from writing the NO_x equation twice, once for each combustion air temperature chosen and subtracting. The explicit excess O_2 concentrations, being the same, drop out as does the constant term A, indicative of burner type. The only things left are the coefficient B and the AFTs for combustion of the fuel with the preheated air and with the reference air. The AFT for the reference condition is actually a constant in the equation.

According to the model, it follows that the relative NO_x from combustion of a particular fuel gas at a given temperature and ambient air conditions is a function of the combustion air temperature and the excess O_2. Relative NO_x depends on the combustion air temperature because the AFT is a function of that temperature. The AFT also provides an excess O_2 functionality (Figure C.3) even though the explicit excess O_2 term drops out in the derivation of Equation (C.4).

A graph of the above equation for combustion of three methane–hydrogen fuel compositions at 2% excess O_2 (wet) is shown in Figure C.10 for a coefficient B of 3.6, its value from Figure C.2 to two significant places. The curves are distinct from one another, but the differences are minor. When plotted on linear coordinates, the curves clearly display an exponential character.

Use of Equation (C.4) with experimental relative NO_x data for preheated combustion air may be another way to determine the coefficient B.

Curves of this type, plotted perhaps without numerical values on the axes, may be familiar from burner manufacturers' literature, for example, Ref. [51], and a rule of thumb quoted from industrial experience is that NO_x can be expected to double as combustion air is preheated from ambient to the range of 500°F (260°C) [51] to 600°F (316°C) [51,52]. Curves for methane containing up to about 75% hydrogen cross the relative $NO_x = 2.0$ line in Figure C.10 within this temperature range.

In addition, a similar set of curves (not shown) for an unspecified fuel and excess O_2 but with numerical values [52] is available showing relative NO_x ranging from 1.85 (min) to 2.3 (max) at 750°F (399°C) with respect to a base temperature of 100°F (37.8°C). Correction between the two temperature bases is about 6%. Relative NO_x from the published curves at 350°F (177°C) is close to 1.2 and between 1.3 (min) and 1.7 (max) at 500°F (260°C) and 600°F (316°C), respectively.

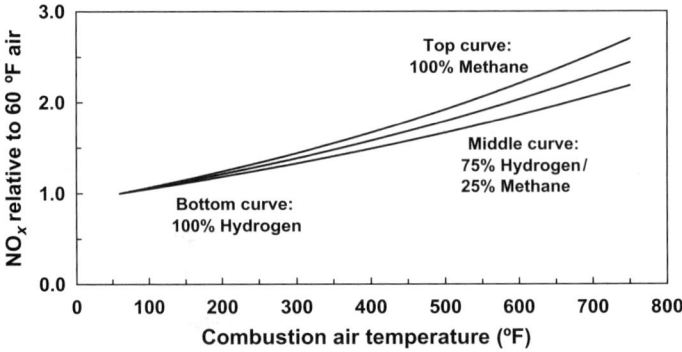

Figure C.10 Relative NO_x at 2% excess O_2 for preheated combustion air, fuels as noted.

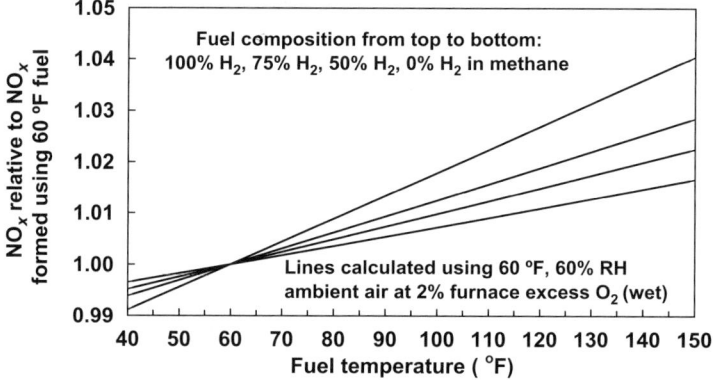

Figure C.11 Relative NO_x from heated fuel at 2% excess O_2 (wet).

Although agreement between those curves and Figure C.10 is not quite exact even on the same basis, Figure C.10 is certainly in the same ballpark while providing more conservative values of NO_x. If desired, the coefficient B can be adjusted to make the relative NO_x curves of the model coincide with one's own experimental data.

C.10.3 Fuel Temperature

Relative NO_x as a function of fuel temperature is plotted in Figure C.11, using 60°F (15.6°C) as the basis. A range of 40°F (4.4°C) to 150°F (65.6°C) is covered. Combustion air at 60°F (15.6°C) and 60% relative humidity (RH) is held constant. Excess O_2 in the furnace is 2% (wet).

Four curves are shown, each nearly linear, for 100% hydrogen fuel and 100% methane fuel at the extremes, 75% H_2 in methane midway between them, and 50% H_2 midway between 75% H_2 and 100% methane. The difference in the temperature range investigated is only 6% for hydrogen fuel (4% between 60°F (15.6°C) and 150°F (65.6°C)) and 2% for methane (1.5% from 60°F (15.6°C) to 150°F (65.6°C)).

Since fuel volume is only about 10% of the combustion air or flue gas at typical excess air conditions for methane/natural gas (about one-third for a pure hydrogen fuel), it is not surprising that the impact of fuel temperature on AFT, and therefore NO_x, is such a small fraction of the effect of combustion air temperature, certainly for fuels containing high concentrations of methane.

C.10.4 Ambient Air Humidity

Figure C.12 shows the effect of humidity in the combustion air intake flow at ambient air temperatures from 40 to 100°F (4.4–37.8°C). The relationship is depicted as a single, nearly linear curve with a negative slope since the miniscule differences in relative NO_x with ambient air temperature are imperceptible at the scale of the figure.

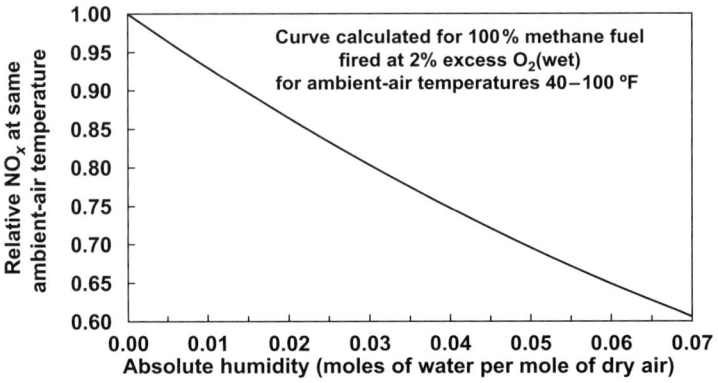

Figure C.12 NO_x relative to NO_x formed using zero-humidity ambient air.

For each temperature, the curve should be used only up to the saturation humidity at that temperature (Table C.4).

Even though the predicted increase in NO_x on days with lower humidity may be minimal compared to some other effects, it is still real. According to the figure, halving the humidity from 0.028 to 0.014 mol of water per mole of dry air (e.g., from 80% RH to 40% RH at 80°F (26.7°C)) results in a relative NO_x increase of 9%, or an absolute increase of about 8 ppm NO_x at 90 ppm, 5.5 ppm at 60 ppm, and 3 ppm at 30 ppm. Such an increase, in general, has been alluded to in testing of burner emissions [15]. This is especially important when ambient conditions cause the operation to approach NO_x permit limits more closely.

C.10.5 Hydrogen Content of Methane/Hydrogen Mixtures

NO_x relative to firing 100% methane for methane/hydrogen fuel blends is plotted in Figure C.13. Fuel–hydrogen content from 0% to 100% is investigated at 2% excess O_2 (wet) in the furnace flue gas. A fuel temperature of 60°F (15.6°C) and combustion air

Table C.4 Saturation Humidity at Indicated Ambient Temperature and Atmospheric Pressure

Temperature (°F/°C)	Absolute humidity (moles of water per mole of dry air)
40/4.4	0.008330
50/10.0	0.012248
60/15.6	0.017735
70/21.1	0.025328
80/26.7	0.035735
90/32.2	0.049894
100/37.8	0.069074

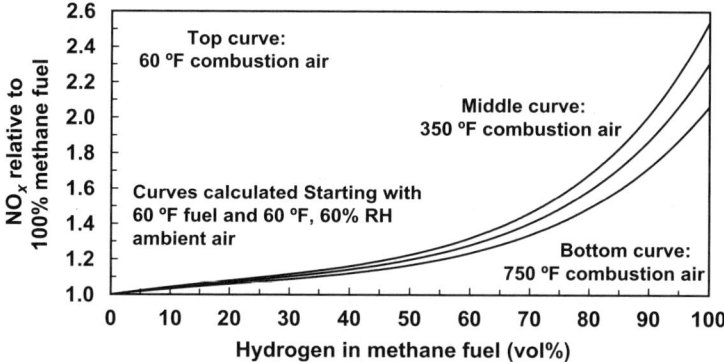

Figure C.13 NO_x with H_2 in fuel relative to firing 100% methane, 2% excess O_2 (wet).

temperatures of 60, 350, and 750°F (15.6, 177, and 399°C) starting with ambient air at 60°F (15.6°C) and 60% RH were employed in the calculations. Basis for comparison is NO_x concentration expressed as ppmd @ 3% O_2 (dry) for 100% methane.

Three distinct curves result, progressing in order from top to bottom with the 60°F (15.6°C) combustion air curve on top. Like Figure C.10, these curves are exponential in character. Relative NO_x increases with increasing hydrogen, gradually at first, but then the curves begin to take off in the range of 50–60% H_2.

Different curves with a lower relative NO_x and a more gradual rise but with a similar takeoff point are obtained when comparing NO_x emissions in lb/MMBtu (HHV) rather than in ppmd @ 3% O_2 (dry). This occurs because the ratio of ppm to lb/MMBtu changes with fuel hydrogen, as shown in Figure C.1. The lb/MMBtu (HHV) relative NO_x curve for 60°F (15.6°C) combustion air is shown as the lowest curve in Figure C.14, along with several others. The shape of this curve has been discussed previously [4], including the experimental observation of no measurable impact of H_2 on NO_x (as reported, tested up to about 50–60 vol% H_2) [17].

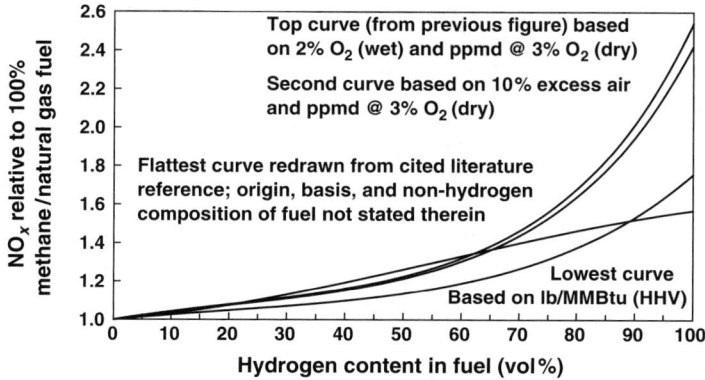

Figure C.14 Relative NO_x from hydrogen content in ethylene furnace fuel—a comparison.

A published graph, termed "the classical and accepted curve for hydrogen influence on NO_x emissions," was found in the literature [52]. However, its origin, basis, and nonhydrogen fuel composition are not clearly stated; the background fuel is assumed here to be natural gas fired with ambient combustion air. This curve, redrawn, appears in Figure C.14 as well. Although it is noted as the flattest curve in the figure, it has been fitted for the plot using a third-order polynomial to capture the rise and inflection point exhibited by the original curve.

The 60°F (15.6°C) combustion air relative NO_x curve at 2% excess O_2 from Figure C.13 and a slightly lower relative NO_x curve based on 10% excess air at the same combustion air temperature are also shown in Figure C.14. Both of these curves deviate from the literature curve by at most 5% up to a concentration of 70% H_2. Beyond that point, the published curve increases slightly from a relative NO_x of 1.4 at 70% H_2 to just less than 1.6 at 100% H_2, compared to a more conservative 2.4–2.6± estimated from the correlation for 100% H_2.

There exists another curve of relative NO_x versus percent H_2 in the fuel [53,54] that is virtually identical to the flattest curve in Figure C.14. The other curve reflects NO_x in conventional and low-NO_x burners when firing refinery fuel gas[4] containing various percentages of H_2 in the mix [54].

With ultralow-NO_x burners, NO_x is said to peak in the range of 60–70% H_2, followed by a decrease at greater hydrogen concentrations (not pictured in Figure C.14) [54]. This phenomenon is attributed to a decrease in prompt NO_x, thought to be predominant in the total NO_x from ultralow-NO_x burners [54]. In other words, as hydrogen in the fuel increases, the concentration of hydrocarbons and their free radicals formed during combustion decreases along with a decrease in the tendency to form prompt NO_x. The influence of prompt NO_x may also explain the deviation beyond 70% H_2 in Figure C.14 between NO_x predicted by the correlation and "the classical and accepted curve for hydrogen influence," as also suggested previously [52].

C.10.6 Acetylene Content of Methane and Hydrogen Fuels

Now that hydrogen has been explored, let us now look at the implications of acetylene in ethylene furnace flue gas. By-product acetylene in the cracked gas would end up in the fuel in the event that it is not being recovered or hydrogenated to ethane and ethylene in the normal processing sequence [55]. Under comparable conditions, acetylene as a pure component is calculated to generate an AFT over 600°F (333°C delta) higher than hydrogen and about 950°F (528°C delta) higher than methane. Firing a significant concentration of acetylene, therefore, has the potential to increase ethylene furnace NO_x emissions substantially.

---

[4]Refinery fuel gas is made up of the leftover gaseous streams in the refining process that is burned for energy recovery in an oil refinery. Refinery fuel gas, in general, contains such ingredients as hydrogen, C_1–C_4 + alkanes with possibly some unsaturated hydrocarbons, sulfur species, and inerts; the composition is highly variable among refineries, over time within the same refinery, and sometimes from one fired unit to the next [54]. (See also Appendix L.)

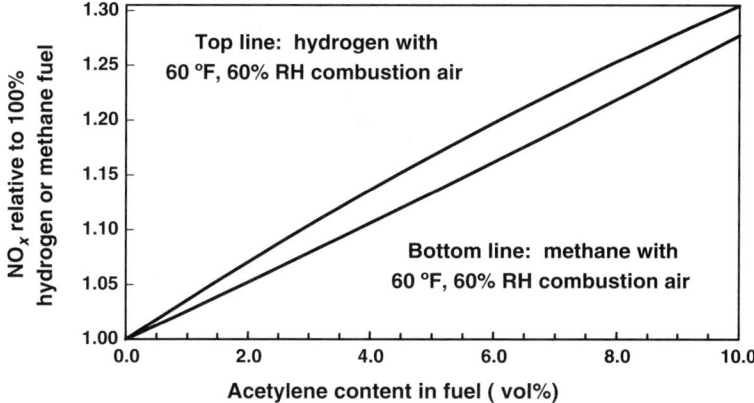

Figure C.15 Relative NO_x from acetylene in ethylene furnace fuel at 2% excess O_2 (wet).

Relative NO_x from combustion of fuel gas containing percentage concentrations up to as high as 10% acetylene is shown in Figure C.15. This range has been exaggerated to magnify differences. Two lines are indicated, one for acetylene mixed with a base fuel of pure hydrogen (upper) and the other for acetylene mixed with a base fuel of pure methane (lower). Both curves were derived using 60°F (15.6°C), 60% RH air for combustion at 2% excess O_2 (wet). The relationships are nearly linear, with a relative NO_x at 10% acetylene of approximately 1.30 compared to pure hydrogen for the hydrogen line, and 1.28 compared to pure methane for the methane line.

With both lines anchored at 1.0 relative NO_x for 0% acetylene, each of the ordinates at 100% acetylene swing down about 0.03 when this same combustion air is heated to 350°F (177°C), and an additional 0.03 for 750°F (399°C) combustion air.

C.11 EXPERIMENTAL VERIFICATION OF NO_X PREDICTIONS

NO_x data for ethylene cracking furnaces in the open literature are few and far between, especially when accompanied by simultaneous operating conditions. Still, it is possible with a little detective work to squeeze out some data for comparison with NO_x predictions by making reasonable assumptions to augment whatever meager information has been reported. Often, the burner NO_x reported is incidental to the main subject under discussion.

A number of usable cases were identified in Refs [4,5], with a synopsis of the data and circumstances for each case, an interpretation of experimental observations, and values predicted by the correlation. These are summarized in a parity plot of predicted versus observed NO_x (Figure C.16). Reported measurements for all cases considered have been converted, where necessary, to a common unit of ppmd @ 3% O_2 (dry).

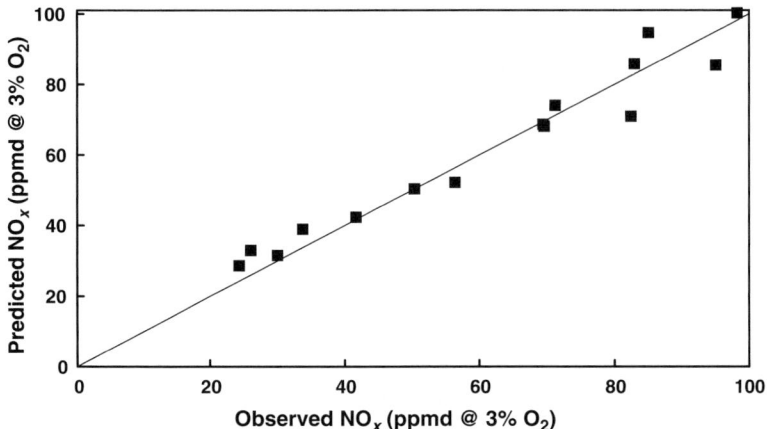

Figure C.16 Parity plot—predicted versus observed NO_x.

One source document [52] also contains NO_x data obtained in the burner manufacturer's pilot test facility. Those data are not representative of full-scale operation, for reasons discussed by the manufacturer, and are not shown here. In another work, the tests of multiple burners have produced significantly higher NO_x levels than single-burner tests in a smaller furnace [56].

The typical NO_x values mentioned in Section C.5 would also plot well in Figure C.16. However, these are not included in the parity plot because they cannot be guaranteed to be from full-scale ethylene cracking furnaces, or even if so, conditions are not specified well enough to validate the model unequivocally. For example, the historical figure of 100–120 ppm NO_x from conventional burners could arise at the maximum point of a NO_x–O_2 curve, according to the correlation, for methane with up to 40% hydrogen in the blend fired with ambient air (Figures C.6 and C.7 and Table C.2), with 100% methane and air preheat up to about 160°F (71°C) (not shown), or from any number of other combinations.

Agreement between the correlated and measured values in Figure C.16 is indeed satisfactory, especially considering that these data were not used in the derivation of the correlation. Regardless of the origin of the proposed equations, their use, as it turns out, appears promising in estimating NO_x for a wide variety of situations encountered in ethylene plants.

C.12 OPPORTUNITIES FOR IMPROVEMENT

The proposed correlating functions appear to check out well enough against spot data. However, the author has not yet come across a complete set of NO_x data in the open literature showing a systematic variation in furnace excess O_2 similar to the SMR data of Figure C.5 to test it further. It is hoped that more complete data sets will become available to provide a definitive check of the NO_x prediction equations for an ethylene cracking furnace.

Data sufficient to correlate and predict NO_x from a plant furnace can often be generated at minimal incremental expense during mandatory stack testing for regulatory purposes. Multiple regression of such data enables one to tailor the constants of the generalized function for a specific furnace, and constants are easily adjusted as more data become available. This can be especially useful when the furnace/burners can be considered the prototype for others of the same model already in service or yet to be built.

Suggested data are listed in Table C.5; concomitant process conditions during emission testing can also be noted and recorded. Target operating conditions are outlined in Figure C.4 of Ref. [2], being careful to remain within one's permit limits at all times.

Table C.5 Suggestions for Ethylene Furnace NO_x Data/Wish List for Data Sets for NO_x Correlation

At furnace

	Fuel (gaseous)	Composition
		Temperature
		Measured heating values (HHV and LHV), if available
		Flow rate (units? wet or dry, actual or standard—basis for std cubic ft, etc., 60°F, 70°F, 68°F) (enables one to calculate mass flow rates of atmospheric contaminants)
	Combustion air	Temperature of ambient air
		Humidity (or enough information, e.g., date and time of testing to get data on ambient air from weather bureau)
		Temperature of preheated air (if employed)
		Flow rate (if available) (units? wet or dry, actual or standard—basis for std cubic ft, etc., 60°F, 70°F, 68°F)
	Flue gas	Excess O_2 (probably wet) at furnace
		Furnace bridgewall or crossover temperature
	Type of burners	Manufacturer and nomenclature
		Standard/conventional
		Low NO_x (type of low NO_x)
		Ultralow NO_x (type)
		Firing orientation
At stack		
	Flue gas	Contaminant concentrations (NO_x and possibly CO) specify wet or dry.
		Composition of major constituents (N_2, CO_2, O_2, and H_2O (moisture), or at least O_2 and H_2O)
		Temperature
		Flow rate (units? wet or dry, actual or standard—basis for std cubic ft, etc., 60°F, 70°F, 68°F)

C.13 CONCLUSIONS

- NO_x correlation extended to ethylene cracking furnace burners.
- Correlation provides reasonable NO_x estimates.
- Correlation can be used to
  - estimate NO_x quickly,
  - evaluate changing conditions,
  - play "What-if" games,
  - complement other techniques such as CFD,
  - interpret experimental data.
- Correlation works for
  - a variety of fuels and fuel temperatures,
  - ambient air for combustion,
  - preheated combustion air,
  - combustion air with varied humidity.
- Correct inputs are required.
- Additional data needed to
  - increase confidence,
  - make modifications, if necessary.

REFERENCES

1. R.G. Kunz, D.D. Smith, N.M. Patel, G.P. Thompson, and G.S. Patrick, Control NO_x from furnaces, *Hydrocarbon Processing*, **71**(8), 57–62, 1992.
2. R.G. Kunz, D.D. Smith, and E.M. Adamo, Predict NO_x from gas-fired furnaces, *Hydrocarbon Processing*, **75**(11), 65–79, 1996.
3. R.G. Kunz, B.R. Keck, and J.M. Repasky, Mitigate NO_x by steam injection, *Hydrocarbon Processing*, **77**(2), 79–84, 1998.
4. R.G. Kunz, Extension of NO_x correlation to ethylene cracking furnaces, ENV-05-197, 2005 NPRA Environmental Conference, Dallas, TX, September 19–20, 2005.
5. R.G. Kunz, Burner NO_x from ethylene cracking furnaces, Proceedings of the 19th Ethylene Producers' Conference, AIChE, New York, 16, 2007, pp. 430–455.
6. W. Gerhartz, et al. editors, *Ullmann's Encyclopedia of Industrial Chemistry*, 5th edn, Vol. **A10**, VCH Publishers, New York, 1987, pp. 45–93.
7. Anonymous, International survey of ethylene from steam crackers—2006, *Oil & Gas Journal*, **104**(12), 51–56, 2006.
8. Texas Commission on Environmental Quality (TCEQ), Revisions to the State Implementation Plan (SIP) for the Control of Ozone Air Pollution: Houston/Galveston/Brazoria Ozone Nonattainment Area, Chapter 1: Executive Summary, adopted December 1, 2004, pp. 1-1–1-2.
9. J.J. McKetta and W.A. Cunningham, editors, *Encyclopedia of Chemical Processing and Design*, Vol. **20**, Marcel Dekker, Inc., New York, 1984, pp. 88–159.
10. R. Patel, B.P. Evans, and W.K. Lam, NO_x reduction technologies for pyrolysis furnaces, Proceedings of the 5th Ethylene Producers' Conference, AIChE, New York, Vol. 2, 1993, pp. 416–431.
11. J.I. Kroschwitz and M. Howe-Grant, editors, *Kirk-Othmer Encyclopedia of Chemical Technology*, 4th edn, Vol. **9**, Wiley, New York, 1994, pp. 877–915.

References 541

12. Texas Commission on Environmental Quality (TCEQ), Emission Inventory Spreadsheet for HGB Ozone Nonattainment Area (circa 2001–2002).
13. 30 TAC Part I, Chapter17, Subchapter B, Division 3, Section 117.206(c)(8)(B), May 19, 2005.
14. R. Just, Flame rollover and other flame shape problems, Proceedings of the 16th Ethylene Producers' Conference AIChE, New York, Vol. 13, 2004, pp. 594–601.
15. R.J. Gartside, P.R. Ponzi, F.D. McCarthy, S.G. Chellappan, P.J. Chapman, and R.T. Waibel, Commercialization of ultra-low NO_x burners for ethylene heaters, Proceedings of the 16th Ethylene Producers' Conference, AIChE, New York, Vol.13, 2004, pp. 618–626.
16. Q. Tang, B. Adams, M. Bockelie, M. Cremer, M. Denison, C. Montgomery, A. Sarofim, and D.J. Brown, Towards comprehensive CFD modeling of lean premixed ultra-low NO_x burners in process heaters, Proceedings of the 17th Ethylene Producers' Conference, AIChE, New York, Vol. 14, 2005, pp. 594–619.
17. G. Stephens and D. Spicer, A low NO_x burner developed for ExxonMobil ethylene furnaces, Proceedings of the 17th Ethylene Producers' Conference, AIChE, New York, Vol. 14, 2005, pp. 487–499.
18. D. Barnett and D. Wu, Flue-gas circulation and heat distribution in reformer furnaces, *Ammonia Plant Safety & Related Facilities: A Technical Manual*, Vol. **41**, AIChE, New York, 2001, pp. 9–16.
19. R.C. Reed and T.K. Sherwood, *The Properties of Gases and Liquids—Their Estimation and Correlation*, McGraw-Hill, New York, 1958, p. 2.
20. R.G. Kunz, D.D. Smith, N.M. Patel, G.P. Thompson, and G.S. Patrick, Control NO_x from Gas-Fired Hydrogen Reformer Furnaces, AM-92-56, 1992 NPRA Annual Meeting, New Orleans, LA, March 22–24, 1992.
21. R.G. Kunz, D.D. Smith, N.M. Patel, G.P. Thompson, and G.S. Patrick, Control NO_x from gas-fired hydrogen reformer furnaces, Emission Inventory Issues—Proceedings of an International Specialty Conference, Durham, NC, October 19–22, 1992, VIP-27, Air & Waste Management Association: Pittsburgh, PA, 1993, pp. 381–392.
22. R.G. Kunz, B.R. Keck, and J.M. Repasky, Mitigate NO_x by steam injection, ENV-97-15, 1997 NPRA Environmental Conference, New Orleans, LA, September 28–30, 1997.
23. R.G. Kunz, D.C. Hefele, R.L. Jordan, and F.W. Lash, Use of SCR in a Hydrogen Plant Integrated with a Stationary Gas Turbine—Case Study: The Port Arthur Steam-Methane Reformer, Paper No. 70093, Air and Waste Management Association (A&WMA) 96th Annual Conference & Exhibition, San Diego, CA, June 22–26 2003.
24. R.G. Kunz, D.C. Hefele, R.L. Jordan, and F.W. Lash, Consider SCR to mitigate NO_x emissions, *Hydrocarbon Processing*, **82**(11), 43–50, 2003.
25. Ya.B. Zeldovich, P.Ya. Sadovnikov, and D.A. Frank-Kamenetskii Oxidation of Nitrogen in Combustion, Academy of Sciences of the USSR, Institute of Chemical Physics, Moscow-Leningrad, translated by M. Shelef of the Scientific Research Staff of the Ford Motor Company, 1947.
26. O.A. Hougen, K.M. Watson, and R.A. Ragatz, *Chemical Process Principles Part I: Material and Energy Balances*, 2nd edn, Wiley, New York, 1954, pp. 354–357, 408–411.
27. G.W. Jones, B. Lewis, J.B. Friauf, and G.St.J. Perrot, Flame temperatures of hydrocarbon gases, *Journal of the American Chemical Society*, **53**(3), 869–883, 1931.
28. A.G. Loomis and G.St.J. Perrot, Measurements of the temperatures of stationary flames, *Industrial and Engineering Chemistry*, **20**(10), 1004–1008, 1928.
29. J.M. Singer, E.B. Cook, M.E. Harris, V.R. Rowe, and J. Grumer, Flame Characteristics Causing Air Pollution: Production of Oxides of Nitrogen and Carbon Monoxide, The U.S. Bureau of Mines R.I. 6958, 1967 33 pp.
30. J. Grumer, M.E. Harris, V.R. Rowe, and E.B. Cook, Effect of Recycling Combustion Products on Production of Oxides of Nitrogen, Carbon Monoxide and Hydrocarbons by Gas Burner Flames, Preprint 37A, 60th Annual Meeting AIChE, New York, November 26–30 1967.
31. H.B. Lange Jr., NO_x formation in premixed combustion, AIChE Symposium Series No. 126, 1972, 17–27.
32. N.M. Patel, R.A. Davis, N. Eaton, D.L. Carlson, F. Kessler, and V. Khurana, 'Across-the-fence' hydrogen plant starts up at California refinery, *Oil & Gas Journal*, **92**(40), 54–61, 1994.
33. J.I. Kroschwitz and M. Howe-Grant, editors, *Kirk-Othmer Encyclopedia of Chemical Technology*, 4th edn, Vol. **13**, Wiley, New York, 1995, pp. 838–894.

34. J.J. McKetta, editor, *Encyclopedia of Chemical Processing and Design*, Vol. **47**, Marcel Dekker, Inc., New York, 1994, pp. 165–203
35. B.M. Tindall and D.L. King, Designing steam reformers for hydrogen production, *Hydrocarbon Processing*, **73**(7), 69–74, 1994.
36. T. Johansen, K.S. Ragharaman, and L.A. Hackett, Trends in hydrogen plant design, *Hydrocarbon Processing*, **71**(8), 119–127, 1992.
37. R.E. Stoll and F. von Linde Hydrogen—what are the costs? *Hydrocarbon Processing*, **79**(12), 42–46, 2000.
38. M. Grayson and D. Eckrorth, editors, *Kirk-Othmer Encyclopedia of Chemical Technology*, 3rd edn, Vol. **12**, Wiley, New York, 1980, pp. 950–982.
39. B. Elvers, S. Hawkins, M. Ravenscroft, and G. Schulz, editors, *Ullmann's Encyclopedia of Industrial Chemistry*, 5th edn, Vol. **A13**, VCH, New York, 1989, pp. 317–328, 435–438.
40. J.H. Gary and G.E. Handwerk, *Petroleum Refining: Technology and Economics*, 4th edn, Marcel Dekker, New York, 2001, pp. 261–285, 313–317.
41. J.R. O'Leary R.G. Kunz, and T.R. von Alten, Selective catalytic reduction (SCR) performance in steam-methane reformer service: the chromium problem, ENV-02-178, 2002 NPRA Environmental Conference, New Orleans, LA, September 9–10, 2002.
42. J.R. O'Leary R.G. Kunz, and T.R. von Alten, Selective catalytic reduction (SCR) performance in steam-methane reformer service: the chromium problem, *Environmental Progress*, **23**(3), 194–205, 2004.
43. R.G. Kunz and W.F. Baade, Predict methanol and ammonia in hydrogen-plant process condensate and deaerator-vent emissions, ENV-00-171, 2000 NPRA Environmental Conference, San Antonio, TX, September 10–12, 2000.
44. R.G. Kunz and W.F. Baade, Predict contaminant concentrations in deaerator-vent emissions, *Hydrocarbon Processing (International Edition)*, **80**(6), 100-A–100-O, 2001.
45. Royal Dutch/Shell, *The Petroleum Handbook*, 6th edn, Elsevier, Amsterdam, 1983, pp. 284, 309, 586, 689.
46. W.F. Bland and R.L. Davidson, *Petroleum Processing Handbook*, Section 14, McGraw-Hill, New York, 1967, pp. 14-1–14-46.
47. K. Funahashi, T. Kobayakawa, K. Ishii, and H. Hata, SCR DeNO$_x$ in new maruzen ethylene plant, Proceedings of the 13th Ethylene Producers' Conference, AIChE, New York, Vol. 10, 2001, pp. 741–755.
48. W.T. Wines, Improve contaminant control in ethylene production, *Hydrocarbon Processing*, **84**(4), 41–46, 2005.
49. C.E. Baukal Jr. and R.E. Schwartz, editors, *The John Zink Combustion Handbook*, CRC Press, Boca Raton, FL, 2001, p. 111.
50. R.G. Kunz and T.R. von Alten SCR Treatment of Ethylene Furnace Flue Gas (A Steam-Methane Reformer in Disguise), Paper presented at Institute of Clean Air Companies (ICAC) Forum '02, Houston, TX, February, 2002.
51. R.T. Waibel, Ultra Low NO$_x$ Burners for Industrial Process Heaters, Paper presented at the Second International Conference on Combustion Technologies for a Clean Environment, Lisbon, Portugal, July 19–22, 1993.
52. K. Krotzer, D. Bishop, and D. Giles, Retrofit application of an ultra low NO$_x$ burner in an ethylene furnaces, (sic) Proceedings of the 9th Ethylene Producers' Conference, AIChE, New York, Vol. 6, 1997, pp. 416–431.
53. Anonymous, *Burners for Fired Heaters in General Refinery Services*, API Publication 535, 1st edn, American Petroleum Institute, Washington, DC, 1995, p. 12.
54. K.A. Modi and R.T. Waibel, NO$_x$ Emissions from process heaters and boilers: improving emissions estimates, ENV-95-163, 1995 NPRA Environmental Conference, San Francisco, CA, October 15–17, 1995.
55. *Ullmann's Encyclopedia of Industrial Chemistry*, 6th Completely Revised Edition, Vol. **12**, WILEY-VCH, Weinheim, Germany, 2003, pp. 531–583.
56. W. Bussman, R. Poe, B. Hayes, J. McAdams, and J. Karan, Low NO$_x$ burner technology for ethylene cracking furnaces, Proceedings of the 13th Ethylene Producers' Conference, AIChE, New York, Vol. 10, 2001, pp. 774–796.

Appendix D

What Is BOD and How Is It Measured?

Never eat anything you make yourself in the laboratory.

D.1 SUMMARY

This appendix provides a summary of dissolved oxygen (DO), biochemical oxygen demand (BOD), and chemical oxygen demand (COD), the important metrics used in pollution control. This summary is specially intended for those individuals who regularly make use of these data but who may not be familiar with how they are measured and the inherent margin of error to be expected. It is divided into three interrelated sections, each helpful in understanding the others. We begin first with DO required in the determination of BOD, continue with BOD itself, and finish up with COD, another nonspecific measure of pollution often used to supplement BOD.

D.2 DISSOLVED OXYGEN (DO) AND ITS MEASUREMENT

To assess the quality of surface waters, for the BOD test (Section D.3), and other purposes, it is important to understand the characteristics of DO contained in water and wastewater.

D.2.1 Overview

Oxygen dissolves in water by a purely physical process, proportional to its partial pressure in the gas in contact with the water. It is dependent on the temperature and the concentration of dissolved salts, notably chlorides. Table D.1 contains a summary of calculations for the saturation values of DO obtained using several methods and

---

Environmental Calculations: A Multimedia Approach, by Robert G. Kunz
Copyright © 2009 John Wiley & Sons, Inc.

Table D.1 Comparison of Calculated Oxygen Solubilities (mg/L)

Temperature (°C)	Henry's law[a] (dry air)	Henry's law[b] (saturated air)	Truesdale[c]	ASCE[d]	Standard methods[e,f]	
10	11.4	11.3	10.9	11.3	11.3	9.0
15	10.2	10.1	9.8	10.0	10.2	8.1
20	9.3	9.1	8.8	9.0	9.2	7.4
25	8.5	8.3	8.1	8.2	8.4	6.7
30	7.8	7.5	7.5	7.4	7.6	6.1
35	7.3	6.9	7.0	7.1[g]	7.1	–
40	6.9	6.4	6.6	6.1[g]	6.6	–

[a] C_s (mg/L) = α (mg/L atm) p_{O_2} (atm); α computed from $(MW_{O_2}/MW_{H_2O})(\rho_{H_2O}/H(\text{atm/mole fraction}))$; H in atm/mole fraction tabulated against temperature in Ref. [1, p. 765]; p is the partial pressure of O_2 in dry air assumed to be 0.209 atm.

[b] Same as in footnote a, except that O_2 partial pressure is reduced by saturation of air with water vapor.

[c] C_s (mg/L) = $14.161 - 0.3943T + 0.007714T^2 - 0.000646T^3$ at zero salt concentration (T in °C) [4, p. 108].

[d] C_s (mg/L) = $14.652 - 0.41022T + 0.0079910T^2 - 0.0000079774T^3$ at zero salt concentration (T in °C) [5, p. 23–10, Table II].

[e] "Standard Methods" tabulation, Table 218(1) in 13th edition [2, pp. 480–481] based on C_s (mg/L) = $A(P - V_p)/B$ [2, p. 479].

where

$$A = 0.678$$
$$B = T(°C) + 35 \quad 0 \leq T(°C) \leq 30$$

and

$$A = 0.827$$
$$B = T(°C) + 49 \quad 30 \leq T(°C) \leq 50$$

[f] All DO values shown are calculated for a zero salt concentration except for column 7. Entries in column 7 correspond to a chloride concentration of 20,000 mg/L, about 5% greater than the chloride level in seawater [3, p. F-203].

[g] Temperature range is 0–30°C.

covering a temperature range from 10 to 40°C. A simple Henry's law [1] relationship was used to calculate the entries in columns 2 and 3. Column 2 is based on the assumption of dry air and a constant atmospheric oxygen partial pressure of 0.209 atm. The results in columns 3–7 are based on water-saturated air in which the water vapor serves to reduce the oxygen partial pressure. Columns 4–7 contain values from several accepted empirical formulations, including the basis for "Standard Methods" Table 218(1) [2]—column 6 for zero salinity and column 7 for a chloride concentration of 20,000 mg/L, about 5% greater than the chloride level in seawater [3, p. F-203]. DO values at saturation for a salt concentration between 0 and 20,000 mg/L chloride ion can be estimated reliably by linear interpolation.

As indicated in the table, the dissolved oxygen level decreases with the increasing temperature. The saturation solubility of oxygen in fresh water ranges from about

11 mg/L at 10°C to just under 7 mg/L at 35°C under atmospheric pressure, and somewhat less in the presence of dissolved salts, as in seawater. If one is able to remember only one number for DO under all conditions in fresh water, 8 mg/L is a good rule of thumb.

These figures indicate the maximum amount that a solution can contain. The actual dissolved oxygen concentration will normally be less than saturation. The rate of solution by aeration depends upon the difference between the saturation level and the actual value [4, p. 23–11].

Special precautions are taken when collecting samples for dissolved oxygen [2, pp. 475–477]. The sample must not remain in contact with air or be agitated since either condition would cause a change in gaseous content. Samples from any depth in natural bodies of water and samples from pressurized vessels such as boilers need special precautions to eliminate changes in temperature and pressure. Sample bottles should be filled from bottom until they overflow several times their volume to ensure collection of representative samples. Care should be taken to prevent turbulence and the formation of bubbles while filling the bottle. The temperature should be recorded to at least the nearest degree Celsius.

D.2.2 DO Measurement

Dissolved oxygen is measured by a number of chemical techniques, the so-called Winkler iodometric method and its several modifications, the choice depending on the type of water/wastewater and the kinds of interferences present [2, pp. 474–484, 488] or by various instrumental methods including use of (portable) membrane electrodes [2, pp. 484–488].

D.2.2.1 Chemical Methods

Originally, the measurement of dissolved oxygen was performed by heating samples to drive out the dissolved gases and by analyzing the collected gas for oxygen by standard gas analysis techniques [6, pp. 388–389; 7, p. 411]. To eliminate this time consuming and cumbersome process, the Winkler technique was developed. The azide modification of the Winkler technique is employed when nitrite interference is possible or present as in most sewage, effluent, and stream samples [2, pp. 477–479]. A description of this method is given in a later section [6, pp. 388–391; 7, pp. 411–413].

The Winkler method depends on the fact that divalent manganese can be oxidized to a higher valence state (+ 4) under alkaline conditions in the presence of oxygen, and manganese in this higher valence state can oxidize iodide ion quantitatively to free iodine under acid conditions. The reactions are as follows:

With no dissolved oxygen present

$$MnSO_4 + 2NaOH \rightarrow Mn(OH)_2\downarrow + Na_2SO_4 \qquad (D.1)$$

white precipitate

If a white precipitate persists, no oxygen was present in the sample, and the test is complete.

With dissolved oxygen present

$$2Mn(OH)_2 + O_2 \rightarrow 2MnO(OH)_2 \downarrow \quad (D.2)$$
$$\text{brown precipitate}$$

$$2NaI + H_2SO_4 \rightarrow Na_2SO_4 + 2H^+ + 2I^- \quad (D.3)$$

$$2H^+ + 2I^- + 2MnO(OH)_2 + H_2SO_4 \rightarrow MnSO_4 + 3H_2O + I_2^0 \quad (D.4)$$

The solution has the I_2^0 brown color. Each molecule of iodine formed is equivalent to a molecule of oxygen in the original sample. By reducing the free iodine back to iodide ion, one obtains a quantitative measure of the oxygen originally present. Sodium thiosulfate previously standardized with potassium dichromate is used for this purpose.

$$2S_2O_3^{2-} + I_2^0 \rightarrow S_4O_6^{2-} + 2I^- \quad (D.5)$$

Starch is added as an indicator, and the disappearance of iodine is accompanied by a change in color from the blue-black of the starch–iodine complex to colorless during titration.

The unmodified Winkler method is subject to interference from a great many substances such as nitrite, divalent and trivalent iron, sulfite, sulfide, and polythionates. Nitrite is one of the most frequent interferences from sewage treatment plants, employing biological processes, in river waters, and in incubated BOD samples. The reactions are as follows:

$$2NO_2^- + 2I^- + 4H^+ \rightarrow I_2^0 + N_2O_2 + 2H_2O \quad (D.6)$$

$$N_2O_2 + \tfrac{1}{2}O_2 + H_2O \rightarrow 2NO_2 + 2H^+ \quad (D.7)$$

Nitrite reacts with iodide to form free iodine and N_2O_2. N_2O_2 reacts with the additional oxygen that enters the sample during the titration procedure and regenerates the nitrite. These reactions are cyclic, making it impossible to obtain a clear end point in the titration. As soon as the blue-black color of the iodine–starch complex disappears, the nitrite reacts with iodide ion to form new free iodine that once again ties up with the starch to give the characteristic blue-black color. Nitrite interference can be overcome by the use of sodium azide (NaN_3). When sulfuric acid (H_2SO_4) is added, the following reactions occur, destroying the interfering nitrite:

$$NaN_3 + H^+ \rightarrow I_2^0 + HN_3 + Na^+ \quad (D.8)$$

$$HN_3 + NO_2^- + H^+ \rightarrow N_2 + N_2O + H_2O \quad (D.9)$$

Other techniques for dissolved oxygen measurement are [2, pp. 474–484]

- Unmodified Winkler method—used when no interferences are present
- Permanganate modification—used in the presence of ferrous ion
- Alum flocculation modification—used with interfering suspended solids present

- Copper sulfate sulfamic acid modification—used in activated sludge mixed liquors

These along with the azide modification constitute the iodometric techniques.

D.2.2.2 Instrumental Methods

Instrumental methods are listed below [2, pp. 484–487]:
The polarographic method using dropping mercury electrode is applicable to clean water only, and it is not always reliable in sewage and industrial wastewaters for strong waste effluents and/or for other situations, containing substances that interfere with the chemical tests.

D.2.2.3 Field Samples

Samples are normally collected in the field where it is impractical to complete the entire procedure when using a chemical analysis technique [6, p. 386; 7, p. 409]. Since oxygen may change radically with time because of the activity of microorganisms, certain reagents are added to "fix" the dissolved oxygen in solution, or other reagents are added to arrest biological activity and maintain the dissolved oxygen in an "unfixed" state. In either case, sealed samples are normally stored at a reduced temperature, perhaps in the dark, to prevent the action of photosynthesis by algae, and so on. Laboratory analysis is then completed without delay within a matter of hours.

D.2.3 Reproducibility

Iodine-titration chemical methods with visual end point detection using a starch indicator can produce a DO measurement precision of $\pm 50\,\mu g/L$ for experienced analysts, and $\pm 5\,\mu g/L$ [2, p. 475] or better [8, p. 535] with electrometric instrumental determination of the end point. Elsewhere in "Standard Methods" [2, p. 385], a range of 20–60 $\mu g/L$ standard deviation for DO measurements is given, with distilled water on the low end and secondary sewage treatment effluents on the high end of the range. Tap water can be expected to fall within this range. In the author's experience, even novice analysts with care are capable of a standard deviation of $\pm 57\,\mu g/L$ for tap water by visual means.

An accuracy of ± 0.1 mg/L is quoted for most commercially available membrane electrode systems [2, p. 474]. The "TENTATIVE" label on the membrane electrode method in the 13th edition of "Standard Methods" [2, p. 484] has been removed in a later edition [9, p. 450].

D.2.4 Wrap-Up

The dissolved oxygen test is an important technique in water pollution control [6, pp. 385–386; 7, p. 408; 11, pp. 217–244, 240–243]. Dissolved oxygen determinations are used in maintaining aerobic conditions in natural waters that receive polluted matter and in aerobic treatment processes intended to purify domestic and industrial wastewaters. Self-purification of a stream depends on having enough dissolved oxygen present [5, pp. 33–16 to 33–29; 10, pp. 184–193]. As long as oxygen is not depleted too rapidly, water quality can be expected to improve. If, however, the oxygen utilization rate exceeds its rate of replenishment, the opposite will occur and the condition of the stream will get worse. In the extreme, the stream will become septic and anaerobic bacteria will take over, producing noxious gases, black sludges, and odors as end products of their metabolism.

If nuisance conditions are to be avoided in a stream or if certain types of fish are to be maintained, it is essential that the dissolved oxygen level in the stream does not drop below certain critical levels. The metabolic rate and therefore the oxygen required by microorganisms and fish increase with temperature, but the oxygen level in a stream goes down as temperature increases (Table D.1). The minimum dissolved oxygen for maintaining fish in healthy condition is 5 mg/L at 20°C or about 57% of saturation [4, pp. 111–112]. Water becomes hazardous or lethal to fish when the dissolved oxygen falls to about 3 mg/L or less. The exact figure depends upon the species and age of the fish and their degree of acclimation, the temperature, the composition of the water, and the presence of toxic substances. A more complete account of the sensitivity of fish to dissolved oxygen is given by Klein [11, pp. 38, 169–170, 226–268].

All aerobic treatment processes depend on the presence of dissolved oxygen. The determination of dissolved oxygen serves as the basis of the BOD test, which is used to assess the strength of a waste and to design and operate treatment equipment. Oxygen is also significant in the corrosion of iron and steel, particularly in water distribution systems and in stream boilers. Removal of oxygen from boiler feed water is monitored by means of the dissolved oxygen test. Although the presence of dissolved oxygen in drinking water is not absolutely necessary and is not mentioned in the U.S. Drinking Water Standards [40 CFR Parts 141 and 143] and no health-based guideline is recommended by the World Health Organization [12], absence of DO causes a characteristic flat taste.

D.3 BIOCHEMICAL OXYGEN DEMAND (BOD) AND ITS MEASUREMENT

D.3.1 Theoretical Considerations

The BOD is the amount of DO required by microorganisms, mainly bacteria, for the oxidation of organic material in a waste under aerobic conditions. The BOD test is a bioassay technique involving the measurement of oxygen consumed by the bacteria while stabilizing the organic matter in the waste as they would normally do in nature but under normal laboratory conditions. By convention, the test is conducted for a period of 5 days at 20°C (68°F).

The 5-day BOD test (BOD_5) was introduced to the United States from England where the time-of-travel of that country's freshwater streams to reach the ocean is not more than 5 days [10, p. 172]. A 5-day time period was therefore chosen as the basis for the maximum oxygen required by the microorganisms to stabilize the organic matter in sewage/wastewater. This basis survived when the test was brought to the United States, despite the lack of significance of the 5-day period *per se* since many streams here exhibit a greater transit time.

BOD measured at different times is in fact a continuum, increasing from a zero value initially to an asymptotic value for complete bio-oxidation. For domestic sewage, BOD_5 is reported as 65–70% [10, p. 172] or 70–80% [6, p. 396; 7, p. 417] of the ultimate oxygen demand. The 70–80% is said to apply to many industrial wastewaters as well [6, p. 396; 7, p. 417]. The theory and mechanics of BOD measurement follow. Details of dissolved oxygen measurement are contained in Section D.2.

The BOD test is a nonspecific analysis that includes all biologically oxidizable species according to the following general chemical reaction [6, p. 395; 7, p. 417]:

$$C_nH_aO_bN_c + \left(n + \frac{a}{4} - \frac{b}{2} - \frac{3c}{4}\right)O_2 \rightarrow nCO_2 + \left(\frac{a}{2} - \frac{3c}{2}\right)H_2O + cNH_3 \quad \text{(D.10)}$$

The relative ease of biological oxidation of various classes of organic compounds is discussed in the cited reference [8, pp. 472–274]. In brief, alcohols, phenols, cresols, simple aldehydes, esters, organic acids, amino acids, carbohydrates, and some surfactants and chlorinated compounds are included among the biodegradable species.

The level of dissolved oxygen present in a sample is limited to the saturation value of 9.2 mg/L at 20°C, the temperature at which the test is normally run. However, the strength of typical wastes is such that several hundred mg/L is required for oxidation. In nature, this is accomplished by constant reaeration of the stream into which the waste is discharged. In the laboratory, a portion of the waste is diluted with oxygen-saturated water to such an extent that the oxygen requirement is less than this saturation value, and reaeration is prevented. For wastes of unknown strength, several dilutions are necessary.

To grow, reproduce, and consume organic waste material, bacteria require favorable temperature, pH, accessory nutrients, proper osmotic pressure, freedom from toxic substances, and sufficient oxygen. Therefore, the temperature is maintained at 20°C (within the favorable range for bacterial growth); a solution of inorganic salts is added to the dilution water to provide a pH buffer, adequate osmotic pressure, and accessory nutrients; external sources of toxic substances are excluded; and the test is run between specified dissolved oxygen levels. It has been found that for the test to be valid, at least 2 mg/L oxygen must be utilized (indicating that oxygen is not limiting) [2, p. 492]. Another source states that at least 0.5 mg/L remaining is adequate [6, p. 403; 7, p. 424]. To ensure sufficient population of a variety of bacteria that will be able to utilize the heterogeneous organic compounds present in the sample, 1–2 mL of a bacterial seed acclimated to the particular waste is also added to the dilution water. A sample of seeded dilution water is incubated for the same period of time as the samples to serve as a blank. The seed can usually be obtained downstream of the point of discharge of the waste [2, p. 490].

When the preceding conditions are met, the rate of biochemical oxidation of the organic matter is proportional to the remaining concentration of unoxidized material. This leads to the exponential relationship of a first-order reaction for BOD at any time:

$$\text{BOD (mg/L)} = L(1-10^{-kt}) \tag{D.11}$$

where L is the "ultimate" (asymptotic) BOD, k is the deoxygenation constant (reciprocal days) and t is the incubation time measured in days. For this relationship, it is customary to employ base 10 exponentials rather than those to the base e (e = 2.718281828...) in the equivalent expression:

$$\text{BOD (mg/L)} = L(1-e^{-k't}) = L(1-\exp(-k't)] \tag{D.12}$$

where $k' = k/\log_{10}(e) = (2.3026...)(k)$

This reaction with the overall appearance of first order is, however, complex since at various times bacteria consume the waste, utilize oxygen, and grow, and protozoa increase in numbers, utilize oxygen, and consume the bacteria [10, pp. 173–175; 6, p. 409; 7, p. 430]. In addition, the biological oxidation rate for a mixed wastewater is the sum of the individual component rates and thus exhibits a smoothing effect [6, p. 410; 7, p. 431]. For sewage, the value of k is often quoted as 0.1 although it may vary from less than half to more than two times this value, and it is said that one should not be surprised if a good exponential fit is not obtained [2, p. 489].

In practice, the BOD curve follows the exponential relationship for only the first 8–10 days [6, p. 398; 7, p. 420]. After 10 days, the curve departs from an exponential shape because nitrifying bacteria build up a sufficient population and begin to utilize oxygen to oxidize ammonia and nitrites to nitrates [11, p. 324; 6, p. 398; 7, p. 420].

Nitrosomonas (nitrite-forming bacteria) $\quad 2NH_3 + 3O_2 \rightarrow 2NO_2^- + 2H^+ + 2H_2O$
$$\tag{D.13}$$

Nitrobacter (nitrate-forming bacteria) $\quad 2NO_2^- + O_2 + 2H^+ \rightarrow 2NO_3^- + 2H^+$
$$\tag{D.14}$$

The ammonia arises from the wastewater itself, nutrient solution added for the BOD test, and degradation of organic compounds such as proteins and amino acids decomposed during the carbonaceous-BOD phase. Beyond 10 days, the measured BOD is a combination of the first-stage or carbonaceous BOD and the second stage of nitrifying BOD, as depicted in Figure D.1 and Figure 24.2 of Ref. [6], Figure 22.2 of Ref. [7], and Figure 16-8 of Ref. [10]. Numerical values on the axes of Figure D.1 refer to a BOD experiment discussed in a later section of this appendix.

To avoid this difficulty, the 5-day BOD is usually free from interference by nitrifying bacteria and is another reason why it has been selected as the standard [6, p. 399; 7, p. 420]. This 5-day BOD is a typically reported value, but caution must be exercised in its interpretation, especially when drawing inferences from one waste to another of a different character.

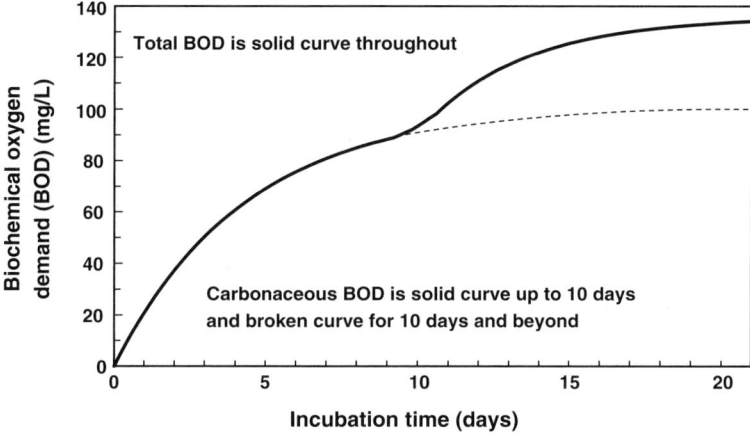

Figure D.1 BOD versus incubation time sewage sample.

The BOD curve may also exhibit a lag period as shown in Figure D.2, caused by a low initial bacterial population, bacteria not acclimated to the wastewater in question, or the presence of toxic substances [8, p. 471; 10, pp. 175–176].

D.3.2 Mechanics of the BOD Test

Serial dilutions of wastewater are mixed with aerated dilution water and a nutrient solution, along with an acclimated microbiological seed in a 250–300 mL BOD incubation bottle [2, pp. 489–492]. Dilutions are chosen in anticipation that at least one sample per batch will fall within the acceptable range. "Standard Methods" calls for saturated dilution water in the BOD procedure (9.2 mg/L DO at 20°C—Table D.1). Blanks containing seeded and unseeded dilution water are also prepared. Samples and

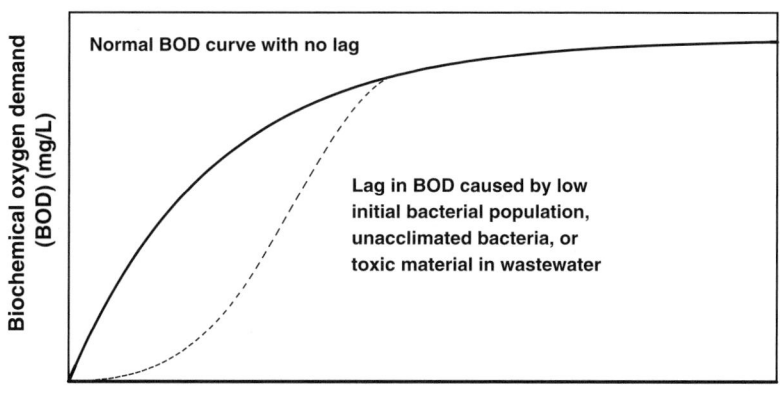

Figure D.2 BOD versus incubation time showing lag.

blanks are incubated in the dark at a constant temperature of 20°C for the specified time period, usually 5 days. The purpose of incubating the seeded blank is to compensate for the utilization of oxygen by microorganisms in the seeded dilution water. The DO depletion in the unseeded blank ought not to be more than 0.1–0.2 mL; it should not be used as a blank correction but rather to serve as a rough check or the dilution water quality. Replicate samples for wastewater and blanks are desirable; at least three blanks are suggested for statistical reliability [6, p. 402; 7, p. 424].

The BOD bottles contain a specifically fitted matching ground glass stopper and a flared mouth/lip around the neck of the bottle to accommodate a water seal. All air bubbles must be excluded from the bottle.

To prevent the involvement of extraneous organic matter and possible microbiological contamination from a previous use, the sample bottles must be scrupulously cleaned of any residual scum, using dichromate cleaning solution, sulfuric acid alone, or detergent, followed by a thorough rinse with water. This is especially important with the blank samples to ensure that the oxygen utilized is due only to the seeded dilution water and not to the BOD bottles themselves.

After completion of the incubation period, DO is determined for each of the sample bottles by a method described in Section D.2. The BOD values are then computed for a given incubation time using the following equation [6, p. 404; 7, p. 426]:

$$\text{BOD (mg/L)} = [(DO_{blank} - DO_{sample})(100/\% \text{ dilution})] - [DO_{blank} - DO_{undiluted\ sample}] \quad (D.15)$$

The second term above drops out when the DO of the dilution water and the undiluted sample is the same and becomes less significant as the value of BOD increases. Dissolved oxygen in domestic sewage at saturation is reported to be about 95% of that of clean water [5, p. 23–10].

Samples are rejected if less than 2 mg/L of O_2 has been consumed or if an inadequate concentration of DO remains in the sample. Only the results for valid samples are reported as BOD.

D.3.3 BOD Experiment

Dissolved oxygen remaining in diluted sewage samples was measured after various incubation periods up to 21 days in order to determine the BOD with respect to time [Equation (D.15)]. The DO in the blank samples also decreased with incubation time, by up to 2 mg/L below the DO saturation value. Values falling outside the range of acceptable DO utilization were discarded. Almost all of the tests that were considered invalid showed that at least 2 mg/L of oxygen had not been used up.

The curve correlating the valid BOD data from this experiment on this very weak domestic sewage is plotted in Figure D.1. A reasonable exponential fit for the carbonaceous BOD for this particular wastewater was obtained using the typical deoxygenation constant of 0.1 reciprocal days.

$$\text{BOD (mg/L)} = (101)(1 - 10^{-0.1 \cdot t \text{ (days)}}) \quad (D.16)$$

Although the individual results are not shown, their agreement with the curve used to correlate them is quite good, considering the semiquantitative nature of the BOD test and the variability in technique between different analysts. However, the total BOD as measured is consistently below that anticipated for a waste that is primarily domestic sewage (BOD typically 150–250 mg/L). (Operating conditions for the sewage treatment plant are unknown, including the possibility for dilution by storm water.)

The stabilization of the wastewater was essentially complete by 14 days, and the results for 14 days and beyond scattered, showing no further trend with time. As determined from the curve fit, the BOD for nitrification is likely on the order of 25% of the total. The variability of the majority of the data obtained on the sample turns out to be about twice the 5% precision obtained by a single trained analyst on a standard glucose–glutamic acid mixture (± 11 mg/L standard deviation at a BOD of 218 mg/L) [2, pp. 492–493] (see Section D.3.4).

The 69 mg/L average BOD determined at 5 days is 68% of the "ultimate" BOD projected for the carbonaceous portion, compared to 65–70% quoted by Ref. [10, p. 172] and 70–80% from Refs [6, p. 396 and 7, p. 417] and is considered reasonable.

D.3.4 Reproducibility

A solution of 300 mg/L glucose ($C_6H_{12}O_6$, MW = 180.16) [3, p. C-317] has a theoretical oxygen demand of $\sim$320 mg/L and a measured 20-day BOD and calculated L values ranging from 250 to 285 mg/L ($\sim$80–90% of theoretical). The difference is explained by the portion of organic waste material converted to cell tissue [6, p. 409; 7, p. 430]. However, glucose has a high and quite variable oxidation rate with simple seeds. When mixed with glutamic acid, the rate is stabilized and behaves similarly to domestic sewage [2, p. 493].

A solution of 150 mg/L each of glucose and glutamic acid ($C_5H_9O_4N$, MW = 147.13) [3, p. C-318] is used as a reference material to validate the reagents and technique of the BOD test [2, pp. 547–548]. By calculation, this solution has a theoretical oxygen demand of $159.85 + 146.81 \cong 307$ mg/L. The BOD_5 of this standard solution has been found to vary, between an average of 207 and 242 mg/L (68–79% of theoretical), with a standard deviation from ± 7 to ± 13 mg/L, depending on the type of seed used [2, p. 548]. From these figures, the precision of a single analyst working in his own laboratory is on the order of 5%.

D.3.5 Wrap-Up

BOD measurements can determine the oxygen required by domestic and industrial wastes discharged into streams under aerobic conditions. It is the only test to provide the rate of biological oxidation of organic waste matter. BOD is used in sizing treatment processes, in assessing their effectiveness, and in evaluating the purification capacity of receiving waters. It is employed by many sewage authorities to calculate sewage treatment charges and is used by regulatory agencies to assure compliance with environmental control laws.

D.4 CHEMICAL OXYGEN DEMAND (COD) AND ITS MEASUREMENT

D.4.1 The COD Test

The COD test determines the equivalent oxygen consumption in the waste of the constituents that are susceptible to oxidation to carbon dioxide and water by a strong oxidizing agent [2, pp. 494–499]. Potassium dichromate ($K_2Cr_2O_7$) in strong acid solution (0.25 N dichromate) is the most commonly used oxidizing agent for the COD test, and the reaction proceeds as follows [6, p. 415; 7, pp. 434–435]:

$$C_nH_aO_b + cCr_2O_7^{2-} + 8cH^+ \rightarrow nCO_2 + \frac{a+8c}{2}H_2O + 2c\,Cr^{3+} \qquad (D.17)$$

where

$$c = \frac{2n}{3} + \frac{a}{6} - \frac{b}{3}$$

The reaction is conducted by refluxing at boiling temperature for a nominal 2 h. During refluxing, the solution changes from the orange of the dichromate ion to the greenish Cr^{3+}.

It would be nice to be able to determine the ultimate BOD, both carbonaceous and nitrification, of a wastewater without having to wait until it is completely stabilized (on the order of 2–3 months incubation time) [13, p. 10]. At that time, virtually all of the organic carbon will have been oxidized to carbon dioxide (CO_2), and the organic nitrogen and ammonia will have been converted to nitrate. This ultimate oxygen demand (UOD) can be estimated from

$$\text{UOD (mg/L)} = 2.67\,C + 4.57\,N \qquad (D.18)$$

where C is the mg/L of organic carbon and N is the total mg/L for each of organic nitrogen and ammonia nitrogen.

Experimental verification would indeed be a tedious exercise and the long-delayed value no longer useful for its intended purpose.

The COD test was originally meant to satisfy this purpose [8, p. 476]. Where a wastewater contains only readily biodegradable organic matter and nothing toxic, the experimental COD can be used to approximate the ultimate carbonaceous BOD [2, p. 495], although it is likely that both this ultimate BOD and the COD will be less than the theoretical oxygen demand, as observed for wastewater of known chemical composition.

Furthermore, many organic species are not completely oxidized under the COD test conditions, some materials that show up as COD are not biodegradable even by acclimated organisms, and biodegradable substances that are picked up as BOD are not attacked by acidified potassium dichromate. For example, both glucose and lignin are oxidized in the COD test [6, p. 413; 7, p. 433]. Lignin is extremely resistant to biological oxidation while glucose is readily oxidized biologically [6, pp. 407–408; 7, p. 428].

D.4 Chemical Oxygen Demand (COD) and Its Measurement

In the absence of a catalyst (silver sulfate), potassium dichromate does not oxidize straight-chain aliphatic hydrocarbons, linear alcohols, or short chain fatty acids [2, p. 496]. For a compound such as acetic acid, the percentage of the theoretical yield from the COD test is negligible without a catalyst and 90% with a catalyst; the yield from the BOD test is 80% [8, p. 476]. Aromatic hydrocarbons (such as benzene and toluene) and pyridine (a heterocyclic ring compound containing nitrogen in the ring in addition to carbon) are not oxidized to any appreciable extent in the COD test even with a catalyst [2, pp. 496, 499]. These types of aromatics also appear to be highly resistant to biological oxidation [8, p. 472]. Loss of volatile organics during initial heating of the COD sample before refluxing also occurs [8, pp. 475–476]. The dichromate solution oxidizes some organic nitrogen compounds [8, p. 475], but it does not react with ammonia originally present in the waste or formed as a degradation product during reaction with nitrogenous compounds [2, p. 295].

The catalyst used is silver sulfate (Ag_2SO_4) dissolved in the concentrated sulfuric acid that is added to the potassium dichromate. In the absence of this catalyst, the aforementioned compounds (e.g., short chain fatty acids) would not be oxidized, and the COD results would be erroneously lower than the true values.

Nitrites, ferrous ions, sulfides, and chlorides constitute possible interferences in the method since they can reduce dichromate and cause the COD to appear higher than its true value. If necessary, sulfamic acid (HSO_3NH_2) can be added to eliminate nitrite interference. However, significant amounts of nitrite seldom occur in wastes or natural waters, and for such samples, nitrite interference is minor. The same is also true for ferrous ions and sulfides, for which a correction determined from a separate analysis can be made, if critical. The most serious interference is from chlorides, which occur at significant levels in most wastewaters [6, p. 417; 7, p. 436].

$$6Cl^- + Cr_2O_7^{2-} + 14H^+ \rightarrow 3Cl_2 + 2Cr^{3+} + 7H_2O \tag{D.19}$$

To eliminate chloride interference, mercuric sulfate is added before the addition of the other reagents. The mercuric ion combines with the chloride ions to form poorly ionized mercuric chloride.

$$Hg^{2+} + 2Cl^- \leftrightarrow HgCl_2 (K_{ip} = 2.6 \times 10^{-15}) \tag{D.20}$$

This method is applicable for eliminating chloride interference up to about 2000 mg/L. Other procedures are recommended for higher chloride concentrations. If the chlorides are not removed, the COD determination will be erroneously higher than its true value.

The dichromate is used in excess. By making the oxidizable compounds the limiting reagents, one ensures that some dichromate is left over at the end of the test. The amount of dichromate used can then be computed from knowledge of the dichromate initially added and that remaining at the test conclusion. In practice, a blank containing all reagents except the sample under test is used as a measure of the initial dichromate. This technique compensates for extraneous foreign matter in the distilled water and other reagents, which it is impractical to exclude from the test.

The blank is expected to retain its orange color with only a limited amount of Cr^{3+} being formed.

The determination of dichromate is made by using a strong reducing agent, ferrous ammonium sulfate $[Fe(NH_4)_2(SO_4)_2 \cdot H_2O]$ of about equal strength (approximately 0.25 N) in the presence of ferroin indicator.

$$6Fe^{2+} + Cr_2O_7^{2-} + 14H^+ \rightarrow 6Fe^{3+} + 2Cr^{3+} + 7H_2O \qquad (D.21)$$

Potassium dichromate is a primary standard [14, p. 127], and the ferrous ammonium sulfate solution is standardized against it. A primary standard is a substance that can be obtained in sufficient purity, can be weighed accurately on an analytical balance, and can be dissolved and diluted to a known volume [14, p. 30]. The endpoint is reached when the solution changes from greenish blue to reddish brown. The dichromate volume or concentration never occurs in the calculation of COD, which is given directly in terms of the ferrous ammonium sulfate used.

$$\text{COD(mg/L)} = \frac{(a-b)(c)(8000)}{\text{mL sample}} \qquad (D.22)$$

where a is the mL of ferrous ammonium sulfate titrant used for blank, b is the mL of ferrous ammonium sulfate titrant used for sample, and c is the normality of ferrous ammonium sulfate solution [2, p. 499].

As with all strong reducing agents, the ferrous ammonium sulfate should be standardized daily since it gradually loses strength because of oxidation on standing. A weaker reducing agent of lower chemical normality will necessitate the addition of a greater volume during titration. The product of the higher volumes used and the erroneously high normality will result in a COD value greater than the true value that would have been obtained by using the true normality.

D.4.2 Reproducibility

In the author's experience, a sewage effluent from the primary settling tank of a sewage treatment tank showed an average COD of 251.2 mg/L, as determined from five aliquots of the same sample by five different analysts. The standard deviation was 2.4 mg/L or less than 1% of the average COD.

A previous determination on a sample from another day for the same plant showed a quite similar COD but with a deviation of about 5% for different aliquots. The difference was probably due to use of a lower normality for ferrous ammonium sulfate in the later case (0.1 N versus the 0.25 N called for), where the titrations consume more reducing agent solution but determination of the end point is less sensitive.

Use of the same standard technique for COD on the final settling tank effluent upstream of chlorination resulted in negative values calculated for COD. Since the value for COD after an expected 90% removal is on the order of 25–30 mg/L, the alternate test procedure in "Standard Methods" for CODs between 10 and 50 mg/L should be followed. In that procedure, lower reagent strengths (0.025 N $K_2Cr_2O_7$ and 0.10 N ferrous ammonium sulfate) are used [2, p. 498].

Potassium acid phthalate ($KHC_8H_4O_4$), another primary standard [14, p. 39], exhibits a COD of 98–100% of its theoretical oxygen demand and is often used as a standard to evaluate technique and reagents [2, p. 499]. "Standard Methods" refers to a set of synthetic unknown samples containing potassium acid phthalate and sodium chloride that was tested by 74 laboratories. At 200 mg/L COD and no chloride, the standard deviation was 10 mg/L (6.5% of COD). At 150 mg/L COD and 1000 mg/L chloride, the standard deviation was 14 mg/L (9.33% of COD) [2, p. 499].

D.4.3 Wrap-Up

COD determinations find extensive application in the analysis of industrial wastewaters and are particularly useful in surveys designed to determine and control losses to sewer systems. In conjunction with BOD, COD is useful in indicating toxic conditions and the presence of biologically resistant organic substances; the COD test is the only means available to determine oxidation requirements when biologically toxic substances are present, as in industrial wastewaters. The COD test may serve as a preliminary index of appropriate dilution for the BOD test. A correlation can be obtained between BOD and COD values for the same wastewater under similar conditions but should not be expected between different wastes. Once the correlation is established, COD can replace BOD as a routine control parameter [4, p. 31]. COD results may be obtained within a relatively short time (2 h reflux time versus 5 days for BOD), and therefore the method is good for treatment plant control for which corrective measures may be instituted immediately. However, the COD test by itself does not provide any rate of oxidation information.

REFERENCES

1. J.H. Perry, editor, *Chemical Engineers' Handbook*, 3rd edn, McGraw-Hill, New York, 1950, 675 pp.
2. *Standard Methods for the Examination of Water and Wastewater*, 13th edn, American Public Health Association, Washington, DC, 1971, 874 pp.
3. R.C. Weast, editor, *CRC Handbook of Chemistry and Physics*, 58th edn, CRC Press, Inc., Cleveland, OH, 1977.
4. L. Klein, *River Pollution I. Chemical Analysis*, Butterworths, London, 1959, 206 pp.
5. G.M. Fair, J.C. Geyer, and D.A. Okun, Water purification and wastewater treatment and disposal. *Water and Wastewater Engineering* Vol. 2, Wiley, New York, 1968.
6. C.N. Sawyer and P.L. McCarty, *Chemistry for Sanitary Engineers*, 2nd edn, McGraw-Hill, New York, 1967, 518 pp.
7. C.N. Sawyer and P.L. McCarty, *Chemistry for Environmental Engineering*, 3rd edn, McGraw-Hill, New York, 1978, 532 pp.
8. K.H. Mancy and W.J. Weber Jr., *Analysis of Industrial Wastewaters*, Wiley-Interscience, New York, 1971, 562 pp.
9. *Standard Methods for the Examination of Water and Wastewater*, 14th edn, American Public Health Association, Washington, DC, 1976, 1193 pp.

10. R.E. McKinney, *Microbiology for Sanitary Engineers*, McGraw-Hill, New York, 1962, 293 pp.
11. L. Klein, *River Pollution II. Causes and Effects*, Butterworths, London, 1962, reprinted 1965, 456 pp.
12. World Health Organization (WHO), Guidelines for Drinking-Water Quality, First Addendum to Third Edition, Vol. 1: Recommendations, Geneva, 2006, 215 pp.
13. L. Klein, *River Pollution III. Control*, Butterworths, London, 1966, 484 pp.
14. R.A. Day Jr. and A.L. Underwood, *Quantitative Analysis*, Prentice-Hall, Englewood Cliffs, NJ, 1958, 465 pp.

Appendix E

Cooling Water Calculations

It's as easy as falling off a bicycle (or riding a log).

E.1 SUMMARY

This appendix outlines a methodology for performing calculations on an open recirculating cooling system and compares calculated versus measured values for certain parameters in several case studies. Different versions of this material have been presented previously at an American Institute of Chemical Engineers Meeting [1] and published in *Chemical Engineering* magazine [2]. It is updated somewhat and included here to make it more accessible along with the other calculations in this book applicable to environmental permitting. The published version has been treated favorably in subsequent presentations by other investigators [3,4], and it is the author's understanding that this material has been used as lecture notes in at least one college/university course.

These procedures show how to calculate evaporation, makeup, and blowdown and the chemical composition of the recirculating water in an open recirculating cooling system. Recommended concentration limits to determine allowable cycles of concentration are given as well as methods to estimate the pH and conductivity of the concentrated water based on theoretically sound correlations of original operating data. The principles are illustrated by a worked example and several case studies for operating plants.

E.2 TOWER PARAMETERS

A cooling tower removes heat by evaporative cooling of a recirculating stream of cooling water; that water picks up heat by passing through the process heat exchangers (Figure E.1). The cooling effect occurs because some of the water evaporates into a stream of air in contact with the water within the cooling tower.

The evaporation can by estimated from a heat balance around the cooling tower

$$E = 0.001(\text{CR})(\Delta T) \tag{E.1}$$

Environmental Calculations: A Multimedia Approach, by Robert G. Kunz
Copyright © 2009 John Wiley & Sons, Inc.

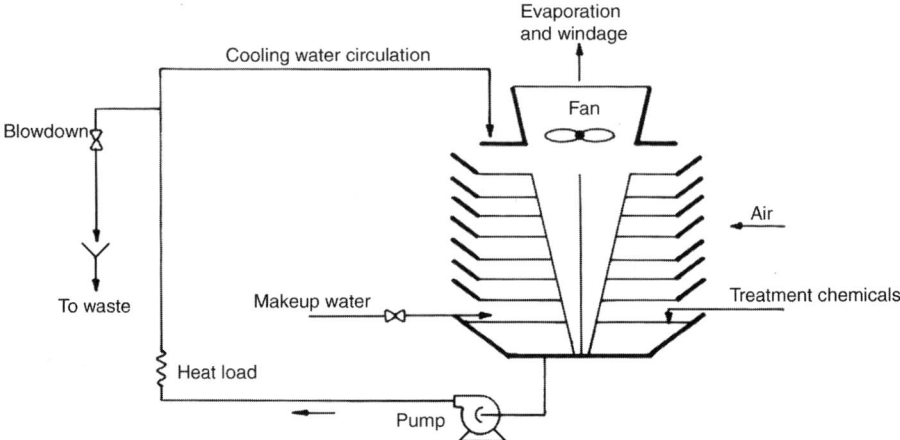

Figure E.1 Cooling system schematic.

in which E and CR are the evaporation and the recirculation rates in gallons per minute (gpm) and ΔT is the temperature difference (°F) between the hot and cold water entering and leaving the tower. This relationship is derived from evaporative cooling with no sensible heat effect, a constant latent heat of evaporation of 1000 Btu/lb, and a heat capacity of the recirculating water at a constant 1 Btu/lb/°F. The evaporation amounts to 1% of the recirculation rate for every 10°F ΔT.

During the evaporation process, the nonvolatile impurities in the makeup water are concentrated. To prevent their buildup to excessive concentrations, some of the recirculating water must be removed, or blown down, from the system. Besides this deliberate blowdown, a relatively small amount of water is lost as fine droplets of liquid entrained by the air stream. This is the so-called windage or drift loss. Unlike evaporation, drift loss carries dissolved impurities with it and reduces rather than concentrates dissolved solids in the recirculating water. (Cooling tower drift is considered to be a source of atmospheric particulate emissions since the water eventually evaporates in the atmosphere, leaving airborne solids. This is discussed in Chapter 3.)

For typical mechanical draft towers that use fans to drive air circulation, 0.1–0.3% of the circulation rate is considered typical [5]. With patented mist eliminator designs, some cooling tower manufacturers warrant as low as 0.008% for windage loss [6]. (The minimum drift loss has changed since the time that this material was originally written; based on the information contained in Problem 3.43, the drift factor can be as low as 0.0005%.)

Makeup water (MU) must be added to replace the evaporation (E), blowdown (B), and drift (W) losses. From a material balance,

$$\text{MU} = E + B + W \tag{E.2}$$

Moreover, the drift/windage is often included in the blowdown term. In that case, the B term represents the upper limit of the amount of water to be removed as blowdown.

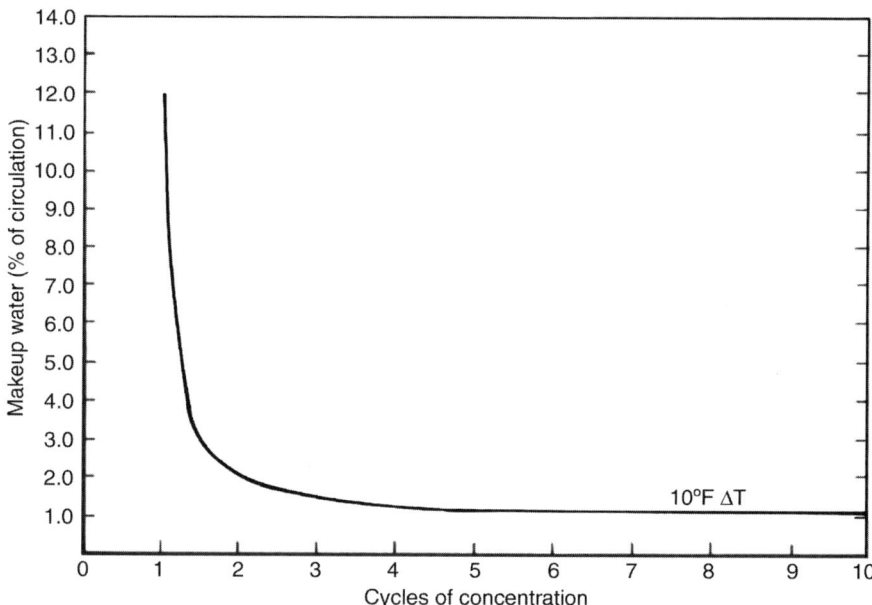

Figure E.2 Makeup decreases with increasing cycles.

Cycles of concentration is a term used to indicate the degree of concentration of the recirculating water with respect to the makeup water

$$\text{Cycles } (C) = X \text{ in recirculating water}/X \text{ in makeup water} \qquad (E.3)$$

where X is any conservative species dissolved in the makeup water; that is, one for which no volatilization, precipitation, or shift in ionic equilibrium occurs.

Makeup (MU), evaporation (E), and cycles of concentration (C) are related by the following equation:

$$\text{MU} = (E)(C)/(C-1) \qquad (E.4)$$

This relationship is shown in Figure E.2 for a 10°F ΔT [5]. It is evident from Figure E.2 that maximizing the number of cycles can minimize makeup water consumption. However, the curve is nearly flat beyond about 3 cycles and reaches a point of diminishing returns at about 6–7 cycles. This provides little incentive to operate at higher cycles for fear of scale formation or enhanced corrosion encountered when the concentrations of certain species are exceeded.

Control of cycles of concentration is discussed below in Section E.4.

E.3 WATER PARAMETERS

E.3.1 Impurities in Cooling Water

Water in its natural state never exists in pure form. It is a universal solvent that dissolves a little of everything and also carries suspended material. The type and amount of foreign matter depends upon the water source and the surrounding environment. The cooling tower is an efficient scrubber of suspended particulates plus gases such as carbon dioxide (CO_2), sulfur dioxide (SO_2), and ammonia (NH_3) from the atmosphere. Other uncontrolled contamination can occur from leakage of process fluids from the heat exchangers into the recirculating water.

The major dissolved impurities in cooling tower makeup are sodium, calcium, magnesium, iron, bicarbonate, chloride, sulfate, and silica. Other mineral impurities occur in lesser amounts. Concentrations are expressed as milligrams/liter (mg/L) or parts per million by weight (ppmw). Analyses of several cooling tower makeup waters are shown in Table E.1. A strict accounting of makeup water impurities is necessary if one is to predict the composition of the recirculating water. These along with flow rates and cycles of concentration are necessary inputs for one's cooling water calculations.

E.3.2 pH, Alkalinity, and Hardness

Among the most important measurements to characterize the impurities in the cooling water are pH, alkalinity, and hardness.

Table E.1 Composition of Four Cooling Tower Makeup Waters

Constituent[a]	Water A	Water B	Water C	Water D[b]
Aluminum	0	0	0.3	0
Calcium	15	147	31	15
Magnesium	6	99	8	0
Sodium[c]	279	602	21	39
Potassium	0	0	0	2
Iron	0	0	0	0
Bicarbonate	334	179	31	31
Carbonate	0	0	15	0
Sulfate	5	137	51	80
Chloride	60	1272	30	9
Fluoride	0	0	1	1
Phosphate	1	0	0	0
Hydroxide	[d]	[d]	1.7	[c]
Carbon dioxide	0	0	0	0
Silica	27	15	7	0
pH (units)	8.0	7.6	10	7.2
P-alkalinity	0	0	25	0
M-alkalinity[e]	274	147	50	25

[a] All units are in mg/L of the constituent, unless indicated otherwise.
[b] Same as water analysis of Table E.4.
[c] Calculated values based on electroneutrality.
[d] Negligible.
[e] Rounded off numbers.

The pH is defined as the negative of the logarithm to the base 10 of the hydrogen ion activity. In dilute solution, the activity is equal to the concentration in g mol/L.

$$\text{pH} = -\log_{10}[\text{H}^+] \tag{E.5}$$

The pH is important because pure water dissociates into hydrogen and hydroxyl ions as follows:

$$[\text{H}^+][\text{OH}^-] = K_w = 10^{-14} \text{ at } 25°\text{C } [7] \tag{E.6}$$

Since the pOH can be similarly defined

$$\text{pOH} = -\log_{10}[\text{OH}^-] \tag{E.7}$$

therefore,

$$\text{pH} + \text{pOH} = pK_w = 14 \text{ at } 25°\text{C} \tag{E.8}$$

Although the hydrated proton, or hydronium ion $[\text{H}_3\text{O}^+]$, is thought to be the actual species present, it is written as the hydrogen ion $[\text{H}^+]$ for simplicity.

This equilibrium is shifted by acids, which donate H^+, or by bases, which contribute OH^-. The acid range is denoted by pH numbers from 0 to 7 and the basic or alkaline range from 7 to 14. The pH of a neutral solution at 25°C is 7.0 and decreases slightly with increasing temperature [8]. In an actual system, the pH is a function of all of the dissolved species in solution. Knowledge of pH is important in predicting the tendency for scaling and corrosion.

Solution pH, the measure of free hydrogen ion activity, should not be confused with alkalinity, which indicates the resistance to change of pH upon addition of an acid. Alkalinity, commonly expressed as mg/L of CaCO_3, is defined as the capacity to neutralize an acid and is measured by titration of a sample with a strong acid. Typical titration curves are depicted in Figure E.3.

Phenolphthalein (P) alkalinity refers to the amount of acid necessary to reach the titration end point for the color change of phenolphthalein indicator from pink to colorless (approximately pH 8.3). Methyl orange (M), or total, alkalinity is the amount necessary to cause a color change in the methyl orange indicator from yellow-orange to a salmon pink (approximately pH 4.3). Alkalinity along the x-axis is an indication of the buffer capacity of the original sample. Here again, the plot reaffirms that alkalinity of the original sample is not synonymous with pH (plotted on the y-axis). Note that in this illustration, the sample with the higher initial pH exhibits the lower alkalinity.

Conversely, acidity is defined as the capacity to neutralize a base. The end points and indicators for the titration (using a strong base such as sodium hydroxide (NaOH)) would be the same, during titration, but the color changes are reversed.

The principal contributors to the alkalinity of natural water are the hydroxide (OH^-), carbonate (CO_3^{2-}), and bicarbonate (HCO_3^-) ions. Alkalinity measurements are often interpreted exclusively in terms of those ions, and the contribution of other species including phosphates, silicates, borates, sulfides, sulfites, the ammonium ion, and organic acids that might also be present is ignored. On this basis, simplified mathematical relationships among the various forms and sources of alkalinity are shown in Table E.2 [9].

Table E.2 Alkalinity Relationships[a]

Condition	Hydroxide alkalinity as $CaCO_3$	Carbonate alkalinity as $CaCo_3$	Bicarbonate alkalinity as $CaCO_3$
$P = 0$	0	0	M
$P < 0.5M$	0	$2P$	$M - 2P$
$P = 0.5M$	0	$2P$	0
$P > 0.5M$	$2P - M$	$2(M - P)$	0
$P = M$	M	0	0

After Ref. [9].

[a] P is the phenolphthalein alkalinity, M is the methyl orange alkalinity, and

$$\text{Hydroxide alkalinity as } CaCO_3^{2-} = [OH^-] \times 50,000$$
$$\text{Carbonate alkalinity as } CaCO_3 = [CO_3^{2-}] \times 100,000$$
$$\text{Bicarbonate alkalinity as } CaCO_3 = [HCO_3^-] \times 50,000$$

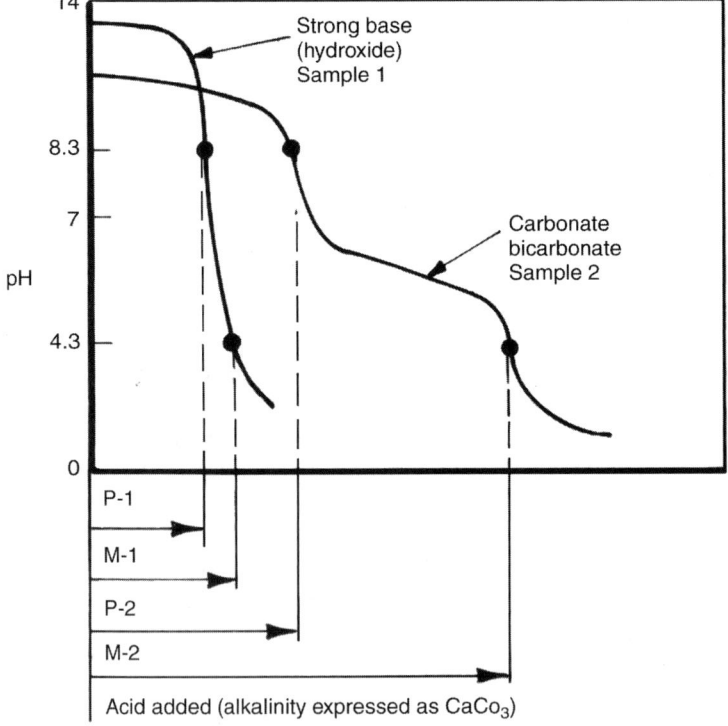

P = Phenolphthalein alkalinity
M = Methyl orange alkalinity
1 = Sample 1
2 = Sample 2

Figure E.3 Typical alkalinity titration curves.

This table presumes the incompatibility of hydroxide and bicarbonate alkalinities, an assumption not valid above pH 9 [10].

The various alkalinity relationships may also be computed from methyl orange (total) alkalinity and an accurately determined pH value by solving Equation E.5 along with the following simultaneous equations:

$$[OH^-] + 2[CO_3^{2-}] + [HCO_3^-] - [H^+] = \text{M-alkalinity}/50,000 \quad (E.9)$$

$$\frac{[H^+][CO_3^{2-}]}{[HCO_3^-]} = K_2 = 4.7 \times 10^{-11} \text{ at } 25°C \text{ [7]} \quad (E.10)$$

Dissolved CO_2 can then be calculated from

$$\frac{[H^+][HCO_3^-]}{[CO_2]} = K_1 = 4.5 \times 10^{-7} \text{ at } 25°C \text{ [7]} \quad (E.11)$$

The P-alkalinity measurement can be used as a check on the sum of hydroxide and carbonate alkalinities. A plot of the CO_2, HCO_3^-, and CO_3^{2-} system is shown in Figure 2.1.

The original definition of hardness included all polyvalent metal ions capable of precipitating soap. The principal contributors to hardness are calcium, magnesium, strontium, ferrous, manganous, and aluminum ions. Of these, only calcium and magnesium are of practical importance in natural waters, and hardness is commonly understood to include only these two ions. Calcium and magnesium can deposit scale in a cooling system by reacting with certain of the anions present.

Calcium, magnesium, and total hardness are also expressed as mg/L of calcium carbonate. Calcium hardness is always determined by measurement. Magnesium hardness is either measured directly or computed by difference of total and calcium hardness. Total hardness can be determined directly by titration with ethylenediamine tetraacetic acid EDTA [9] or from separate measurements of calcium and magnesium. Waters containing up to about 50–75 mg/L hardness are considered soft and above 200–300 are very hard. Waters in the intermediate range are classified as hard.

E.3.3 Milligrams Per Liter as Calcium Carbonate

The term milligrams per liter (mg/L) as $CaCO_3$ is encountered in water treatment calculations. Alkalinity and hardness are invariably expressed in terms of this unit; but concentrations of other constituents such as sodium, chloride, and sulfate are also frequently expressed as calcium carbonate. Care must therefore be exercised in interpreting water analyses.

Milligrams per liter expressed as $CaCO_3$ means the quantity of material having the same number of chemical equivalents as the indicated amount of calcium carbonate and has nothing to do with composition. By definition

$$\text{mg/L CaCO}_3 = \text{mg/L X (EW of CaCO}_3/\text{EW of X)} \quad (E.12)$$

where X represents any species under consideration and EW stands for equivalent weight, the molecular weight (MW) divided by the valence.

Table E.3 Conversion Factors for Calcium Carbonate Equivalent

Constituent	Ion to $CaCO_3$ equiv	$CaCo_3$ equiv to ion
	Multiply by	Multiply by
Aluminum	5.56	0.18
Calcium	2.50	0.40
Iron (Fe^{2+})	1.79	0.56
Iron (Fe^{3+})	2.69	0.37
Magnesium	4.10	0.24
Potassium	1.28	0.78
Sodium	2.18	0.46
Bicarbonate	0.82	1.22
Carbonate	1.67	0.60
Chloride	1.41	0.71
Fluoride	2.63	0.38
Phosphate	1.58	0.63
Sulfate	1.04	0.96
Sulfite	1.25	0.80
Sulfide	3.13	0.32

Since calcium carbonate has a molecular weight of 100 and an equivalent weight of 50, it serves as a convenient common denominator. This artifice enables one to perform calculations using whole numbers of the same magnitude as the original mg/L of each constituent rather than with gram equivalents of the order of 10^{-4}. Conversion factors for several important ions are listed in Table E.3. A more complete listing for both ions and chemical compounds is contained in Ref. [11].

E.3.4 Electroneutrality

The first step in water treatment calculations is to examine the composition of the makeup water for electroneutrality. According to the principle of electroneutrality, the positive and negative ionic charges must be in balance. For example,

$$[H^+] + [Na^+] + 2[Ca^{2+}] + 2[Mg^{2+}] + \cdots = \\ [OH^-] + [HCO_3^-] + 2[CO_3^{2-}] + [Cl^-] + 2[SO_4^{2-}] + 3[PO_4^{3-}] \quad \text{(E.13)}$$

where the bracketed terms are in gram moles per liter. These terms, along with their numerical coefficients, are the chemical equivalents of each ion and can be expressed as (milli)equivalents per liter, or more conveniently, as mg/L as $CaCO_3$.

Hence

$$\text{Total cations (mg/L as } CaCO_3) = \text{Total anions (mg/L as } CaCO_3) \quad \text{(E.14)}$$

In theory, for a complete and correct water analysis, Equation (E.14) will be satisfied exactly. In practice, however, typical water analysis data are seldom complete, and certain deviations from electroneutrality as calculated can be expected.

Table E.4 Ion Balance Calculation

Constituent	Concentration as ion (mg/L)	Cations (mg/L as CaCO$_3$)	Anions (mg/L as CaCO$_3$)
Cations (+)			
Aluminum	0	0	—
Calcium	15	37.5	—
Magnesium	0	0	—
Sodium	—	—	—
Potassium	2	2.56	—
Iron	0	0	—
Anions (−)			
Bicarbonate	31^a	—	25.4
Carbonate	0^a	—	0
Sulfate	80	—	83.2
Chloride	9	—	12.7
Fluoride	1	—	2.63
Neutral species			
Silica	0	—	—
Total	138	40.06^b	123.93

^{a} P-alkalinity = 0; M-alkalinity = 25.4 mg/L as CaCO$_3$.
^{b} Total anions − total cations = (123.93 − 40.06) = 83.87 mg/L as CaCO3. Assuming that the difference is entirely attributable to sodium, sodium concentration as Na will be (83.87)(0.46, from Table E.3) = 39.58 mg/L, or about 39 mg/L as Na.

For example, the sodium ion concentration is often not measured but rather is computed by difference from other constituents. Moreover, in the pH range 6–9, the hydrogen and hydroxide ion concentrations tend to cancel each other and are small compared to the other ions. They are therefore normally neglected in ion balance calculations. Results of an ion balance computation are shown in Table E.4.

E.4 CONTROL OF CYCLES OF CONCENTRATION

Cycles of concentration are controlled by comparing some measurement of concentration in the recirculating water with its value in the makeup water. In practice, chloride, hardness, or conductivity, which is roughly proportional to total dissolved solids (TDS), is used.

Chloride is chosen because it remains soluble even at a high degree of concentration. However, use of the chloride concentration is not valid with less than 5 mg/L in the makeup water [12] and is questionable when the recirculating water has a high chlorine demand and chlorine is being used as a biocide. Chlorine demand is the difference between the amount of chlorine added and the amount of chlorine left in solution at the end of a specified chlorination contact period [13]. Hardness in likewise questionable if precipitation is at all likely. Conductivity is much preferred and is amenable to automatic control. Prediction of recirculating water conductivity is discussed in a subsequent section.

E.5 pH EFFECTS

E.5.1 The pH Increases with Cycles of Concentration

The pH of the recirculating water tends to increase with increasing cycles of concentration and a higher pH intensifies the tendency for scale deposition. As the recirculating water becomes more concentrated, its bicarbonate alkalinity increases, but dissolved CO_2 in equilibrium (Equation E.11) remains more or less constant.

Acid is often added for pH control to allow operation at higher cycles. The added acid will react with the water's alkalinity as follows:

$$2HCO_3^- + H_2SO_4 \rightarrow SO_4^{2-} + 2CO_2 \uparrow + 2H_2O \qquad (E.15)$$
$$CO_3^{2-} + H_2SO_4 \rightarrow SO_4^{2-} + CO_2 \uparrow + H_2O \qquad (E.16)$$

to replace carbonate and bicarbonate ions with sulfate ions and depress the pH. The bulk of the CO_2 formed is stripped from the tower, and the added water is negligible.

Left alone without acid addition, the pH will naturally equilibrate between 8 and 8.5 at the operating cycles of concentration [14]. When high-alkalinity water is used as makeup, the equilibrium pH may lie above 8.5 [15]. With certain Gulf Coast well waters that contain very high alkalinity, the pH may exceed 9.

These generalizations are borne out by the curves of Figure E.4. This figure portrays the expected pH change with cycles of concentration for the four different makeup waters of Table E.1. Without adding acid, the pH of water A would reach 9.5 at 6 cycles. The pH of water B at 6 cycles would be lower because this water contains a lower alkalinity. However, in practice, water B would be limited to less than 2 cycles without acid addition because of hardness and potential scaling. The corresponding pH is 8.3.

Water C is a lime-softened, alum-coagulated water whose alkalinity is low despite its high pH, because of incomplete recarbonation. The pH would first decrease as the water is recirculated and carbon dioxide in solution achieves a more reasonable level. After this initial drop, pH would rise for this water also. Similarly, the initial pH dip for water D indicates an unsaturated condition with respect to CO_2, typical of some well waters. The low alkalinity would permit the pH to increase only to about 8 even at 6 cycles. These curves demonstrate that the makeup water's alkalinity and not its pH is responsible for the pH of the recirculating water.

E.5.2 Prediction of Circulating Water pH

Unless one is considering an existing tower where pH can be measured, the pH of the concentrated water must somehow be predicted. This prediction is a complex problem [16] and is considered here in some detail below.

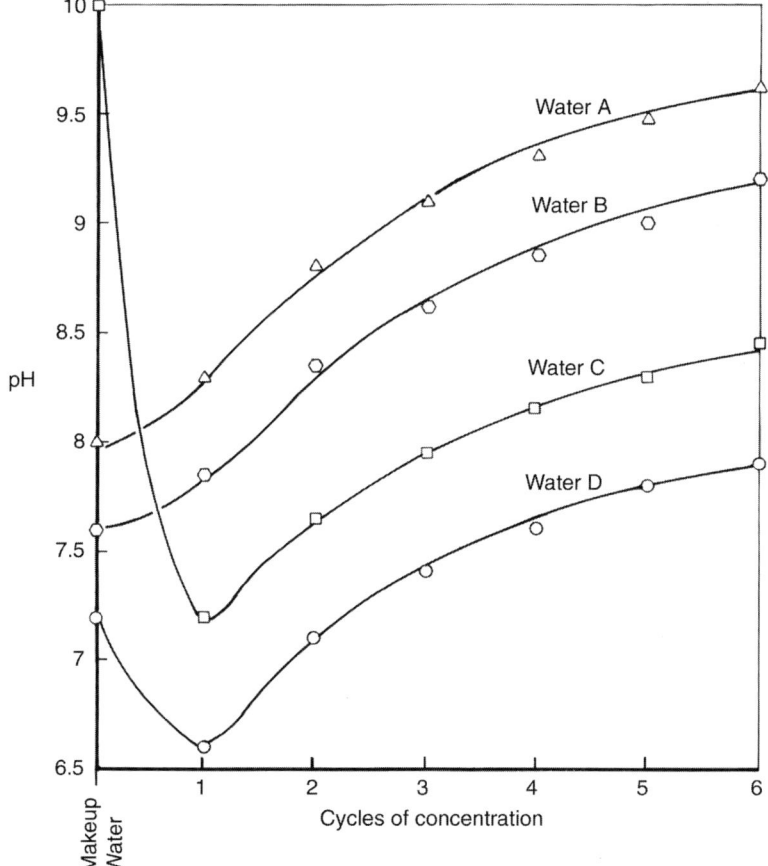

Figure E.4 pH Increases with recirculating water concentration.

The pH of the recirculating water is affected primarily by the carbonate–bicarbonate–CO_2 system. The stripping or scrubbing of carbon dioxide in the cooling tower, the equilibrium distribution between bicarbonate and carbonate, and the pH of the recirculating water are interrelated. Even with adequate input information, to solve the simultaneous equations that describe the relationship for these species is at best cumbersome.

Instead, the empirical correlation of total alkalinity and pH of the recirculating water (Figure E.5) based on routine monitoring data gathered from over 40 cooling towers provides an excellent, simple tool for estimating purposes. These towers represent different types of mechanical construction as well as different makeup water compositions and water treatment programs. Cold water temperatures are typically in the range of 90°F with a 10–20°F ΔT.

Approximately 400 pH–alkalinity data points were subjected to least squares regression in which pH was considered as the dependent variable. The 0.8 correlation coefficient obtained reflects the scatter but indicates a definite correlation between pH

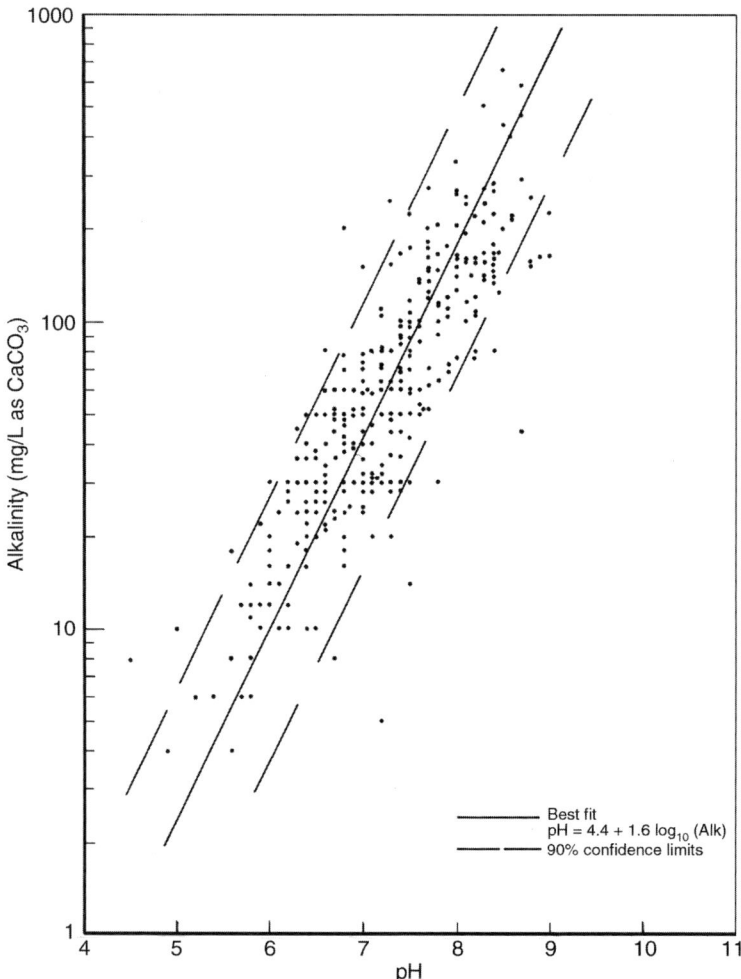

Figure E.5 Cooling tower water alkalinity–pH relationship.

and alkalinity [17]. At the 90% confidence level, pH can be predicted within ±0.75 units when the alkalinity is known. These data approximate the rule of thumb that 20–50 mg/L of methyl orange alkalinity will correspond to a pH of 6.0–7.0 [12]. Water quality, treatment, and aeration in the tower may affect both alkalinity and pH [12].

The scatter is caused primarily by differences in the concentration of dissolved CO_2. A 1 mg/L increase in CO_2 in the range of 1–10 mg/L produces a decrease of 0.1–0.2 pH units at constant alkalinity. The CO_2 concentration maintained in solution would be expected to vary both from tower to tower and with liquid and gas rates in a single tower. Analytical results depend on sample handling.

Variability in temperature and dissolved solids also contributes slightly to scatter. An increase in temperature from 50 to 140°F would result in a decrease of 0.2 pH units; for an increase in TDS of 3000 mg/L, the pH would decrease by only 0.1.

Changes in the ambient atmosphere from time to time and from place to place may also alter the pH–alkalinity characteristics. For example, Sussman [14] points out the difference in pH between two cooling towers in the same immediate vicinity caused by the pickup of SO_2 from a plant stack by one of them but not the other.

Scatter in the pH-alkalinity relationship is also caused by the presence of various water treatment chemicals. Detailed consideration of additives to control corrosion, scaling, and biological fouling is outside the scope of this discussion; an excellent treatment is contained elsewhere [18].

E.5.3 Dissolved CO₂ Is Not Exactly Constant

In Figure E.6 are shown the line of best fit correlating the cooling water pH-alkalinity data of Figure E.5 plus several lines computed from theory with dissolved CO_2 as parameter.

Since the majority of the alkalinity between a pH of about 4.3–8.3 is caused by bicarbonate ion in equilibrium with CO_2 (Table E.2 with $P=0$), then from Equation (E.11) in logarithmic form, a semilogarithmic plot of alkalinity versus pH should yield a straight line with a slope of unity.

$$pH = \log_{10}[HCO_3^-] + (pK_1 - \log_{10}[CO_2]) \tag{E.17}$$

provided that CO_2 is constant.

However, the observed slope of the best-fit line implies that the dissolved CO_2 is not exactly constant, but is indeed a function of pH. To attain low pH values, a good deal of acid must be added and much CO_2 is formed in solution (see Equations E.15 and E.16). The empirical fit in the middle pH range from 6.8 to 7.6 corresponds to a dissolved CO_2 concentration between 5 and 10 mg/L. Applebaum [19] indicates concentrations of 5–10 mg/L CO_2 attainable with typical wooden degasifiers having geometries similar to cooling towers. This range of CO_2 concentrations is considerably in excess of the 0.4 mg/L in equilibrium with the CO_2 in the atmosphere [8]. Higher than anticipated dissolved CO_2 values imply a mass transfer limitation in stripping this gas from the recirculating water. This point is expanded upon in the subsequent Ref. [3], in that the greater amount of CO_2 produced in the water by acid addition to achieve the pH and alkalinity at the lower end of the plot results in higher than equilibrium CO_2 concentrations that must (but cannot) be stripped off in the tower.

Lower CO_2 values at the upper end of the experimental line are to be expected since little or no acid would be added for control in this region, with minimal formation of dissolved CO_2. The only other sources of CO_2 are the makeup water, thermal breakdown of bicarbonate ion,

$$2HCO_3^- \rightarrow CO_3^{2-} + CO_2 \uparrow + H_2O \tag{E.18}$$

and pickup from the atmosphere. Finally, the presence of carbonate above pH 8.3 should cause the line to curve upward, when total methyl orange alkalinity is plotted against pH, but the anticipated curvature in the empirical relationship is obscured by the scatter of the data in Figure E.5.

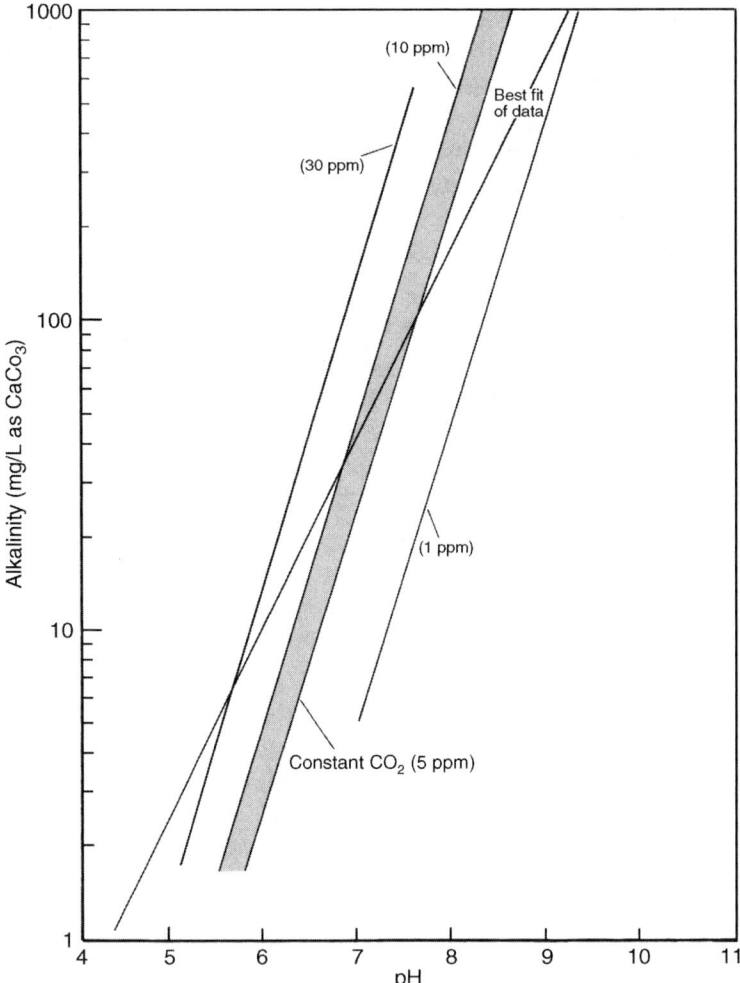

Figure E.6 pH Prediction using constant CO_2.

E.5.4 Quantity of Acid Required

To control the pH at the desired level, sulfuric acid is added to the recirculating water (Equations E.15 and E.16). Since one molecular weight of sulfuric acid will destroy two equivalents of alkalinity, the quantity of sulfuric acid required can be formulated as follows:

$$W = \frac{(\Delta \text{Alk})}{50,000} \times 0.5 \times 98 \times B \times 1440 \times \frac{1}{10^3} \times \frac{8.34}{0.9319} \quad (E.19)$$
$$= 0.0126 \times \Delta \text{Alk} \times B$$

where W is the 66° Baumé sulfuric acid required (lb/day), ΔAlk is the alkalinity to be reduced in the recirculating water (mg/L as $CaCO_3$), and B is the blowdown water flow rate (including drift/windage loss) (gpm).

Towers employing chromate ions as the corrosion inhibitor normally operate slightly below a pH of 7 (6.2–6.8 is a common control range), and nonchromate inhibitor programs are usually accompanied by a somewhat higher pH.

Chromate (hexavalent chromium) programs have fallen into environmental disfavor. They have either been replaced by nonchromate programs, or the chromate is removed before discharge of blowdown. Two such removal processes are reduction to insoluble trivalent chromium using a sacrificial iron electrode [20,21] followed by sedimentation and centrifuging of the sludge produced [21] or addition of sulfur dioxide (SO_2) [22] followed by lime precipitation [23].

E.6 TOTAL DISSOLVED SOLIDS AND CONDUCTIVITY

Total dissolved solids and conductivity are two of the most critical variables in cooling tower control. The ability to estimate these two quantities permits the consideration of design limits and allows the selection of a conductivity meter with an appropriate range. Conductivity is often used to control the cycles of concentration.

E.6.1 TDS

TDS can be calculated by summation of all the dissolved species in the water including silica (expressed as $Si(OH)_4$) but not including dissolved gases. A calculated TDS value should be higher than the measurement of filterable residue. In such an analysis, a sample is usually dried at 103–105°C, resulting in loss of carbon dioxide and conversion of bicarbonate to carbonate (Equation E.18). According to the stoichiometry, up to about 51% of the bicarbonate originally present will not be measured as TDS [9].

E.6.2 Prediction of Conductivity

For a rough estimate, conductivity can be calculated by dividing the filterable residue by a factor in the range 0.55–0.7 [9]. A more rational estimate is obtainable from the conductivity factor technique. Conductivity of the recirculating water can be predicted within ±25% of measured values from the concentrations of its major ions, their activity coefficients, and a factor specific for each ion [7,9] (Table E.5), followed by summation of the products, according to the following equations:

$$\text{Ionic strength } I = 0.5 \sum_{i=1}^{n} C_i Z_i^2 \tag{E.20}$$

$$\text{Activity coefficient } \log f_i = -AZ_i^2 \left(\frac{I^{1/2}}{1+I^{1/2}} - 0.2I \right) \quad (I < 0.5) \tag{E.21}$$

$$\text{Activity } a_i = f_i c_i \tag{E.22}$$

Table E.5 Conductivity Factors for Ions Commonly Found in Water

Ion	Conductivity @ 25°C, μmho/cm per mg/L of ion
Aluminum	3.44
Bicarbonate	0.715
Calcium	2.6
Carbonate	2.82
Chloride	2.14
Fluoride	2.86
Iron (2+)	1.93
Iron (3+)	3.65
Magnesium	3.82
Potassium	1.84
Phosphate	2.18
Sodium	2.13
Sulfate	1.54

where I is the ionic strength (g mol/L), C_i is the concentration for the ith ion (g mol/L), Z_i is the charge of the ith ion, A is a constant $\cong 0.5$ (tabulated as a function of temperature in Ref. [7]), f_i is the activity coefficient for the ith ion, a_i is the activity for the ith ion (mg/L), and c_i is the concentration for the ith ion (mg/L).

Activity coefficients are necessary to account for the effect of concentration of the recirculating water. The Davies equation [24], Equation (E.21), has been found to produce activity coefficients adequate for conductivity predictions. The method is valid in the pH range 6–9, where the concentrations of the highly conductive H^+ or OH^- ions are not significant, and for conductivities above 90 μmho/cm [9].

In Figure E.7 are plotted the predicted versus measured conductivities at 18 or 25°C for the available 136 recirculating water data points and 116 makeup water points. Ionic strength varied between 0.01 and 0.12 for the cooling waters and as low as 0.001 for fresh makeup. The 1:1 45° angle parity line appears to correlate the data quite well, even for low values of conductivity. Conductivity increases (or decreases) 2% for every 1°C above (or below) 25°C [9]. This relationship permits the adjustment of conductivity from ambient to tower temperatures.

E.7 ALLOWABLE CYCLES OF CONCENTRATION

Several criteria for determining the maximum allowable cycles of concentration are summarized in Table E.6 and are discussed briefly below, along with some others. These guidelines given in the literature should be tempered by one's own experience and/or the recommendations of a reputable water treatment services company. Unless otherwise stated, quoted limits refer to the recirculating water.

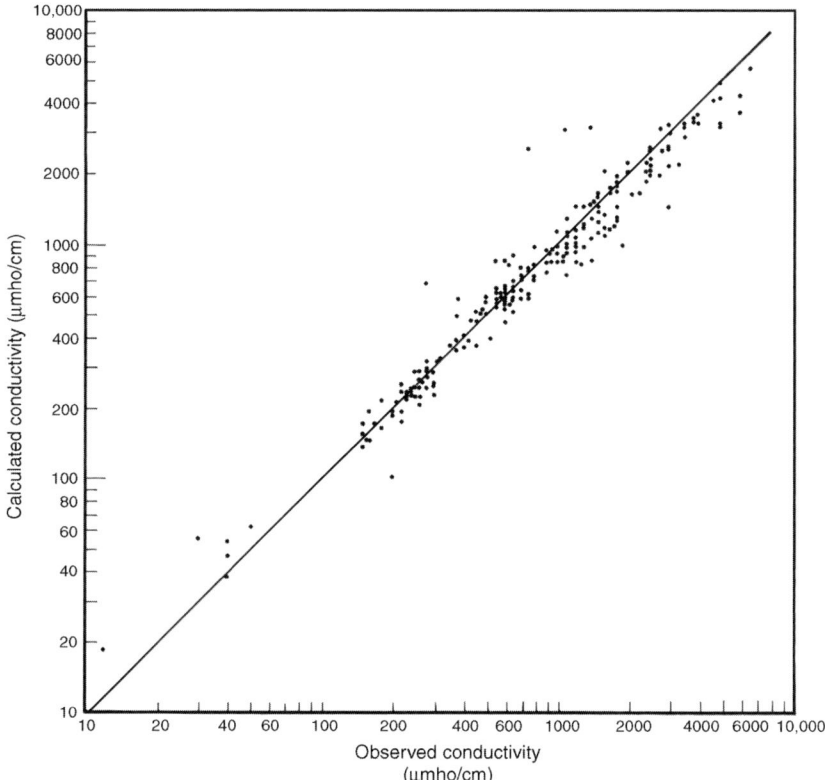

Figure E.7 Results of conductivity prediction.

E.7.1 Langelier Saturation Index

The theoretically based Langelier index is defined as the difference between the actual pH and the pH at which a given water would be saturated with calcium carbonate ($CaCO_3$):

$$pH - pH_s = pH - (pK_2 - pK_s + pCa + pAlk) \quad (E.23)$$

The terms in the equation are negative logarithms of the second ionization constant for carbonic acid (Equation E.10), the solubility product for calcium carbonate ($K_s = 4.5 \times 10^{-9}$ at 25°C [7], and the molar and equivalent concentrations of calcium ion and methyl orange alkalinity, respectively. The constants are a function of temperature and total dissolved solids [8]. Nomographs are available to simplify this calculation [5,25].

When $pH - pH_s$ is positive, the system has a tendency to deposit $CaCO_3$ scale. When the difference is negative, the system tends to dissolve $CaCO_3$ scale and, by inference, is corrosive. When $pH - pH_s = 0$, the system is theoretically in balance. However, an index of -0.5 to $+0.5$ is said to be unreliable for predicting scaling tendency [14].

Table E.6 Guidelines for Cooling Tower Operation

Parameter	Limits (min)	Limits (max)	Remarks
Langelier saturation index	+0.5	+1.5	Nonchromate programs
Ryznar stability index	+6.5	+7.5	Nonchromate programs
pH	6.0	8.0	
Calcium (mg/L as $CaCO_3$)	20–50	300	Nonchromate programs
		400	Chromate program
Total iron (mg/L)		0.5	
Manganese (mg/L)		0.5	
Copper (mg/L)		0.08	
Aluminum (mg/L)		1	
Sulfide (mg/L)		5	
Silica (mg/L)		150	For pH < 7.5
		100	For pH > 7.5
(Ca)(SO_4) (product)		500,000	Both calcium and sulfate expressed as mg/L $CaCO_3$
Total dissolved solids (mg/L)		2500	
Conductivity (μmho/cm)		4000	
Suspended solids (mg/L)		100–150	

Where still in use, chromate treatments operate at negative values of the Langelier index and treat the water to prevent corrosion. Nonchromate treatments operate at positive values to take advantage of the corrosion protection offered by a thin film of calcium salts; gross precipitation of $CaCO_3$ is inhibited by the addition of deposit control chemicals including chelating agents and crystal growth modifiers.

Nonchromate treatments are operated between 0.5 and 1.0 [26,27], at 1.0 [14,26], or at a maximum of 2.5 for calcium carbonate precipitation and 1.5 for calcium phosphate precipitation [28]. These values vary according to the maximum temperature in the system [27]. A range of about 1.0–1.5 appears to be in order.

E.7.2 Ryznar Stability Index

This is an empirical index, defined as $2pH_s - pH$, and is based on a study of operating data with water having various saturation indices. For this index, 6.5 is the nominal neutral point. Values of 6 or less indicate scaling [12,14,26], and 7 or greater are corrosive [14]. The 6 to 7 range is inconclusive. The range 5.5–6 corresponds to a Langelier index of about +1.0.

E.7.3 Corrosion Ratio

This corrosion ratio (R), defined as follows

$$R = ((1.4\,\text{mg/L}\,(\text{Cl}^-) + 1.04\,\text{mg/L}\,(\text{SO}_4^{2-}))/(\text{alkalinity mg/L as CaCO}_3) \quad (\text{E.24})$$

was developed from work done on the corrosion of steel and cast iron in the Great Lakes Waters [29] and is an attempt to quantify the effect of certain ions in the water. The numerical coefficients are conversion factors to change from chemical equivalents per million (epm) to mg/L. For a pH of 7–8, a value of R of 0.1 or less for untreated water indicates relative freedom from corrosion [18]. The higher the value, the more corrosive is the water. High chloride, high alkalinity waters treated with sulfuric acid for pH control can result in index values approaching 100.

E.7.4 Allowable pH Range

The pH should be kept within the control range for the corrosion inhibitor program in use. In general, chromate programs should be kept above a minimum of about pH 6 [18], with somewhat higher values recommended for other types of treatment.

At high pH [12,26,30], especially with high alkalinity low hardness water [31], chlorination for biological control oxidizes the lignin in cooling tower wood to aldehydes and acids [18]. Delignification reportedly commences at a pH of 7.8 [18]. Moreover, chlorine's biocidal properties are weak at pH 8 and negligible at pH 9 [18]. (This is caused by ionization of hypochlorous acid resulting from chlorine addition and is discussed further in Chapter 5). The effectiveness of certain other biocides is also diminished in the same pH range.

E.7.5 Calcium Salts

Solubility of calcium carbonate, 20–50 mg/L at cooling tower temperature [14,27] is accounted for in the Langelier index. Magnesium carbonate requires a pH of 9 or greater for precipitation and is generally not encountered in cooling systems [12].

With a nonchromate corrosion inhibitor, calcium hardness should be at least 20–50 mg/L as CaCO_3 [27,32] but less than 300 mg/L as CaCO_3 [18]; for a chromate program, the upper limit can be extended to 400 [18]. The minimum concentration is to enable the formation of a protective film [32], and the maximum is to ensure that calcium can be kept from precipitating using normal concentrations of treatment chemicals [18].

Calcium also forms a precipitate with sulfate, $\text{CaSO}_4 \cdot 2\text{H}_2\text{O}$ (gypsum), at much higher concentrations, 1200–2000 mg/L [13,27]. To avoid scaling, the sum of calcium and sulfate (both expressed as mg/L CaCO_3) should be below 1500, and their product should be less than 400,000–500,000 [14]. This corresponds to a gypsum concentration of about 1200 mg/L and a solubility product of the same order of magnitude as $K_{sp} = 2.4 \times 10^{-5}$ quoted by McCoy [18]. Precipitation is possible whenever the actual product of the ionic species exceeds the solubility product constant, K_{sp}. However, this approach appears to be highly conservative at ionic strengths well above the range normally encountered in a cooling tower because of the effect of activity coefficients on the apparent solubility product [33].

$$K_{\text{sp apparent}} = \frac{K_{\text{sp actual}}}{f_{\text{Ca}} f_{\text{SO}_4}} = [\text{Ca}^{2+}][\text{SO}_4^{2-}] \quad (\text{E.25})$$

Where f_{Ca} and f_{SO_4} are activity coefficients, approaching unity in dilute solution.

E.7.6 Calcium and Zinc Phosphate Saturation Indexes

Phosphate, whether originating in the makeup water or from added treatment chemicals, forms sticky phosphate sludges with calcium, zinc, and other metals. Calcium originates in the makeup water, and zinc ion is sometimes added as a water treatment chemical. The tendency to precipitate is accentuated with increasing metal concentrations and pH. For calcium and zinc phosphates, the tendency can be expressed semiquantitatively by saturation indexes.

Like the Langelier index, the calcium phosphate saturation index (CPSI) is the algebraic difference of the cooling water pH and the equilibrium pH at which calcium phosphate $Ca_3(PO_4)_2$ is saturated:

$$\text{CPSI} = \text{pH} - (\text{pH}_s)_{\text{cp}} \quad (\text{E.26})$$

Similarly, the zinc phosphate saturation index (ZPSI) is

$$\text{ZPSI} = \text{pH} - (\text{pH}_s)_{\text{zp}} \quad (\text{E.27})$$

Although an orthophosphate determination is expressed as PO_4^{3-}, only a fraction actually exists in the PO_4^{3-} form and only the PO_4^{3-} form precipitates with calcium. Other species, HPO_4^{2-}, $H_2PO_4^-$, and H_3PO_4 also contribute to the total, depending on the pH. Mathematically,

$$[\text{Ca}^{2+}]^3 [\text{PO}_4^{3-}]^2 = (K_s)_{\text{cp}} \quad (\text{E.28})$$

$$\frac{[\text{H}^+][\text{H}_2\text{PO}_4]}{[\text{H}_3\text{PO}_4]} = K_1' \quad (\text{E.29})$$

$$\frac{[\text{H}^+][\text{HPO}_4^{2-}]}{[\text{H}_2\text{PO}_4^-]} = K_2' \quad (\text{E.30})$$

$$\frac{[\text{H}^+][\text{PO}_4^{3-}]}{[\text{HPO}_4^{2-}]} = K_3' \quad (\text{E.31})$$

$$\text{TP} = [\text{H}_3\text{PO}_4] + [\text{H}_2\text{PO}_4^-] + [\text{HPO}_4^{2-}] + [\text{PO}_4^{3-}] \quad (\text{E.32})$$

Simultaneous solution, rearrangement of terms, and taking logarithms yield

$$[\text{Ca}^{2+}]^3 ((\text{TP}) K_1' K_2' K_3')^2 / (K_1' K_2' K_3' + [\text{H}_s^+]^3_{\text{cp}} + K_1' [\text{H}_s^+]^2_{\text{cp}} + K_1' K_2' [\text{H}_s^+]_{\text{cp}})^2 = (K_s)_{\text{cp}} \quad (\text{E.33})$$

Details of the derivation are contained elsewhere [34]. Solution for $(H_s^+)_{cp}$ and ultimately $(pH_s)_{cp}$ is by trial and error.

It should be noted that tricalcium phosphate is not the sole species that forms calcium phosphate scale. For example, magnesium may also participate in the precipitation reaction. Tricalcium phosphate was selected as the basis for the above calculations because it is the least soluble of the calcium phosphates for which the necessary solubility data are available. Once deposited, the calcium phosphate precipitate slowly transforms to the crystalline hydroxyapatite, $Ca_{10}(PO_4)_6(OH)_2$.

In a similar manner, one obtains an analogous expression for zinc with the value for $(K_s)_{zp}$ of 9.1×10^{-33} [18]. This figure is in excellent agreement with a value of 10^{-37} based on field experience [35].

The interested reader is referred elsewhere for further details [18,34].

E.7.7 Silica

Silica in natural waters results from the degradation of silica-containing rocks in the earth's crust. This silica occurs as suspended particles, in a colloidal or polymeric state, and as dissolved material [9]. Dissolved silica in natural water exists in the form of $Si(OH)_4$ orthosilicic acid; silicate anions are formed only above a pH of about 9 [24]. For these reasons, silica concentrations are not used in computing the cation–anion balance.

In a cooling system, silica may deposit as a scale on heat transfer surfaces and may therefore be the limiting parameter in a cooling water treatment program. The history and form of silica is important [28]. The scaling potential of magnesium silicate increases as the pH rises above 8 [28], but magnesium silicate is generally not actually encountered in cooling systems unless the pH is above 9 [12]. Dissolved silica can coprecipitate with iron and manganese hydroxides and may be adsorbed on various hydroxides including aluminum, manganese, and magnesium. Examples of silica scale precipitated with calcium salts are pictured in Ref. [36]. Deposit control agents containing acrylate polymers reduce the tendency of silica scale to form.

Levels in the makeup of 40–50 mg/L, expressed as SiO_2, are a source of silica scale and may limit cycles of concentration [27]. Recommended maximum values of silica in the recirculating water, expressed as mg/L SiO_2 are 150–175 [18], 150–200 [12,27], and 200 maximum [28]. If the silica level is above 100, a pH of 7.5 or less is preferred [28]. Based on this information, a maximum silica concentration of 100–150 mg/L in the recirculating water appears to be conservative, with values at the lower end of the range being indicated for a pH of 7.5 and above.

E.7.8 Other Dissolved Materials

Heat-insulating deposits can also form by oxidation of soluble ferrous ion in the water supply followed by precipitation of a hydrated ferric oxide sludge [26], which is converted by heat to iron oxides [12]. Iron deposits can also originate from corrosion products, and precipitation of iron compounds can be aggravated by iron bacteria [12].

Iron should be limited to 0.5 mg/L [27]. Above 0.5 mg/L, a phosphonate/polymer treatment is required to control iron deposition [28]. Concentrations greater than 3 mg/L require higher phosphonate levels because of adsorption of the phosphonate on iron [28]. With levels approaching 10 mg/L, iron removal should be considered [28]. Likewise, manganese should be limited to 0.5 mg/L. [27].

Acceptable concentrations of copper range from 0.04 to 0.08 mg/L; concentrations greater than 0.1 mg/L cause significant galvanic corrosion [18]. High values of total dissolved solids will accelerate copper corrosion with nonchromate treatments [28].

Aluminum causes deposition problems. The recommended level is less than 1 mg/L [28]. Fluoride added as sodium fluoride or sodium fluorosilicate complexes aluminum ion but not suspended alumina floc carryover from a clarifier [12]. Sulfides should be limited to less than 5 mg/L because of their corrosive effects [27].

Zinc hydroxide solubility should be checked to determine if zinc added in a treatment program is being utilized effectively or is being wasted by precipitation and sludge formation. In one study [37], greater amounts of zinc hydroxide solids were encountered with increasing pH in the pH range of 7–8.5. At typical orthophosphate concentrations and pH values in a cooling tower, zinc will be more or less soluble as the phosphate salt than as the hydroxide, and it is well to check both possibilities.

Magnesium is an important ion in water softening, in which it is removed as the hydroxide at high pH. Its solubility may also be considered according to the solubility product principle, and others have presented an index for magnesium hydroxide stability [38]. In any event, at typical magnesium levels in a cooling tower, magnesium hydroxide does not begin to cause trouble unless the pH is somewhere between 9 and 10, at which point other problems will probably already have arisen. Although the solubility of magnesium phosphate is said to be relatively high [12], certainly relative to that of calcium phosphate, magnesium phosphate will likely pose more of a limitation than the hydroxide.

Other candidates for sludge and scale formation can be similarly evaluated. The list of compounds is seemingly limited only by one's own imagination.

E.7.9 TDS and Conductivity

High values of total dissolved solids increase conductivities, cause galvanic corrosion, interfere with inhibitor film formation, accelerate the corrosion of copper in nonchromate programs, and destroy effectiveness of certain biocides. An upper limit of 2000 mg/L has been recommended to minimize galvanic corrosion [27]. Conductivities above 4000 μmho/cm indicate extremely corrosive conditions and can cause marginal to poor results in a nonchromate program [28]. A conductivity of 4000 μmho/cm would correspond to a TDS between about 2000 and 3000 mg/L. A good cutoff point for TDS therefore appears to be approximately 2500 mg/L.

E.7.10 Suspended Solids

Suspended solids enter the cooling tower with the makeup water, with the air passing through the cooling tower, and from debris within the cooling tower circuit. They can be inorganic or biological in origin.

Suspended solids in a cooling tower are abrasive [26] and can lead to under-deposit corrosion [26,30]. Suspended solids also consume certain biocides [31,39], reducing the effectiveness of the microbiological control program. Recommended limits are 100 mg/L [14,27] up to a maximum of 150 mg/L [27]. Suspended solids can be controlled by chemicals that tend to disperse or agglomerate the particles, or they can be removed to the desired level by installing a sidestream filter.

Sidestream filters are frequently employed to control suspended solids in the recirculating water [12,14,25–27,30,40] by filtering from 1% to 5% of the total circulation. This technique has obtained as much as 80% reduction in suspended solids 10 μm and greater [12]. The smaller, colloidal particles can be dispersed with a suitable silt dispersant [39]. Additional design information and operating experience on sidestream filters can be found elsewhere [41].

E.8 EXAMPLE PROBLEM

Step-by-step calculation procedures are illustrated below for 5 cycles, using water D as makeup (Table E.1).

(1) Compute evaporation, makeup, and blowdown: for a circulation rate of 50,000 gpm and a 15°F cooling range, evaporation is 750 gpm (Equation E.1), makeup is 937.5 gpm (Equation E.4), and the sum of blowdown and windage losses is 187.5 gpm (Equation E.2 rearranged). The windage contribution will amount to 100 gpm at 0.2% of the circulation rate.

(2) Check electroneutrality of makeup water: a comparison of total cation and anion concentrations, all expressed as mg/L $CaCO_3$, shows anions to be in excess (Table E.4). To satisfy electroneutrality, 39 mg/L (as Na) must be added to account for the difference.

(3) Cycle it up: for all constituents except carbonate, bicarbonate, and carbon dioxide, the concentration at 5 cycles (Table E.7) is simply 5 times that of the makeup (Table E.4). Total alkalinity expressed as $CaCO_3$ will be five times 25.4 mg/L, or 127 mg/L. The corresponding pH is 7.8 (Figure E.5). The concentrations of carbonate and bicarbonate can be calculated using Equations (E.9) and (E.10), and the CO_2 concentration from Equation (E.11).

For 5 cycles and above, the pH is near the upper limit in Table E.6, and provisions for acid addition may be justified. Concentrations of bicarbonate, carbonate, carbon dioxide, and sulfate would change in adding acid. To achieve an arbitrarily selected pH of 7, the alkalinity must be reduced to 42 mg/L (Figure E.5). At this pH, the bicarbonate and carbonate concentrations are 51 and 0 mg/L of the ion, respectively. Carbon dioxide concentration is about 8 mg/L.

To reduce the alkalinity from 127 to 42 mg/L takes $(127 - 42) \times 1/50 \times 0.5 \times 98 = 83.3$ mg/L of H_2SO_4 or 82 mg/L as SO_4^{2-}, requiring 201 lb/day of 66° Baumé acid (Equation E.19). Total sulfate in the recirculating water at 5 cycles (482 mg/L) is the sum of the sulfate added (82 mg/L) plus that concentrated from the makeup water ($5 \times 80 = 400$ mg/L).

Table E.7 Calculation Procedure Illustration for Water D at 5 Cycles

Constituent	Concentration (mg/L)	$0.5 C_i Z_i^2$ (10^3 mol/L)	f_i	A_i (mg/L)	Conductivity (μmho/cm)
Aluminum	0	0	0.300	0	0
Calcium	75	3.74	0.586	44.0	114
Magnesium	0	0	0.586	0	0
Sodium	195	4.24	0.875	171	364
Potassium	10	0.128	0.875	8.75	16
Iron	0	0	0.586	0	0
Bicarbonate	155	1.28	0.875	137	98
Carbonate	0.45	0.015	0.586	0.264	1
Sulfate	400	8.33	0.586	234	360
Chloride	45	0.636	0.875	39.4	84
Fluoride	5	0.13	0.875	4.38	13
Carbon dioxide	4	0	0	0	0
Silica	0	0	0	0	0
Sum	890	18.5			1050

TDS (calculated)(mg/L) = 890
Theoretical filterable residue (mg/L) = 810
pH (units) = 7.8

(4) Calculated TDS: summing concentrations of all constituents yields a TDS of 890 mg/L before adding acid (Table E.7). Similarly, TDS after acid addition is 870 mg/L (Table E.8).

(5) Estimated conductivity: compute ionic strength (Equation E.20) and activity coefficients (Equation E.21). The product of concentration and activity coefficient is activity (Equation E.22). Multiply the activity of each ion by the appropriate conductivity factor from Table E.5 to yield the conductivity contributed by that ion and sum all contributions to obtain 1050 μmho/cm, the conductivity of the water (Table E.7).

(6) Compute parameters for other cycles of concentration: these results are listed in Table E.8.

(7) Determine the approximate cycles allowable for water D. A comparison of the entries of Table E.8 with the guidelines in Table E.6 reflects the good quality of water D. Six to perhaps 7 cycles appear reasonable.

(8) Determine allowable cycles for other waters. Similarly, for waters A, B, and C, the calculated values of limiting water quality parameters at the critical cycles of concentration are presented in Table E.9.

Water A—because of high alkalinity, the natural pH at 2 cycles already far exceeds 8 with a Langelier index of +1.85. Acid should be added in order to lower pH

Table E.8 Detailed Calculation Results for Water D (All units are in mg/L of the constituent unless otherwise indicated)

Constituent	Adjusted MU	2X BA	2X AA	3X BA	3X AA	4X BA	4X AA	5X BA	5X AA	6X BA	6X AA
Calcium	15	30	30	45	45	60	60	75	75	90	90
Magnesium	0	0	0	0	0	0	0	0	0	0	0
Sodium	39	78	78	117	117	156	156	195	195	234	234
Potassium	2	4	4	6	6	8	8	10	10	12	12
Bicarbonate	31	62	51	93	51	124	51	155	51	185	51
Carbonate	0	0	0	0	0	0	0	1	0	1	0
Sulfate	80	160	178	240	272	320	368	400	482	480	586
Chloride	9	18	18	27	27	36	36	45	45	54	54
Fluoride	1	2	2	3	3	4	4	5	5	6	6
Carbon dioxide	0	7	8	6	8	5	8	4	8	4	8
pH (units)	7.2	7.1	7.0	7.4	7.0	7.6	7.0	7.8	7.0	7.9	7.0
P-alkalinity (as $CaCO_3$)	0	0	0	0	0	0	0	0	0	2	0
M-alkalinity (as $CaCO_3$)	25.4	51	42	76	42	102	42	127	42	152	42
LSI @ 100°F (units)	−1.92	−0.84	−1	−0.2	−0.85	+.024	−0.74	+.063	−0.65	+.089	−0.57
RSI @ 100°F (units)	+10.44	+8.78	+9.00	+7.80	+8.70	+7.12	+8.48	+6.54	+8.3	+6.12	+8.14
TDS (calculated)	180	350	360	530	520	710	700	890	870	1060	1040
Theoretical filterable residue	160	320	340	480	500	650	660	820	840	970	1010
Conductivity @ 25° (μmho/cm)	240	460	470	650	660	840	840	1050	1050	1200	1200

Note: BA, before adding acid; AA, after adding acid.
X is cycles of concentration.

Table E.9 Values of Critical Parameters for Various Waters

Constituent	Water A, 2X BA	Water A, 2X AA	Water A, 4X BA	Water A, 5X AA	Water B, 2X BA	Water B, 2X AA	Water B, 3X AA	Water C, 3X BA	Water C, 3X AA	Water C, 5X BA	Water C, 5X AA
Langelier saturation index @ 100°F (units)	+1.85	−1.05	+2.92	−0.68	+2.04	−0.14	+0.02	+1.09	−0.45	+1.86	−0.25
Ryznar stability index @ 100°F (units)	+5.1	+9.1	+3.47	+8.36	+4.27	+7.28	+6.96	+5.77	+7.90	+4.58	+7.50
pH (units)	8.8	7.0	9.31	7.0	8.35	7.0	7.0	7.95	7.0	8.3	7
Calcium (mg/L as CaCO$_3$)	75	75	150	187.5	735	735	1102.5	232.5	232.5	387.5	387.5
Total iron (mg/L)	0	0	0	0	0	0	0	0	0	0	0
Manganese (mg/L)	0	0	0	0	0	0	0	0	0	0	0
Copper (mg/L)	0	0	0	0	0	0	0	0	0	0	0
Aluminum (mg/L)	0	0	0	0	0	0	0	0.9	0.9	1.5	1.5
Sulfide (mg/L)	0	0	0	0	0	0	0	0	0	0	0
Silica (mg/L)	54	54	108	135	30	30	45	21	21	35	35
(Ca)·(SO$_4$) (product) (mg/L as CaCO$_3$)^2	780	38,690	3120	280,860	209,445	389,080	919,580	36,995	66,770	102,770	189,810
TDS (mg/L)	1200	1070	2400	2670	4740	4820	7350	670	640	1120	1070
Conductivity (µmho/cm)	1270	1260	2450	2810	6900	6830	9650	775	795	1200	1210

Key: BA, before adding acid; AA, after adding acid.
X is cycles of concentration.

and Saturation Index. At 4 cycles, silica barely exceeds 100 mg/L. TDS is marginally above 2000 mg/L and the natural pH is too high. At 5 cycles or above, TDS is too high. The maximum cycles of concentration for this poor quality water would be limited to about 4 for a nonchromate program. Field experience has confirmed this prediction.

Water B—this water contains a high level of calcium hardness, exceeding the allowable calcium level even at 2 cycles. Its natural pH is also too high. The Langelier index is exceeded at 2 cycles without acid, and calcium sulfate controls at 3 cycles with acid. TDS and conductivity are high; without proper treatment, galvanic corrosion is likely to occur. In practice, this water is limited to about 1.5 cycles with acid addition.

Water C—without adding acid, the maximum allowable cycles would probably be 3. The pH and Langelier and Ryznar Indices are approaching their limits. Scaling is likely to occur at higher cycles. Decreasing the pH to 7, or slightly lower, and employing a chromate program, if allowable, would be able to increase the allowable cycles to about 5. At this pH, the corrosive tendency could be controlled by chromate. Calcium exceeds the guideline for a nonchromate program. However, the negative Langelier Saturation Index suggests that this would be academic. Aluminum is also a potential problem. The current control range for this water is 4–6 cycles.

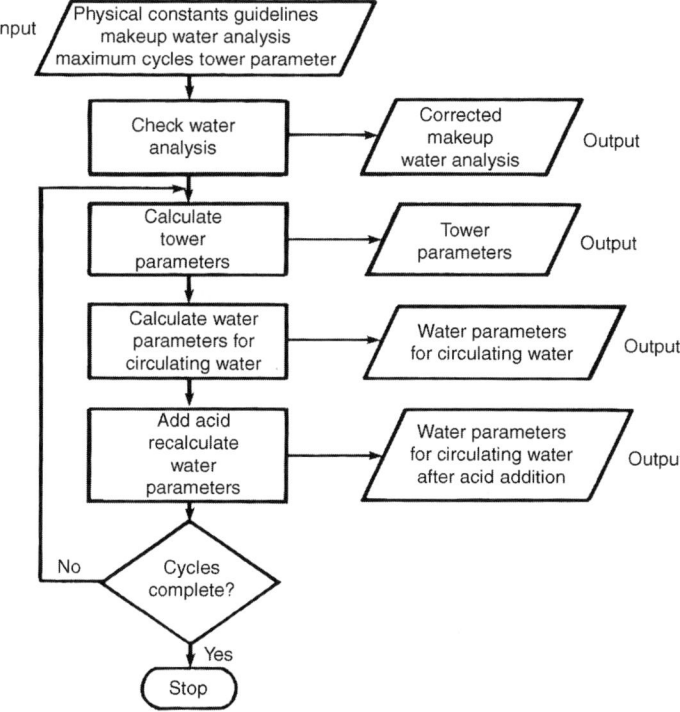

Figure E.8 Cooling water program flow sheet.

Table E.10 Inputs and Outputs

Tower parameter inputs
  Hot and cold water temperatures—data
  Wet bulb temperature—data
  Circulation rate—data
  Cycles of concentration—data

Tower parameter outputs
  Evaporation, blowdown, and makeup rates
  Chemical usage, acid addition, etc.

Water parameter inputs
  Species present (Ca^{2+}, HCO_3^-, SiO_2, etc.)—data
  Physical constants (ionization constants, solubility products)—program
  Desired pH—data
  Guidelines—program

Water parameter outputs
  Balanced water analysis
  Total dissolved solids
  Ionic strength
  Naturally occurring pH
  Conductivity
  Activity coefficients (not printed)
  Water quality indexes (Langelier, Ryznar, CPSI, ZPSI)
  Solubilities ((Ca^{2+})(SO_4^{2-}), (Zn^{2+})(OH^-)^2, (Mg^{2+})(OH^-)^2)

E.9 COMPUTERIZED CALCULATIONS

The calculations of the previous section are straightforward, but tedious. Alternatively, use of the generalized flowchart (Figure E.8) with the required inputs (Table E.10) produces recirculating water compositions at integral cycles of concentration up to any value desired. By comparison with recommended concentration limits, the maximum allowable cycles of concentration can be determined. These procedures are readily implemented on a digital computer, subject to one's own needs and preferences. Many such computer programs of a proprietary nature have been reviewed in an international survey [4].

E.10 CASE STUDIES

The computations described above have been performed for several operating plants to compare the calculated results with operating data in the field. According to the calculations, several of the plants are operating at or near the optimum point; conditions at others could be changed slightly, while still others appeared to be far

Table E.11 Comparison of Conductivity and Acid Predictions

Plant	Operating pH	Cycles based on Mg	Cycles based on SiO_2	Cycles based on Conductivity	Measured conductivity at 25°C	Calculated conductivity at 25°C	Measured sulfate in tower water	Calculated from sulfate measurement[a]	Sulfuric acid requirement gal/day 66° Bé @ 60°F Calculated by program[b]
1	7.6	12.6[c]	—	12	360	556–593	73	0	0.15–0.16
2	7.2	3.0	3.8[c]	3.9	680	549–693	232	2.7	3.0–2.9
3	6.9	8.1[c]	—	6.3	1725	1321–1444	607	10.9–2.6	7.8
4	8.7	7.6	4.9[c]	5.2	2420	1955–2312	475	8.6–2.7	3.4–6.0
5	8.0	4.5[c]	—	4.0	1840	1601–1930	181	0	0.5–1.2
6	6.4	6.6	5.5[c]	5.1	460	491–574	139	0.57	1.8
7	5.3	2.8[c]	2.8[c]	2.9	3200	1675–2243	1653	40.0	43.1–32.4
8	6.6	—	4.7[c]	3.4	3735	2908–3414	2000	87.3–46.3	60.3–56.8

[a] Calculated at best estimate of cycles unless a range is given.
[b] Range corresponds to one cycle below to one cycle above best estimate.
[c] Best estimate of cycles. Range corresponds to the range of measured cycles.

removed from predicted optimal conditions. Modifications have been made, and all towers are being continually monitored to improve the predictive techniques through field experience and vice versa.

For example, a comparison of conductivity and acid predictions is contained in Table E.11. It is important to note that these more recent data were not used in establishing either the conductivity or pH correlations.

Except for plants 1 and 7, agreement for conductivities is within the advertised 25% and much better in the majority of the cases. Plan 1 compares less favorably because of accumulated errors in predicting concentrations of individual ions and activity coefficients at the high cycles of concentration. Plant 7 is not a fair test because the low pH of 5.3 is outside the range of validity of the conductivity correlation.

Comparison of acid requirements is even more gratifying. In all cases, the computed figure is very much "in the ballpark" of values calculated from sulfate measurements and other operating data. Based on the sulfate measurements, plants 1 and 5 show zero acid addition. In actuality, plant 1 does not add any acid, and plant 5 adds only a very small amount on an irregular basis from time to time as the makeup water alkalinity changes.

Such calculations are a useful tool to monitor both cooling tower operation and the composition of the blowdown for environmental control.

REFERENCES

1. R.G. Kunz, A.F. Yen, and T.C. Hess, Basic cooling water calculations: general algorithm for cooling water chemistry, in water—1979, AIChE Symposium Series No. 197, 76, 1980, 174–183.
2. R.G. Kunz, A.F. Yen, and T.C. Hess, Cooling-water calculations, *Chemical Engineering*, **84**(16), 61–71, 1977.
3. D.A. Johnson and K.E. Fulks, Computerized water modeling in the design and operation of industrial cooling systems, Paper No. IWC-80-42 presented at the 41st Annual International Water Conference, Pittsburgh, PA, October 20–22, 1980.
4. G. Caplan, Cooling water computer calculations: do they compare? Paper No. 100 presented at National Association of Corrosion Engineers (NACE) Meeting Corrosion 90, Las Vegas, NV, April 23–27, 1990.
5. *Betz Handbook of Industrial Water Conditioning*, 6th edn, Betz Laboratories, Trevose, PA, 1962, Chapters 31–33.
6. D. Furlong, The cooling tower business today, *Environmental Science & Technology*, **8**(8), 712–716, 1974.
7. J.A. Dean, editor, *Lange's Handbook of Chemistry*, 11th edn, McGraw-Hill, New York, 1973, Section 5.
8. G.M. Fair, J.C. Geyer, and D.A. Okun, *Water and Wastewater Engineering*, Vol 2, Wiley, New York, 1968, Chapters 28–29.
9. *Standard Methods for the Examination of Water and Wastewater*, 14th edn, American Public Health Association, Washington, DC, 1976, 1193 pp.
10. C.N. Sawyer and P.L. McCarthy, *Chemistry for Sanitary Engineers*, 2nd edn, McGraw-Hill, New York, 1967, p. 331.
11. F.N. Kemmer, editor, *Water: The Universal Solvent*, Nalco Chemical Company, Oak Park, IL, 1977, pp. 50–51.

12. *Cooling Water Treatment Manual*, National Association of Corrosion Engineers, Houston, TX, 1971, 35 pp. Technical Practices Committees, Publication No. 1.
13. G.C. White, *Handbook of Chlorination*, Van Nostrand Reinhold Company, New York, 1972, p. 295.
14. S. Sussman, Facts on water use in cooling towers, *Hydrocarbon Processing*, **54**(7), 147–153, 1975.
15. A.J. Freedman and J.E. Shannon, Alkaline cooling water treatments, Paper presented at the International Water Conference sponsored by the Engineers' Society of Western Pennsylvania, Pittsburgh, PA, October, 1972.
16. M.R. Beychock, *Aqueous Wastes from Petroleum and Petrochemical Plants*, Wiley, London, 1967, p. 153.
17. C. Lipson and N.J. Sheth, *Statistical Design and Analysis of Engineering Experiments*, McGraw-Hill, New York, 1973, pp. 387–390.
18. J.W. McCoy, *The Chemical Treatment of Cooling Water*, Chemical Publishing Company, New York, 1974, 237 pp.
19. S.B. Appelbaum, *Demineralization by Ion Exchange*, Academic Press, New York, 1968, pp. 110–113.
20. A.J. Reitano Jr. and R.R. Lessard, Hexavalent chromium reduction in cooling tower blowdown—evaluation of the electrolytic process, Paper presented at the 1977 Annual Cooling Tower Institute Meeting, Houston, TX, January 31–February 2, 1977.
21. J.H. Haggenmacher and S.B. Gale, Electrochemical treatment of cooling water blowdown, Paper presented at the 38th International Water Conference, Pittsburgh, PA, November 2, 1977.
22. R.G. Kunz, T.C. Hess, A.F. Yen, and A.A. Arseneaux, Kinetic model for chromate reduction in cooling tower blowdown, *Journal Water Pollution Control Federation*, **52**(9), 2327–2329, 1980.
23. W.M. Skinner and G.L. Porter, Chromium removal from cooling-water blowdown, Presented at the 86th National AIChE Meeting, Houston, TX, Paper No. 79b, April 1–5, 1979.
24. W. Stumm and J.J. Morgan, *Aquatic Chemistry*, Wiley-Interscience, New York, 1970, pp. 83, 395–396.
25. F. Caplan, Is your water scaling or corrosive, *Chemical Engineering*, **82**(18), 129, 1975.
26. E.W. Brower, Corrosion and water treatment, *ASHRAE Handbook & Product Directory 1973 Systems*, American Society of Heating, Refrigerating, and Air Conditioning Engineers, New York, 1973, Chapter 36, pp. 36.1–36.20.
27. S. Sussman, Treatment of water for cooling, heating, and steam generation, *Water Quality and Treatment*, 3rd edn, American Water Works Association Handbook of Public Water Supplies, McGraw-Hill, New York, 1971, pp. 499–525.
28. J.R. Rue, Non-chromate treatment in cooling water systems, Paper No. 47e presented at the American Institute of Chemical Engineers 82nd National Meeting, Atlantic City, NJ, September 1, 1976.
29. T.D. Larson and R.V. Skold, *Corrosion*, **14**, 285t, 1958.
30. C.B. Capper, The protection of open recirculating cooling systems, *Effluent and Water Treatment Journal*, **14**(10), 577–583, 1974.
31. J.C. Grier and R.J. Christensen, Biocides give flexibility in water treatment, *Hydrocarbon Processing*, **54**(11), 283–286, 1975.
32. W.L. Harpel and J.M. Donahue, Effective phosphate/phosphonate treatments replace chromate-based programs, Paper No. TP 117A presented at the Cooling Tower Institute Annual Meeting Houston, TX, January 29–31, 1973.
33. E.F. Klen and D.A. Johnson, Paper presented at National Association of Corrosion Engineers Annual Meeting and Industrial Pollution Conference and Exposition, Houston, TX, March–April, 1976.
34. J. Green and J.A. Holmes, Calculation of the pH of saturation of tricalcium phosphate, *Journal of American Water Works Association*, **39**, 1090–1096, 1947.
35. R.V. Comeaux, *Cooling Towers: a CEP Technical Manual*, American Institute of Chemical Engineers, New York, 1972, pp. 78–82.
36. Midkiff W.S. and H.P. Foyt, Amorphous silica scale in cooling towers, Paper No. TP 148A presented at the Cooling Tower Institute Annual Meeting Houston, TX, January 19–21, 1976.
37. R.E. Badger, Paper presented at the Cooling Tower Institute Annual Meeting, Houston, TX, February, 1977.

38. T.E. Larson, R.W. Lane, and C.H. Neff, *Journal of American Water Works Association*, **51**, 1551–1558, 1959.
39. W.J. Ward, Cooling water treatment chemicals, *Cooling Towers, a CEP Technical Manual*, Vol. 2, American Institute of Chemical Engineers, New York, 1975, pp. 60–63.
40. J.M. Donahue and W.W. Hales, Improve cooling water treatment, *Hydrocarbon Processing*, **47**(6), 101–106, 1968.
41. J.W. Hayes Jr., Current practice in sidestream filtration for cooling towers, Paper No. TP 25A presented at the Cooling Tower Institute Annual Meeting, Houston, TX, January 23, 1967.

Appendix F

Increase in Runoff from Industrial/Commercial/Urban Development: The Telltale Bridge

Never try to teach a pig to sing.
You just waste your time,
...and it annoys the pig.

F.1 SUMMARY

This appendix relates from the author's personal experience, the sad story of the cumulative impact of incremental development on storm water runoff flow, flooding, erosion, and water quality for a stream in the Northeastern United States. On paper, this creek's high quality water is designated for special protection as a cold water trout stream. However, it is subject to pollution from suspended solids, soluble contaminants, and the occasional release of raw sewage. Wetlands exist along the creek and its flood plain. Downstream it serves as public water supply for a large city.

In many localities, this *creek* (20 miles long, drainage area 68.5 square miles) would be of sufficient size to be called a *river* as, for example, the small river discussed in Appendix G. Unlike Appendix G, names and places here have been disguised to avoid further controversy.

F.2 INTRODUCTION

As development proceeds, structures, driveways, sidewalks, and parking lots are constructed on what was once bare ground. Rain falling on such developed areas is then no longer able to percolate into the ground to the same extent as formerly, and a

---

Environmental Calculations: A Multimedia Approach, by Robert G. Kunz
Copyright © 2009 John Wiley & Sons, Inc.

greater proportion of the precipitation runs off. The increased surface flow is felt on the site itself and on adjacent and downstream properties. In addition, the water quality of the runoff in the "after" case cannot be guaranteed to be the same as runoff "before."

Typical standard design practice is to collect, detain, and release runoff flow at no higher a rate than that experienced before development [1]. Responsible design is also concerned with controlling whatever sediment and chemical species are picked up by the runoff. A further concern is to mitigate any increase in runoff temperature from its contact with blacktop paving warmed by the sun. Since the higher peak flows generated by the addition of impervious surface area flow for longer times at predevelopment rates, the increased erosion of the banks of the receiving stream can also be an issue. Furthermore, with increased surface flow, less infiltration and groundwater recharge occurs. Nonetheless, not all of these considerations are legal requirements everywhere in this governmental jurisdiction.

F.3 THE CASE IN POINT

The specific case under consideration concerns a broad, shallow natural depression in the landscape that drains storm water runoff by gravity from several square miles of tributary area into the aforementioned creek. Historical use of the land in and around this natural geological feature (termed a *swale*) has been agricultural, including a predominance of cornfields. Over the years, commercial, residential, and public development has proceeded on the edges of the swale and within the swale itself. So far, the land in the immediate vicinity is not known to have been used for heavy industrial/manufacturing purposes.

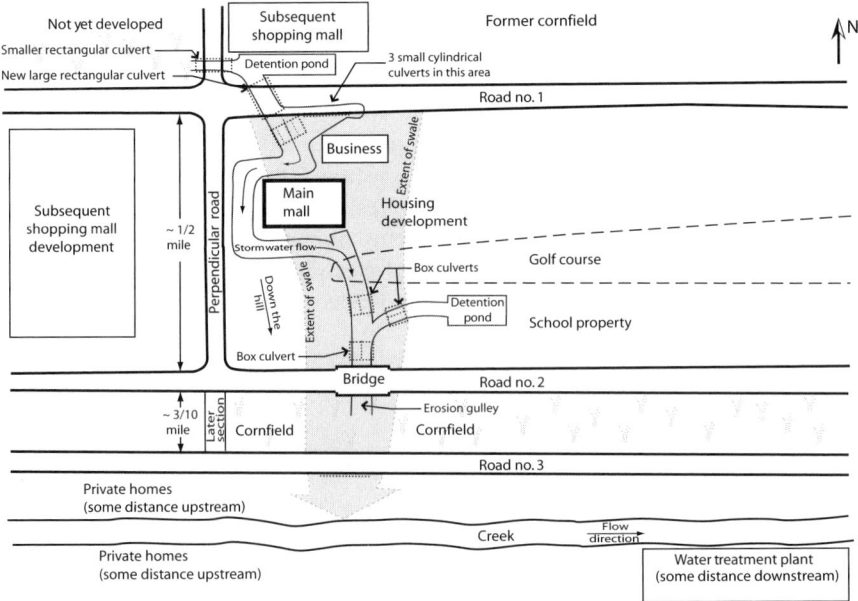

Figure F.1 Sketch of area surrounding swale.

The swale in general extends from north to south (Figure F.1). It is traversed nearest the creek by three east–west roads. They are denoted here, from north to south, as road no. 1, road no. 2, and road no. 3. Road no. 1 in this area is flat and once carried a U.S. route designation; road no. 2 is a two-lane state road, which dips to follow the contour of the swale. Road no. 3, also two-lane, is a local road. Distance between road no. 1 and road no. 2 is 0.5 miles approximately; distance between road no. 2 and road no. 3 is about 0.3 miles. Road no. 3 is located all the way down the hill in the flood plain of the creek, often in close visual proximity.

F.4 THE BRIDGE

In 1935 or thereabouts, a bridge was constructed to allow road no. 2 to cross the natural swale at its low point at this location. At that time, the opening under the bridge was of ample size to handle the runoff from the largely undeveloped countryside between adjacent farm fields. The open area for flow as found in 1998 was 15 ft^2. It could be enlarged to as much as 42 ft^2, depending on how much the accumulated silt and clay and the native soil and rocks were to be subsequently excavated. From a distance, or when one drives over it quickly, the structure might appear to be a culvert, but it is in reality an arch bridge built of stone according to a classic design, as evidenced by its foundation of rocks and masonry.

Sometime later, in the 1940s–1960s perhaps, responsibility for this bridge, along with others statewide, was "inherited" from local government by the state highway department according to an act of the state legislature. Original information such as drawings and calculations, if indeed these ever existed, was lost. It is understood that the subsequent file indicates repointing and installation of a concrete face coat on the sides and underneath the bridge in 1989 and a guardrail or rails in 1992.

A 1998 close-up of the north face of the bridge showing an earthen mound presumably for sedimentation control during construction of a public school project is depicted in Figure F.2. Within a week after this photograph was taken, the trees that

Figure F.2 Close-up of bridge (north face).

Figure F.3 Close-up of north face of bridge (later photograph).

had grown up adjacent to the bridge were removed, and the earthen mound was replaced by crushed stones (Figure F.3).

F.5 STORM WATER FLOW IN THE SWALE

Storm water originating in the tributary area north of road no. 1 at one time passed through three cylindrical culverts at different locations under that road (diameters of 18, 24, and 42 in., for a total flow area of 14.5 ft^2). It proceeded through four parallel box culverts 8.55 ft wide by 3 ft high each (total flow area of 102.6 ft^2), under the parking lot of a business immediately south of road no. 1. The flow path then headed through a largely undeveloped field. A housing development occupies the eastern fringe of the swale, out of the normal flow path.

When a major retail shopping mall and road were constructed on the field, storm water flow was diverted around the parking lot by means of a shallow concrete channel of rectangular cross section several feet wide below the new grade. Provision was made for heavier storm water runoff to overflow into a wider grass-bottomed auxiliary channel in the landscaped open trench. At the south end of the parking lot, the storm water passing through is joined by runoff of unknown abatement characteristics from the mall. Other shopping centers have been built on adjacent properties, and additional development is under consideration there.

The runoff, pre- or postmall, then crosses one corner of a golf course (an interesting water hazard) and enters the school property. It traverses this land via a manmade channel lined with closely spaced concrete tiles set in grass (western branch). This main flow then mixes with the flow from another similar channel draining the rest of the school property (eastern branch) before it reaches the bridge. The eastern branch channel originates from a detention pond some distance away on the school site. Before it was developed, the school property was farmland and woodland. Pairs of box culverts are installed to permit foot traffic and other passage over the school property channels at several locations (two each of 6 ft depth × 9 ft width = 108 ft^2).

Flow continues southward past the bridge at road no. 2 onto undeveloped farmland and then to road no. 3, where it pools up on the north side of the road. There storm water flows of any consequence regularly overwhelm the two small culverts, each no larger than 12 in. in diameter (1.6 ft^2 total, if clean and open). The water overtops the pavement on its way to the creek, often swollen onto its flood plain. In the all too common extreme case, the rain-swollen creek covers the road.

F.6 BOTTLENECKS

One notices immediately the disparity between the sizes of the conduit structures along the watercourse. The box culverts in the swale at road no. 1 and in the channels on school property upstream of the bridge are obviously designed to accommodate greater flows that can be handled beneath the bridge and through the previously existing culverts at road no. 1.

F.6.1 At the Bridge

Indeed, heavy rainfall generates runoff that backs up at the north face of the bridge, completely covering the opening in the photograph of Figure F.4, when the storm brought only about 2/3 of an inch of rain. On two or three prior occasions photographed, storm runoff came close to submerging the open area.

The bridge functions as a sort of gauge or telltale such that one cannot fail to notice increases in runoff. In other words, flooding of the bridge is a dead giveaway that runoff has increased from predevelopment levels. Like the coal miner's canary, it

Figure F.4 Flooded bridge January 1999.

ought to provide a warning to be ignored at one's own peril. This picture (Figure F.4) was taken in January 1999, before the construction of the shopping mall and with the three cylindrical culverts in place under road no. 1.

High-velocity water surges through the narrow bridge aperture onto the cultivated farmland downstream. During the period when these pictures were taken, this resulted in an erosion gulley several hundred feet long, evident in the farm field during dry weather by the exposure of the underlying rocks and the stunted growth of corn stalks. When the runoff totally inundated the opening in Figure F.4, a continuous stream of major dimensions not normally encountered during lesser rains formed downstream of the bridge.

The dirty, muddy water covered road no. 3, as usual, and made its way to the trout stream. Following the increased siltation from heavy rains, the public water utility downstream would regularly close its intake gates until the surge of suspended solids passed by.

Repeated scouring action from high-velocity water also erodes the underside of the bridge, possibly undermining its structural integrity. Authorities were notified of the situation, and a repair crew was subsequently spotted performing maintenance at the bridge. It was rumored over 7 years ago that there was money in the highway department's budget to replace the bridge within 2 years of that time, but the same bridge is still in place as the time of this writing.

A contemporary dry weather photograph of the bridge is shown in Figure F.5. The bottom of the channel under the bridge is now concrete. Some excavation has occurred, and the height of the opening has been increased to about 4.5 ft. Concrete shoulders have been poured extending partway upward from the floor of the channel, and a new coat of masonry has been applied to the rest of the underside of the bridge. The open area is now estimated at 30–35 ft^2, increased from the 15 ft^2 in 1998 as a result of greater headroom but decreased by the cross-sectional area of the shoulders.

Figure F.5   North face of bridge contemporary photograph.

The erosion gully downstream of the bridge as seen from the road is deeper, especially right at the southern side of the bridge. Large rocks are either newly exposed or have been placed in the gully to break up the flow. Riprap (bunches of large rocks to prevent erosion) and gabions (riprap contained in a strong chicken-wire mesh) have been placed along both sides of the channel on the south side of the bridge, apparently to contain the flow and prevent it from expanding laterally too close to the road.

F.6.2 At the Culverts under Road No. 1

Storm water has flooded road no. 1 at least once within recent memory. With the three small cylindrical culverts in place, this occurred during a hurricane event, closing the road to traffic. At least one longtime local resident cannot recall this ever happening in the previous 25 years.

Development of the former cornfield in the swale north of road no. 1 has already begun. The road perpendicular to road no. 1 has been extended northward, and a number of stores and restaurants with a single huge continuous parking lot plus a detention pond are open for business. In conjunction with this development, there have been installed a large rectangular culvert under road no. 1 and a somewhat smaller one north of road no. 1 under the perpendicular road. The larger culvert feeds directly into the four box culverts at the business south of road no. 1 and is of comparable size to their total area.

Runoff from this and future development activities north of road no. 1 is no longer held up by the former bottleneck under road no. 1, and it is now possible for a greater surge to be felt at the bridge at road no. 2. It is the author's firm belief that one of these days, the additional surge during a severe storm will be sufficient to submerge the bridge completely and flood that road. The water level needs to rise only another 3.5 ft above the top of the arch of the bridge before reaching the grade of the road. It is unlikely that road no. 1 will flood again now that the new larger rectangular culvert has been installed under that road.

F.7 CONTINUED FLOODING

Control of storm water permitting during the development process at this location is fragmented among various public agencies, and clearly mistakes were made in arriving at the present situation. Even without mistakes, each governmental body involved appears to be taking a narrow view of its responsibilities in following procedures and simply passing on its work product to the next agency. There seems to be no single entity responsible for an overview of an individual project, much less an overview of the cumulative impacts of multiple projects.

The result is that the area is not being properly protected. It is small consolation to the concerned citizens who warned time and again of the upcoming debacle to be able to say, "We told you so."

Flooding still takes place regularly in the vicinity of road no. 3, both upstream and downstream of the confluence of the swale's discharge into the creek, closing that road. Some would contend that flooding occurs now every time it rains.

Flooding is said to be getting worse every year; in addition to closing the road, the flooding also affects nearby homes [2]. The local newspaper goes on to state that two 500-year flood events have occurred within a 16-month period. According to the newspaper, many residents in the flooded areas along the creek attribute the repeated flooding to overdevelopment upstream and predict a worsening with additional development.

In other stories, the Army Corps of Engineers is reported to have been requested to help solve the flooding problem. Local government has appropriated 50,000 to fund such a study, but additional approvals, authorizations, and funding are required to conduct the study and implement its findings in competition with other priorities. Time will tell.

REFERENCES

1. T.A. Seybert, *Storm Water Management for Land Development: Methods and Calculations for Quality Control*, Wiley, Hoboken, NJ, 2006, pp. 2–4.
2. Local Newspaper Articles (February–December 2007).

Appendix G

Water Quality Improvement for a Small River

When making friends with an unfamiliar cat, let the cat come to you.

This appendix is a condensation of a previously unpublished original investigation of water quality for the Whippany River in Northern New Jersey. It was conducted in 1973 as an engineering research project (not a thesis) in partial fulfillment of the requirements for the Degree of Master of Science in Environmental Engineering from Newark College of Engineering (now New Jersey Institute of Technology).

Its original title was "Whippany River Water Quality—Past and Present"

G.1 SUMMARY AND CONCLUSIONS

- Information necessary for the evaluation of Whippany River water quality was obtained from a number of sources available to the public.
- Two 18-month periods were studied—July 1964–December 1965 ("before") and July 1971–December 1972 ("after").
- Water quality in the river, as characterized by dissolved oxygen (DO), biochemical oxygen demand (BOD), and to a lesser extent suspended solids (SS), improved dramatically between the two periods.
- The positive change in water quality is due largely to an upgrading of secondary treatment facilities, completed in 1967, at the Whippany Paper Board Company's Eden Mill. During the period investigated, this industrial source was the largest single user of the Whippany for discharge of wastewater.
- In the "before" case, septic conditions in the river frequently occurred during the summer months downstream of the paperboard company; "after," significant positive DO values were achieved at that location all year long, approaching the before and after upstream data.

Environmental Calculations: A Multimedia Approach, by Robert G. Kunz
Copyright © 2009 John Wiley & Sons, Inc.

- The DO sag curve follows, in general, the theoretical Streeter–Phelps relationship, both before and after.
- With the improvement in water quality, the effect on DO caused by the two municipal wastewater treatment plant discharges, Morristown and Hanover, becomes apparent. Previously this condition was masked by the paperboard company's industrial discharge.
- On the average, the unchanging BODs from the sewage treatment plants then become dominant versus the lower effluent BOD from the improved paper mill wastewater treatment.
- Reduction of BOD from the sewage treatment plants could conceivably improve water quality even further, as well as control of other pollutants such as phosphates and oil, which is not considered here.

Details from the original report as written at the time are given below.

G.2 INTRODUCTION

This report deals with the water quality of the Whippany River. Its scope is limited, being concerned primarily with DO and BOD, although some reference is made to SS. By pulling together information from diverse sources, it is possible to make a meaningful comparison between the water quality of several years ago and more recent data. The dramatic improvement in the Whippany River between 1964–1965 and 1971–1972 can be traced directly to the upgrading of the waste treatment facilities in the Whippany Paper Board complex. After major improvements to treatment, which were completed in 1967, the paper mill effluent regularly meets the state regulations for BOD and suspended solids. However, there may still be room for improvement in river water quality to be achieved by a reduction of BOD from the sewage treatment plants that also discharge into the Whippany.

G.3 FLOW OF WHIPPANY RIVER

The Whippany River rises in the hills west of Morristown, NJ, and flows in a general northeasterly direction to Hatfield Swamp where it joins the Rockaway River before confluence with the Passaic. The Whippany drops 140 ft in elevation in some 20 miles, and the stream velocity is moderate. A sketch of the Whippany River system is shown in Figure G.1 [1,15].

The Whippany starts out in a downright pastoral setting. Then along its route, the Whippany passes several population centers—Morristown, Cedar Knolls, Whippany, and East Hanover. In addition to natural and urban runoff, the river receives the wastes from sewage treatment plants and many of the industries located along its banks, also augmenting its flow. The main stream is joined by several tributaries along its course as well, and the size of the channel increases to accommodate the increased flow.

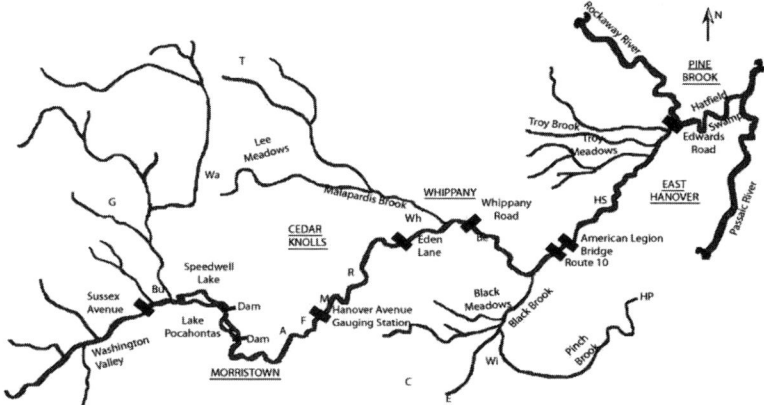

Figure G.1 Simplified sketch of Whippany river system.

Sources having a permit to discharge wastewater into the Whippany and its tributaries are listed in Table G.1 [9]. The approximate location of each discharge along the river is identified in Figure G.1 by the appropriate symbol in the table. This table also describes the type of waste treatment and the average effluent rate. The chief contributor to wastewater flow, and therefore stream quality, is the Eden Mill of the Whippany Paper Board Company, followed by the sewage treatment plants at Morristown and Hanover (Table G.1).

The river flow is regularly determined at the United States Geological Survey (USGS) gauging station near Hanover Avenue in Morristown and upon occasion at several other locations along the river [14,16,18]. The gauging station is located on the left bank at the Morristown sewage treatment plant (STP), 0.8 miles northeast of Morristown and 9.0 miles upstream from the mouth of the Whippany River. The average flow of the Whippany at Morristown is about 48 cfs. Complete records for flow at Morristown are contained in Refs [14] and [16].

Measured flow rates at various points along the river are tabulated for the record in Table G.2. These can be related to the flow at the Morristown gauging station. Although there is some scatter in the data, in general, the flow at Sussex Avenue upstream of Morristown (14.2 miles upstream of the mouth of the river) is approximately 60% of gauging station flow, and the flows at the two downstream locations, Route 10 and Edwards Road, are roughly 1.5 and 2 times the flow at Morristown. Refer again to Figure G.1 for the relative locations of these flow-measuring stations.

G.4 WATER QUALITY OF RIVER

Assessment of the quality of Whippany River water is based upon fragmentary measurements reported by the USGS [16,18] and more complete unpublished data on file at both the Passaic Valley Water Commission [10] and the New Jersey Department of Environmental Protection [9].

Table G.1 Discharges by Permit into Whippany River System

Permit holder	Symbol in Figure G.1	Type of treatment	Flow (MGD)
Airtron Div. of Litton Industries (industrial)	A	Settling High-rate trickling filter Digestion Sludge drying Chlorination	–
Bell Telephone Laboratories (industrial)	Be	Mixed settling Sludge digestion Industrial waste treatment	0.15
Butterworth Farms STP (domestic)	Bu	Contact stabilization Settling Chlorination Sludge holding beds	2.0
College of St. Elizabeth (domestic)	C	Chemical flocculation Settling Sand filters Digestion Sludge drying beds	0.15
Delbarton School (domestic)	D	Extended aeration Settling Chlorination Sludge holding	0.037
Esso Research & Engineering Co. (domestic)	E	Clarigester Std. rate trickling filter Final settling Sand filters Chlorination Sludge drying	0.29
Flintcote Company Research Lab	F	Settling tanks Sand filters Chlorination	0.01
Greystone Park	G	—	0.8–1.3
Hanover Park Regional High School (domestic and cafeteria)	HP	Secondary settling Chlorination Sludge holding Sludge drying beds	0.06

Table G.1 (*Continued*)

Permit holder	Symbol in Figure G.1	Type of treatment	Flow (MGD)
Hanover Sewage Authority (domestic)	HS	Settling High-rate trickling filter Digestion Sludge drying Chlorination	1.5
Morristown Sewer Dept. (domestic)	M	Primary and final setting Contact stabilization Chlorination Digestion Sludge drying beds	3.0
Rayonier, Inc., Eastern Research Division (domestic and industrial)	R	Neutralization Imhoff tank High-rate trickling filter Final settling Chlorination	0.259
Texas U.S. Chem. Co. (domestic and industrial) (industrial)	T	Primary settling High-rate trickling filter Digester Chlorination Final settling	0.0105 0.0107
Warner Lambert Pharmaceutical Co. (domestic and industrial)	Wa	Clarigester and primary settling High-rate trickling filter Final settling Sand filters Chlorination	0.33
Whippany Paper Board Company (industrial)	Wh	Primary settling Biological oxidation Sludge thickening Sludge stabilization Purifax treatment of sludge to settling lagoons Sludge trucked away	4.5–5.0
Wilbur Driver Company	Wi	Extended aeration Chlorination	0.001

Table G.2 Whippany River Flow Rates at Various Locations[a,b]

Date	Bridge on Sussex Avenue 14.2 miles upstream of mouth	Morristown Gauging Station	Route 10 Bridge	Edwards Road near Pine Brook
06/25/1963	—	19.0	23.6	29.1
07/30/1963	—	45.0	101.0	123.0
08/27/1963	—	10.0	17.4	24.3
09/24/1963	—	8.8	16.2	22.1
03/17/1964	—	49.0	72.5	112.0
06/17/1964	—	24.0	—	39.1
08/11/1964	—	12.0	16.8	25.8
08/28/1964	2.32	9.2	—	—
08/30/1964	4.00	9.6	–	–
09/22/1964	—	6.9	13.0	20.2
04/15/1965	10.20	29.0	—	—
04/30/1965	—	29.0	39.4	62.7
07/07/1965	3.25	9.8	—	—
09/07/1965	—	11.0	—	24.9
09/08/1965	2.08	8.8	19.2	—
04/12/1966	11.20	26.0	33.5	44.3
07/14/1966	3.75	10.0	—	—
08/10/1966	—	9.6	11.5	23.4
09/09/1966	2.01	8.8	—	—
06/07/1967	14.10	32.0	—	55.6
09/11/1967	9.07	24.0	—	—
09/12/1967	—	20.0	—	25.3
03/04/1968	23.20	25.0	—	—
08/29/1968	—	17.0	—	29.9
09/16/1969	8.40	14.0	—	—
04/28/1970	28.20	49.0	—	—
09/04/1970	4.58	16.0	—	—
04/14/1971	30.90	62.0	—	—

[a] All flows in cfs.
[b] Data from Refs [14,18].

A major diversion of the Passaic River is taken for public water supply at Little Falls by the water commission, and therefore the Passaic and its upstream tributaries are constantly monitored as a public health measure. The streams are regularly sampled once a month at several locations, and special daily measurements are made at other stations. These records are considered to be among the most comprehensive water quality data available anywhere. Measurements made by the other agencies appear to have been made with equal care, but until recently lack the frequency of the water commission's measurements.

G.4 Water Quality of River

Locations of the sampling stations for water quality are also shown in Figure G.1. These sampling stations were selected to monitor the effect on water quality of the major discharges previously tabulated (Table G.1). Thus, provisions have been made for sampling the Whippany upstream and downstream of the Morristown Sewage Plant, the Whippany Paper Board Company, and the Hanover Sewage Plant. Therefore, in addition to the flow measurement stations already identified where water quality data are also taken, there is a station at Eden Lane upstream of the Whippany Paper Board Company's Eden Mill and another at Whippany Road downstream of the mill. This station at Whippany Road replaced the Route 10 location between 1965 and 1971. For reasons to be explained later, the Passaic Valley Water Commission has singled out the paperboard company for special surveillance and monitors daily their effluent along with the river downstream at Edwards Road.

G.4.1 Dissolved Oxygen

Values of dissolved oxygen for water samples obtained at the Hanover Avenue Bridge upstream of the Morristown Sewage Plant are plotted in Figure G.2 against time of the year. A complete listing of dissolved oxygen values with the temperature of measurement for this and downstream stations is recorded in Tables G.3–G.5. All of these DO values were recorded during daylight and do not reflect the significant diurnal variations expected because of the presence of algae. Such diurnal fluctuations in dissolved oxygen varying between 1.5 mg/L for darkness and 3.0 mg/L for daylight during

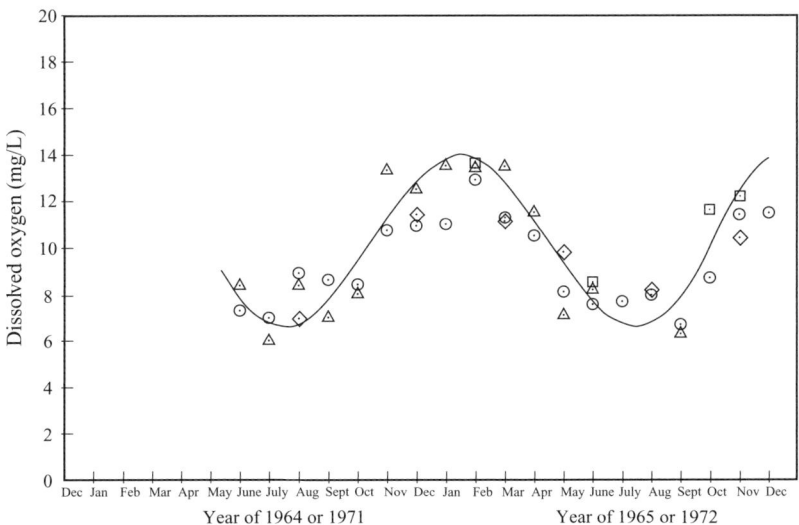

Figure G.2 Dissolved oxygen measurements at Hanover Avenue Bridge Morristown. △ Passaic Valley Water Commission Data 1964–1965; ⊙ Passaic Valley Water Commission Data 1971–1972; ☐ United States Geological Survey Data 1971–1972; ◇ New Jersey Department of Environmental Protection Data 1971–1972.

Table G.3 Dissolved Oxygen Levels in Whippany River[a,b]

Date	Hanover Avenue		Eden Lane		Route 10		Edwards Road	
	DO	Temperature (°F)	DO	Temperature (°F)	DO	Temperature (°F)	DO	Temperature (°F)
06/16/1964	8.4	67	5.7	66	3.2	72	0.2	70
07/29/1964	6.0	75	2.0	76	0.2	84	0.8	79
08/11/1964	8.4	68	3.0	68	0.0	80	0.0	74
09/22/1964	7.0	61	1.9	60	0.0	65	1.0	62
10/20/1964	8.0	52	3.1	53	0.2	67	0.4	56
11/12/1964	13.3	48	3.7	49	0.0	68	2.2	54
12/15/1964	12.5	38	10.3	38	9.8	40	6.7	40
01/19/1965	13.5	33	12.0	34	6.1	48	5.6	38
02/24/1965	13.4	37	12.8	35	10.0	48	9.8	41
03/24/1965	13.5	42	11.1	42	9.5	48	7.5	46
04/27/1965	11.5	52	9.2	52	7.7	56	6.7	55
05/26/1965	7.1	74	4.5	74	0.3	74	0.8	77
06/23/1965	8.2	75	4.0	78	0.0	87	0.0	77
08/18/1965	—	—	6.2	76	0.6	84	0.2	78
09/21/1965	6.3	72	4.2	74	1.1	80	0.8	75
10/27/1965	—	—	8.4	52	2.4	66	3.6	56
11/27/1965	—	—	8.6	46	4.5	57	2.4	53
12/07/1965	—	—	5.3	40	10.8	47	8.7	40

Date	Hanover Avenue		Eden Lane		Whippany Road		Edwards Road	
	DO	Temperature (°F)	DO	Temperature (°F)	DO	Temperature (°F)	DO	Temperature (°F)
06/22/1971	7.3	72	5.6	72	5.4	78	2.4	76
07/20/1971	7.0	68	6.5	68	5.4	68	3.7	72
08/24/1971	8.9	66	7.0	66	2.8	79	2.8	70
09/28/1971	8.6	58	8.5	59	7.6	64	3.8	62
10/19/1971	8.4	56	8.4	56	8.0	62	3.2	60
11/16/1971	10.7	45	10.3	46	9.2	54	5.3	49
12/14/1971	10.9	42	11.0	42	10.6	48	7.5	44
01/25/1972	11.0	43	10.0	44	9.8	42	8.0	48
02/23/1972	12.9	33	13.1	33	11.4	43	10.7	36
03/21/1972	11.3	46	11.0	46	10.6	48	9.7	47
04/25/1972	10.5	52	10.3	52	9.8	56	7.8	53
05/23/1972	8.1	62	8.2	64	7.9	64	4.3	66
06/21/1972	7.6	64	7.8	65	8.0	67	2.9	68
07/19/1972	7.7	78	6.8	78	5.5	88	2.0	96
08/22/1972	8.0	70	6.7	70	6.0	76	0.0	73
09/19/1972	6.7	70	7.1	70	7.1	72	4.6	72
10/17/1972	8.7	52	8.7	52	6.5	64	3.6	58
11/29/1972	11.4	42	11.4	42	11.1	46	6.3	43
12/13/1972	11.5	42	11.4	44	11.3	46	7.6	42

[a] All DO measurements in mg/L.
[b] Data from Ref. [10].

Table G.4 USGS Water Quality Data [18]

Date	Average discharge at Morristown (cfs) [16]	Temperature (°C)	DO (mg/L)	BOD mg/L
Whippany river at Morristown – Latitude 40°48′21″, Longitude 74°27′22″				
07/14/1971	25	—	—	4.6
09/27/1971	56	—	—	3.8
02/01/1972	37	3.6	13.6	—
03/03/1972	309	4.5	—	—
		4.0	—	—
03/14/1972	85	1.0	—	—
05/04/1972	261	16.0	—	—
05/09/1972	177	—	—	—
06/13/1972	66	—	—	>10.0
		16.9	8.5	—
06/21/1972	214	—	—	—
06/22/1972	485	—	—	—
09/18/1972	23	23.0	—	—
10/26/1972	—	10.7	11.6	6.0
11/20/1972	—	4.4	12.2	2.0
Whippany River at Pine Brook – Latitude 40°50′42″, Longitude 74°20′51″				
07/14/1971	25	—	—	10.0
09/24/1971	67	—	—	6.0
02/01/1972	37	—	—	—
06/12/1972	66	—	—	16.0
07/20/1972	—	27.2	—	6.6
09/18/1972	23	24.5	—	—
10/25/1972	—	14.6	4.3	12.0
11/20/1972	—	3.8	9.8	5.0

September 1968 have been reported in the Passaic [20]. A summary of microorganisms, including algae, identified in the Whippany River is listed in Table G.6 [9].

Passaic Valley Water Commission measurements during 1964–1965 and all measurements for 1971–1972, as shown in Figure G.2, appear to cluster around the same anticipated periodic function, with no significant difference between the data for the early years and the more recent results. Passaic Valley data for 1971–1972 are corroborated by the data from the other two agencies. The reason for the periodic nature of the dissolved oxygen readings is the seasonal variation in temperature. Almost without exception, dissolved oxygen levels varied from 70% to 100% of saturation values [13] at the temperature of measurement. No better correlation could be found using the degree of saturation. Neither could any significant trends in temperature be detected between any data for the same time of year. The values are

Table G.5 Water Quality Data from NJ State Department of Environmental Protection

Date	Average discharge at Morristown (cfs) [16]	Temperature (°C)	DO (mg/L)	BOD mg/L
Whippany River at Morristown				
08/04/1971	45	21.0	6.9	2
12/01/1971	—	2.0	11.4	2
03/02/1972	193	3.0	11.2	3
05/02/1972	71	15.0	9.8	2
08/01/1972	42	21.0	8.2	2
11/28/1972	—	4.5	10.4	1
Whippany River at Route 10 Crossing				
08/04/1971	45	23.0	3.1	5
12/01/1971	—	3.0	9.7	6
03/02/1972	193	4.0	11.4	5
05/02/1972	71	17.0	7.6	5
08/01/1972	42	20.0	4.6	15
11/28/1972	—	5.5	9.3	6
Whippany River at Rockaway Neck Edwards Road				
08/04/1971	45	22.0	1.3	3
12/01/1971	—	2.0	7.4	2
03/02/1972	193	3.0	9.4	3
05/02/1972	71	17.0	5.0	4
08/01/1972	42	20.0	7.0	6
11/28/1972	—	5.0	6.4	3

also representative of the dissolved oxygen measured farther upstream on several occasions during 1972 by the New Jersey Department of Environmental Protection [9]. Those data are shown in Table G.7.

In contrast, dissolved oxygen values measured at the Eden Lane station downstream of the Morristown STP but upstream of the paperboard company show a decided difference from 1964–1965 to 1971–1972, the major difference occurring during the summer months (Figure G.3). Although some of the difference can be accounted for by lower river flows in 1965 compared to 1972, such is not the case in 1964 versus 1971. The principal reason for the difference is an upgrading in operation of the sewage treatment plant during the intervening period. Nevertheless, the treatment plant is currently overloaded (as of 1973), operating at about double its design capacity. For the past several years, Morristown has been under orders to make about $3.6 million worth of improvements to its sewage system but has been awaiting federal assistance before proceeding. Thus, the dissolved oxygen in the river was already significantly depressed during 1964–1965 before it reached the paper mill.

Table G.6 Microorganisms Identified in Whippany River Upstream and Downstream of Whippany Paper Board Company 12/29/1972 [9]

Type	Subcategory
Green Algae	Chlorococcum
Diatoms	Cocconus
	Cyclotella
	Cymbella
	Diatoma
	Gomphonema
	Melosira
	Navicula
	Nitzschia
Green rods and spheres	
Green and yellow-brown filaments	
Protozoa	Ciliate
	Rhizapoda
Ruptured cells	

This, in combination with the potent waste discharged by the mill at that time, served to lower downstream oxygen levels still further, often to septic conditions (Figures G.4–5). Values of zero dissolved oxygen in August 1964, 1965, and 1972 resulted in a blackening of the Passaic River and gigantic fish kills. Zero DO in August 1972 occurred because of an upset at the paper mill.

The more recent data indicate significant improvement in DO levels. With the one exception noted above, a comparison of the recent data plotted in Figures G.3 and G.4 indicates only a small decrease in dissolved oxygen as the river passes by and through the paper mill, but this minor decrease is almost unnoticeable compared to the corresponding decrease in dissolved oxygen for the earlier years, irrespective of the lowered upstream DO for 1964–1965. A further discussion of the effect of the paperboard company's effluent upon the river will be presented in a subsequent section.

In Figure G.5, further changes in dissolved oxygen can be noted. For the 1971–1972 data, the level of DO continues to drop. For the earlier period, the wintertime data show the same trend. However, the summer values were so bad that some net reaeration took place despite the added load placed on the river by the intervening Hanover STP.

It appears that the levels of dissolved oxygen in the Whippany River are now generally higher than those during previous times. However, the Whippany is classified as an FW-2 stream under the Rules and Regulations of the New Jersey Department of Environmental Protection [8] by reason of the eventual use of its water as public water supply at Little Falls. This classification implies that not less than

Table G.7 Upstream Water Quality [9]

Location	Date	Temperature (°C)	DO field (mg/L)	DO laboratory (mg/L)	BOD (mg/L)
Above Butterworth STP	04/20/1972	12.0	9.0	8.8	3
	08/24/1972	19.0	8.0	7.5	4
At Effluent of Butterworth STP	04/20/1972	10.5	8.1	6.5	<3
	08/24/1972	20.0	7.7	5.9	<3
Below Butterworth STP	04/20/1972	12.0	7.6	7.9	5
	08/24/1972	19.0	7.5	7.7	<2
Tributary from Greystone Park above confluence with Whippany	04/20/1972	12.5	8.7	7.5	9
	08/24/1972	21.0	9.0	8.8	3
Speedwell Lake Inlet	03/09/1972	1.1	12.8	12.4	2
	04/20/1972	13.1	9.1	8.9	3
	08/24/1972	19.5	7.5	7.6	2
Speedwell Lake Outlet (above dam)	04/20/1972	15.5	10.0	7.7	4
	08/24/1972	22.0	10.6	10.6	3
Speedwell Lake Outlet (below dam)	03/09/1972	0.9	12.5	12.1	2
	08/24/1972	21.5	8.6	8.8	4
Pocahontas Inlet	03/09/1972	1.5	11.7	12.3	2
	04/20/1972	14.5	6.8	8.0	3
	08/24/1972	21.0	7.7	9.2	3
Pocahontas Outlet (above dam)	04/20/1972	15.0	–	7.6	3
	08/24/1972	22.0	9.5	9.3	4
Pocahontas Outlet (below dam)	03/09/1972	3.4	12.5	12.1	2
	08/24/1972	22.0	9.0	9.1	4

4.0 mg/L of dissolved oxygen is to be present at any time. Applying this criterion to the Whippany at Edwards Road during the summer months (Figure G.5) indicates that further improvement is necessary.

G.4.2 Biochemical Oxygen Demand

The reason for the drop in dissolved oxygen along the river is the addition of organic matter having a biochemical oxygen demand. Values of BOD upstream of major pollution sources are 3–4 mg/L (Table G.7). This level appears reasonable in light of the swampy, grassy areas through which the river flows even in its upstream reach. After passing through the Morristown urban area, these levels are sometimes exceeded and have been measured at as high as about 10 mg/L during the periods under investigation. Data for these periods at the Morristown sampling station are

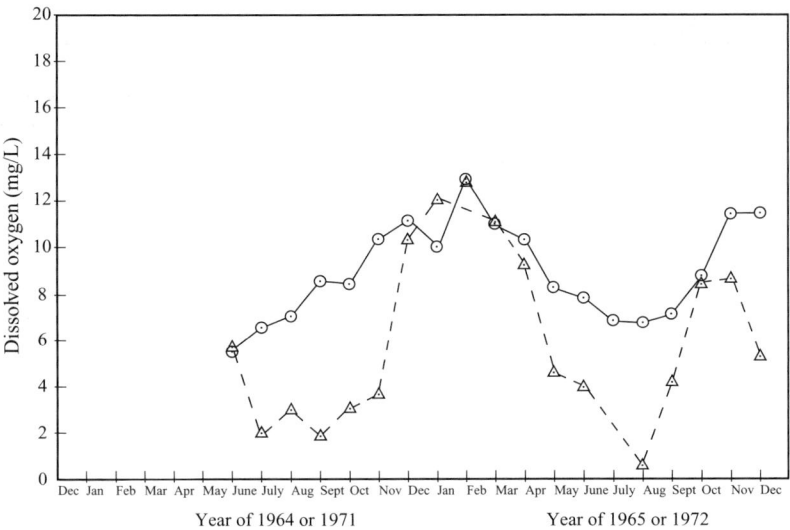

Figure G.3 Dissolved oxygen measurements at Eden Lane. △ Passaic Valley Water Commission Data 1964–1965; ⊙ Passaic Valley Water Commission Data 1971–1972.

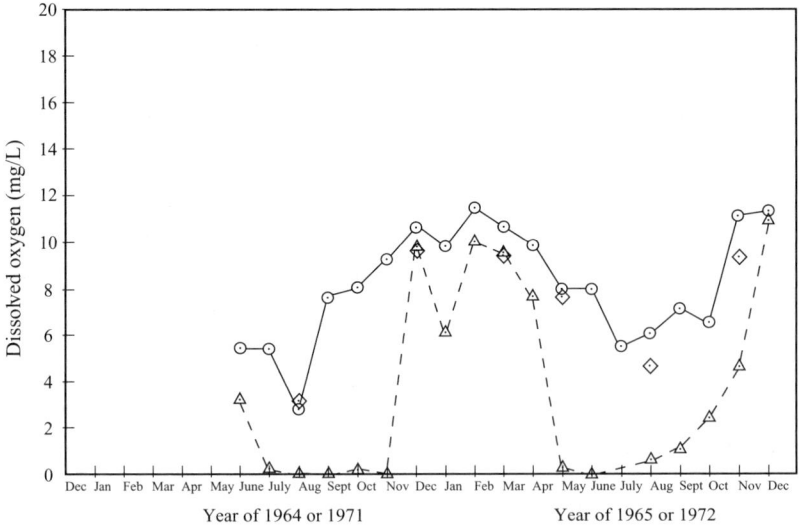

Figure G.4 Dissolved oxygen measurements at Route 10 (1964–1965) or Whippany Road (1971–1972). △ Passaic Valley Water Commission Data 1964–1965; ⊙ Passaic Valley Water Commission Data 1971–1972; ◇ New Jersey Department of Environmental Protection Data 1971–1972.

G.4 Water Quality of River 613

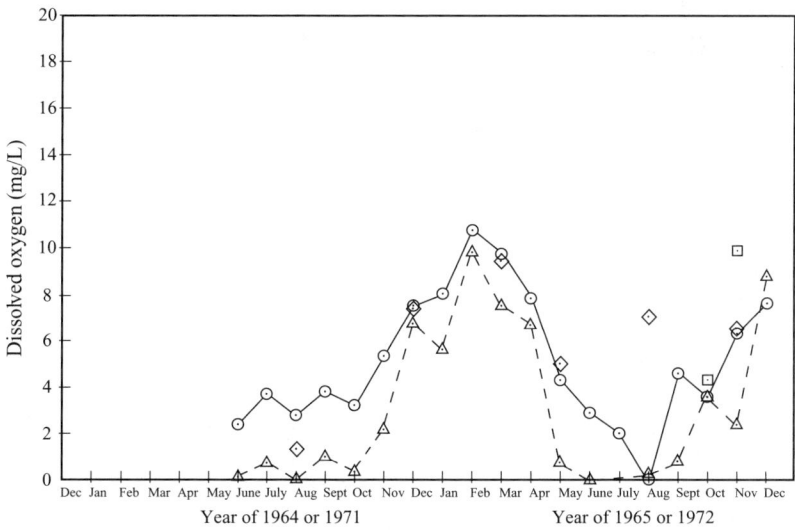

Figure G.5 Dissolved oxygen measurements at Edwards Road. △ Passaic Valley Water Commission Data 1964–1965; ⊙ Passaic Valley Water Commission Data 1971–1972; ▢ United States Geological Survey Data 1971–1972; ◇ New Jersey Department of Environmental Protection Data 1971–1972.

shown in Figure G.6 plotted against the river's flow rate at Morristown. A complete listing of BOD data from all Whippany River stations for these periods is recorded in Tables G.4, G.5, G.7, and G.8. In spite of appreciable scatter, the data plotted in Figure G.6 show an apparent gradual decreasing trend with river flow at Morristown

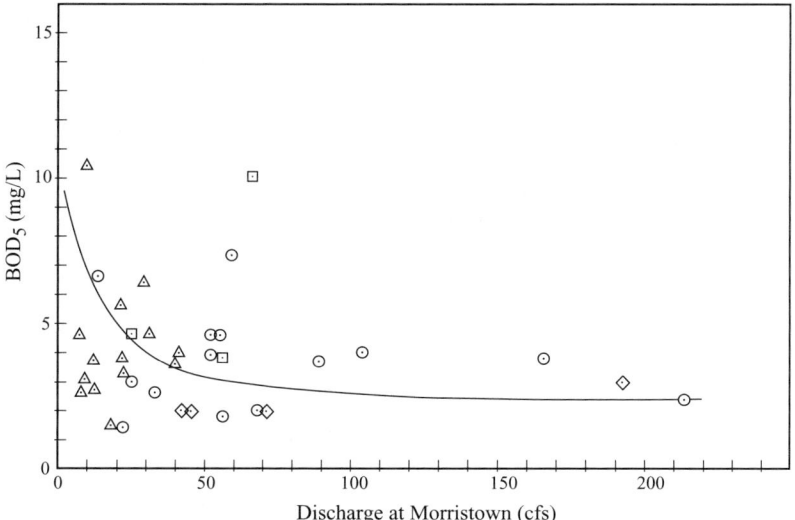

Figure G.6 BOD_5 measurements at Hanover Avenue Bridge Morristown. △ Passaic Valley Water Commission Data 1964–1965; ⊙ Passaic Valley Water Commission Data 1971–1972; ▢ United States Geological Survey Data 1971–1972; ◇ New Jersey Department of Environmental Protection Data 1971–1972.

Table G.8 Biochemical Oxygen Demand in Whippany River[a,b]

Date	Flow at Morristown (cfs)	Hanover Avenue	Eden Lane	Route 10	Edwards Road
06/16/1964	29.0	6.4	6.4	50.0	56.0
07/29/1964	22.0	3.3	5.8	56.0	31.2
08/11/1964	12.0	3.7	8.6	110.0	62.0
09/22/1964	6.9	4.6	4.2	82.0	46.0
10/20/1964	8.8	3.1	6.8	68.0	80.0
11/12/1964	9.6	10.4	12.0	132.0	40.0
12/15/1964	21.0	5.6	19.8	54.0	60.0
01/19/1965	31.0	4.6	5.6	120.0	82.0
02/24/1965	22.0	3.8	5.8	21.2	20.2
03/24/1965	40.0	3.6	6.8	28.0	38.8
04/27/1965	41.0	4.0	5.6	48.8	25.0
05/26/1965	18.0	1.5	7.0	32.0	12.0
06/23/1965	12.0	2.7	6.8	126.0	75.0
08/18/1965	9.3	—	4.4	93.4	45.0
09/21/1965	7.3	2.6	5.2	24.0	11.0
10/27/1965	15.0	—	10.0	35.0	18.0
11/27/1965	45.0	—	10.8	74.0	18.0
12/07/1965	13.0	—	12.5	4.2	72.0

Date	Flow at Morristown (cfs)	Hanover Avenue	Eden Lane	Whippany Road	
06/22/1971	25.0	3.0	7.0	9.9	8.8
07/20/1971	55.0	4.6	6.3	19.8	9.6
08/24/1971	13.0	6.6	5.0	10.5	15.0
09/28/1971	56.0	1.8	2.7	15.0	5.4
10/19/1971	33.0	2.6	5.2	8.4	8.0
11/16/1971	—	2.3	4.6	12.9	10.0
12/14/1971	59.0	7.3	9.2	24.8	20.0
01/25/1972	52.0	3.9	7.0	10.0	5.6
02/23/1972	52.0	4.6	2.8	11.1	9.6
03/21/1972	89.0	3.7	2.7	4.5	4.5
04/25/1972	68.0	2.0	1.9	4.8	4.2
05/23/1972	104.0	4.0	2.5	5.2	1.8
06/21/1972	214.0	2.4	1.8	0.9	1.0
07/19/1972	—	1.2	3.2	9.9	7.2
08/22/1972	22.0	1.4	5.4	8.0	18.4
09/19/1972	166.0	3.8	5.2	10.4	7.0
10/17/1972	—	1.4	8.7	22.0	10.0
11/29/1972	—	2.7	3.6	6.0	3.6
12/13/1972	—	2.3	2.4	3.0	2.1

[a] All flows in cfs; all BODs in mg/L.
[b] Flow data from Refs [16,18]; BOD data from Ref. [10].

with no significant difference in data from 1964–1965 and 1971–1972 or any difference among the various sources of 1971–1972 data. The single correlating curve shown exhibits a general hyperbolic shape, with concentrations of BOD higher at low flows and *vice versa*. This is in contrast to an earlier study [19], which reported that Whippany River BOD–discharge data were greatly scattered with no definite trend.

This same trend of BOD_5 decreasing with river flows at Morristown also applies to data from the sampling stations downstream of Morristown (although not shown here). Since the flows at these other stations are nearly proportional to the more readily available flow at Morristown, using the reference flow at Morristown merely compresses the horizontal axis in such plots.

In summary, at Eden Lane the 1971–1972 BOD_5 data are lower than the 1964–1965 data where comparable flows exist. Lower BODs are the result of improved operation in the Morristown treatment plant. At higher river flows, only BOD data from the later period are available to show the effect of dilution in lowering BOD.

At Route 10 (1964–1965) or Whippany Road (1971–1972), downstream of the Whippany Paper Board Company, there is a decided difference between 1964–1965 data and 1971–1972 data. Most of the early data were measured at high levels (some over 100 mg/L in the river), but all of the more recent data lie below 25 mg/L. All sources of the 1971–1972 data are in agreement and at low levels regardless of flow; the high 1964–1965 data all occurred at relatively low flow rates.

The same comments also apply to the downstream BOD_5 data at Edwards Road.

G.5 WASTE TREATMENT AT THE WHIPPANY PAPER BOARD COMPANY

The Whippany River Paper Board Company is the key to water quality on the Whippany River. Effectiveness of the waste treatment facilities at this plant determines to a large extent the dissolved oxygen and BOD levels in the river all the way to the Passaic. Therefore, this source deserves special consideration here, and the development of its waste treatment facilities is discussed in this section.

G.5.1 Nature of the Waste

The Whippany Paper Board Company produces various grades of paperboard, linerboard, and corrugating medium from recycled newspaper and Kraft corrugated board on cylinder type and Fourdrinier machines. The manufacturing process also uses alum, caustic, starch, rosin, and a biocide for slime control. The aqueous effluent from this process constitutes a typical paper mill waste high in settleable and suspended solids and biodegradable organic materials.

G.5.2 Treatment Before 1964–1965

The paperboard complex in Whippany, NJ, consists of three mills, Eden, Hanover, and Stony Brook. These mills were originally built as separate enterprises around 1941 but were consolidated under single ownership before the three in 1947. Between 1947 and 1956, Whippany installed primary clarifiers at each of the three mills, and in 1959 after an expansion of the Eden Mill, built secondary biochemical treatment facilities there. In 1961 and 1962, the Hanover and Stony Brook primary effluents were linked to the Eden secondary treatment system, necessitating an enlargement of that system in 1963 to handle the combined primary-treated effluent of the three mills [5–7].

This biological waste treatment system consisted of mixing the overflow from primary sedimentation in a 12-million-gallon capacity cooling tower to cool 7–8 million gallons per day (MGD) from 140 to 85°F upstream of a 1-million-gallon equalization tank. Flow at that time was in parallel through two activated sludge units, each with its own separate thickener. The smaller 110 ft diameter, 1.3-million-gallon unit was installed in 1959; the larger 175 ft diameter, 4.2-million-gallon unit was completed in 1963 to augment the overloaded smaller unit. Ammonia (1 lb nitrogen/20 lb BOD) and phosphoric acid (1 lb phosphorous/80–100 lb BOD) were added to balance the high carbohydrate diet of the bacteria. The liquid effluent was chlorinated before discharge into the river or reuse in the mill.

Sludge, formed at the rate of 0.5 lb/1 lb BOD was discharged to the thickeners and the supernatant was recycled to the activated sludge units. The thickened sludge was lagooned before disposal as landfill. Some trouble with sludge consistency was reported but was expected to be rectified with the installation of five 16-cell filter units. Further details of the $2.5 million addition to the treatment system in 1963 can be found in Ref. [3]. Although the range of 7–13 mg/L quoted by this source as "best values" for effluent BOD may be technically correct, it seems a bit misleading in light of overall sustained treatment performance.

G.5.3 Effect of Early Treatment on Water Quality

This low level is indeed optimistic when one considers the effect of the mill's effluent on the river during 1964–1965. In addition to these monthly data for the river, discussed previously, there exist other data collected daily by the Passaic Valley Water Commission to monitor both the river and the plant effluent. These data were taken because of this plant's status as the largest single pollution source upstream of the Little Falls water treatment plant.

The monthly maximum, average, and minimum values for these daily samples by the Passaic Valley Water Commission are shown in Figures G.7–G.9 for 1964–1965. The single-sample DO and BOD values discussed previously fall within the extremes as shown in Figure G.7. The effluent BOD values in Figure G.9 are particularly informative. While the effluent BOD was recorded in the vicinity of 7–13 mg/L on several occasions, those periods were rare. A more realistic range would be 100–200 mg/L with excursions to levels above 300. There can therefore be no doubt of the mill's responsibility for the poor river quality during 1964–1965 despite the very sizable expenditure made for pollution control before this period.

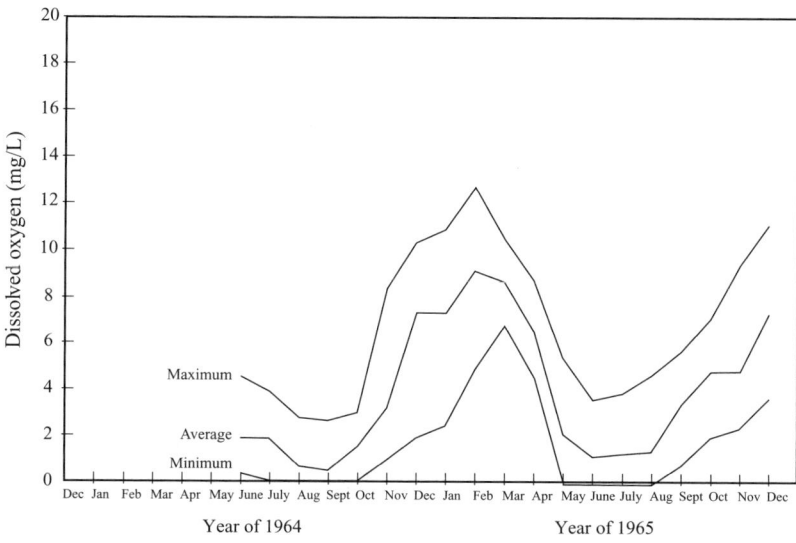

Figure G.7 Monthly maximum, average, and minimum values of daily dissolved oxygen measurements at Edwards Road 1964–1965.

G.5.4 Upgrading of Treatment Facilities

In early 1966, the Whippany Paper Board Company entered into a consent judgment with the water commission and the State Department of Health, agreeing to limit its discharge to 4.7 MGD and to curtail production if the river dissolved oxygen at

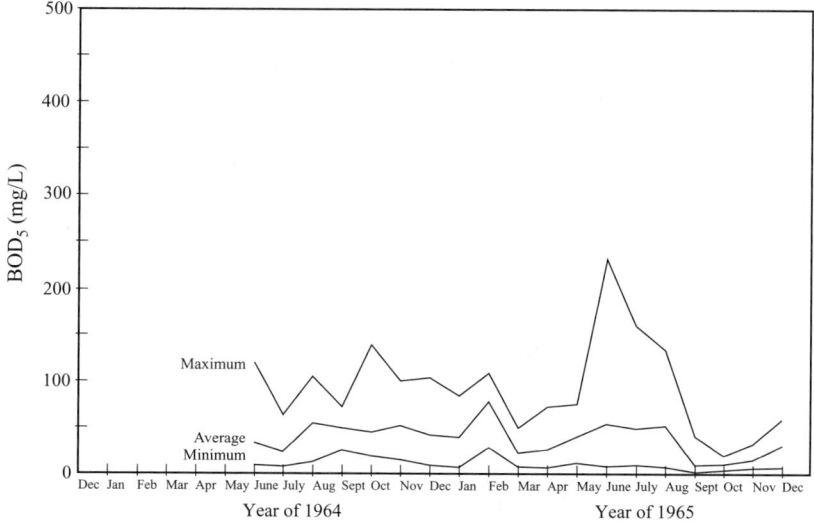

Figure G.8 Monthly maximum, average, and minimum values of daily BOD_5 measurements at Edwards Road 1964–1965.

618 Appendix G Water Quality Improvement for a Small River

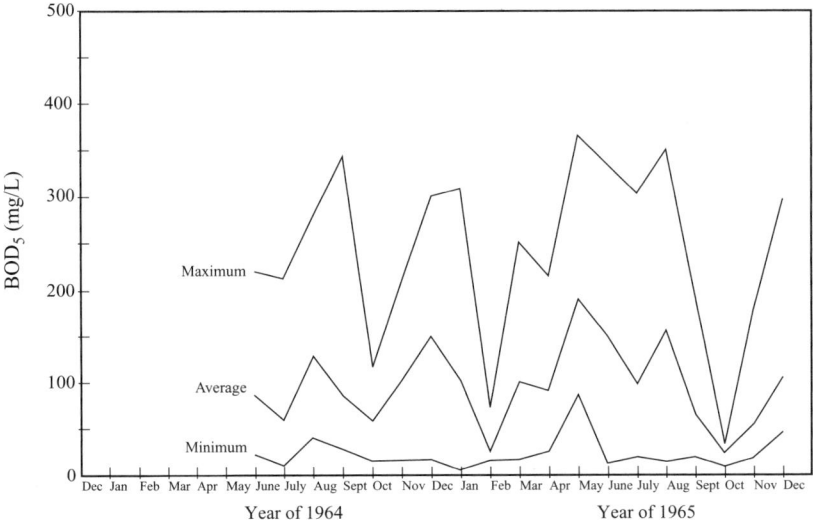

Figure G.9 Monthly maximum, average, and minimum values of daily BOD$_5$ measurements for Whippany Paper Board Effluent 1964–1965.

Edwards Road dropped below 2 mg/L or if the effluent contained more than 2% settleable solids by volume. At that time, the company also agreed to undertake an engineering study to upgrade the plant to produce an effluent quality of 25 mg/L BOD or less.

The study followed a classical approach of problem definition, wastewater volume reduction, and equipment modifications. The major culprit responsible for the poor performance was identified as the predominance of filamentous organisms causing a bulking of the activated sludge. The accelerated growth of these organisms was in turn traced to the carbohydrate nature of the waste materials, low pH, and high temperatures. Additional problems, such as poor performance of the primary clarifier treating the Fourdrinier white water from Eden machines 3 and 4, deficiencies in sludge wasting and disposal, and high solids carryover to the activated sludge units, were also identified.

The wastewater volume was cut in half from more than 8 MGD to approximately 4 MGD primarily by recirculation and in-mill reuse of primary-treated effluent wastewater. Although BOD and suspended solid loadings in lb/day remained constant, the retention time in the biological system was doubled and the overflow rate in the secondary clarifiers was reduced to half the former rate. (Since that time, wastewater volume has increased to 4.5–5 MGD as listed in Table G.1.)

Once the wastewater volume was reduced, modifications were made to the existing primary treatment systems. The primary treatment facilities handling the cylinder board machines (Hanover, Stony Brook, and Eden machines 1 and 2) were performing properly, but the clarifier treating the white water from the two Fourdrinier machines (Eden 3 and 4) was ineffective for a number of reasons. These included a high proportion of fines, thermal gradients leading to complete turnover of the bottom sludge layer, severe surges in flow, and generation of solid–buoyant gases and volatile acids caused by anaerobic conditions in the sludge layer. To eliminate these problems, a disk-type vacuum save-all was installed in place of the clarifier, and the clarifier was converted into an additional thickener for the secondary sludge.

Further equipment modifications were made. Four 7.5 hp mixers were installed in the 1-million-gallon equalization tank to prevent undesired settling and periodic discharge of the septic sludge to the biological system. Provision was made for wasting the sludge from the activated sludge units to any one of the several sludge thickeners. Formerly, each of the two parallel-activated sludge units was piped to only its own sludge thickener. As an interim measure, a unit was installed to chlorinate the wasted sludge before lagooning, while pilot studies of sludge dewatering (centrifugation and evaporation) were begun. However, later information was that the chlorination unit was still operating and the sludge was being trucked away after lagooning (Table G.1).

Perhaps the most important modification made was in the configuration of the originally parallel activated-sludge units. As a result of pilot- and full-scale studies, the two units were changed to series operation with the smaller (1.3 MG) unit ahead of the larger (4.2 MG) unit. With this change to a first stage at high loading followed by a second stage with lower loading, the sludge bulking because of the filamentous growth was greatly reduced. Production of filamentous sludge could then be controlled by a combination of dissolved oxygen regulation, pH control, sludge wasting and interchange between the units, and chlorination techniques. Nitrogen and phosphorous levels were unchanged. Additional details concerning these treatment facilities, including the process flowsheet that is too complex to reproduce here, are contained in Ref. [12].

G.5.5 Effect of Improvements on Water Quality

For the six months following these changes completed in late April 1967, the chlorinated plant effluent was stated to average 14 mg/L and was reported to be less than the State standard of 25 mg/L BOD between this date and early 1971, or up until the time that the article was published. Similarly, suspended solids decreased from the order of 100 mg/L to less than 25–50 mg/L. Settleable solids, high before 1967, dropped to 0.03–0.4 mg/L.

These figures appear quite reasonable when compared to independent measurements [10] during 1971–1972 (Figures G.10–G.12). Figures G.10 and G.11 depict the dissolved oxygen and BOD levels at Edwards Road, and Figure G.12 shows the effluent BOD. The river values are much better than their pre-expansion 1964–1965 counterparts (Figures G.7 and G.8) as are the effluent BODs compared to their previous values (Figure G.9). The scales on corresponding before and after figures have been kept the same to facilitate these comparisons. The effluent BODs average about 20 mg/L during the more recent period. Maximum values, however, have been above 25 mg/L upon occasion.

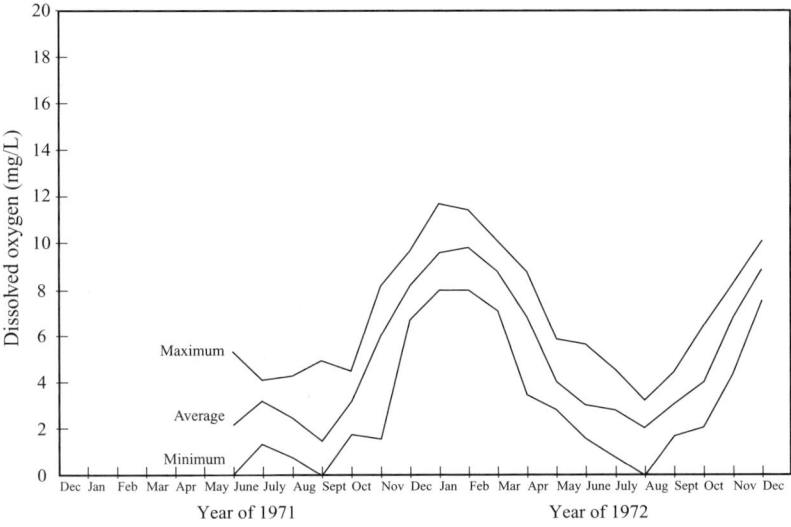

Figure G.10 Monthly maximum, average, and minimum values of daily dissolved oxygen measurements at Edwards Road 1971–1972.

In addition to these data showing the effect of high-effluent BODs on the river, the effect of completely eliminating the effluent BOD during periods of strike or summer vacation shutdown at the mill is even more dramatic. River BOD values dropped to their upstream levels, and low or zero dissolved oxygen reverted to normal upstream levels [10].

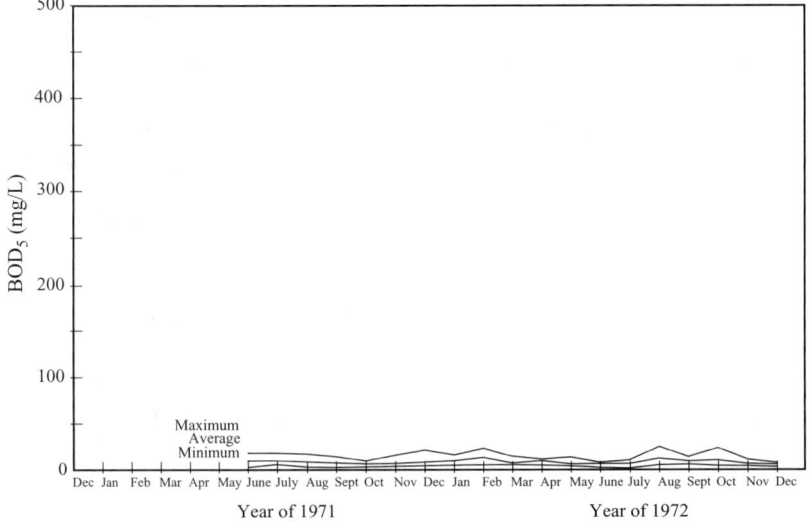

Figure G.11 Monthly maximum, average, and minimum values of daily BOD_5 measurements at Edwards Road 1971–1972.

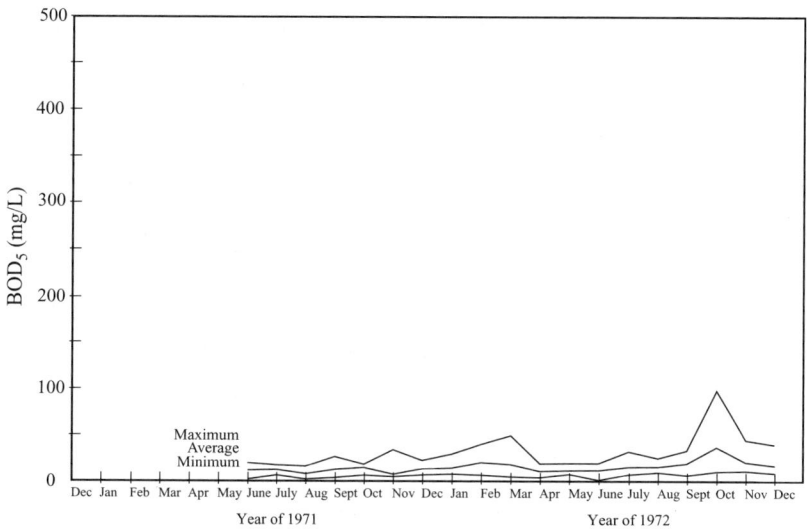

Figure G.12 Monthly maximum, average, and minimum values of daily BOD$_5$ measurements for Whippany Paper Board Effluent 1971–1972.

G.5.6 Summary

In conclusion, the purpose of this section is not to point out the seemingly obvious conclusion that treating waste lowers BOD, but rather to contrast the difference between unsuccessful and effective treatment on the receiving body. Expenditure of funds *per se* does not guarantee effective treatment or statutory compliance. Unfortunately, in the case of the paper mill, construction of treatment facilities that turned out to be inadequate delayed the sought-after improvement in river water quality by some 4–5 years.

G.6 MODELING THE WHIPPANY RIVER

The dissolved oxygen sag along the last 9 miles of the Whippany River was evaluated in terms of a simple Streeter–Phelps model:

$$D = \frac{kL_a}{r-k}[\exp(-kt) - \exp(-rt)] + D_a \exp(-rt) \quad (G.1)$$

where D is the dissolved oxygen deficient ($DO_{sat} - DO_{actual}$) (mg/L), D_a is the dissolved oxygen deficit at point of pollution (mg/L), L_a is the ultimate first-stage BOD at point of pollution (mg/L), k is the deoxygenation constant (days^{-1}), r is the reaeration constant (days^{-1}), t is the stream flow time from point of pollution (days).

Along this reach are located the important sources of pollution – the Morristown STP, the Whippany Paper Board Company, and the Hanover STP.

Besides the initial DO–BOD data, necessary inputs to this equation are values for the constants and a relationship to convert distance along the river into time of flow. The DO–BOD data have already been presented in previous sections. The evaluation of the deoxygenation and reaeration constants and the stream velocity to relate distance and time are discussed in the next sections.

G.6.1 Selection of Constants

The value for the deoxygenation constant has been selected as 0.293 at 20°C. This has been reported [19] as the mean of eight annual values varying from 0.190 to 0.342 during the period from 1957 to 1967 in the Passaic River immediately downstream of its confluence with the Whippany and Rockaway. Variation with temperature has been taken as

$$k_T = k_{20} 1.045^{(T-20)°C} = k_{68} 1.0247^{(T-68)°F} \tag{G.2}$$

The reaeration constant has been assumed to be

$$r = 5V/R^{5/3} \tag{G.3}$$

Where V is the stream velocity and R is the hydraulic radius [4]. The hydraulic radius is the wetted area of flow divided by the wetted perimeter, not including the free surface of the stream exposed to atmospheric air. Values computed for the Whippany according to the flow–velocity–depth relationship developed in subsequent sections are plotted in Figure G.13. According to this figure, typical values of the reaeration constant vary between 1.0 and 2.6. The temperature dependence for the reaeration constant has been taken as

$$r_T = r_{20} 1.024^{(T-20)°C} = r_{68} 1.0133^{(T-68)°F} \tag{G.4}$$

G.6.2 Theoretical Calculation of Stream Velocity Impractical

The Manning equation

$$V = \frac{1.486}{n} R^{2/3} S^{1/2} \tag{G.5}$$

comes immediately to mind to obtain the velocity of flow in an open channel. In this equation, S is the slope of the terrain measured along the river, ft/ft, and the so-called Manning's n is a dimensionless coefficient of roughness. The other symbols have been defined earlier. With the proper elevations and horizontal distance to compute the slope and the channel cross section to calculate the hydraulic radius, estimation of the stream velocity is straightforward, provided that the appropriate value of n is used. The roughness coefficient, n, is normally assumed to depend upon surface roughness, vegetation, channel irregularity, obstructions, hydraulic radius, stage, and discharge. In general, any situation tending to increase the extent of surface contact between the

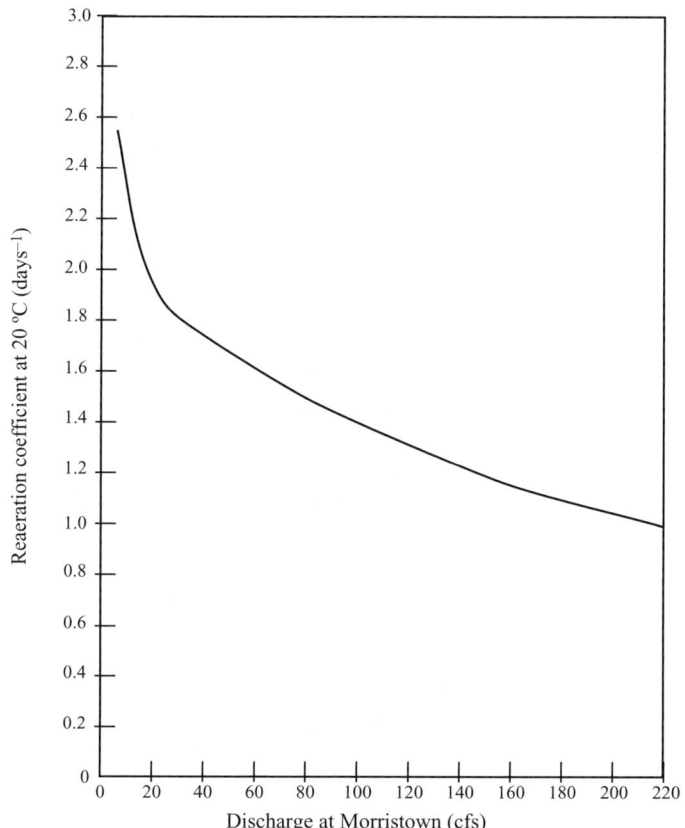

Figure G.13 Assumed Whippany river reaeration coefficient versus discharge.

stream and its bed results is an increase in n, and *vice versa*. When the river is shallow and the water is forced to flow over many irregularities, increased frictional resistance is reflected in higher n values. Conversely, for flows at high discharge, n becomes lower and nearly constant.

There is a procedure given in Ref. [2], permitting one to build up a roughness coefficient from the observed condition of a stream and its banks. Applying this method to the Whippany gives estimated n values varying between 0.033 minimum to 0.040 normal and 0.045 maximum. However, a value of 0.121 has been previously back-calculated from data using the Manning equation [19]. The flow used in the calculation was low; approximately 10 cfs, and therefore a high n value would be anticipated. Pictures of typical channels [2] with typical roughness coefficients confirm that n can easily vary between these limits.

In light of this possible wide variation in n values, it was therefore decided to determine the Whippany's velocity experimentally and to relate it empirically to the discharge recorded daily at the Morristown gauging station. Details of the measurements used in generating the desired relationship are given below.

G.6.3 Velocity—Time of Travel Measurements

Using a dye injection technique, the USGS has conducted several time-of-travel measurements along the Whippany River at flows up to 55 cfs measured at Morristown. These data are summarized in Table G.9. Fifty-five cfs is slightly above average flow, but still well below some of the measured discharges corresponding to BOD data during 1972 (Table G.8). The USGS data were therefore supplemented with independent measurements by the author during periods of higher flow.

These latter measurements were not nearly so sophisticated. Time of travel and hence velocity were obtained by observing floating objects between the Route 10 bridge and the bridge leading to American Legion Post No. 421, 0.35 miles downstream. This stretch of river is almost parallel to Ridgedale Avenue, providing easy access to both stations, and is approximately midway along the reach to be modeled. The distance was estimated using an automobile odometer and confirmed by scale measurement on the USGS Morristown quadrant map [15].

Depth of water was measured by dropping a plumb line from the American Legion Bridge, the width of the river was approximated using 50-ft steel tape along the bridge, and the discharge was computed by multiplying the velocity by the area of flow. This reach appears to be of uniform width, and it has been assumed that its depth is uniform also.

Measurements were made during mid-April 1973 to take advantage of a period of high spring flow and again at the end of June 1973 when the river was expected to be somewhat lower (Table G.10). The initial June measurement was followed by others during the next several days when the opportunity presented itself to observe the river during and after a period of extremely heavy rainfall. Some comments concerning the individual measurements are given below.

On April 14, the river had just receded from an auxiliary channel near the American Legion Bridge, but the water was still high. The depth recorded in the table was virtually uniform across the channel with some sloping near the banks. On June 28, the river was lower, and the banks on both sides were exposed and partially overgrown with vegetation. At this time, the river was only 39 ft wide, corresponding to a drop in water level from 34 to 25 in. The very next day, similar measurements were made in the midst of a violent rainstorm. During and immediately after the storm, there were streaks of oil on the water, apparently due to runoff from the nearby roads.

Table G.9 USGS Time-of-Travel Measurements[a]

Date	Mean Discharge at Morristown (cfs)	Reach Length (miles)	Travel Time of Peak (h)	Velocity (fps)
08/09–10/1966	9.6	6.2	27.7	0.32
08/09–10/1966	9.6	3.9	16.2	0.35
09/02–03/1968	28.0	6.2	14.2	0.65
04/21/1969	55.0	6.2	10.5	0.86

[a] Time of travel determined using 40% Rhodamine BA dye injection technique [17].

Table G.10 Author's Time-of-Travel Measurements[a]

Date	Time (min)	Velocity[b] (fps)	Depth of water (in.)	Width of channel (ft)	Discharge at location (cfs)	Discharge[c] at Morristown (cfs)
04/14/1973	19:05	1.34	34	44	168.0	118.0
06/28/1973	24:00	1.06	25	39	86.5	60.5
06/29/1973	12:55	1.97	55	44	398.0	279.0
06/30/1973	11:00	2.13	69	44 (?)	540.0	378.0

[a] Time of travel measured by observing floating objects from Route 10 Bridge to American Legion Bridge. Reach length in all cases 0.35 miles. (See Figure G.1).

[b] Calculated by dividing reach length by 1.2 times the travel time. The factor 1.2 corrects the surface velocity in an open channel to the mean velocity [11].

[c] Computed from observed discharge at location and the flow factor extrapolated from the data of Table G.2 (approximately 1.5 at Route 10 versus Morristown).

The main channel of the river had again widened to 44 ft, partially overflowing into the auxiliary channel. The depth increased to 55 in.

It had continued to rain all night, and the river was near flood stage on the next morning. The depth was found to be 69 in. across the main channel. Although the auxiliary channel near the bridge was also filled to near overflowing, the width for flow measurement was still taken as 44 ft, the estimated width of the single channel upstream of the point of measurement. This may result in an underestimation of the discharge, but this flow rate is well beyond the range of flows encountered with the 1972 BOD data and is therefore of little consequence to this study.

The data from Tables G.9 and G.10 are plotted against discharge in Figure G.14. Two lines are shown, one referring to the discharge at the location of measurement and

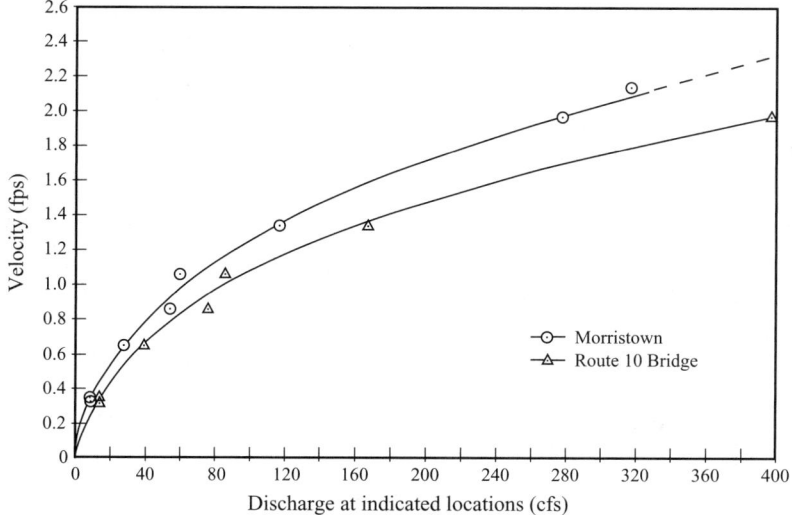

Figure G.14 Whippany river velocity versus discharge.

the other to the discharge at the Morristown reference station. Discharges at the two locations have been related using the flow factor extrapolated from the data of Table G.2 (approximately 1.5 at Route 10 in Morristown).

These curves are of the form $V = Q/A$ plotted against Q, where Q is the discharge and A is the cross-sectional area. If A were constant, the relationship would be linear, passing through the origin because a zero velocity condition corresponds to zero discharge. But since A increases with increasing Q, the curves have the indicated shape. Furthermore, it can be shown to follow from the Manning equation that the velocity is proportional to the 2/5 power of the discharge for constant n and a rectangular cross section whose width is much greater than its depth. These conditions are approximated at high flows in the Whippany, and the data of Figure G.14 exhibit the theoretical variation of velocity with discharge.

Purely as a matter of interest, a value of n was estimated from the data and distances and elevations scaled from the USGS maps [15]. For the low flows, n was found to be in the range 0.1–0.2, approximately 0.095 for the intermediate flows (55 and 86.5 cfs at Morristown), and falling off to 0.07 for flows in excess of 118 cfs at Morristown.

Based on the foregoing considerations, the velocity data are concluded to be valid for use in the modeling study. Time-of-travel for various stretches of the river was obtained by dividing distance by velocity. USGS reports the location of the Morristown gauging station and the Morristown STP as 9.0 miles upstream from the river's mouth. This and the other distances were scaled from the USGS maps [15] and are recorded in Table G.11.

G.6.4 Assembling the Model

Using the parameters developed in previous sections, the lower reach of the river was modeled in segments, and the predicted versus measured values of DO and BOD were compared whenever possible. For the river, only the initial DO–BOD data at Hanover

Table G.11 Distances Along Whippany River

Location	Miles upstream from mouth [15]
Hanover Avenue Bridge (Gauging Station, Morristown STP)	9.00
Eden Lane Bridge (Whippany Paper Board Company)	6.50
Whippany Road Bridge	5.55
Route 10 Bridge	4.05
American Legion Bridge	3.70
Troy Road Bridge (Hanover STP)	2.60
Edwards Road Bridge	0.30

Notes: Distance for the Morristown Gauging Station and the Morristown STP is given as 9.0 miles upstream from mouth by the USGS. Other distances scaled from USGS maps [15].

Table G.12 Plant Effluents used in Modeling Whippany River

Source	Flow (MGD)	BOD (mg/L)	Data
Morristown STP	3.0	25	Monthly BOD range: 7–44, April 1972; 15–40, Aug 1972, Ref. [9]
Whippany Paper Board Company	8.0, 1964–1965	Variable, high	Figure G.9
	4.0, 1971–1972	Variable, low	Figure G.12
Hanover STP	1.5	65	Monthly BOD range: 40–84, May 1972; 42–88, July 1972, Ref. [9]

Avenue, the river flow at Morristown, and the river temperatures were used as inputs (Tables G.3 and G.4). The effluent flows and BOD were taken as shown in Table G.12. Values of r at 20°C were calculated from the flow by means of Figure G.13, and both r and k were adjusted for temperature using Equations (G.2) and (G.4). Zero DO was assumed in the effluents.

The river was modeled in segments: the first from Hanover Avenue (Morristown STP) to Eden Lane, the second from the paperboard company effluent downstream of Eden Lane to the Hanover STP at Troy Road, and the third from Troy Road to Edwards Road. Intermediate checks were also made at Route 10 (for 1964–1965 data) or Whippany Road (1971–1972). Other flows, either natural or man-made, that enter the main stream between Hanover Avenue and Edwards Road have been disregarded.

G.6.5 Results and Discussion

A comparison of the predicted and measured DOs is plotted in Figure G.15. The modeling is generally poorer for the period 1964–1965 both for BOD and DO. This is illustrated for DO by the upper three parity plots of Figure G.15. There are two reasons for this lack of agreement: first, the modeling is not too good for the extremely low flows encountered during 1964–1965 even upstream of the Whippany Paper Board Company. Perhaps the value of the reaeration constant for these low flows is overly optimistic. Secondly, the high and highly variable BODs from the paperboard operation in 1964–1965 strongly influence the calculation downstream of the paper mill.

Modeling of dissolved oxygen is better for the later period when the measured BODs are lower, more constant, and in better agreement with the calculated values. This better agreement in DO values is shown in the lower three plots of Figure G.15.

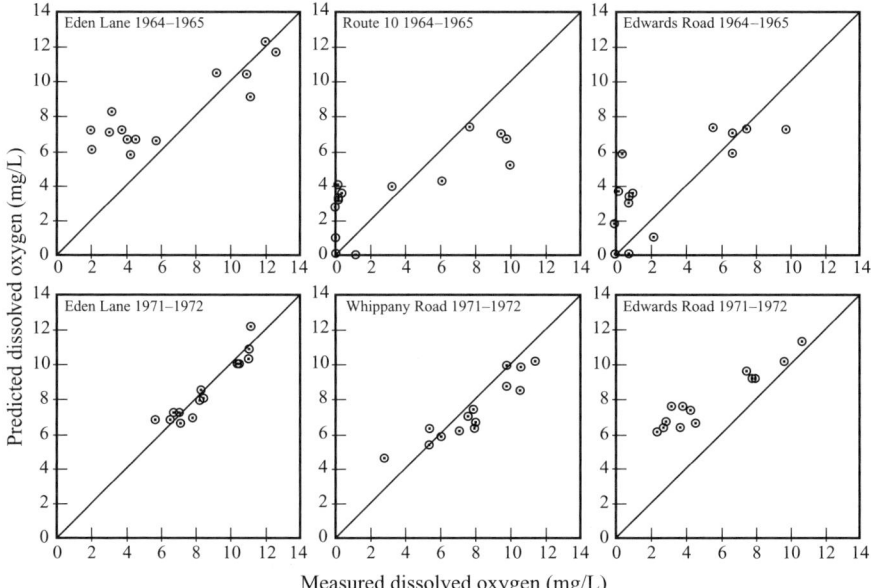

Figure G.15 Predicted versus measured dissolved oxygen levels in the Whippany River.

Even for the later period, the predicted DO values at Edwards Road are above the measurements. The BODs, however, are still in reasonable agreement. These values are the result of modeling the stretch of river from Eden Lane to Edwards Road with an intermediate check shortly downstream of the paper mill. This reach is 6.2 miles, the longest single segment modeled. Accumulated errors from the upstream calculations could be contributing to the discrepancy. It is also possible that the magnitude of flow and/or BOD from the Hanover STP is actually higher than that used in the calculations. This would not influence the river BODs too much but would have a marked effect on the DO downstream. Again, the reaeration constant may also be overly optimistic.

The data used for comparison are good but are of a routine nature for control purposes and are not research data on a single slug of water followed from point to point. This would be necessary for the practical case of an unsteady flow. Even the effluent quantities and qualities vary with time. In addition, sources of unrecorded BOD [19] were not considered. Finally, the input velocity, time, and depth of river relationship were based on a few measurements at a single location on the river. Nevertheless, despite these minor shortcomings, this model taken as a whole appears to be an adequate representation of the Whippany River, and further refinements are not considered warranted at the present time.

G.7 POSTSCRIPT

This newly written section is intended to provide a sort of closure to the investigation. The technology of the Internet, not available at the time of the study, was employed to update the information to the present day. As it turns out, the river is still there, but not all of the circumstances are the same.

The Whippany Paper Board Company closed down the Hanover, Stony Brook, and Eden Mills over a period of 3 years [21] sometime around 1979 [22] and declared bankruptcy [21]. The company filed for bankruptcy in 1980 apparently to complete its sale [23], and the case was decided in 1981 [24].

This in turn adversely affected the business fortunes of the local freight-hauling railroad, itself operating in receivership; freight volume was said to have dropped suddenly from thousands of cars per year to a few cars per week [21]. The railroad was reorganized in 1982 and subsequently expanded operations. However, the historical commentary through 1998 makes no mention of any resumption of paper mill freight traffic or reopening of the mill under new management.

Before the paper mill shutdown and bankruptcy, and independent of the author's Whippany River study conducted in 1973, the United States Attorney's Office in Newark, NJ filed a 20-count indictment against the Whippany Paper Board Company for "illegal discharge pollution into the navigable Whippany River" [25]. (Navigable means that if you can float a rowboat, canoe, raft, or some other such thing in it under any circumstance, it's navigable). Unbeknownst to the author at the time, water samples and photographic evidence gathered between 1971 and 1973 by two undergraduate students at another local university were submitted to the U.S. Attorney's Office.

The company pleaded guilty on 14 counts, and the U.S. District Court in 1974 applied both the 1899 Refuse Act (Section 13 of the Rivers and Harbors Act, 33 USC 407) and the recently enacted Federal Water Pollution Control Act (FWPCA) of 1972 simultaneously, and awarded the students half the amount of the fine levied against the company. The Refuse Act prohibits depositing refuse matter of any kind into navigable waterways and contains a provision entitling citizens providing such evidence to 50% of the fine imposed by the court upon conviction of a polluter. This case sets the precedent that the 1972 FWPCA does not supersede the 1899 Refuse Act and its bounty paid to environmental activists.

Annual average flow of the river at the Morristown gauging station is now somewhat higher than the 48 cfs noted for the study period—62.7 cfs for Calendar Year 2005, 72.1 cfs for Water Year (October 2005–September 2006), and 54.7 cfs averaged over Water Years 1972–2006 [26]. Morristown's wastewater/sewage treatment plant capacity has increased from its Table G.1 value of 3.0 MGD to 3.45 MGD [27]. The Hanover Sewage Authority wastewater treatment plant, serving the unincorporated communities of Whippany and Cedar Knolls, NJ, has shown an annual average flow just over 2.0 MGD covering the years 1999–2006, up from the 1.5 MGD figure of Table G.1 [28]. In addition, it is likely that other entries in Table G.1, including the list of dischargers itself, have changed over the years.

Finally, the Whippany River at Morristown is identified by the New Jersey Department of Environmental Protection (NJDEP) as "impaired" for phosphorous at total phosphorous concentrations greater than 0.1 mg/L in the river [29]. This had been identified as an opportunity for further improvement in the author's original report.

One final postscript: the author's personal reconnaissance from the public roads in the Whippany River area revealed no contemporary trace of paper mill operations along the river and railroad line, except perhaps a boarded-up abandoned building in the trees behind locked chain-link fences and a rusting water tower off in the distance beyond. When questioned about the whereabouts of the Whippany Paper Board Company, a local resident pointed in the general direction of the water tower and indicated that the place had closed down many years ago.

REFERENCES

1. Board of Chosen Freeholders, Morris County, N.J., *Map of Morris County, N.J.*, General Drafting Co., Convent Station, NJ, 1969.
2. Ven Te Chow, *Open Channel Hydraulics*, McGraw-Hill, New York, 1959.
3. A.W.J. Dyck, Whippany lets the 'Bugs' do it, *Paper Industry*, **45**, 374–375, 1963.
4. G.M. Fair, J.C. Geyer, and D.A. Okun, *Water and Wastewater Engineering*, Vol. 2, Wiley, New York, 1968.
5. L.L. Klinger, Whippany completing final link in tri-mill waste treatment system, *Paper Trade Journal*, **146** (20), 36–43, 1962.
6. L.L. Klinger, Whippany expands tri-mill treatment complex, *Water Works and Wastes Engineering*, **1** (1), 60–65, 94–95, 1964.
7. L.L. Klinger, Dynamic activated sludge treatment in paperboard mills, *Paper Industry*, **47**(1), 76–80, 1965.
8. New Jersey Department of Environmental Protection, Rules and Regulations Establishing Surface Water Quality Criteria, June 30, 1971.
9. New Jersey Department of Environmental Protection, Trenton, NJ, unpublished data (1971–1972).
10. Passaic Valley Water Commission, Little Falls, NJ, unpublished data (1964–1972).
11. J.G. Rabosky and D.L. Koraido, Gauging and sampling industrial wastewaters, *Chemical Engineering*, **80**(1),111–120, 1973.
12. R.K. Shaw, Wastewater treatment experiences at Whippany Paper Board Co., *Water and Sewage Works*, **118**, 114–120, 1971.
13. *Standard Methods for the Examination of Water and Wastewater*, 13th edn, Table 218(1), American Health Association, Washington, DC, 1971, pp. 480–481.
14. State of New Jersey, Special Report 31 Stream Flow Records October 1960 – September 1965.
15. United States Geological Survey Quadrant Maps Caldwell, NJ, N 4045–W 7415/7.5 AMS 6165 IV SE-Series V 822; Mendham, NJ, N 4045–W 7430/7.5 AMS 6065 I SE-Series V 822; Morristown, NJ, N 4045–W 7422.5/7.5 AMS 6165 IV SW-Series V 822 USGS, Washington, DC (1954–Photorevised 1970).
16. United States Geological Survey, Trenton, NJ, provisional flow and water quality data (1971–1972).
17. United States Geological Survey, Trenton, NJ, unpublished time-of-travel measurements Whippany River (1966–1969).
18. United States Geological Survey, Water Resources Data for New Jersey: Part I Surface Water Records; Part II Water Quality Records, Prepared in cooperation with the State of New Jersey and with other agencies, Trenton, NJ (1966–1971).
19. W. Whipple Jr., Preliminary Mass Balance of BOD on Three New Jersey Rivers, New Jersey Water Resources Research Institute, Rutgers—The State University of New Jersey, 1969.
20. W. Whipple Jr., J. V. Hunter, B. Davidson, F. Dittman, and S. Yu, *Instream Aeration of Polluted Rivers*, New Jersey Water Resources Research Institute, Rutgers—The State University of New Jersey, August, 1969.

Supplemental References for Postscript to Appendix G

21. S.P. Helper, The Morristown & Erie Railway, article dated 1990, available at http://jcrhs.org/m&e.html, accessed on October 1 2007.
22. J. Cheslow, If You're Thinking of Living In Whippany, NJ; Where Houses Are In High Demand, The New York Times, August 8, 1999, available on New York Times Archives http://query.nytimes.com, accessed on October 1, 2007.
23. J.P. Sheppard, Beautifully broken benches: a typology of strategic bankruptcies and the opportunities for positive shareholder returns, *Journal of Business Strategies*, **12**(1), 99–134, 1995.
24. Whippany Paper Board Co. (15 B.R. 312, 314, 315 Bkrtcy. NJ 1981).
25. Refuse Act, Wikipedia, the free encyclopedia, available on http://en.wikipedia.org/wiki/Refuse_Act, last modified September 22, 2007—accessed on October 1, 2007.
26. USGS Water-Data Report 2006 01381500 Whippany River at Morristown, NJ Passaic River Basin, available on http://wdr.water.usgs.gov/wy2006/pdfs/01381500.2006.pdf, accessed on August 31, 2008.
27. Department of Public Works Sewer Utility, available on Town of Morristown Internet: http://www.townofmorristown.org/pw_sewer.html, accessed on October 1, 2007.
28. New Jersey Department of Environmental Protection, Municipal/Sanitary NJDEP/DSW Permit Flow Data, Spreadsheet dated January 27, 2006, available on NJDEP website http://www.state.nj.es/dep/er/, accessed on October 1, 2007.
29. New Jersey Department of Environmental Protection (NJDEP), Amendment to the Northeast, Upper Raritan, Sussex County and Upper Delaware Water Quality Management Plans Total Maximum Daily Load Report for the Non-Tidal Passaic River Basin Addressing Phosphorous Impairments Watershed Management Areas 3, 4, and 6, NJDEP Division of Watershed Management P.O. Box 418 Trenton, NJ 08625-0418 (proposed May 7, 2007, adopted April 24, 2008), available on www.state.nj.us/dep/watershedmgt/passaic_tdml_adopt_12_6_07.pdf, accessed August 31, 2008.

Appendix H

Experimental Determination of Coefficient for Draining of Tank

There are three kinds of people in this world—those who can do math and those who can't.

H.1 SUMMARY

This appendix presents original, unpublished data from a typical laboratory-scale experiment in which an open-top tank of uniform cross section at atmospheric pressure is allowed to drain by gravity. Its purpose is to show with real data the type of data obtained in such an experiment and the treatment of those data necessary to estimate the discharge coefficient and the time to drain the tank. Results of this laboratory-scale experiment show a discharge coefficient (C_D) on the order of 0.35 for the exit valve and tubing configuration, or a K-factor of 7.2 calculated by rearrangement of Equation (7.7) in Chapter 7.

A series of single measurements were made using nonprecision equipment and measuring devices. Despite this limitation, the results suited the purpose of the investigation. However, certain phenomena encountered on a laboratory scale are unlikely to arise in full-scale equipment, for reasons to be discussed in this appendix. When representative of the full-scale operation, data reduction methods are valid, but the reader is cautioned against using the numerical results themselves directly for scale-up.

H.2 DESCRIPTION OF EXPERIMENT—EQUIPMENT

An insulated beverage cooler of nominal 5 gal capacity and 1 in. (2.54 cm) wall thickness was used as the tank in this experiment. The standard push-button dispenser

---

Environmental Calculations: A Multimedia Approach, by Robert G. Kunz
Copyright © 2009 John Wiley & Sons, Inc.

at the bottom of the tank was replaced by a ball valve and tubing assembly. The outlet structure inside the tank immediately surrounding the outlet tube (a circular sink nut on a threaded fitting) was about 25 mm in diameter. It extended inside the tank about 15 mm beyond the inside wall surface; the inside-the-tank end of the outlet tube was flush with the outlet structure.

The push-button dispenser was changed in favor of a plastic ball valve with inlet and outlet polyethylene tubing of 0.170 in. (0.4318 cm) stated inside diameter. This allowed shutoff of water flow from the tank if need be without having to hold the outlet device closed manually on a continuous basis. Also, the valve and tubing slowed down the normal flow from the beverage cooler to allow enough time for manual readings of depth of liquid in the tank and volume collected.

H.2.1 Equipment Used

- Tank with modified outlet structure (as above)
- Wristwatch with sweep-second hand for time measurements
- Tape measure, calibrated in centimeter (cm) and inches (in.) to ascertain the tank dimensions and gauge liquid depth
- Multiple (labeled) jars to collect water samples
- Graduate cylinders for measuring sample volumes
- Small funnel to aid in transfer of each water sample to graduate cylinder
- Dial thermometer to determine water temperature

H.3 DESCRIPTION OF EXPERIMENT—PROCEDURE

- Measure the height of the outlet drain above the flat bottom of the tank (2.8 cm). (When filled with liquid, this heel of liquid below the outlet is undrainable by gravity through the outlet without tipping the tank.)
- Fill tank to brim with outlet valve closed.
- Measure the initial depth/height of liquid (cm) from the water surface to the bottom/floor of the tank; measure the water temperature.
- Turn the valve wide open at time $t = 0$ and allow the tank to drain by gravity.
- Measure water depth every 60 s while an assistant switches the sample-collection container from the one that has just finished collecting liquid from the previous minute to an empty container starting the collection for the next minute.

A preliminary run was first conducted without measurements to get an idea of what to expect for the speed at which the tank drained, the ease of making depth measurements and collecting samples, and the total time to drain the tank. For the measured run, height/depth data were measured (point valves) and volumetric grab samples (1 min samples), were switched "on the fly" every minute in one continuous operation, without stopping from start to finish until the last drops of liquid flowing from the outlet tubing eventually dribbled to a halt at 68 min and 40 s into the run.

The sample containers, previously labeled, were set aside for subsequent measurements of volume after the run. After the end of the active experiment, the valve was closed. Liquid volume for each sample was measured using graduate cylinders and recorded. The liquid from each measured sample was returned to the tank, and the final depth of accumulated water was measured at 42.0 cm, accounting for 99% of the water initially in the tank. Temperature of the composite collected samples, which had been sitting in uninsulated containers, was then measured in the tank at 60°F (15.6°C).

H.4 EXPERIMENTAL DATA

Dimensional data for the tank and outlet assembly are given in Table H.1. Table H.2 lists the elapsed time (column 1), and the depth of volume collected at 1 min intervals during the experimental run (columns 2 and 4). The third column contains the height of liquid above the outlet after subtraction of the outlet height (2.8 cm) above the tank bottom.

Table H.1 Tank and Outlet Dimensions, Conditions, and Physical Constants

Diameter of tank	Height of tank	Height of outlet	Area of tank (A_t)
26.0 cm	39.6 cm	2.8 cm	530.929 cm^2
10.236 in.	15.59 in.		82.29417 in.^{2}
10.25 in. (approx.)	1.30 ft		
Volume of heel	Tank volume without heel	Total tank volume	Volume of 1 cm height
1486.60 cc	21024.79 cc	22511.39 cc	530.929 cc
Outlet diameter	Outlet area (A_0)	A_t/A_0	Gravity constant (g)
0.4318 cm	0.146438 cm^2	3625.613	32.17 ft/s^2
0.170 in.	0.022698 in.^{2}	(dimensionless)	9.80665 m/s^2
			980.665 cm/s^2
			3530394 cm/m in.^{2}
Length of outlet		Water temperature	Viscosity of water
Tube tank to valve	5.4 cm	50°F	0.01310 cp at
Ball valve	3.0 cm	10°C	stated temperature
Discharge tube	2.4 cm		
Total length	10.8 cm		

Table H.2 Experimental Height and Volume Readings

Elapsed time (t, min)	Liquid height above bottom (h raw, cm)	Height above outlet (h, cm)	Volume collected for 1 min (V, cc)
0.0	42.4	39.60	0
1.0	40.8	38.00	833
2.0	39.0	36.20	820
3.0	37.7	34.90	850
4.0	35.9	33.10	839
5.0	34.2	31.40	802
6.0	33.0	30.20	795
7.0	31.4	28.60	766
8.0	29.9	27.10	738
9.0	28.3	25.50	728
10.0	27.0	24.20	708
11.0	25.8	23.00	703
12.0	24.8	22.00	638
13.0	23.5	20.70	641
14.0	22.2	19.40	611
15.0	21.1	18.30	596
16.0	19.9	17.10	567
17.0	18.8	16.00	553
18.0	17.8	15.00	524
19.0	16.7	13.90	504
20.0	15.8	13.00	473
21.0	15.3	12.50	451
22.0	14.4	11.60	452
23.0	13.8	11.00	392
24.0	13.4	10.60	378
25.0	12.5	9.70	368
26.0	11.0	8.20	340
27.0	10.0	7.20	316
28.0	10.0	7.20	309
29.0	9.3	6.50	280
30.0	8.9	6.10	264
31.0	8.1	5.30	238
32.0	7.6	4.80	234
33.0	7.0	4.20	210
34.0	6.8	4.00	192
35.0	6.1	3.30	182
36.0	5.8	3.00	168
37.0	5.8	3.00	147
38.0	5.6	2.80	138
39.0	5.2	2.40	128

(continued)

Table H.2 (*Continued*)

Elapsed time (*t*, min)	Liquid height above bottom (*h* raw, cm)	Height above outlet (*h*, cm)	Volume collected for 1 min (*V*, cc)
40.0	4.9	2.10	116
41.0	4.8	2.00	102
42.0	4.6	1.80	95
43.0	4.4	1.60	82
44.0	4.2	1.40	76
45.0	4.0	1.20	73
46.0	3.9	1.10	65
47.0	3.8	1.00	60
48.0	3.7	0.90	55
49.0	3.5	0.70	52
50.0	3.5	0.70	48
51.0	3.4	0.60	39
52.0	3.3	0.50	40
53.0	3.2	0.40	29
54.0	3.2	0.40	25
55.0	3.0	0.20	20
56.0	3.0	0.20	18
57.0	2.9	0.10	16
58.0	2.9	0.10	14
59.0	2.9	0.10	13
60.0	2.8	0.00	13
61.0	2.8	0.00	12
62.0	2.8	0.00	10
63.0	2.8	0.00	10
64.0	2.8	0.00	10
65.0	2.8	0.00	6
66.0	2.8	0.00	7
67.0	2.8	0.00	6
68.0	2.8	0.00	5
68.7	2.8	0.00	2

Liquid height/head data above the outlet at each minute are plotted against the elapsed time on linear coordinates in Figure H.1.

The points lie on a smooth curve except for some scatter midway through the plot. This is caused by an increased difficulty in reading the liquid depth without parallax error as the water depth recedes. The tail end of the curve, where depth measurement is both difficult and imprecise, is reconstructed (smoothed) from the cumulative volumes collected. This is a much better measurement than small values of liquid depth until accumulated errors in cumulative volume propagate to an unacceptable level.

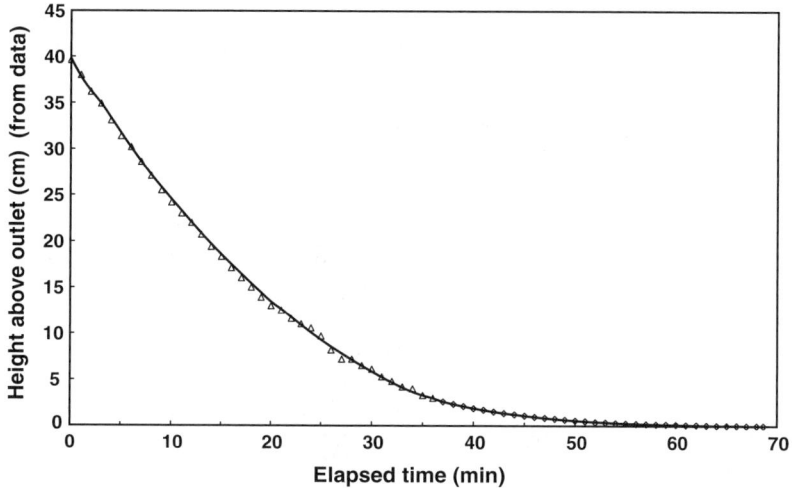

Figure H.1 Height of liquid remaining in vertical cylindrical tank, laboratory-scale experiment (linear scale).

Average volumetric flow over each minute and the average velocity derived from it by dividing the flow by the circular cross-sectional area of the outlet tube (with no observable *vena contracta*) are shown in Figure H.2. These are also plotted on linear coordinates but at the half-minute, the midpoint of each range. The flow at the end of the run is defined as zero, and the velocity at $t = 0$ (y-intercept) is generally consistent with what would be calculated from the initial height of liquid above the drain. There is some scatter for the first few readings at lowest times, where the instantaneous height and therefore the velocity are changing most rapidly and the averaging

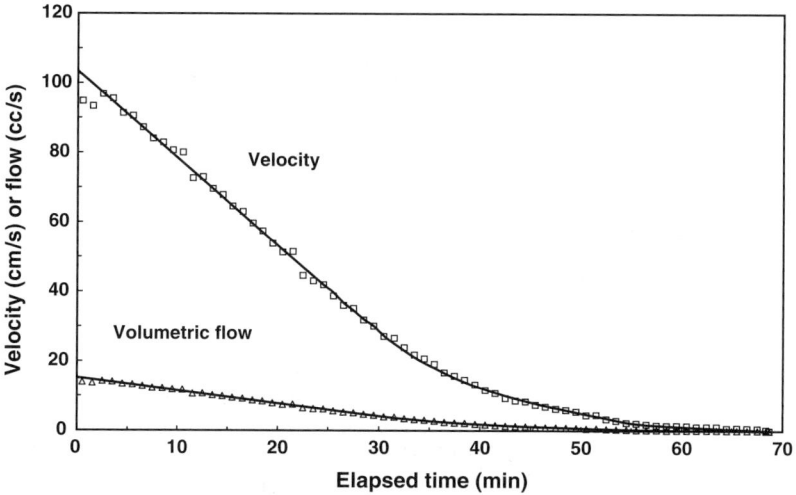

Figure H.2 Velocity and flow rate derived from experimental liquid volumes collected.

H.5 DATA ANALYSIS—FALLING HEAD/UNSTEADY-STATE EXPERIMENT

For Bernoulli's Equation (Equation (7.1)) and therefore Torricelli's theorem/law, Equation (7.2), to apply, the integrated form of the relationship is given by Equation (7.14)

$$H^{1/2} - h^{1/2} = (1/2)(A_0/A_T)(2g)^{1/2} C_D t \qquad (7.14)$$

This plots as a straight line of $(H^{1/2} - h^{1/2})$ versus t with slope given by the coefficient of t in Equation (7.14). Symbols are as defined in Chapter 7. The discharge coefficient (C_D) is related to this straight line slope by

$$C_D = [(A_T/A_0)(2/g)^{1/2}] \text{ (slope)} \qquad (7.15)$$

A graph of $(H^{1/2} - h^{1/2})$ versus t is shown in Figure H.3. The points plot as a straight line up to perhaps the 40 min point, or for 95% of the active volume drained from the tank. At this point, the data and the straight line part ways. A curved line then follows the actual data in the plot beyond the point where the data deviate from the straight line fit. This straight line slope, determined by least-squares analysis, can be used to calculate a discharge coefficient (C_D) of 0.35.

Drainage of the last 5% of the active tank volume deviates from this straight line because of a change in flow characteristics arising from the hydraulic scale of the experiment, correlated with the Reynolds number (Re). Because of the size of typical

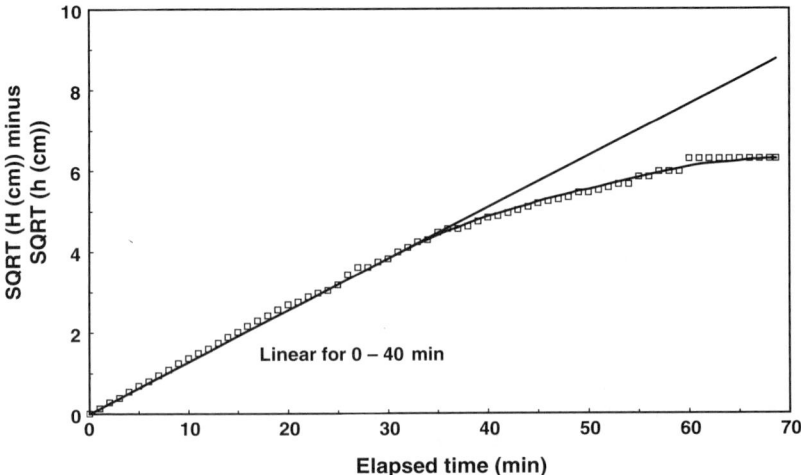

Figure H.3 Height of liquid remaining in vertical cylindrical tank (laboratory-scale experiment) (square root function).

industrial equipment, this situation is not likely to apply when draining aqueous solutions from full-scale tanks with their associated piping.

The dimensionless Reynolds number is defined as

$$Re = Dv\rho/\mu \tag{H.1}$$

where D is the pipe or tube diameter, v is the velocity, ρ is the fluid density, and μ is the liquid viscosity, all in consistent units.

Viscosity is commonly denoted by the Greek letters mu (μ) or eta (η). An example of a Reynolds number calculation in cgs units is illustrated in Ref. [1, pp. 43–44].

Using values here for the full tank

$$Re = \frac{(0.4318 \text{ cm})(97.5 \text{ cm/s})(0.99971 \text{ g/cm}^3)}{(0.01310 \text{ g/cm s} = \text{poise})} \cong 3200$$

Selected values for various times and velocities are shown in Table H.3.

For pipe flow, flow is turbulent above a Reynolds number of 10,000 and laminar, or streamline, below about 2000–2100 but might be laminar at Re as low as 1000 [1, pp. 43–44; 2, pp. 162–163; 3, p. 49]. A transition region, where the flow may be either turbulent or laminar, exists between 2100 and 4000–10,000 [1, pp. 43–44; 3, p. 49]. Flow is not likely to be laminar when draining a full-scale tank even down to the last several inches. For example, in the case of a tank with a 1 in. diameter outlet and a C_D of 0.6, the Reynolds number would be above 10,000.

Although correspondence is not exact, flow is in the turbulent or the transition region down to perhaps at least the 30 min point of the experiment, and clearly laminar at 47 min and beyond. As observed during the experiment, the outlet flow changed from a vigorous stream to a lazy discharge about midway through.

Table H.3 Reynolds Numbers for Laboratory-Scale Experiment

Time (min)	Approximate velocity (cm/s)	Approximate Reynolds number (Re)	Active % of tank
0	98	3200	100
10	81	2700	61
20	54	1800	33
30	30	1000	15
35	21	700	8
36	19	600	7.5
38	16	500	7
40	13	400	5
47	7	200[a]	2
59	2	60[a]	0.4
60	1	30[a]	0.3

[a] Clearly laminar.

Appendix H Experimental Determination of Coefficient

If flow is laminar in the outlet tube of length (L), the Hagen–Poiseuille Law [3, p. 88] will apply

$$\Delta p = 32\, Lv\mu/(gD^2) \tag{H.2}$$

with symbols as defined previously; v is the average velocity across the tube cross section. For laminar flow in the outlet tube, the governing equation for draining the tank transforms into the form of

$$dh/dt = -\alpha h\ [4] \tag{H.3}$$

where α is theoretically

$$\alpha = D_0^4 \rho g/(32\mu L D_T^2) \tag{H.4}$$

The solution of the differential equation above is

$$h = H\exp[-\alpha t] \tag{H.5}$$
$$\ln h = \ln H - \alpha t \tag{H.6}$$
$$\text{or } \ln(h/H) = -\alpha t \tag{H.7}$$

A plot of the data on semilogarithmic coordinates versus time for times between 47 and 59 min is shown in Figure H.4

It is a dead-on straight line. Slope determined from a least-squares fit is 0.143.

Outside this range, a distinct curvature is visible when plotted on these coordinates (not shown). The points between 0 and 40 min are depicted above in the linear section of the ($H^{1/2} - h^{1/2}$) versus t plot of Figure H.3. The semilogarithmic plot with a slope of 0.143 is used to generate the curve that hugs the data between 47 and 59 min in Figure H.3. In both plots, the points between 40 and 47 min (exclusive) represent a transition; the points beyond 59 min (less than 0.5% of the active volume of

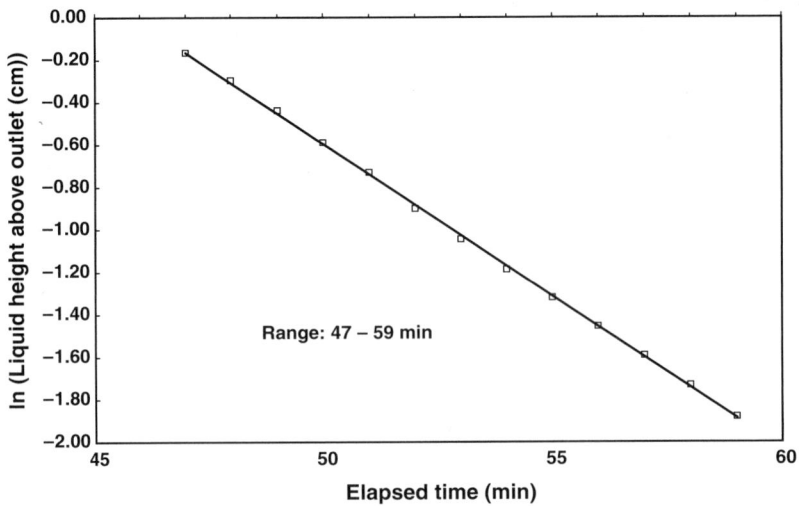

Figure H.4 Comparison of experimental and calculated drainage times.

the tank) are undoubtedly affected by assorted nonidealities not considered in the models, such as surface tension effects.

The theoretical slope from the Hagen–Poiseuille Law is given by Equation (H.4). Inserting numbers

$$\alpha = (0.4318)^4 (0.999712299)(980.665)/[(32)(0.01310)(10.8)(26)^2]$$
$$= 0.01114 \text{ s}^{-1} \times 60 = 0.668 \text{ min}^{-1}$$

The absolute value of the experimentally determined least-squares slope from Figure H.4 is 0.14336 min^{-1}, of the same order of magnitude as the theoretical value of the experimental coefficient.

The most readily explained source of difference is the outlet tube diameter, which is raised to the fourth power in Equation (H.4). If the effective diameter of the outlet tube were actually only 1 mm less, the experimental and theoretical coefficients would agree. There are in addition, entrance and exit losses within the valve between the tubing diameter and the valve diameter and an obstruction to flow within the valve.

Flow through small capillary tubes is used for the measurement of liquid viscosity [4]. The measurements here are not precise enough to determine viscosity, nor were they intended to.

In the final figure from this experiment, the exit velocities of Figure H.3 are plotted against $(2gh)^{1/2}$ (Figure H.5) as an alternate means of determining C_D. In this graph, low values of $(2gh)^{1/2}$ plotted at the left-hand end correspond to long times and low liquid depths. Values at the right-hand end correspond to short experimental times and the upper range of liquid depths, where the straight line section of Figure H.3 applies. Implicit in the velocity "data" is the outlet diameter chosen to convert flow rate into velocity.

The straight line slope in Figure H.5 as determined from the 0–40 min range of data was found to be 0.41 by least-squares analysis of the data. Agreement in this case

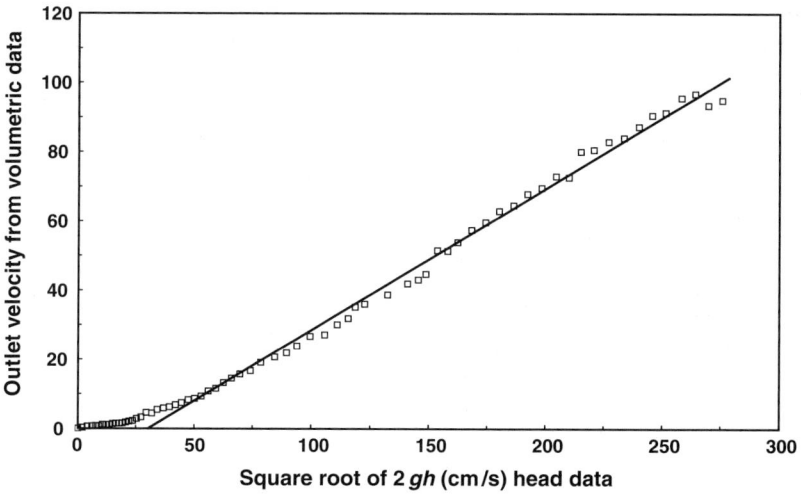

Figure H.5 Determination of discharge coefficient from 1 min average experimental data.

is close to 0.35, confirming the general consistency of the independent measurements of liquid height/depth and collected sample volumes, but this is not the whole story as explained below.

H.6 TIME TO DRAIN THE TANK

Time to drain the tank to various liquid heights/depths excerpted from the listing of Table H.2 can be found from

$$t = (A_T/A_0)(2/g)^{1/2}[H^{1/2} - h^{1/2}]/C_D \qquad (H.8)$$

the form of the Equation (7.16) with $h = 0$. Equation (H.8) applies when h and therefore $h^{1/2} \neq 0$. Results are shown in Table H.4.

In the range where the turbulent flow model is valid, a discharge coefficient (C_D) of 0.35 in conjunction with an outlet diameter of 0.170 in. (0.4318 cm) results in a difference between calculated drainage times and the observed values within experimental error for C_D and its roundoff to two significant figures. The deviations become greater at longer times and lower liquid depths where the $H^{1/2} - h^{1/2}$ data begin to deviate from the straight line in Figure H.3. Emptying the entire active volume of the tank involves a large interval where the turbulent flow model does not apply.

Agreement using a C_D of 0.41 is not quite so good throughout. This is believed to be caused by errors in the independent set of sample collection data and the process of averaging the derived time-varying outlet velocity over each 1 min time interval. The initial data from the shorter times, greater liquid height/depths, and higher velocities show scatter in Figure H.3. The derived velocities also exhibit the most change in this area, leading to more error in determining "average" values, in contrast to the instantaneous values of liquid height above the outlet. In addition to greater error, the short-time data plot to the right, away from the origin of coordinates, in Figure H.5, where they have more influence to swing the line of least-squares fit ever so slightly to a slope of $C_D = 0.41$.

Be that as it may, this difference in C_D estimated from flow data is largely of academic interest. Use of the field data would require the collection and processing of enormous volumes of liquid samples and is not expected to be the method of choice to estimate the discharge coefficient in a full-scale industrial setting.

Table H.4 Comparison of Experimental and Calculated Drainage Times

Observed time (min)	Final height (cm)	% Active volume drained	Calculated Time to drain (min) $C_D = 0.35$	$C_D = 0.41$
30	6.1	85	29.8	25.4
35	3.3	92	34.9	29.8
36	3.0	92.5	35.6	30.4
40	2.1	95	37.8	32.3
68.87	0	100	49.1	35.8

H.7 STEADY-STATE EXPERIMENT

Another laboratory-scale experiment for obtaining the discharge coefficient is a steady-state study in which the tank is allowed to drain, while keeping the head constant by adding a constant makeup flow [5]. In theory, it would provide a check on the results of the falling head, unsteady-state experiment previously discussed. It was not performed but is described briefly for the sake of completeness. Because of the additional logistical problems involved, it is also doubtful that an experiment at this scale would be used to estimate C_D for full-scale equipment.

To use a simple analogy, this can be called the "bathtub problem." An appropriately chosen inlet flow with the drain open will result in a final level within the tub. Too low a flow, and the tub never fills; too high a flow, and the tub overflows. The final depth of water in the tub can be made to vary by adjusting the inlet flow.

This situation can be analyzed mathematically as follows:

Rate of change in liquid volume in the tank = rate of flow in − rate of flow out

$$dV/dt = Q_{in} - Q_{out}$$

Since tank volume (V) is equal to tank cross-sectional area (A_T) times height (h) and substituting for $Q_{out} = C_D A_0 (2gh)^{1/2}$

$$\frac{d}{dt}(A_T h) = A_T \frac{dh}{dt} = Q_{in} - C_D A_0 (2/gh)^{1/2} \tag{H.9}$$

$$dh/dt = (Q_{in}/A_T) - C_D(A_0/A_T)(2g)^{1/2} h^{1/2} \tag{H.10}$$

At steady state, $dh/dt = 0$ and

$$Q_{in}/A_T = C_D (A_0/A_T)(2g)^{1/2} h^{1/2} \tag{H.11}$$

from which

$$h = (Q_{in})^2 / [C_D^2 A_0^2 (2g)] \tag{H.12}$$

For a practical solution to occur, h must lie between 0 and H. At steady state, C_D can be determined by rearrangement of Equation (H.12) and measurements for all the other variables

$$C_D = (Q_{in})^2 / [A_0 (2gh)^{1/2}] \tag{H.13}$$

Note that C_D in Equation (H.13) does not depend on the cross-sectional area of the tank.

As an example of filling the tank, consider a flow (Q_{in}) of 600 cc/min (10 cc/s), a C_D of 0.35 as found in the unsteady-state experiment, and dimensions as listed in Table H.1.

$$h = \{10/[(0.35)(0.1464)(44.2869)]^2\} = 19.4+ \text{ cm above the outlet drain}$$

The flow was chosen to achieve a liquid height/depth about halfway up the tank.

Changing the inlet flow (Q_{in}) repeatedly and waiting for the system to stabilize each time will enable a curve of h versus Q_{in}^2 to be drawn. Better yet, a plot of h versus

Q_{in}^2 will yield a straight line with the slope of $1/(C_D^2 A_0^2 2g)$, and from it, a value for C_D (both curves not shown). Flow would be measured by collecting a theoretically constant volume or mass of liquid during a known time period ("bucket and stopwatch" method). The steady-state experiment has the advantage of not having to average a time-varying flow as in the falling head experiment previously discussed. Flow data can be checked using the principle of continuity for incompressible flow at steady state

$$Q_{in} = Q_{out} \tag{H.14}$$

Exit velocity can be derived from Q_{out} (or even Q_{in}) by dividing by $C_D A_0$, and the relationship $v = C_D(2gh)^{1/2}$ can be verified using the experimental data.

H.7.1 Transient Solution to Fill and Drain Problem

The unsteady-state transient to the steady-state case is provided by the solution of Equation (H.10) with $dh/dt \neq 0$. With substitutions

$$a = Q_{in}/A_T \tag{H.15}$$
$$b = (A_0/A_T)C_D(2g)^{1/2} \tag{H.16}$$

and rearrangement, Equation (H.10) becomes

$$dh/dt = a - bh^{1/2} \tag{H.17}$$

Further substitution of $u = h^{1/2}$ and $dh = 2u\,du$, which follows from it, yields upon rearrangement

$$\frac{2u\,du}{a - bu} = dt \tag{H.18}$$

Starting with an empty tank, the time to achieve a given height or depth of liquid above the outlet drain for the fill and drain case, is found by the integration of Equation (H.18), subject to the initial condition of $h = 0$ at $t = 0$.

$$t = (2/b^2)(a\ln[a/(a - bu)] - bu) \tag{H.19}$$

However, the relationship of Equation (H.19), derived from an analytical solution of the defining differential equation, is backward. It predicts the dependence of the independent variable (t) on the dependent variable (h), rather than *vice versa*. Nonetheless, it can be plotted in the conventional manner as h versus t, as shown in Figure H.6 for the example above with an inlet flow rate of 600 cc/min. This curve exhibits the same general form as a true exponential buildup curve, $y = \beta[1 - \exp(-\alpha t)]$. The curve in Figure H.6 approaches the final steady-state h of just over 19.4 cm at long times. This asymptote occurs at $a = bu$, making the denominator in Equation (H.19) equal to zero and the time to reach >19.4 cm theoretically infinite.

In performing this experiment, one would want to start out close to the point of "equilibrium" rather than from an empty tank. Otherwise, it would be about 2.5 h just to reach 99% of the desired final level before being able to make any steady-state

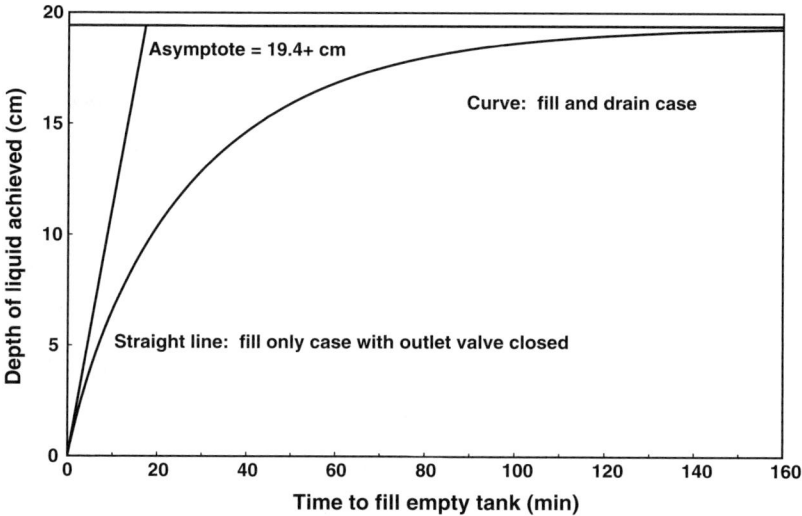

Figure H.6 Transient for steady-state experiment tank of Table H.1 and 600 cc/min fill rate.

measurements. The straight line in Figure H.6 depicts the course of filling the tank at 600 cc/min with the outlet valve closed (~17 min to the 19.4+ cm level).

Strictly speaking, Equation (H.19) and the resulting curve in Figure H.6 assume that the turbulent flow regime is valid throughout the entire time of filling the tank. However, using the alternate, laminar flow model for the lowest 5% of the active volume of the tank (0–2.1 cm) where it applies makes little difference in the time to achieve a given height in this region and in the overall shape of the entire curve that results.

Other considerations for such an experiment would be (1) to add the liquid makeup in such a way so as not to disturb a calm surface to measure a valid height. (2) Since such an experiment has the potential to consume huge quantities of water even on a laboratory scale, one may be faced with having to recycle the inlet flow. (3) One would also prefer to study heights/depths and effluent velocities corresponding to the turbulent flow linear range of Figure H.3. Each condition should be observed for a long enough time to convince oneself that the system has indeed equilibrated well enough for meaningful measurements before moving on to the next experimental condition.

Finally, although this may be an interesting laboratory-scale experiment, it is doubtful that the steady case would ever be examined experimentally on a full scale in an industrial setting, as previously stated.

REFERENCES

1. J.P. Reynolds, J. S. Jeris, and L. Theodore, *Handbook of Chemical and Environmental Engineering Calculations*, Wiley–Interscience, Hoboken, NJ, 2002, 948 pp.

2. R.L. Daugherty and A.C. Ingersoll, *Fluid Mechanics with Engineering Applications*, 5th edn, McGraw-Hill, New York, 1954, 472 pp.
3. W.L. Mc Cabe, J. C. Smith, and P. Harriott, McGraw-Hill, New York, 1993, 1130 pp.
4. Anonymous, Physics 3 Summer 1990 Lab 7—Hydrodynamics, available at http://www.dartmouth.edu/~physics.labs/writeups/hydrodynamics.pdf, last accessed on May 31, 2009.
5. J.B. Calvert, Coefficient of Discharge, available at http://www.du.edu/~jcalvert/tech/fluids/orifice.htm, June 15, 2003, last accessed on May 25, 2004.

Appendix I

Noise Case Studies

Better to be criticized for spending too much money than for failure to get the job done.

I.1 CASE 1—SOUND METER READINGS BEHIND A HIGHWAY NOISE BARRIER

I.1.1 Theory

Sound waves incident on a wall are either reflected, absorbed by the material, or transmitted through it [1, p. 13]. Behavior for different frequencies is different for the same material. High frequencies are more easily reflected than low frequencies [1, p. 58]. They also pass directly through holes or breaks in the structure but do not diffuse around its edges.

A highway barrier decreases the sound level at a receptor in the acoustic shadow behind the barrier through a combination of reflection away from the barrier and absorption of the sound by the barrier material. Like light waves at the boundary of a different medium, sound waves are also able to be diffracted around acoustic barriers. This diffraction is more pronounced for the lower frequencies. A highway noise barrier will preferentially block the upper frequencies of traffic noise from appearing behind it until distance and whatever air absorption might occur to dissipate their intensity; however, the less offensive low frequency noise appears to "crawl over" much closer to the wall.

I.1.2 Synopsis of the Measurements

Original unpublished sound pressure data were taken by the author during a brief field exercise of the Rutgers University New Jersey Community Noise Control Course. Measurements were made behind the noise barrier wall running along the west side of the nominally north–south New Jersey Turnpike in East Brunswick, NJ. The barrier

---

Environmental Calculations: A Multimedia Approach, by Robert G. Kunz
Copyright © 2009 John Wiley & Sons, Inc.

Table I.1 NOAA Weather Data, Newark, NJ 05/29/1991 [2]

Time EDT[a]	Temperature (°F)	Relative humidity (%)	Barometric pressure (in.Hg)
3:50 p.m.	89.0	39.1	29.89
4:50 p.m.	90.0	39.3	29.88

[a] Eastern daylight time.

shields a residential area and school from traffic noise. Data were obtained on a clear day at the start of the weekday evening rush hour in late spring of 1991.

On-site temperature was measured at 92°F (33.3°C); wind measured there was calm at less than 2 mph (miles per hour) (3.2 kph (kilometers per hour)). NOAA measurements of temperature, relative humidity, and barometric pressure from Newark, NJ [2], are listed in Table I.1. Instantaneous readings of dBA and dBC were recorded manually using a type II handheld sound meter with windscreen in the meter's slow response mode. The meter was properly calibrated.

Many years later in the fall of 2006, the site was found again by means of satellite photographs of the turnpike and was revisited to verify some ancillary measurements and to take pictures before publication.

I.1.3 Turnpike Noise Barriers

The post-and-panel barrier is constructed from masonry containing embedded stones. On the turnpike side, the surface is fluted with vertical grooves an inch or so (a few centimeters) deep (Figures I.1 and I.2). Vertical spacers/columns/supports made of concrete separate adjacent panels. Except for a break in the wall

Figure I.1 Photograph of turnpike side of wall.

I.1 Case 1—Sound Meter Readings Behind a Highway Noise Barrier

Figure I.2 Close-up photograph of turnpike side of wall.

(Figure I.3) to allow a local street and its sidewalks to pass over the turnpike via a bridge to the right (north) of the section of wall shown in Figure I.1, the barrier is continuous up and down the turnpike as far as the eye can see. A similar wall runs along the east side of the turnpike as well.

The turnpike runs in a depressed corridor below the grade of the neighborhood with enough clearance beneath the bridge for large trucks to pass (perhaps at least 14 ft).

Figure I.3 Photograph of street opening in the wall.

At this point of the turnpike's route, there are 10 traffic lanes [3] (each 12 ft wide) in the so-called dual–dual configuration. The two outer lanes in each direction are open to all vehicles; three inner lanes in each direction are designated for cars only. Each group of lanes is separated from the others by median strips/dividers. Shoulders on either side of the road are 12 ft wide. The outer two-lane strip of the northbound roadway increases to three lanes an estimated 50 ft (~15 m) north of the bridge. The analogous outer southbound roadway necks down from three to two lanes at 2500 ft ± (760 m) north of the bridge (as estimated from satellite photographs). Pavement is asphalt. Posted speed limit was 55 mph (88.5 kph) at the time of the noise measurements [4]. (It had been increased to 65 mph but was later reduced to 55 mph once again.) Current published average traffic volume for this section of the turnpike is 130,000 vehicles per day [3], or about 1 vehicle per lane every 7 s. This is a correct order of magnitude for the measurement, but frequency of vehicles as observed passing under the bridge in 1991 was somewhat higher.

I.1.4 Turnpike Noise Measurements

Turnpike noise levels as measured from the bridge at various locations amid the deafening roar and rumble of the steady traffic ranged from 74 to 84 dBA and from 78 to 89 dBC, low frequency noise as expected, with dBC > dBA on the average.

I.1.5 Neighborhood Side of the Noise Barrier

The street passing over the turnpike (Sullivan Way) intersects at nearly right angles a second local street (Corona Road), which parallels the sound barrier wall (Figure I.4). The southeast corner of the intersection contains a somewhat rectangular grassy park-like area, mowed at the time of the noise measurements. It lies between the wall and the second street. It is bounded on the south by a nearby residential property. Other residential properties farther away (not shown) abut the wall. Many of the homes in the neighborhood are split levels with a one- or two-car garage. A school building (Lawrence Brook Elementary School) is located across Corona Road.

The grassy area stretches for eight noise barrier panels more or less (Figure I.5) or slightly more than 120 ft (37 m) along the wall and an estimated 200 ft (61 m) on a perpendicular from the wall to the edge of the near sidewalk across the street from the school. Four more panels (plus or minus) are visible behind the neighboring residential property.

During the intervening years, vines and other vegetation have grown up next to the wall and in the grassy field. A walking path cutting diagonally across the field from the east end of the crosswalk at the school to the bridge crossing the turnpike has worn away the grass cover. Erosion of the underlying soil by storm water flow is clearly visible first hand and in the more recent satellite photographs.

The wall panels are each 14 ft wide (4.3 m). Concrete columns between them and at the bridge end are 14 in. (35.6 cm) wide. The column at the bridge end is 18 in.

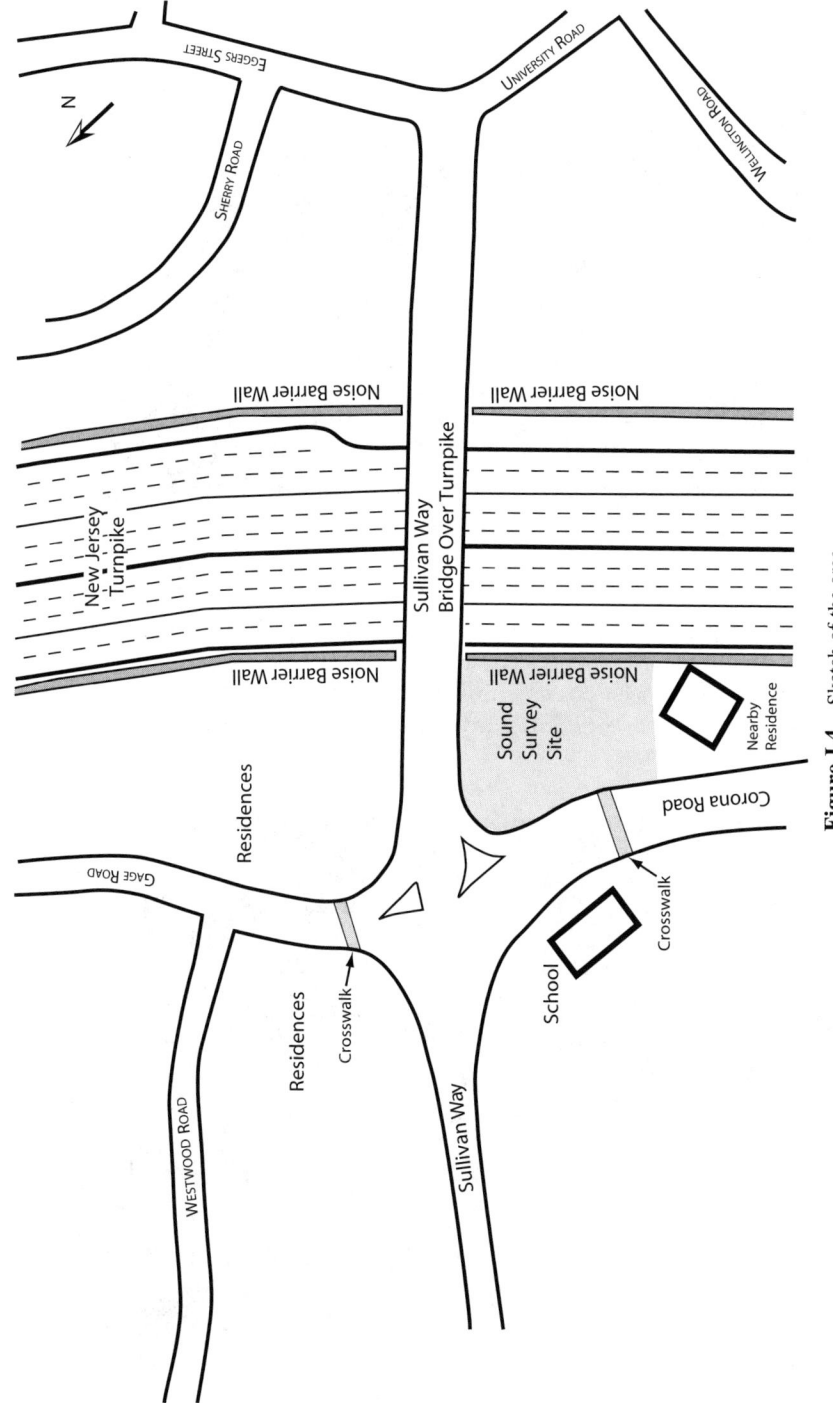

Figure I.4 Sketch of the area.

Figure I.5 Photograph of neighborhood side of wall.

(45.7 cm) deep, and the wall thickness at that location was estimated as 6 in. (15.2 cm) at the time of the measurements. The neighborhood (sheltered) side of the wall (Figure I.6) may have been coated subsequently with a 1 in. (2.5 cm) curd stucco material, making its surface different from that of the turnpike side.

Figure I.6 Close-up photograph of neighborhood side of wall.

At the bridge, the wall is an estimated 16 ft (4.9 m) tall (four horizontal strips at an estimated 4 ft (1.2 m) height each). The grade of the land along the wall on the neighborhood side drops down at first as one proceeds southward from the bridge, but then appears to remain constant. The top of the wall remains even with the ground. Height from the grade of the field at the south end of the fourth panel from the bridge is 18 ft (5.5 m). This is the approximate center of the wall behind the grassy field. The next four panels step down to 16 ft (4.9 m) above the local grade.

I.1.6 Noise Measurements behind the Barrier

Noise measurements were made approximately 48 in. (1.2 m) above the ground on a perpendicular from the wall along the centerline of the grassy field, roughly between panels 4 and 5 counting from the bridge end. Noise measurements were made at the wall and at 10 ft (~3 m) intervals moving away from the wall up to 70 ft (21.3 m). Measurements were also made along the 50 ft (15.2 m) contour line parallel to the wall at even intervals (say 20 ft (6.1 m)). Data are shown schematically in Tables I.2 and I.3.

According to the data obtained, the quietest location was immediately behind the wall. At this location, turnpike traffic noise was reduced by up to 30 dBA but only 15 dBC, indicative of high-frequency abatement. However, in contrast to previous observations of point-source plant noise, where both the noise readings dropped and the plant noise completely disappeared in the acoustical shadow of a distant building, traffic noise here is still audible although at a much lower sound level. There were and are no other distinctive continuous noise sources in the area.

The dBC readings in Table I.2 reach a maximum at somewhere around 20 ft (6.1 m) from the wall. Beyond this point, geometric spreading and attenuation over distance take over. It is not known whether the second, smaller bump at 60 ft (18.3 m) is real or is caused by an aberration in the data. A maximum in dBA readings appears to be located in this same region but is more difficult to detect. The dBA readings seem to have achieved background levels (55 dBA) some 60–70 ft (18.3–21.3 m) from the barrier. This is the reason that the walkaway from the wall was terminated.

Both the dBA and dBC readings, taken parallel to the wall on the 50 ft (15.2 m) contour line (Table I.3), decline gradually as one proceeds from north to south away from the bridge and the gap in the otherwise continuous noise barrier. Nevertheless, the readings must extrapolate to somewhat higher values at the north end, where the street crossing the turnpike interrupts the continuity of the wall and noise from the turnpike diminished by only a short distance is felt directly.

The dBA at the southernmost data point appears to have achieved background (55 dBA); while the 65 dBC reading at this same point is lower than the lowest dBC reading taken perpendicular to the wall (Table I.2), one cannot say with certainty whether 65 dBC is indicative of the true background at this location.

The goal of obtaining practice in conducting sound/noise measurements in the field was certainly achieved, and it is hoped that the results presented here are informative.

Table I.2 Measurements Behind Traffic Noise Barrier Perpendicular to the Wall—New Jersey Turnpike

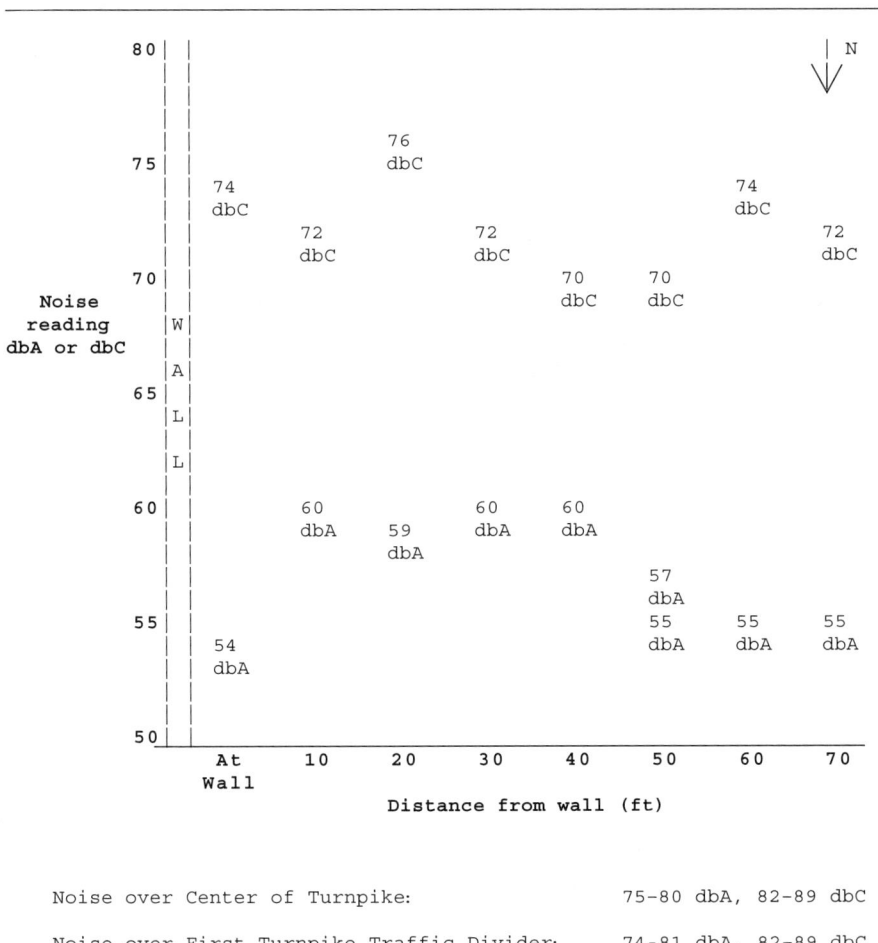

Noise over Center of Turnpike:	75-80 dbA, 82-89 dbC
Noise over First Turnpike Traffic Divider:	74-81 dbA, 82-89 dbC
Noise over Near Side of Turnpike:	75-84 dbA, 78-86 dbC

I.2 CASE 2—ANOTHER NOISE BARRIER STUDY

Data from the brief study of case 1 are, in general, consistent with results from a science project done by a seventh grade student in Washington state during 1999–2000 and posted on the Internet [5]. He studied the ability of a concrete wall, a wooden fence, an earthen berm, and vegetation versus an open field to abate highway noise. He is to be congratulated on his good work. Results from his investigation of the concrete barrier are summarized and discussed below.

Table I.3 Measurements Behind Traffic Noise Barrier Parallel to the Wall, 50 ft away—New Jersey Turnpike

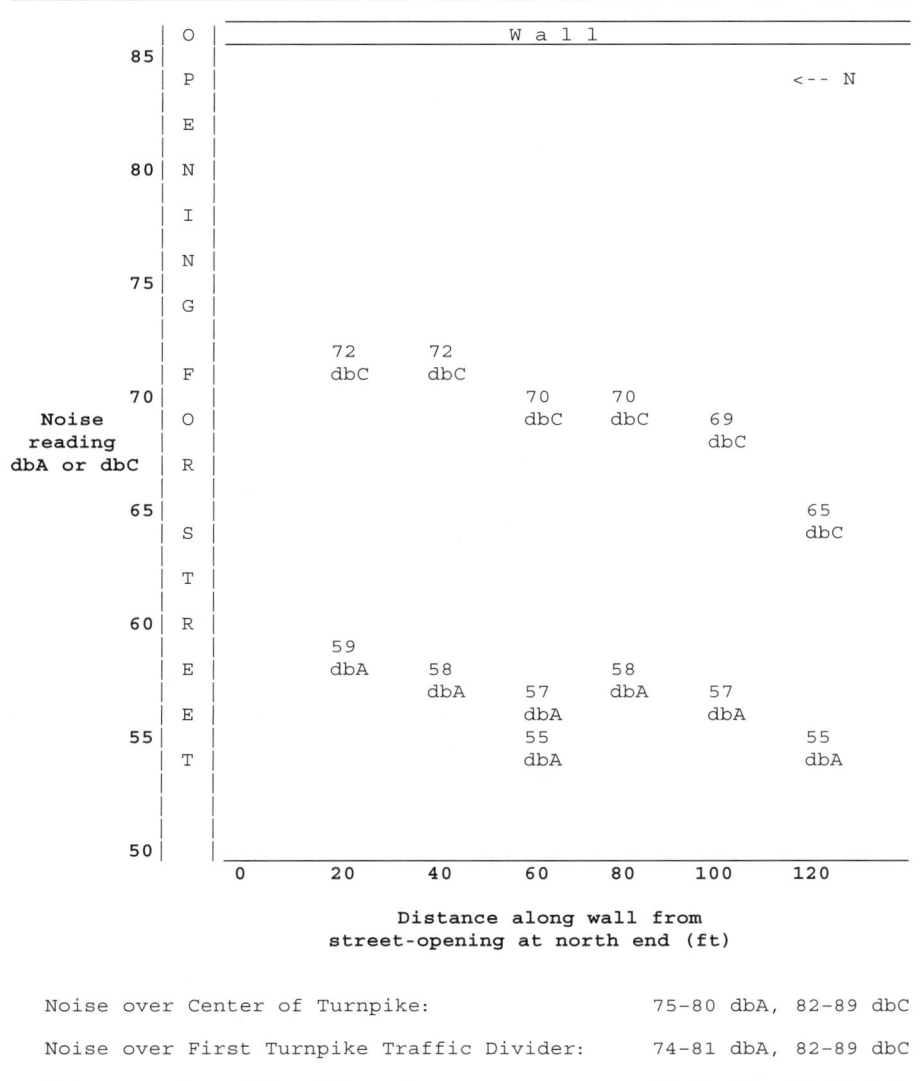

```
Noise over Center of Turnpike:                    75-80 dbA, 82-89 dbC
Noise over First Turnpike Traffic Divider:        74-81 dbA, 82-89 dbC
Noise over Near Side of Turnpike:                 75-84 dbA, 78-86 dbC
```

I.2.1 Highway Noise Data

This part of the experiment was conducted along a highway with a concrete barrier adjacent to an open field. Primary data consisted of 10 min L_{eq} values determined manually from dBA readings every 30 s using handheld sound meters. Maximum dBA readings were also recorded. Only A-weighted data were obtained, but it was stated that the frequency of traffic noise is about 550 Hz.

Noise data were obtained at a variety of locations back from the road and in front of and behind the concrete barrier, apparently for some data using as many as 3 sound meters at the same time. A simultaneous traffic count was taken by another person during each 10 min period, counting the number of light, medium, and heavy vehicles passing by the location. However, traffic counts were not reported.

Noise measurement points included "Road Side," 2 m in front of the wall, and cumulative distances of 3, 20, 37, and 43 m "back" without a completely clear specification of the reference point in the case of the open field. However, the terms road side and 2 m in front of barrier are equated in the caption of one of the investigator's graphs reporting L_{eq} values.

The following information was not reported:

- Type of highway, number of lanes, and speed limit.
- Distance of the wall from the highway.
- Dimensions of concrete barrier.
- Transverse distance along the wall to the behind-the-wall measurement points.
- Distance between the end of the wall and the open-field measurement points.

In addition, any sampling point closer to the road than 2 m in front of the concrete wall was not defined.

Some data are reported on graphs comparing point dBA readings and L_{eq} values at Road Side with a point 43 m behind the wall and similar results at 37 m for the concrete barrier versus reading in the open field. Other data are reported in a narrative form comparing results at various points with one another.

I.2.2 Summary of Results

The entries in Table I.4 have been pieced together from the report, with notes as to how they were determined. One "data" point had to be estimated. The data behind the barrier show the same sort of dip in the acoustic shadow behind the wall followed by a local maximum point as in case 1 above. The open-field data simply decline with distance from the road.

I.3 CASE 3—SUCCESSFUL NOISE PERMITTING PROCEDURE

A case study, prepared by a consultant, is presented in the literature [6,7]. The step-by-step technique employed secured a favorable outcome for the project described therein. The method is straightforward yet elegant. Its salient points are summarized in the next sections.

Table I.4 Measurements Behind Concrete Traffic Noise Barrier and in Adjacent Open Field—Washington State Highway

Distance from road (m)	Distance behind barrier (m)	L_{eq} without barrier (dBA)	L_{eq} with barrier (dBA)	Remarks
0 (Roadside/2 m in front of barrier)	−2	81	81	Read from graph
5	3	73 ← Δ = 9 dBA →	↑ Δ = 17 dBA → 64	Statements 17 dBA reduction at 3 m behind wall 9 dBA reduction compared to open-field readings
25	23	71 ← Δ = 1 dBA → (estimated by interpolation)	70	Statement 1 dBA reduction compared to open-field readings
37	35	68 ← Δ = 5 dBA →	↑ 63–66 (avg = 64.5) 15 dBA Δ from 81	68 dBA read from graph 63 dBA read from graph *Statements* 5 dBA reduction compared to open field = 63 dBA 15 dBA reduction behind wall = 66 dBA
45	43	—	↑ 20 dBA Δ from 81 → 61	Read from graph *Statement* 20 dBA reduction at 43 m behind wall

I.3.1 The Project

A grocery company wished to build a new supermarket in an urban area of Massachusetts during the late 1980s. Control of noise emissions (as an air contaminant) was required by the Massachusetts Department of Environmental Protection (DEP) regarding permissible noise from the proposed facility. Written approval from the department had to be obtained before "construction, installation, modification, or operation" of the proposed equipment. Sources of noise were the refrigeration fans, condensers, compressors, and HVAC unit that would be mounted on the roof of the supermarket.

I.3.2 Method of Execution

Project execution proceeded according to the following stepwise methodology, embellished only slightly from the consultant's itemized list.
- Identify conditions of any required permits/approvals.
- Conduct baseline noise monitoring at the appropriate receptor locations.
- Determine noise-generating characteristics of equipment to be installed in the new facility.
- Predict future noise levels.
- Compare with allowable limits.
- Select mitigation measures, if necessary, and make new prediction. (Repeat if necessary until the mitigation is adequate to achieve compliance.)
- Install controls and start up the equipment.
- Conduct postconstruction ambient monitoring to verify compliance.

Many of these items are driven by Massachusetts DEP requirements [8].

I.3.3 The Background Noise Survey

In brief, conditions of approval were that the background sound pressure level at the property line and at the closest inhabited residence (110 ft away) not be increased by more than 10 dBA above ambient levels and that a pure tone not be produced. A pure tone is defined in Massachusetts as the condition where the sound pressure level in any octave band exceeds the sound pressure level in the two adjacent octave band frequencies by 3 dB or more [6–8] (rather than the 1/3 octave band definition indicated in Chapter 9 for some locations). Further, the DEP requested that all sound level monitoring be conducted around 2 a.m. The greatest change in ambient sound levels arising from the supermarket's 24 h operation of the refrigeration equipment could be expected to occur during the period of minimal human activity, typically from 2–3 a.m. on a weekday.

The baseline noise level at the nearest residence was typical of a residential area at night (42.6 dBA). The low-frequency nature of the background noise and pure tone condition identified for the 63 Hz octave band appeared to originate from mechanical equipment at a local factory. Octave band frequency spectrum data for the initial survey and follow-up are reported in Refs [6,7].

I.3.4 Estimate of Noise from New Equipment

In lieu of the requisite sound-level information difficult to obtain from the refrigeration equipment manufacturers, noise design data were derived from another source. Similar equipment of equal or larger size being operated by this same grocery company at a newly constructed store elsewhere were used to predict off-site noise levels for the proposed facility, first without and then with appropriate abatement. A rooftop noise barrier and a fan shroud were designed to mitigate off-site effects to acceptable levels, since unabated off-site noise was predicted at about 60 dBA (including a 3 dBA safety factor).

I.3.5 Postconstruction Noise Survey Confirmed Compliance

A 2 a.m. postconstruction noise survey was conducted within days of the 1-year anniversary of the preconstruction baseline test. It confirmed compliance for even worst-case noise levels. Residential noise was an acceptable 47.0 dBA (in agreement with the predicted level); the pre-existing pure tone from an extraneous source remained.

The project was successfully executed with a single preconstruction and a single postconstruction noise survey. Changes in the ambient level without the store's contribution in the intervening year must have been minimal.

However, if substantial time has elapsed since the original survey, it is only common sense that an additional baseline noise survey be considered/conducted before start-up to confirm that no significant changes have occurred in the ambient background noise level.

REFERENCES

1. Anonymous, *Noise Control Principles and Practice*, 1st edn, Brüel & Kjaer, Naerum, Denmark, 1982, 156 pp.
2. U.S. Department of Commerce, National Oceanic and Atmospheric Administration (NOAA) National Weather Service, Surface Weather Observations Newark, NJ for May 29, 1991, National Climatic Data Center, Asheville, NC, printed December 15, 2006.
3. New Jersey Turnpike Historic Overview, http://www.nycroads.com/roads/nj-turnpike/, last accessed on September 2, 2008.
4. J. Preston, Speed Limit to Hit 65 on Major New Jersey Highways, The New York Times, January 13, 1998, http://www.nytimes.com, accessed on September 2, 2008.

5. Steven P., The Effects of Sound Barrier Design on Highway Noise Attenuation, available at http://www.selah.k12.wa.us/soar/sciproj2000/StevenP.html, accessed on November 5, 2006.
6. R.D. O'Neal, Noise barrier insertion loss: a case study in an urban area, Paper 90-156.3 presented at the 83rd Annual Air & Waste Management Association (A&WMA) Annual Meeting & Exhibition, Pittsburgh, PA, June 24–29, 1990.
7. R.D. O'Neal, Predicting potential sound levels: a case study in an urban area, *Journal of Air & Waste Management Association*, **41**(10), 1355–1359, 1991.
8. Massachusetts Department of Environmental Protection (DEP), Noise Pollution Guidance Interpretation and Permit Fact Sheet, available at http://www.mass.gov/dep/air/laws/noisepol.htm, accessed on September 12, 2008.

Appendix J

Air Pollution Aspects of the Fluid Catalytic Cracking Process[1]

Always carry enough pens in your pocket protector; you never know when you'll need two green ones.

J.1 SUMMARY

Petroleum refineries, among other industrial sources of pollution, are mandated by federal, state, and local environmental agencies to meet certain air quality emission limits, as well as discharge criteria for other media. This appendix describes the fluid catalytic cracking (FCC) process employed in a high conversion oil refinery to produce gasoline and other fuels. The focus is on the generation and treatment of environmental contaminants emitted to the atmosphere. Liquid effluents, such as those produced by flue gas scrubbing, and/or solid wastes, such as spent cracking catalyst or collected airborne catalyst fines, are separate issues and are not considered here.

J.2 INTRODUCTION

The fluid catalytic cracking unit (FCCU), also known as the cat cracker, is the central process at the heart of a typical high-conversion refinery producing gasoline. It cracks large gas oil molecules into smaller molecules within the gasoline range plus light gases, other liquid products, and petroleum coke. The term gas oil denotes a

---

[1] Adapted and updated from Ref. [1].

petroleum distillate fraction with an atmospheric boiling range between kerosene and lubricating oil.

A recent refining survey [2] lists slightly over 130 petroleum refineries operating in the Continental U.S., Alaska, and Hawaii. Approximately 100 of these employ some form of fluid catalytic cracking in the refining process. Total cat cracking feed at a given refinery ranges from 2300 to 227,000 b/cd (barrels per calendar day) (0.37–36.1 million L/cd). Median and average values are 54,000 and 56,000 b/cd, respectively, or 8.6 and 8.9 million L/cd in metric units.

Unfortunately, without abatement, a refinery's cat cracker also generates atmospheric contaminants as undesired by-products. The FCCU is the biggest single source of atmospheric pollution in an oil refinery, primarily from sulfur oxides and particulates [3]. Although on a lesser scale, half of the NO_x (oxides of nitrogen = nitric oxide (NO) plus nitrogen dioxide (NO_2)) in a refinery is estimated to originate from the FCCU.

J.3 FCC PROCESS DESCRIPTION

The FCC process [4–11] is conducted in a pair of fluidized-bed vessels containing catalytic cracking catalyst, which circulates continuously from one vessel to the other through a set of transfer lines (Figure J.1). Those vessels are referred to as the reactor and regenerator, respectively. Conditions are summarized in Tables J.1 and J.2. Somewhat different configurations of the basic hardware are currently in use [8,12,13], having evolved separately from the joint cooperative effort at the time of World War II to develop the process [8–10,12–22]. Reactor and regenerator may sit side by side or one atop the other.

J.3.1 FCC Cracking Catalyst

Fresh cat cracking catalyst is a porous silica–alumina material, white in color and finer than most beach sand. It was originally produced from naturally occurring minerals but is now made from synthetic crystalline zeolites and amorphous materials. Individual FCC particles exhibit a microspheroidal shape. Its particle-size distribution approximates lognormal, with a range from close to 0 to over 100 µm and a mass median diameter (50% point) of approximately 50–80 µm [23,24] (Figur 2.13).

When a gas or vapor is passed upward through a bed of FCC catalyst at a velocity in excess of the pressure drop needed to support the bed, it becomes "fluidized" and exhibits the properties of a liquid. Fluidization occurs when the pressure drop across the bed is sufficient to support the weight of the particles in the stream of flowing fluid. For FCC catalyst, the minimum, or incipient, fluidization velocity is on the order of 0.001–0.01 ft/s (0.0003–0.003 m/s) [25,26]. It is two to three orders of magnitude below normal operating velocities [22,27–39]. Further information on fluidization is available elsewhere [30–35].

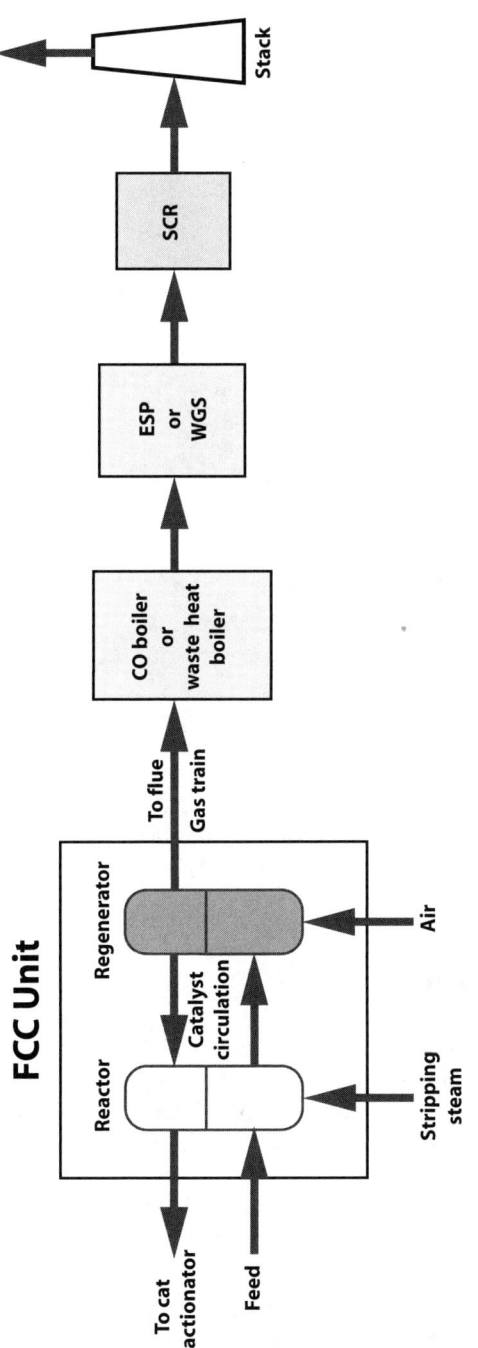

Figure J.1 Generic process flow diagram for FCCU and ancillary flue gas treatment.

Table J.1 Summary of Typical FCCU Conditions[a]

Condition	Reactor	Regenerator
Temperature, °F (°C)	Bed or riser: 900–1050 (480–565)	Partial-burn unit (with CO boiler): 1100–1250 (595–675)[b]
	Oil feed: 500–800 (260–425)	Full-burn unit: 1250–1500 (675–815)[b]
Pressure, psig	10–35[c]	8–40[c]
Catalyst Condition:		
Entering	Regenerated	Spent
Leaving	Spent	Regenerated
Fluidizing gas	Hydrocarbon vapors Steam (for feed atomization and steam stripping)	Air Flue gas Some steam
Atmospheric pollutants at exit of regenerator or CO boiler:		
CO	—	Partial burn: in % range Full burn or out of CO boiler: <500 ppm
Particulates	—	See Table J.2.
SO_x	—	Low to hundreds of ppm or more, depending on feed, hydrodesulfurization, and additives
NO_x	—	<100–500+ ppm

[a] Entries in the table were compiled from the references cited in the text.
[b] Temperature leaving CO boiler or waste heat boiler can be as low as 600°F (315°C).
[c] Range of cited references.

J.3.2 The Reactor

In the FCC process, a hydrocarbon feed, usually in the gas oil range, is vaporized upon contact with hot catalyst in the transfer line to the reactor. Cracking of large molecules takes place on the surface and in the pores of the catalyst in the transfer line or in the reactor itself. The desired products from cat cracking pass upward in the vaporized state through the reactor bed and out of the reactor to be separated into gas and various liquids (gasoline components and middle distillates) in the cat fractionator distillation column downstream. The chemistry of hydrocarbon cracking reactions is discussed at length in Refs [37,38].

One (undesired) product of cracking is coke, a condensed hydrocarbon somewhat deficient in hydrogen. Coke has a black color and contains carbon, hydrogen, sulfur, nitrogen, and metals. Sulfur, nitrogen, and metals originating in crude oil tend to

Table J.2 Summary of FCCU Particulate Matter (PM) at Various Locations

Location	Collection efficiency (%)	Particulate loading at exit (mg/Nm³)	Particle size at exit (μm)	Remarks[a]
Regenerator bed	—	See table below	Average 50–80 in dense bed	At some point above 90 μm average particle size (APS), fluidization becomes poor
Regenerator cyclones (first and second stages combined)	99.995 (combined)	≥200	Average 20 or less	See Section J.4.2.1.
Third-stage cyclones	70–90 (conventional)	50, 75, or 100–150	Average 5 (98% <10 μm)	Usually located in pressure vessel outside regenerator
	40–50 (multiclones)			Works best for particle sizes >10 μm. Do not depend on to meet a permit limit of <50 mg/Nm³
Fourth-stage cyclones	—	—	—	Used for solids separation when transporting the underflow from collection hopper in vessel containing third-stage cyclones
Electrostatic precipitator	90–98	10–50	2–5	Probably the most commonly employed technology downstream of the regenerator cyclones. Effective on particles down to 1 μm
Wet gas scrubber	90–95	≤50	—	Also removes SO_x
Baghouse	Virtually 100%	<25	—	Baghouses have not been used except on small slipstreams for testing or as a substitute for fourth-stage cyclones noted above (see Section J.4.2.5)

(continued)

Table J.2 (*Continued*)

Entrainment from regenerator bed			
Regenerator superficial velocity (ft/s)[b]	2	3	4
Particulate loading at cyclone inlet (lb/ft³)[b]	0.2	1.0	2.0

Source: Information compiled largely from NPRA Q&A Transcripts (1990–1999) [36].
[a]Data are presumed not to include upset conditions.
[b]1 ft/s = 0.3048 m/s; 1 lb/ft³ ≅ 1.6 × 10⁷ mg/Nm³, or 16 kg/Nm³.

accumulate in the heavier liquid fractions and coke produced during refining. Metals include nickel, which enhances coke formation, and vanadium, which catalyzes oxidation of sulfur dioxide (SO_2) to sulfur trioxide (SO_3) during combustion of sulfur compounds. The sulfur and nitrogen in the coke are in the form of large, heavy organic molecules. Buildup of coke and its accompanying metals on the catalyst interferes with the desired cracking reactions and must be continuously removed to some low level. This takes place in the regenerator.

Catalyst from the reactor riser or reactor vessel is constantly moving downward, into, and through the transfer line to the regenerator. Simultaneously, catalyst continues to circulate from the regenerator to the top of the reactor. Before entering the transfer line, catalyst leaving the reactor first encounters a stripping section, where steam is injected to strip the hydrocarbon vapors remaining in, on, and around the catalyst particles. Catalyst leaving the reactor on its way to the regenerator is said to be spent, and is termed regenerated when leaving the regenerator (Table J.1).

J.3.3 The Regenerator

In the regenerator, the coke is burned off to prepare the catalyst for its return to the reactor. Compressed air is fed to the bottom of the regenerator for this purpose. Coke on regenerated catalyst (CRC) attains a very low percentage but is not removed all the way to zero, and regenerated catalyst is a light gray to black color.

The heat generated by the combustion reaction vaporizes the feed oil and supplies the energy necessary for the endothermic cracking reactions taking place in the reactor vessel plus whatever heat losses occur. If possible, the unit operates naturally in heat balance through control of the temperatures and circulation rate. Otherwise, heat can be added by preheating the feed, and it can be removed by making steam via heat exchange with the hot flue gas leaving the regenerator or by means of a catalyst cooler [39].

J.4 ATMOSPHERIC CONTAMINANTS FROM THE REGENERATOR—ORIGIN AND TREATMENT

Atmospheric contaminants are formed in the regenerator, either chemically during the combustion process or physically from the carryover of catalyst particles entrained in the regenerator flue gas. These include carbon monoxide (CO), catalyst particles (particulate matter (PM) or total suspended particulates (TSP)), sulfur oxides (SO_x = sulfur dioxide plus sulfur trioxide), and oxides of nitrogen. Origin and treatment of CO, PM, SO_x, and NO_x are considered below. Additional information is contained in the literature [40].

EPA emission limitations for CO, PM, and SO_x and opacity from FCCU regenerators are contained in 40 CFR 60, Subpart J [41]; if one is affected, the latest version should be consulted for all pertinent details. With certain specifics of applicability and exceptions, CO is limited to 500 ppm by volume on a dry basis (ppmvd), and PM is limited to 1 kg/1000 kg (1 lb/1000 lb or 2.0 lb/ton) of coke burnoff. For sulfur oxides (SO_x), SO_2 is limited as follows to: the least stringent of 50 ppm by volume (vppm) or 90% removal with an add-on control device, SO_x calculated as SO_2 9.8 kg/1000 kg (20 lb/ton) of coke burnoff without an add-on control device, or a fresh feed to the FCCU of 0.03% by weight sulfur (S). The limit for PM in FCCU regenerator flue gas passing through an incinerator or waste heat boiler firing supplemental liquid or solid fossil fuel may be increased by an incremental rate of up to 43.0 g/MJ (0.10 lb/MMBtu) attributable to the supplemental fuel. Opacity, related to PM concentration, is limited to 30%, except for one 6 min average opacity reading in any 1 h period. Any applicable state or local regulations may not be less stringent than EPA regulations, but may indeed be stricter.

A number of refiners in the United States are operating under consent agreements with EPA to reduce air emissions from their cat crackers, including NO_x emissions. As indicated in these decrees, EPA believes that a NO_x emission concentration of 20 ppmvd (annual average at 0% oxygen, O_2) is achievable by an selective catalytic reduction (SCR) system [40].

J.4.1 Carbon Monoxide (CO)

Carbon monoxide is formed in the regenerator from incomplete combustion of the carbon content of the coke on catalyst. There are, in general, two types of regenerators: the so-called partial-burn units built up to about 1972, and the full-burn units constructed thereafter. In a partial-burn unit, coke is burned to CO_2 plus CO in the percent range with little or no excess O_2. These older units produce a regenerator flue gas with a CO/CO_2 ratio of around 1.0 and containing up to 10% CO. Since combustion of CO to CO_2 liberates about 2.5 times as much heat as combustion of carbon to CO, this allowed a lower regenerator temperature and easier control but resulted in an excess of CO emitted to the atmosphere (Table J.1).

To meet an EPA mandated effluent limit of 500 ppm, the flue gas is processed in an external combustion device, known as a CO boiler, in which auxiliary fuel is fired to incinerate the excess CO in the regenerator effluent to within acceptable limits. Auxiliary fuel is necessary because of the small heating value (20–40 Btu/SCF) of the regenerator flue gas [42]. Heat is recovered from the CO boiler flue gas by making steam.

Full-burn units operate with a higher excess O_2 in the regenerator itself and produce CO within the parts per million (ppm) range. These run hotter, burn more coke off the catalyst, but are more difficult to control. Where a CO boiler is not necessary to control CO emissions, the regenerator flue gas is cooled in an unfired heat exchanger, termed a waste heat boiler. Full-burn units may be built to employ a two-stage combustion in separate zones [43].

Certain additives can also be blended in with the cracking catalyst to facilitate CO combustion in the regenerator. These employ a platinum-based CO oxidation catalyst, or promoter, to enhance CO combustion in the regenerator dense phase and are also used to prevent excessive temperatures from the so-called afterburning of CO remaining in the dilute phase.

J.4.2 Particulate Matter (PM)

In both the reactor and regenerator vessels, the fluidizing gas (hydrocarbon vapors in the reactor and air/flue gas in the regenerator) carries the smaller catalyst particles, the so-called fines, out of the bed and upward into the dilute phase above the bed. The process of expelling particles from the bed is known as elutriation, the phenomenon of solids being carried by the gas entrainment, and the height between the expanded bed and the exit of the vessel the freeboard. The particle- and gas-mixture in the freeboard is termed the dilute, or lean phase, to distinguish it from the dense phase within the bubbling bed below. Elutriation and entrainment increase with increasing gas superficial velocity, the volumetric flow rate divided by the cross-sectional area of the empty vessel. Superficial velocity in the regenerator is nominally in the range from 1 to more than 6 ft/s [27–29]. Entrainment data for regenerator superficial velocities between 2 and 4 ft/s at the inlet to the cyclones are shown at the bottom of Table J.2 [27].

The larger entrained particles typically settle out on their own. The smaller particles, or fines, are captured mechanically and returned to the bed in cyclone separators to prevent excessive loss of bed material. Fines are contained in the original FCC catalyst particle-size distribution and are created during operation by particle breakdown, fracture of larger particles, and/or abrasion and erosion of their outer surface—processes known as attrition.

J.4.2.1 Cyclones in the Regenerator

The smaller of the fines expelled from the regenerator bed enter the two stages of inertial separators in series to affect their capture. These separators are known as cyclones. Gas containing suspended particulates enters the first stage (primary), passes on to the second stage (secondary), and vents out the top of the second stage. Multiple sets of such cyclones are arranged in parallel to handle the total gas flow.

Some of the particles entering the cyclones are captured and are returned by gravity to the bed. The others escape from the regenerator, and unless further abated downstream, are emitted to the atmosphere via the regenerator flue gas stack. Collection efficiency is higher in the first stage than in the second stage since the larger particles are collected preferentially. Typical combined collection efficiency from both stages is 99.995%. Typical operating conditions, solids loadings, and particle sizes are shown in Table J.2.

The reactor vessel is also provided with two stages of cyclones. The catalyst particles escaping from the reactor cyclones are caught in the cat fractionator and form slurry in the bottom of that column. These do not constitute an emission to the atmosphere.

J.4.2.2 Third-Stage Cyclones [44,45]

In some cases, the exhaust gas from the regenerator is let down in pressure through a turbo-expander for power recovery. The resulting high-pressure steam can also be used to drive a steam turbine. To protect the blades in the turbo-expander from the sandblasting effect of the entrained catalyst, a third stage of cyclones (tertiary) is added. These cyclones are usually contained in a separate pressure vessel external to the regenerator.

The extremely fine particles captured here may or may not be returned to the circulating catalyst inventory. Emissions from transporting the collected catalyst fines in the underflow of the external collection vessel are treated with a fourth stage of cyclones or another particulate-collection device, such as an electrostatic precipitator (ESP) or a baghouse.

Third-stage cyclones cannot be used by themselves for environmental control except in those jurisdictions where the regulations are sufficiently lenient. The usual devices for particulate control to meet regulatory requirements are electrostatic precipitators or wet scrubbers.

J.4.2.3 Electrostatic Precipitators

These are a commonly employed particulate clean-up technology downstream of the regenerator cyclones. They are high-efficiency collection devices and are effective on particles down to 1 μm (Table J.2). They cause minimal pressure drop and can operate over a wide range of temperatures but appear to be limited to a maximum of 650°F (345°C) when placed upstream of a CO boiler [46].

In an ESP, flue gas flows between large vertical parallel plates carrying an electrical charge. Finely divided solids in the gas acquire the opposite charge and migrate toward the plates, where they are collected in solid sheets. Periodically, the plates are gently rapped to dislodge the solids into a collection hopper below. As with third-stage cyclones, some or all of the precipitator catch may or may not be recycled to the regenerator.

Collection efficiency for FCC catalyst fines is enhanced by ammonia and/or SO_3 either already occurring in the flue gas or injected upstream.

J.4.2.4 Wet Gas Scrubbers

This device contacts the flue gas with an atomized spray of fine droplets, which encapsulate suspended solids and absorb sulfur oxides and any other water-soluble contaminants [47–49]. The droplets themselves, much larger than catalyst fines, are then separated from the gas by cyclonic action in a second vessel. Wet gas scrubbing is commercially demonstrated on FCCU regenerator flue gas, and particulate collection of 90–95% is regularly achieved (Table J.2) along with SO_x removals of 90–99%. An alkali, such as caustic, soda ash, or lime, is added to the recirculating scrubbing liquor to control pH.

J.4.2.5 Baghouses

A baghouse contains multiple filter bags or socks through which a gas must flow. The bag material must be able to withstand the temperature of the gas. Solids in the gas impinge upon the mesh surface of the bag and form a filter cake. This cake then acts as the collecting surface.

As solids build up on the surface, pressure drop increases. Periodically, triggered either by time on line or by pressure drop, the bags are cleaned by shaking or by a reverse pulse of air. Groups of bags are arranged in separate chambers to allow the service flow to continue, while some of the bags are taken off-line for cleaning. It is also important to be able to isolate and replace bags that have sprung a leak, without having to shut down the entire system.

Baghouses can collect solids at virtually 100% efficiency. However, they are not known to have been used on FCCUs except on small slipstreams for testing or as a substitute for fourth-stage cyclones on the third-stage cyclone catch (Table J.2); a baghouse on an FCCU would adversely affect the delicate regenerator pressure balance and is said to increase the risk of upset and possibly catastrophic loss of control [50].

J.4.3 Sulfur Oxides (SO_x) [3]

Sulfur in the cat-cracker feed makes its way to several locations: 70–95% leaves the reactor overhead in the liquid products or as H_2S in the product gas, and the remaining 5–30% ends up in organic thiophenes and similar structures in the coke. When the coke is burned off in the regenerator, up to hundreds or even thousands of ppm of SO_x can result. Sulfur trioxide can constitute up to about 10% of the total SO_2 (sulfur dioxide) plus SO_3 [51], compared to a typical combustion effluent with SO_3 at a nominal 1–3%, discussed in Chapter 3. Vanadium and other metals in the coke or in a NO_x-removal SCR catalyst (described in later section) tend to increase the SO_3/SO_2 ratio.

There are four options to lower SO_x to acceptable levels: use of low-sulfur feeds to the cat cracker; hydrodesulfurization, or hydrotreating, of higher sulfur feedstocks; blending of sulfur-removal additives into the FCC catalyst; and wet gas scrubbing (WGS) of the regenerator flue gas [52].

Having to employ naturally *low sulfur materials* as cat-cracker feed is restrictive, and *hydrotreating* to bring the sulfur level down to the same range is the most capital intensive option [53] but is done under certain circumstances.

Catalyst additives function by enhancing the fraction of SO_3 formed upon combustion in the regenerator and then tying it up with the catalyst until the catalyst reaches the reactor. The sulfur content is then liberated as H_2S in the reactor, to be separated from the overhead product gas, rather than becoming an atmospheric pollutant emitted from the regenerator. The viability of this method in any specific case depends on the amount of sulfur removal possible at reasonable addition rates.

A *WGS* can remove the sulfur oxides in regenerator flue gas at the high efficiencies discussed above. Disadvantages include the following: (1) a waste scrubbing liquor containing suspended and dissolved solids requiring treatment and disposal plus (2) a water-saturated, corrosive exhaust gas at relatively low temperature forming a visible steam plume and possibly containing traces of highly corrosive sulfuric acid [54].

For a regenerator flue gas at 600°F (315°C) with 10% moisture entering the scrubber, calculations with simplifying assumptions show the adiabatic saturation temperature to be approximately 150°F (65°C) as the gas is cooled by the evaporating water, the water is heated by the hot flue gas, and the scrubber water evaporates up to its saturation concentration in the flue gas. (See also Problem 2.37.)

In addition to the problem with suspended particulate matter, SO_2 dissolved in the scrubber water becomes sulfite ion (SO_3^{2-}) and bisulfite (HSO_3^-) ion when an alkaline pH control agent such as sodium hydroxide (NaOH) is present. These sulfite species exert a chemical oxygen demand (COD), which must be satisfied by oxidation to sulfate (SO_4^{2-}) before wastewater from the WGS can be disposed of. (See Appendix D for COD.)

J.4.4 Oxides of Nitrogen (NO_x)

J.4.4.1 Source of NO_x

NO_x levels in the flue gas from a commercial FCCU regenerator are typically on the order from 100 or less to 500 or more ppm [55–59] and predominantly in the form of nitric oxide [60]. Organic nitrogen compounds constitute between about 0.05% and 0.5% of FCCU feed. Although the majority of this nitrogen appears in the FCCU products, nearly half of the nitrogen ends up in the coke on catalyst. When the coke is burned, the nitrogen is liberated in various forms.

During regeneration, 70–90% of the nitrogen in the coke is reduced to N_2, and the other 10–30% makes nitric oxide. This is so-called fuel NO_x [61], coming from the condensed organic nitrogen compounds in the coke. These include aromatic ring structures containing nitrogen atoms, amines, pyridine compounds, pyrrole derivatives, and amides. Temperatures in the regenerator are too low to produce any appreciable thermal NO_x, from the combination of the N_2 and O_2 in the combustion air; a *maximum* of 10–30 ppm of thermal NO_x is estimated even if chemical equilibrium were to be achieved [55,56].

The organic nitrogen compounds in the coke break down into such species as ammonia (NH_3), hydrogen cyanide (HCN), nitrous oxide (N_2O), nitric oxide (NO), and molecular nitrogen (N_2). Ammonia from commercial FCCUs has been measured by others in a few cases at levels of 400–1000 ppm [62]. The presence of HCN in the flue gas from a commercial regenerator has also been reported [63]. Molecular nitrogen is thought to arise from the reaction of NO with carbon (C) or CO, thereby decreasing the amount of NO.

Increasing the amount of oxygen in the regenerator decreases both the carbon on regenerated catalyst (i.e., the coke) and the carbon monoxide. It also causes the remaining coke to become richer in nitrogen species, which tend to burn last after the carbon. Both effects lead to an increase in regenerator-exit NO with increasing O_2 [64]. For a full-burn unit, NO_x is formed at excess O_2 in one overall step. Full-burn units operate in an oxidizing environment at higher excess O_2 to produce lower concentrations of NH_3 and HCN and to oxidize these intermediates to NO.

Under the reducing conditions of a partial-burn unit with an excess CO concentration, the organic nitrogen compounds in the coke react to elemental nitrogen (N_2), reduced nitrogen compounds such as ammonia and HCN, and some NO_x. When combustion is completed in a CO boiler, the reduced nitrogen compounds form more NO_x [65], and thermal NO_x plus perhaps some fuel NO_x are added from the CO boiler fuel [66]. Low-NO_x burners are available for CO boilers and will lower the thermal NO_x output but not the fuel NO_x [67].

In summary, a NO_x concentration up to hundreds of ppm can be expected in either case, depending on the overall excess O_2 concentration in the regenerator and the nitrogen level in the cat-cracker feedstock. NO_x from the regenerator may also be affected by higher local O_2 concentrations there caused by poor mixing.

J.4.4.2 NO_x-Abatement Technology

Prospective techniques for NO_x removal fall into several categories: hydrodesulfurization/hydrotreating, catalyst additives, scrubbing, and chemical reaction such as selective noncatalytic reduction (SNCR) and selective catalytic reduction (SCR).

Hydrotreating of FCCU feed will decrease feed nitrogen, nitrogen in the coke on catalyst, and consequently NO_x from the regenerator, but it is costly and unlikely to be done for NO_x abatement alone.

Additives. The amount of NO is enhanced by the presence of a platinum-based CO combustion promoter and depressed by an additive specifically engineered to decrease NO_x [59]. Such NO_x-control additives can also be used for fine-tuning of other NO_x-abatement techniques.

Ozone Scrubbing. Since regenerator NO_x is primarily NO with its low solubility in aqueous scrubbing solutions, NO_x removal in a conventional WGS for abatement of sulfur oxides and particulates is limited [60]. However, a scrubbing process with ozone (O_3) in the water [68] converts the NO_x (NO + NO_2) to nitric acid (HNO_3). Ozone is generated on-site as needed.

Chemical reactions in the scrubbing process are summarized by equations such as those exemplified below [69,70]:

$$NO + O_3 \rightarrow NO_2 + O_2 \tag{J.1}$$
$$NO_2 + O_3 \rightarrow NO_3 + O_2 \tag{J.2}$$
$$2NO_2 + O_3 \rightarrow N_2O_5 + O_2 \tag{J.3}$$
$$NO_2 + NO_3 \rightarrow N_2O_5 \tag{J.4}$$
$$2N_2O_5 + H_2O \rightarrow 2\,HNO_3 \tag{J.5}$$

The ozone to NO_x ratio is 1.2–2.2 to 1 [68], depending on such variables as temperature of the flue gas and residence time [71].

The nitric acid is then neutralized by the alkali, sodium hydroxide, present (NaOH). (A calcium- or magnesium-based reagent can also be used [68].)

$$HNO_3 + NaOH \rightarrow NaNO_3 + H_2O \tag{J.6}$$

The sodium nitrate ends up in the wastewater from the scrubber [71].

Sulfur dioxide and particulate (PM) are captured and removed along with NO_x in a single step [72]. Furthermore, ozone does not react with gaseous SO_2, nor with CO, to any appreciable extent [69–71]. This technology offers the possibility of retrofitting a WGS originally installed to control SO_2 and particulate alone.

The maximum temperature for ozone reaction is stated as 300°F (~150°C) [69], comfortably above the adiabatic saturation temperature for wet scrubbing of FCCU regenerator flue gas.

The ozone technology has been shown to work on several types of emission sources, either during slipstream pilot testing or on a few commercial installations [70,71]. It was tested on a refinery FCCU regenerator in 2002 using a small-scale scrubber [69,71,73,74]. No unusual conditions were observed with this new application, and an outlet NO_x from 10–20 ppm was demonstrated. Optimal temperature was observed to be 140–150°F (60–66°C), removal of SO_2 remained at over 99%, and nitrate concentration was found to be small compared to the sulfate resulting from scrubbing the SO_2 [71,73].

A full-scale ozone scrubbing process was installed at this refinery and is operating successfully [74]. Four other commercial units are scheduled for start-up later in 2007 [74]. Some of these ozone-scrubbing units are retrofits of scrubbers already designed for removal of SO_2 and particulate; some incorporate NO_x removal from the start.

SNCR reacts ammonia or urea (H_2NCONH_2) with NO_x to produce molecular nitrogen and water vapor. No catalyst is employed, and reaction takes place in a narrow high-temperature window, compared with SCR [75]. NO_x-removal efficiencies are typically much lower than that with SCR. The advantages of SNCR are that there is no catalyst to plug, foul, or blind, and the price is cheaper. The disadvantage is a much lower NO_x-removal efficiency. This technology has been demonstrated on various boilers, heaters, furnaces, and incinerators as well as on a coal-fired fluidized-bed boiler [76].

An application on a refinery FCCU in the Los Angeles, California area is described in a presentation given in 1994 [77]. Information is contained there with other sources in the public domain [2,8,12,13,40,78]. Hydrotreated FCC feed is preheated in an external furnace before entering the reactor; the regenerator is followed by a third-stage separator, expander turbine, CO boiler, and ESP, with a waste heat boiler at the tail end [77]. Results of a commercial-scale test of a CO combustion promoter were reported in 1979 [78], and a promoter may still be in use.

As others have found [55–59], inlet NO_x was said to originate from the organically bound nitrogen in the FCCU gas oil feed combined with the thermal NO_x produced by combustion of CO and auxiliary fuel in the CO boiler. An SNCR system utilizing urea was installed on the FCCU CO boiler (760–1010°C, 1400–1850°F) in the 1988 time frame [77]. The temperature window quoted in the paper for the urea-based SNCR is 760–1149°C (1400–2100°F), compared to 760–972°C (1400–1780°F) [75] for an SNCR process utilizing ammonia.

For this particular plant, the SNCR technology provided a day-to-day NO_x-removal efficiency between 40% and 55%, with performance test results under controlled conditions as high as 66%; original NO_x-reduction goal was 24% [77].

The SNCR was replaced by an SCR unit starting up in April 2000 [79]. Replacement of SNCR by SCR enabled the refinery to comply by 2003 with refinery-wide NO_x emission limitations mandated by the South Coast Air Quality Management District (SCAQMD), the governing air quality jurisdiction. SCR is capable of reducing NO_x from an FCCU by over 90% [1,79].

In the *SCR* process, the oxides of nitrogen NO and NO_2, commonly known as NO_x, are reacted with ammonia in the presence of a flow-through honeycomb catalyst containing titanium dioxide, vanadium, and other base metals to give nitrogen and water vapor. Reaction stoichiometry with ammonia, injected upstream, depends on the relative amount of each oxide and whether or not oxygen is present. For applications such as FCCU flue gas containing excess O_2 and parts per million concentrations of NO in excess of NO_2 [60], the equations given below apply. In the absence of competing side reactions, the theoretical molar ratio of NH_3 reacted to NO_x destroyed is 1.0.

$$4NO + 4NH_3 + O_2 \xrightarrow{\text{Catalyst}} 4N_2 + 6H_2O \tag{J.7}$$

$$NO + 2NO_2 + 2NH_3 \xrightarrow{\text{Catalyst}} 2N_2 + 3H_2O \tag{J.8}$$

Ammonia beyond that required to participate in the SCR reactions appears in the SCR effluent and is designated as ammonia slip. Sufficient catalyst must be present to provide the required degree of NO_x removal at an acceptable level of ammonia slip [80,81]. NO_x-removal efficiency depends ultimately on the amount of catalyst; the NH_3 to NO_x ratio; and the local distribution of ammonia, NO_x, and flow across the SCR inlet.

In the absence of SO_3 in the flue gas, the operating temperature range for SCR is 400–750°F (204–399°C); however, as the SO_3 concentration rises, so does the minimum operating temperature, and it may be as high as 650°F (343°C) with sufficient SO_3 in FCC flue gas [1]. The increase in minimum temperature is to inhibit the formation of the "sticky" ammonium sulfate (($NH_4)_2SO_4$) and/or bisulfate (NH_4HSO_4) salts and their potential for fouling of the flow-through SCR catalyst and downstream equipment. In the general case, a typical temperature range quoted for base-metal SCR catalysts composed of titanium dioxide (TiO_2), vanadium pentoxide (V_2O_5), and other ingredients is 600–750°F (316–399°C) [82,83].

Additional SO_3 can also lead to the formation of sulfuric acid (H_2SO_4) with flue gas moisture; corrosion of downstream equipment may result if the flue gas temperature falls below the sulfuric acid dew point (303–347°F, 150–175°C) [54]. (See also the sulfuric acid dew point problems in Chapter 3.)

SCR treatment of nitrogen oxides in FCC regenerator flue gas must also account for the other atmospheric contaminants generated simultaneously in the FCCU process as well as the type and location of pollution control devices installed to abate those contaminants [1]. For example, upstream of a WGS, one must also address high particulate loading; this leads to increased catalyst pitch (along with soot blowers or sonic horns) required to minimize plugging/blinding of the catalyst flow passages. Downstream of a WGS, one must be aware of possible poisoning by the metal ions in pH control agents carried over in any liquid-droplet entrainment that might occur. Whatever the flue gas treatment configuration, upset conditions in the FCCU may lead to carryover of flammable oil mist to the SCR catalyst, possibly causing high temperatures and permanent catalyst damage plus a possible safety issue for personnel and property.

Despite all of these potential problems, performance of SCR treating FCCU regenerator flue gas has been demonstrated on a commercial scale for eight FCC units identified by name, including the one mentioned above in the discussion of SNCR [1,79]. (See also Ref. [84].) Flue gas flows in Ref. [1] range from about 75,000 to over 530,000 Nm^3/h (47,000–330,000 SCFM (standard ft^3/min @ 60°F, 1 atm)). Additional installations have started up since the time of the cited [1] presentation.

J.5 SUMMARY

- The fluid catalytic cracker (FCCU) produces much needed gasoline and fuel oil.
- However, the FCCU is the biggest single source of air pollution in an oil refinery.
- Atmospheric contaminants from the FCCU include CO, PM, SO_x, and NO_x.
- A number of refiners in the United States are operating under consent agreements with EPA to reduce emissions from the FCCU.
- Abatement devices for various pollutants must be compatible with one another.
- Designs must also address upset conditions.

REFERENCES

1. G.D. Bouziden, J.K. Gentile, and R.G. Kunz, Selective catalytic reduction of NO_x from fluid catalytic cracking—case study: BP Whiting Refinery, Paper ENV-03-128 presented at the 2003 NPRA National Environmental & Safety Conference, New Orleans, LA, April 23–24, 2003.
2. J. Stell, editor, 2002 Worldwide Refining Survey, *Oil & Gas Journal*, **100**(52), 68–111, 2002.
3. K.R. Gilman, H.B. Vincent, and T.F. Walker, The cost of controlling air emissions generated by FCCUs, Paper AM-98-15 presented at the 1998 NPRA Annual Meeting, San Francisco, CA, March 15–17, 1998.
4. J.H. Gary and G.E. Handwerk, *Petroleum Refining: Technology and Economics*, 4th edn, Marcel Dekker, New York, 2001, pp. 93–135.
5. J.I. Kroschwitz and M. Howe-Grant, editors, *Kirk-Othmer Encyclopedia of Chemical Technology*, 4th edn, Vol. 5, Wiley, New York, 1993, pp. 419–448.
6. J.J. McKetta and W.A. Cunningham, editors, *Encyclopedia of Chemical Processing and Design*, Vol. 13, Marcel Dekker, Inc., New York, 1981, pp. 1–132.
7. ENERGETICS Incorporated, Columbia, Maryland for U.S. Department of Energy Office of Industrial Technologies, Energy and Environmental Profile of the U.S. Petroleum Refining Industry, pp. 55–59, U.S.D.O.E., Washington, DC, December 1998.

References

8. W.F. Bland and R.L. Davidson, editors, *Petroleum Processing Handbook*, Section 3, McGraw-Hill, New York, 1967, pp. 3-1–3-17.
9. Royal Dutch/Shell, *The Petroleum Handbook*, 6th edn, Elsevier, Amsterdam, 1983, pp. 284–294.
10. D. Kunii and O. Levenspiel, *Fluidization Engineering*, Wiley, New York, 1969, pp. 16–20.
11. F.A. Zenz and D.F. Othmer, *Fluidization and Fluid-Particle Systems*, Reinhold, New York, 1960, pp. 7–15.
12. J.H. Gary and G.E. Handwerk, *Petroleum Refining: Technology and Economics*, 4th edn, Marcel Dekker, New York, 2001, pp. 95–96.
13. D. Kunii and O. Levenspiel, *Fluidization Engineering*, Wiley, New York, 1969, Figure 3 on p. 20.
14. L.E. Carlsmith and F.B. Johnson, Pilot plant development of fluid catalytic cracking, *Industrial and Engineering Chemistry*, **37**(5), 451–455, 1945.
15. J.H. Gary and G.E. Handwerk, *Petroleum Refining: Technology and Economics*, 4th edn, Marcel Dekker, New York, 2001, pp. 93–135.
16. E.J. Gohr, Background, history and future of fluidization, in: D.F. Othmer, editor, *Fluidization*, Reinhold, New York, 1956, pp. 102–116.
17. C.E. Jahnig, H.Z. Martin, and D.L. Campbell, *The Development of Fluid Catalytic Cracking*, in: J.R. Grace and J.M. Matsen, editors, *Fluidization*, Plenum Press, New York, 1980, pp. 3–24.
18. C.E. Jahnig, D.L. Campbell, and H.Z. Martin, History of fluidized solids development, *Chemtech*, **14**(2), 106–112, 1984.
19. J.J. McKetta and W.A. Cunningham, editors, *Encyclopedia of Chemical Processing and Design*, Vol. 13, Marcel Dekker, Inc., New York, 1981, pp. 1–16.
20. P.K. Niccum, M.F. Gilbert, M.J. Tallman, and C.R. Santner, Future Refinery—FCC's role in refinery/petrochemical integration, Paper AM-01-61 presented at the 2001 NPRA Annual Meeting, New Orleans, LA, March 18–20, 2001.
21. A.D. Reichle, Fluid cat cracking—fifty years ago and today, Paper AM-92-13 presented at the 1992 NPRA Annual Meeting, New Orleans, LA, March 22–24, 1992.
22. Royal Dutch/Shell, *The Petroleum Handbook*, 6th edn, Elsevier, Amsterdam, 1983, pp. 284–288.
23. F.A. Zenz and D.F. Othmer, *Fluidization and Fluid-Particle Systems*, Reinhold, New York, 1960, Figure 3.20 on p. 125 and Figure 7.17 on p. 251.
24. J.J. McKetta and W.A. Cunningham, editors, *Encyclopedia of Chemical Processing and Design*, Vol. 13, Marcel Dekker, Inc., New York, 1981, pp. 100–102.
25. J.F. Frantz, Minimum fluidization velocities and pressure drop in fluidized beds, *Chemical Engineering Progress Symposium Series No. 62*, **62**, 21–31, 1966.
26. R.G. Kunz, Minimum Fluidization Velocity: Fluid Cracking Catalysts and Spherical Gas Beads, *Powder Technology*, **4**, 156–162, 1970/1971.
27. NPRA Q&A Transcript, 1998, p. 51.
28. J.J. McKetta and W.A. Cunningham, editors, *Encyclopedia of Chemical Processing and Design*, Vol. 13, Marcel Dekker, Inc., New York, 1981, pp. 8, 11.
29. J.I. Kroschwitz and M. Howe-Grant, editors, *Kirk-Othmer Encyclopedia of Chemical Technology*, 4th edn, Vol. 5, Wiley, New York, 1993, pp. 419–448.
30. F.A. Zenz, and D.F. Othmer, *Fluidization and Fluid-Particle Systems*, Reinhold, New York, 1960, Chapters 3, 7, 8, and 9, pp. 94–135, 230–312.
31. J.I. Kroschwitz and M. Howe-Grant, editors, *Kirk-Othmer Encyclopedia of Chemical Technology*, 4th edn, Vol. 11, Wiley, New York, 1993, pp. 138–171.
32. M. Leva, *Fluidization*, McGraw-Hill, New York, 1959, 327 pp.
33. J.F. Frantz, Design for fluidization: Parts 1, 2, and 3, *Chemical Engineering*, **69**(19), 161–178, (September 7, 1962); **69**(20), 89–96 (October 1, 1962); **69**(22), 103–110 (October 29, 1962).
34. J.F. Davidson and D. Harrison, *Fluidised Particles*, Cambridge University Press, London, 1963, 155 pp.
35. D. Kunii and O. Levenspiel, *Fluidization Engineering*, Wiley, New York, 1969, 534 pp.
36. NPRA Question & Answer (Q&A) Transcripts, 1990–1999.
37. J.J. McKetta and W.A. Cunningham, editors, *Encyclopedia of Chemical Processing and Design*, Vol. 13, Marcel Dekker, Inc., New York, 1981, pp. 19–26.

38. J.H. Gary and G.E. Handwerk, *Petroleum Refining: Technology and Economics*, 4th edn, Marcel Dekker, New York, 2001, pp. 106–108.
39. J.I. Kroschwitz and M. Howe-Grant, editors, *Kirk-Othmer Encyclopedia of Chemical Technology*, 4th edn, Vol. 5, Wiley, New York, 1993, pp. 444–445.
40. P.K. Niccum, E. Gbordzoe, and S. Lang, FCC flue gas emission control options, Paper AM-02-27 presented at the 2002 NPRA Annual Meeting, San Antonio, TX, March 17–19, 2002. See also Optimize FCC Flue-Gas Emission Control—Part 1, *Hydrocarbon Processing*, **81**(9), 71–76 (September 2002), and Optimize FCC Flue-Gas Emission Control—Part 2, *Hydrocarbon Processing*, **81**(10), 85–91 (October 2002).
41. 40 CFR 60, Standards of Performance for New Stationary Sources, Subpart J—Standards of Performance for Petroleum Refineries, United States Government Printing Office, Washington, DC (electronic version accessed at http://ecfr.gov/on Mar. 5, 2008).
42. D.G. Hammond, M.R. Parrish, and V.A. Citarella, Fluid coking: a competitive option for heavy feed processing, Paper AM-96-72 presented at the 1996 NPRA Annual Meeting, San Antonio, TX, March 17–19, 1996.
43. R.B. Miller, T.E. Johnson, C.R. Santler, A.A. Avidan, and J.H. Beech, Comparison between single and two-stage FCC regenerators, Paper AM-96-48 presented at the 1996 NPRA Annual Meeting, San Antonio, TX, March 17–19, 1996.
44. R. Hypes and E. Tenney, Methods for control of particulate matter (PM) generated in the FCCU process, Paper ENV-00-197 presented at the 2000 NPRA Environmental Conference, San Antonio, TX, September 10–12, 2000.
45. E.D. Tenney, Third stage cyclone separators for FCC regenerator gases, Paper 66d presented at the 2000 AIChE Spring National Meeting, New Orleans, LA, March 5–9, 2000.
46. NPRA Q&A Transcript, 1994, p. 85.
47. J.D. Cunic and A.S. Feinberg, Innovations in FCCU wet gas scrubbing, Paper AM-96-47 presented at the 1996 NPRA Annual Meeting, San Antonio, TX, March 17–19, 1996.
48. J.D. Cunic and E.M. Roundtree, Control technology selection with MACT II FCC particulate emission standards, Paper AM-99-17 presented at the 1999 NPRA Annual Meeting, San Antonio, TX, March 21–23, 1999.
49. J.A. Herlevich Jr., S.T. Eagleson, A.H. Roth, and E.H. Weaver, Wet scrubbing for FCCUs—a case study examining site specific design considerations, Paper AM-01-12 presented at the 2001 NPRA Annual Meeting, New Orleans, LA, March 18–20, 2001.
50. NPRA Q&A Transcript, 1992, 69 pp.
51. J.I. Kroschwitz and M. Howe-Grant, editors, *Kirk-Othmer Encyclopedia of Chemical Technology*, 4th edn, Vol. 5, Wiley, New York, 1993, 432 pp.
52. D.P. McArthur, H.D. Simpson, and K. Baron, Catalytic control of FCC SO_x emission looking good, *Oil & Gas Journal*, **79**(8), 55–59, 1981.
53. A. Vierheilig and M. Evans, The role of additives in reducing fluid catalytic cracking SO_x and NO_x emissions, *Petroleum and Coal*, **45**(3–4), 147–153, 2003.
54. J.I. Kroschwitz and M. Howe-Grant, editors, *Kirk-Othmer Encyclopedia of Chemical Technology*, 4th edn, Vol. 5, Wiley, New York, 1993, p. 443.
55. A.W. Peters, G.D. Weatherbee, and X. Zhao, Origin of NO_x in the FCCU regenerator, *Fuel Reformulation*, **5**(3), 45–50, 1988.
56. A.W. Peters, X. Zhao, and G.D. Weatherbee, The origin of NO_x in the FCCU regenerator, Paper AM-95-59 presented at the 1995 NPRA Annual Meeting, San Francisco, CA, March 19–21, 1995.
57. A.W. Peters, X. Zhao, G. Yaluris, S. Davey, and A. Kramer, Catalytic NO_x control strategies for the FCCU, Paper AM-98-43 presented at the 1998 NPRA Annual Meeting, San Francisco, CA, March 15–17, 1998.
58. S.W. Davey, Environmental fluid catalytic cracking technology, Paper presented at the European Refining Technology Conference, 19 pp. (undated), www.gracedivision.com/custpubs/fccother/ptqpaper.pdf assessed on January 24, 2002.
59. W.-C. Cheng, G. Kim, A.W. Peters, X. Zhao, K. Rajagopalan, M.S. Ziebarth, and C.J. Pereira, Environmental fluid catalytic cracking technology, *Catalysis Reviews Science and Engineering*, **40**(1&2), 39–79, 1998.

60. J.I. Kroschwitz and M. Howe-Grant, editors, *Kirk-Othmer Encyclopedia of Chemical Technology*, 4th edn, Vol. 5, Wiley, New York, 1993, pp. 434–435.
61. R.G. Kunz, D.D. Smith, N.M. Patel, G.P. Thompson, and G.S. Patrick, Control NO_x from gas-fired hydrogen reformer furnaces, Paper AM-92-56 presented at the 1992 NPRA Annual Meeting, New Orleans, LA, March 22–24, 1992.
62. H.B. Lange, J.K. Arand, M.N. Mansour, and R.E. Hall, Laboratory evaluation of NO_x reduction techniques for refinery CO boilers, Paper No. 84-42.4 presented at the 77th Annual Meeting of the Air Pollution Control Association, San Francisco, CA, June 24–29, 1984.
63. W.-C. Cheng, G. Kim, A.W. Peters, X. Zhao, K. Rajagopalan, M.S. Ziebarth, and C.J. Pereira, Environmental fluid catalytic cracking technology, *Catalysis Reviews Science and Engineering*, **40**(1 & 2), 73, 1998.
64. W.-C. Cheng, G. Kim, A.W. Peters, X. Zhao, K. Rajagopalan, M.S. Ziebarth, and C.J. Pereira, Environmental fluid catalytic cracking technology, *Catalysis Reviews Science and Engineering*, **40**(1&2), Figure 17 on p.72, 1998.
65. J.O.L. Wendt and C.V. Sternling, Effect of ammonia in gaseous fuels on nitrogen oxide emissions, *Journal of the Air Pollution Control Association*, **24**(11), 1055–1058, 1974.
66. NPRA Q&A Transcript, 1998, p. 44.
67. K.A. Modi and R.T. Waibel, NO_x emissions from process heaters and boilers: improving emissions estimates, Paper ENV-95-163 presented at the 1995 NPRA Environmental Conference, San Francisco, CA, October 15–17, 1995.
68. Wet Scrubbing Apparatus and Method for Controlling NO_x Emissions, United States Patent, U.S. patent 7214356, issued to Belco Technologies Corporation, May 8, 2007.
69. E.H. Weaver, N. Confuorto, and M.J. Barrasso, FCCU NO_x control with a wet scrubbing system—state of the art technology, Paper ENV-03-127 presented at the 2003 NPRA National Environmental & Safety Conference, New Orleans, LA, April 23–24, 2003.
70. N. Confuorto, M. Barasso, and N. Suchak, Clean generation, Reprint *Hydrocarbon Engineering*, June 2003, 5pp.
71. Anonymous, Assessment of NO_x Emissions Reduction Strategies for Cement Kilns—Ellis County, Attachment C (ERG Report on LoTOx™ Application to Refineries), Final Report to Texas Commission on Environmental Quality, ERG, Inc., July 14, 2006.
72. DuPont™Belco® Clean Air Technologies—LoTOx™ Technology, http://www.belcotech.com/products/nox.html accessed on March 6, 2008.
73. J. Sexton, N. Confuorto, M. Barasso, and N. Suchak, LoTOx™ technology demonstration at marathon ashland petroleum LLC's refinery in Texas City, Texas, Paper AM-04-10 presented at the 2004 NPRA Annual Meeting, San Antonio, TX, March 21–23, 2004.
74. N. Confuorto and J. Sexton, Wet scrubbing based NO_x control using LoTOx™ technology—first commercial FCC start-up experience, Paper ENV-07-100 presented at the 2007 NPRA Environmental Conference, Austin, TX, September 24–25, 2007.
75. B.E. Hurst, Improved thermal $DeNO_x$ process verified by field testing, Paper No. 84-42.1 presented at the 77th Annual Meeting of the Air Pollution Control Association, San Francisco, CA, June 24–29, 1984.
76. T.C. Hess, Advanced NO_x and SO_2 emission control performance on a CFB, *Power Engineering*, **93**(7), 47–49, 1989.
77. M.P. Younis, Implementation of NO_x control technologies in petroleum refining applications, Mobil Torrance refinery, Conference Proceedings from Clean Air '94—First North American Conference & Exhibition, Emerging Clean Air Technologies and Business Opportunities: Meeting Global Air Challenges through Partnerships, Toronto, Ontario, Canada, September 26–30, 1994, pp. 301–314.
78. F.D. Hartzell and A.W. Chester, CO burn promoter produces multiple fluid catalytic cracker benefits, *Oil & Gas Journal*, **77**(16), 83–86, 1979. See also FCCU gets a catalyst promoter, *Hydrocarbon Processing*, **58**(7), 137–140, 1979.
79. K. Gentile, BACT/LAER Technology for Tier II, Slides from Panel Discussion at the 2000 NPRA Annual Environmental Conference, San Antonio, TX, September 10–12, 2000.

80. R.G. Kunz, SCR performance on a hydrogen reformer furnace—a comparison of initial and first- and second-year anniversary emissions data, Paper No. 96-RA 120.01 presented at the 89th Annual Meeting Air & Waste Management Association, Nashville, TN, June 23–28, 1996.
81. R.G. Kunz, SCR performance on a hydrogen reformer furnace, *Journal of the Air & Waste Management Association*, **48**, 26–34, 1998.
82. J. Czarnecki, C.J. Pereira, M. Uberoi, and K.P. Zak, Put a lid on NO_x emissions, *Pollution Engineering*, **26**(12), 26–29, 1994.
83. D. Fusselman and D. Lipsher, Several technologies available to cut refinery NO_x, *Oil & Gas Journal*, **90**(44), 45–50, 1992.
84. M. Yamamura and K. Suyama, Operation experience of selective catalytic NO_x removal systems for miscellaneous flue gas, Paper 88-83.2 presented at the 81st Annual Meeting of APCA, Dallas, TX, June 19–24, 1988.

Appendix K

Case Studies in Air Emission Control

Never live west of where you work and drive back and forth to your job on the day shift. (Think about it.)

K.1 SUMMARY

This appendix presents a synopsis of several cases within the author's experience resulting in the reduction of NO_x emissions to the atmosphere. Further details are contained in the cited references.

K.2 ADDITION OF STEAM TO REDUCE BURNER NO_X

This case study of burner NO_x reduction shows how selection of a NO_x control system depends on many factors, including a window for shutdown. In this case history [1,2], an existing steam methane reformer (SMR) had to expand production to meet increased customer demand. However, the proposed modifications required that the existing air-quality permit then be changed to reflect a higher firing rate to support greater production. Furthermore, even without a production increase, some emission reduction was necessary to comply with new regulatory initiatives.

The burners in the reformer furnace were first-generation staged-air low-NO_x burners operating on natural gas and preheated combustion air. These burners were sufficient to control NO_x at the maximum firing rate in accordance with the original permit limit and existing regulations at the time the plant was built.

Several NO_x control options were considered. These included reduced air preheat, selective catalytic reduction (SCR), better performing low-NO_x burners, and steam injection into the new or existing burners. Limiting production or doing nothing was not deemed acceptable. The design options were limited because of the desire to minimize plant shutdown, thus favoring those strategies that could be implemented within the customer's coincident 14-day annual turnaround.

Environmental Calculations: A Multimedia Approach, by Robert G. Kunz
Copyright © 2009 John Wiley & Sons, Inc.

Reducing air preheat to lower NO_x was quickly eliminated as an option because of the extensive plant-wide modifications required plus the additional ongoing operating expense of burning additional fuel because of less efficient combustion.

Construction of an SCR posttreatment device could be accomplished on the run, except for a brief outage for final tie-ins. However, its estimated price tag to achieve NO_x control well beyond the needs of this location was high.

Replacement of the existing burners with a later model (with or without the capability to add steam to them) would have controlled the NO_x emissions at a much lower capital cost than SCR but with a negative impact on scheduling considerations.

Adding steam to the existing burners would result in some NO_x reduction, and like the SCR, external steam lines could be added while the plant continued to operate. This was the option finally decided upon, with steam to be added to the combustion air. Steam addition to the fuel typically achieves twice the NO_x removal as adding the same amount of steam to the combustion air [3], and addition of steam to the fuel would be the choice if adding steam in a grassroot project. However, since the requisite steam flow is appreciable compared to that of the fuel, this often causes bottlenecks, whereas steam injection into the much larger combustion airflow does not. This strategy was implemented, along with a parameter monitoring system authorized by the environmental regulatory agency.

Following start-up, stack testing to demonstrate environmental compliance within an enforceable operating envelope was conducted in conjunction with plant performance tests for typical operating scenarios at various production rates. The reformer furnace was fired solely on pipeline natural gas at 90°F (32°C), average combustion air temperature was 530°F (277°C), and saturated steam at 50 psig was added for the steam-addition runs. The average ambient air humidity measured was close to the EPA Method 19 default humidity of 0.027 mol H_2O per mol of wet air [4]. Operations were held steady during a given run.

The stack testing confirmed experimentally the burner manufacturer's prediction [3] of the effect of steam on NO_x emissions. At the same excess O_2 in the furnace, relative NO_x upon adding steam was shown during plant testing to decrease linearly from 1.0 (no effect) at zero steam addition to 0.8 for steam at 0.3 lb steam/lb fuel, the operational limit set for this plant to avoid excessive steam usage and an unstable flame.

To ensure ongoing compliance, online parameter monitoring was incorporated into the plant's distributed control system (DCS). This allowed the DCS to follow the process parameters on a real-time basis, calculate the uncontrolled NO_x emission rates for comparison with the permit limit, actuate the steam injection system, if necessary, and compute the final NO_x emission. This then enabled the parameters to be recorded, reported, and tracked for regulatory purposes.

K.3 ADDITION OF SCR TO REDUCE BURNER EMISSIONS

SCR, an add-on device used to control emissions, is shown in this case study to reduce NO_x from a high air preheat case to acceptable levels. Selective catalytic reduction is a process in which oxides of nitrogen (NO_x) are reacted with ammonia to form nitrogen (N_2) and water vapor (H_2O) in the presence of a suitable catalyst. The nitrogen and water vapor are discharged to atmosphere in the flue gas, along with any remaining NO_x and unreacted ammonia (NH_3), the so-called ammonia slip.

As required by the air-quality permit-to-construct, an SCR was installed to abate the NO_x from a hydrogen plant reformer furnace located at a California refinery [5–7]. The permit further required a continuous emissions monitoring system (CEMS) on the flue gas stack to monitor NO_x, carbon monoxide (CO), oxygen (O_2), volumetric flow rate, and temperature, plus periodic stack sampling to certify the CEMS measurements and to determine the ammonia slip. Concentration limits for outlet CO, NO_x, and NH_3 in the permit are based on rolling averages of hourly data—50, 10, and 25 ppm by volume, respectively. These limits are specified on a dry basis, expressed at 3% O_2 (dry) by volume (ppmvd).

Pressure swing adsorption (PSA) purge gas [8], supplemented with natural gas or refinery fuel gas, was used as fuel, and combustion air was preheated to a high temperature to maximize recovery of flue-gas waste heat and minimize export steam.

As reported for annual compliance runs, excess O_2 at the SCR inlet ranged from 651 to 673°F (344–356°C), excess O_2 at the SCR inlet from 3.0% to 3.7% (dry), and furnace flue gas moisture from 14% to 18%. NO_x-removal efficiency for the compliance demonstration runs was about 85%. The measured outlet NO_x and ammonia slip were in compliance with their permit limits. CO was found to be unaffected across the SCR and was measured at the stack at well below its permit limit.

K.4 INTEGRATION OF A FURNACE WITH GAS TURBINE EXHAUST

This particular case study shows how integration of a furnace with gas turbine exhaust can reduce furnace emissions, with a single SCR treating the combined effluent. This case study describes an SMR integrated with gas turbine technology at a Gulf Coast location to produce hydrogen, steam, and power [9,10]. In this plant, the combined furnace flue gas is treated using SCR. The hydrogen plant is capable of operating on its own without the turbine. Conversely, hot turbine exhaust gas can be discharged for short periods through its own stack upstream of the SCR. However, this mode is thermally inefficient and is minimized for economic reasons.

The hydrogen plant furnace is fired on PSA purge gas [8], a low-Btu fuel, supplemented by natural gas or refinery fuel gas as the trim fuel. Firing of PSA purge gas in the fuel mixture produces lower NO_x in an SMR reformer furnace compared to burning 100% trim fuel. The gas turbine is fired on natural gas.

Firing natural gas in a combustion gas turbine generates electricity and a hot exhaust gas containing about 15% O_2 on a dry basis (~13% O_2, wet). Discharge of the hot turbine exhaust directly to the atmosphere without cooling is termed *simple cycle*. When the hot exhaust is cooled by making steam in a heat recovery steam generator (HRSG) before discharge to atmosphere, the process is called *combined cycle*. The turbine can be integrated with furnace operation by replacing all or part of the furnace combustion air by hot gas turbine exhaust and using the furnace to perform the function of the HRSG. In doing so, the NO_x is further reduced.

This is seen by comparing NO_x formation during integrated operation with NO_x from the SMR furnace standalone operation at the time of an extended period of gas turbine downtime, both at the same maximum hydrogen production rate. NO_x from the integrated operation is some 40% lower on a ppm basis and lower on a lb/h basis as well. It was also noted that NO_x in the furnace flue gas upstream of the SCR during integrated operation was lower than expected, based on the burner NO_x for a similar hydrogen plant.

These observations are explained as follows: higher concentrations of inerts nitrogen (N_2), carbon dioxide (CO_2), and water vapor (H_2O) in the turbine exhaust give rise to a lower furnace burner flame temperature in contrast to using ambient air for combustion, in spite of the elevated temperature of the turbine exhaust ($\sim 1000°F$, $538°C$). In addition, more fuel was being fired with ambient air to make up for the lack of "air preheat" usually provided by the turbine exhaust, and a greater proportion of trim fuel was being fired since the amount of PSA purge gas is fixed and limited at a given production by process conditions.

Provided that one can make cost-effective use of the generated power, integration of gas turbine technology may be a viable option for reducing NO_x from other process heaters as well. Similar gas turbine integration is practiced for pyrolysis cracking furnaces in the ethylene industry [11].

REFERENCES

1. R.G. Kunz, B.R. Keck, and J.M. Repasky, Mitigate NO_x by steam injection, Paper ENV-97-15 presented at the 1997 NPRA Environmental Conference, New Orleans, LA, September 28–30, 1997.
2. R.G. Kunz, B.R. Keck, and J.M. Repasky, Mitigate NO_x by steam injection *Hydrocarbon Processing*, **77**(2), 79–84, 1998.
3. R.T. Waibel, John Zink Co., Personal Communication, March 7, 1997.
4. U.S. Environmental Protection Agency, Standards of Performance for New Stationary Sources – Test Methods, 40 CRF Part 60 Appendix A Method 19, 1995, p.812.
5. N.M. Patel, R.A. Davis, N. Eaton, D.L. Carlson, F. Kessler, and V. Khurana, Across-the-fence' hydrogen plant starts up at California refinery, *Oil & Gas Journal*, **92**(40), 54–61, 1994.
6. R.G. Kunz, SCR performance on a hydrogen reformer furnace—a comparison of initial and first- and second-year anniversary emissions data, Paper No. 96-RA 120.01 presented at the 89th Annual Meeting Air & Waste Management Association, Nashville, TN, June 23–28, 1996.
7. R.G. Kunz, SCR performance on a hydrogen reformer furnace, *Journal of the Air & Waste Management Association*, **48**, 26–34, 1998.
8. R.G. Kunz, D.D. Smith, and E.M. Adamo, Predict NO_x from gas-fired furnaces, *Hydrocarbon Processing*, **75**(11), 65–79, 1996.
9. R.G. Kunz, D.C. Hefele, R.L. Jordan, and F.W. Lash, Use of SCR in a hydrogen plant integrated with a stationary gas turbine—case study: the Port Arthur steam-methane reformer, Paper No. 70093 presented at the Air and Waste Management Association (A&WMA) 96th Annual Conference & Exhibition, San Diego, CA, June 22–26, 2003.
10. R.G. Kunz, D.C. Hefele, R.L. Jordan, and F.W. Lash, Consider SCR to mitigate NO_x emissions, *Hydrocarbon Processing*, **82**(11), 43–50, 2003.
11. J.I. Kroschwitz and M. Howe-Grant, editors, *Kirk-Othmer Encyclopedia of Chemical Technology*, 4th edn, Vol. 9, Wiley, New York, 1994, p.888.

Appendix L

Combustion of Refinery Fuel Gas

When all is said and done, there is nothing left to say or do.

L.1 REFINERY FUEL GAS

According to 40 CFR 63.641, refinery fuel gas (RFG) means a gaseous mixture of methane, light hydrocarbons, hydrogen, and other miscellaneous species that is produced in the refining of crude oil and/or petrochemical processes and that is separated for use in boilers and process heaters throughout the refinery [1]. With certain exceptions, EPA considers any gas generated and combusted at a petroleum refinery as *fuel gas*, including natural gas combined with RFG in any proportion [2]. Natural gas is a commercially available fuel consisting of a high percentage of methane (generally greater than 85%) and varying amounts of ethane, propane, butane, and so on and inerts such as nitrogen, carbon dioxide, and possibly helium [3].

RFG is typically a blend of gas streams produced as by-products in various refinery units including catalytic reforming, hydrotreating and hydrocracking, catalytic cracking, and coking [4]. On a dry basis, major components in RFG can consist of percent levels of numerous hydrocarbons (C_1–C_6^+, saturated and unsaturated), hydrogen (H_2), carbon oxides (CO and CO_2), nitrogen (N_2), and possibly some oxygen (O_2). Saturated hydrocarbons in RFG come from catalytic reforming, hydrotreating, hydrocracking, and the like; unsaturates arise from such refining processes as catalytic cracking and coking [4].

Hydrogen in RFG may vary from 10% up to about 60% [5,6], if not separated for use in hydrotreating operations for sulfur removal from refinery feeds and products [7]. Hydrogen recovery from fuel gas streams is but one method of meeting an increased hydrogen demand brought about by clean fuels initiatives [8].

Environmental Calculations: A Multimedia Approach, by Robert G. Kunz
Copyright © 2009 John Wiley & Sons, Inc.

RFG also contains various sulfur species, including hydrogen sulfide (H_2S) and mercaptans (RSH), at concentrations expressed in units of parts per million (ppm). The sulfur species originate from the organic sulfur compounds contained in the crude oil processed by the refinery.

RFG is in fact a leftover stream containing those gas-phase constituents deemed not economically recoverable for use as a feedstock or for sale as products. Burning of the RFG to recover its energy content at the same time eliminates having to dispose it of as a waste stream. It is consumed in the refinery's boilers, furnaces, and fired heaters to make steam or raise the temperature of refinery process streams. A single source of refinery fuel gas may be burned locally in the unit where it is generated, or it may be sent to one or more refinery fuel headers to be combined with other fuel gas streams. To meet strict environmental regulations, the RFG must be treated to reduce its sulfur content before combustion, or the sulfur oxides (SO_x) generated must be scrubbed out of the resulting flue gas.

Compared to natural gas, combustion of RFG results in greater emissions of sulfur oxides because of its higher fuel–sulfur content even after treatment, and in somewhat greater NO_x emissions because of the higher resulting flame temperatures [9]. But those are topics for a different discussion.

L.2 COMBUSTION CALCULATIONS FOR REFINERY FUEL GAS

There will be occasions when combustion calculations must be performed for refinery fuel gas, often with a minimum of information on even an average fuel composition. Since this uncertain analysis of a mixture of gases from multiple sources typically changes from minute to minute, characterization of RFG chemical composition is meaningful only on a long-term basis. In many cases, only its heating value is routinely measured.

Since a component-by-component chemical reaction mass balance is difficult at best, this situation readily lends itself to the use of the techniques of EPA Method 19, explained in a later section.

L.3 EPA METHOD 19 COMBUSTION CALCULATIONS

Method 19 in 40 CFR 60 Appendix A [10] provides data reduction procedures for the results of stack testing. In addition, one can make use of F factors of Method 19 as a shortcut in performing combustion calculations. Use of F factors, and the relationships among them, for this purpose is illustrated in the example appearing later in this appendix.

Method 19 relates the volume of combustion products to the higher heating value (HHV) of the fuel, a quantity which is denoted there as the *gross calorific value* (GCV). The factor F_d is defined as the ratio of the *dry* flue gas volume for stoichiometric combustion at 0% excess O_2 to the GCV (HHV). Flue gas volume is

Table L.1 Tabulated F Factors for Natural Gas (English Units)

F_d SCF (dry) per 10^6 Btu (MMBtu)	F_w SCF (wet) per 10^6 Btu (MMBtu)	F_c SCF CO_2 per 10^6 Btu (MMBtu)
8710	10,610	1040

computed at standard conditions of 68°F (20°C) and 1 atm (760 mmHg, 29.92 in Hg, or 14.696 psia). At these conditions, the molar volume is 385.3 SCF/lb mol.

Conversion of flue gas volume to different conditions of temperature and pressure can be accomplished via the ideal gas law (Chapter 2). Conversion of flue gas volume to other conditions of flue gas oxygen is accomplished by means of Equation (3.19) and repeated here for convenience.

$$Q_{dry}/(Q_{dry} @ 0\% \, O_2) = [20.9/(20.9 - \% \, O_2 dry)] \quad (3.19)$$

The dry basis F_d factor appears to be the most useful and the most frequently used in normal practice.

Two other F factors, F_w and F_c, are also defined in Method 19. F_w relates the *wet basis* stoichiometric flue gas flow to the heating value. F_w reflects only the water of combustion derived from the hydrogen atoms contained in fuel constituents and does not include moisture/humidity in the combustion air or the water added by a wet scrubber, steam injection, and so on. [10].

The factor F_c denotes the amount of carbon dioxide in the flue gas per quantity of fuel heat input. Carbon dioxide is generated from the carbon-based materials in the fuel. F_c is completely independent of excess O_2 in the flue gas as long as the ambient combustion air contains zero CO_2, a very close approximation to the few hundred parts per million of CO_2 actually present.

The advantage of Method 19 is that one need not have a detailed fuel composition to proceed. Some F factor values are tabulated in Method 19 for various general classes of fuels. These include natural gas; crude, distillate, and residual oil; anthracite, bituminous, and lignite coal; wood and wood bark; and municipal solid waste. Numerical values are different for each type of fuel. F factors for natural gas in English units of SCF (@ 0% O_2) per 10^6 Btu or MMBtu (HHV) are given in Method 19 (Table L.1).

However, no F factors are listed specifically for refinery fuel gas, and the F factors for natural gas have been used in their absence.

L.4 *F* FACTORS FOR REFINERY FUEL GAS

To test the validity of that assumption, a brief investigation was conducted using refinery fuel gas compositions from public sources [5,9,11–16]. The study is not exhaustive but is thought to be adequate for its purpose. Combustion calculations were performed for a number of refinery fuel gases fired with dry atmospheric air having a simplified dry basis composition of 79.1% N_2 and 20.9% O_2. That composition is

688 Appendix L Combustion of Refinery Fuel Gas

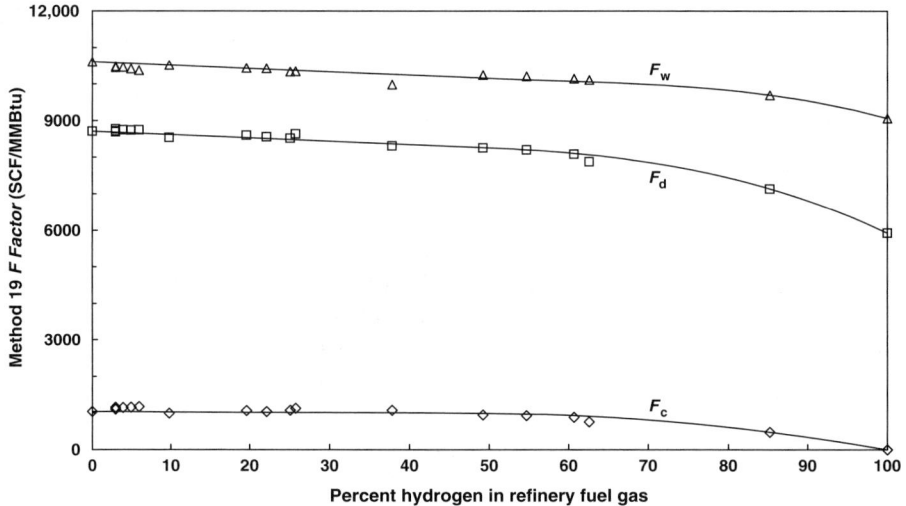

Figure L.1 Proposed correlation of calculated F factors for combustion of refinery fuel gas.

used in Method 19 to correct for changes in flue gas volume at differing values of excess flue gas oxygen (see Equation (3.19)). Results are cast in the form of stoichiometric SCF/h divided by the HHV firing rate of the fuel in MMBtu/h to produce F_d (dry), F_w (wet), and F_c (CO_2 generated). The heating value for each fuel gas was computed from the heating values of its component compounds [17].

Calculated F-factor point values are plotted in Figure L.1 against percent H_2 in the fuel, the major constituent in the fuel with the greatest likelihood to influence combustion characteristics. Three correlating curves are also shown, from top to bottom, for F_w, F_d, and F_c. They are anchored on the left-hand end at 0% H_2 by the F factor values for natural gas and on the right-hand end at 100% H_2 by the F factor values computed for pure hydrogen. The curves are linear up to the 50–60% H_2 range, approaching the values for natural gas as percent H_2 approaches zero. The straight-line sections were curve-fitted empirically by the method of least squares, with a power-series function tacked on in the upper range of H_2.

Furthermore, the linear portions are relatively flat, showing a decline in both F_w and F_d of approximately 9 F factor units for every percent H_2 in the fuel. The F_c relationship is even flatter, decreasing by approximately 1 unit for each H_2 percentage point. Constants of the linear curve fits are listed in Table L.2.

Table L.2 Values of Constants in Linear Curve Fit Equations (English Units)

Constant	F_w SCF (wet) per 10^6 Btu (MMBtu)	F_d SCF (dry) per 10^6 Btu (MMBtu)	F_c SCF CO_2 per 10^6 Btu (MMBtu)
Intercept	10,610	8710	1040
Slope	−9.13491	−9.03155	−0.98519

The correlation was derived from and is valid for typical refinery fuel gas compositions. It does not work for low-Btu fuel gas [13], for pressure swing adsorption (PSA) tail gas [9], for Flexicoker offgas [9], or for any other gaseous fuel whose percentage composition of CO, CO_2, or N_2 lies outside the norm for refinery fuel gas. Those readers who have access to proprietary RFG compositions may wish to check the correlating curves against their data.

Use of the generalized natural gas F factors in combustion calculations for refinery fuel gas provides a fairly close approximation, even up to a hydrogen percentage of 50–60%. For F_d and F_w, the maximum deviation between the actual individually calculated values and the corresponding Method 19 F factors for natural gas was found to be on the order of 5–6% at the upper range of applicability of the linear correlation. The deviation can be expected to decrease linearly with the percentage of H_2 in the fuel. For example, at 10% H_2, F_d from Figure L.1 or from the equation is about 8620, a 1% deviation. At 50% H_2, F_d from the correlation calculates to 8258, or a difference in F_d of only 5%, assuming that the F_d value for natural gas is absolutely correct and known exactly. The difference between the calculated and observed F_d and F_w factors is considerably less. Use of the published natural gas F factors as a substitute for the unspecified refinery fuel gas factors would produce a higher estimate of flue gas flow than would be calculated using individually computed F_d and F_w factors or the correlating lines.

Relative differences in the F_c factor are greater because it is lower in magnitude than F_d and F_w. The percentage difference between individual points and the correlating line shows a greater variation than F_d and F_w. This can be as high as 12% compared to an actual individual value. The difference between an individual F_c for refinery fuel gas and the Method 19 F_c for natural gas has been found to be from as high as $+13\%$ to as low as -10% anywhere within the range of applicability of the correlation. This may be adequate for routine work, but may not be good enough to calculate greenhouse gas emissions, requiring an individually determined F_c for a particular gas in question. Estimation and use of F_c should therefore be approached with caution, depending on its purpose.

L.4.1 Calculated *F* Factors for Published Natural Gas Compositions

A side study was conducted using readily available sources of published data on hand for 34 natural gas compositions including pure methane [15,16,18,19]. Calculated results for F_d, F_w, and F_c in SCF/MMBtu are shown in Table L.3. Relevant statistics for these sample distributions are tabulated at the bottom of each column.

The average F_d factor found is 8671 as opposed to 8710, an average F_w of 10,634 versus 10,610, and an average F_c of 1034 compared to 1040. The calculated ranges of F_d and F_w are less than 3% of the mean for the Method 19 F factors. Once again, as with the refinery fuel gases, the variability of the F_c factor is higher.

The relative closeness of the mean and median for each of the F factors indicates that the distributions are reasonably normal. As such, 95% of the calculated results should lie within plus or minus two times the sample standard deviation, and for the most

Table L.3 Method 19 F Factors Calculated for Published Natural Gas Analyses

Analysis no.	F_d	F_w	F_c
1	8614	10,625	1005
2	8624	10,635	1005
3	8657	10,645	1024
4	8800	10,780	1032
5	8657	10,588	1045
6	8679	10,618	1060
7	8720	10,557	1094
8	8656	10,635	1026
9	8677	10,662	1018
10	8633	10,624	1025
11	8725	10,699	1034
12	8657	10,594	1041
13	8641	10,622	1026
14	8661	10,644	1030
15	8645	10,622	1032
16	8649	10,630	1028
17	8679	10,622	1045
18	8642	10,607	1031
19	8841	10,777	1046
20	8677	10,623	1042
21	8674	10,625	1036
22	8641	10,620	1028
23	8664	10,613	1041
24	8653	10,612	1034
25	8662	10,617	1039
26	8638	10,623	1024
27	8656	10,667	1014
28	8664	10,592	1046
29	8683	10,564	1069
30	8677	10,662	1018
31	8676	10,626	1037
32	8727	10,702	1031
33	8655	10,593	1041
34	8627	10,619	1015
Mean	8671	10,634	1034
Standard deviation	45	47	17
% Standard deviation	0.5	0.4	1.6
Maximum	8841	10,780	1094
3rd Quartile	8677	10,644	1041
Median	8659	10,623	1032
1st Quartile	8645	10,613	1025
Minimum	8614	10,557	1005

Table L.3 (*Continued*)

Analysis no.	F_d	F_w	F_c
Range	227	223	89
% Range	2.6	2.1	8.6
Method 19	8710	10,610	1040

part they do. The few outlying results are responsible for the slightly larger range indicated. The factor F_d tabulated for natural gas in Method 19 would fall within the top quartile of this sample distribution, with the Method F_w and F_c factors for natural gas close to the quartile boundary between the lowest and next lowest quartiles.

The overall conclusion from the side study is that the variability in the calculated F factors for refinery fuel gas is of the same order of magnitude as the variability of individually calculated F factors for various natural gases compared to the official Method 19 F factors for natural gas. Here also, F_d and F_w are better behaved than F_c.

L.5 EXAMPLE—USE OF *F* FACTORS IN COMBUSTION CALCULATIONS

In this section, combustion calculations will be performed for a refinery fuel gas containing 50% H_2 using F factors estimated from the correlation, compared with a generic natural gas whose F factors are given in Method 19. The 50% H_2 content has been chosen at the upper end of the linear range to magnify any differences between RFG and natural gas. Assume a firing rate of 100 MMBtu/h (HHV), the natural gas F factors in Table L.1, and RFG factors (F_d, F_w, and F_c) from the correlation −8258, 10,153, and 991 SCF/MMBtu (HHV), respectively. Atmospheric air is at 60°F, with 60% relative humidity (RH).

L.5.1 Stoichiometric Case

Dry flue gas flow ((F_d) (firing rate))

$$\text{Natural gas } (8710)(100) = 871,000 \text{ SCFH}$$
$$\text{RFG } (8258)(100) = 825,800 \text{ SCFH}$$

Wet flue gas flow (including only moisture of combustion) ((F_w) (firing rate))

$$\text{Natural gas } (10,610)(100) = 1,061,000 \text{ SCFH}$$
$$\text{RFG } (10,153)(100) = 1,015,300 \text{ SCFH}$$

Water of combustion ((F_w) (firing rate) − (F_d) (firing rate))

$$\text{Natural gas } 1,061,000 - 871,000 = 190,000 \text{ SCFH}$$
$$\text{RFG } 1,015,300 - 825,800 = 189,500 \text{ SCFH}$$

Carbon dioxide (CO_2) ((F_c) (firing rate))

$$\text{Natural gas } (1040)(100) = 104{,}000 \text{ SCFH}$$
$$\text{RFG } (991)(100) = 99{,}100 \text{ SCFH}$$

Nitrogen (N_2) in the flue gas is calculated by difference of all dry constituents minus CO_2 and O_2. At stoichiometric conditions, O_2 is zero.

Nitrogen (N_2) in flue gas (total dry flue gas − CO_2)

$$\text{Natural gas } 871{,}000 - 104{,}000 = 767{,}000 \text{ SCFH}$$
$$\text{RFG } 825{,}800 - 99{,}100 = 726{,}700 \text{ SCFH}$$

Nitrogen in the combustion air is equivalent to nitrogen in the flue gas, assuming that there is no nitrogen in the fuel. At low fuel N_2 concentrations, equating the nitrogen in the flue gas and the combustion air becomes an approximation, but still a good one.

Nitrogen (N_2) in combustion air (N_2 in flue gas)

$$\text{Natural gas } 767{,}000 \text{ SCFH}$$
$$\text{RFG } 726{,}700 \text{ SCFH}$$

Dry combustion air ((N_2) (100/79.1))

$$\text{Natural gas } (767{,}000)(100/79.1) = 969{,}659 \text{ SCFH}$$
$$\text{RFG } (726{,}700)(100/79.1) = 918{,}710 \text{ SCFH}$$

Since F_w includes only the water of combustion, the total flue gas moisture must be computed from the sum of the water of combustion and the moisture in the combustion air. The combustion air moisture can be determined from the moisture content of the air and the total air used.

Moisture/humidity (H) in combustion air ((air) (H))

$$H = p_w(\%RH/100)/[P_T - p_w(\%RH/100)]$$

where p_w is the vapor pressure of water at a given temperature and P_T is the total atmospheric pressure in the same units [20].

From steam tables [21], p_w at $60°F = 0.25611$ psia

$$H = (0.25611)(60/100)/[14.696 - 0.25611(60/100)]$$
$$= 0.010567 \text{ SCF water/SCF dry air}$$

$$\text{Natural gas } (969{,}657)(0.010567) = 10{,}246 \text{ SCFH water in combustion air}$$
$$\text{RFG } (918{,}710)(0.010567) = 9708 \text{ SCFH water in combustion air}$$

Wet combustion air (dry air + H_2O in combustion air)

$$\text{Natural gas } 969{,}659 + 10{,}246 = 979{,}905 \text{ SCFH}$$
$$\text{RFG } 918{,}710 + 9708 = 928{,}418 \text{ SCFH}$$

L.5 Example—Use of *F* Factors in Combustion Calculations

Table L.4 Calculated Combustion Flows (SCFH) Stoichiometric Case

Constituent	Combustion air		Flue gas	
	Natural gas[a]	RFG[a]	Natural gas[a]	RFG[a]
Nitrogen (N_2)	767,000	726,700	767,000	726,700
Carbon dioxide (CO_2)	0	0	104,000	99,100
Oxygen (O_2)	202,659	192,010	0	0
Total dry flow	969,659	918,710	871,000	825,800
Water in air	10,246	9708	10,246	9708
Water of combustion	—	—	190,000	189,500
Total water	10,246	9708	200,246	199,208
Total wet flow	979,905	928,418	1,071,246	1,025,008

[a] Fuel.

Total flue gas moisture (water of combustion + water in combustion air)

$$\text{Natural gas } 190{,}000 + 10{,}246 = 200{,}246 \text{ SCFH}$$
$$\text{RFG } 189{,}500 + 9708 = 199{,}208 \text{ SCFH}$$

Total wet flue gas flow (dry flue gas flow + total flue gas moisture)

$$\text{Natural gas } 871{,}000 + 200{,}246 = 1{,}071{,}246 \text{ SCFH}$$
$$\text{RFG } 825{,}800 + 199{,}208 = 1{,}025{,}008 \text{ SCFH}$$

Flows for the stoichiometric case are summarized in Table L.4. Wet- or dry-basis composition for the major flue gas constituents can be obtained by dividing the flow of an individual constituent by the total wet or dry flow. This is exemplified for water (on a wet basis) below.

$$\% \, H_2O = (100)(200{,}246)/(1{,}071{,}246) = 18.69\% \text{ for natural gas}$$
$$= (100)(199{,}208)/(1{,}025{,}008) = 19.43\% \text{ for RFG}$$

Stoichiometric case flue gas compositions are shown in Table L.5. Within roundoff, flue gas flows (Q) in Table L.4 conform to the following relationships from Chapter 3:

$$Q_{\text{wet}} = Q_{\text{dry}}[100/(100-\% \, H_2O)] \qquad (3.23)$$

$$Q_{\text{dry}} = Q_{\text{wet}}[(100-\% \, H_2O)/100] \qquad (3.24)$$

The non-H_2O concentration entries (C) in Table L.5 are related by

$$C_{\text{dry}} = C_{\text{wet}}[100/(100-\% \, H_2O)] \qquad (3.29)$$

$$C_{\text{wet}} = C_{\text{dry}}[(100-\% \, H_2O)/100] \qquad (3.30)$$

Table L.5 Calculated Flue Gas Compositions Stoichiometric Case

Constituent	Natural gas[a]		RFG[a]	
	% wet	% dry	% wet	% dry
Nitrogen (N_2)	71.60	88.06	70.90	88.00
Carbon dioxide (CO_2)	9.71	11.94	9.67	12.00
Oxygen (O_2)	0.00	0.00	0.00	0.00
Water (H_2O)	18.69	—	19.43	—
Total dry	81.31	100.00	80.55	100.00
Total wet	100.00	—	100.00	—

[a] Fuel.

L.5.2 3% Excess O_2 Case

At 3% excess O_2 on a dry basis in the flue gas, total dry flue gas volume increases in accordance with Equation (3.19).

Dry flue gas flow

$$\text{Natural gas} \quad (871,000)[20.9/(20.9-3)] = 1,016,978 \text{ SCFH}$$
$$\text{RFG} \quad (825,800)[20.9/(20.9-3)] = 964,202 \text{ SCFH}$$

The difference between the dry flue gas flow at the excess O_2 condition and stoichiometric flow is the dry excess air in the flue gas. This excess air is composed of 20.9% O_2 and 79.1% N_2.

Excess O_2 in flue gas ((flue gas with excess O_2 − stoichiometric flue gas) (20.9/100)]

$$\text{Natural gas} \quad (1,016,978-871,000)(20.9/100) = 30,509 \text{ SCFH}$$
$$\text{RFG} \quad (964,202-825,800)(20.9/100) \quad = 28,926 \text{ SCFH}$$

Added N_2 in flue gas ((flue gas with excess O_2 − stoichiometric flue gas)(79.1/100))

$$\text{Natural gas} \quad (1,016,978-871,000)(79.1/100) = 115,469 \text{ SCFH}$$
$$\text{RFG} \quad (964,202-825,800)(79.1/100) \quad = 109,476 \text{ SCFH}$$

Total N_2 in flue gas (stoichiometric N_2 + added N_2)

$$\text{Natural gas} \quad 767,000 + 115,469 = 882,469 \text{ SCFH}$$
$$\text{RFG} \quad 726,700 + 109,476 = 836,176 \text{ SCFH}$$

As in the stoichiometric case, N_2 in the combustion air equals N_2 in the flue gas, and the dry combustion air equals this N_2 times the factor (100/79.1).

L.5 Example—Use of F Factors in Combustion Calculations

Dry combustion air ((N_2 in flue gas or in combustion air)(100/79.1))

$\quad$ Natural gas $(882,469)(100/79.1) = 1,115,637$ SCFH

$\quad$ RFG $(836,176)(100/79.1) = 1,057,113$ SCFH

Moisture/humidity (H) in combustion air ((air) (H))

$\quad$ Natural gas $(1,115,637)(0.010567) = 11,789$ SCFH water in combustion air

$\quad$ RFG $(1,057,113)(0.010567) = 11,200$ SCFH water in combustion air

The ambient humidity is unchanged from the stoichiometric case, as are the CO_2 generated and the water of combustion.

Wet combustion air (dry air + water in combustion air)

$\quad$ Natural gas $1,115,637 + 11,789 = 1,127,426$ SCFH

$\quad$ RFG $1,057,113 + 11,200 = 1,068,313$ SCFH

Total flue gas moisture (water of combustion + water in combustion air)

$\quad$ Natural gas $190,000 + 11,789 = 201,789$ SCFH

$\quad$ RFG $189,500 + 11,200 = 200,700$ SCFH

Total wet flue gas flow (dry flue gas flow + total flue gas moisture)

$\quad$ Natural gas $1,016,978 + 201,789 = 1,218,767$ SCFH

$\quad$ RFG $964,202 \quad + 200,700 = 1,164,902$ SCFH

Check: total wet flue gas flow (sum of constituents $N_2 + O_2 + CO_2 + H_2O$)

Natural gas $882,469 + 30,509 + 104,000 + (190,000 + 11,789) = 1,218,767$ SCFH

$\quad$ RFG $836,176 + 28,926 + 99,100 + (189,500 + 11,200) \ = 1,164,902$ SCFH

Flows for the 3% excess O_2 case are summarized in Table L.6, and flue gas composition is shown in Table L.7. At the same excess O_2, the calculated flue gas

Table L.6 Calculated Combustion Flows (SCFH) 3% Excess O_2 Case

	Combustion air		Flue gas	
Constituent	Natural gas[a]	RFG[a]	Natural gas[a]	RFG[a]
Nitrogen (N_2)	882,469	836,176	882,469	836,176
Carbon dioxide (CO_2)	0	0	104,000	99,100
Oxygen (O_2)	233,168	220,937	30,509	28,926
Total dry flow	1,115,637	1,057,113	1,016,978	964,202
Water in air	11,789	11,200	11,789	11,200
Water of combustion	—	—	190,000	189,500
Total water	11,789	11,200	201,789	200,700
Total wet flow	1,127,426	1,068,313	1,218,767	1,164,902

[a] Fuel.

Table L.7 Calculated Flue Gas Compositions 3% Excess O_2 Case

Constituent	Natural gas[a]		RFG[a]	
	% wet	% dry	% wet	% dry
Nitrogen (N_2)	72.41	86.78	71.78	86.72
Carbon dioxide (CO_2)	8.53	10.22	8.51	10.28
Oxygen (O_2)	2.50	3.00	2.48	3.00
Water (H_2O)	16.56	—	17.23	—
Total dry	81.31	100.00	80.55	100.00
Total wet	100.00	—	100.00	—

[a] Fuel.

compositions in Tables L.5 and L.7 are nearly the same on a dry basis for combustion of natural gas and refinery fuel gas. Wet basis compositions are different because the moisture content of RFG flue gas is somewhat higher. This is consistent with step-by-step combustion calculations done with chemical reaction equations, where the increased presence of molecular H_2 in the fuel increases the combustion moisture in the flue gas. Use of natural gas F factors for combustion of RFG does not capture this phenomenon.

L.6 CONCLUSIONS

- F factors for combustion of RFG are not specified in 40 CFR 60, Appendix A, Method 19 [10].
- The proposed correlation of refinery fuel gas F factors against percent hydrogen in the fuel is a good start.
- The deviations between the correlating lines and individual point values are in the low percent range. Percentage deviations of the smaller F_c factor for CO_2 emissions are larger than that for the factors F_d and F_w, used to estimate overall flue gas flow.
- The observed deviations, percentage-wise, between individual points for RFG and the correlating curves are of the same order of magnitude as the differences between the F factors derived for a specific natural gas compared to the generic natural gas F factors listed in Method 19.
- At 0% H_2 in RFG, the RFG F factor correlations correspond to the Method 19 F factors listed for natural gas. They decrease in value by about 1% for every 10% increase in RFG H_2 concentration up to 50–60% H_2. Beyond that range, they are nonlinear with few data points and should be used only with extreme caution.
- The Method 19 F factors for natural gas have been applied to RFG, and there is merit in employing officially sanctioned factors for combustion calculations

in light of the small errors noted above. Nevertheless, use of the natural gas factors for refinery fuel gas will overestimate the flue gas flow and underestimate its moisture content.

- The F_c factor for carbon dioxide, either from the correlation or by using the factor for natural gas, may be adequate for routine work. However, it may prove necessary to come up with case-specific F_c factors for more precise calculations of greenhouse gas emissions.

REFERENCES

1. U.S. Environmental Protection Agency, National Emission Standards for Hazardous Air Pollutants for Source Categories, Subpart AA National Emission Standards for Hazardous Air Pollutants from Phosphoric Acid Manufacturing Plants, Section 63.641—Definitions, p. 43, 40 CFR 63.641, U.S. Government Printing Office, Washington DC, revised as of July 1, 2005, available at http://edocket.access.gpo.gov/cfr_2005/julqtr/40cfr63.641.htm, accessed on March 29, 2009.
2. U.S. Environmental Protection Agency, Standards of Performance for Petroleum Refineries, 40 CFR 60, Subpart J, Section 60.101(d), U.S. Government Printing Office, Washington, DC, Electronic Code of Federal Regulations, http://ecfr.gpoaccess.gov/cgi/t/text/text, current as of June 12, 2008.
3. Compilation of Air Pollutant Emission Factors, AP-42, Volume I, Fifth Edition, Supplement D, Chapter 1: External Combustion Sources, Section 1.4 Natural Gas Combustion, U.S. EPA, Washington, DC, July, 1998.
4. B. Grover and P. Di Zanno, Study examines use of refinery fuel gas for hydrogen production, *Oil & Gas Journal*, **105**(24), 2007.
5. J. Colannino, *Modeling of Combustion Systems: A Practical Approach*, CRC Press, Taylor & Francis Group, Boca Raton, FL 2006, pp. 104, 406.
6. State of the Art (SOTA) Manual for Petroleum Refineries, p. 3.3–6, State of New Jersey Department of Environmental Protection Air Quality Permitting Program, Trenton, NJ (July 1997), available at www.nj.gov/dep/aqpp/downloads/sota/sota3.doc, accessed on March 30, 2009.
7. H. Isalski, Hydrogen recovery from fuel gas can help meet refinery demand, *Oil & Gas Journal*, **91**(12) 1993.
8. K.A. Simonsen, L.F. O'Keefe, and W.F. Wong, Changing fuel formulations will boost hydrogen demand, *Oil & Gas Journal*, **91**(12), 1993.
9. C.E. Baukal, Jr., and R.E. Schwartz, editors, *The John Zink Combustion Handbook*, CRC Press, Boca Raton, FL, 2001, pp. 158–163, 437, 446.
10. U.S. Environmental Protection Agency, Determination of Sulfur Dioxide Removal Efficiency and Particulate Matter, Sulfur Dioxide, and Nitrogen Oxide Emission Rates, 40 CFR 60, Appendix A, Method 19, July 1, 2007.
11. J.J. McKetta, and W.A. Cunningham, editors, *Encyclopedia of Chemical Processing and Design*, Vol. **28**, Marcel Dekker, Inc., New York, 1988, 230 pp.
12. Application Note Refining, Measuring Hydrogen Sulfide in Refinery Fuel Gas with a Simple TCD-based Gas Chromatograph, p. 2, Rosemount Analytical, Emerson Process Management (2008), available at www.emersonprocess.com/, accessed on March 28, 2009.
13. K.A. Kramer, N. Patel, S. Sekhri, and M.G. Brown, Flexible hydrogen plant utilizing multiple refinery hydrocarbon streams, Paper AM-96-59 presented at the 1996 NPRA Annual Meeting, San Antonio, TX, March 17–19, 1996.
14. D.R. Bartz, K.W. Arledge, J.E. Gabrielson, L.G. Hays, and S.C. Hunter (KVB Engineering, Inc.), Control of Oxides of Nitrogen in the South Coast Air Basin (of California), Appendix C (Oil Refineries), PB-237 688, National Technical Information Service (NTIS), Springfield, VA, September, 1974.
15. Phased Project Requested Amendment, Table 3, Typical Gaseous Fuel Properties, Document 2006-19-06.pdf available at www.efsec.wa.gov/bpcogen/SCA, accessed on March 29, 2009.

16. R.H. Perry, C.H. Chilton, and S.D. Kirkpatrick, editors, *Chemical Engineers' Handbook*, 4th edn, McGraw-Hill, New York, 1963, p. 9–8 and p. 12–32.
17. ASTM Method D 3588-81, Standard Method of Calculating Calorific Value and Specific Gravity (Relative Density) of Gaseous Fuels, American Society for Testing and Materials, Philadelphia, PA, January 1982.
18. Perry, R.H. and C.H. Chilton, *Chemical Engineers' Handbook*, 5th edn, McGraw-Hill, New York, 1973, p. 9–12.
19. *North American Combustion Handbook,* 1st edn, 2nd printing, The North American Manufacturing Co., Cleveland, OH, 1957, 34 pp.
20. R.G. Kunz, Calculate water in combustion air, *Hydrocarbon Processing*, **77**(9), 119–122, 1998.
21. C.A. Meyer, R.B. McClintock, G.J. Silvestri, and R.C. Spencer, Jr., *1967 ASME Steam Tables*, 2nd edn, American Society of Mechanical Engineers (ASME), New York, 1968, p. 88.

Index

acetylene, 536–537
activity coefficient, 46–58
adiabatic flame temperature (AFT), 131–136, 143–144, 520–522
adiabatic saturation temperature, 68–70, 166, 672
air composition, 87–90, 96–98, 513, 687
air infiltration, 180–182
air preheat, 145, 530–533, 681–684
alkalinity, 271–273, 562–565, 569–571, 581–586
alpha radiation, 381, 470–471, 476
Alta Dena Dairy, 229–232
Amagat's Law, 16
ammonia, 35, 38, 211–218, 264–265, 276–277, 550, 672–675, 682–683
ammonia ionization, 35, 38, 276–277
ammonia slip, 211–218, 675, 683
antineutrino, 471
Antoine Equation, 25–30, 92–94, 197
atmospheric dispersion modeling, 238–245
atom, 6–9
atomic number, 7–8, 471
atomic weight/atomic mass number, 7–9, 471
average molecular weight (AMW), 88–90, 96–98, 119, 122–123, 130, 171–173
Avogadro's Number, 14, 482
azeotrope, 49–52

baghouse (See fabric filter)
beat phenomenon, 465–466
Becquerel, 468–469, 482, 484, 495–496, 500–501, 503
benzene, toluene, xylenes (BTX), 524
Bernoulli's Equation, 340, 638
beta radiation, 381, 469–471, 482
biochemical oxygen demand (BOD), 260–268, 548–557, 599–600, 608–621, 627–628
blowdown, 293–296, 560, 581–586
BOD lag curve, 551
boiler, 84, 145, 214, 216, 229, 291–300, 512, 685
boiler horsepower, 231

boiler operations, 291–300
bridgewall temperature, 524
buddy system, 466
burner No$_x$, 511–542, 681–684
butane, 85, 120
B$_{wa}$ moisture fraction in ambient air, 111–115
B$_{ws}$ moisture fraction in flue gas, 113–114

carbon-14, 470, 475
carbonaceous BOD, 550–551
carbon dioxide (CO$_2$), 34–38, 261, 272–273, 402, 405–408, 412, 521–523, 569–572, 684, 687–688, 696–697
carbonate-bicarbonate-CO$_2$ system, 34–38, 272–273, 569
carbon monoxide (CO), 132, 138, 223–224, 410–411, 522–526, 668, 672–676, 683
carbon monoxide (CO) boiler/waste heat boiler, 216, 663–664, 673–674
case studies, 586–588, 647–660, 681–684
cesium-137, 475, 495, 498, 500, 503–504
chlorine/chlorination-dechlorination, 277–285, 567
chemical oxygen demand (COD), 260–268, 554–557
chemical tracers, 315–317, 323–324, 624
Cipolletti weir formula, 306
CNEL, 431–432, 435–437, 441
coal, 85, 157, 216, 238, 376, 512
cobalt-60, 475, 496, 498, 500
coefficient of area (C$_A$), 341
coefficient of discharge (C$_D$), 341, 638, 642–644
coefficient of velocity (C$_V$), 340–341
coke, 523, 527, 665–668
combined cycle, 223, 683
combustion calculations, 12–13, 83–186, 192–194, 511–516, 686–697
community noise survey, 459–466, 647–660
compressibility factor (Z), 17

Environmental Calculations: A Multimedia Approach, by Robert G. Kunz
Copyright © 2009 John Wiley & Sons, Inc.

699

concentration, 19–21, 137–140, 511–515
conductivity, 257, 295, 568, 573–575, 580–588
compressor noise, 441, 445–446
computational fluid dynamics (CFD), 518
conventional burners, 519–521, 527
continuous emission monitoring system (CEMS), 180–185, 214, 683
conversion factors, 142–143, 386–390
cooling tower drift, 190–191, 560
cooling tower noise, 441, 444–445
cooling water, 291, 559–590
corrected source strength, 425–426
Crater Lake water quality, 253–257
crickets, 418, 465
Curie, 468, 482, 496, 500–501, 503
current meter, 314–315
curriculum, 1, 507–510
cycles of concentration, 561, 567, 574–576, 581–588
cyclone separators, 232–237, 665–666, 669–670

dBA, 428–430, 460, 462, 648–657
$dB_{all\ pass}/dB_{flat}$, 428–430, 462
dBB, 428–430
dBC, 428–430, 462, 648, 653–655
Dalton, John, 7
Dalton's Law, 25–16
dating of antiquities, 470
daughter species, 468, 471, 486, 493–494
discharge coefficient (C_D), 341, 638, 642–644
discharge monitoring report (DMR), 303
dissolved oxygen (DO), 252, 257–259, 269–271, 543–548, 599–600, 605–617, 621–622, 627–628
dissolved oxygen (DO) sag, 269–271, 600, 621–622, 628
destruction and removal efficiency (DRE), 393–400
decibel (dB), 417, 441–442
decibel addition, 422–427
de Pitot, Henri, 173
dew point, 92–94
disaster, environmental, 2, 495, 503
distributed control system (DCS), 682
domestic sewage/wastewater, 268, 591
do's and don'ts, 4–5, 462
draining of tank, 339–374, 632–646
dry flue gas, 107–110, 687–688
dry scrubbing, 238
duration of noise event, 417

electron, 7–9, 469, 472, 482
electroneutrality, 39–41, 253–257, 286–288, 566–567, 581–586

electrostatic precipitator (ESP), 235–237, 663, 665, 670
element, 7–9
emission factor, 21–22, 137–138, 511–516
entrainment, 91, 666, 669
environmental disaster, 2, 495, 503
EPA Method 19, 110–115, 682, 686, 696
ethane, 84–85, 120
ethylene cracking furnace, 143–144, 213, 517–542
evaporation, 559–561, 581–586
excess air, 13, 99, 101–105, 118, 123, 126, 131, 153, 169, 404–409, 512–515, 535–536
excess oxygen (O_2), 13, 100–107, 116–118, 123–126, 130–131, 395–409, 512–514, 521, 526–539, 668, 682–683, 686–688, 693–694

F Factors, 110–116, 119, 123–129, 224–226, 230–231, 686–697
fabric filter, 665, 670–671
Federal Water Pollution Control Act of 1972 (FWPCA), 629
film badge, 469
fish and DO, 270–271, 548
fish kill, 270, 610
flare, 207–211
flashing factor, 67–68
floating objects, 317–318, 624–625
flooding, 591–598
flue gas, 84, 107–110, 687–688
flue gas moisture, 91, 111, 157, 512, 683, 687
flue gas recirculation (FGR), 218–223
fluid catalytic cracking catalyst, 80, 662
fluid catalytic cracking unit (FCCU), 216, 661–676
flume, 308–313
Francis weir formula, 304–305
frequency/pitch, 416–417, 457–459
fuel, 84
fuel sulfur, 142, 149, 224, 524, 536, 686
fugitive emissions, 186–192
furnace, 84, 145, 179–180, 229, 512, 517–542
Fyrite analysis, 169–170

GEP stack height, 240–241
Galileo, 340
gamma radiation, 469–471
gas turbine, 140, 223–225, 512, 683–684
Geiger, Hans, 478
Geiger-Müller Counter, 477
general license, 499–500
grade efficiency curve, 232
gray, 469
groundwater, 324–326, 591–592
gross calorific value (GCV), 110

Index **701**

Hagen-Poisseuille Law, 640–641
half life, 474–479, 482–483, 486–487, 492, 495–496
Hanover Township, NJ Sewage Treatment Plant, 603, 610, 629
hardness, 252, 254, 562, 565
heat capacity, 60–66, 133–137, 404, 409–410
heat recovery steam generator (HRSG), 223–224, 683
heat transfer, 62, 71–72
height of dike, 372–374
Henry's Law, 31–41, 52–60, 258, 544
Hertz (Hz), 416, 419–421, 428–431, 440, 444–446, 449–454, 458–459, 465, 656, 659
hexane, 85, 120
higher heating value (HHV), 22–24, 85–87, 118, 146–147, 158, 513–515, 519, 535, 686
highway noise, 424, 454–456, 464–465, 647–657
highway noise barrier, 454, 464, 647–656
horizontal cylindrical tank – partially filled, 355
humidity, 29–30, 83, 91–96, 114, 402, 512, 533–534, 682
hydraulic elements graph, 332–334
hydraulic radius, 318–321
hydrogen chloride (HCl), 166, 264–266, 394–398
hydrogen in fuel, 132, 519–520, 522–526, 529–537, 685, 688, 696
hydrogen plant, 213, 517, 520–525, 681–684
hypochlorous acid ionization, 278–279

ideal gas, 13–19, 61–62
ideal gas molar volume, 15
impulse sound, 417, 439–440, 463–464
incinerator/incineration, 84, 400–412
incinerator temperature, 404–409
intensity/loudness, 416–417

K value for valve, 341
Kunz's Maxims, xxi, 2

L_{dn}, 431–433, 435–437, 441
L_{eq}, 431–433, 435, 655
L_{01}, L_{10}, L_{50}, L_{90}, 431, 462
latent heat, 61, 66
lead, 470–471, 475, 486–488, 491–492
least squares, 480, 484–485, 569, 688
log-mean temperature difference, 71–72
Log-Normal distribution, 80
Log Pearson Type III distribution, 323
low-flow conditions (e.g., $7Q_{10}$), 322–324
lower heating value (LHV), 22–24, 85–87, 118, 152, 686
low-NO_x burners, 145, 214, 519–521, 527, 681–682

Low Temperature Oxidation (LTO), 226–232

makeup rate, 293–296, 560–561, 581–586
manifest, 375
Manning Equation, 318–322, 331–336, 622–623
Margules Equations, 47
mass flow rate, 19–21, 24–25
mean life, 474
membrane electrode method for DO, 258, 547
Mendeleev, Dmitri, 7
methane, 22–23, 84–85, 98–107, 115–116, 120, 132–137, 519, 524–526, 685, 689
Meyer, Lothar, 7
mg/L as $CaCO_3$, 252–254, 565–566
microorganisms, 259, 274, 610
mole/mol, 6, 9, 14
molecule, 7–9
Morristown, NJ Sewage Treatment Plant, 601, 603, 605, 609, 626, 629
municipal solid waste (MSW), 380–381, 400–412
municipal solid waste (MSW) composition, 401–402

nappe, 304
nastygram, 183
natural gas, 84–85, 119–123, 129–131, 139–143, 215, 223–224, 469, 515, 519, 521–526, 681–697
neutrino, 471
neutron, 7–9, 468
New Jersey Community Noise Control Course, 439–440, 647
New Jersey Department of Environmental Protection, 601, 609–613
New Jersey Noise Regulations, 439–440
New Jersey Turnpike, 647–654
nitric oxide (NO) to nitrogen dioxide (NO_2) ratio, 140–141, 521
nitrifying reactions 274–276, 550–551
No. 2 Fuel Oil, 142, 151–161
noise barrier, equipment, 457, 464, 647–656, 659
noise permitting procedure, 461–462, 466, 656–659
noise regulations, 415, 438–441, 461, 463
notice of violation (NOV), 183
nuclear power/weapons, 470, 479, 499. 502
nuclear/radioactive waste
Nuclear Regulatory Commission, 497–499
nucleon, 468
nucleus, 468

octave, 419–421
octave band measurements, 418, 421, 460, 658–659
oil, 84–85, 142, 151–161, 214, 223, 512, 624
oil/water separator, 288–291

702 Index

Ontario Ministry of the Environment (MOE), 455–456
opacity, 167, 668
open channel flow, 318–322, 331–337
Orsat analysis, 169–175
oxides of nitrogen (NO_x), 138, 211–232, 410–411, 511, 513, 517–542, 662–664, 667–668, 672, 675–676, 681–684, 686
ozone, 227–232, 519, 673–674

Palmer-Bowlus flume 311–313
parameter modeling (See predictive emission monitoring system (PEMS))
parent nucleus, 468, 471, 482, 487, 492–495
Parshall flume, 308–311
particulate matter (PM), 138, 190–191, 232, 410–411, 667–671, 676
Passaic River, 601, 604, 610
Passaic Valley Water Commission, 601, 608
penetration, 207, 218
pentane, 85, 120
period, 9, 416–417
Periodic Table, 7–10
permanganate oxygen demand test, 260–261
permit application procedure, 4, 461–462, 466, 656–659
petroleum refining, 80, 149–151, 216, 232, 524, 536, 661–676, 685
pH, 34–41, 237, 272–273, 276–279, 285–288, 562–563, 568–572, 581–586
phosphate rock, 469, 502
pitot tube, 173–176, 178, 314
pollutant emissions, 137–140
positron, 472, 482
potassium-40, 469, 475, 481–483
preapplication meeting, 4, 658
predictive emission monitoring system (PEMS), 183, 682
pressure swing adsorption (PSA), 229, 682–684, 689
principal organic hazardous constituent (POHC), 393–400
process heater, 84, 229, 512, 685
product of incomplete combustion (PIC), 394, 398
propane, 84–85, 116–120, 123–125, 146–149
proton, 7–9, 468, 471
public meeting/hearing caricature, 180
public water supply, 591, 604, 610
pure tone, 439–440, 444, 465, 658–659
pyrolysis fuel oil, 523–524
pyrolysis gasoline (pygas), 523

qualified observer, 167

rad, 469
radioactive decay/radioactivity, 468–469, 472–481
radioactive decay series, 471, 486–495
radioactive material categories, 497–504
radionuclide/radioisotope, 468, 479–480, 483, 503
radium, 381, 468, 470, 475–476, 502
radon, 469, 475–476, 502
Raoult's Law, 41–46
rational method, 326–331
recirculation rate, 559–561, 581
reduction efficiency, 207, 218
refinery fuel gas, 149–151, 524, 536, 685
reflection of sound, 464
regenerative cycle, 223
regulation of radioactive sources, 497
relative accuracy test audit (RATA), 183–185
relative biological effectiveness (RBE) factor, 469
relative humidity, 91–94
rem, 469, 496, 503
retrograde solubility, 294
Reynolds Number, 638–639
Rivers and Harbors Act, 629
Rockaway River, 600–601
Roentgen, Wilhelm, 469
rolling average, 76–77, 683
Rutherford, Ernest, 478

salinity and dissolved oxygen (DO), 543–545
selective catalytic reduction (SCR), 211–214, 217–218, 517, 663, 675, 668, 681–685
selective noncatalytic reduction (SNCR), 214–218, 674–675
septic tanks, 268–269
sievert, 469, 496
simple cycle, 223, 683
sine wave, 417–418
sound attenuation, 443–454
sound meter, 418, 460, 463–464, 648, 656
sound/noise survey, 459–466, 647–660
sound power, 441–443
sound pressure, 417, 441
sound survey report, 461–462, 647–656
speech interference, 430–431, 440
speed of sound, 457–459
spherical tank –partially filled, 346, 358, 360
stack draft, 84, 246–247
stack test wish list, 168, 539
stack testing, 166–186
Standard Normal Probability Curve, 77–80
statistics, 72–81, 184–185, 322–324, 390–393, 431–434, 480, 484–485, 569–572, 688–691
steam methane reformer (SMR), 143–144, 213, 517, 520–525, 681–684

Index

steam plume, 166, 671
steam/water injection, 91, 111, 223–224, 681–682, 687
stoichiometric flue gas, 98–99, 101, 104, 120–121, 231, 404–405, 512, 686–687, 691
storm water, 326–337, 591–598, 624–625
storm water drainage channels (various shapes), 331–336
Streeter-Phelps Equation, 268–271, 600, 621–622
strontium-90, 475, 479, 498, 500
sublimation, 25
sulfur dioxide/sulfurous acid system, 238, 285–288
sulfur oxides (SO_x), 139, 151–161, 411, 664–668, 671–672, 686
sulfuric acid, 572–573, 581–588
sulfuric acid dew point, 157–159, 165–166, 675
surrogate compound, 394, 399–400
suspended solids, 591, 619

theoretical oxygen demand, 262–268
thorium, 469–471, 475, 488, 498, 502
timbre, 418
time of concentration, 329–331, 335–337
time of travel, 317–318, 624–626
Torricelli's Law, 340, 346, 638
total dissolved solids (TDS), 570, 573, 581–586
total organic carbon (TOC), 261–268
traffic noise frequency, 454, 655
tramp air, 180
transformer noise, 441, 443–444, 465
trajectory of liquid jet, 367–374
trout stream, 591
typical sound pressure levels, 418

ultimate BOD, 550
ultimate oxygen demand (UOD), 554
ultralow-NO_x burners, 519, 527, 536
universal gas constant (R), 15, 61
uranium, 469–471, 498, 502

urea, 214–216, 674
U.S. EPA, 497
U.S. Federal Highway Administration (FHWA) Noise Model, 454, 456
USGS Gauging Station Morristown, NJ, 601, 626, 629
USGS Surface Water Records, 322–324, 604, 608–609, 614, 624–625, 629
USGS Water Quality Data, 253, 257, 605, 608, 613

valence, 9–10
van Laar Equations, 47–52
vena contracta, 341, 637
vapor-liquid equilibrium ((VLE), 41–60, 196–202
vapor pressure, 25–30, 91–94, 196–202
volatile organic compounds (VOCs), 138, 191–202
volatile organic sampling train (VOST), 398–399

waste categories, 375–385
wastewater discharge permit, 303
water of combustion, 91, 111, 687
water quality limits, 322
wavelength, 416, 457–459
weather effects, 460, 462–463
weir, 304–308
wet- and dry-bulb temperatures, 94–96
wet flue gas, 107–110, 687–688
wet gas scrubber (WGS), 236–237, 663, 665, 670–672
Whippany Paper Board Company, 599–605, 609–610, 615–621, 627–630
Whippany River, 599–631
Winkler Technique for DO, 545–548
wipe test, 501

X-Ray, 469, 471, 498

Zeldovich Kinetics, 521